Introduction to Planetary Science

Introduction to Planetary Science

The Geological Perspective

GUNTER FAURE
*The Ohio State University,
Columbus, Ohio, USA*

TERESA M. MENSING
*The Ohio State University,
Marion, Ohio, USA*

 Springer

A C.I.P. Catalogue record for this book is available from the Library of Congress.

ISBN-13 978-1-4020-5233-0 (HB)
ISBN-13 978-1-4020-5544-7 (e-book)

Published by Springer,
P.O. Box 17, 3300 AA Dordrecht, The Netherlands.

www.springer.com

Cover art: The planets of the solar system.
Courtesy of NASA.

A Manual of Solutions for the end-of-chapter problems can be found at the book's homepage at
www.springer.com

Printed on acid-free paper

*In memory of Dr. Erich Langenberg,
David H. Carr, and Dr. Robert J. Uffen
who showed me the way.*

Gunter Faure

*For Professor Tom Wells and
Dr. Phil Boger who taught me to
reach for the stars.*

Teresa M. Mensing

Table of Contents

The Earth and the Moon. Montage of separate images recorded in 1992 by the spacecraft Galileo on its way to the Jupiter system (Image PIA 00342, courtesy of NASA/JPL-Caltech)

Preface

This textbook is intended to be used in a lecture course for college students majoring in the Earth Sciences. Planetary science provides an opportunity for these students to apply a wide range of subject matter pertaining to the Earth to the study of other planets of the solar system and their principal satellites. As a result, students gain a wider perspective of the different worlds that are accessible to us and they are led to recognize the Earth as the only oasis in space where we can live without life-support systems. In addition, planetary science tends to unify subjects in the Earth Sciences that are customarily taught separately: geophysics, volcanology, igneous petrology, mineralogy, geomorphology, geochemistry, hydrogeology, glacialogy, tectonics, economic geology, historical geology, as well as meteoritics, microbiology, physics, astronomy, atmospheric science, and even geopolitics and international relations. Clearly, planetary science is well suited to be taught as a **capstone course** for senior undergraduates and beginning graduate students at universities everywhere. The exploration of space is by necessity a cooperative enterprise in which national borders, cultural differences, and even language barriers fade. Therefore, we propose that planetary science should be incorporated into the Earth Science curriculum even at institutions that do not have an active research program in planetary science. Students will benefit from such a course because it will expand their intellectual horizons and will prepare them to contribute to the on-going exploration of the solar system.

This book evolved from a popular course we taught from 1988 until 2005 in the Department of Geological Sciences of The Ohio State University. The course combined lectures with class discussions and fieldtrips to the planets via videotapes. In addition, students solved numerical homework problems and wrote short essays on assigned topics which required them to use their own resources and to form their own opinions based on factual information. As a result of the progress in the exploration of the solar system that has already been made, the planets of the solar system and their satellites are no longer faint points of light in the night sky. Instead, they have become real **worlds** with surfaces that contain records of their histories and of their present environmental conditions. The data that are pouring in from robotic spacecraft that are exploring different parts of the solar system need to be interpreted by Earth scientists who are accustomed to constructing realistic hypotheses from large sets of numerical data and visual information.

Our current knowledge about the Earth and the solar system is expressed in the form of theories that explain the interactions of matter and energy in a way that is consistent with all relevant observations available to us at the present time. Accordingly, we emphasize the importance of the scientific method in the exploration of the solar system. In addition, we demonstrate the consistency of the physical and chemical properties of the planets and their satellites with the current theory of the origin and evolution of the solar system. In this way, the differences in the properties of planets and satellites can be viewed as a consequence of the processes that produced the solar system and that continue to cause the evolution of its many components. Instead of treating each body in the solar system as a unique entity, we emphasize the patterns of variation among them. In this way, we demonstrate that the solar system is a unit that has many parts with properties that vary in predictable ways based primarily on their distances from the Sun.

For this reason, we begin this presentation with the Sun which dominates the solar system and exemplifies the processes that cause stars in the Milky Way galaxy to continue to form and evolve. In addition, we relate the evolution of stars to the synthesis of chemical elements by the nuclear reactions which release the energy that stars radiate into interstellar space. We simplify this complex subject by emphasizing the main events in the evolution of stars and stress that the origin of the Earth and of the other planets of the solar system is part of an ongoing process that started with the creation of the Universe and that will continue until the end of time.

Next, we present Kepler's Laws and demonstrate how they govern the motions of all objects that orbit the Sun or any other massive body in space. The motions of the Earth around the Sun are the basis for the way we measure time and for the seasonal changes in the climate of the Earth. The discussion of celestial mechanics also introduces students to basic information about the geometries of ellipses and circles. In addition, we demonstrate that the bulk densities of the terrestrial planets decrease with increasing distance from the Sun in agreement with the prevailing theory of the formation of the solar system.

The next six chapters deal with the terrestrial planets starting with meteorites, followed by the Earth-Moon system, Mercury, Venus, Mars, and the asteroids. Meteorites and the Earth-Moon system are discussed first because they are familiar to us and because they contain information about the early history of the solar system. For example, the cratered surface of the Moon records a period of intense bombardment by solid objects like those that still reside in the asteroidal belt. In addition, the gravitational interaction between the Moon and Earth gives rise to the tides in the oceans, generates heat, and causes the period of rotation of the Moon to be locked to the period of its revolution around the Earth. Such gravitational interactions occur also between Mercury and the Sun and between all of the planets and their principal satellites including Jupiter and its Galilean satellites.

The descriptions of the terrestrial planets provide information on their orbital parameters, their masses and volumes, the presence or absence of magnetic fields and atmospheres, and their bulk chemical compositions. These properties can be used to deduce their internal structure, their geologic history and present activity, and the presence or absence of water or other volatile compounds in solid, liquid, or gaseous form. We emphasize that the present surfaces of the terrestrial planets are shaped by the geological processes that have occurred in the past or are still occurring at the present time.

The so-called gaseous planets and their satellites are alien worlds with properties that are radically different from those of the Earth and the other terrestrial planets. We pay particular attention to Jupiter and its Galilean satellites because they exemplify the bizarre properties of Saturn, Uranus, and Neptune and their major satellites. Each one of these giant planets is unique and yet all of these planets and their satellites fit the mold provided by the theory of the solar system.

Until quite recently, very little information was available about the Pluto-Charon system on the outer fringes of the solar system. Pluto is now known to be a member of a newly-defined group of objects composed primarily of ices of several volatile compounds. The largest members of this group, including Pluto, are now classified as dwarf planets. In addition, the entire solar system is surrounded by a "cloud" of icy bodies whose highly eccentric orbits take them far into interstellar space. When these small solar-system bodies return to the inner solar system, they may either collide with the gaseous and terrestrial planets or they may swing around the Sun as comets.

The discussion of comets returns our attention to the inner solar system and thus to the Earth, which is the only planet we know well. Therefore, the next to last chapter contains a review of the dominant planetary processes and of the way these processes are manifested on Earth. The principal point of these comparisons is to recognize the uniqueness of the Earth as "an abode fitted for life". In the final chapter we take note of the existence of planets in orbit around other stars in our galactic neighborhood. Most of the planets that have been discovered so far are more massive than Jupiter and some of them orbit close to their central star. Although none of the 160 extrasolar planets known at the present time are likely to harbor life, we are beginning to consider the possibility that alien organisms may occur on other planets in the galaxy. We are at a critical juncture in the history of humankind because our scouts are about to leave the "cradle of life" to venture forth to other planets of the solar system and beyond.

The teaching aids used in this textbook include black-and-white diagrams that illustrate and clarify the text. We also provide numerous and carefully chosen color pictures of planetary surfaces and extraterrestrial objects. All of the diagrams and color pictures have extensive explanatory captions that complement the information provided in the text. This textbook also includes end-of-chapter problems by means of which students can test their understanding of the principles of planetary science and can verify statements made in the text. In addition, we provide subject and author indexes, and appendices containing relevant formulae and lists of the properties of the planets and their satellites. In conclusion, we reiterate our basic premise that the Earth scientists in training today will benefit from a capstone course that unifies what they have learned about the Earth and that broadens their intellectual horizon to include the solar system in which we live.

Our interest in and knowledge of Planetary Science has been stimulated by the positive feedback we have received from many former students, including especially Rene Eastin, Edson Howard, Nancy Small, Tim Leftwich, Jeffrey Kargel, Stuart Wells, Christopher Conaway, Brad Cramer, Kate Tierney, Allison Heinrichs, Benjamin Flaata, Andy Karam, and many others. In addition, we have enjoyed the friendship of Gerald Newsom, Professor of Astronomy, and of Ralph von Frese, Professor of Geology, at The Ohio State University. We have also benefitted from our association with colleagues of the ANSMET program (Antarctic Search for Meteorites) including W.A. Cassidy, Ralph Harvey, John Schutt, Christian Koeberl, Kunihiko Nishiizumi, Ghislaine Crozaz, Robert Walker, Peter Wasilewski, and David Buchanan. However, we alone are responsible for any errors of omission or comission.

We are pleased to acknowledge the excellent working relationship we have enjoyed with Petra D. van Steenbergen, Senior Publishing Editor (Geosciences) of Springer and with Ria Balk, Publishing Assistant. We thank the individuals who gave us permission to use copyrighted images and text: Ron Miller, Peter Brown, Bevan French, Paolo Candy, Kris Asla, John Gleason, Chinellato Matteo, Robert Postma, Joe Orman, and Rick Scott. We also thank Scot Kaplan for the photograph of the Ahnighito meteorite and we gratefully acknowledge the support of The Ohio State University of Marion. Betty Heath typed the manuscript including all of its numerous revisions and corrections. Her skill and dedication helped us to complete this project for which we are most grateful.

Gunter Faure and Teresa M. Mensing

The Urge to Explore

On May 25, 1961, President *John F. Kennedy* of the USA (Figure 1.1) made a speech before a joint session of Congress concerning urgent national needs. In this speech, he proposed to send an American astronaut to the Moon before the end of the decade and to bring him back alive:

"...I believe that this nation should commit itself to achieving the goal, before this decade is out, of landing a man on the moon and returning him safely to the earth. No single space project in this period will be more impressive to mankind, or more important for the long-range exploration of space; and none will be so difficult or expensive to accomplish..." <http://www.cs.umb.edu/jfklibrary/j05256/.htm>

This bold proposal immediately captured the support of the American people who were concerned at the time because the USA had fallen behind the Soviet Union in launching spacecraft into Earth orbit (Sorensen, 1965).

The accomplishments of the Soviet Union in the exploration of space at that time included the launching of Sputnik 1 into Earth orbit on October 4, 1957, followed by Sputnik 2 on November 2 of the same year. During this period, several American rockets exploded on the launch pad until the Explorer 1 satellite was launched on January 31, 1958, by a rocket built by *Wernher von Braun* and his team of engineers at the Redstone Arsenal near Huntsville, Alabama (Ward, 2005). Subsequently, the USA launched seven Pioneer spacecraft between October 11, 1958 and December 15, 1960 from Cape Canaveral, Florida, for planned flybys of the Moon, but all of them either missed the Moon or malfunctioned after launch (Weissman et al., 1999).

The Soviet Union also had its share of failures with the Luna program designed to explore the Moon. However, Luna 3, launched on October 4, 1959, did reach the Moon and photographed the farside, which had never before been seen by human eyes. The Soviet Union achieved another important success when *Yuri Gagarin* orbited the Earth on April 12, 1961, and returned safely.

In the meantime, the National Aeronautics and Space Administration (NASA) had designed and built a spacecraft for Project Mercury whose goal it was to put humans into space. On April 1, 1959, seven military test pilots were selected to be the first Mercury astronauts, including *Alan Shepard* who carried out a 15 minute suborbital flight on May 5, 1961, followed by a second suborbital flight on July 21, 1961, by *Virgil "Guss" Grissom*. Finally, on February 20, 1962, *John Glenn* orbited the Earth three times in a Mercury capsule (Glenn, 1999).

When viewed in this perspective, President Kennedy's dramatic proposal of May 25, 1961, to put a man on the Moon was a giant leap forward. It was intended to rouse the American people from their depression and to demonstrate to the world that the USA possessed the technology to achieve this goal and that it had the collective will to carry it out. Therefore, President Kennedy's proposal was primarily motivated by political considerations related to the Cold War with the USSR. He did not say that the astronauts should explore the Moon by taking photographs and by collecting rocks, nor did he mention how many flights should take place. These matters were advocated later by geologists at academic institutions and by members of the United States Geological Survey and were adopted by NASA with some reluctance (Wilhelms, 1993).

The American effort to put a man on the Moon succeeded brilliantly when *Neil Armstrong* and

1.1 The Exploration of Planet Earth

The exploration of the Earth began more than one million years ago as groups of human hunter-gatherers (*Homo erectus*) roamed across the plains of Africa and Asia in search of food. More recently, at the end of the last ice age between about 20,000 and 10,000 years ago, when sealevel was about 120 meters lower than it is today, hunting parties from Siberia crossed the Bering Strait on foot to Alaska. These people subsequently explored the new territory they had discovered and eventually populated all of North and South America. When the continental ice sheet started to recede about 20,000 years ago, sealevel increased and eventually flooded the Bering Strait about 10,000 years ago (Emiliani, 1992, Figure 24.9).

In a much more recent period of time, starting in 1415 AD, Portuguese mariners began a sustained and state-supported effort to sail south along the coast of West Africa (Bellec, 2002). Their goal was to find a direct route to the Far East, which was the source of silk, spices, and other high-priced goods that reached the Mediterranean region primarily by land via slow-moving caravans. In the process, the Portuguese navigators disproved much misinformation about the perils of crossing the equator of the Earth, they constructed accurate maps, improved navigational techniques, and designed more seaworthy sailing ships. After 83 years of effort by many successive expeditions, the Portuguese explorer *Vasco da Gama* in 1498 AD reached the city of Calicut (Kozhikode) in India. Fourteen years later, in 1512 AD, the Portuguese arrived in the Spice Islands (Moluccas) and, in 1515 AD, they "discovered" China. Actually, the Portuguese were by no means the only explorers at that time because Chinese navigators had already reached the east coast of Africa, *Marco Polo* had lived in China from 1276 to 1292 AD, and *Alexander the Great* crossed the Himalayas into India in 326 BC, or 1824 years before Vasco da Gama sighted the Malabar coast of India.

While the Portuguese were seeking a direct route to China by sailing east, *Christopher Columbus* proposed to *Queen Isabella of Spain* that China could be reached more directly by sailing west. In 1486 the Queen appointed a Commission to evaluate this proposal. After

Figure 1.1. John F. Kennedy, 35th President of the USA from 1961 to 1963. (John F. Kennedy Presidential Library and Museum, Boston, MA)

Edwin Aldrin landed in the Sea of Tranquillity on July 20, 1969, in accordance with President Kennedy's challenge (Collins, 1974; Aldrin and McConnell, 1991; Hansen, 2006). The lunar landing of Apollo 11 was followed by six additional flights, the last of which (Apollo 17) returned to Earth on December 19, 1972 (Cernan and Davis, 1999). The Soviet space program also succeeded by landing robotic spacecraft on the Moon starting with Luna 9 (launched on January 31, 1966) followed later by several additional flights, some of which returned samples of lunar soil (e.g., Luna 20, 1972; Luna 24, 1976).

These landings on the surface of the Moon achieved a long-standing desire of humans to explore the Moon and the planets whose motions in the night sky they had observed for many thousands of years. Some people considered the exploration of the Moon by American astronauts and by Soviet robotic landers to be a violation of the sanctity of a heavenly body, but most viewed it as an important milestone in the history of mankind.

deliberating for four years, the Commission declared that Columbus' proposal was nonsense. However, the defeat of the Moors by the Spanish armies in the battle of Granada in 1492 caused the Queen to reconsider. Columbus was recalled to the Spanish court and his proposal was approved on April 17 of 1492. Six months later, on October 12, Columbus sighted land after sailing west across the Atlantic Ocean six years before Vasco da Gama reached India.

Christopher Columbus had, in effect, redis-covered North America which had been populated thousands of years earlier by people from Siberia and which had also been visited by Vikings led by *Leif Eriksson*, son of *Erik the Red,* who had established a settlement in Greenland in 986 AD. In fact, all of the conti-nents (except Antarctica) and most of the large islands contained human populations in the 15th century when the Portuguese began their quest to find a way to India, China, and the islands in the Molucca Sea of Indonesia.

In spite of the discoveries Christopher Columbus had made, he had not actually reached China or the Moluccas. Therefore, *Emperor Charles V* of Spain authorized the Portuguese nobleman *Ferdinand Magellan* to reach the Moluccas by sailing around the southern end of the landmass that Columbus had discovered. A small fleet of five ships sailed on September 20, 1519, with 200 people on board. One year later, on November 28, 1520, the three remaining ships sailed through a narrow passage (now known as the Straits of Magellan) and entered a new ocean on a calm and sunny day. Accord-ingly, Magellan and his surviving associates named it the *Pacific Ocean.* However, the voyage across the Pacific Ocean was a nightmare because of the lack of fresh food and water. Many crew members died of scurvy and Magellan himself was killed in a battle with natives in the Philippines. The 18 survivors completed the first circumnavigation of the Earth and returned to Seville in 1522 AD after three years at sea.

All of the expeditions of the fifteenth to the seventeenth centuries were motivated primarily by the desire for wealth and power. The explo-ration of the Earth that resulted from these voyages was an unintended byproduct. Not until the eighteenth century did scientific exploration become a factor in motivating the expeditions

organized by England and France. Noteworthy among the leaders of these extended voyages of discovery were *Captain Louis Antoine de Bougainville* of France (1729–1811) and *Captain James Cook* of Britain (1728–1779).

Captain de Bougainville had an illustrious career in the service of France. He first served as aide-de-camp to General Louis-Joseph Montcalm in Canada during the French and Indian War (1756–1763) described by Anderson (2005). In 1763, he led an expedition to the South Atlantic Ocean and set up a French colony on the Falkland Islands. Next, he sailed around the world between 1766 and 1769 accompanied by a group of scientists. In 1772, de Bougainville became the secretary to King Louis XV of France and served as the Commodore of the French fleet which supported the American Revolution (1779–1782). After the French Revolution in 1792, he settled on his estate in Normandy. Later, Napoleon I appointed him to be a Senator, made him a Count, and a Member of the Legion of Honor. He died on August 31, 1811, in Paris at the age of 82.

James Cook learned his craft as a seaman on wooden sailing ships that hauled cargo along the North-Sea coast of England. In 1752 he enlisted in the Royal Navy and participated in the siege of Louisburg in Nova Scotia during the Seven-Years War (1756–1763) between Great Britain and France (Anderson, 2000). He also helped General Wolfe to cross the St. Lawrence River for an attack on the city of Quebec. The British army commanded by General Wolfe scaled the rocky cliff along the northern bank of the river under cover of darkness and defeated the French army of General Montcalm on the Plain of Abraham on September 13, 1759.

In 1768 the Royal Society of England selected James Cook to command the first scientific expedition to the Pacific Ocean. His orders were to transport a group of British scientists led by Joseph Banks to the island of Tahiti in order to observe the transit of Venus across the face of the Sun on June 3, 1769. Subsequently, he was ordered to sail south to explore the so-called Terra Australis. During this leg of the expedition Cook discovered and circumnavigated New Zealand. He then crossed the Tasman Sea and, on April 19,1770, discovered the east coast of Australia. On this voyage he mapped not only the coast of New Zealand but also the Great Barrier Reef which is still considered to be a serious navigational hazard. On his way back to England, Cook passed through the Torres Strait between Timor and the York peninsula of Australia, stopped briefly in Batavia (i.e., Jakarta) and arrived in England in 1771 without losing a single member of his crew to scurvy, although 30 men died of fever and dysentery contracted in Batavia.

James Cook led two additional scientific expeditions to the Pacific Ocean. During the second voyage (1772–1775) he entered the Pacific Ocean from the west and approached the coast of East Antarctica before turning north. He mapped Easter Island and Tonga and discovered New Caledonia before continuing to cross the Pacific Ocean. By sailing east around the tip of South America he entered the Atlantic Ocean where he discovered the South Sandwich Islands and South Georgia Island. During this voyage Cook concluded that the hypothetical Terra Australis existed only in the form of New Zealand and Australia and that Antarctica was not the hoped-for oasis of the southern hemisphere.

During his third and last cruise (1776–1779) Cook again entered the Pacific Ocean from the west, passed New Zealand and Tahiti before turning north to the Hawaiian Islands. From there he sailed north along the west coast of North America into the Bering Strait and entered the Chukchi Sea. On his return to Hawaii Captain Cook was killed on February 14, 1779, during a skirmish with native Hawaiians on the beach at Kealakekua Bay on the Big Island.

During his three expeditions between 1768 and 1779 Cook mapped large areas of the Pacific Ocean from the coast of Antarctica to the Bering Strait. In addition, the scientists on his ships contributed greatly to the advancement of botany, zoology, anthropology, and astronomy. Cook's expeditions were motivated primarily by the desire to explore the Earth and thereby set a precedent for the current exploration of the solar system.

We may have inherited the urge to explore from our distant ancestors (i.e., Homo erectus) who had to keep moving to survive. Those who were bold and took risks were occasionally rewarded by finding new places to hunt and

to gather food. The more recent voyages of discovery by Vasco da Gama, Christopher Columbus, and Ferdinand Magellan, and many others were motivated primarily by the hope of becoming rich and famous. These explorers endured great hardships and many lives were lost. Nevertheless, the voyages did stimulate world trade and increased the wealth and power of certain European nations. The voyages also led to advances in science and to improvements in the technology of the time as exemplified by the expeditions of Louis Antoine de Bougainville of France and of Captain James Cook of Great Britain.

When the 20th century ended a few years ago, the exploration of the Earth had been successfully concluded. In the meantime, advances in science and technology had prepared the way for the exploration of the space around the Earth and of the Moon, our nearest celestial neighbor. In addition, robotic spacecraft are now providing increasingly detailed views of the planets of the solar system and their satellites. As a result, we are contemplating a future when humans may live in space stations orbiting the Earth and in settlements on the Moon, on Mars, and perhaps elsewhere. This giant leap for mankind has become possible because of the development of powerful rocket engines and of sophisticated electronic computers.

1.2 Visionaries and Rocket Scientists

One of the technological prerequisites for the exploration of the solar system was the development of the rocket engine. In addition, the people of the world had to be made aware of the possibility that space travel by humans is feasible. Therefore, the exploration of the solar system was first popularized by visionaries and then realized by rocket scientists.

The first proposal for space travel was written in the seventeenth century by *Johannes Kepler* (1571–1630 AD) who was both a scientists and a visionary. He published a story called Somnium (The Dream) in which he described a trip to the Moon by means of his mother's spells. The authorities were not amused and accused his mother of being a witch. She was imprisoned for six years before Kepler was able to secure her release. Although Kepler's science fiction did not catch on, he did explain the motions of the planets in their orbits around the Sun and thereby provided the foundation for modern astronomy. However, two centuries passed before the people of the world were ready to think about space travel again.

The science fiction novels of *Jules Verne* (1828–1905) and later those of *H. G. Wells* (1866–1946) generated interest in space travel and inspired some of the pioneers whose scientific contributions have made the exploration of the solar system a reality. In addition, the paintings of Chesley Bonestell, Ron Miller, W. K. Hartmann, Pamela Lee, and other space artists have let us see places in the solar system and in the Milky Way galaxy we have not yet been able to reach. Chesley Bonestell's images, which appeared in numerous books, magazine articles and movies starting in the early 1940s, inspired an entire generation of future explorers of the solar system including Wernher von Braun, Arthur C. Clarke, Carl Sagan, Fred Whipple, and Neil Armstrong (Miller and Durant, 2001; Hartmann et al., 1984; Miller and Hartmann, 2005). The individuals to be considered here include Konstantin E. Tsiolkovsky (1857–1935), Robert H. Goddard (1882–1945), and Wernher von Braun (1912–1977).

Konstantin   Eduardovitch   Tsiolkovsky (Figure 1.2) became interested in space travel as a young man after reading novels by Jules Verne (e.g., De la Terre à la Lune; From the Earth to the Moon, published in 1865). In 1880 Tsiolkovsky took a job as a mathematics teacher in Borovsk, Kaluga Province of Russia, where he began to think seriously about how humans could live in space. In 1892 he moved to the town of Kaluga where he designed liquid-fueled, steerable rockets, and spacecraft that included gyroscopes for attitude control and airlocks through which astronauts could exit for extra-vehicular activity (EVA) in free space. He also designed an orbiting space station which contained a greenhouse to provide food and oxygen for the occupants. These plans were included in a scientific paper which Tsiolkovsky wrote in 1903: "Investigations of space by means of rockets".

Tsiolkovsky was convinced that humans would eventually colonize the solar system and then move on to planets orbiting other stars in the Milky Way galaxy:

Figure 1.2. Konstantin Eduardovitch Tsiolkovsky (1857–1935) was a Russian rocket scientist and pioneer of cosmonautics

"Men are weak now, and yet they transform the Earth's surface. In millions of years their might will increase to the extent that they will change the surface of the Earth, its oceans, the atmosphere, and themselves. They will control the climate and the Solar System just as they control the Earth. They will travel beyond the limits of our planetary system; they will reach other Suns and use their fresh energy instead of the energy of their dying luminary." (i.e., our Sun).

Although Tsiolkovsky did not actually build rockets or spacecraft, he influenced many young Russian engineers including *Sergey Korolev* who later became the "Chief Designer" of the Soviet space program. Korolev led the team of engineers that designed and built the rockets that launched Sputniks 1 and 2 in 1957 and which put Yuri Gagarin in orbit around the Earth in 1961. <http://www.informatics.org/museum/tsiol.html>.

Robert Hutchings Goddard (Figure 1.3), like Tsiolkovsky, became interested in space flight as a young man after reading the science fiction novel: "War of the Worlds" by H.G. Wells. He studied physics at Clark University in Worcester, Massachusetts, and received his doctorate in 1908. Subsequently, he joined the faculty of Clark University and started to experiment with

Figure 1.3. Robert H. Goddard (1882–1945). (Courtesy of NASA)

could launch his rockets in secret. In the 1930s his rockets first broke the sound barrier and reached an altitude of 1.6 km (i.e., one mile) above the surface of the Earth. Goddard's inventions were used after his death in 1945 to build intercontinental ballistic missiles (ICBMs) as well as the multi-stage rockets that put a man on the Moon in 1969.

The contributions Goddard made to rocketry were officially recognized in 1959 when the Congress of the USA authorized the issuance of a gold medal in his honor and established the Goddard Space Flight Center in Greenbelt, Maryland. In addition, the Federal Government agreed to pay $1,000,000 for the use of Goddard's patents. (<http://www.gfsc.nasa.gov/gsfc/service/gallery/fact-sheets/general/goddard/goddard.htlm>)

Wernher von Braun (Figure 1.4) grew up in Germany and, like Tsiolkovsky and Goddard, became an enthusiastic advocate of space travel after reading the science fiction novels of Jules Verne and H.G. Wells (Ward, 2005). He also studied a seminal scientific report by *Hermann Oberth* published in 1923: "Die Rakete zu den Planetenräumen" (The rocket into interplanetary space). The desire to understand the technical details of Oberth's report motivated von Braun to learn physics and mathematics when he was still in highschool. In 1930, he joined the German Society for Space Travel and assisted Oberth in testing liquid-fueled rocket engines. After receiving a B.S. degree in mechanical engineering in 1932 from the Technical Institute of Berlin, von Braun enrolled in the University of Berlin and received his Ph.D. degree in physics in 1934. His dissertation was a theoretical and experimental study of the thrust developed by rocket engines.

When the Society for Space Travel experienced financial difficulties during the worldwide economic depression in the early 1930s, von Braun went to work for Captain *Walter R. Dornberger* who was in charge of solid-fuel rockets in the Ordnance Department of the German Army (Reichswehr). Within two years, von Braun's team successfully launched two liquid-fueled rockets that achieved an altitude of more than 2.4 kilometers (1.4 miles). Because the original test site was too close to Berlin, the Army built a new facility near

rockets in a small basement laboratory. His results demonstrated that rockets can generate forward thrust in a vacuum and he invented a rocket engine that burned liquid oxygen and gasoline. Finally, on March 16, 1926, he successfully tested such a rocket in Auburn, Massachusetts.

Goddard's research was funded by the Smithsonian Institution of Washington, D.C., which published a report he had submitted in 1919 in which he mentioned the possibility that rockets could be used to reach the Moon. Much to Goddard's dismay, this idea was ridiculed by newspaper reporters, which caused him to avoid publicity of any kind in the future. However, copies of Goddard's report were read with great interest in Europe and may have encouraged the founding of the German Rocket Society in 1927.

In spite of the unfavorable publicity in the USA, Goddard continued to improve the rockets he was building with new funding arranged for him by *Charles Lindbergh*. He moved his operation to Roswell, New Mexico, where he

Figure 1.4. Wernher von Braun (1912–1977), Director of the NASA Marshall Space Flight Center. (Courtesy of NASA)

the village of Peenemünde on the coast of the Baltic Sea in northeastern Germany. Dornberger was the military commander and von Braun was the civilian technical director of the new rocket-research center. One of the rockets that was developed there was the A-4, which became known as the V-2 when it was used by the German Army to strike targets in Great Britain during World War II (1939–1945). At that time, the German rockets developed in Peenemünde were superior to those of any other country, although von Braun acknowledged the importance of Robert Goddard's work in the development of rocket technology.

When the Russian Army approached Peenemünde during the winter of 1944/45, von Braun moved his entire staff and a large number of rockets to western Germany and surrendered to the advancing American forces. Within a few months, he and about 100 of his engineers were taken to Fort Bliss in Texas. From there they traveled to White Sands, New Mexico, where they launched V-2 rockets for high-altitude research. In 1952, von Braun and his team

moved to Huntsville, Alabama, where von Braun was the technical director and later chief of the U.S. Army ballistic-weapons program.

Even though he was designing and testing rockets for the US Army, von Braun published several popular books between 1952 and 1956 in which he advocated space travel and developed plans for landing astronauts on the Moon and Mars (See Section 1.7, Further Reading). After the Soviet Union launched the Sputniks in October and November of 1957, von Braun was asked to use one of his rockets to launch an American satellite into orbit. The launch of Explorer 1 by a Redstone rocket took place on January 31, 1958, and thereby initiated the exploration of space by the USA. As a result, von Braun who had become a naturalized citizen of the USA in 1955, became a national hero and leading advocate of the exploration of the solar system.

The organization that von Braun had built up in Huntsville was later transferred to NASA and was renamed the C. Marshall Space Flight Center with von Braun as its first director. The Center developed the Saturn rockets, which were used between 1969 and 1971 to launch spacecraft to the Moon in response to President Kennedy's directive. These rockets performed flawlessly and enabled astronauts to set foot upon the Moon. As a result, we now have the technological capability to explore the solar system and, in time, to establish human settlements in space and on other planets, as Konstantin Tsiolkovsky envisioned more than 100 years ago.

1.3 Principles of Rocketry and Space Navigation

Before a spacecraft built on Earth can travel anywhere in the solar system it must have a powerful engine that can lift it far enough above the surface so that it can escape from the gravitational force the Earth exerts on all objects on its surface. The engine that is required for this purpose must also be able to function in the upper atmosphere of the Earth where the air gets very "thin" and where the amount of oxygen is insufficient to operate internal combustion engines powered by gasoline. The only propulsion system that can provide the necessary power and that can operate in the vacuum of space is the rocket engine. Therefore, space travel and the exploration of the solar system were not possible until after Robert Goddard, Wernher von Braun, and Sergey Korolev designed and tested rocket engines that could lift a spacecraft, the rocket engine itself, and the necessary fuel into orbit around the Earth.

The principle of operation of the rocket engine is derived from a law of nature discovered by *Isaac Newton* (1642–1727), which states that to every action there is an equal and opposite reaction. In the rockets in Figure 1.5 a jet of hot gas is expelled through a nozzle at the rear which causes the rocket itself to move forward. The hot gas is generated by combustion of either a solid or a liquid fuel. The solid fuel burns even in the vacuum of space because the fuel contains the oxygen that is needed for its combustion. Liquid-fueled rockets carry oxygen and fuel in liquid form. Both are pumped into the combustion

chamber where they burn to form the hot gas that is expelled through the nozzle of the engine.

The force of gravity exerted by the Earth on all objects decreases with increasing distance from its center in accordance with Newton's law of gravity (Science Brief 1.5.1). According to this law, the force exerted by the Earth on an object on its surface decreases when that object is moved farther away from the Earth (i.e., into the space between the planets). Figure 1.6 demonstrates that the force of gravity exerted by the Earth decreases from "one" when the object is located on the surface of the Earth to only 0.01 when the object is moved to a point located ten Earth radii from its center. At that distance, the object is nearly weightless and therefore could remain in free space without falling back to the surface of the Earth.

Figure 1.6 also demonstrates that a spacecraft launched from the surface of the Earth has to climb out of the gravitational hole into free space before it can start its journey to the Moon, or Mars, or to other more distant locations. At the present time, rocket engines burning solid or liquid fuel are the only way a spacecraft can escape from the gravitational force of the Earth.

Rocket Using Solid and Liquid Fuel

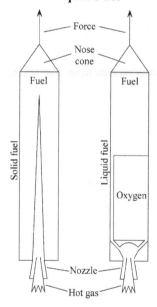

Figure 1.5. Schematic diagrams of rockets powered by solid and liquid fuel. Both derive forward thrust from the jets of hot gas expelled through the nozzles in the rear. The nose cone contains the payload, which is small compared to the weight of the body of the rocket and the fuel it contains. The arrows extending from the nose cones represent the magnitude and direction of the force exerted by the rocket engines. (Adapted from Figure 5.2 of Wagner, 1991)

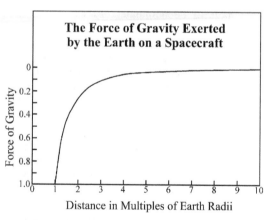

Figure 1.6. This diagram illustrates that the force of gravity acting on an object of constant mass decreases from 1.0 on the surface of the Earth to near zero when that object is transported into interplanetary space. The distance between the center of the Earth and the object is expressed in multiples of the radius of the Earth, which has a value of 6378 kilometers. The scale of the vertical axis was inverted in order to make the point that a rocket must "climb" out of a gravitational hole in order to reach free space where it becomes weightless because the force of gravity exerted on it by the Earth approaches zero (See Science Brief 1.5.1)

After a spacecraft has arrived in free space, a comparatively small push from a rocket motor can move it in the direction of its desired destination.

When a spacecraft approaches its destination, it is acted upon by the force of gravity exerted by the target planet which pulls it into a gravitational hole like that which surrounds the Earth. To avoid a crash landing, the spacecraft must go into orbit around the planet and then descend gradually for a controlled landing.

A spacecraft in a stable circular orbit around a planet is acted upon by two forces identified in Figure 1.7. The gravitational force (F_1) exerted by the planet is pulling the spacecraft toward itself. This is opposed by the centrifugal force (F_2) that is generated by the velocity (v) of the spacecraft in its circular orbit around the planet.

A Spacecraft in Orbit Around a Planet

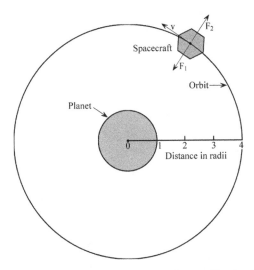

Figure 1.7. Forces acting on a spacecraft or satellite that is "in orbit" around a planet. The velocity (v) of the spacecraft generates a centrifugal force (F_2) which must be equal in magnitude but opposite in direction to the gravitational force (F_1). When $F_1 = F_2$, the spacecraft is in a stable circular orbit above the surface of the planet. When the velocity (v) decreases, F_2 decreases, whereas F_1 remains constant. Consequently, the radius of the orbit decreases which causes F_1 to increase as the spacecraft approaches the surface of the planet in accord with Figure 1.6. When a spacecraft descends to a lower circular orbit, its velocity must increase in order to raise the centrifugal force sufficiently to balance the increase in the gravitational force. (See Science Brief 1.5.2 and 1.5.3)

When the two opposing forces are equal, the spacecraft is weightless and can remain in orbit unless its velocity is decreased by an external process (i.e., the firing of a "retrorocket"). When that happens, the centrifugal force (F_2) decreases while the force of gravity that acts on the spacecraft remains constant. As a result, the altitude of the spacecraft begins to decrease. Therefore, a spacecraft in orbit around a planet can be made to land by gradually reducing its orbital velocity (See also Science Brief 1.5.2).

1.4 Summary

The modern era of space exploration can be traced to the work of four individuals: Konstantin E. Tsiolkovsky and Sergey Korolev of Russia, Robert H. Goddard of the USA, and Wernher von Braun of Germany and the USA. These rocket scientists were inspired by the science-fiction novels of Jules Verne and H.G. Wells to design the rocket engines that are required to place artificial satellites in orbit around the Earth and to propel robotic or manned spacecraft from the Earth to the Moon and to the planets of the solar system.

The exploration of the solar system actually began thousands of years ago with the exploration of the Earth. Early humans (Homo erectus) in Africa and Asia were hunter/gatherers who needed to keep moving in order to survive. Some groups of hunters from Siberia walked across the Bering Strait during the last ice age more than about 10,000 years ago and discovered new land on the other side which was uninhabited by humans.

Still more recently, in 1415 AD, the rulers of Portugal on the Iberian peninsula of Europe attempted to find a sea-route to India, China, and the so-called spice islands in order to gain wealth and political power. Many hundreds of sailors died on the expeditions that attempted to sail around the southern tip of Africa and across the Indian Ocean. In spite of the losses, the Portuguese mariners ultimately established a highly profitable trade in silk, spices, and other valuable commodities.

The success of the Portuguese merchants motivated other countries to participate in the search for wealth: Spain, The Netherlands, Britain, and France. Expeditions sent by Britain and France included scientists who wrote reports about their discoveries, which substantially increased knowledge of the Earth. During the 20th century, the exploration of the Earth progressed rapidly with the result that all continents were mapped and thoroughly investigated from the air and on the ground.

The exploration of the Earth has now evolved to the exploration of the other planets and their satellites in the solar system made possible by the development of the rocket engine. Seven of the eight planets of the solar system have been examined by remote sensing from robotic spacecraft. Only Pluto has not yet been visited by a spacecraft from Earth, but a robotic spacecraft called New Horizons will arrive in its vicinity in 2015. In addition, several spaceships from Earth have landed on Mars and Venus and twelve humans have walked on the Moon. The modern era of exploration of the solar system has begun and will continue with the active participation of many nations including the USA, Russia, the European Union, Japan, and China. In addition, India, Pakistan, Iran and North Korea have rocket-powered ballistic missiles that could be used for space exploration.

1.5 Science Briefs

1.5.1 Force of Gravity

The force of gravity exerted by a planet of mass (M) on another body of mass (m) is expressed by Newton's Law of gravitation:

$$F = G\frac{M \times m}{r^2} \qquad (1.1)$$

where F = force of gravity in newtons (N)
  M = mass of the planet in kilograms (kg)
  m = mass of the other object in kilograms
  r = distance between the object and the center of the planet in meters (m)
  G = gravitational constant
    $= 6.67 \times 10^{-11}$ N × m^2/kg^2
  Let M $= 5.98 \times 10^{24}$ kg (mass of the Earth)
    r $= 6.378 \times 10^6$ m (one Earth radius)
    m = 1.0 kg

$$F = \frac{6.67 \times 10^{-11} \times 5.98 \times 10^{24} \times 1.0}{(6.378 \times 10^6)^2} = 9.805\,N$$

Let r $= 10 \times 6.378 \times 10^6$ m (10 Earth radii)

$$F = \frac{6.67 \times 10^{-11} \times 5.98 \times 10^{24} \times 1.0}{(63.78 \times 10^6)^2}$$

$$F = 0.09805\,N$$

The results indicate that, if the force of gravity acting on the object on the surface of the Earth is equal to 1.0, the force acting on the same object 10 radii from the center of the Earth is equal to $0.09805/9.805 = 0.01$ and the units cancel.

1.5.2 Stable Circular Orbit

A spacecraft of mass (m) in orbit about a planet of mass (M) is acted upon by the gravitational force (F_1) expressed by equation 1.1

$$F_1 = G\frac{Mm}{r^2}$$

A spacecraft in circular orbit around a planet is also acted upon by the centrifugal force (F_2) depending on its velocity (v):

$$F_2 = \frac{mv^2}{r} \qquad (1.2)$$

If the spacecraft is in a stable circular orbit:

$$F_1 = F_2 \qquad (1.3)$$

In case the velocity of a spacecraft in a stable circular orbit is decreased by external means (e.g., by firing an on-board rocket), the centrifugal force (F_2) in equation 1.2 decreases, whereas the gravitational force (F_1) in equation 1.1 is unaffected. Therefore, the gravitational force (F_1) becomes stronger than the centrifugal force (F_2) and the spacecraft begins to spiral toward the surface of the planet.

In case the velocity of a spacecraft in a stable orbit is increased (e.g., by firing an on-board rocket), the centrifugal force (F_2) becomes stronger than the gravitational force (F_1). As a result, the orbit of the spacecraft changes from a circle to an ellipse. If the spacecraft is accelerated sufficiently, its orbit ultimately changes from an ellipse into a parabola, which causes the spacecraft to move away from the planet without returning.

1.5.3 Average Velocity of an Object in Circular Orbit

The condition for a circular orbit expressed by equation 1.3 is: $F_1 = F_2$:
Substituting equations 1.1 and 1.2:

$$\frac{GMm}{r^2} = \frac{mv^2}{r}$$

$$v = \left(\frac{GM}{r}\right)^{1/2} \qquad (1.4)$$

This result indicates that the velocity of a spacecraft in a circular orbit is inversely proportional to the square root of the radius of the orbit. Alternatively, equation 1.4 can be solved for r:

$$r = \frac{GM}{v^2} \qquad (1.5)$$

Note that equations 1.4 and 1.5 apply only to spacecraft in stable circular orbits as required by equation 1.3.

1.6 Problems

1. Exploration has always been a risky business that has caused great hardship and death. The polar regions of the Earth, in some respects, resemble the surface environment of the planet Mars. Select one of the several polar explorers and write a short report about him (2 pages double-spaced). Identify what aspect of the planning and execution led to success or failure of the expedition:

   1. Roald Amundsen   4. Fridtjof Nansen
   2. Robert F. Scott   5. Lauge Koch
   3. Ernest Shackleton   6. Peter Freuchen

2. Science Brief 1.5.3 contains the equation:

$$\frac{GMm}{r^2} = \frac{mv^2}{r}$$

Solve this equation first for the velocity (v) and then for the orbital radius (r).

3. Calculate the velocity of the Earth in its orbit around the Sun given that:

   $M = 1.99 \times 10^{30}$ kg (mass of the Sun)
   $r = 149.6 \times 10^9$ m (radius of the orbit of the Earth)
   $G = 6.67 \times 10^{-11}$ Nm^2/kg^2 (gravitational constant)

   Express the orbital velocity in km/s. (Answer: $v = 29.78$ km/s).

4. What important assumption underlies the calculation in problem 3 above?

5. Calculate the average orbital velocity of the Earth given that:

   $r = 149.6 \times 10^6$ km (radius of the orbit of the Earth)
   $p = 365.25$ days (time required for the Earth to complete one orbit around the Sun)
   $C = 2\pi r$ (circumference of a circle).
   $\pi = 3.14$

   Express the velocity in km/s. (Answer: $v = 29.77$ km/s).

6. Does the good agreement of the orbital velocities obtained in problems 3 and 5 prove that the orbit of the Earth is a circle?

1.7 Further Reading

Aldrin B, McConnell M (1991) Men from Earth. Bantam Falcon Books, New York

Anderson F (2000) Crucible of war: The Seven-Years' War and the fate of empire in British North America (1754–1766). Knopf, New York

Anderson F (2005) The war that made America: A short history of the French and Indian War. Viking, New York

Bellec F (2002) Unknown lands: The log books of the great explorers (Translated by Davidson L, Ayre E). The Overlook Press, Woodstock and New York

Bergaust E (1960) Reaching for the stars (An authoritative biography of Wernher von Braun). Doubleday, New York

Cernan E, Davis D (1999) The last man on the Moon. St Martin's Griffin, New York

Collins M (1974) Carrying the fire: An astronaut's journeys. Ballantine Books, New York

Emiliani C (1992) Planet Earth: Cosmology, geology and the evolution of life and environment. Cambridge University Press, Cambridge, UK

Ferris T (1988) Coming of age in the Milky Way. Anchor Books, Doubleday, New York

Glenn J (1999) John Glenn: A memoir. Bantam Books, New York

Hansen JR (2006) First man: The life of Neil Armstrong. Simon and Schuster, New York

Hartmann WK, Miller R, Lee P (1984) Out of the cradle: Exploring the frontiers beyond Earth. Workman, New York

Kosmodemyansky AA (1956) Konstantin Tsiolkovsky: His life and works. Foreign Languages Publishing House, Moscow, Russia

Lehman M (1963) This high man: The life of Robert H. Goddard. Farrar-Straus, New York

Miller R, Durant III, FC (2001) The art of Chesley Bonestell. Paper Tiger, Collins and Brown, London, UK

Miller R, Hartmann WK (2005) The grand tour: A traveler's guide to the solar system. Workman, New York

Seeds MA (1997) Foundations of astronomy, 4th edn. Wadsworth Pub. Co

Sorensen TC (1965) Kennedy. Harper and Row, New York

Tsiolkovsky KE (1960) Beyond planet Earth (Translated by Kenneth Sayers). Press, New York

Von Braun W (1953a) Man on the Moon. Sidgwick and Jackson, London, England

Von Braun W (1953b) The Mars project. University of Illinois Press, Urbana, IL

Von Braun W, et al. (1953c) Conquest of the Moon. Viking, New York

Von Braun W (1971) Space frontier, Revised edition. Holt, Rinehart, and Winston, New York

Wagner JK (1991) Introduction to the solar system. Saunders College Pub., Philadelphia, PA

Ward B (2005) Dr. Space: The life of Wernher von Braun. Naval Institute Press, Annapolis, MD

Weissman PR, McFadden L-A, Johnson TV (eds) (1999) Encyclopedia of the solar system. Academic Press, San Diego, CA

Wilhelms DE (1993) To a rocky Moon: A geologists history of lunar exploration. The University of Arizona Press, Tucson, AZ

From Speculation to Understanding

Since time beyond reckoning, humans have observed the daily motions of the Sun and the Moon and wondered about this important spectacle. They were also vitally concerned about the weather, the growth of plants and the ripening of fruits and seeds, the migration of animals, and about many other environmental phenomena that affected their survival. The early humans lived in constant fear because they did not understand and therefore could not predict natural phenomena in their environment. For this reason, humans everywhere invented stories that explained where the Sun went at night, what caused the face of the Moon to change in a regular pattern, and why it rained at certain times and not at others. These stories were passed by the parents to their children and gradually evolved into a coherent body of *mythology* that gave these people a sense of security because it explained the natural phenomena that governed their lives. Mythology explained not only the movement of the Sun and the Moon, but also provided answers about life and death and established rules of conduct which allowed the people to live in harmony with their environment.

2.1 The Geocentric Cosmology of Ancient Greece

Science began during the early Stone Age (Paleolithic Period: 2,500,000 to 200,000 years ago) when humans picked up stones and considered whether they could be used to make a hand ax, or blades for scraping and cutting. Good stones were valuable and therefore were collected and traded. Individuals who knew where good stones could be found and who could make tools of good quality gained the respect of their less knowledgeable companions and thus

rose in the hierarchy of their Stone-Age clans (Faul and Faul, 1983).

From these ancient roots evolved a new way of explaining natural phenomena that was based on observations and deductive reasoning. In time, the accumulated information was interpreted by gifted individuals who passed their insights to their students. The scholars among the Sumerians, the Chinese, the Babylonians, the Egyptians, and the Greeks studied the stars and planets as a basis for tracking the progress of the seasons and as an aid to navigation. For example, *Homer* (9th or 8th century BC) stated that the Bear never bathes, meaning that the constellation Ursa Major (Big Dipper) in Figure 2.1 never sinks into the water of the Mediterranean Sea, which means that it does not set and therefore can be relied upon to identify the North star Polaris.

Later, Greek philosophers developed rational explanations about the apparent motions of heavenly bodies based on their knowledge that the Earth is a sphere and on *Plato's* requirement that the explanations be geometrically pleasing. For example, the Greek geometer *Eudoxus* (440-350 BC) proposed in 385 BC that the Universe consists of 27 concentric translucent spheres that surround the Earth. This model provided rational explanations for the observable motions of the planets. However, the model of Eudoxus failed to explain the more extensive astronomical observations of the Babylonians, who were defeated by Alexander the Great in 330 BC.

Nevertheless, *Aristotle* (384 to 322 BC) adopted Eudoxus' model and elaborated it by increasing the number of transparent spheres to 55 with the Earth positioned at the center. This geocentric model of the Universe was later refined by the astronomer *Claudius Ptolemy* who

Figure 2.1. The constellation Ursa Major (Big Dipper) is a well known feature of the northern sky at night. The stars that form this constellation are located at different distances from the Earth within the Milky Way galaxy. (Reproduced by permission of Paolo Candy)

incorporated it into his book entitled: *Almagest* published about 150 AD (Science Brief 2.5.1). The final version of this theory included epicycles and eccentrics as well as off-center motions of the spheres with the result that the entire contraption reproduced all of the available data concerning stars and planets, not perfectly, but pretty well. A rival theory, which placed the Sun at the center of the Universe, advocated by *Aristarchus* (310 to 230 BC) and a few other Greek astronomers, failed to convince the geocentrists and was forgotten.

This episode in the history of astronomy illustrates how science was done at the time of the Greeks between 2000 and 2500 years ago. Eudoxus had gone to Egypt to study geometry and applied his knowledge to the study of the stars. He set up an astronomical observatory where he mapped the night sky. His observations were crude by present-day standards, but they were state-of-the art about 2500 years ago. He then used these observations and those of his predecessors to construct a hypothesis about the way the Universe is organized. The hypothesis was subsequently tested and was found to be inadequate, especially when the astronomical observations of the Babylonians

became available. Nevertheless, Aristotle and others improved the hypothesis of Eudoxus, which caused Claudius Ptolemy to adopt it (Science Brief 2.5.1). In this way, the geocentric model of the Universe became the leading cosmological theory until the 16th century AD when *Nicolaus Copernicus* (1473 to 1543) reinstated the heliocentric (Sun-centered) model of the Universe. Although the geocentric theory of Eudoxus, Aristotle, and Ptolemy turned out to be wrong, it was originally based on a set of factual observations and measurements that were available in their lifetimes. Similarly, the cosmological theories we accept today may be replaced by better theories in the future if new observations about the Universe indicate that the present theories are inadequate (Gale, 1979; Ferris, 1988).

2.2 The Scientific Method

The process by means of which Eudoxus devised his hypothesis is an early example of the scientific method, which was later demonstrated even more clearly by *Johannes Kepler* (1571–1630 AD) who is widely regarded as the first true scientist. The scientific method is based on the accumulation of precise and accurate measurements concerning the phenomenon under investigation. These measurements must be reproducible and their reliability must be beyond question. The measurements are then used to invent an explanation (called a *hypothesis*) that must take into account all of the available information, but it may differ from the scientific knowledge existing at the time. A given set of data may be explainable by two or more hypotheses, all of which satisfy the existing data. For example, light can be treated as an electromagnetic wave or as a stream of particles called photons (Science Brief 2.5.2, 3).

A hypothesis concerning a particular phenomenon must be tested to determine whether it is an acceptable explanation of the relevant facts. These tests generally arise from predictions made by extrapolating the hypothesis. The predictions are tested either by experiments in the laboratory or by additional measurements, or both. The testing of a hypothesis is carried out in an adversarial manner in order to demonstrate that the hypothesis is incorrect. This phase of the investigation may last several years or even decades and ends only when a general consensus is reached that the hypothesis is, in fact, a valid explanation of the available information. At this point in the process, a hypothesis becomes a *theory* that is added to the fabric of scientific knowledge (Carey, 1998, Cleland, 2001).

The theories that constitute scientific knowledge are judged to be acceptable because they are successful in explaining all known facts concerning natural phenomena even if previously accepted theories pertaining to related phenomena have to be discarded. The beauty of scientific theories is their internal consistency and their success in explaining the world around us, including the nuclei of atoms, the Earth, the solar system, the galaxy in which we live, and the Universe at large. In spite of their success, all scientific theories are subject to change as new observations become available. Therefore, they represent our present understanding concerning natural phenomena rather than the ultimate truth about them (Sagan, 1996; Wong, 2003).

Theories that appear to explain fundamental properties of matter are elevated to the stature of natural *laws*. For example, Isaac Newton expressed the movement of bodies under the influence of gravitational forces by means of three statements that are universally applicable and are therefore known as *Newton's laws of motion*. Similarly, Johannes Kepler described the movement of the planets orbiting the Sun by three laws that govern such motions everywhere in the entire Universe. However, even Newton's and Kepler's laws need to be modified by the theory of relativity of *Albert Einstein* (1879–1955) in cases where the velocities approach the speed of light or where very large masses are present.

The scientific method demands a high degree of honesty from scientists engaged in research. Observations may not be altered to fit a preconceived hypothesis and all relevant facts must be disclosed (Powell, 1998). Similarly, the hypothesis that is derived from a body of data must follow logically from these data even when it contradicts some strongly held convictions of the investigator, or of the sponsor, or of contemporary society. In some cases, scientists have suffered the consequences of expressing unpopular ideas arising from the results of their research.

An example of the conflict between the formal teachings of the Catholic Church of the 16th century and astronomy is the execution of *Giordano Bruno* (1548–1600 AD) by burning at the stake on February 17 of 1600 in Rome. Bruno had openly advocated the heliocentric model of the Universe proposed by Copernicus and had stated that the Earth moves around the Sun. In addition, he claimed that the Universe is infinite. During his trial, which lasted seven years, he refused to retract these and other heretical ideas he had published in his turbulent life. Finally, *Pope Clement VIII* declared him to be an "impenitent and pertinacious heretic". When he was sentenced to death on February 8 of 1600, he is reported to have told the judges of the inquisition: "Perhaps your fear in passing judgement on me is greater than mine in receiving it."

Even the great *Galileo Galilei* (1564–1642 AD) was not spared (Figure 2.2). He invented the astronomical telescope and used it to make observations in 1610 that proved that the Earth moves around the Sun as Copernicus had stated. Subsequently, he wrote a book in which he presented persuasive arguments in favor of Copernican cosmology. Although he

Figure 2.2. Galileo Galilei (1564–1642) was prosecuted for having taught the Copernican doctrine that the Earth and the planets of the solar system revolve around the Sun. (http://www.gap.dcs.st-and.ac.uk/~historyPictDisplay/Galileo.html)

also discussed Ptolemy's geocentric cosmology and although his final conclusions favored Ptolemy (as the Pope had ordered), Galileo's arguments in favor of Copernicus were clearly more persuasive than his token consideration of Ptolemy. Consequently, Galileo was prosecuted in 1633 on suspicion of heresy. He was convicted of having taught the Copernican doctrine and sentenced to serve time in jail, even though he had recanted his teachings, and even though he was 69 years old at the time. Fortunately, *Pope Urban VIII* commuted his sentence to house arrest at his small estate at Arcetri near Florence. Galileo continued his work on physics and astronomy while under house arrest for eight more years until his final illness and death on January 8, 1642 (Hester et al., 2002; O'Connor and Robertson, 2003). Galileo's heretical views on the solar system were partially vindicated on October 31 of 1992 when *Pope John Paul II* officially stated that errors had been made in his trial. The Vatican now has an active astronomical observatory staffed by scientists who are free to publish the results of their research.

2.3 Units of Measurement

Before we can properly describe the Sun and the planets of the solar system, we need to define the units of distance, mass, time, and temperature we will use. These units are in use in most countries in order to facilitate international commerce, standardize manufacturing, simplify global communication, and permit scientific research. The USA is one of the few remaining countries that still uses the traditional measures of distance, mass, volume, and temperature. A summary of units and equations used in astronomy is provided in Appendix Appendix 1.

2.3.1 Distance

The standard international (SI) unit of distance is the *meter* (m). The meter was originally defined in 1791 by the French Academy of Sciences as one ten millionth of the length of the meridian (line of longitude) which extends from the North Pole to the South Pole of the Earth and passes through the city of Paris. Subsequently, the International Bureau of Weights and Measures in 1889 redefined the meter as the distance between two lines on a bar composed of an alloy of platinum (90%) and iridium (10%). In the SI system of units the meter is defined as the distance traveled by light in a vacuum in a time interval of one 299 792 458 th of one second. In spite of the different ways in which the meter has been defined, its length has not changed significantly.

The meter is subdivided or expanded in multiples of ten defined by standard prefixes:

$$\text{kilo (k)} = 10^3 \qquad \text{milli (m)} = 10^{-3}$$
$$\text{deci} = 10^{-1} \qquad \text{micro } (\mu) = 10^{-6}$$
$$\text{centi (c)} = 10^{-2} \qquad \text{nano (n)} = 10^{-9}$$

Accordingly, the hierarchy of units of length based on the meter (m) consists of the following derived units:

$$1 \text{ kilometer (km)} = 10^3 \text{ m} \qquad 1 \text{ millimeter (mm)} = 10^{-3} \text{ m}$$
$$1 \text{ decimeter} = 10^{-1} \text{ m} \qquad 1 \text{ micrometer } (\mu\text{m}) = 10^{-6} \text{ m}$$
$$1 \text{ centimeter (cm)} = 10^{-2} \text{ m} \qquad 1 \text{ nanometer (nm)} = 10^{-9} \text{ m}$$

The conversion of statute miles, feet, and inches used in the USA into SI units of length is based on the relation between inches and centimeters: 1 inch $= 2.54$ cm, 1 foot (12 inches) $= 2.54 \times 12 = 30.48$ cm, and

$$1 \text{ mile} (5280 \text{ feet}) = 30.48 \times 5280 = 160,934 \text{ cm}$$
$$= 1609.34 \text{ m} = 1.60934 \text{ km}.$$

The basic unit of distance in studies of the solar system is the *Astronomical Unit* (AU), which is defined as the average distance between the center of the Earth and the center of the Sun. The astronomical unit is equal to 149.6 million kilometers (i.e., 1 AU $= 149.6 \times 10^6$ km). (See also Glossary). Distances from the Earth to the stars of the Milky Way galaxy and to the other galaxies of the Universe are expressed in units of *lightyears* (ly). One light year is the distance that light travels through the vacuum of interstellar and intergalactic space in one sidereal year at a speed (c) of 2.9979×10^8 m/s.

$$1 \text{ ly} = 2.9979 \times 10^8 \times 365.25 \times 24 \times 60 \times 60$$
$$= 9460 \times 10^{12} \text{ m}$$
$$1 \text{ ly} = 9.46 \times 10^{12} \text{ km. (See Glossary).}$$

When distances are expressed in units of lightyears, they indicate how long the light we see has traveled through space to reach the Earth. Therefore, we see stars and galaxies how they were in the past. In other words, when we look at stars and galaxies in the sky at night, we are looking into the past. For example, the distance to the Andromeda galaxy is 2.3×10^6 ly. Therefore we see it today the way it was 2.3×10^6 years ago.

2.3.2 Time

The basic unit of time in the SI system (as well as in the cgs and mks systems) is the *second* defined as the duration of 9,192,631,770 periods of the radiation corresponding to the transition between the two hyperfine levels of the ground state of the atom of cesium -133. This definition was adopted by the 13th Conference on Weights and Measures in order to ensure that the second defined in this way is indistinguishable from the so-called *ephemeris second* which is based on the motions of the Earth in its orbit around the Sun (Science Brief 2.5.4). The ephemeris second is defined as 1/31,556,925.9747 of the tropical year January 0, 1900 AD, 12h ET. Therefore, 1 tropical year $= 31,556,925.9747$ s. The tropical year is the time that elapses between two passes of the vernal equinox by the Earth in its orbit. The precession of the vernal equinox causes the tropical year to be slightly less than the time it takes the Earth to complete one orbit about the Sun.

In everyday life, time is measured in years, days, hours, and minutes which are defined as follows (Science Brief 2.5.4):

$$\begin{aligned}
1 \text{ minute (min)} &= 60 \text{ s} \\
1 \text{ hour (h)} &= 60 \text{ min} = 3600 \text{ s} \\
1 \text{ day (d)} &= 24 \text{ h} = 86,400 \text{ s}
\end{aligned}$$

Therefore, if $1 \text{ d} = 86,400$ s, one tropical year has:

$$\frac{31,556,925.9747}{86,400} = 365.242199 \text{ days}.$$

By adding one extra day every four years and by other manipulations the calender (Section 7.4), the year is held to 365 days.

The length of the sidereal year, during which the Earth completes one orbit around the Sun and returns to the same place in its orbit as determined with reference to fixed background stars, is:

$$1 \text{ sidereal year} = 365.256366 \text{ days}.$$

The periods of revolution of planets around the Sun and of satellites around their planets are expressed in sidereal years.

2.3.3 Velocity and Speed

The rate of motion of a body can be described by the distance traversed in one second, which is a statement of its velocity or speed. The two words do not have the same meaning because "velocity" has both magnitude and direction, whereas "speed" has magnitude but not direction. For example, a bullet fired out of a gun has velocity because it has direction, while light emitted by a lightbulb travels in all directions at the same speed (Science Brief 2.5.2).

The units of velocity or speed are distance divided by time (e.g., miles per hour, centimeters per second, or meters per second). The rates of motion of bodies in space are expressed in terms of kilometers per second (km/s).

Albert Einstein determined that the speed of light (c) in a vacuum cannot be exceeded by any body traveling in space because the mass of the body increases and becomes infinite when its velocity approaches the speed of light (Section 24.4.3). Therefore, the speed of light is an important fundamental constant of nature. It has been measured carefully and is known to be:

Speed of light (c) $= 2.99792458 \times 10^8$ m/s

In other words, light travels 299.79 million meters in one second and nothing can move faster than this. We begin to appreciate the vast distances between stars in the Milky Way Galaxy from the fact that the distance to the star Proxima Centauri, the Sun's nearest neighbor, is 4.2 lightyears (Section 24.4.3).

2.3.4 Mass

All objects composed of matter posses a property called *mass*. The mass of an object on the surface of the Earth is expressed by its weight, which depends on the magnitude of the gravitational force the Earth exerts on it. Figure 1.6 demonstrates that the weight of an object *decreases* when it is moved away from the Earth, whereas its mass remains constant regardless of where the object is located.

The basic unit of mass in the cgs system is the *gram* (g), which is the weight of one cubic centimeter of pure water at the temperature of 4 °C at a place where the acceleration due to gravity is 980.655 cm/sec/sec. In the SI system of units, mass is expressed in kilograms (kg). The Bureau of Weights and Measures in Paris actually maintains an object composed of platinum whose weight is one kilogram. The gram and kilogram are related by the statement: $1 \text{ kg} = 10^3$ g. In addition, the metric tonne is defined as : $1.0 \text{ tonne} = 10^3$ kg.

The gram is subdivided into:

milligrams (mg) $= 10^{-3}$ g   nanograms (ng) $= 10^{-9}$ g

micrograms (μg) $= 10^{-6}$ g   picograms (pg) $= 10^{-12}$ g

These weights are used in geochemistry to express the concentrations of trace elements in solids such as meteorites (e.g., $5 \mu\text{g/g}$). By the way, one pound (lb) avoirdupois is equal to 453.59 g, whereas one metric pound is defined as 500.00 g.

2.3.5 Temperature

Under normal circumstances, such as on the surface of the Earth, the temperature is a measure of the amount of heat an object contains. In thermodynamics heat is defined as a form of energy that is convertible into work. Heat behaves like an invisible fluid that always flows from a point of higher to a point of lower temperature. Temperature is expressed on the *Celsius scale* (°C), which has been adopted in virtually all countries except in the USA where the *Fahrenheit scale* (°F) is still in use.

On the Celsius scale, the temperature of pure water in contact with ice is set equal to 0 °C (i.e., the freezing temperature of water) and the temperature of pure boiling water at sealevel is 100 °C. The boiling temperature of pure water increases with increasing pressure and it decreases with decreasing pressure (Section 12.8.2). This relationship is the reason why food cooks faster in a pressure cooker than in an open pot (i.e., boiling water reaches temperatures above 100 °C when the pressure is greater than one atmosphere). Similarly, on the summits of high mountains, water boils at less than 100 °C because the pressure exerted on it by the atmosphere is lower than it is at sealevel.

The conversion of temperatures measured in Fahrenheit to the Celsius scale is expressed by the equation:

$$C = \frac{(F - 32)5}{9} \qquad (2.1)$$

For example, the temperature on the Celsius scale of 95 °F is: 95–32 = 63; 63/9 = 7; 7 × 5 = 35 °C.

A third temperature scale called the *Kelvin scale* (K) is used in physical chemistry. The conversion of kelvins to °C is based on the equation:

$$C = K - 273.15 \qquad (2.2)$$

This equation indicates that if K = 0, C = −273.15°. In other words, "absolute zero"

(K = 0) occurs at −273.15 °C. The equation also indicates that C = 0° is equivalent to 273.15 K.

A very different situation occurs in the measurement of temperatures in dilute gases that exist in the space around the Sun and other stars. In such environments the temperature is related to the velocities of the molecules that make up the gas. When the density of the gas is very low, the molecules reach high velocities before colliding with another molecule. The high average velocities of the molecules cause the dilute gas to have a high temperature even though the total amount of heat energy of the gas is low. The so-called kinetic theory of gases also implies that the molecules of a gas (or of any other form of matter) at absolute zero stop moving (i.e., their velocities are equal to 0 m/s).

2.3.6 Pressure

The pressure that is exerted by the atmosphere on the surface of the Earth is measured in *atmospheres* (atm), where 1 atm is the atmospheric pressure at sealevel at the equator. Actually, pressure is expressed more precisely in *bars* where:

$$1 \text{ bar} = 0.987 \text{ atm} \qquad (2.3)$$

One atmosphere is also equal to 759.968 mm of mercury (Hg) or 29.92 inches of Hg, which is the unit of pressure used in barometers in the USA.

The unit of pressure in the SI system of units is the pascal (Pa) which is defined by the relation:

$$1 \text{ Pa} = 10^{-5} \text{ bar} \qquad (2.4)$$

The pascal is the preferred unit of pressure in science, but pressure is still expressed in atmospheres in planetary science because it is easy to visualize what it means. The nature of light and its wavelength spectrum are presented in Science Briefs 2.5.2 and 2.5.3.

2.4 Summary

Science has evolved from the efforts of some of the oldest human civilizations to explain the world by developing natural philosophies about important aspects of their environment. Geometry and astronomy were among the first subjects to be investigated because they were useful to the people at the time. For example, the philosophers of ancient Greece, more than 2000 years ago, developed the geocentric model of the Universe. This model was later adopted by the Catholic Church, which thereby stifled research in astronomy for nearly 1500 years.

The scientific method, by means of which knowledge is generated, evolved in the 16th century AD from the work of Nicolaus Copernicus, Tycho Brahe, Johannes Kepler, and Galileo Galilei who derived conclusions by the rational interpretation of their own measurements, or of those of their contemporaries, instead of relying on the teachings of Aristotle and Ptolemy, as was customary at that time. The reinterpretation of the movement of the planets of the solar system by the heliocentric model of Copernicus caused a revolution that led to the rebirth of science and of the arts (i.e., the Renaissance).

The scientific method is based on factual observations and accurate measurements of natural phenomena. These observations and measurements concerning a particular phenomenon must be reproducible and therefore are not in dispute. The explanations that are proposed to explain the observations (hypotheses) are tested by means of experiments, or by additional measurements, or both. The goal of testing hypotheses is to falsify them by demonstrating that they are wrong. This process continues until a consensus develops in the community of scientists that one of the original hypotheses cannot be falsified and therefore is acceptable as a theory. Such theories may be revised later when new observations or measurements become available. Therefore, scientific knowledge consist of a body of theories each of which is an acceptable explanation of a set of factual observations and measurements at any given time.

The observations from which new hypotheses and theories are derived are expressed in terms of units of measurements. Scientists in all countries use a set of carefully defined standard international (SI) units which enable scientists from all countries to communicate their results unambiguously.

The journals in which reports about the solar system are published include: Meteoritics and Planetary Science, Journal of Geophysical

Research, Reviews of Geophysics and Space Physics, Nature, Geochimica et Cosmochimica Acta, Earth and Planetary Science Letters, Space Science Reviews, Icarus, Science, as well as Astronomy and Sky and Telescope.

2.5 Science Briefs

2.5.1 The Cosmology of Ptolemy

Ptolemy was born during the second century AD in Ptolemais on the river Nile. He became an astronomer and mathematician at the astronomical observatory in Canopus located about 24 km east of the city of Alexandria in Egypt. Ptolemy published a book in which he presented an elaborate model of the motions of the Sun, the Moon, and the planets based on the geocentric cosmology of the Greek astronomers. In the 9th century AD the Arab astronomers who used Ptolemy's book referred to it by the Greek word "megiste", which means "greatest". Therefore, the book became known as "al megiste" which was eventually corrupted to "Almagest", meaning "The Greatest." Ptolemy's statement that the Earth is fixed at the center of the Universe was adopted as dogma by the Catholic Church.

2.5.2 The Nature of Light

Electromagnetic waves have certain properties such as wavelength, period, and speed, where the wavelength is the distance between two successive peaks in the wave, the period is the time required for two successive peaks to pass a fixed point, and the velocity of the wave is the ratio of the wavelength divided by the period. All kinds of waves transport energy as they travel away from their source. The energy of a ray of light is equivalent to a certain quantity of mass in accordance with Einstein's famous equation:

$$E = mc^2$$

where (E) is the energy, (m) is the mass, and (c) is the speed of light. Therefore, a quantum of energy of an electromagnetic wave can be expressed as the equivalent mass of a photon.

2.5.3 Electromagnetic Wavelength Spectrum

The wavelengths of electromagnetic waves range continuously from hundreds of meters to fractions of one nanometer (nm). Light that is visible to human eyes has a narrow range of wavelengths between 400 (blue) and 750 (red) nanometers. The characteristics of electromagnetic waves whose wavelengths are longer and shorter than those of visible light are indicated in Table 2.1 The energy of electromagnetic radiation increases in proportion to the reciprocal of the wavelength. In other words, the shorter the wavelength the higher is the energy. Therefore, gamma rays are much more energetic than x-rays, and ultraviolet light is more energetic than visible light.

2.5.4 Origin of the Divisions of Time

About 4400 years ago, the Sumerians who lived in Mesopotamia (presently southern Iraq) determined that a year had approximately 360 days. Consequently, they divided the year into 12 months of 30 days each for a total of 360 days. About 900 years later, the Egyptians divided the day into 24 hours. Still later, between 300 and 100 BC, the Babylonians divided hours into 60 minutes and minutes into 60 seconds.

The sexagesimal system of numbers used by the Babylonians has the advantage that the number 60 is divisible by 12 different numbers: 1, 2, 3, 4, 5, 6, 10, 12, 15, 20, 30, and 60. Therefore, when 60 is divided by any of these numbers, the result is a whole number: $60/2 = 30$, $60/3 = 20$, $60/4 = 15$, $60/5 = 12$, etc.

Table 2.1. The wavelength spectrum of electromagnetic waves

Wavelength	Character
$> 10^3 - 1$ m	radio
$1 - 0.1$ m	television
$10 - 0.1$ cm	microwave
$100 - 0.75\,\mu$m	infrared
$750 - 400$ nm	visible
$400 - 10$ nm	ultraviolet
$10 - 0.1$ nm	x-rays
$0.1 - 0.001$ nm	gamma rays

One nanometer is equal to one billionth of a meter ($1\,\text{nm} = 10^{-9}\,\text{m}$).

Accordingly, the roots of the units of time in use today extend about 4400 years into the past (i.e., 2400 BC) to the culture of the Sumerians and to later civilizations of the Egyptians, Babylonians, and Greeks who introduced these units to the peoples living in the Mediterranean region. Subsequently, the Romans continued to use the ancient units of time throughout their empire.

2.6 Problems

1. Express 1 kilometer in terms of millimeters.
2. Convert 1 light year into the corresponding number of astronomical units. (Answer: 63,235 AU).
3. How long would it take a spacecraft to reach the star Proxima Centauri assuming that the speed of the spacecraft is 1000 km/s and that the distance to Proxima Centauri is 4.2 ly? Express the result in sidereal years. (Answer: 1259 years).
4. Derive an equation for the conversion of temperatures on the Fahrenheit to the Kelvin scale.
5. Write a brief essay about the life and scientific contributions of one of the following pioneers of astronomy.

a. Galileo Galilei   d. Albert Einstein
b. Johannes Kepler   e. Edwin Hubble
c. Isaac Newton   f. Carl Sagan

2.7 Further Reading

Carey SS (1998) A beginner's guide to the scientific method, 2nd edn. Wadsworth Pub. Co., Belmont, CA

Cleland CE (2001) Historical science, experimental science, and the scientific method. Geology 29(11):987–990

Faul H, Faul C (1983) It began with a stone: A history of geology from the Stone Age to the age of plate tectonics. Wiley, New York

Ferris T (1988) Coming of age in the Milky Way. Anchor Books, Doubleday, New York

Gale G (1979) Theory of science: An introduction of the history, logic, and philosophy of science. McGraw-Hill, New York

Hester J, Burstein D, Blumenthal G, Greeley R, Smith B, Voss H, Wegner G (2002) 21st century astronomy. Norton, New York

O'Connor JJ, Robertson EF (2003) Galileo Galilei. http://www-gap.dcs.stand.ac.uk/~history/Mathematicians/Galileo.html

Powell JL, (1998) Night comes to the Cretaceous. Harcourt Brace and Co., San Diego, CA

Sagan C (1996) The demon-haunted world: Science as a candle in the dark. Random House, New York

Wong K (2003) Stranger in a new land. Sci Am 289(5):74–83

The Planets of the Solar System

The solar system is dominated in all respects by the Sun, which actually is a fairly ordinary star among the hundreds of billions of stars that make up the Milky Way galaxy. Our galaxy is one of untold billions of galaxies and star clusters that populate the Universe. These statements make clear that our Sun and the planets of the solar system comprise only a very small part of the Milky Way galaxy and that our galaxy itself is like a grain of sand on a beach that stretches as far as we can see, and beyond. These statements also tell us that the Earth and all of us who live here are a part of the Universe. In other words, the Universe is not just an immense entity composed of space and galaxies of stars, but it also includes our living rooms and backyards were we live. In fact, the Universe includes us and all living organisms on the Earth.

When viewed from the Earth, the Sun is a radiant disk that rises every morning across the eastern horizon, moves in an arc across the sky, and sets in the evening by appearing to sink below the western horizon. These motions give the mistaken impression that the Sun orbits the Earth. In reality, the Earth and all of the other planets of the solar system orbit the Sun. The familiar daily motions of the Sun across the sky are caused by the rotation of the Earth about a virtual axis that is located at the North and South geographic poles. Many other aspects of the solar system that are apparent to a casual observer are also not the way they appear to be.

Our experiences on the Earth have given us a sense of scale regarding size, distance, and time that is completely misleading in the context of the solar system, not to mention the Milky Way galaxy, and the Universe as a whole. For example, we consider the Earth to be a large planet, but its volume is more than one million times smaller than the volume of the Sun. The average distance between the Sun and the Earth is 149.6 million kilometers, which sounds like a long distance, but it actually places the Earth very close to the Sun within the solar system which extends to more than 50 AU from the Sun, not to mention the Oort cloud which extends to 100,000 AU. We also think that 100 years is a long time and celebrate the fact that we now live in the twenty-first century. Even if we consider that recorded history began about 10,000 years ago, only 100 centuries have passed. Actually, one century is an insignificant length of time compared to the age of the solar system, which is 46 million centuries old (i.e., 4.6 billion years). Even our distant ancestors (Homo erectus), mentioned in Section 1.1, lived 18,000 centuries ago.

Although the 20 centuries of the most recent human history represent only a tiny fraction of time in the history of the solar system, the population of the Earth has increased to more than six billion during this period of time. In addition, we have explored the Earth and are now exploring the solar system. Humans have certainly accomplished a lot in the past 2000 years! Visionaries like Konstantin Tsiolkovsky of Russia predicted that in the fullness of time humans will not only explore and occupy the solar system, but will colonize other planets that orbit "nearby" stars located several lightyears from the Sun. (See Chapter 24).

3.1 The Sun and the Planets of the Solar System

The celestial bodies that populate the solar system have been classified into three categories that were defined on August 24, 2006, by the

International Astronomical Union (IAU) at an assembly in Prague:

Planets are celestial bodies that orbit the Sun, have sufficient mass for their self-gravity to overcome rigid-body forces so that they assume a spherical shape, and they have "cleared the neighborhood around their orbits."

Dwarf planets are less massive than planets although they do have spherical shapes, they have not "cleared the neighborhood around their orbits," and they are not satellites.

Small Solar-System Bodies orbit the Sun but do not have sufficient mass to achieve spherical shapes and they have not "cleared the neighborhood around their orbits."

The practical consequences of these new definitions are that the solar system has only eight planets: Mercury, Venus, Earth, Mars, Jupiter, Saturn, Uranus, and Neptune.

Pluto and the asteroid Ceres are now classified as dwarf planets, which is a new category that will eventually include the largest of the spherical ice bodies that occur in the Edgeworth-Kuiper belt in the space beyond the orbit of Neptune (e.g., 2003 UB 313 described in Chapter 21).

All of the asteroids (except Ceres), centaurs, comets, and the small ice objects that occur in the Edgeworth-Kuiper belt and in the Oort cloud are now classified as small solar-system objects.

The new definitions adopted by the IAU do not specifically address objects that orbit the planets of the solar system except to exclude

them from the category of dwarf plants because they do not orbit the Sun. However, several satellites of Jupiter and Saturn have larger diameters than the planet Mercury. Therefore, we will continue to use the term "*satellite*" to describe celestial bodies of any size and shape that orbit any of the eight regular planets of the solar system and we reserve the name "Moon" for the satellite of the Earth. The orbits of the planets and asteroids are confined to a narrow region of space aligned approximately with the equator of the Sun and with the plane of the orbit of the Earth. The orientations of the orbits of the comets range more widely than those of the planets such that comets may enter the solar system both from above and from below the planetary disk.

Mercury and Venus are the only planets that do not have at least one satellite. The number of known satellites in the solar system has increased from 59 in 1991 (Wagner, 1991) to more than 120 because many additional satellites of Jupiter, Saturn, Uranus, and Neptune have been discovered recently with modern astronomical telescopes and by robotic spacecraft (e.g, the Voyagers, Galileo, and Cassini). The newly discovered satellites have small diameters, are far removed from their home planets, and have irregular orbits.

The best way to describe the solar system is by examining certain physical properties (Table 3.1) that characterize each of the planets. The physical properties to be considered here are the distances of the planets from the Sun, the bulk densities

Table 3.1. Physical properties of the planets (Beatty et al., 1999; Hartmann, 2005; Hester et al., 2002)

Planet	Distance from Sun, AU	Radius, km	Mass, 10^{24} kg	Density, g/cm^3	Average surface temperature, °C
Sun	–	695,510	1,989,000	1.410	+5507
Mercury	0.3871	2440	0.3302	5.43	+167 (−173 to +452)
Venus	0.7233	6052	4.865	5.20	+464
Earth	1.0000	6378	5.974	5.52	+15 (−90 to +58)
Mars	1.5237	3396	0.6419	3.91	−33 (−140 to +20)
Ceres	2.768	457	0.0012	2.3	
Jupiter	5.2026	71,492	1898	1.33	−123 to −153
Saturn	9.5549	60,268	568.5	0.69	−113 to −153
Uranus	19.2184	25,559	86.83	1.318	−195*
Neptune	30.1100	24,766	102.4	1.638	−204*
Pluto	39.5447	1150	0.0132	2.0	−236

* Temperature where the atmospheric pressure is 1.0 bar. Pluto and Ceres are included in this table even though they are dwarf planets.

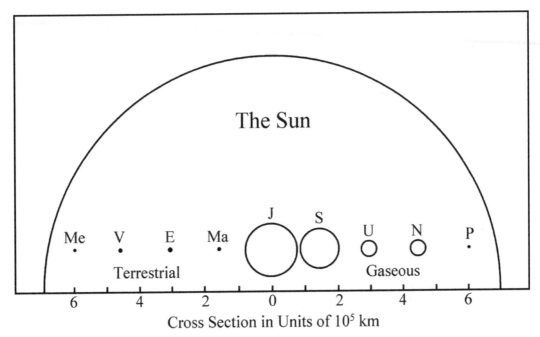

Figure 3.1. Comparison between the diameter of the Sun and those of the planets of the solar system (including Pluto) drawn to scale. The planets and their diameters expressed in kilometers are : Me = Mercury (4880); V = Venus (12,104); E = Earth (12,756); Ma = Mars (6792); J = Jupiter (142,984); S = Saturn (120,536); U = Uranus (51,118); N = Neptune (49,532); P = Pluto (2,300). The diameter of the Sun is 1,391,020 km

of the planets, their average orbital velocities, and their surface temperatures. The existence of patterns in these properties implies the operation of a process that has affected all of the planets in the solar system. In addition, these patterns provide a basis for recognizing deviations from the norm, which are clues to the occurrence of events that may have locally disturbed the patterns.

The data in Table 3.1 indicate that the radius of the Sun is 695,510 km, that its mass is $1,989,000 \times 10^{24}$ kg, its bulk density is 1.410 g/cm^3, and its surface temperature (photosphere) is 5507 °C. Although the Earth is the most massive terrestrial planet (5.974×10^{24} kg), its mass is only a tiny fraction of the mass of the Sun (i.e., 0.00030 %). Even Jupiter, the largest of all the planets in the solar system with a mass of 1898×10^{24} kg, pales by comparison with the Sun (i.e., 0.0954%). The sum of the masses of all of the planets of the solar system (2667.5×10^{24} kg) comprises only 0.134 % of the mass of the Sun. Evidently, the planets of the solar system are

mere crumbs left over during the formation of the Sun. This point is illustrated in Figure 3.1 which shows the planets in relation of the Sun drawn to the same scale.

3.2 The Titius-Bode Rule

The Titius-Bode rule generates a series of numbers that appear to match the average distances of the planets from the Sun expressed in astronomical units. The rule was published in 1772 by two German astronomers named Johann Titius and Johann Bode. The procedure for generating this series of numbers is:

1. Write a string of numbers starting with zero: 0, 3, 6, 12, 24, 48, 96, 192, 384, 768...
2. Add 4 to each number and divide by 10.
3. The result is: 0.4, 0.7, 1.0, 1.6, 2.8, 5.2, 10.0, 19.6, 38.8, and 77.2.

The actual distances of the planets (i.e. the average radii of their orbits) are compared to the Titius-Bode predictions in Table 3.2 and Figure 3.2.

Table 3.2. Average radii of planetary orbits compared to values predicted by the Titius-Bode rule published in 1772

Planet	Avg. radius of orbit, AU	Titius-Bode radius, AU	Year of discovery
Mercury	0.39	0.4	antiquity
Venus	0.72	0.7	antiquity
Earth	1.00	1.0	antiquity
Mars	1.52	1.6	antiquity
Ceres	2.77	2.8	1801
Jupiter	5.20	5.2	antiquity
Saturn	9.56	10.0	antiquity
Uranus	19.22	19.6	1781
Neptune	30.11	38.8	1846
Pluto*	39.54	77.2	1930

* Pluto was reclassified as a dwarf planet in August of 2006.

The Titius-Bode rule predicted the existence of a planet at a distance of 2.8 AU from the Sun between the orbits of Mars and Jupiter.

At the time the rule was published, no such planet was known to exist in this region of the solar system. However, when the asteroid *Ceres* was discovered on January 1, 1801, by Giuseppe Piazza at the predicted distance from the Sun, the Titius-Bode rule appeared to be confirmed. However, the solar distances of Uranus, Neptune, and Pluto, which were discovered later (e.g., Uranus in 1781) deviate significantly from their predicted values.

The good agreement between the predicted and observed solar distances of the inner planets in Figure 3.2 suggests that they formed by a process that caused the spacing of their orbits to follow a pattern that is duplicated by the Titius-Bode rule. However, the rule does not identify this process. The discrepancies between the actual radii of the orbits of Uranus, Neptune, and Pluto and their radii predicted by the Titius-Bode rule may indicate that the orbits of these planets were altered after their formation.

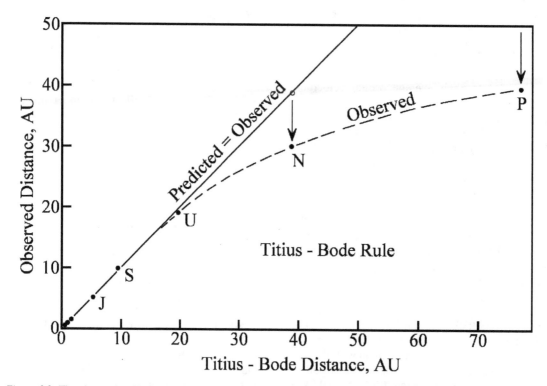

Figure 3.2. The observed radii of the orbits of the terrestrial planets (Mercury, Venus, Earth, Mars, and Ceres) as well as those of Jupiter and Saturn agree with the distances predicted by the Titius-Bode rule. However, the observed average radii of the orbits of Uranus, Neptune, and of the dwarf planet Pluto are progressively shorter than the radii predicted by the Titius-Bode rule

However, it is probably best not to read too much into the deviations from the Titius-Bode rule because the rule is not considered to be a valid description of the spacing of planets in the solar system.

3.3 Average Orbital Velocities

A planet in a stable circular orbit around the Sun must move with a velocity which is sufficiently large to generate a centrifugal force that is equal to the force of gravity exerted by the Sun (Section 1.3). Since the gravitational force acting on a planet located close to the Sun is greater than the force acting on a more distant planet, the average orbital velocity of planets close to the Sun must be higher than the orbital velocity of more distant planets. An equation based on this principle was derived in Science Brief 1.5.3.

The average orbital velocities of the planets of the solar system in Figure 3.3 vary as expected because Mercury (0.3871 AU) has the highest average orbital velocity of 47.87 km/s, whereas the dwarf planet Pluto (39.5447 AU) moves at the lowest velocity of only 4.75 km/s. The average orbital velocity of the Earth (1.0 AU) is 29.79 km/s, which is equivalent to 107,244 km/h or 67,027.5 miles per hour. That is more than one thousand times faster than the speed limit for cars on the Interstate Highways of the USA (65 miles/h).

The observed pattern of variation of the average orbital velocities of the planets in the solar system in Figure 3.3 applies to all objects in stable circular orbits around the Sun. The velocity of any object orbiting the Sun depends only on the average radius of the orbit and on the mass of the Sun but not on the mass of the orbiter. For example, the average velocity of Ceres, which has a much smaller mass than Jupiter, is 17.87 km/s as expected for an object orbiting the Sun at an average distance of 2.768 AU. (Science Brief 3.7.1).

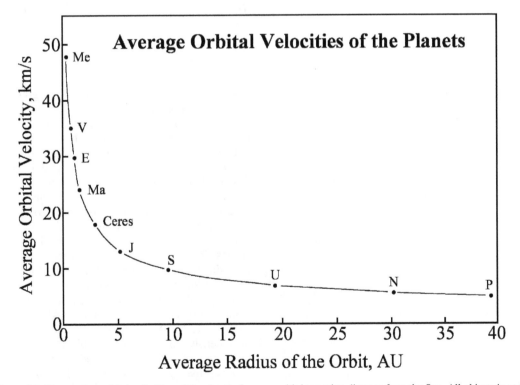

Figure 3.3. The average orbital velocities of the planets decrease with increasing distance from the Sun. All objects in stable circular orbits around the Sun obey this relationship demonstrated in Section 1.3

3.4 Surface Temperatures

The surface temperature of a planet depends primarily on its distance from the Sun and on the presence of an atmosphere containing so-called greenhouse gases such as carbon dioxide (CO_2), water vapor (H_2O), and methane (CH_4). In addition, minor amounts of heat originate in the interiors of some of the planets and are brought to the surface by volcanic activity (e.g., Earth) and by convection of the atmosphere (e.g., Jupiter).

The average surface temperatures of the terrestrial planets in Table 3.1 and in Figure 3.4A decrease with distance from the Sun as expected and range from $+167\,°C$ on Mercury to $-33\,°C$ on Mars. The average surface temperature of Venus ($+464\,°C$) is anomalously high because this planet has a dense atmosphere composed primarily of carbon dioxide which absorbs infrared radiation. The pattern of temperature variations suggests that, without greenhouse warming, the temperature of Venus should only

be about $+75\,°C$. The surface temperature of Mercury has an extreme range of 625 degrees (from $+452°$ to $-173\,°C$) because Mercury lacks an atmosphere and rotates slowly on its axis allowing extreme heating and cooling. The observed range of surface temperatures on Earth is 148 degrees between extremes of $+58°$ and $-90\,°C$. The range of temperatures on Mars is 160 degrees ranging from $+20°$ to $-140\,°C$. The average temperature of Ceres can be inferred from the average radius of its orbit (2.768 AU) in Figure 3.4B to be about $-75\,°C$.

The surface temperatures of the gas planets in Table 3.1 and in Figure 3.4B refer to the level in their atmospheres where the pressure is 1.0 bar. These temperatures decrease with increasing distance from the Sun starting with $-138\,°C$ on Jupiter, $-139\,°C$ on Saturn, $-195\,°C$ on Uranus, and $-204\,°C$ on Neptune. The temperature on the surface of the dwarf planet Pluto is $-236°$.

3.5 Bulk Densities

The density of a solid object is defined by the relation:

$$\text{Density} = \frac{\text{Mass}}{\text{Volume}}\ kg/m^3 \text{ or } g/cm^3. \quad (3.1)$$

The density of an object is the weight in grams of one cubic centimeter of the material of which the object is composed. Consequently, the density of celestial object depends on its chemical composition and on the minerals of which it is composed. The specific gravity is defined in Science Brief 3.7.4 and in the Glossary.

For example, igneous, sedimentary, and metamorphic rocks on the Earth are composed of certain minerals having a wide range of densities depending on their chemical compositions. The minerals listed in Table 3.3 have densities that range from $0.917\,g/cm^3$ for water ice to $5.26\,g/cm^3$ for hematite (Fe_2O_3). The densities of metals range even more widely from $2.69\,g/cm^3$ for aluminum (Al) to 19.3 for gold (Au). Accordingly, a planet having a bulk density of $3.91\,g/cm^3$ could be composed of a core of metallic iron (Fe) surrounded by a rocky mantle of peridotite rocks (augite + olivine + magnetite). The significant aspect of this deduction is that the planet in question has a

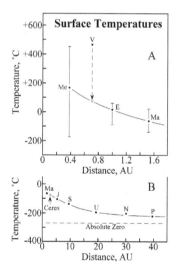

Figure 3.4. A. The average surface temperatures of the terrestrial planets decrease with increasing distance from the Sun. The surface temperature of Venus ($+464\,°C$) deviates from the pattern because the carbon dioxide of its atmosphere absorbs infrared radiation (i.e., extreme greenhouse warming). B. The temperatures at a pressure of 1.0 bar in the atmospheres of the gas planets likewise decrease with distance from the Sun and reach very low values. The surface temperature of the dwarf planet is $-236\,°C$ which is only 37 degrees above absolute zero

Table 3.3. Densities of common terrestrial minerals and selected metals

Mineral	Composition	Density, g/cm^3
Minerals		
Ice	H_2O	0.917
Quartz	SiO_2	2.65
Orthoclase	$KAlSi_3O_8$	2.57
Albite	$NaAlSi_3O_8$	2.62
Anorthite	$CaAlSi_2O_8$	2.76
Olivine	$(Mg, Fe)_2SiO_4$	3.27–3.37
Augite	Ca,Mg,Fe,Al silicate	3.2–3.4
Pyrite	FeS_2	5.02
Magnetite	Fe_3O_4	5.18
Hematite	Fe_2O_3	5.26
Metals		
Aluminum	Al	2.69
Iron	Fe	7.87
Nickel	Ni	8.90
Copper	Cu	8.96
Lead	Pb	11.35
Gold	Au	19.3

core of metallic iron (or of iron sulfide) because its density exceeds that of peridotite. This line of reasoning indicates that the Moon in Table 3.4 can only have a small iron core because its bulk density of $3.34 \, g/cm^3$ is within the range of the minerals that form peridotite (i.e., augite + olivine + minor amounts of magnetite). The volume and radius of the rocky core of Pluto are estimated in Science Briefs 3.7.2 and 3 based on the planetary bulk density of $2.0 \, g/cm^3$ and on some plausible assumptions.

Table 3.4. Comparison between observed bulk densities and uncompressed densities of the terrestrial planets (Seeds, 1997, p. 418; Freedman and Kaufmann, 2002)

Planet	Density, g/cm^3	
	Bulk	Uncompressed
Mercury	5.44	5.4
Venus	5.24	4.2
Earth	5.52	4.2
Moon	3.34	3.35
Mars	3.93	3.75

The density of matter depends not only on its chemical composition but also on the temperature and pressure to which it is subjected. An increase in temperature causes the volume of most materials to increase, whereas pressure reduces the volume of even seemingly incompressible materials such as rocks and metallic iron. Therefore, an increase in pressure causes the density of matter to rise while an increase in temperature causes it to decrease. The net increase of the density of the interior of the Earth in Figure 3.5 reflects the increases of both pressure and temperature as well as changes in the chemical composition of the material. In addition, the minerals that make up the rocks of the mantle of the Earth recrystallize under the influence of pressure to form new minerals that have smaller volumes and greater densities than the minerals that are stable at the surface of the Earth.

Figure 3.5 also demonstrates the abrupt change in density at the boundary between the rocks of the mantle and the liquid iron of the outer core of the Earth. The pressure exerted by the weight of the overlying mantle and crust causes the density of iron in the core of the Earth to rise to more than $12 \, g/cm^3$ compared to only $7.87 \, g/cm^3$ at $20°C$ at the surface of the Earth. Consequently, the bulk density of the Earth $(5.52 \, g/cm^3)$ calculated from its mass and its volume is higher than its so-called uncompressed density $(4.2 \, g/cm^3)$, which has been corrected for the compression of the mantle and core. For the same reason, the uncompressed densities of the terrestrial planets and the Moon in Table 3.4 are all lower than their bulk densities. Although the uncompressed densities are more representative of the internal structure and chemical composition of planets, the bulk densities of the planets are based on direct observations of their masses and volumes.

The bulk densities of the terrestrial planets (Mercury, Venus, Earth, Mars, and Ceres) in Figure 3.6A decrease in a regular pattern with increasing distance from the Sun expressed in astronomical units (AU). The differences in the bulk densities of these planets are caused by corresponding differences in the chemical compositions of the terrestrial planets. In addition, the pattern of density variation suggests that the terrestrial planets originally formed in orbit around the Sun (i.e., they were

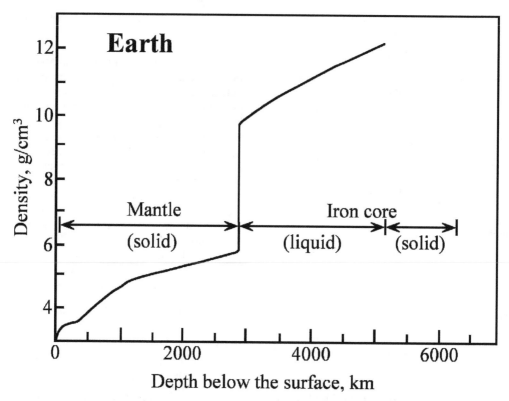

Figure 3.5. The density of the mantle and outer core of the Earth are caused by the increasing pressure as well as by changes in the chemical composition (e.g., rocky mantle and metallic iron core) and by pressure-induced recrystallization of minerals in the mantle to form new minerals having lower volumes and higher densities than the common rock-forming minerals at the surface of the Earth (e.g., plagioclase feldspar, olivine, and pyroxene). Adapted from Bullen (1963)

not captured by the Sun after having originated elsewhere in the solar system).

The pattern of density variations of the terrestrial planets in Figure 3.6A also reveals the existence of three anomalies:

1. The bulk density of the Earth ($5.52\,g/cm^3$) is greater than expected for a planet located at a distance of 1.0 AU from the Sun.
2. The density of the Moon ($3.34\,g/cm^3$) is less than that of the Earth and lower than expected for a body that formed in orbit around the Earth.
3. The densities of the satellites of Mars (Phobos, $1.9\,g/cm^3$ and Deimos, $1.8\,g/cm^3$) are less than those of Mars ($3.91\,g/cm^3$) and deviate from the pattern of planetary densities in Figure 3.6A, but their densities are consistent with those of stony asteroids.

The anomalously high bulk density of the Earth and the comparatively low density of the Moon imply a significant difference in their chemical compositions. A reasonable hypothesis for the low density of the Moon would be that it formed in the space between the orbits of Mars and Jupiter (as suggested in Figure 3.6A) and was later captured by the Earth. This hypothesis has been rejected because the Earth and the Moon would have broken up by gravitational interaction (i.e., tides). Although the resulting fragments could have reassembled to reform the Earth, it is by no means certain that the reassembly of the Earth would have allowed a satellite to form, or that such a satellite, if it formed, would have had the chemical composition and internal structure of the Moon as we know it. The question concerning the origin of the Moon and the reason for the anomalously high density of the Earth will be considered in Chapter 9.

The satellites of Mars consist of rocks composed of silicate minerals, they have small

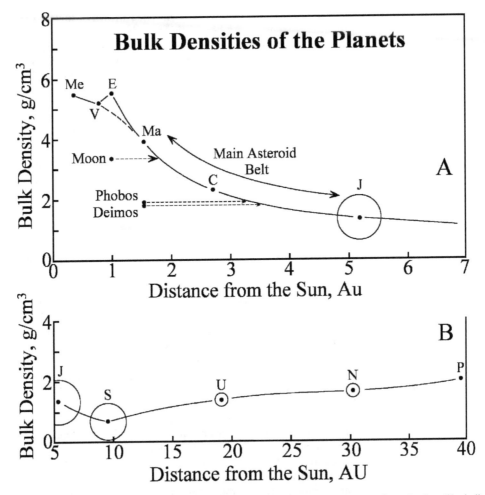

Figure 3.6. A. The bulk densities of the terrestrial planets decrease with increasing distance from the Sun. The bulk density of Earth is anomalously high, whereas that of the Moon is anomalously low as are the densities of the satellites of Mars (Phobos and Deimos). The bulk density of the gas planet Jupiter is noticeably lower than that of the terrestrial planets. B. The bulk densities of the gas planets are similar to that of Jupiter and vary only within narrow limits. Pluto is a solid object with a bulk density of about $2\,g/cm^3$ and has been reclassified as a dwarf planet

masses compared to Mars (Phobos: $1.08 \times 10^{16}\,kg$; Deimos: $1.80 \times 10^{15}\,kg$; Mars: $6.419 \times 10^{23}\,kg$), and they have irregular shapes. Their low masses and irregular shapes combined with their low densities can be explained by the hypothesis that they are asteroids that were captured by Mars.

The bulk densities of Jupiter, Saturn, Uranus and Neptune in Figure 3.6B are all less than $1.64\,g/cm^3$ and vary only from $0.69\,g/cm^3$ (Saturn) to $1.638\,g/cm^3$ (Neptune). The low bulk densities arise from the thick atmospheres

composed of hydrogen (H) and helium (He) that surround the comparatively small cores of rocky material of these planets. The low densities and thick atmospheres of the gas planets distinguish them from the terrestrial planets which have high bulk densities characteristic of the presence of iron cores.

Pluto, located at an average distance of $39.54\,AU$ from the Sun, differs from the planets of the solar system because it is a solid body composed primarily of water ice (H_2O). Its density of about $2.0\,g/cm^3$ indicates that it has a core, which is presumed to be composed of rocky material having a density of about $3.0 \, g/cm^3$ (Table 3.1). The radius of the rocky core of Pluto can be estimated from the densities of ice $(0.917 \, g/cm^3)$ and rocks $(3.0 \, g/cm^3)$ and from the known bulk density of Pluto. (Science Brief, 3.7.2 and 3).

3.6 Summary

The masses and volumes of the planets of the solar system are very small compared to the mass and volume of the Sun. All of the planets are under the gravitational influence of the Sun even though some of the planets are very far from it. The spacing of the orbits of the planets is represented by a series of numbers generated by the Titius-Bode rule. The match between the predicted and observed radii for the terrestrial planets is remarkably good, but discrepancies of increasing magnitude occur for Uranus, Neptune, and Pluto because the observed orbital radii are shorter than the predicted values. Therefore, the Titius-Bode rule is probably not valid although it suggests that the spacing of the planets in the solar system reflects the process by which they formed.

The average orbital velocities of the planets likewise decrease with increasing average radii of their orbits as required by dynamical considerations in Section 1.3. This relationship results from the requirement that an object in a stable circular orbit must move with a velocity that generates a centrifugal force large enough to balance the gravitational force of attraction exerted by the Sun.

The surface temperatures of the planets in the solar system also decrease with increasing distance from the Sun. This relationship is modified in the case of the terrestrial planets (except Mercury) by the presence of atmospheres and by greenhouse warming. The temperatures of the gas planets refer to the surface where the atmospheric pressure is equal to 1.0 bar. These temperatures decline to low values and reach $-204 \, °C$ in the atmosphere of Neptune.

The bulk densities of the planets decrease with increasing distance from the Sun. This pattern of variation indicates that the chemical compositions of the planets differ because of the chemical differentiation of the solar system during its formation. The terrestrial planets as a group are enriched in heavy metals (e.g., iron), whereas the gas planets have thick atmospheres composed primarily of hydrogen and helium. The bulk density of the dwarf planet Pluto $(2 \, g/cm^3)$ indicates that it contains a core of rocky material (about $3 \, g/cm^3$) surrounded by a thick mantle of water ice $(0.917 \, g/cm^3)$.

The patterns of variations of the parameters considered in this chapter relate the planets of the solar system to each other and to the Sun and thereby promote the thesis that the solar system is a coherent unit rather than a collection of individual objects.

3.7 Science Briefs

3.7.1 Average Orbital Velocity of Ceres

Before we demonstrate how the average orbital velocity of Ceres (or of any other planet in the solar system) is calculated, we briefly review the geometry of circles: A circle is a geometric figure consisting of a line (the circumference) that maintains a constant distance (r) from a point located at its center. The length of the circumference of a circle divided by its diameter (2r) is a constant known as π (pi) whose value is 3.14159.... (π = circumference/diameter).

Consequently, the circumference of a circle is $2r\pi$ and the area (A) of a circle is $A = r^2\pi$.

In order to calculate the orbital velocity of Ceres, we make the simplifying assumption that the orbit of Ceres is a circle. Therefore, the average orbital velocity (v) is:

$$v = \frac{2r\pi}{t} \qquad (3.2)$$

where r = radius of the orbit in kilometers
       t = period of revolution in seconds
The radius (r) of the orbit of Ceres is 2.768 AU or $414.09 \times 10^6 \, km$ and its period of revolution (t) is 4.61 years or $4.61 \times 365.256 \times 24 \times 60 \times 60 = 1.4548 \times 10^8 \, s$.

Therefore, the average velocity (v) of Ceres in its orbit around the Sun is:

$$v = \frac{2 \times 414.09 \times 10^6 \times 3.14}{1.4548 \times 10^8} = 17.87 \, km/s$$

The average orbital velocity of Ceres is consistent with the pattern of variation of the average orbital velocities of the planets in the solar system in Figure 3.3.

3.7.2 The Volume of the Rocky Core of Pluto

A sphere is a three-dimensional body whose surface maintains a constant distance (r) from its center. The volume (V) of a sphere is $V = (4/3)\pi r^3$ and its surface (S) is $S = 4\pi r^2$. (Appendix Appendix 1).

The mass of Pluto (M_p) is the sum of the masses of the core (M_c) and the mantle (M_m):

$$M_p = M_c + M_m \qquad (3.3)$$

The mass of an object is related to its density (d) and volume (V) by equation 3.1.

Therefore, the mass of an object is equal to its volume multiplied by its density:

$$\text{Mass} = \text{Volume} \times \text{Density}$$

Let V_p = volume of Pluto
V_c = volume of the core of Pluto,
d_p = bulk density of Pluto ($2.0\,g/cm^3$),
d_c = density of the core ($3.0\,g/cm^3$),
d_m = density of the ice mantle ($0.917\,g/cm^3$),
and $M_p = V_p d_p$, $M_c = V_c d_c$, and $M_m = V_m d_m$.

Substituting into equation 3.3 :

$$V_p d_p = V_c d_c + V_m d_m \qquad (3.4)$$

Note that $V_m = V_p - V_c$, which allows us to eliminate V_m from equation 3.4:

$$V_p d_p = V_c d_c + (V_p - V_c) d_m$$

Solving for V_c yields:

$$V_c = \frac{V_p (d_p - d_m)}{(d_c - d_m)} \qquad (3.5)$$

The volume of Pluto (V_p) can be calculated from its radius (r = 1150 km, Table 3.1):

$$V_p = \frac{4}{3}\pi r^3 = \frac{4 \times 3.14 \times (1150)^3}{3}$$

$$= 6.367 \times 10^9\,km^3$$

Substituting this value into equation 3.5 yields:

$$V_c = \frac{6.367 \times 10^9 (2.0 - 0.917)}{(3.0 - 0.917)} = 3.31 \times 10^9\,km^3$$

Therefore, the volume of the rocky core of Pluto is $V_C = 3.31 \times 10^9\,km^3$

3.7.3 The Radius of the Rocky Core of Pluto

The volume of Pluto's core ($V_c = 3.31 \times 10^9\,km^3$) can be set equal to the volume of a sphere:

$$V_c = \frac{4}{3}\pi r_c^3 \qquad (3.6)$$

from which it follows that:

$$r_c = \left(\frac{3V_c}{4\pi}\right)^{1/3} = \left(\frac{3 \times 3.310 \times 10^9}{4 \times 3.14}\right)^{1/3}$$

$$= (0.790 \times 10^9)^{1/3}$$

$$\log r_c = \frac{\log 0.790 \times 10^9}{3} = 2.9658; r_c = 924\,km$$

Therefore, the radius of the core of Pluto is 924 km and the radius of its ice mantle (r_m) is:

$$r_m = r_p - r_c = 1150 - 924 = 226\,km$$

The numerical results of these calculations depend primarily on the densities of the core and mantle of Pluto and hence on their assumed chemical compositions. Stern and Yelle (1999) considered a three-component model of Pluto in which the mantle is assumed to be composed of a mixture of water and methane ices.

3.7.4 Definition of Specific Gravity

The specific gravity of a body is its density in g/cm^3 divided by the density of pure water at 4 °C which is $1.00\,g/cm^3$. Therefore, the specific gravity is the ratio of two densities and it is dimensionless because the units cancel. (See Glossary).

3.8 Problems

1. Calculate the volumes of the Earth (V_E) and the Sun (V_S) and compare them to each other by dividing the volume of the Sun by the volume of the Earth. Express the result in words.

   Radius of the Sun: 695,700 km

   Radius of the Earth: 6378 km

   Volume of a sphere: $\frac{4}{3}\pi r^3$

   (Answer: $V_S/V_E = 1.3 \times 10^6$)

2. Express the average distance between the Sun and the Earth as a percent of the average distance between the Sun and Pluto. State the result in words. Orbital radii: Earth: 149.6×10^6 km; Pluto: 39.53 AU. (Answer: 2.5 %).

3. Calculate the travel time for light emitted by the Sun to reach the Earth. Express the result in seconds, minutes, hours, and days. The speed of light (c) is 2.99×10^{10} cm/s. Look up the length of one AU in Appendix Appendix 1. (Answer: t = 500.3 s = 8.33 min, etc).

4. Halley's comet was first observed in 240 BC by Chinese astronomers and was also in the sky in the year 1066 AD at the Battle of Hastings. Calculate the number of years that elapsed between these dates and determine how many times this comet reappeared in this interval of time, given that its period of revolution is 76 years but excluding the appearances at 240 BC. (Answer: 1306 y, 16 times).

5. Calculate the density of a rock composed of the four minerals listed below together with their abundances and densities.

Mineral	Abundance, % by volume	Density, g/cm³
Olivine	15	3.30
Augite	50	3.20
Plagioclase	25	2.68
Magnetite	10	5.18

Hint: The total mass of the rock is the sum of the masses of its minerals. (Answer: $d_r = 3.28$ g/cm³).

3.9 Further Reading

Beatty JK, Petersen CC, Chaikin A (eds) (1999) The new solar system, 4th edn. Sky Publishing Corp., Cambridge, MA

Bullen KE (1963) An introduction to the theory of seismology, 3rd edn. Cambridge University Press, Cambridge, UK

Freedman RA, Kaufmann III WJ (2002) Universe: The solar system. Freeman, New York

Hartmann WK (2005) Moons and planets, 5th ed. Brooks/Cole, Belmont, CA

Hester J, Burstein D, Blumenthall G, Greeley R, Smith B, Voss H, Wegner G (2002) 21st century astronomy. Norton, New York

Seeds MA (1997) Foundations of astronomy, 4th edn. Wadsworth, Belmont, CA

Stern SA, Yelle RV (1999) Pluto and Charon. In: Weissman P.R., McFadden L.-A., Johnson T.V. (eds) Encyclopedia of the Solar System. Academic Press, San Diego, CA, pp 499–518

Wagner JK (1991) Introduction to the solar system. Saunders, Philadelphia

Life and Death of Stars

An acceptable theory of the origin of the Earth must also explain the origin of the other planets of the solar system and of the Sun. Moreover, such a theory must account for the origin of the different kinds of stars that exist in the Milky Way galaxy. When we attempt to explain the origin of galaxies, we enter the realm of *cosmology*, which is the study of the Universe including its history, its present state, and its evolution in the future. In modern cosmology the origin of the Earth is intimately related to the origin of the Sun and hence to the origin of the Milky Way galaxy and to the origin of the Universe itself.

We are naturally preoccupied with objects in the Universe that consist of the same kind of matter of which stars, planets, and humans are composed. Actually, the Universe contains far more space than galaxies and stars and the kind of matter that forms the stars and planets of the galaxies makes up only a small fraction of the total amount of matter and energy in the Universe. Modern cosmology indicates that galaxies are embedded within *dark matter* that is invisible and the Universe also contains *dark energy* which opposes the force of gravity. The most recent estimates in Table 4.1 indicate that ordinary matter contributes only 5% of the total amount of matter and energy contained in the Universe, dark matter makes up 25%, and dark energy 70%. The properties of dark matter and dark energy can only be determined indirectly from the effects they have on the objects in the Universe that are visible to us. Even though we are just beginning to appreciate the complexity of the Universe as a whole, the origin and evolution of stars like the Sun are well understood.

4.1 The Big Bang

During the first two decades of the 20th century the Universe was assumed to be static. For example, when the equations of Einstein's general theory of relativity indicated that the Universe could expand, he modified them by adding a "cosmological constant" in order to maintain the Universe in a steady state. The theory of the steady-state Universe was challenged when the American astronomer *Edwin P. Hubble* (1889–1953) concluded from his observations of distant galaxies that the Universe is actually expanding. Hubble discovered the existence of galaxies located far beyond the limits of the Milky Way galaxy and demonstrated by use of the Doppler effect that these galaxies are receding from the Earth at high velocities. (Science Brief 11.7.3). Moreover, in 1929 he showed that the recessional velocities of these distant galaxies are proportional to their distances from the Earth. In other words, the higher the recessional velocity of a distant galaxy, the greater the distance between it and the Earth.

This proportionality is expressed by the so-called Hubble equation:

$$\text{velocity} = \text{H} \times \text{distance} \qquad (4.1)$$

where H = Hubble constant. The Hubble equation implies that the distances between the Milky Way galaxy and other galaxies in the Universe have increased with time in such a way that galaxies with high recessional velocities have moved farther away than galaxies that have low velocities. This means that five billion

Table 4.1. Types of matter and energy contained within the Universe (Cline, 2003)

Type of matter	Abundance, % (mass)
Ordinary matter: (stars, planets, etc.)	5
Cold dark matter:	25
Hot dark matter:	0.3
Dark energy:	70
Radiation (photons)	0.005

years ago, these distant galaxies were closer to the Milky Way galaxy than they are at the present time. Ten billion years ago the distances between galaxies were still smaller. The farther back in time we go, the smaller the Universe becomes until ultimately, at a certain time in the past, all of the galaxies come together in a point.

The consequence of the Hubble equation is that the Universe originated at a certain time in the past when all of its mass and space were contained in a small particle. The expansion of the Universe presumably started when this particle exploded, thereby creating a fireball that subsequently cooled as it expanded to form the Universe we know today. The British cosmologist *Sir Fred Hoyle*, who supported the rival steady-state hypothesis of the Universe, referred to the explosion that started the expansion of the Universe as the "*Big Bang*." He actually meant to ridicule the new hypothesis, but the name was quickly adopted by cosmologists. The Big Bang hypothesis has been confirmed by several independent lines of evidence and has become the accepted theory of cosmology (Riordan and Zajc, 2006).

The age of the Universe, which is the time that has elapsed since the Big Bang, can be calculated from the Hubble equation:

$$\text{velocity} = H \times \text{distance}$$

The velocity (v) is related to the distance (d) by the equation that defines velocity:

$$v = \frac{d}{t}$$

Substituting into the Hubble equation yields:

$$v = \frac{d}{t} = H \times d$$

By dividing both sides of this equation by d we obtain:

$$t = \frac{1}{H} \qquad (4.2)$$

Evidently the expansion age (t) of the Universe is equal to the reciprocal at the Hubble constant. The numerical value of this constant is the slope of the straight line that is defined by the recessional velocities of galaxies (y-coordinate) and the corresponding distances from the earth (x-coordinate).

The recessional velocities of more than 50 galaxies and their distances from the Earth define the straight line in Figure 4.1. The slope of this line yields a value of $22 \, \text{km/s}/10^6$ ly for the Hubble constant (ly = lightyear). The calculation of the corresponding expansion age of the Universe is illustrated in Science Brief 4.6.1. The best presently available estimate of the expansion age of the Universe is:

$$t = 13.6 \times 10^9 \text{ years}$$

with an uncertainty of about $\pm 1.4 \times 10^9$ y. Therefore, the Universe came into existence $(13.6 \pm 1.4) \times 10^9$ years ago by the explosion of an unimaginably small particle and it has evolved from this beginning to its present state.

The Big Bang hypothesis was confirmed in 1964 when *Arno A. Penzias* and *Robert W. Wilson* discovered a microwave radiation that originates from space surrounding the Earth and which was identified by *Robert H. Dicke* as the remnant of the fireball of the Big Bang. The wavelength of this background radiation is equivalent to a temperature of about 2.73 kelvins, which is the present temperature of the Universe.

The third piece of evidence that supports the Big Bang theory is the abundance of hydrogen and helium in the Universe. The expansion of the Universe proceeded so rapidly that only atoms of hydrogen and helium (with minor amounts of lithium, beryllium, and boron) could form. The observed abundances of hydrogen and helium in stars agree with the predicted abundances of these elements (Anders and Grevesse, 1989):

Hydrogen = 76% by weight
Helium = 24% by weight

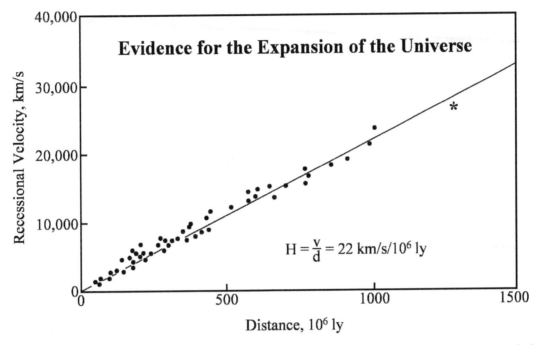

Figure 4.1. The measurements of the recessional velocities (v) and the distances (d) between the Earth and distant galaxies define a straight line whose slope is the Hubble constant (H) defined by the relation: $H = v/d$ or $v = H \times d$ as stated in equation 4.1. A statistical interpretation of these data yields a value of $22 \pm 2.2 \, km/s/10^6$ ly for the Hubble constant. The asterisk identifies the most distant galaxy included in this survey, which is located at almost 1300 ± 10^6 ly from Earth (1.22×10^{22} km) and which has a recessional velocity close to 26,500 km/s (8.8% of the speed of light). These modern measurements extend and confirm the results Edwin Hubble published in 1929. Adapted from Hester et al. (2002, Figure 19.6)

The agreement between the predicted and observed abundances of hydrogen and helium supports the hypothesis that the Universe did start with a Big Bang (Science Brief 4.6.2).

The physical properties of the Universe immediately following the Big Bang have been investigated by the application of certain theories of nuclear physics. All matter in the Universe immediately following the Big Bang was in its most primitive state consisting of a mixture of free *quarks* and *leptons*, which are identified by name in Table 4.2. When the Universe cooled because of the expansion following the Big Bang, the quarks combined with each other to become *protons* and *neutrons*, which subsequently formed the nuclei of hydrogen and helium. The timing of these events was controlled by the temperature, which has been calculated based on the expansion of the fireball (Weinberg, 1977).

The resulting evolution of the Universe is depicted in Figure 4.2 by a line which traces the decrease of the temperature of the Universe as a function of its age expressed in seconds. The values of both temperature and age in Figure 4.2 were plotted in powers of 10 in order to encompass the enormous ranges of these parameters. As a result of this choice of scales, the first second of time in the history of the Universe is greatly expanded in Figure 4.2 compared to all of the time that followed. Nevertheless, the diagram illustrates the conclusion of cosmologists that atomic nuclei of hydrogen and helium (with minor amounts of lithium, beryllium, and boron) formed about three minutes after the Big Bang. Subsequently, these nuclei became electrically neutral atoms by attracting electrons to themselves when the temperature had decreased to about 4000 K and the Universe was about 100,000 years old. Another milestone in the evolution of the Universe occurred several hundred million years later when stars began to form from large clouds of hydrogen and helium gas, which eventually evolved into

Table 4.2. The basic constituents of ordinary matter. (Liss and Tipton, 1997)

Types of particles	Quarks			Leptons		
Names of particles	Up	Charm	Top	Electron	Muon	Tau
	Down	Strange	Bottom	Electron neutrino	Muon neutrino	Tau neutrino

galaxies. For example, some of the stars of the Milky Way galaxy are about 13 billion years old and presumably formed about 600 million years after the Big Bang.

The data in Table 4.3 indicate that the density of the Universe also decreased from an estimated value of 10^{96} g/cm^3 at 10^{-43} seconds to 10^{-30} g/cm^3 at the present time. The decrease of the density of the Universe is primarily caused by the increase in the volume of space between the galaxies. Space and time increase together and are referred to as the "space-time continuum." In other words, the Universe is not expanding into existing space, but it is actually creating space and time as it expands. Evidence obtained recently indicates that the expansion of the Universe is accelerating.

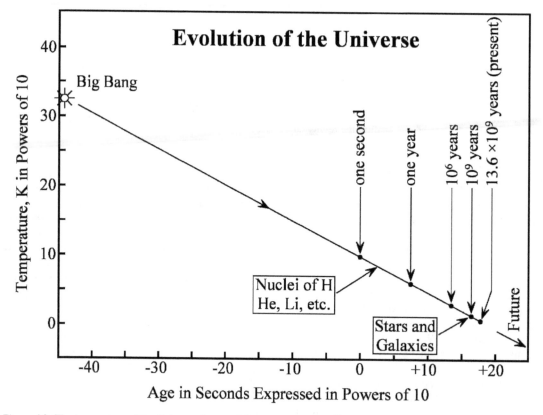

Figure 4.2. The temperature of the Universe decreased from greater than 10^{32} kelvins only a tiny fraction of a second after the Big Bang to 2.73 kelvins at the present time about at 13.6×10^9 years later. The age of the Universe and its temperature are both expressed in powers of 10 meaning that the age "−20" is equal to 10^{-20} seconds and that a temperature of "20" represents 10^{20} kelvins. Several milestones in the evolution of the Universe are identified for reference. Data from Hester et al. (2002, Figure 20.14)

Table 4.3. Decrease of temperature and density of the Universe as a result of expansion (Hester et al., 2002, Figure 20.14)

Time, as indicated	Temperature, K	Density, g/cm^3
10^{-43} s	10^{32}	10^{96}
10^{-35} s	10^{28}	10^{80}
10^{-13} s	10^{16}	10^{32}
15 s	3×10^9	10^4
3 min[1]	10^9	10^2
10^5 y[2]	4000	10^{-20}
10^9 y[3]	18	10^{-27}
13.6×10^9	2.73	10^{-30}

[1] Nuclei of hydrogen, deuterium, helium, lithium, beryllium, and boron atoms formed in about 2 minutes between 3 and 5 minutes after the Big Bang.

[2] Atomic nuclei combined with electrons to form neutral atoms of hydrogen and of the other elements listed above.

[3] Several hundred million years after the Big Bang, clouds of hydrogen and helium atoms began to form stars within galaxies.

As a result, the Universe will become colder, darker, and emptier in the course of billions of years in the future (Ferris, 1997; Christianson, 1999).

4.2 Stellar Evolution

The Big Bang and its immediate aftermath caused clouds of hydrogen and helium to be scattered into the space that was unfolding with the expansion of the Universe. A few hundred million years after the Big Bang, stars began to form within these clouds, which later evolved into clusters of galaxies each of which contained hundreds of billions of stars. The early-formed first-generation stars were composed only of hydrogen and helium (and small amounts of lithium, beryllium, and boron) because, at that time, the Universe did not yet contain any of the other elements of which the planets of our solar system are composed. Typical stars in the Milky Way galaxy today are still composed primarily of hydrogen and helium but also contain elements having higher atomic numbers (numbers of protons) such as carbon (Z = 6), nitrogen (Z = 7), oxygen (Z = 8), fluorine (Z = 9), and neon (Z = 10), where (Z) is the atomic number.

The properties of the light that stars radiate into space are used to determine certain important properties of these stars, such as their chemical composition, their surface temperature, and the "color" of the light they emit, as well as their *luminosity*, which is defined as the total amount of energy a star radiates into space in one second (Glossary). These measurable properties are used to construct the *Hertzsprung-Russell (H-R) diagram* that is used to categorize stars on the basis of their masses and to track their evolution. The stars in a relatively young galaxy are represented by points within a band called the *main sequence*. The luminosity of these stars relative to the luminosity of the Sun ranges over seven orders of magnitude from 10^3 to 10^{-4}. The most luminous stars emit blue light, which implies that they have a high surface temperature ($\sim 40,000$ K), whereas the least luminous stars emit red light because they have low surface temperatures (~ 2000 K). Stars having a high luminosity and a high surface temperature are more massive than stars with low luminosity and low surface temperature. Therefore, the high-luminosity stars are called *blue giants* and the low-luminosity stars on the main sequence are the *red dwarfs*. The very smallest stars called *brown dwarfs* are described in Section 24.2.

The light radiated by stars on the main sequence is generated by nuclear fusion reactions by means of which two protons (nuclei of hydrogen) and two neutrons are combined to form the nucleus of helium. This process releases large amounts of energy in the form of heat that is transported by radiation and convection from the core of the star to its surface where some of it is converted into electromagnetic radiation (light) that is radiated into the space. Stars can also lose energy by ejecting hot ionized gas in the form of flares. Such flares are primarily composed of protons (hydrogen nuclei), alpha particles (helium nuclei), and electrons.

The very first stars of the Milky Way galaxy formed by the contraction of clouds of hydrogen and helium under the influence of gravity and other forces. When the core temperature of such a H-He star reached about 12 million kelvins, the hydrogen-fusion reaction started and a star was born. The luminosity and surface temperature of the new star placed it on the main sequence at a point depending on the mass of the star. For example, the Sun is located roughly in the middle

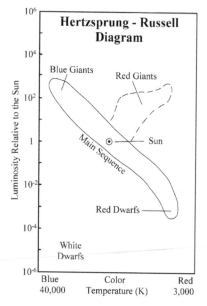

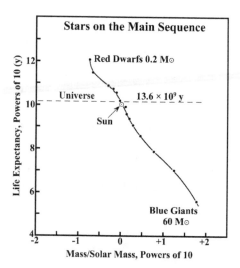

Figure 4.3. The Hertzsprung-Russell diagram is constructed by plotting the luminosity of stars (divided by the luminosity of the Sun) in powers of 10 versus their surface temperature (or color). Most of the stars of an active galaxy plot in the area labeled main sequence. The blue giants are the largest stars with about 60 solar masses, whereas red dwarfs have less mass than the Sun. Stars on the main sequence generate energy in their cores by a nuclear reaction that fuses two protons (nuclei of hydrogen atoms) with two neutrons to form a nucleus of helium. After the hydrogen in the core is exhausted, stars leave the main sequence and become red giants which ultimately eject their gas envelopes leaving their cores, which take the form of white dwarfs, neutron stars, or a black holes depending on the original mass of the star. Adapted from Hester et al. (2002, Figure 12.15)

interiors is continually replenishing the hydrogen in their cores. Blue giants are short-lived because convection in their interiors does not allow the hydrogen in their cores to be replaced.

The relationship between the masses of stars and their life expectancies on the main sequence in Figure 4.4 indicates that the most massive stars (60 solar masses) spend only about 360,000 years on the main sequence, whereas the Sun (1 solar mass) and other stars of similar mass stay on the main sequence for about 10 billion years. Stars with masses less than that of the Sun remain on the main sequence longer than the age of the Universe. Therefore, red dwarfs that formed in the Milky Way galaxy during the initial burst of star formation are still residing on the main sequence, whereas stars more massive than the Sun died earlier and completed their evolution long before the present. The Sun, being a star of modest size, will remain on the main sequence

of the main sequence in Figure 4.3 at a relative luminosity of 1.

The masses of stars on the main sequence range widely from about 0.2 to 60 solar masses. Consequently, the massive stars (blue giants) contain more hydrogen fuel for the H-He fusion reaction than the less massive stars (red dwarfs). We might expect therefore, that blue giants can sustain the H-fusion reaction much longer than red dwarfs because they contain so much more fuel. However, the opposite is true because blue giants "burn" the hydrogen in their cores far more rapidly than red dwarfs do and therefore they spend much *less* time on the main sequence than red dwarfs. The longevity of red dwarfs arises from the fact that convection in their

Figure 4.4. The life expectancies of stars on the main sequence depend strongly on their masses. The most massive stars (60 solar masses) "burn" the hydrogen in their cores in less than one million years and complete their life cycle soon thereafter. The least massive stars (less than one solar mass) spend up to one thousand billion years on the main sequence. As a result, the red dwarfs are the oldest and most abundant stars in the Milky Way and other galaxies, whereas blue giants are the youngest and least abundant stars. The life expectancy of the Sun on the main sequence is ten billion years. We also know that the Sun is 4.6 billion years old which means that it will continue to shine for about another 5.4 billion years before it becomes a red giant. Data from Hester et al. (2002, Table 12.3)

for an additional 5.4 billion years, based on its present age of the 4.6 billion years.

4.3 Nucleosynthesis

Stars on the main sequence can only convert hydrogen in their cores into helium. All of the other chemical elements are formed by nuclear reactions that occur after a star has left the main sequence and has evolved into a red giant on the H-R diagram in Figure 4.3. When most of the hydrogen in the core of a star on the main sequence has been converted to helium, the rate of energy production initially decreases, which causes the star to contract. The resulting compression of the interior of the star raises the temperature of the core and of the shell around the core. Therefore, the site of the hydrogen-fusion reaction shifts from the core to a shell outside of the core, thereby causing the luminosity of the star to increase. As a result, the star moves off the main sequence and into the region in Figure 4.3 occupied by red giant stars.

The heat generated by hydrogen-fusion in the shell also increases the temperature of the core. When the core temperature reaches about 100 million K, the helium in the core begins to form nuclei of carbon atoms by the so-called triple-alpha process, which occurs in two steps. First, two helium nuclei collide with each other to form a nucleus of beryllium $\left(^{8}_{4}\text{Be}\right)$:

$$^{4}_{2}\text{He} + ^{4}_{2}\text{He} \rightarrow ^{8}_{4}\text{Be} + \text{energy} \qquad (4.3)$$

This beryllium nucleus is unstable and must assimilate a third alpha particle almost immediately to produce a nucleus of carbon:

$$^{8}_{4}\text{Be} + ^{4}_{2}\text{He} \rightarrow ^{12}_{6}\text{C} + \text{energy} \qquad (4.4)$$

The additional energy generated by the helium fusion reaction causes the star to expand into a *red giant*. The surface temperature of red giant stars actually decreases because the energy is spread over a larger surface area than before.

The transformation of a main-sequence star into a red giant is an important event in the evolution of stars because, at this stage, stars can generate energy by synthesizing nuclei of carbon and many other elements. As a result, the chemical composition of red giants changes as they synthesize elements containing up to 92 protons in their nuclei. The elements that contain more than two protons in their nuclei are called "metals" by cosmologists even though some of these elements are actually non-metals such as carbon, nitrogen, oxygen, fluorine, sulfur, and many others. Nevertheless, the synthesis of nuclei of chemical elements by red giants is said to increase their "*metallicity.*" The synthesis of chemical elements in red giant stars takes place by several different kinds of nuclear reactions that occur at different stages in their evolution. These reactions were originally described by Burbidge et al. (1957) and have since become an integral part of the theory of stellar evolution and energy generation in stars (Bethe, 1968).

The nucleosynthesis reactions occurring in red giant stars are caused by the bombardment of nuclei by alpha particles (alpha process) and by the capture of neutrons (neutron-capture process). For example, if the nucleus of carbon-12 $\left(^{12}_{6}\text{C}\right)$ collides with an alpha particle $\left(^{4}_{2}\text{He}\right)$, it forms the stable nucleus of oxygen-16:

$$^{12}_{6}\text{C} + ^{4}_{2}\text{He} \rightarrow ^{16}_{8}\text{O} + \text{energy} \qquad (4.5)$$

This process continues until the positive electrical charges of the product nuclei repel the positively charged alpha particles, because particles having electrical charges of the same polarity repel each other. The alpha process becomes ineffective when the target nuclei contain 26 or more protons, which characterizes the nuclei of the atoms of iron and elements of higher atomic number.

In addition to the alpha process described above, nucleosynthesis in red giant stars occurs by the capture of neutrons. This process is not limited by electrostatic repulsion because neutrons do not have an electrical charge. Consider for example, the neutron capture by the stable nucleus of iron-56 $\left(^{56}_{26}\text{Fe}\right)$:

$$^{56}_{26}\text{Fe} + ^{1}_{0}\text{n} \rightarrow ^{57}_{26}\text{Fe}(\text{stable}) \qquad (4.6)$$

$$^{57}_{26}\text{Fe} + ^{1}_{0}\text{n} \rightarrow ^{58}_{26}\text{Fe}(\text{stable}) \qquad (4.7)$$

$$^{58}_{26}\text{Fe} + ^{1}_{0}\text{n} \rightarrow ^{59}_{26}\text{Fe}(\text{unstable}) \qquad (4.8)$$

$$^{59}_{26}\text{Fe} \rightarrow ^{59}_{27}\text{Co} + \text{beta particle} + \text{neutrino} + \text{energy} \qquad (4.9)$$

The unstable nucleus of $^{59}_{26}$Fe decays by the spontaneous emission of a *beta particle* (an electron emitted from the nucleus) to form stable cobalt-59, which has 27 protons and $59 - 27 = 32$ neutrons. The *neutrino* is a lepton listed in Table 4.2 which interacts sparingly with matter but carries kinetic energy (Faure and Mensing, 2005).

The decay of the nucleus of $^{59}_{26}$Fe is an example of *radioactivity* which occurs by the spontaneous emission of nuclear particles from an unstable nucleus and of electromagnetic radiation from the resulting product nucleus. The nuclei of atoms in the cores of red giants and main-sequence stars have positive electrical charges (arising from the presence of protons) because the nuclei are unable to interact with electrons at the high temperatures (i.e., > 100 million K in red giants and > 12 million K in main-sequence stars).

The hypothesis explaining the origin of the chemical elements by nucleosynthesis in stars, proposed by Burbidge et al. (1957), was fine-tuned by comparing the predicted abundances of the elements to the chemical composition of the solar system revealed by optical spectroscopy of sunlight and by chemical analyses of meteorites which originated from the asteroids. The abundances of the chemical elements derived from these sources are displayed in Figure 4.5.

The abundances of the chemical elements in the solar system are expressed in terms of numbers of atoms of an element divided by 10^6 atoms of the element silicon (Si). For example, the abundance of hydrogen is 2.79×10^{10} atoms of H per 10^6 atoms of Si, which can be restated in terms of powers of 10:

$$2.79 \times 10^{10} = 10^{10.44}$$

Therefore, the abundance of hydrogen in Figure 4.5 is expressed by a point whose coordinates are: 10.44 and 1 (i.e., the nucleus of H contains only one proton).

The abundances of the elements included in Figure 4.5 illustrate several important observations:

1. Hydrogen and helium, which formed during the Big Bang, are more abundant than all other elements.
2. Lithium, beryllium, and boron have low abundances because only small amounts of these elements formed during the initial expansion of the Universe.
3. Elements with even atomic numbers are more abundant than elements with odd atomic numbers. (Oddo-Harkins rule).
4. The abundances of the elements generally decline as their atomic numbers increase.
5. Iron stands out by being more abundant than expected from the pattern of abundances.

The chasm in Figure 4.5 caused by the low abundances of Li, Be, and B is bridged by the triple-alpha process, which permits all of the other elements to form by nucleosynthesis reactions starting with carbon-12. Iron is the end product of the alpha process, which may explain the anomalously high abundance of this element. Elements having higher atomic numbers than iron formed by neutron capture and other nuclear reactions during the red-giant stage and its aftermath. The present theory of nucleosynthesis successfully explains the abundances of most of the elements in the solar system. The good agreement between the theoretical predictions and the measured abundances of chemical elements in the solar system provides confidence in the reliability of the theory of stellar nucleosynthesis.

The nuclear reactions that continue to synthesize chemical elements enable red giant stars to maintain their large size until eventually all nuclear fuel is used up. Therefore, red giants stars become increasingly unstable with age and are ultimately destroyed in gigantic explosions that occur after the central core collapses. The resulting shockwave blows away the outer layers of gas that contains not only residual hydrogen and helium, but also the new elements whose nuclei were synthesized by nuclear reactions during the red giant stage. When a star more massive than the Sun reaches the end of the red giant stage it explodes as a *supernova*. The gas ejected by a supernova into interstellar space forms a *planetary nebula*, which in some cases has the shape of a balloon with the remnant of the stellar core at its center. The fate of the stellar core depends on its mass. Stars whose mass is similar to that of the Sun become *white dwarfs* which are characterized by low luminosities but high surface temperatures (Figure 4.3). These stellar corpses gradually cool and ultimately fade from view

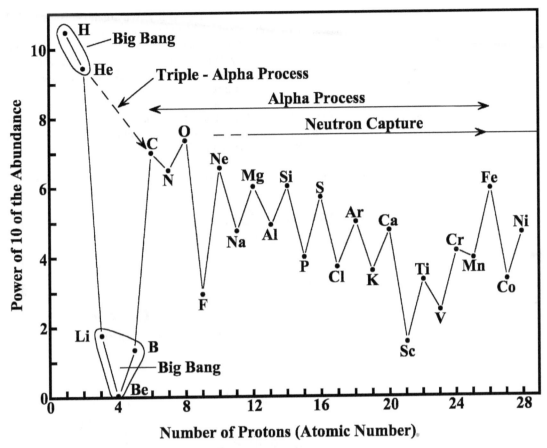

Figure 4.5. The abundances of the chemical elements in the solar system are expressed in terms of numbers of atoms per million atoms of silicon and were plotted in this diagram in powers of 10. The abundances of hydrogen and helium are up to 1000 times higher than those of carbon and other elements of higher atomic number (i.e., number of protons). The chemical symbols used to identify the elements are: H = hydrogen, He = helium, Li = lithium, Be = beryllium, B = boron, C = carbon, N = nitrogen, O = oxygen, F = fluorine, Ne = neon, Na = sodium, Mg = magnesium, Al = aluminum, Si = silicon, P = phosphorus, S = sulfur, Cl = chlorine, Ar = argon, K = potassium, Ca = calcium, Sc = scandium, Ti = titanium, V = vanadium, Cr = chromium, Mn = manganese, Fe = iron, Co = cobalt, and Ni = nickel. Only the first 28 of the 92 naturally occurring chemical elements are included in this diagram. Data from Anders and Grevesse (1989)

when their surface temperature drops below about 750 K.

The cores of stars that are significantly more massive than the Sun collapse until all atomic nuclei disintegrate and the protons are converted into neutrons by interacting with electrons. The resulting *neutron stars* are as dense as atomic nuclei and hence are relatively small. The contraction also increases their rate of rotation up to about 60 rotations per second although a few rotate much faster than this. As a result, the rapidly spinning neutron stars emit electromagnetic radiation in pulses, which

is why neutron stars are also called *"pulsars"* (Irion, 1999).

The cores of the most massive stars collapse until their gravitational force prevents even photons of electromagnetic radiation to escape. Therefore, they have the appearance of *black holes* in the galaxy in which they occur. Black holes can grow in mass by pulling matter from passing stars. Before falling into the black hole, the material forms a disk that orbits the hole and emits an intense beam of x-rays as it disappears into the black hole forever. The Milky Way and other galaxies each contain a supermassive black hole in their central bulge. The mass of a superheavy black hole can reach millions of solar masses.

4.4 The Milky Way Galaxy

Our galaxy in Figure 4.6 has the shape of a flat disk formed by spiral arms and is surrounded by a halo containing numerous globular clusters of stars and small satellite galaxies. The central region of the Milky Way galaxy is occupied by a spherical bulge containing a large number of bright stars. These features identify the Milky Way galaxy as a *barred spiral*. Other types of galaxies include ellipticals, spirals, and irregulars.

The diameter of the disk of the Milky Way galaxy is about 100 thousand light years and that of the halo is 300 thousand light years. The Sun is located within the galactic disk at a distance of

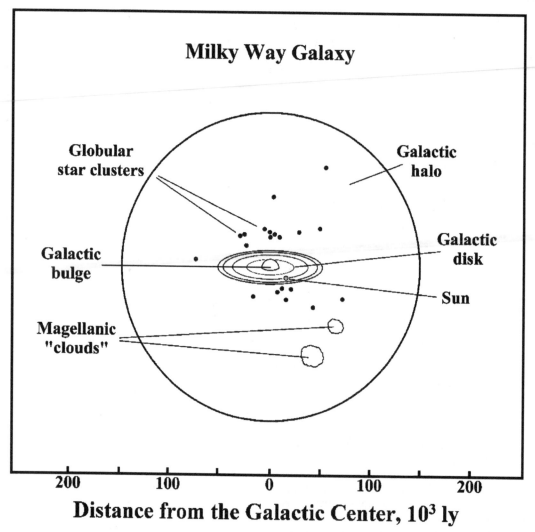

Milky Way Galaxy

Distance from the Galactic Center, 10^3 ly

Figure 4.6. Schematic diagram of the Milky Way galaxy. The galactic disk consists of spiral arms which are composed of stars and clouds of gas and dust. The galactic bulge contains a large number of luminous stars as well as a supermassive black hole. The galactic halo includes small satellite galaxies such as the Magellanic "clouds" and numerous globular clusters of stars. The galactic halo also contains dark matter, which makes up most of the mass of the galaxy. Adapted from Hester et al. (2002, Figure 18.6)

27 thousand light years from the central bulge. The entire Milky Way galaxy is rotating about its central bulge with a radial velocity of 250 km/s at a distance of 27 thousand lightyears from the galactic center, which corresponds to a period of revolution of 200 million years (Science Brief 4.6c).

The rotation of the Milky Way galaxy causes it to have the structure of a two-armed spiral and triggers star formation by compressing the clouds of gas in the spiral arms. Therefore, the spiral arms are illuminated by the light emitted by massive young stars whose life expectancy on the main sequence is so short that they do not have time to drift far from the spiral arm in which they formed.

The rotation of the Milky Way and other spiral galaxies causes stars like the Sun to revolve around the galactic center. If most of the mass of the galaxy is concentrated in the central bulge, as suggested by the amount of light that is generated there, then the orbital velocities of stars should decrease with increasing distance from the galactic center like those of the planets of the solar system in Figure 3.5. In actual fact, recent measurements of the orbital velocities of stars in spiral galaxies in Figure 4.7 indicate that their orbital velocities are close to being constant and independent of their distance from the galactic center. Therefore, the mass of a spiral galaxy is *not* concentrated in the central bulge as initially assumed. The radial distribution of orbital velocities (called the *rotation curve*) of spiral galaxies indicates that a large fraction of the total mass of a galaxy resides in its halo in the form of *dark matter*. The matter that is responsible for the mass of the halo does not emit electromagnetic radiation and therefore is not detectable by optical telescopes or other kinds of radiation detectors.

The halo of the Milky Way galaxy in Figure 4.6 contains several small galaxies that are being cannibalized by the host galaxy (e.g., the Magellanic "clouds") or will collide with it in the future (e.g., the Andromeda galaxy). As a result, the Milky Way galaxy is gaining large amounts of gas and dust from which new stars will continue to form in the future (Wakker and Richter, 2004).

Evidently, the Milky Way and other galaxies in the Universe are growing in mass by assim-

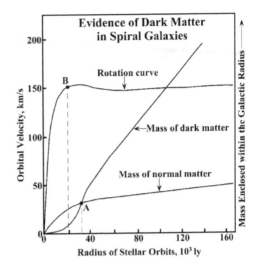

Figure 4.7. The rotation curve of the spiral galaxy NGC 3198 is flat indicating that the orbital velocities of stars do not decrease with increasing distance from the galactic center. The explanation is that the mass of this and other spiral galaxies is not concentrated in its central bulge, but resides primarily in the halo surrounding the disk of spiral arms. The curves demonstrate that normal luminous matter dominates only in the inner part of the galaxy within a radius of about 32×10^3 ly (point A) and that dark nonluminous matter becomes increasingly abundant at greater distances from the galactic center. The presence and distribution of dark matter explains why the orbital velocities of stars in this spiral galaxy remain constant at distances greater than about 20×10^3 ly (point B) from the galactic center. Adapted from Hester et al. (2002, Figure 17.14)

ilating smaller galaxies and clouds of gas from intergalactic space. Although the oldest known stars in our galaxy formed about 13 billion years ago and although billions of stars have formed and died since then, the Milky Way galaxy continues to grow and to form new stars with no end in sight.

4.5 Summary

The observation that the Universe is expanding implies that the distances between galaxies were shorter in the past than they are at present and that at some time in the past all of the matter and energy of the Universe was concentrated in a small particle. This particle exploded spontaneously thereby starting the Universe with a Big Bang.

About 10^{-43} seconds later, the Universe was composed of a mixture of elementary nuclear particles which were interacting with each other in accordance with theories of nuclear physics and depending on the rapidly decreasing pressure and temperature. About three minutes after the Big Bang at a temperature of one billion kelvins, nuclei of hydrogen and helium formed, as well as much smaller amounts of lithium, beryllium, and boron.

The first stars which began to form several hundred million years later were initially composed only of the elements that were synthesized during a brief interval of time following the Big Bang (i.e., hydrogen and helium). These stars generated energy by the nuclear fusion of hydrogen to helium while they resided on the main sequence of the H-R diagram. The largest stars consumed most of the hydrogen in their cores in less than one million years and subsequently sustained themselves for even shorter periods of time by the energy derived from the synthesis of atomic nuclei containing more than two protons. These newly synthesized nuclei were ejected when the aging stars died by the collapse of their cores and the explosive loss of the surrounding shells of gas (i.e., supernova).

The resulting planetary nebulae allowed new stars to form that were composed not only of hydrogen and helium, but also contained elements whose nuclei contained more than two protons. Consequently, the Milky Way and all other galaxies in the Universe have evolved by the birth and death of stars and by the production of chemical elements within these stars. The theory of stellar evolution and nucleosynthesis also leads to the insight that the stars of the Milky Way galaxy have a range of ages and chemical compositions. However, all stars are composed of the same chemical elements that exist in the Sun and in the planets of our solar system.

Another important insight is that the Sun, which formed only 4.6 billion years ago, is a relatively young star in the Milky Way galaxy. We know this not only from the age of the solar system, but also because of the presence in the Sun of elements that did not form during the Big Bang. In other words, the Sun is a second-generation star that formed from the gas and dust that had accumulated in our galactic neighborhood by the explosion of ancestral red giant stars.

The presence of these "metallic" elements in the cloud of gas and dust from which the Sun formed, made possible the accumulation of planets composed of solid compounds. These planets include the Earth on which we live. We see now that the Earth is composed of chemical elements that were synthesized by nuclear reactions in red giant stars that have since exploded. Human beings and all other forms of life on the Earth are likewise composed of the "ashes" of the nuclear "fires" like those that are still burning in all of the hundreds of billions of stars in the Milky Way and all other galaxies of the Universe.

The Milky Way galaxy is classified as a barred spiral because it contains a bar-shaped central bulge connected to two arms composed of stars and clouds of gas and dust. The arms are being wound into a trailing spiral by the rotation of the central bulge. The rotation curve of the Milky Way galaxy indicates that the orbital velocities of stars are constant and do not decrease with increasing distance from the galactic center. This anomalous property of spiral galaxies is attributable to the presence of large amounts of dark matter in the halo surrounding the galactic disk. The halo also contains numerous dwarf galaxies and globular clusters which are being assimilated by the Milky Way galaxy. The clouds of gas and dust entering the Milky Way galaxy are causing new stars to form and are thereby rejuvenating the galaxy.

4.6 Science Briefs

4.6.1 The Hubble Constant and the Age of the Universe

The data points in Figure 4.1 define a straight line in coordinates of the distance to other galaxies and their recessional velocities. The distance is expressed in units of millions of lightyears while the velocity is expressed in kilometers per second. According to equation 4.1, the Hubble constant is equal to the slope of this line (Hester et al., 2002).

$$H = \frac{velocity}{distance} = 22 \frac{km/s}{10^6 \, ly}$$

The expansion age (t) of the Universe expressed by equation 4.2 is the reciprocal of the Hubble constant:

$$t = \frac{1}{H} = \frac{10^6 \, ly}{22 \, km/s}$$

We calculated in Section 2.3.1 that one lightyear is equal to 9.460×10^{12} km.

Therefore:

$$t = \frac{9.460 \times 10^{12} \times 10^6 \, km}{22 \, km/s}$$

Note that the distance in kilometers cancels, leaving:

$$t = \frac{9.460 \times 10^{18}}{22} \, s$$

Converting to years:

$$t = \frac{9.460 \times 10^{18}}{22 \times 60 \times 60 \times 24 \times 365.26}$$
$$= 0.01362 \times 10^{12} \, y$$
$$t = 13.62 \times 10^9 \, y$$

This result is the best available estimate of the age of the Universe. The scatter of data points above and below the straight line in Figure 4.1 allows the value of the Hubble constant to vary from 20 to 25 km/s/10^6 ly. The corresponding values of the age of the Universe range from 15 billion (H = 20 km/s/10^6 ly) to 12 billion years (H = 25 km/s/10^6 ly). Astronomers are still working to reduce the uncertainty of the best estimate of the age of the Universe.

4.6.2 The Atomic Nuclei of Hydrogen and Helium

The atoms of hydrogen consist of only one proton, which forms the nucleus, and one electron. The nuclei of a small fraction of the hydrogen atoms (0.015 % by number) contain both a proton and a neutron giving them a mass number of two. Therefore, this atomic variety of hydrogen is known as *deuterium* after the Greek word for "two." In contrast to the nucleus of hydrogen atoms, all nuclei of helium atoms contain two protons and most of them

also contain two neutrons, although a few helium atoms (0.000138 % by number) contain only one neutron. The make-up of the atomic nuclei of hydrogen and helium is indicated by the way these atoms are identified in atomic physics:

$$\text{Hydrogen} : {}^1_1 H \text{ and } {}^2_1 H$$

$$\text{Helium} : {}^4_2 He \text{ and } {}^3_2 He$$

where the subscript identifies the number of protons (atomic number) and the superscript indicates the number of protons and neutrons (mass number).

4.6.3 The Period of Revolution of the Sun

Information provided by Hester et al. (2002) indicates that the Sun is located at a distance of 27×10^3 ly from the central bulge of the Milky Way galaxy and that the radial velocity at that distance is 250 km/s. Therefore, the circumference (d) of the orbit of the Sun around the center of the galaxy is:

$$d = 2r\pi = 2 \times 27 \times 10^3 \times 9.460 \times 10^{12} \times 3.14$$
$$= 1604 \times 10^{15} \, km$$

where $1 \, ly = 9.460 \times 10^{12}$ km (Section 2.3.1). The period (t) of revolution of the Sun is:

$$t = \frac{d}{v} = \frac{1604 \times 10^{15}}{250} = 6.416 \times 10^{15} \, s$$

Converting from seconds to years:

$$t = \frac{6.416 \times 10^{15}}{60 \times 60 \times 24 \times 365.26} = 203 \times 10^6 \, y$$

The period of revolution of the Sun around the center of the Milky Way galaxy is about 200 million years.

4.7 Problems

1. Calculate the diameter of the Milky Way galaxy in kilometers assuming that its radius is 50 ly. (Answer: 9.46×10^{14} km).
2. Calculate the period of revolution of a star that is located 35 ly from the center of the galaxy assuming that its radial velocity is 150 km/s. (Answer: 4.39×10^5 y).

3. Calculate the age of an expanding universe characterized by a Hubble constant $H = 15\,km/s/10^6\,ly$. (Answer: $19.98 \times 10^9\,y$).

4. According to the Oddo-Harkins rule, chemical elements with even atomic numbers are more abundant than elements that have odd atomic numbers. Examine Figure 4.5 and note the elements Al, Si, and P. Determine whether these elements obey the Oddo-Harkins rule and suggest an explanation for this phenomenon. (See Faure, 1998).

5. Write an essay about one of the following scientists who contributed significantly to the theory of cosmology and stellar evolution.

  a. George Gamow   c. Hans Bethe
  b. Fred Hoyle   d. Edwin Hubble

4.8 Further Reading

Anders E, Grevesse N (1989) Abundances of the elements: Meteoritic and solar. Geochim Cosmochim Acta 53:197–214

Bethe H, (1968) Energy production in stars. Science 161:541–547

Burbidge EM, Burbidge GR, Fowler WA, Hoyle F (1957) Synthesis of the elements in stars. Rev Modern Phys 29:547–650

Christianson G (1999) Mastering the Universe. Astronomy 27(2):60–65

Cline DB (2003) The search for dark matter. Sci Am 288(3):50–59

Faure G (1998) Principles and applications of geochemistry, 2nd edn. Prentice Hall, Upper Saddle River, NJ

Faure G, Mensing TM (2005) Isotopes: Principles and applications, 3rd edn. Wiley, Hoboken, NJ

Ferris T (1997) The whole shebang: a state-of-the-Universe report. Touchstone Books, Simon and Schuster, New York

Hester J, Burstein D, Blumenthal G, Greeley R, Smith B, Voss H, Wegner G (2002) 21st century astronomy. Norton, New York

Irion R (1999) Pursuing the most extreme stars. Astronomy 27(1):48–53

Liss TM, Tipton PL (1997) The discovery of the Top Quark. Sci Am 276(9):54–59

Riordan M, Zajc WA (2006) The first few microseconds. Sci Am 294(5):34–41

Wakker BP, Richter P (2004) Our growing and breathing galaxy. Sci Am 290(1):38–47

Weinberg S (1977) The first three minutes. Bantam Books, Basic Books, New York, 177 p

Origin of the Solar System

The Milky Way galaxy contains clouds of gas and dust that occupy large regions of interstellar space. These clouds are composed primarily of hydrogen and helium, but they also contain all of the other chemical elements that were synthesized by red giant stars prior to and during their final explosions as novae or supernovae. Therefore, the clouds of gas and dust in the galaxy are the remnants of *planetary nebulae* that formed when the envelopes of hot gas of red giant stars were blown into interstellar space following the collapse of their cores.

The gases of planetary nebulae are initially hot enough to cause the atoms to lose some or all of their extranuclear electrons. Consequently, the planetary nebulae are initially composed of electrically charged atoms called *ions* (Science Brief 5.5.1). Gases composed of ions and electrons represent a state of aggregation called *plasma*. Matter in the form of plasma in interstellar space responds to differences in the electrical potential and can contain magnetic fields inherited from the ancestral red giant stars in..

The hot and ionized gas of a planetary nebula produced by a supernova expands at high velocities approaching the speed of light until it collides and mixes with interstellar gas of previous supernovae. The Sun and the planets of the solar system originated by contraction of such a cloud of gas and dust called the *solar nebula* which contained not only hydrogen and helium but also chemical elements that were ejected by ancestral red giant stars in our neighborhood of the Milky Way galaxy.

5.1 The Solar Nebula

The gases of planetary nebulae cooled as a result of expansion and by emitting electromagnetic radiation. Consequently, the ions that were present when the gas was still hot attracted electrons to themselves and thus became neutral atoms of the chemical elements. These atoms continued to move along random paths at high velocities depending on the temperature. Consequently, the atoms of different chemical elements collided with each other, which permitted covalent bonds to form by "sharing" of electrons. Chemical bonds that result from the electrostatic attraction between positively and negatively charged ions are called "*ionic*" bonds. In reality, covalent bonds between atoms in a molecule in most cases have a certain amount of ionic character because the electrons are not shared equally by the atoms that form a particular molecule.

The collisions among atoms in a cloud of gas composed of neutral atoms produced a large number of different molecules. The most abundant molecule in the solar nebula was diatomic hydrogen (H_2), which formed by covalent bonding of two hydrogen atoms. The solar nebula also contained individual atoms of helium, which is a *noble gas* and therefore does not form bonds with other atoms. The other noble gases (neon, argon, krypton, xenon, and radon) in the solar nebula also did not form bonds for the same reason as helium.

The atoms of the other elements, some of which are listed in Table 5.1, formed covalently bonded molecules when their atoms in the solar nebula collided with each other. For example, an atom of carbon (C) that collided with an atom of oxygen (O) produced the molecule CO, which we know as carbon monoxide:

$$C + O \rightarrow CO + energy \tag{5.1}$$

Table 5.1. The periodic table of the chemical elements having atomic numbers from 1 to 18

	Groups							
	1	2	3	4	5	6	7	8
Period 1	H(1)							He(2)
Period 2	Li(3)	Be(4)	B(5)	C(6)	N(7)	O(8)	F(9)	Ne(10)
Period 3	Na(11)	Mg(12)	Al(13)	Si(14)	P(15)	S(16)	Cl(17)	Ar(18)

Notes: The elements are arranged in order of increasing atomic number (number of protons in the nuclei of their atoms). Elements in the same group have similar chemical properties. The horizontal sequences (periods) reflect the progressive filling of the available electron orbitals with increasing atomic numbers. The elements are identified by their chemical symbols: H = hydrogen, He = helium, Li = lithium, Be = beryllium, B = boron, C = carbon, N = nitrogen, O = oxygen, F = fluorine, Ne = neon, Na = sodium, Mg = magnesium, Al = aluminum, Si = silicon, P = phosphorus, S = sulfur, Cl = chlorine, Ar = argon. The complete periodic table contains 92 chemical elements.

The energy released by this process is called the "heat of the reaction" which caused the temperature of the solar nebula to rise. However, the heats of chemical reactions are much lower than those of nuclear reactions.

Molecules of CO may have collided with another atom of oxygen to form CO_2, which is carbon dioxide (CO_2):

$$CO + O \rightarrow CO_2 + energy \qquad (5.2)$$

Carbon monoxide (CO) and carbon dioxide (CO_2) are gases at the temperature and pressure that prevail on the surface of the Earth. However, at the low temperature of interstellar space, CO and CO_2 form solids we call ice.

Another example of compound formation involved atoms of hydrogen (H) and oxygen (O):

$$H + O \rightarrow HO + energy \qquad (5.3)$$

$$H + HO \rightarrow H_2O + energy \qquad (5.4)$$

H_2O is the chemical formula for water ice which melts at $0\,°C$ on the surface of the Earth. Liquid water can be converted into a gas by raising its temperature to $100\,°C$ at a pressure of 1 atm. The important point here is that water molecules formed spontaneously in the solar nebula as a result of collisions of atoms of H and O and that the water molecules existed in the form of small particles of ice.

Reactions between atoms of carbon (C) and hydrogen (H) in solar nebula formed so-called *organic molecules*, such as methane (CH_4) and other hydrocarbons. In addition, the solar nebula, like other clouds in interstellar space, contained a large number of different complex organic compounds such as methanol (CH_3OH), acetone [$(CH_3)_2CO$], formaldehyde (H_2CO), and about one hundred others. Some of these compounds have been detected in *comets* where they have been preserved at low temperature for 4.6 billion years (Hartmann, 2005).

The atoms of metallic elements such as calcium (Ca), magnesium (Mg), and iron (Fe) interacted with atoms of oxygen in the solar nebula to form the oxides CaO, MgO, and FeO. These compounds formed small solid particles at the low pressure and temperature of the solar nebula. In this way, the hot plasma ejected by the ancestral supernovae evolved into the solar nebula which was composed of molecular hydrogen, helium atoms, organic molecules, as well as solid particles of ice, oxides of metals, and other kinds of solids (e.g., graphite).

5.2 Origin of the Sun and Planets

The Sun and the planets of the solar system formed as a result of the contraction of gas and dust of the solar nebula which took place in accordance with the laws of physics and chemistry. However, the sequence of events and the present make-up of the solar system are, in general, unpredictable because the process is strongly affected by feedback loops which magnify minor perturbations into major differences in the final outcome. For example, attempts to reproduce the formation of the solar system by computational methods do not necessarily yield the solar system we know. In some cases, two stars can form instead of one and the number of planets and their satellites may vary unpredictably.

The solar nebula was part of a much larger cloud of gas and dust which had accumulated in our galactic neighborhood during a long period of time that preceded the comparatively rapid formation of our solar system. The Elephant's Trunk nebula in Figure 5.1 is an example of such a cloud of gas and dust in which stars are forming at the present time. Another example is the Orion nebula described by O'Dell (2003). Therefore, other stars may have originated from the same cloud of gas and dust in which the Sun formed and these sister stars may still reside in our galactic neighborhood.

Figure 5.1. The Elephant's Trunk nebula in the Milky Way galaxy, seen here in false-color infrared light by the Spitzer Space-Telescope. The nebula is a stellar nursery in which several young stars glowed brightly when the light was emitted about 2450 years ago. This nebula also contained dust particles and molecules of hydrogen and of complex polycyclic aromatic hydrocarbons (PAHs). (Courtesy of NASA/JPL-Caltech/W. Reach (SSC/Caltech). (http://ipac.jpl.nasa.gov/media_images/ssc2003-06b1.jpg)

The contraction of the solar nebula under the influence of gravity may have been initiated by the passage of a shockwave caused by the explosion of a nearby red giant star, or by the pressure of photons emitted by stars in the galaxy, or by random motions of atoms and dust particles, or by other causes. The resulting contraction of the solar nebula caused most of its mass to be concentrated in the form of the *protosun*. The subsequent compression of the gas and dust caused the temperature of the interior of the protosun to increase. As a result, all solid dust particles and molecules that were incorporated into the protosun disintegrated into atoms and ultimately into ions. When the core temperature of the *protosun* reached about 12 million kelvins, the hydrogen-fusion reaction started to produce energy, thereby causing the newly-formed Sun to radiate light. The contraction of the protosun until ignition of the hydrogen-fusion reaction lasted only about 30 million years. Protostars that are more massive than the Sun contract faster, whereas less massive protostars contract more slowly than the Sun (e.g., 0.16 million years for 15 solar masses and 100 million years for 0.5 solar masses). The resulting star then takes its place on the main sequence on the H-R diagram depending on its mass and hence on its luminosity and surface temperature or color (Section 4.2).

The contraction of the solar nebula also increased its rate of rotation and caused the left-over gas and dust to be spun out into the *protoplanetary disk* surrounding the protosun. This disk extended 50 AU or more from the center of the protosun. The temperature of the protoplanetary disk nearest the young Sun increased substantially primarily because of the transfer of heat by infrared radiation from the Sun. As a result, the ice particles in the protoplanetary disk closest to the Sun sublimated and only particles composed of refractory compounds survived, such as the oxides of metals. At a distance of greater than about 5 AU from the Sun, the protoplanetary disk remained cold enough for ice particles to survive. In this way, the temperature gradient altered the chemical composition of the protoplanetary disk because only particles of refractory compounds (i.e., those having high melting temperatures) survived near the Sun, whereas all kinds of particles and molecules of the solar nebula survived in the disk at distances greater than about 5 AU (i.e., the radius of the orbit of Jupiter).

During the final stages of contraction of the Sun, the solar wind (to be discussed in Section 5.3.2 and 5.3.4) blew away most of the gas of the protoplanetary disk close to the Sun. As a result, the refractory dust particles in the inner part of the solar system could stick together and thereby formed larger particles which had a wide range of diameters from one millimeter or less to several hundred kilometers or more. The resulting objects called *planetesimals* continued to grow in mass by colliding with each other to form the terrestrial planets in the inner part of the solar system and the rocky cores of the large gas planets.

The planetesimals near the outer edge of the protoplanetary disk were composed of mixtures of refractory dust particles and ices (e.g., water, carbon dioxide, methane, ammonia, etc.), whereas the planetesimals in the inner (hot) part of the protoplanetary disk contained only small amounts of ice. A large number of ice planetesimals have survived in the outer fringe of the present solar system in the form of Kuiper-belt objects (Chapter 21) whose orbits occasionally take them into the inner part of the solar system as comets (Chapter 22).

The scenario outlined above explains the existence of the terrestrial planets in the inner part of the solar system. In addition, the systematic decrease of the bulk densities of these planets (Figure 3.6A) is a consequence of the chemical fractionation of the protoplanetary disk by the temperature gradient that was imposed on it. The high temperature nearest the Sun not only caused ice particles to sublimate but also enriched the remaining solid particles in iron and other refractory metal oxides such that the planetesimals that ultimately formed the planet Mercury contained more iron than the planetesimals that formed Mars. However, the theory of the origin of the solar system does not explain the anomalously high density of the Earth and the low density of the Moon discussed in Section 3.5 and documented in Table 3.4. These anomalies will be considered in Chapter 9 on the Earth-Moon system.

The bulk densities of the gas planets in Figure 3.6B, considered in the context of the theory of the origin of the solar system, indicate that these planets have rocky cores surrounded by thick atmospheres of hydrogen, helium, and molecular compounds like CO, CO_2, CH_4, H_2O, NH_3, and others. The chemical compositions of the atmospheres of the gas planets are similar to the chemical composition of the Sun because these gases originally existed in the solar nebula. However, the Sun does not contain molecular compounds because they have disintegrated at the high temperatures that prevail in the Sun.

On the other hand, the atmospheres of the terrestrial planets: Venus, Earth, and Mars contain only trace amounts of hydrogen and helium because these gases were expelled from the inner part of the protoplanetary disk by the solar wind. In addition, the masses of these planets are not large enough to hold hydrogen and helium (Appendix 1). Consequently, the atmospheres of the terrestrial planets had a different origin than the atmospheres of the gas planets. Another point to be made here is that Jupiter may have captured a lion's share of the gas expelled from the inner part of the solar system, which may explain why it is the most massive planet in the solar system (Freedman and Kaufmann, 2002; Beatty et al., 1999; Taylor, 1992).

5.3 The Sun

An obvious property of the Milky Way and other galaxies is that the stars are separated from each other by large distances. Consequently, the stars in our galactic neighborhood appear as mere points of light in the sky at night. The resulting isolation of the solar system from neighboring stars protects life on the Earth from the dangerous explosions that can occur during the evolution of stars and from the ultraviolet light and x-rays they emit.

The remoteness of the solar system from the nearest stars in the Milky Way galaxy also increases the difficulties astronomers experience who try to study stars at different stages in their evolution in order to determine their internal structure and to understand the processes that occur in their interiors. Much of what is known about stars on the main sequence of the H-R diagram is based on observations of the Sun, which rules the solar system. The very existence

of the Earth and all other objects in the solar system resulted from the origin of the Sun, which also holds the planets in their orbits and warms their surfaces by irradiating them with light (Section 6.1). If the Sun were to be replaced by a black hole of equal mass, the planets would revolve around it in total darkness and their surface temperatures would be far too low for life to exist anywhere in the solar system. The Sun has sustained life on Earth for close to four billion years and has enabled sentient beings on the Earth to ask questions about how it all works. (See also Chapter 23).

5.3.1 Internal Structure

The data in Table 3.1 indicate that the Sun has a radius of 695,510 km, a mass of 1.989×10^{30} kg, a bulk density of $1.410 \, g/cm^3$, and a surface temperature of 5507 °C. We also know that the average distance between the Earth and the Sun is 149.6×10^6 km, which allows us to observe the Sun with a delay of only 8.3 minutes caused by the time it takes light emitted by the Sun to reach the Earth. However, observers must be careful *never to look directly at the Sun* in order to avoid permanent damage to the eyes.

The interior of the Sun in Figure 5.2 consists of a core from which prodigious amounts of energy are released by the hydrogen-fusion reaction. The temperature at the center of the core exceeds 15×10^6 K, the pressure is greater than 225×10^9 atmospheres, and the density is about $150 \, g/cm^3$ even though the core of the Sun is composed primarily of hydrogen and helium. The graphs in Figure 5.3 demonstrate that temperature, pressure, and density of the interior of the Sun decrease toward the surface.

The energy released by hydrogen fusion in the core is transported toward the surface in the form of radiation rather than by conduction because gas is an inefficient conductor of heat. However, when the temperature decreases to about 2.2 million K at 0.71R from the center of the Sun, the gas becomes opaque which reduces the efficiency of heatflow by radiation. For this reason, the heat is transported the rest of the way to the surface by convection, which occurs when hot gas rises to the surface, cools, and then sinks back into the interior of the Sun.

The Sun

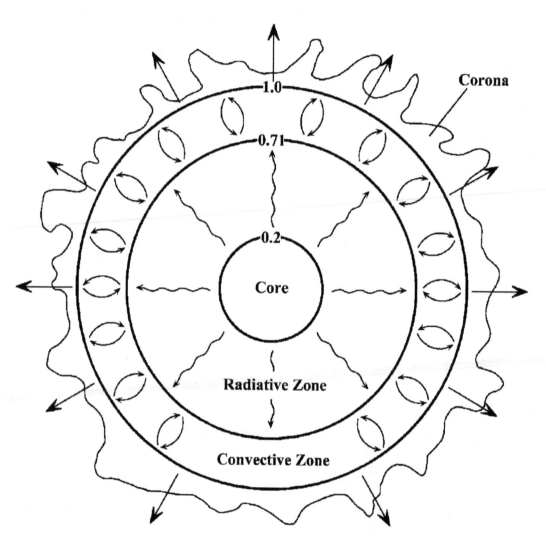

Figure 5.2. The interior of the Sun contains the core (0 to 0.2R) where energy is released by the hydrogen-fusion reaction. The energy is transported towards the surface in the form of radiation (0.2R to 0.8R) and by convection (0.8R to 1.0R), where R is the radius of the Sun (695,510 km). The energy is then radiated into space through the corona, which extends irregularly for millions of kilometers into space. Adapted from Hester et al. (2002, Figure 13.6) and from Wagner (1991, Figure 6.12)

The hot gas at the top of the convective zone radiates light into the atmosphere of the Sun which consists of the basal *photosphere*, the overlying *chromosphere*, and the *corona*. The density and temperature of matter in these layers of the solar atmosphere vary systematically with height above the base of the photosphere. The graphs in Figure 5.4 indicate that the density decreases from less than 10^{-6} g/cm^3 at the base of the photosphere to less than 10^{-14} g/cm^3 at the base of the corona about 2250 km above the top of the convective zone. The temperature of the Sun's atmosphere rises from about 5780 K in the photosphere to about 10,000 K in

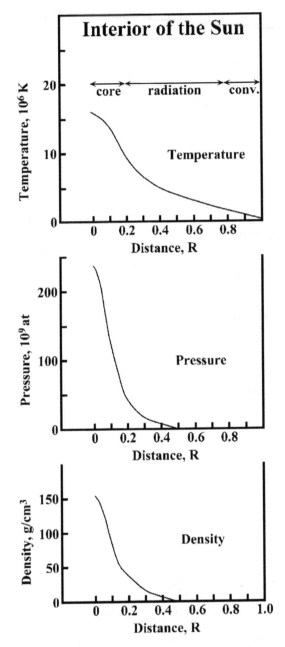

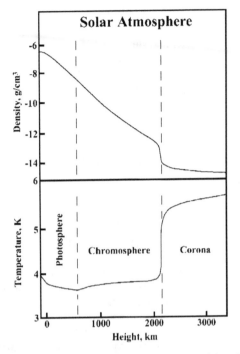

Figure 5.4. Profiles of density and temperature of the solar atmosphere expressed in powers of 10. For example, "−10" on the scale of densities is 10^{-10} g/cm^3 and "5" on the scale of temperatures is 10^5 K. The light emitted by the surface of the Sun originates from the photosphere. Adapted from Hester et al. (2002, Figure 13.11)

Figure 5.3. Variation of temperature, pressure, and density in the interior of the Sun, which is subdivided into the core (0 to 0.2R), the radiation zone (0.2R to 0.8R) , and the convection zone (0.8R to 1.0R) where R is the radius of the Sun (695,510 km) and 0.2R = 139, 102 km. All three parameters have maximum values in the center of the core (R = 0) and decline toward the surface. Adapted from Hester et al. (2002, Figure 13.2)

the chromosphere. It then rises abruptly to values between 2 and 3 million kelvins in the overlying corona. The temperature increase of the corona which coincides with the decrease of its density, has been attributed to heating by the magnetic field of the Sun (Lang, 1999). The high temperature of the corona causes atoms to lose electrons and to acquire positive electrical charges. The resulting ions emit ultraviolet light and x-rays when electrons enter vacant orbitals in ions such as Fe^{13+} and Ca^{14+} (Hester et al., 2002). The corona becomes visible when the Sun is eclipsed by the Moon or when the direct light of the Sun is blocked by a coronagraph. The *solar wind*, which consists of protons, electrons, and ionized atoms, originates deep within the corona and blows continually at 350 to 700 km/s throughout the solar system to the outermost boundary of the heliosphere (Schrijver, 2006; Human, 2000).

5.3.2 Energy Production

The luminosity and surface temperature place the Sun on the main sequence in the H-R diagram as shown in Figure 4.3. Like all other stars on the main sequence, the Sun generates energy by the nuclear fusion of hydrogen to helium called the proton-proton chain and represented by the equations:

$$^1_1H + ^1_1H \rightarrow ^2_1H + positron(e^+) + neutrino(\nu) \tag{5.5}$$

$$e^+ + e^- \rightarrow 2 \text{ gamma rays}(\gamma) \tag{5.6}$$

$$^2_1H + ^1_1H \rightarrow ^3_2He + \gamma \tag{5.7}$$

$$^3_2He + ^3_2He \rightarrow ^4_2He + ^1_1H + ^1_1H \tag{5.8}$$

The first step in the process (equation 5.5) requires the collision of two hydrogen nuclei (protons) with sufficient energy to overcome the electrostatic repulsion of the two protons and thus to form the nucleus of deuterium composed of one proton and one neutron as well as a positron (positively charged electron) and a neutrino which are released from the product nucleus (Science Brief 5.5.2). The positron is annihilated by interacting with a negatively charged electron as shown in equation 5.6. This antimatter-matter reaction converts the positron and the electron into two gamma rays, which have equal energy but travel in opposite directions. The deuterium nucleus collides with another proton (equation 5.7) to form the nucleus of stable helium-3 $\left(^3_2He\right)$, which de-excites by emitting a gamma ray. The final step in the process (equation 5.8) occurs when two helium-3 nuclei collide with sufficient energy to form the nucleus of stable helium-4 $\left(^4_2He\right)$ thereby releasing two protons, which return to the hydrogen reservoir in the core of the Sun. The fusion of four protons into one nucleus of ^4_2He releases energy because the four protons have more mass than the product nucleus of ^4_2He and the excess mass is converted into energy in accordance with Einstein's equation:

$$E = mc^2 \tag{5.9}$$

where E = energy, m = mass, and c = speed of light. The total amount of energy released by this process is 19.794 million electron volts

(MeV) per atom of helium produced. (Note that $1\,MeV = 1.60207 \times 10^{-6}\,erg$). The rate of conversion of hydrogen into helium is approximately equal to 600 million metric tons per second (Hester et al., 2002). These facts justify the statement that the Sun generates prodigious amounts of energy by the nuclear reactions in its core.

The energy released by the proton-proton chain explains why stars on the main sequence are able to radiate energy in the form of light. However, when hydrogen fusion was first proposed in 1920 by Sir Arthur Eddington, some of his colleagues complained that the temperature inside stars is not high enough to overcome the electrostatic repulsion between two protons. Sir Arthur refuted this argument with the statement:

"We do not argue with the critic who urges that the stars are not hot enough for this process; we tell him to go and find a hotter place."

The proton-proton chain actually contributes only about 85% of the total amount of energy released by hydrogen fusion in the Sun. The remaining 15% arises from the production of helium by the CNO cycle discovered by Hans Bethe. The CNO cycle depends on the presence of nuclei of carbon-12 in the Sun and other second-generation stars. The sequence of reactions that constitute the CNO cycle starts with the addition of a proton $\left(^1_1H\right)$ to the nucleus of carbon-12 $\left(^{12}_6C\right)$:

$$^{12}_6C + ^1_1H \rightarrow ^{13}_7N + \gamma(\text{unstable}) \tag{5.10}$$

$$^{13}_7N \rightarrow ^{13}_6C + e^+ + \nu(\text{stable}) \tag{5.11}$$

$$^{13}_6C + ^1_1H \rightarrow ^{14}_7N + \gamma(\text{stable}) \tag{5.12}$$

$$^{14}_7N + ^1_1H \rightarrow ^{15}_8O + \gamma(\text{unstable}) \tag{5.13}$$

$$^{15}_8O \rightarrow ^{15}_7N + e^+ + \nu \tag{5.14}$$

$$^{15}_7N + ^1_1H \rightarrow ^{12}_6C + ^4_2He \text{ (stable)} \tag{5.15}$$

where N = nitrogen, O = oxygen, γ = gamma ray, e^+ = positron, and ν = neutrino.

The nucleus of $^{13}_7N$ produced in reaction 5.10 is unstable and decays to stable $^{13}_6C$ by emitting a positron (equation 5.11). Carbon-13 is stable and absorbs an additional proton to form stable $^{14}_7N$ (equation 5.12). The process continues when $^{14}_7N$ absorbs a proton to form unstable $^{15}_8O$ (equation 5.13), which decays by positron emission to stable $^{15}_{7}N$ (equation 5.14). The last reaction in the CNO cycle (equation 5.15) occurs when $^{15}_{7}N$ absorbs a proton and then disintegrates into $^{12}_{6}C$ and $^{4}_{2}He$.

In this way, the CNO cycle converts four hydrogen nuclei $\left(^{1}_{1}H\right)$ into one nucleus of helium $\left(^{4}_{2}He\right)$ and releases the same amount of energy as the proton-proton chain (i.e., 19.794 MeV). The positrons (e^{+}) are annihilated by interacting with extranuclear electrons (e^{-}) and the neutrinos escape from the Sun without interacting with it in any way. (Science Briefs, 5.5.2). The nucleus of $^{12}_{6}C$ acts as the catalyst that facilitates the fusion of four protons into a helium nucleus. It is released at the end of the process and is available for another passage through the CNO cycle.

5.3.3 Magnetism and Sunspots

All stars act like giant magnets because they extend lines of magnetic force into the space around them (Science Brief 5.5.3). The lines of magnetic force that surround a star constitute its *magnetic field*. The magnetism of stars is induced by electrical currents that are activated when hot plasma in their interiors is made to flow in closed loops. The flow of hot plasma is probably related to the fact that all stars rotate about a virtual axis. For example the period of rotation of the photosphere of the Sun ranges from 25.7 days at the equator to 33.4 days at a solar latitude of 75°. In addition, the rates of rotation of the Sun change with depth below the surface but converge to about 28 days at the boundary between the top of the core and the bottom of the radiative zone. The magnetic field of the Sun is probably generated in a layer about 20,000 km thick that is located between the radiative and convective zones (Lang, 1999).

The lines of magnetic force extend from the surface of the Sun into space in the form of closed loops. The geometry of the solar magnetic field in Figure 5.5 reflects the fact that the Sun has magnetic poles of opposite polarity located close

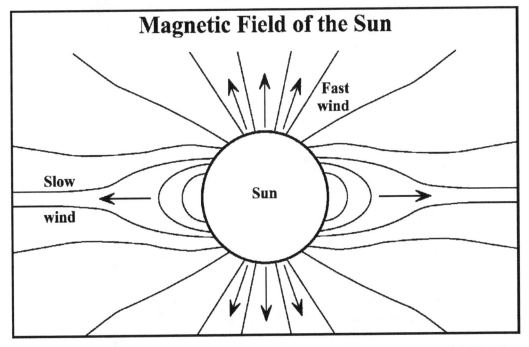

Figure 5.5. The Sun acts like a bar magnet by extending lines of magnetic force into the space around it. These lines of force form closed loops that are anchored within the Sun. Recent data obtained by the spacecraft Ulysses indicate that the lines of magnetic force are spaced fairly evenly over the surface of the Sun. The fast solar wind (700 km/s) follows the field lines emanating from the poles, whereas the low-velocity particles (350 km/s) escape from the equatorial region of the Sun. Adapted from Lang (1999, Figure 21)

to the axis of rotation. The pattern of lines of magnetic force is disturbed by the rotation of the Sun which causes the loops of the magnetic field to be twisted. In addition, the magnetic field of the Sun interacts with the solar wind and forces the ions to flow along lines of magnetic force. When large amounts of hot plasma called *solar flares* erupt from the surface of the Sun as in Figure 5.6, the magnetic field shapes the flares into arches and loops.

In the seventeenth century AD, astronomers studying the Sun observed the occurrence of dark spots in the photosphere. These *sunspots* are cooler than the surrounding photosphere by about 1500 K and are caused by localized increases of the strength of the magnetic field caused by the convective currents of plasma in the Sun (Byrne, 2005). Individual sunspots have a dark center called the *umbra*, which is surrounded by

a region that is less dark called the *penumbra*. Sunspots occur in pairs connected by arches of magnetic lines of force. The existence of these arches indicates that the sunspots that form a pair have opposite magnetic polarity. Although individual sunspots fade in a few days, the intensity of sunspot activity and its location on the surface of the Sun in Figure 5.7 vary in cycles of about 11 years. At the start of a new cycle, sunspots first appear at about 30° north and south latitude on the surface of the Sun. In the following years, the sites of sunspot activity move towards the equator where the activity ends 11 years after it started.

The sunspot cycle has been observed for about 400 years. The records indicate that the period of the sunspot cycles has actually varied from 9.7 to 10.8 years and that their intensity has also fluctuated. For example, during the Little Ice

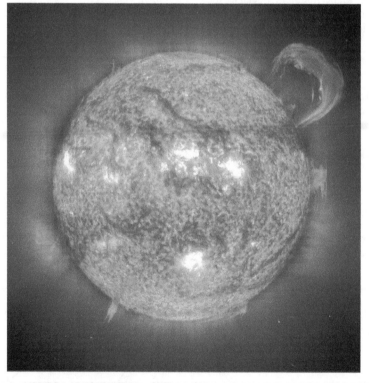

Figure 5.6. Image of the Sun taken on September 14, 1999, with the Extreme Ultraviolet Imaging Telescope (EIT). This view shows the upper chromosphere/lower corona of the Sun at a temperature of about 60,000 K. The large prominence is composed of protons and electrons that are being ejected by the Sun into the corona where the temperature typically exceeds one million kelvins. When a solar flare like this is directed toward the Earth, it can disturb its magnetic field and cause a spectacular aurora in the polar regions. Courtesy of the SOHO/EIT consortium. SOHO is a project of international cooperation between ESA (European Space Agency) and NASA. (http://soho.nasc om.nasa.gov/)

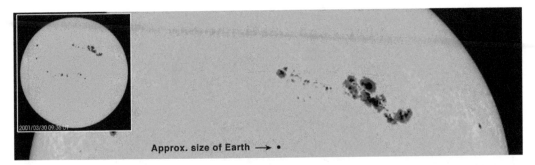

Approx. size of Earth ➝ •

Figure 5.7. A large group of sunspots was observed by SOHO on March 30, 2001. The area encompassed by these sunspots is equivalent to 13 times the surface of the Earth. Several large flares and coronal mass ejections occurred in this area of the Sun, including one on April 2, 2001, which was the largest flare recorded in 25 years. The number of sunspots varies with a period of about 11 years. The sunspots in this image consist of dark centers (umbra) surrounded by lighter rims (penumbra). Time-dependent variations in the number of sunspots correlate with variations in the frequency and magnitude of solar flares which disrupt the reception of radio and television broadcasts on the Earth and affect global weather. The images of the Sun in Figures 5.6 and 5.7 reinforce the point that the surface environment of the Earth is affected by the activity of the Sun. Courtesy of SOHO (ESA & NASA)

Age (Maunder Minimum) from 1645 to 1715 AD almost no sunspots were observed. In addition, the polarity of the magnetic field associated with sunspots changes from cycle to cycle with a periodicity of 22 years.

5.3.4 Effect on Space

The Sun affects the surrounding interstellar space by its gravitational and magnetic fields, as well as by the solar wind. The strength of the two force fields and of the solar wind decreases with increasing distance from the Sun by the reciprocal of the distance squared. Therefore, at a distance of about 150 AU from the Sun the solar wind is no longer able to displace the gas and dust in the space between the stars and therefore it piles up against the *interstellar medium*. This phenomenon defines the boundary of a bubble of space in Figure 5.8 within which the solar wind blows unhindered. This boundary is called the *termination shock*. The bubble of space, defined by the termination shock, forces the gas and dust of the interstellar medium to flow around it and thereby defines the *heliopause* which is the surface of the *heliosphere*. The heliosphere is not a perfect sphere because the front is compressed by the resistance of the interstellar medium against the forward motion of the Sun in its orbit around the center of the Milky Way galaxy and the backside of the heliosphere trails behind the Sun as shown in Figure 5.8. The

heliopause is not the outer limit of the solar system because the gravitational field of the Sun extends to about 150,000 AU, which includes the ice bodies of the Oort cloud to be discussed in Section 21.3 (Weissman, 1999, 25–26).

The Voyager 1 spacecraft, which was launched from Earth in 1977, approached the termination shock boundary in August of 2002. It will eventually cross the boundary of the heliopause and will then become the first man-made object to leave the heliosphere and thus to enter interstellar space. Present estimates place the distance between the Sun and the heliopause at between 153 and 158 AU. Voyager 1 may reach this final frontier at about 2020 AD to be followed by Voyager 2, its sister spacecraft, which is 29 million kilometers behind Voyager 1.

5.3.5 Life and Death of the Sun

The theory of stellar evolution can be used to explain the past history of the Sun and to predict its future. For example, the luminosity of the Sun soon after its formation 4.6 billion years ago was only about 70% of its present value and has been increasing gradually. The luminosity will continue to increase until about 1.4 *billion* years from now, when the Sun will be hot enough to cause the oceans to boil thus turning the Earth into a lifeless desert (Hayden, 2000). Four billion years later, when the age of the Sun approaches 11 billion years, the Sun will leave the main

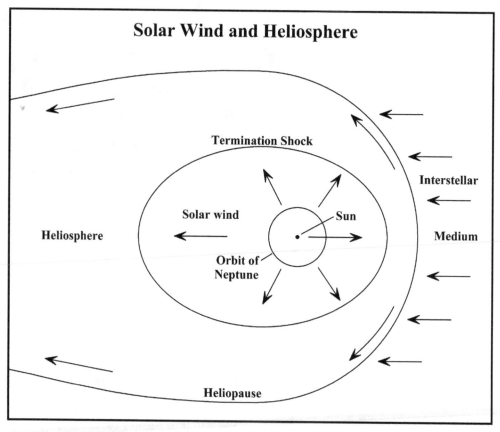

Solar Wind and Heliosphere

Figure 5.8. The solar wind forms an asymmetric bubble around the Sun. The boundary surface of this bubble is defined by the termination shock which occurs about 150 AU from the Sun when the nuclear particles of the solar wind pile up against the gas and dust of the interstellar medium through which the Sun is moving (large arrow). The bubble of solar wind forces the interstellar medium to flow around it, thereby defining the heliosphere which is bounded by the heliopause. Adapted from Hester et al. (2002, Figure 13.17)

sequence and will evolve into a red giant. At an age of 12.25 billion years (about 7.7 billion years from now), the Sun's luminosity will be about 2000 times its present value and its diameter will increase until it reaches the orbit of Venus. At this time, the planet Mercury will have been assimilated into the body of the red giant Sun. The mass of the Sun will actually decline at this stage, which will cause the radii of the orbits of Venus, Earth, and Mars to increase and thereby prevent their assimilation by the red giant Sun. After the Sun's age reaches 12.33 billion years, it will become increasingly unstable because of repeated explosive helium-fusion events in shells around the core. Finally, 12.3655 billion years after its arrival on the main sequence, the Sun's

core will collapse into a white dwarf and the remaining envelope of gas will be blown into interstellar space as a planetary nebula (Lang, 1999). The evolution of the Sun follows the same pattern as that of other stars of similar mass and chemical composition which start on the main sequence and eventually become red giants that last less than one billion years before they explode as supernovae or die as white dwarfs, depending on their masses.

In this way, the fate of the planets of the solar system is directly linked to the evolution of the Sun. The planets originated during the formation of the Sun, the conditions of their existence is governed by the Sun, and they will be destroyed when the Sun dies in a gigantic explosion. This drama has occurred billions of times in the Milky Way and other galaxies since the Universe came into existence.

5.4 Summary

The Milky Way galaxy contains large clouds of gas and dust which formed from mixtures of primordial hydrogen and helium with other chemical elements that were ejected into interstellar space during the violent explosions of red giant stars. Parts of these interstellar clouds contract to form new generations of stars, some of which may have planets in orbit around them. The Sun is such a second-generation star that formed from a cloud of interstellar gas and dust by a process that has caused stars to form in the Milky Way galaxy throughout its history.

The gas of a planetary nebula ejected by a supernova is initially composed of positively-charged ions and electrons because the atoms of the chemical elements do not retain all of their electrons at the high temperature the gas inherits from the supernova. As the gas cools, the ions attract electrons to themselves until their positive charges are neutralized. The resulting atoms of different elements then form covalently bonded molecules of a large number of compounds, including: water, carbon dioxide, methane, hydrocarbons, oxides of metals, and many others. Some of these compounds form small crystals of ice (e.g., water, carbon dioxide, methane, and others) or solid dust particles (e.g., FeO, MgO, CaO, graphite, and others). As a result, the cloud of hot gas ejected by a supernova cools and becomes a mixture of gas (H_2, He) and dust.

The solar nebula, which contracted to form the Sun and the planets of the solar system, was part of a much larger cloud from which several stars in our galactic neighborhood may have formed. The compression of the solar nebula increased the temperature of the resulting protosun. When the core temperature of the protosun reached 12 million kelvins, the nuclear fusion reaction by the proton-proton chain and by the CNO cycle started. The luminosity and surface temperature of the newly-formed star placed it on the main sequence of the Hertzsprung-Russell diagram. The Sun has resided on the main sequence for the past 4.6 billion years and is expected to remain there for a total of about 10 billion years.

The contraction of the solar nebula resulted in an enormous decrease of its volume and caused a corresponding increase in the rate of rotation of the protosun. As a result, small amounts of gas and dust that were left behind were spun into the protoplanetary disk which surrounded the protosun. The temperature of the disk closest to the protosun increased primarily by heat radiated by the protosun and caused ice particles to evaporate. More distant parts of the protoplanetary disk remained cold enough to allow ice particles to survive. The refractory dust particles that remained in the inner part of the disk agglomerated into solid planetesimals composed primarily of refractory compounds of iron, magnesium, and calcium. The planetesimals in the inner part of the protoplanetary disk were composed of refractory particles and formed the terrestrial planets. The planetesimals in the outer part of the protoplanetary disk were composed of both ices and refractory particles and formed the rocky cores of the gas planets.

When the protosun began to generate heat by nuclear reactions in its core, the resulting Sun began to emit a strong wind, which pushed the hydrogen and helium gas in the protoplanetary disk to the outer part of the disk where it was captured by the gravitational fields of the large rocky cores of Jupiter, Saturn, Uranus, and Neptune. The terrestrial planets (Mercury, Venus, Earth, Mars, and Ceres) inherited very little of the primordial hydrogen and helium because their gravitational fields are not strong enough to hold the atoms of these elements. The evolution of the protoplanetary disk into a set of planets and their satellites occurred in less than one hundred million years.

The ultimate fate of the Earth is closely tied to the future evolution of the Sun whose luminosity will continue to increase even while it continues to reside on the main sequence. About 1.4 billion years from now, the surface temperature of the Earth will have increased sufficiently to cause the water in the oceans to boil. Still later, about 5.4 billion years from now, the luminosity and the diameter of the Sun will increase dramatically as it evolves into a red giant. The planet Mercury will be assimilated by the Sun at this time, but Venus, Earth, and Mars may be spared because the radii of their orbits will increase as the Sun loses mass by a series of explosions. Finally, the red giant Sun will explode and its core will collapse into a white dwarf star. The origin of the Sun and of the Earth, their evolution through time, and their ultimate destruction are all part of the ongoing evolution of stars in the Milky Way galaxy.

The Sun is the only star that is accessible to close observation and therefore has been studied intensively. The internal structure of the Sun arises from the way in which heat is transported from the core towards the surface. The major units are: the core (0.20 R), the radiative zone (0.20–0.71)R, and the convective zone (0.71–1.0) R, where R is the radius of the Sun. The heat is then radiated from the photosphere through the chromosphere and the corona into interstellar space in the form of electromagnetic radiation (i.e., visible, ultraviolet, and infrared light). In addition, the Sun emits a flux of ions and electrons called the solar wind which follows lines of magnetic force emanating from the Sun.

The magnetic field of the Sun arises by induction resulting from electrical currents caused by currents of plasma that flow in its interior. The plasma currents may locally compress the magnetic field of the Sun and cause the appearance of sunspots. The number and location of sunspots varies in cycles with a period of about 11 years.

The so-called solar wind forms a bubble around the solar system whose boundary surface (termination shock) is located about 150 AU from the Sun. As the Sun moves through interstellar space, the gas and dust of the interstellar medium are forced to flow around it, thereby defining a volume of space called the heliosphere. The two Voyager spacecraft, which were launched from Earth in 1977, approached the termination shock boundary in 2002 and will leave the heliosphere around 2020 when they pass through the heliopause.

5.5 Science Briefs

5.5.1 Atoms and Ions

Atoms consist of positively charged nuclei surrounded by a sufficient number of negatively charged electrons to make the atoms electrically neutral. We know from the study of electrical charges that opposite charges (positive and negative) attract each other, whereas identical charges repel each other. When neutral atoms are exposed to heat energy (as in the Sun), some of the electrons may break away from their atoms. As a result, atoms that have lost electrons have a positive charge. The magnitude of the charge depends on the number of electrons that were lost. If two electrons are lost by a neutral atom, the resulting ion has a charge of +2.

5.5.2 Neutrinos

Neutrinos are low-mass nuclear particles that are released by unstable nuclei undergoing beta decay. They carry varying amounts of kinetic energy but interact sparingly with matter. Recent research indicates that neutrinos can assume three different forms identified in Table 4.2. In spite of their elusive properties, neutrinos can be detected and are being used to monitor the nuclear reactions in the Sun. In addition, neutrinos were released by the supernova 1987A which occurred in the Large Magellanic Cloud (Figure 4.6).

5.5.3 Magnetic Fields

Magnetism is a force that is exerted by certain types of materials such as the mineral magnetite or objects composed of metallic iron or certain ceramic materials that were magnetized during manufacture. Magnetism has polarity similar to electrical charges. Therefore, the ends of a magnetized iron bar (i.e., a bar magnet) have opposite magnetic polarity, which causes a magnetic force field to form between them as shown in Figure 5.5 (See also Science Briefs 6.7.3, 6.7.4, and 6.7.5).

5.6 Problems

1. The average speed of the solar wind is 850,000 miles per hour. Convert that speed to kilometers per second (Answer: v = 380 km/s).
2. Interstellar space contains about one atom (or ion) per cubic centimeter of space. Calculate the number of atoms (or ions) in one cubic kilometer of space. (Answer: 10^{15} atoms).

3. The fusion of four protons into one nucleus of $^4_2$He releases energy in the amount of 19.794 MeV. Convert this amount of energy into the equivalent energy expressed in joules ($1 \, eV = 1.6020 \times 10^{-19}$ J). (Answer: 31.709×10^{-13} J).

4. Calculate the number of $^4_2$He atoms that must be produced in the Sun each second to provide enough energy to equal the solar luminosity of 3.9×10^{26} J/s. (Answer: 1.229×10^{38} atoms of $^4_2$He per second).

5. Calculate the number of hydrogen nuclei that exist in the Sun based on the following information:

Mass of the Sun	$= 1.99 \times 10^{30}$ kg
Concentration of hydrogen	$= 74\%$ by mass
Atomic weight of hydrogen	$= 1.00797$
Number of atoms per mole	$= 6.022 \times 10^{23}$

(Answer: 8.79×10^{56} atoms)

6. Given that the rate of helium production is 1.229×10^{38} atom/s and that the number of hydrogen nuclei in the Sun is 8.79×10^{56} atoms, estimate how much time in years is required to convert all of the hydrogen in the Sun into helium. Remember that four hydrogen atoms form one helium atom. (Answer: 57×10^9 y).

7. Refine the calculations of the life expectancy of the Sun by considering that hydrogen fusion in the Sun occurs only in its core and that the density of the core is greater than the bulk density of the Sun as a whole.

   a. Calculate the mass of the core of the Sun given that its radius is 140,00 km and that its density is $150 \, g/cm^3$. (Answer: 1.72×10^{33} g)

   b. Calculate the number of atoms of hydrogen in the core of the Sun assuming that its concentration is 74 % (mass). (See problem 5) (Answer: 7.62×10^{56} atoms).

   c. Estimate how much time in years is required to convert all of the hydrogen in the core of the Sun into helium. (Answer: 49×10^9 y).

   d. How does this result compare to the accepted life expectance of the Sun and consider possible reasons for the discrepancy.

5.7 Further Reading

Beatty JK, Petersen CC, Chaikin A (1999) The new solar system, 4th edn. Sky Publishing, Cambridge, MA

Byrne G (2005) Cycle of the Sun. Astronomy 33(6):40–45

Cowen R (2003) Cool cosmos: Orbiting telescope views infrared universe. Science News 164 Dec 20 and 27:164

Freedman RA, Kaufmann III WJ (2002) Universe: The solar system. Freeman and Co., New York

Hartmann WK (2005) Moons and planets, 5th edn. Brooks/Cole, Belmont, CA

Hayden T (2000) The evolving Sun. Astronomy, 28(1):43–49

Hester J, Bursten D, Blumenthal G, Greeley R, Smith B, Voss H, Wegner G (2002) 21st century astronomy. W.W. Norton, New York

Human K (2000) When the solar wind blows. Astronomy 28(1):56–59

Lang KR (1999) The Sun. In: Beatty J.K., Petersen C.C., Chaikin A (eds), The new solar system. 4th edn. Sky Publishing, Cambridge, MA, pp. 23–38

O'Dell CR (2003) The Orion nebula: Where stars are born. Belknap, New York

Schrijver CJ (2006) The science behind the solar corona. Sky and Telescope 111(4):28–33

Taylor SR (1992) Solar system evolution: A new perspective. Cambridge University Press, New York

Wagner JK (1991) Introduction to the solar system. Saunders College Publishing, Philadelphia, PA

Weissman PR (1999) The solar system and its place in the galaxy. In: Weissman P.R., McFadden L.-A., Johnson T.V. (eds), Encyclopedia of the solar system. Academic Press, San Diego, CA, pp 1–33

Earth: Model of Planetary Evolution

The Earth in Figure 6.1 is the most massive of the four terrestrial planets that formed in the inner part of the solar system from planetesimals composed primarily of refractory compounds of oxygen, silicon, aluminum, iron, calcium, magnesium, sodium, potassium, and titanium. These planetesimals contained only small amounts of ice composed of volatile compounds such as water, carbon dioxide, methane, and ammonia because the ice particles in the inner part of the protoplanetary disk sublimated when heat emanating from the protosun caused the temperature in its vicinity to rise. In spite of the loss of these volatile compounds from the inner part of the solar system, the Earth has a large volume of water on its surface. Water is also present on Mars in the form of ice in the polar regions. Therefore, questions arise about the sources of water on Earth and Mars and about whether Mercury, Venus, and the Moon may also have contained water at the time of their formation.

Other significant properties that characterize the Earth include the presence of an atmosphere containing molecular oxygen, the presence of a magnetic field, and a narrow range of surface temperatures within which water can exist in liquid form. Even more important is the fact that some of the surface features of the Earth are still controlled by dynamic processes that are energized by heat that continues to be generated in its interior by the radioactivity of the atoms of uranium, thorium, and potassium.

6.1 Growth from Planetesimals

The solid dust particles in the inner region of the protoplanetary disk (i.e., less than 5 AU from the Sun) were composed primarily of refractory compounds because the elevated temperature in this region caused ice particles to sublimate. These refractory dust particles accreted to form the terrestrial planets of the solar system. The details of this process remain obscure because it was dominated by random events that are not reproducible. Initially, a large number of small bodies may have formed which continued to grow by attracting dust particles from the disk. The rate of growth at this stage depended on the masses of the growing bodies, such that the most massive bodies grew faster than less massive objects. The resulting gravitational interactions eventually caused the smaller objects to impact on the largest bodies, thereby accelerating the growth and raising the internal temperatures of the large bodies that were destined to become the earthlike planets of the solar system (Kenyon, 2004).

The rate of accretion of the terrestrial planets was slow at first but accelerated and became violent at the end when the protoplanetary bodies were bombarded by the impacts of solid objects that had not grown as rapidly as the protoplanets. At this stage, the protoplanets were heated by the impacts, by the continuing decay of radioactive atoms that had been produced in the ancestral stars, and by the compression of the increasing amount of matter of which they were composed.

The accretion of dust particles in the outer region of the protoplanetary disk proceeded similarly, except that both refractory and ice particles were present. The largest bodies that formed became the rocky cores of the giant planets: Jupiter, Saturn, Uranus, and Neptune. The strong force of gravity exerted by the rocky cores of these planets attracted hydrogen, helium, and other volatile compounds from the protoplanetary disk to form the atmospheres that characterize these giant gas planets. The extent of the "outer region" of the

Figure 6.1. This picture of the "Visible Earth" is one of the most important images ever recorded in human history because it confirms that the Earth is a spherical body suspended in space by the force of gravity exerted by the Sun (Science Brief 1.5.2,3; Appendix Appendix 1). The water vapor of the atmosphere condenses to form clouds which reveal the movement of air masses that determine the local weather. Most of the surface of the Earth is covered by an ocean of liquid water that surrounds the continents and islands where we live. This view of the Earth shows the continent of North America, including the peninsulas of Florida, Yucatán, and Baja, as well as most of the southwestern region of the USA and Canada. The unique feature of this view is that the Earth is revealed to be an oasis in the vastness of space. Although the Earth is affected by extraterrestrial forces and processes (e.g., the amount of solar energy), its surface environment is largely controlled by the dynamics of its own mantle, crust, hydrosphere, and atmosphere. Courtesy of NASA/visible Earth. (http://visibleearth.nasa.gov/)

solar system is far greater than that of the "inner region" both in terms of the volume of space and in terms of the mass it contains (Problem 1, Section 6.8). The smaller objects that formed in this vast region of space either impacted on the protoplanets that were forming, or were captured by them into orbit, or were deflected into the inner region of the solar system, or were ejected to the outermost reaches of the protoplanetary disk.

All of these bodies were initially composed of random mixtures of refractory particles and ice consisting of volatile compounds (e.g., water, oxides of carbon, methane, and ammonia). The largest of these bodies generated enough internal heat to differentiate into a rocky core surrounded by an ice mantle and crust. The heat required for this internal differentiation originated from the decay of radioactive atoms, from the compression of the material of which they were composed, and as a result of tidal interactions with the large planets in this region and with each other.

The ice-bearing bodies that were deflected into the inner region of the solar system eventually collided with the terrestrial protoplanets and released water and other volatile compounds they contained. As a result, all of the terrestrial planets initially developed atmosphere composed of the volatile compounds that were delivered by ice-bearing impactors from the outer region of the solar system. In addition, ices of volatile compounds were occluded in the impactors that had formed in the inner region and had survived within these bodies.

The protoplanetary disk and all of the objects of whatever size and composition continued to orbit the Sun. The contraction of the solar nebula into the Sun and the protoplanetary disk and the subsequent evolution of that disk into the planets and their satellites occurred in less than about 100 million years, which is rapid on the timescale of the Universe (i.e., 13.6×10^9 y). The accretion of the Earth took only about 10 million years or less which is also rapid compared to its age of 4.56×10^9 y.

The processes that caused the formation of the planets of the solar system described above are an extension of the theory of star formation supplemented by information derived from the study of meteorites and the Moon. In addition, the sequence of events is consistent with the applicable laws of physics and chemistry with due allowance for the chaotic nature of this process. By adopting this hypothesis of the origin of the planets, we can recognize the unique set of circumstances that affected not only the origin of each planet but also shaped its subsequent evolution. For example, the chemical differentiation of the protoplanetary disk is a plausible explanation for the differences between the masses and compositions of the planets in the inner and outer solar system. Even the regular decrease of the bulk densities of the terrestrial planets with increasing distance from the Sun fits into this framework. However, it does not account for the anomalously high density of the Earth and for the equally anomalous low density of the Moon. Therefore, we conclude from this example that the properties of the planets are not necessarily attributable to the process by which they formed and that special circumstances in their subsequent history have intervened to produce the planets in their present form. This subject was discussed in complete detail and with excellent clarity by Taylor (1992) and in an issue of the journal "Elements" (vol. 2, No. 4, 2006).

6.2 Internal Differentiation

The impact of planetesimals upon the Earth released large amounts of energy primarily in the form of heat. The planetesimals impacted at a rapid rate such that the heat generated in this way was sufficient to melt the Earth. Therefore, the Earth was initially composed of a mixture of molten silicates, molten sulfide, and molten iron. In geology a body of molten silicate that may form in the crust and upper mantle of the Earth is called a *magma*. Therefore, when the Earth was completely or even partially molten, it is said to have had a *magma ocean* shown in Figure 6.2 (Miller et al., 2001). Metallurgists who smelt ores in order to recover the metals they contain have known for many centuries that silicate melts, sulfide melts, and liquid iron are immiscible in each other and separate from each other based on their densities. Accordingly, globules of molten iron coalesced into larger masses which sank through the magma ocean towards the center of the Earth where they collected to form its iron core. The globules of molten sulfide likewise sank through the magma ocean but moved more slowly than the molten iron because iron sulfide has a lower density than metallic iron (Table 3.3).

At the end of the period of accretion, the Earth began to cool which caused silicate minerals to crystallize from the silicate magma. The sequence of minerals that crystallize from magma at decreasing temperatures depends on environmental factors (e.g., temperature and pressure) as well as on the chemical composition of the magma. In general, minerals can crystallize from a melt of the appropriate

Figure 6.2. This painting by Chesley Bonestell shows the Earth at an early stage of its evolution about 4.6 to 4.5 billion years ago. At that time, minerals were crystallizing in a global ocean of molten silicate as the Earth differentiated into a core composed of liquid iron and nickel, a mantle consisting of olivine and pyroxene crystals, and a primordial crust of anorthite crystals that floated in the magma ocean. At that time, the Earth already had a primordial atmosphere containing clouds that formed when water vapor condensed as the temperature of the atmosphere was decreasing even though the surface of the Earth was still being bombarded by impacts of solid objects. The Moon appears large on the horizon because it was closer to the Earth when it first formed than it is today (Section 9.6). This view of the early Earth is a realistic visualization of a stage in the evolution of the Earth. This and other images by Bonestell popularized the exploration of the solar system many years before the technology became available that has enabled the current exploration of the planets and their satellites (Miller et al., 2001). (Reproduced by permission of Ron Miller)

chemical composition when the temperature of the melt has decreased to the melting temperature of the minerals under consideration. This simple rule must be modified in case the magma contains water and other compounds that can lower the melting temperatures of minerals. Nevertheless, in the simplest case, the minerals crystallize in the order of their decreasing melting temperatures as expressed by Bowen's reaction series in Figure 6.3. According to this scheme, olivine $[(Mg, Fe)_2SiO_4]$ and the Ca-plagioclase anorthite $(CaAl_2Si_2O_8)$ crystallize at temperatures between 1880 and 1535 °C. The early-formed olivine crystals may react with the residual magma to form pyroxenes which are silicates of magnesium, iron, and calcium with varying amounts of aluminum. The important point is that olivine and pyroxene have high densities of about $3.4 \, g/cm^3$ that exceed the density of basalt magma (i.e., $3.0 \, g/cm^3$). Consequently, the early-formed crystals of olivine and pyroxene *sank* in the magma ocean of the early Earth and were deposited around the

iron core thereby forming the peridotite rocks of the mantle of the Earth.

The anorthite that crystallized from the magma at about the same high temperature as olivine and pyroxene has a low density of only $2.76 \, g/cm^3$ and therefore *floated* toward the top of the magma ocean where it accumulated as a solid crust composed of the rock anorthosite. This crust continued to be broken up by violent explosions caused by the impact of planetesimals that collided with the Earth for about 600 to 800 million years after its formation. In addition, the surface of the Earth was disrupted by volcanic activity which occurred more frequently and was more violent than at present because the Earth contained more heat, because the crust was thinner than it is today, and because the impacts of planetesimals triggered large-scale melting in the mantle. The evolution of the Earth at this time in the past was discussed by Stevenson (1983), Ernst (1983), Lunine (1999), Wilde et al. (2001), Moorbath (2005), and Valley (2005). During this time, conditions on the surface of

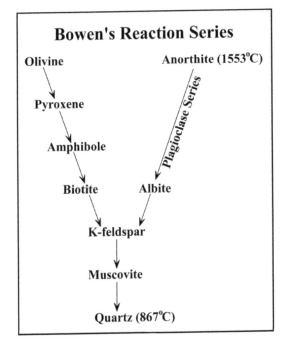

Figure 6.3. Bowen's reaction series indicates the sequence of silicate minerals that crystalize from a cooling magma of basaltic composition. Olivine [$(Mg, Fe)_2SiO_4$] and the calcium-rich plagioclase anorthite ($CaAl_2Si_2O_8$) crystallize first at the highest temperature of 1553 °C. Olivine has a high density between 3.27 and 3.37 g/cm^3 depending on its iron content, whereas anorthite has a low density of 2.76 g/cm^3. Therefore, olivine crystals forming from basalt magma (density = 3.0 g/cm^3) sink, while anorthite crystals float. The removal of early-formed crystals of olivine, pyroxene, and Ca-rich plagioclase depletes the remaining magma in magnesium, iron, and calcium and enriches it in sodium, potassium, and silica (SiO_2). Adapted from Press and Siever (1986, Figure 15.8) and Emiliani (1992, Tables 10.4 and 10.5)

the Earth and of the other terrestrial planets were quite different than the tranquillity of the present solar system. Therefore, geologists refer to the time interval between 4.6 and 3.8 billion years in the past as the Hadean Eon (i.e., hades = hell).

The separation of immiscible iron and sulfide liquids and the crystallization of the silicate liquid can account for the large-scale features of the internal structure of the Earth in Figure 6.4. The iron core at the center of the Earth consists of an inner solid core (radius = 1329 km) and an outer liquid core giving the combined core a radius of 3471 km. The core is overlain by

the mantle which is composed of peridotite and its high-pressure equivalent called *pyrolite* for a combined thickness of 2883 km.

The mantle of the Earth is subdivided into the *asthenosphere* and the *lithosphere*. The two subdivisions of the mantle differ primarily in their mechanical properties which depend largely on the temperature. The rocks of the asthenosphere can deform plastically when they are subjected to stress because the minerals are so close to their melting temperatures that a small amount of melt may be present. The lithosphere is rigid and forms a hard shell about 120 km thick around the underlying asthenosphere. In addition, the Earth has continents which are about 40 km thick and are composed primarily of silicate minerals that have low densities, such as: orthoclase, the mica minerals biotite and muscovite, and quartz. The average density of the rocks of the continental crust is about 2.7 g/cm^3, which means that the continents are actually floating in the underlying lithospheric mantle whose density is 3.2 g/cm^3 in accordance with Archimedes' Principle (Science Brief 6.7.1).

The differentiation of the Earth as a result of the segregation of immiscible liquids caused the chemical elements to be partitioned into those with an affinity for metallic iron (siderophile), those which associate with sulfur (chalcophile), and the elements that enter the silicate melt (lithophile). The preferential association of the chemical elements with the three immiscible liquids is the basis for the geochemical classification of the elements developed by the geochemist *V.M. Goldschmidt (1888–1947)*. (See Faure, 1998)

6.3 Atmosphere

Goldschmidt's geochemical classification also includes the *atmophile* elements: hydrogen, nitrogen, helium, neon, argon, krypton, xenon, and radon. These elements as well as certain volatile compounds (e.g., water, oxides of carbon, methane, ammonia, and hydrogen sulfide) were present in the planetesimals that formed the Earth. When the planetesimals exploded on impact or fell into the magma ocean, the gases were released and formed an atmosphere that enveloped the Earth. Any residual hydrogen and helium leaked away

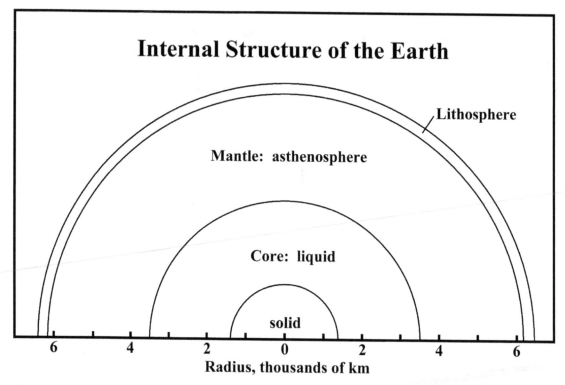

Internal Structure of the Earth

Lithosphere

Mantle: asthenosphere

Core: liquid

solid

6   4   2   0   2   4   6

Radius, thousands of km

Figure 6.4. The interior of the Earth consists of an iron core, the outer part of which is liquid, a solid mantle composed of silicate rocks, and a thin oceanic and continental crust (not shown). The upper part of the mantle and the overlying crust form a rigid shell called the lithosphere which is about 120 km thick. The rest of the mantle called the asthenosphere is solid but can deform plastically when subjected to stress acting over long periods of time. The variation of the density of the mantle and core at increasing depth is illustrated in Figure 3.5. Adapted from Faure (1998, Figure 4.1)

because the force of gravity of the Earth is not strong enough to retain these elements in the atmosphere. The violent explosions caused by the impacts of large planetesimals disrupted the primordial atmosphere, but the resulting losses were offset by the impact of late-arriving ice planetesimals from the outer fringes of the solar system which released water and the other volatile compounds into the primordial atmosphere of Earth. In addition, the vigorous volcanic activity that occurred during the period of intense bombardment allowed gases to escape from the interior of the Earth.

As the main phase of the accretionary period of the Earth ended, the atmosphere contained most of the water and other volatile compounds listed above as well as diatomic nitrogen molecules (N_2) and the noble gases because the Earth was still too hot to contain them. However, molecular oxygen (O_2) was present only in trace amounts because most of the oxygen in the planetesimals was bonded to other chemical elements. Small amounts of oxygen and hydrogen were released by the decomposition of water molecules by ultraviolet light emitted by the Sun. The hydrogen molecules escaped into interplanetary space, but the oxygen reacted with methane (CH_4) to form carbon dioxide (CO_2):

$$CH_4 + O_2 \rightarrow CO_2 + 2H_2 \qquad (6.1)$$

Carbon dioxide, water vapor, and methane are so-called greenhouse gases because they absorb infrared radiation and thereby cause the temperature of the atmosphere to rise. Therefore, the presence of these gases in the atmosphere may have prevented the Earth from lapsing into a deep freeze at a time when the luminosity of the Sun was only 70% of its present value. However, after the anorthosite crust insulated the atmosphere from the hot interior of the Earth, the temperature of the atmosphere decreased, which caused water vapor to condense and rain started falling all over the Earth (Science Brief 6.7.2). As the atmosphere continued to cool, most of the water it contained was transferred to the surface of the Earth in the form of rain and became a global ocean. The raindrops that formed in the atmosphere dissolved carbon dioxide and other volatile compounds and thus removed these greenhouse gases from the atmosphere, which moderated the global climate of the Earth and prevented the water in the global ocean from boiling, as it may have done on Venus.

Although the details of the evolution of the atmosphere are still uncertain, geological evidence proves that liquid water existed on the surface of the Earth more than 3.8 billion years ago and may have been present much earlier (Moorbath, 2005). In addition, evidence derived from the study of the Moon indicates that the terrestrial planets, including the Earth, continued to be bombarded by objects from space until about 3.8 billion years ago. Nevertheless, geological processes began to shape the surface of the Earth as soon as liquid water could exist on its surface.

The geological activity during the Hadean Eon (4.6 to 3.8 billion years) included volcanic eruptions and the formation of volcanic islands in the global ocean. The lava flows that were extruded were composed of pyroxene, olivine, calcium-rich plagioclase, and magnetite (Fe_3O_4) which identifies them as varieties of *basalt* known as komatiite and tholeiite (Faure, 2001). The water in the ocean, which was not yet salty at this time, dissolved these minerals in a process known as *chemical weathering*. The ions of magnesium, iron, and calcium that entered the ocean reacted with the dissolved carbon dioxide to form solid precipitates of sedimentary carbonate minerals:

$$Fe^{2+} + CO_2 + H_2O \rightarrow FeCO_3 + 2H^+ \quad (6.2)$$
(siderite)

$$Mg^{2+} + CO_2 + H_2O + \rightarrow MgCO_3 + 2H^+$$
(magnesite)
$$(6.3)$$

$$Ca^{2+} + CO_2 + H_2O \rightarrow CaCO_3 + 2H^+ \quad (6.4)$$
(calcite)

The deposition of carbonate minerals removed carbon dioxide dissolved in the water and transferred it into the rock reservoir, from which it could only escape when the rocks were later exposed to chemical weathering at the surface of the Earth. This was the start of geological activity and the beginning of the continuing modulation of the global climate by the storage of carbon dioxide in rock reservoirs, including not only carbonate rocks but also fossil fuel (coal, petroleum, and natural gas), amorphous carbon particles, and solid hydrocarbon molecules (kerogen) in shale (Holland, 1984; Lewis and Prinn, 1984).

The chemical composition of the present atmosphere in Table 6.1 indicates that dry air consists primarily of diatomic molecular nitrogen (78.084 %) and oxygen (20.946 %), and that the concentration of carbon dioxide (CO_2) is only 0.033 %. These data demonstrate that geological and biological processes have profoundly changed the chemical composition of the primordial atmosphere by removing carbon dioxide and by releasing oxygen. The removal of carbon dioxide by chemical weathering of silicate minerals followed by the precipitation of solid carbonates in the oceans is augmented by photosynthesis of green plants which combine carbon dioxide and water molecules to form glucose ($C_6H_{12}O_6$) and molecular oxygen:

$$6CO_2 + 6H_2O \rightarrow C_6H_{12}O_6 + 6O_2 \quad (6.5)$$

The reaction is energized by sunlight which is absorbed by chlorophyll (i.e., the molecule that

Table 6.1. Chemical composition of dry air (Faure, 1998)

Constituent	Concentration by volume	
	%	µL/L
Nitrogen (N_2)	78.084	—
Oxygen (O_2)	20.946	—
Carbon dioxide	0.033	—
Argon (Ar)	0.934	—
Neon (Ne)	—	18.18
Helium (He)	—	5.24
Krypton (Kr)	—	1.14
Xenon (Xe)	—	0.087
Hydrogen (H_2)	—	0.5
Methane (CH_4)	—	2
Nitrogen oxide	—	0.5

72

CHAPTER 6

colors green plants). Accordingly, some of the carbon dioxide of the primordial atmosphere was later converted into molecular oxygen after photosynthetic lifeforms had evolved in the ocean prior to about 3.5 billion years ago.

6.4 Interior of the Earth

The continuing tectonic activity of the Earth is driven by the convection of the asthenospheric part of the mantle in response to differences in the temperature. In other words, the basic cause for the tectonic activity of the Earth is the unequal distribution of heat in the mantle. At the time of its formation, the Earth received increments of heat from several sources:

1. Conversion of the kinetic energy of planetesimals into heat at the time of impact upon the Earth.
2. Decay of naturally occurring radioactive elements which were more abundant 4.6 billion years ago, but which still generate heat in the Earth at the present time.
3. Compression of the Earth as its mass continued to rise.
4. Release of gravitational potential energy by molten iron sinking toward the center of the Earth.
5. Heat generated by the gravitational interactions between the Earth and the Moon.

The last-mentioned source of heat (lunar tides) has contributed to the heat budget of the Earth only after the Moon formed following the initial accretion of the Earth. The heat sources listed above apply to all of the terrestrial and gaseous planets of the solar system.

The variation of density in the interior of the Earth was discussed in Section 3.5 and is illustrated in Figure 3.5 which shows that the density of rocks and minerals increases when they are compressed by rising pressure with increasing depth in the Earth. The information we have about the variation of density of the interior of the Earth is derived largely from the interpretation of earthquake records because the velocity of seismic waves depends on the mechanical properties of the rocks, which in turn are affected by the density and the temperature of the rocks in the interior of the Earth.

6.4.1 Temperature

The present temperature distribution in the mantle of the Earth is not known with certainty, although seismic data do indicate that the rocks that compose the mantle are solid, which constrains the temperature to be less than the melting temperatures of peridotite and pyrolite. Limited experimental data indicate that the melting temperature of dry peridotite rises with increasing pressure from about 1090 °C at the surface of the Earth (pressure = 0.001 kilobars) to 1340 °C at a depth of 100 km (pressure = 34 kilobars). Extrapolation of these data to a depth of 400 km (pressure = 132 kilobars) in Figure 6.5 yields an estimated melting temperature for dry peridotite of about 1700 °C (Press and Siever, 1986). Additional estimates referred to by Wyllie (1971) suggest a temperature of 1770 °C at a depth of 700 km. This means that the rocks of the mantle of the Earth are "white hot" but remain solid as indicated by the fact that they transmit seismic waves (i.e., S-waves).

The iron of the outer part of the core is liquid because the temperature at that depth exceeds the melting temperature of iron in spite of the high pressure. However, the pressure close to the center of the Earth causes the melting point of iron to rise above the prevailing temperature of about 5000 to 6000 °C. Therefore, the inner core of the Earth is solid.

6.4.2 Melting of Peridotite

The melting curve of dry peridotite in Figure 6.5 is significant because it illustrates how hot peridotite in the mantle of the Earth can melt to form basalt magma. Consider a volume of hot peridotite at point P where the temperature is 1500 °C and the pressure is 140 kilobars. In this environment, dry peridotite does not melt, which is indicated by the location of point P in the field labeled "solid." In order to cause the rock to melt, either the temperature must *increase* to 1720 °C at point R or the pressure must *decrease* to 70 kilobars at point Q. The peridotite at point P in Figure 6.5 can also melt in case water or other volatile compounds are present, because such compounds act as a flux that lowers the melting temperature of solid rocks. However, for the purpose of the present discussion, the

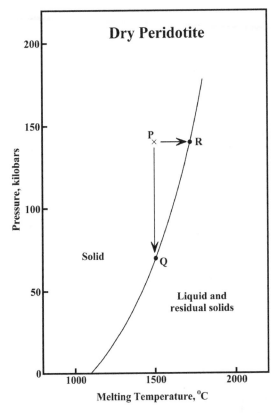

Figure 6.5. The melting curve of dry peridotite constrains the temperature of the mantle of the Earth, which remains solid even though its temperature rises to 1700 °C at a depth of 400 km where the pressure is 132 kilobars. In addition, the melting curve illustrates how peridotite at point P can melt either by increasing the temperature to 1740 °C at point R or by lowering the pressure from 140 to 70 kilobars at point Q. Melting of rocks in the mantle by decompression is an important process that leads to the formation of magma, which can either cause volcanic eruptions on the surface of the Earth or can result in the formation of plutonic igneous rocks in the continental crust. Adapted from Press and Siever (1986, Figure 15.12)

Table 6.2. Production rates of heat by the decay of radioactive atoms of uranium (U), thorium (Th), and potassium (K) in different kinds of plutonic igneous rocks (Emiliani, 1992, Table 11.8)

Rock type	Heat production,			10^{-6} cal/g/y
	U	Th	K	Total
Granite	3.4	4	1.08	8.48
Diorite	1.9	1.8	0.67	4.37
Gabbro	0.66	0.5	0.12	1.28
Peridotite	0.011	0.002	0.008	0.021

granite (8.48×10^{-6} cal/g/y) because peridotites have much lower concentrations of radioactive elements than granite. Consequently, peridotite in the mantle of the Earth is more likely to melt as a result of a decrease in pressure than because of an increase in temperature.

Rocks composed of silicate minerals do *not* melt completely at a unique temperature the way ice does. In contrast to ice which melts *congruently*, silicate minerals such as olivine, plagioclase, and pyroxene melt *incongruently* over a range of temperature to form a melt and residual solids. The chemical composition of the melt depends on the extent of melting and differs from the chemical composition of the rock before any melting occurred. Therefore, the partial melting of peridotite in the mantle can produce a wide range of melt compositions depending on the extent of melting. Small degrees of melting (e.g., 1% or less) yield melts that are enriched in alkali metals such as sodium, potassium, and rubidium, whereas large degrees of partial melting (e.g., 20 to 30%) produce melts of basaltic composition with elevated concentrations of iron, magnesium, and calcium,

6.4.3 Mantle Plumes

The volcanic activity on oceanic islands, along mid-ocean ridges, and along continental rift zones is caused by decompression melting that occurs when plumes or diapirs of hot rock rise through the asthenospheric mantle until they encounter the underside of the rigid lithosphere (Faure, 2001). Plumes arise at depth in the asthenosphere where the irregular distribution of radioactive elements (explained in Section 6.4.4) causes localized increases of the temperature. As a result, the density of the affected rocks

peridotite at point P is considered to be "dry" which means that it can only melt in response to an increase in temperature or a decrease in pressure.

The rocks of the mantle of the Earth do contain radioactive elements (uranium, thorium, and potassium) that generate heat when the nuclei of their atoms decay spontaneously. However, the data in Table 6.2 demonstrate that peridotite generates only small amounts of heat (0.021×10^{-6} cal/g/y) by this process compared to

decreases which causes them to become buoyant relative to the surrounding rocks. Therefore, these rocks start to move upward through the asthenospheric mantle. Such mantle plumes can develop a head and a tail having diameters of several tens to hundreds of kilometers as they slowly rise to the top of the asthenosphere.

When plumes encounter the rigid lithospheric layer at the top of the asthenospheric mantle, their heads spread radially and can reach diameters of up to 1000 km. Such large plumes exist under Iceland, under the island of Tristan da Cuñha in the South Atlantic Ocean, and under the Afar depression at the entrance to the Red Sea and at many other places on the Earth. The impact of an asthenospheric plume on the underside of the lithosphere has significant geological consequences especially in case the plume is located beneath continental crust:

1. The outward flow of the head of the plume causes fractures to form in the overlying lithosphere. These fractures can propagate to the surface and cause rifts in the oceanic or continental crust.
2. The development of fractures reduces the pressure in the head of the plume and in the lithospheric mantle directly above it.
3. The resulting decompression causes large-scale partial melting of the rocks in the head of the plume and in the overlying lithospheric mantle.
4. The basalt magma produced by decompression melting follows the fractures upward and may cause volcanic eruptions on the surface of the Earth.
5. Magma that crystallizes before reaching the surface forms sills and dikes or large-scale bodies of plutonic igneous rocks in the oceanic and continental crust.
6. The upward flow of hot magma transports heat from the lower asthenosphere into the oceanic and continental crust and thereby makes the crustal rocks more susceptible to partial melting to form magmas ranging in composition from andesite to rhyolite.
7. The increase of the temperature of the lithospheric mantle adjacent to a plume and of the overlying crust causes expansion of the rocks, which manifests itself by uplift at the surface.
8. The rifts that form in the continental crust may widen in the course of geologic time to form ocean basins exemplified by the Atlantic Ocean and the Red Sea.

The theory of mantle convection presented above explains how heat generated by radioactivity causes the upward movement of plumes of hot peridotite and how the interaction of these plumes with the rigid lithospheric layer at the top of the mantle causes many of the geologic phenomena we observe at the surface of the Earth. In order to complete the presentation of this theory we next explain why the radioactive elements are irregularly distributed in the asthenospheric mantle.

6.4.4 Plate Tectonics

The lithosphere of the Earth consists of about 20 plates, some of which are quite small (e.g., the Gorda/Juan de Fuca, Caroline, Bismarck, and Solomon plates in the Pacific Ocean). The plates are bounded by spreading ridges, rift valleys, strike-slip faults, and subduction zones. Some plates consist entirely of suboceanic lithosphere with the corresponding overlying oceanic crust (e.g., the Pacific, Scotia, Philippine, Cocos, and Nazca plates), while others contain both continental and oceanic crust underlain by the corresponding lithosphere (e.g., the Antarctic, African, Somali/Eurasian, North American, Caribbean, South American, and Indian/Australian plates). A few of the smaller plates include only continental crust and lithosphere (e.g., Arabian and Anatolian plates).

Plates form by volcanic activity at oceanic spreading ridges (e.g., the Mid-Atlantic Ridge and East-Pacific Rise) and are transported away from these ridges by convection currents in the underlying asthenosphere caused by the motions of plumes. For example, the Pacific plate in Figure 6.6 moves in a northwesterly direction at a rate of up to 11 cm/y and carries along with it all of the chains of volcanic islands that have formed on it (e.g., the Hawaiian islands). The Pacific plate is consumed in subduction zones that extend from the west coast of North America across the North Pacific Ocean and along the eastern margin of Asia all the way south to the South-East Indian Ridge.

The down-going Pacific plate consists of a thin layer of terrigenous sediment, basaltic rocks

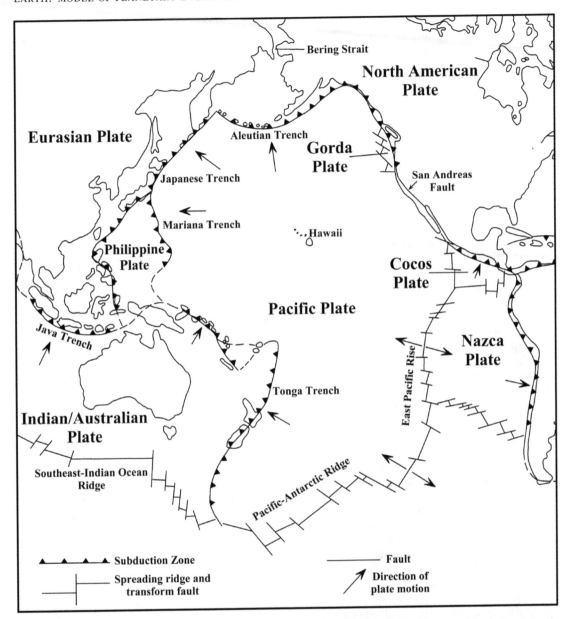

Figure 6.6. The Pacific plate originates by volcanic activity on the seafloor along the East Pacific rise and the Pacific-Antarctic ridge. The plate moves in a northwesterly direction at up to 11 cm/y and is subducted along the system of deep-sea trenches along the northern and western margin of the Pacific basin. The subducted marine sediment and underlying basalt of the oceanic crust are dehydrated, chemically altered, and recrystallized. Several hundred million years after subduction, these rocks form plumes that manifest their presence by volcanic activity within the Pacific basin. The Hawaiian Islands are an examples of this phenomenon. Adapted from Press and Siever (1986)

of the oceanic crust, and the underlying litho-spheric mantle. The sediment and the basaltic rocks contain uranium, thorium, and potassium derived by weathering of crustal rocks and by the alteration of sediment and basalt by seawater. When the sediment and basaltic rocks are subducted into the mantle along deep-sea trenches, most of the water escapes into the wedge of mantle rocks above the down-going plate. The addition of water to the peridotite of the mantle wedge decreases its melting temperature and therefore causes partial melting. The resulting magmas of basaltic and andesitic composition rise to the surface and are erupted by volcanoes located behind the deep-sea trenches that mark the subduction zone. The resulting island arcs along the northern and western margin of the Pacific Ocean include the Aleutian Islands of North America and the Kuril, Japanese, Ryukyu, and the Philippine islands. The feature continues as the Bismarck Archipelago and the Solomon, Vanuatu, and the Tonga-Kermadec islands, as well as the North and South Islands of New Zealand. This system of deep-sea trenches (subduction zones) in Figure 6.6 and volcanic island arcs is one of the largest topographic features on the surface of the Earth. The theory of plate tectonics, which was developed between about 1950 and 1970, explains this and many other aspects of global geology. Examples of volcanic activity caused by rising plumes and descending plates are illsturated in Figure 6.7 and 6.8.

The theory of plate tectonics also explains the origin of plumes in the asthenospheric mantle because the subducted sediment and basalt recycle the heat-producing elements back into the mantle. The rocks that pass through the process of subduction are dehydrated, their chemical composition is altered, and the minerals recrystallize at the elevated pressure and temperature of the asthenosphere. These changes also increase the density of the subducted rocks allowing them to sink to a level where they are in buoyant equilibrium.

The radioactive elements in the subducted rocks produce heat as they continue to decay for hundreds of millions of years. The heat so generated accumulates and causes the temperature of the subducted rocks to rise above that of the surrounding rocks of the asthenosphere. The

increase in temperature causes expansion, which lowers the density, and thereby creates a buoyant force that is directed upward. As a result, a body of subducted rocks that has been incubated for a sufficient period of time begins to move upward through the surrounding asthenosphere.

The rising body of hot rock becomes a plume that slowly forces its way upward through the solid rocks of the asthenosphere, which yield by plastic deformation. Similarly, the rocks of the plume itself also remain solid because the temperature of the plume initially is only one or two hundred degrees higher than that of the asthenosphere. When the plumehead reaches the underside of the overlying lithospheric shell of the Earth, it spreads laterally and thereby causes fractures to form which may propagate upward into the continental or oceanic crust. The resulting decompression causes partial melting of the rocks in the head of the plume and in the overlying lithosphere as indicated previously in Section 6.4.3.

The arrival of a large plume under a continent can cause uplift, rifting, volcanic activity, earthquakes, and may eventually lead to the formation of a new ocean. The width of the ocean increases as the crustal fragments move apart. In this way, plumes explain continental drift (e.g., the increasing separation between Africa and South America). Continental drift may move land masses into different climatic zone, which affects the flora and fauna they carry. In addition, the break-up of large continents may lead to the dispersal of the resulting fragments. Subsequent changes in the movement of asthenospheric convection currents may reassemble previously dispersed continental fragments into larger continents. For example, the granitic crust of the present continents is a mosaic of microcontinents of differing ages, compositions, and geologic histories that were swept together by the convection currents of the asthenospheric mantle.

6.4.5 Magnetic Field

The Earth, like the Sun, acts like a bar magnet in Figure 6.9 and extends a magnetic field into the space around it. However, the Earth cannot actually be a permanent bar magnet because all materials lose their magnetic properties at

Figure 6.7. The shield volcano Kilauea on the island of Hawaii has been erupting basalt lava from vents on its flank. The braided lava flows in this image originated from the fountains on the slope of the Pu'u O'o cone. The lava erupted at this site has a low viscosity and moves at an average speed of about 0.5 km/h as it flows to the coast where it is quenched by the water of the Pacific Ocean. The magma originates by decompression melting of rocks in the head of a plume that is currently located beneath Kilauea volcano. The volcanic eruptions will end in the future when the movement of the Pacific plate (Figure 6.6) displaces Kilauea from its magma source and a new island called Loihi will emerge from the ocean south of the present island of Hawaii. The entire chain of the Hawaiian islands and the related Emperor seamounts consists of extinct volcanoes that originally formed above the "hot spot" beneath the present island of Hawaii starting about 65 million years ago (Carr and Greeley, 1980; Rhodes and Lockwood, 1995; Faure, 2001). Courtesy of the U.S. Geological Survey, Department of the Interior. (http://hvo.wr.usgs.gov/hazards/dds24167_photocaption.html)

temperatures greater than about 500 °C, which is their Curie temperature, named after Pierre Curie.

The existence of the magnetic field of the Earth was discovered during the 12th century AD and led to the construction of the magnetic compass, which has been an important aid to navigation ever since (Science Brief 6.7.3). The magnetic compass points in the direction of the magnetic pole located near the geographic north pole. The magnetic poles of the Earth wander randomly in the polar region and do not coincide with the geographic north and south poles (Science Brief 6.7.4). In addition, the magnetic field of the Earth has varied in terms of its strength and has actually changed its polarity on a timescale of millions of years. The history of polarity reversals during the past five million years has been used to construct a chronology presented in Table 6.3. The polarity of the magnetic field of the Earth throughout its history has been and continues to be recorded by volcanic rocks during their crystallization from cooling lava flows.

The magnetic field of the Earth is attributed to *induction* caused by electrical currents flowing in the liquid part of its iron core. However, the details of the process that generates and sustains the magnetic field of the Earth are not yet completely understood. The simplest explanation we can offer is that the magnetic field of the Earth is generated by electrical currents that flow around the circumference of the core, in which case the Earth could be regarded as an *electromagnet* (Science Brief 6.7.5). Unfortunately, this

Figure 6.8. Mt. Fuji-san (also called Fujiyama) is a typical stratovolcano near the Pacific coast of the island of Honshu, Japan. Its summit rises to an elevation of 3,776 m, making it the highest mountain in Japan. Mt. Fuji is composed of interbedded flows of andesite lava and layers of volcanic ash that were erupted explosively through the crater at the summit. Andesite lava has higher concentrations of SiO_2 and is more viscous than basalt. Note that "SiO_2" is a chemical component of lava and does not represent the mineral quartz whose chemical formula is also SiO_2. The andesite magma forms by partial melting of ultramafic rocks of the lithospheric mantle above subduction zones that surround the Pacific basin in Figure 6.6. Melting is initiated when water is released by the rocks of the Pacific plate that is being subducted into the Japan deep-sea trench. The addition of water (and other volatile compounds) to the rocks of the mantle wedge above the subduction zone lowers their melting temperature and causes magma to form of basaltic to andesitic composition (Wilson, 1989; Faure, 2001). The image of Mt. Fuji was taken at sunrise from Kawaguchiko, Yamanashi Prefecture, Japan. (http://commons.wikimedia.org)

hypothesis fails to explain the origin of the postulated electrical current.

An alternative hypothesis is to compare the Earth to a dynamo, which is a device for generating an electrical current by rotating magnets within coils of insulated copper wire (Science Brief 6.7.6). The problem with this hypothesis is that dynamos are designed to generate electrical current rather than magnetism. Therefore, the Earth cannot be a conventional dynamo. Instead, Walter Elsasser and Edward Bullard proposed that the magnetic field of the Earth is generated by a *self-exciting dynamo* operating in the liquid part of the iron core.

The hypothesis of Elsasser and Bullard proposes that liquid iron, which is a good electrical conductor, is stirred by thermal convection, or by the rotation of the Earth, or both. As a result, electrical currents are induced in the streams of liquid iron by the presence of a stray magnetic field. The electrical currents in the liquid part of the iron core then induce the magnetic field of the Earth. A small fraction of the magnetic field of the Earth continues

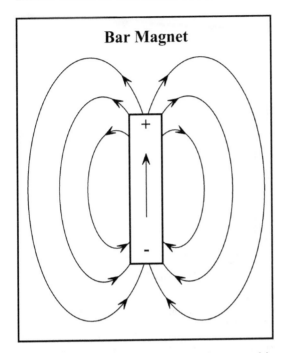

Bar Magnet

Figure 6.9. The polarity of a bar magnet is expressed in terms of plus (+) and minus (−), where plus means "north" and minus means "south" (See also Science Brief 6.7.7). The lines of magnetic force outside of the bar magnet extend from the positive pole to the negative pole but from minus to plus inside the magnet. The magnetic field of the Earth closely resembles that of a bar magnet. Adapted from Emiliani (1992, Figure 11.28)

which cause and effect are not closely related. However, the process is feasible and does explain why the magnetic poles are located in the vicinity of the geographic poles of the Earth (Press and Siever, 1986).

Accordingly, the necessary conditions for the existence of planetary magnetic fields include but may not be limited to:

1. The presence of an electrically conducting fluid, which, in the case of the Earth, is the molten iron in the outer core of the Earth.
2. The flow of the electrical conductor must be activated either by a rapid rate of rotation of the planet, or by thermally driven convection currents, or both.
3. A magnetic field of unspecified origin must exist in order to induce the electrical current in the liquid part of the core.

Accordingly, planets that do not rotate as rapidly as the Earth, or whose core has solidified, are not expected to have a magnetic field. These statements provide a frame of reference that will enable us to explain why some planets in the solar system have magnetic fields and why some do not. As we progress in the exploration of the planets and their satellites, we will discover that magnetic fields can be generated by quite different processes, which we cannot anticipate from a consideration of the Earth alone.

to induce the electrical currents in the streams of liquid iron, thereby creating a self-exciting dynamo in the core of the Earth.

The proposed process is highly unstable which explains why the strength and polarity of the magnetic field of the Earth vary spontaneously. In other words, the magnetic field of the Earth is generated by a chaotic process in

6.5 Interactions with the Solar Wind

The magnetic field of the Earth resembles that of a bar magnet depicted in Figure 6.9 and described in Science Brief 6.7.7. However, the resemblance is only superficial because the interior of the Earth is too hot to be permanently magnetized and because a bar magnet cannot spontaneously reverse its polarity as the magnetic field of the Earth has done many times throughout geologic time. In addition, the magnetic field of the Earth in Figure 6.10 is not symmetrical like that of the bar magnet in Figure 6.9. Instead, the magnetic field is distorted by interacting with the solar wind which blows with a speed of about 450 km/s as it approaches the Earth. The speed of the solar wind decreases abruptly at the boundary called the *shock wave* where the motion of the particles becomes turbulent as they are deflected by the magnetic field of the Earth. The *magnetopause* in Figure 6.10 is a second boundary where the

Table 6.3. Chronology of the reversals of the polarity of the magnetic field of the Earth (Faure and Mensing, 2005; Mankinen and Dalrymple, 1979)

Epoch	Polarity	Age, Million years
Brunhes	Normal	0 to 0.72
Matuyama	Reversed	0.72 to 2.47
Gauss	Normal	2.47 to 3.40
Gilbert	Reversed	3.17 to 5.41

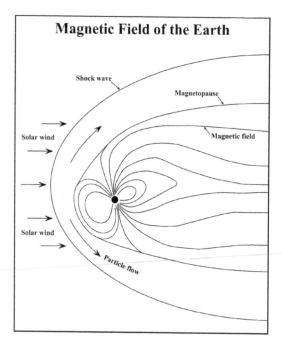

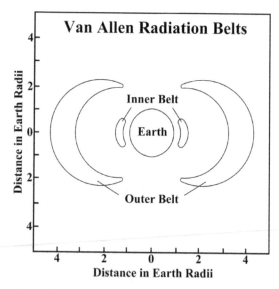

Figure 6.10. The magnetic field of the Earth is distorted by the solar wind composed of protons and electrons. As the Earth revolves around the Sun, the magnetic field facing the Sun is compressed by the solar wind while the magnetic tail trails away from the Sun. Adapted from Freedman and Kaufmann (2002, Figure 8.20)

Figure 6.11. The Van Allen radiation belts were discovered in 1958 during the flight of Explorer I, which was the first American satellite launched into space by Wernher von Braun (Section 1.2). James Van Allen, a physics professor at the University of Iowa, insisted that the satellite should carry a detector of energetic charged particles (Geiger counter) and subsequently discovered the radiation belts that bear his name. Adapted from Wagner (1991, Figure 9.6) and from Freedman and Kaufmann (2002, Figure 8.20)

pressure exerted by the solar wind is exactly balanced by the strength of the magnetic field. In this way, the magnetic field shields the Earth from the protons and electrons of the solar wind by forcing them to flow around the *magnetosphere*, which is the volume of space within the magnetopause.

Some of the charged particles of the solar wind leak across the magnetopause and are then trapped by the magnetic field within the magnetosphere in the form of two donut-shaped equatorial rings known as the *Van Allen belts* in Figure 6.11. The inner belt is located between about 2000 and 5000 km above the equator of the Earth and is composed primarily of protons. The second belt consists mostly of electrons and extends farther north and south above the equator than the first (lower) belt. The second belt lies about 16,000 km above the surface of the Earth and is about 6,000 km thick. These so-called radiation belts are hazardous to astronauts and to electronic equipment in spacecraft that are outward bound

to the Moon or to other destinations in the solar system (Van Allen and Bagenal 1999).

When the Sun ejects unusually large amounts of protons and electrons in the form of solar flares, the magnetosphere of the Earth is overloaded with nuclear particles that follow the lines of the magnetic field to the region above the magnetic poles of the Earth where they descend towards the surface. As they enter the atmosphere, the particles collide with molecules of nitrogen and oxygen and thereby transfer their energy to them. These energized atoms lose the excess energy by emitting electromagnetic radiation with wavelengths in the visible part of the spectrum (Science Brief 2.5.3, Table 2.1). The resulting displays of light in Figure 6.12 seen at night in the polar regions are known as the northern lights (*aurora borealis*) or southern lights (*aurora australis*) and originate about 80 to 160 km above the surface of the Earth. Auroral displays in the northern hemisphere are accompanied by simultaneous displays in the southern hemisphere. The most common colors

Figure 6.12. Spectacular view of the aurora borealis over the frozen Mackenzie River at Fort Simpson in the Northwest Territories of Canada. The red dot on the distant horizon is a microwave tower. The view is to the north with the Mackenzie River in the foreground. Image courtesy of Robert Postma. < http://www.distanthorizons.ca >

are red and green. Auroral displays are visual evidence that the atmosphere of the Earth is affected by energetic electrons (and protons) of the solar wind that are trapped by the magnetic field of the Earth. (Jago, 2001).

6.6 Summary

The dust particles in the protoplanetary disk accreted to form planetesimals having diameters that ranged from a few centimeters or less to hundreds of kilometers or more. The largest planetesimals in the inner part of the disk grew into the terrestrial planets by impacts of smaller bodies. Computer models of this process indicate that the time required to form the Earth was between one and ten million years.

The rate of accretion of the Earth by impacts of planetesimals was sufficiently rapid to cause the Earth to melt. The resulting magma ocean contained three immiscible components: liquid iron, liquid sulfides of iron, and molten silicates of iron, magnesium, and calcium. Consequently, the metallic iron sank to the center of the Earth to form the core, the liquid sulfide also sank but did not form a continuous layer around the iron core, and the silicate melt began to crystallize in accordance with Bowen's reaction series.

Early-formed crystals of olivine and pyroxene sank in the magma ocean, whereas crystals of Ca-rich plagioclase floated toward the top of the magma ocean. The olivine and pyroxene crystals formed the peridotite of which the mantle of the Earth is composed. The crystals of anorthite accumulated in a layer at the surface of the magma ocean to form a crust consisting of anorthosite. The gases and volatile compounds that were released by the impacting planetesimals formed a dense atmosphere.

Most of the water and other volatile compounds that had been delivered to the Earth by the planetesimals formed the primordial atmosphere. As time passed, the atmosphere cooled sufficiently to become saturated with water vapor which caused the water to rain out to form a global ocean. The transfer of water from the atmosphere to the ocean also caused a major fraction of the carbon dioxide in the atmosphere to dissolve in the water. The removal of greenhouse gases such as water vapor and carbon dioxide allowed the atmosphere to cool further and prevented the water in the ocean from boiling as it may have done on Venus. The presence of liquid water on the surface of the Earth initiated chemical weathering of volcanic rocks in the ocean and resulted in the deposition of marine carbonate rocks which sequestered carbon dioxide and thereby stabilized the global temperature on the surface of the Earth.

The interior of the Earth at the present time includes an iron core, the outer part of which is liquid, whereas the inner core is solid. The mantle, composed of peridotite and pyrolite, is solid even though it is very hot because the high pressure prevents the rocks from melting. The outer layer of the mantle forms a brittle shell about 120 km thick called the lithosphere, whereas the underlying mantle, called the asthenosphere, responds to stress by deforming plastically.

The subduction of lithospheric plates with the associated oceanic crust and marine sediment recycles the radioactive elements back into the asthenospheric mantle where these elements release heat as their atomic nuclei decay. After a period of incubation lasting hundreds of millions of years, the subducted rocks become buoyant and form a plume that slowly rises toward the top of the mantle. When a large plume reaches the underside of the lithosphere, the diameter of the head increases. As a result, fractures form in the overlying lithospheric plate which leads to a decrease of the lithostatic pressure and thereby causes basalt magma to form by partial melting of the rocks in the head of the plume and in the adjacent lithospheric mantle. The geological phenomena that result from the inter-action of mantle plumes with the lithospheric plate include: uplift, volcanic activity, earthquakes, rifting, seafloor spreading, continental drift, and climate change.

The magnetic field of the Earth resembles that of a bar magnet but actually originates by electro-magnetic induction associated with convection of liquid iron in the outer core. This process is unstable, which explains why the strength of the magnetic field and its polarity have changed unpredictably throughout geologic time. The magnetic field deflects the protons and electrons of the solar wind except during episodes of intense solar flares. At these times, some of the charged particles leak into the magnetosphere and descend toward the surface of the Earth in the polar regions. The energy of the protons and electrons is transferred to molecules of oxygen and nitrogen which re-radiate the excess energy in the form of visible light. The resulting display gives rise to the northern and southern lights.

The theories of the origin of the solar system (Chapter 5) and of the Earth (Chapter 6) are direct extensions of the Big Bang cosmology and of stellar evolution. Therefore, planets and their satellites, like those of our solar system, can form every time a star arises by contraction of a large volume of diffuse gas and dust. The only constraints on the development of solar systems are that elements able to form refractory compounds were present and that the process was allowed to run its course without disruption by the violent activities of nearby stars.

6.7 Science Briefs

6.7.1 Archimedes' Principle

According to the Law of Buoyancy, discovered by the Greek mathematician Archimedes (290-212 BC), a body that is completely or partially submerged in a fluid at rest is acted upon by an upward-directed or buoyant force. The magnitude of this force is equal to the weight of the fluid displaced by the body. Another way of expressing this principle is to say that a body will float in a liquid if the weight of the body is less than the weight of the liquid the body displaces. For example, a ship constructed of steel sinks into the water until the weight of the water the ship displaces is equal to the weight of the ship. If the ship weighs less than the water it displaces, then the ship will have freeboard (i.e., it will stick out of the water). If the weight of the ship is greater than the weight of an equal volume of water, the ship will have no freeboard and therefore will sink.

Archimedes' Principle applies to blocks of continental crust floating in the mantle of the Earth. Even though the rocks of the mantle are solid, they are able to yield to long-term stress by deforming plastically. A volume of continental crust weighs less than the same volume of the mantle because the density of the continental crust ($d = 2.9\,g/cm^3$) is less than the density of the mantle ($d = 3.2\,g/cm^3$). Therefore, blocks of continental crust have freeboard (i.e., they stick out of the mantle) as illustrated in Figure 6.13.

If the crust is floating in the mantle (i.e., is in isostatic equilibrium), the weight of the crustal block is equal to the weight of the mantle it displaces. If f = freeboard and the average thickness of the crust is 40 km, the ratio of the weight of the crustal block (W_c) divided by the weight of the mantle displaced by the crustal block (W_m) is equal to one:

$$\frac{W_c}{W_m} = \frac{40 \times 2.9}{(40 - f)3.2} = 1.0$$

Solving for the freeboard (f):

$$f = 3.75\ km$$

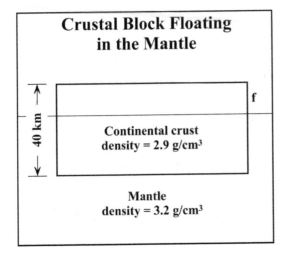

Crustal Block Floating in the Mantle

40 km

f

**Continental crust
density = 2.9 g/cm³**

**Mantle
density = 3.2 g/cm³**

Figure 6.13. Blocks of continental crust float in the upper mantle because crustal rocks are less dense than the rocks of the mantle. In this illustration, the crust is assumed to be 40 km thick, and (f) is the "freeboard" in kilometers. The result of the calculation in the text (f = 3.75 km) agrees with the average depth of the oceans

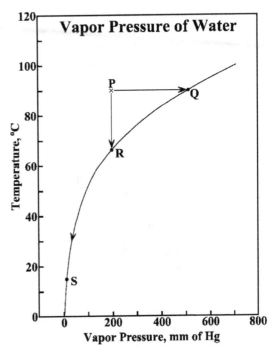

Figure 6.14. The curve represents the increase of the vapor pressure of water with increasing temperature. This curve is used to illustrate the effect of cooling an air mass at point P whose coordinates are a vapor pressure of 200 mm of Hg and a temperature of 90 °C. As shown in the text, this air mass is undersaturated with respect to water and has a relative humidity of 39.2%. As the temperature decreases to about 66 °C, the relative humidity rises to 100% and therefore raindrops start forming by condensation. The air remains at 100% relative humidity as the temperature continues to decline to 15 °C at point S. Data from Emiliani (1992, Table 13.6)

According to this calculation, the average depth of the ocean basins is 3.75 km which is in good agreement with the actual average depth of 3.73 km (Emiliani, 1992, Table 14.1).

6.7.2 Relative Humidity

Liquid water and ice form vapor composed of water molecules. When the rate of evaporation of water molecules from the surface of liquid water or ice is equal to its rate of condensation, the vapor is said to be saturated, which means that a state of equilibrium exists between the vapor and the liquid water or ice.

The concentration of water vapor in air is expressed in units of grams of water per cubic meter (g/m³). Alternatively, the concentration of water vapor can also be expressed in terms of the pressure it exerts measured in mm of Hg. The concentration of water vapor or its pressure in Figure 6.14 increases with rising temperature regardless of whether the vapor exists in a vacuum or in a mixture of other gases such as air in accordance with Dalton's Law.

The *relative humidity* of air is the ratio of the measured concentration or pressure of water vapor at a specified temperature divided by the concentration or pressure of saturated vapor at

the same temperature. For example, point P in Figure 6.14 represents air at 90 °C having a vapor pressure of 200 mm of Hg. The air represented by point P is undersaturated with respect to water vapor because saturated vapor at 90 °C (point Q) has a pressure of about 510 mm of Hg. Therefore, the relative humidity (RH) of the air at point P is:

$$RH = \frac{200}{510} \times 100 = 39.2\%$$

If the air is cooled from 90 °C at point P to 66 °C at point R, the air becomes saturated even though the vapor pressure of water has not changed. Therefore, water vapor condenses at point R to form rain drops that fall out of the air mass. If the temperature continues to decline, the vapor pressure of water remaining in the air must decrease along the curve by continued condensation of water. When the temperature reaches 15 °C at point S, the vapor pressure of the water remaining is only about 10 mm of Hg. As the air mass cooled from 66 °C (point R) to 15 °C (point S), the relative humidity of the air remained at 100%, provided that equilibrium between the vapor and drops of liquid water was maintained.

The example in Figure 6.14 illustrates how the primordial atmosphere reached saturation by cooling and how most of the water it contained rained out as the temperature decreased to the present average global temperature of 15 °C.

The graph in Figure 6.14 also demonstrates that the vapor pressure of water at 100 °C is 760 mm of Hg, which is equal to one atmosphere. Consequently, water boils at 100 °C because its vapor pressure is equal to the total atmospheric pressure of one atmosphere. On the summit of Mt. Everest, where the pressure is only 236 mm of Hg, water boils at 70 °C and potatoes take a lot longer to cook than they do at sealevel where water has a higher boiling temperature. (Emiliani, 1992, p. 274.)

6.7.3 Magnetic Compass

The magnetic compass consists of a magnetized iron or steel needle which is suspended to allow it to move freely in a horizontal plane. The needle therefore aligns itself with the magnetic lines of force such that one end points toward the magnetic pole located close to the geographic north pole of the Earth, whereas the other end points to the magnetic pole near the geographic south pole. The magnetic poles do not coincide exactly with the geographic poles whose positions are fixed by the axis of rotation of the Earth. Therefore, magnetic compasses must be adjusted to compensate for the *declination*, which is the angle between a line pointing to "true" north and the direction indicated by the magnetic compass. The magnitude of the declination depends on the locations of the observer and also varies with time on a scale of centuries because the magnetic poles wander randomly in the polar regions.

6.7.4 Magnetic Poles

Magnetism is associated with polarity similar to the polarity of electrical charges. Objects with opposite magnetic polarity attract each other, whereas objects having the same magnetic polarity repel each other. The polarity of the magnetic poles of the Earth is determined by the needle of a magnetic compass. The end of the needle that points to geographic north is identified as the north magnetic pole of the needle. Therefore, the magnetic pole, which is associated with geographic north pole of the Earth, must actually be the south magnetic pole because it attracts the north magnetic pole of the compass needle. Similarly, the magnetic pole in Antarctica is actually the north magnetic pole because it attracts the south magnetic pole of the compass needle. Accordingly, the *south magnetic pole of the Earth* is located close to the geographic north pole in the Canadian Arctic at 77°18′N and 101°48′W. The *north magnetic pole* is located in East Antarctica at 65°48′S and 139°0′E (Emiliani, 1992).

6.7.5 Electromagnet

An electromagnet consists of an iron core around which insulated copper wire has been wound. When a direct electrical current (i.e., DC rather than AC) flows through the windings of the copper wire, a magnetic field is induced in the iron core which thereby becomes an electromagnet. The strength of the magnetism of an electromagnet can be adjusted by selecting the number of copper windings and by controlling the magnitude of the electric current that passes through them.

6.7.6 Dynamo

A dynamo is a device that generates electric power by induction, which occurs when a magnet is rotated within windings of insulated copper wire. The energy required to rotate the magnet is provided by turbines that are driven either by steam, or by the flow of water, or by diesel engines. The steam is made by heating water either by burning fossil fuel (e.g., coal) or by fissioning nuclei of atoms of ^{235}U. Regardless of the design of the dynamo and the power source, dynamos generate electricity rather than magnetism.

6.7.7 Bar Magnets

The polarity of bar magnets is subject to the convention that north is "plus" and south is "minus". The positive (north) pole of a bar magnet repels the north magnetic pole of a compass needle, whereas the negative (south) pole of a bar magnet attracts the north magnetic pole of a compass needle. The lines of magnetic force that emerge from the poles of a bar magnet in Figure 6.9 extend from the positive (north) pole to the negative (south) pole.

6.8 Problems

1. Calculate the surface area (A_1) of a disk having a radius of 50 AU and compare it to the area (A_2) of a disk whose radius is only 5 AU. Express the areas in units of square kilometers and state the results in words with reference to Section 6.1. (Answer: $A_1 = 1.75 \times 10^{20}$ km^2; $A_2 = 1.7568 \times 10^{18}$ km^2.

2. Compare the volume (V_1) of the solid part of the core of the Earth to the volume (V_2) of the whole core both by means of the percent difference and by a factor (e.g., the volume of the inner core is "x" % of the volume of the whole core and is "x" times larger than the inner core).

Radius of the inner (solid) core:	1329 km	Answer: $V_1 = 0.0561\,V_2$
Radius of the whole core:	3471 km	or: $V_1 = 5.612\,\%$ of V_2
Volume of a sphere:		$V = 4/3\pi r^3$

3. Calculate the mass of the core of the Earth and of its mantle (without the core) and interpret the results by expressing their masses as a percent of the total mass of the Earth.

4. Estimate the depth of the hypothetical global ocean in kilometers before any continents or oceanic islands had formed. Assume that the total volume of water on the surface of the Earth was 1.65×10^9 km^3. (Answer: 3.23 km).

Radius of the whole core:	3471 km	Answer: Core = 32.2%
Radius of the Earth:	6378 km	Mantle = 67.8% of the mass of the Earth.
Density of iron:	7.87 g/cm^3	
Density of peridotite:	3.2 g/cm^3	

6.9 Further Reading

Carr MH, Greeley R (1980) Volcanic features of Hawaii: A basis for comparison to Mars. NASA, Sci Tech Inform Branch, Washington, DC

Emiliani C (1992) Planet Earth; Cosmology, geology, and the evolution of life and environment. Cambridge University Press, New York

Ernst WG (1983) The early Earth and the Archean rock record. In: Schopf J.W. (ed) Earth's Earliest Biosphere; Its Origin and Evolution. Princeton University Press, Princeton, NJ, pp 41–52

Faure G (1998) Principles and applications of geochemistry, 2nd edn. Prentice Hall, Upper Saddle River, NJ

Faure G (2001) Origin of igneous rocks: The isotopic evidence. Springer-Verlag, Heidelberg

Faure G, Mensing TM (2005) Isotopes; Principles and applications. Wiley, Hoboken, NJ

Freedman RA, Kaufmann III WJ (2002) Universe: The solar system. Freeman, New York

Holland HD (1984) The chemical evolution of the atmosphere and oceans. Princeton University Press, Princeton, NJ

Jago L (2001) The northern lights. Knopf, Random House, NY

Kenyon S (2004) Cosmic snowstorm. Astronomy 32(3):43–53

Lewis JS, Prinn RG (1984) Planets and their atmospheres: Origin and evolution. Academic Press, Orlando, FL

Lunine JL (1999) Earth: Evolution of a habitable world. Cambridge University Press, Cambridge, UK

Mankinen EA, Dalrymple GB (1979) Revised geomagnetic polarity time scale for the interval 0–5 m.y. BP. J Geophys Res 84(B2):615–626

Miller R, Durant III FC, Schuetz MH (2001) The art of Chesley Bonestell. Paper Tiger, Collins and Brown, London, UK

Moorbath S (2005) Oldest rocks, earliest life, heaviest impacts, and the Hadean-Archaean transition. Applied Geochem 20:819–824

Press F, Siever R (1986) Earth, 4th edn. Freeman and Co., New York

Rhodes JM, Lockwood JP, (eds) (1995) Mauna Loa revealed: Structure, composition, history, and hazards. Geophys. Monograph 92, Amer Geophys Union, Washington, DC

Schopf JW (1983) Earth's earliest biosphere: Its origin and evolution. Princeton University Press, Princeton, NJ

Stevenson DJ (1983) The nature of the Earth prior to the oldest known rock record: The Hadean Earth. In:

Schopf JW (ed) Earth's Earliest Biosphere: Its Origin and Evolution. Princeton University Press, Princeton, NJ, pp. 32–40

Taylor SR (1992) Solar system evolution: A new perspective. Cambridge University Press, New York

Van Allen JA, Bagenal F (1999) Planetary magnetospheres and the interplanetary medium. In: Beatty JK, Petersen CC, Chaikin A. (eds) The New Solar System, 39–58, 4th edn. Sky Publishing Corp., Cambridge, MA

Valley JW (2005) A cool early Earth? Sci Am 293:57–65

Wagner JK (1991) Introduction to the solar system. Saunders College Pub., Orlando, Florida

Wilde SA, Valley JW, Peck YH, Graham CM (2001) Evidence from detrital zircons for the existence of continental crust and oceans on the Earth 4.4 billion years ago. Nature 409:175–178

Wilson M (1989) Igneous petrogenesis: A global tectonic approach. Unwin Hyman, London, UK

Wyllie PJ (1971) The dynamic Earth: Textbook in geosciences. Wiley, New York

The Clockwork of the Solar System

The motion of the Earth in its orbit around the Sun is profoundly important to all forms of life including humans. Even the most primitive people on Earth observed the daily rising and setting of the Sun, the changing phases of the Moon, and the motions of the planets in the sky at night. In addition, most plants and animals respond to the daily sequence of light and dark periods and to the passage of the seasons by automatic adjustments of their metabolic functions. Throughout history, humans have made careful measurements in order to understand these matters and to predict celestial events before they occurred. The contributions of the ancient civilizations to astronomy have already been mentioned in Section 2.1 in connection with cosmology. Section 2.3.2 and Science Brief 2.5.4 relate the units of time to the period of rotation of the Earth and to the period of its orbit around the Sun.

The people of the Middle Ages also had problems with the calendar they inherited from the Romans. The problems with the calendar became so serious that in 1514 AD the astronomer Nicholaus Copernicus was asked by the Lateran Council of the Catholic Church to advise it on the reform of the calendar. Copernicus declined to get involved at that time because he thought that the motions of the Sun and Moon were not understood well enough. However, even at that time Copernicus had become dissatisfied with Ptolemy's cosmology.

7.1 The Pioneers of Astronomy

Nicolaus Copernicus (1473–1543) was a nineteen-year-old student at the University of Krakov in 1492 when Christopher Columbus embarked on his first voyage west across the Atlantic Ocean. Copernicus was also a contemporary of Ferdinand Magellan whose crew successfully circumnavigated the Earth and returned to Spain in 1522 (Section 1.1).

Copernicus devoted much of his professional life to the improvement of the cosmological model of Ptolemy and eventually came to the conclusion that all of the planets of the solar system revolve around the Sun rather than around the Earth. He presented a detailed description of the heliocentric model of the solar system in his book entitled: De revolutionibus orbium coelestium (On the revolutions of the celestial spheres) but did not allow his work to be published until 1543 when he was on his deathbed because he did not want to incur the sanctions of the Catholic Church (Gingerich, 2004). Copernicus proposed that the Sun is located at the center of the solar system and that the orbits of the six planets known at that time are perfect circles. Unfortunately, the heliocentric model of Copernicus did not predict the motions of the planets any better than Ptolemy's model did because the orbits of the planets are not circular and the Sun is not located at the exact center of the solar system. Copernicus tried to make adjustments similar to those Ptolemy had used, but in the end he could not reconcile the differences between his model and astronomical observations of the motions of the planets. Nevertheless, the resurrection of the heliocentric concept by Copernicus initiated a revolution in astronomy that contributed to the liberation of the human spirit known as the Renaissance (Ferris, 1988).

After Copernicus died in 1543, his model was refined by the work of *Tycho Brahe* (1546–1601) even though Brahe initially adhered to the cosmology of Ptolemy. However, when his own work demonstrated the deficiencies of Ptolemy's cosmology, Brahe adopted a hybrid model in which the Sun orbits the Earth (as proposed by Ptolemy), but all other planets orbit the Sun (as postulated by Copernicus). More important for the future of astronomy was Brahe's resolve to map the stars and to track the planets with the highest precision possible. Fortunately, Tycho Brahe was a very wealthy man thanks to grants he received from King Frederick II of Denmark. He used these funds to build two astronomical observatories (Uraniborg and Stjerneborg) on the island of Ven in Oresund between Denmark and Sweden. The data that Brahe and his assistants accumulated at these observatories between 1576 and 1597 were ultimately analyzed by Johannes Kepler after Brahe had been forced to leave Denmark upon the death of his royal patron.

Johannes Kepler (1571–1630) was educated at a Lutheran seminary in Germany but failed to be ordained. Instead, he became a mathematics teacher at a highschool in Graz, Austria. Kepler was an ardent Copernican and fervently believed that the orbits of the planets are related to each other in a simple and beautiful way. However, his idealism was constrained by his rational mind and by his intellectual honesty. Kepler knew that Tycho Brahe possessed the data he needed to test his ideas about the relationships among the planetary orbits. Therefore, he contacted Brahe at Benatek Castle near Prague where he lived following his expulsion from Denmark. After Kepler joined Brahe at Benatek Castle, they repeatedly quarreled because Brahe hesitated to give Kepler his data and because Kepler disapproved of Brahe's excessive lifestyle. But when Brahe died unexpectedly on October 24 in 1601, Kepler took possession of his data and began a lengthy series of calculations to determine the shapes of planetary orbits. After many trials and errors he eventually arrived at three conclusions known to us as *Kepler's laws*:

1. The orbit of a planet about the Sun is an ellipse with the Sun at one focus (Figure 7.1).
2. A line joining a planet in its orbit and the Sun sweeps out equal areas in equal intervals of time (Figure 7.2).
3. The square of the period of an object orbiting the Sun is equal to the cube of the average radius of its orbit.

Kepler was not pleased to acknowledge that planetary orbits are ellipses rather than circles as he and most astronomers of his time assumed. However, his laws solved the problems that afflicted the model of Copernicus. In addition, we now know that Kepler's laws apply to orbiting bodies throughout the Universe including spacecraft orbiting a planet, to binary star systems revolving around their common center of gravity, and even to pairs of galaxies that orbit each other (Freedman and Kaufmann, 2002).

When Kepler published his first two laws in 1609, *Galileo Galilei* (1564–1642) was 45 years old. Although the two pioneers of astronomy did not meet, Galileo supported the heliocentric model of the solar system by his telescopic observations starting in 1610. For example, Galileo noticed that the apparent size of Venus is related to the phases of Venus as seen from Earth (to be discussed in Chapter 11). He also discovered that Jupiter has four satellites that revolve around it like a miniature solar system (Chapter 15).

The final chapter in the quest to understand the motions of the planets in the solar system was written by *Sir Isaac Newton* (1642–1727) who was born in the year Galileo Galilei passed away. Newton experimented with the motions of bodies when acted upon by a force and expressed his conclusions in the form of the three laws of motion that bear his name (Science Brief 7.6a). Newton also recognized that the motions of the planets around the Sun are caused by the force of gravity described by the equation (Section 1.3):

$$F = G\frac{m_1 \times m_2}{r^2} \qquad (7.1)$$

where F = gravitational force between two objects expressed in newtons (N),
m_1 = mass of the first object (e.g., the Sun) expressed in kilograms,
m_2 = mass of the second object (e.g., a planet) expressed in kilograms,
r = distance between the centers of the two objects expressed in meters,
G = universal constant of gravitation.
The value of the constant of gravity in the mks system of units is:

$$G = 6.67 \times 10^{-11} \text{ newton m}^2/\text{kg}^2$$

Newton demonstrated that Kepler's laws follow directly from his three laws of motion and from

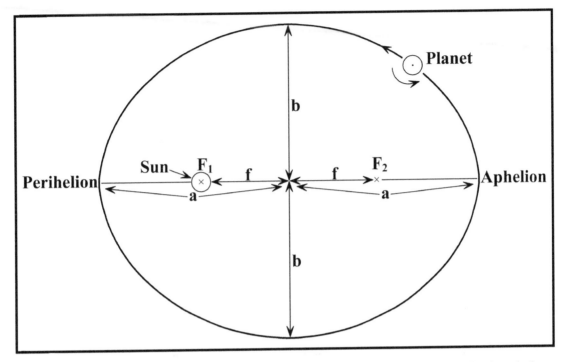

Figure 7.1. The orbits of the planets are ellipses with the Sun at one of the focal points: a = semi-major axis; b = semi-minor axis; f = focal length; F_1 and F_2 = focal points; perihelion is the point on the orbit that is closest to the Sun; aphelion is the farthest point on the orbit from the Sun. The planets revolve around the Sun in the prograde (counterclockwise) direction when viewed from a point in space located above the plane of the orbit of the Earth. The planets also rotate about imaginary axes and most do so in the prograde direction. Only Venus, Uranus, and Pluto have retrograde (clockwise) rotation

the equation describing the force of gravity. In fact, he extended Kepler's third law:

$$p^2 = a^3 \qquad (7.2)$$

where p = period of revolution of the planet expressed in years and

a = average distance between the planet and the Sun measured in astronomical units, to make it applicable to any two bodies that orbit each other:

$$p^2 = \frac{4\pi^2}{G(m_1 + m_2)} a^3 \qquad (7.3)$$

When this equation is used, time must be expressed in seconds (s), the distance between the objects must be in meters (m), and the masses must be in kilograms (kg).

7.2 Elliptical Orbits of Planets

The geometry of an ellipse is defined by certain parameters identified in Figure 7.1. These include the two focal points (F_1 and F_2), the focal length (f), the semi-major axis (a), the semi-minor axis (b), as well as the perihelion and aphelion (Science Brief 7.6.2).

7.2.1 Eccentricity

Ellipses differ from circles by their eccentricity (e), which is defined by the equation:

$$e = \frac{f}{a} \qquad (7.4)$$

The eccentricity of circles is equal to zero because circles have one center rather than two focal points. Therefore, the focal length (f) of circles is zero, which causes the eccentricity (e) in equation 7.4 to be zero. With the exception of

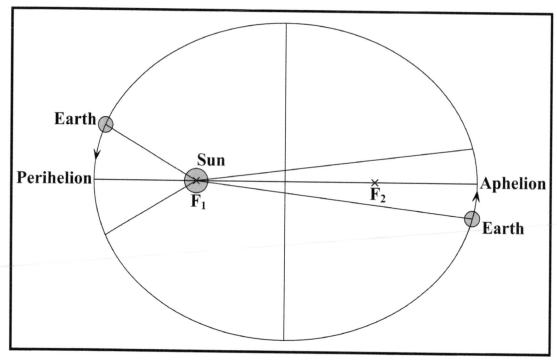

Figure 7.2. According to Kepler's second law, the areas of the triangles swept out in equal intervals of time by a planet in its orbit are equal. The diagram illustrates the fact that a planet at perihelion moves faster in its orbit than it does at aphelion. This law applies to all bodies that revolve around a second body because of the ellipticity of such orbits

Mercury and Pluto, the eccentricities of the orbits of the planets in Table 7.1 are generally low and range only from 0.0068 (Venus) to 0.093 (Mars). The orbit of Mercury has an eccentricity of 0.206 whereas that of Pluto is 0.250. The eccentricity of the orbit of the Earth is 0.017. Consequently, the orbits of most planets are very nearly circular.

7.2.2 Average Distance

The distance between a planet moving in an elliptical orbit and the Sun changes continuously depending on the location of the planet. Reference to Figure 7.1 demonstrates that at perihelion the distance (d_1) between the Sun and a planet is:

$$d_1 = a - f$$

whereas at aphelion the distance (d_2) is:

$$d_2 = a + f$$

Therefore, the average distance between the Sun and a planet in its orbit is:

$$\frac{d_1 + d_2}{2} = \frac{a - f + a + f}{2} = \frac{2a}{2} = a \qquad (7.5)$$

In other words, the semi-major axis (a) of the orbit of a planet is equal to the average distance between the planet and the Sun.

7.2.3 Revolution and Rotation

As the planets *revolve* around the Sun in their elliptical orbits in Figure 7.1, they also *rotate* about imaginary axes. The direction of the revolution and rotation of the planets is determined from a vantage point in space *above* the plane of the orbit of Earth (plane of the ecliptic defined in Section 7.2.4). The direction of revolution and rotation of a planet is *prograde* if it is counterclockwise when viewed from above. Similarly, if the direction is clockwise, it is called *retrograde*. All of the planets of the solar system in Table 7.1 revolve around the Sun in the prograde (counterclockwise) direction. Most also rotate in the prograde direction. The only exceptions to this generalization are the planets Venus, Uranus, and Pluto whose rotation is retrograde. The prevalence of prograde revolution and rotation of the planets is consistent with the prograde rotation of the Sun (Section 5.3.3) and implies that the protoplanetary disk also rotated in the prograde direction. We will find out later that some satellites of Jupiter, Saturn, Uranus, and Neptune revolve in the retrograde direction.

7.2.4 Plane of the Ecliptic

The plane of the orbit of the Earth, called the plane of the ecliptic, serves as a reference for the orbits of the other planets in the solar system. This relationship is expressed in Figure 7.3 by the angle of inclination (i) between the plane of the planetary orbit and the ecliptic. The inclinations of the planetary orbits to the ecliptic in Table 7.1 range from zero (Earth) to 3.39° (Venus) but rise to 7.00° for the orbits of Mercury and 17.12° for Pluto. The evident alignment of the orbits of the planets with the plane of the ecliptic is consistent with the theory that the planets of the solar system formed within the protoplanetary disk that surrounded the protosun following the contraction of the solar nebula (Section 5.2).

7.2.5 Axial Obliquity

The theory of the origin of the planets within the protoplanetary disk leads us to expect that their axes of rotation should be oriented at right angles to the planes of their orbits. However, this expectation is not fulfilled as illustrated in Figure 7.4 because the inclinations of the axes of rotation (axial obliquity) relative to the vertical vary widely from 0° (Mercury) to 178° (Venus). Uranus and Pluto also have highly inclined axes of rotation of 97.96° and 122.5°, respectively (Hartmann, 2005).

The deviations of the axes of rotation of the planets from the vertical direction may have been caused by impacts of large bodies during the early history of the solar system. Even the axes of rotation of the gaseous planets may have been affected by impacts on their rocky cores before these planets accumulated their large atmospheres of hydrogen and helium. The present inclination of the axis of Uranus (97.9°) indicates that this planet was actually "tipped over" and that its north pole is now pointing in a southerly direction. In that case, the direction of rotation of Uranus, viewed from above, changed from prograde to retrograde. Similar explanations apply to Venus and Pluto whose present retrograde rotation may be due to "tipping" of their axes of rotation, thus causing their former geographic north poles to point south.

Table 7.1. Physical properties of planetary orbits (Freedman and Kaufmann, 2002)

Planet	Semi-major axis(AU)	Sidereal period of revolution (d,y)	Period of rotation (d,h)	Eccentricity	Inclination of orbit to the ecliptic (°)	Average orbital speed (km/s)
Mercury	0.3871	87.969 d	58.646 d	0.206	7.00	47.9
Venus	0.7233	224.70 d	243.01^R d	0.0068	3.39	35.0
Earth	1.0000	365.256 d	1.000 d	0.017	0.00	29.77
Mars	1.5236	686.98 d	1.026 d	0.093	1.85	24.1
Jupiter	5.2026	11.856 y	9.936 h	0.048	1.30	13.1
Saturn	9.5719	29.369 y	10.656 h	0.053	2.48	9.64
Uranus	19.194	84.099 y	17.232^R h	0.043	0.77	6.83
Neptune	30.066	164.86 y	16.104 h	0.010	1.77	5.5
Pluto*	39.537	248.60 y	6.387^R d	0.250	17.12	4.7

R = retrograde
* Pluto has been reclassified as a dwarf planet (Section 3.1).

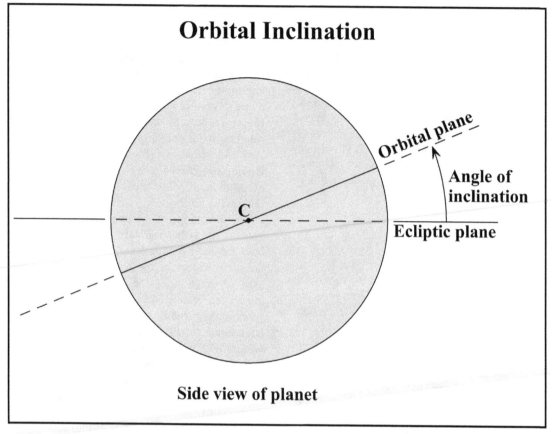

Figure 7.3. The orbital inclination (i) of a planet (or satellite) is the angle between the plane of the planetary orbit and the plane of the ecliptic, which is the plane of the orbit of the Earth. In cases where the orbital inclination is greater than 90°, the direction of revolution of the planet (or satellite) changes from prograde (anti-clockwise) to retrograde (clockwise)

7.2.6 Conjunctions and Oppositions

The systematic decrease of the average orbital velocities of the planets with increasing distance from the Sun (Figure 3.5) causes them to constantly change position relative to each other. For example, the Earth moves faster in its orbit than Mars and therefore regularly passes Mars. Similarly, Venus catches up to the Earth and passes it regularly because the average orbital velocity of Venus is greater than that of the Earth. The resulting alignments of the planets with the Earth and the Sun in Figure 7.5 are referred to as *conjunctions* or *oppositions*. A conjunction occurs when a planet is positioned on a straight line between the Earth and Sun or its extension on the other side of the Sun. If one of the so-called

inferior planets (Mercury and Venus) is located between the Earth and the Sun, the resulting alignment is called an *inferior conjunction*. If one of the above-mentioned planets is positioned on the other side of the Sun, a *superior conjunction* results.

Conjunctions of the superior planets (i.e., Mars, Jupiter, Saturn, Uranus, Neptune, and Pluto) occur when one of these planets is located behind the Sun relative to the Earth. However, when the alignment of a superior planet occurs on the same side of the Sun as the Earth, that kind of a conjunction is referred to as an *opposition*. Note that only the "inferior" planets have what we call "inferior" and "superior" conjunctions and that "superior" planets have either conjunctions or oppositions depending on whether they occur

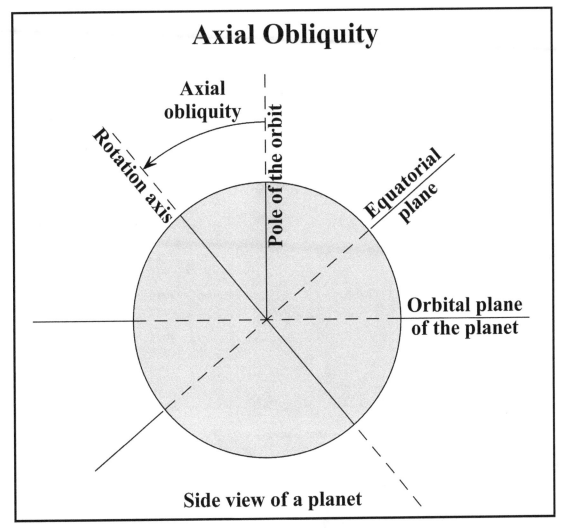

Axial Obliquity

Side view of a planet

Figure 7.4. The axial obliquity of a planet (or satellite) is the inclination of its axis of rotation relative to the pole of the orbit, which is the imaginary line drawn from the center of the planet at 90° to the plane of its orbit. The obliquity determines the amount of sunlight received by the northern and southern hemispheres of a planet and thus causes seasonal changes of the weather. In cases were the axial obliquity is greater than 90°, the sense of rotation changes from prograde (anti-clockwise) to retrograde (clockwise). For example, Venus (178°), Uranus (97.9°), and Pluto (122.5°) rotate in the retrograde direction because their obliquities are greater than 90°

behind the Sun as seen from the Earth or on the same side of the Sun as the Earth. The reader can verify by reference to Figure 7.5 that the inferior planets cannot have oppositions.

When Mercury or Venus pass between the Earth and the Sun, their images may be visible from the Earth as dark disks that move across the face of the Sun. Such an event, called a *transit*, is not visible from the Earth in all cases because

of the inclinations of their orbits relative to the plane of the ecliptic (i.e., the orbit of the Earth). Similarly, superior planets that are in conjunction are also not visible from the Earth because the conjunctions take place behind the Sun. However, when a superior planet is in opposition in Figure 7.5, it is visible from the Earth at night.

The movement of the planets relative to the Sun can also be expressed by the magnitude of

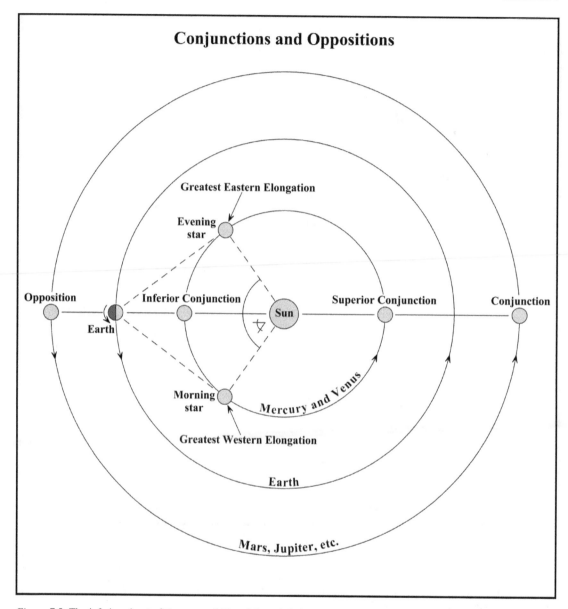

Figure 7.5. The inferior planets (Mercury and Venus) have inferior and superior conjunctions when the Earth and the Sun are aligned with either Mercury or Venus. The superior planets (Mars, Jupiter, Saturn, etc.) are in opposition when they are aligned on the same side of the Sun as the Earth and they are in conjunction when they are located behind the Sun. The elongation is the angle formed by a planet with the Sun. For example, the elongation of a planet at inferior conjunction or opposition is equal to zero. When Mercury or Venus are at their greatest eastern elongation, they are visible in the evening in the western sky. When either of these planets is at its greatest western elongation, it is visible in the morning in the eastern sky. Adapted from Figure 4.6 of Freedman and Kaufmann (2002)

the angle between the lines joining a planet to the Sun and the Earth to the Sun. Astronomers call this angle the *elongation* of the planet. The elongation of an inferior planet (Mercury or Venus) increase as it moves in its orbit from 0° at inferior conjunction to the position of its greatest elongation. When Mercury or Venus are in the position of greatest western elongation, they are visible from Earth in the morning. Hence, Venus is called the "morning star." However, when Mercury or Venus are in the position of greatest eastern elongation, they are visible in the evening, which explains why Venus is also called the "evening star". The same could be said about Mercury, but it is much closer to the Sun than Venus and is therefore more difficult to see. In addition, Mercury and Venus are planets rather than stars, but a casual observer on Earth may have difficulty distinguishing planets from stars.

7.2.7 Sidereal and Synodic Periods of Revolution

The position of a planet in its orbit can also be fixed with reference to background stars in the Milky Way galaxy. The resulting reference point is used to determine the sidereal period of revolution of a planet, which is defined by the statement: The *sidereal period* of revolution of a planet (or any celestial object) is the time that elapses during one revolution about the Sun with respect to the stars.

An alternative method of describing the revolution of planets is by means of their *synodic period*, which is the time that elapses between successive conjunctions with the Earth. The synodic periods of revolution of Mercury, Venus, and Mars in Table 7.2 are longer than their sidereal periods whereas the synodic periods of Jupiter, Saturn, Uranus, Neptune, and Pluto are shorter. The reason for the shortness of the synodic periods of these planets is that they move more slowly in their orbits than the Earth (Table 7.1), which causes the Earth to overtake them before they have moved very far in their own orbits.

The relation between the sidereal and synodic periods of planets is expressed by equations that are derived in Science Brief 7.6.3. For inferior

Table 7.2. Sidereal and synodic periods of revolution of the planets (Freedman and Kaufmann, 2002)

Planet	Sidereal period	Synodic period
Mercury	88 d	116 d
Venus	225 d	584 d
Earth	365.256 d	–
Mars	686.98 d	780 d
Jupiter	11.9 y	399 d
Saturn	29.5 y	378 d
Uranus	84.0 y	370 d
Neptune	164.8 y	368 d
Pluto*	248.5 y	367 d

* Pluto has been reclassified as a dwarf planet.

planets (Venus and Mercury):

$$\frac{1}{P} = \frac{1}{E} + \frac{1}{S} \qquad (7.6)$$

For superior planets (Mars, Jupiter, etc.):

$$\frac{1}{P} = \frac{1}{E} - \frac{1}{S} \qquad (7.7)$$

Where P = sidereal period of a planet
E = sidereal period of the Earth
S = synodic period of the planet
and the periods must be expressed in the same units of time.

For example, the synodic period of revolution of Jupiter in Figure 7.6 can be calculated from equation 7.7 given that the sidereal period of Jupiter (P) is 11.856 y:

$$\frac{1}{11.856} = \frac{1}{1.0} - \frac{1}{S}$$

$$-0.9156 = -\frac{1}{S}$$

$$S = \frac{1}{0.9156} = 1.092\,y = 398.9\,d$$

Rounding to three significant digits yields a value of 399 days for the synodic period of Jupiter as in Table 7.2.

The synodic period of revolution of inferior and superior planets indicates the length of time between successive alignments with the Earth when the distance between them is at a

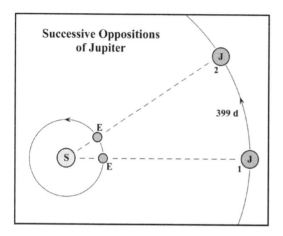

Figure 7.6. The synodic period of Jupiter is 398.9 days which is the time that elapses between successive alignments of Jupiter, the Earth, and the Sun. While Jupiter moves from opposition 1 to 2, the Earth has completed one complete revolution of 360° and is starting on the second revolution. Adapted from Box 4.1 of Freedman and Kaufmann (2002). Not drawn to scale

minimum. For example, the Earth and Mars are in conjunction every 780 days (2.13 years) which defines the window of opportunity for travel from the Earth to Mars and for the return to Earth.

7.2.8 Solar and Lunar Eclipses

The motion of the Moon around the Earth occasionally places it between the Earth and the Sun. During these occurrences, observers on the day-side of the Earth see that a dark object (i.e., the Moon) passes across the face of the Sun and causes temporary darkness. Such *solar eclipses* greatly disturbed humans who lived thousands of years ago. The fear caused by the darkening of the Sun may have motivated them to build solar observatories (e.g., Stonehenge in England) in order to predict such solar eclipses and thereby to lessen the fear that these events aroused.

The alignment of the Earth, Moon, and Sun in Figure 7.7 demonstrates that a total solar eclipse occurs only in those parts of the Earth that pass through the dark shadow (umbra) which the Moon casts upon the Earth. The diameter of the Moon's shadow is 269 km when the Earth-Moon distance is at a minium and even less when it is at a maximum caused by the eccentricity of the Moon's orbit. The shadow of the Moon sweeps

across the surface of the Earth at a rapid rate because of the orbital velocity of the Moon and the rotational velocity of the Earth. As a result, a total solar eclipse cannot last more than 7.5 minutes and, in many cases, takes even less time than that (Hester et al., 2002).

The eclipse starts when only a part of the Sun is obscured by the Moon (position 1 in Figure 7.7), totality occurs at position 2, and the phenomenon ends when the Sun re-emerges from behind the Moon (position 3). When the distance between the Earth and the Moon is at its maximum, the area of the Moon as seen from the Earth is less than the area of the Sun. On these occasions, we see a bright ring of sunlight around the shadow of the Moon in what is called an *annular eclipse* (not shown).

The revolution of the Moon around the Earth can also place it in the shadow of the Earth, which results in a lunar eclipse visible at night (not shown). The shadow of the Earth (umbra) at the distance of the Moon is nearly 9200 km wide or more than 2.5 times the diameter of the Moon. Therefore, a total lunar eclipse can last up to 100 minutes as the Moon slowly moves across the shadow of the Earth before it re-merges into sunlight.

7.2.9 Orbital Velocities

The velocity of a planet in a circular orbit around the Sun was derived from basic principles in Science Brief 1.5.2 and is expressed by the equation:

$$v_0 = \left(\frac{GM}{r} \right)^{1/2} \qquad (7.8)$$

where M = mass of the Sun in kilograms
  r = distance between the center of the planet and the center of the Sun in meters.
G = gravitational constant (6.67 × 10^{-11} Nm^2/kg^2)
v_0 = orbital velocity of the planet in m/s.
Equation 7.8 applies to any body in the solar system in orbit around another body provided that the shape of its orbit is a circle.

In reality, Kepler's first law states that the orbits of the planets are ellipses rather than circles, which means that equation 7.8 does not strictly apply to the planets of the solar

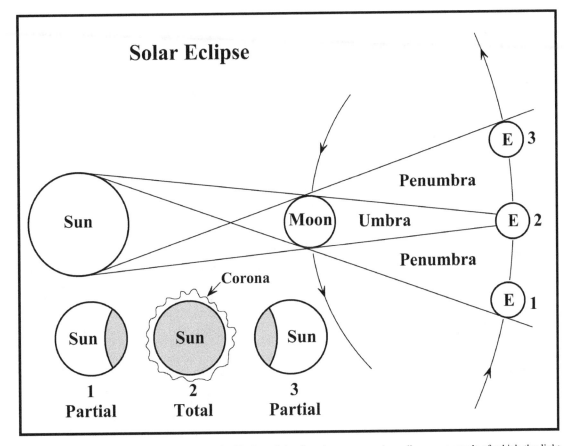

Figure 7.7. When the Moon passes between the Earth and the Sun, it causes a solar eclipse, as a result of which the light of the Sun is partially or completely blocked from reaching the Earth. A total eclipse occurs only in that part of the Earth that actually passes through the umbra whose width decreases with increasing distance between the Moon and the Earth. The diagram is not to scale and was adapted from Figure 2.24 of Hester et al. (2002)

system. Nevertheless, the orbital velocity of Mars (24.19 km/s), calculated by means of equation 7.8 in Science Brief 7.6.4, is in good agreement with the observed period of its orbit (Table 7.1) because the eccentricity of its orbit (0.093) is low, which means that the orbit of Mars is very nearly circular.

According to Kepler's second law, a planet in Figure 7.2 sweeps out triangular areas of equal area in equal intervals of time. This statement implies that the velocity of a planet at perihelion is greater than it is at aphelion. Therefore, the velocity of a planet in its orbit varies continuously depending on its position between perihelion and aphelion. However, if the orbit is nearly circular (i.e., it has a low eccentricity), the average orbital velocity can be calcu-

lated by dividing the circumference of its orbit by the period of its revolution.

For example, the length of the semi-major axis (a) of the orbit of the Earth (i.e., its average distance from the center of the Sun) is a = 149.6×10^6 km (Section 2.3.1) and its period of revolution is p = 365.256 days (Table 7.1). The circumference of a circle is equal to $2r\pi$ where r is the radius of the circle and $\pi = 3.14$. Therefore, the average orbital velocity (v_o) of the Earth is:

$$v_o = \frac{2a\pi}{p} = \frac{2 \times 149.6 \times 10^6 \times 3.14}{365.256 \times 24 \times 60 \times 60} \text{ km/s}$$

$$v_o = 29.77 \text{ km/s}.$$

Nothing in our every-day experience approaches velocities of this magnitude. For example, the velocity of a rifle bullet is only about one kilometer per second. Nevertheless, we are unaware of the high orbital velocity of the Earth because we experience velocities only with respect to fixed points in our frame of reference. The average orbital velocities of the planets in the solar system are listed in Table 7.1 and are illustrated in Figure 3.3.

The decrease of the average orbital velocities with increasing distance of the planets from the center of the Sun is explained in Science Brief 1.5.3 by the equation:

$$r = \frac{GM}{v_o^2} \qquad (7.9)$$

where all symbols have the same meaning as before. The equation indicates that the radii of planetary orbits depend on the reciprocal of the square of their orbital velocities, provided that the orbits are circular. In other words, a planet having a small circular orbit (small value of r) must have a high velocity. Conversely, a planet with a large circular orbit must have a low orbital velocity, as shown in Figure 3.3.

7.2.10 Launching Satellites into Orbit

The laws that determine the average velocities of the planets in their orbits around the Sun also apply to artificial satellites launched from the Earth or from any other body in the solar system. The critical parameter that determines whether a satellite launched from the Earth will achieve a stable orbit is the so-called *escape velocity* which is defined by the equation:

$$v_e = \left(\frac{2GM}{r} \right)^{1/2} \qquad (7.10)$$

and is derived in Science Brief 7.6e. In this equation:

M = mass of the Earth in kilograms,
G = universal gravitational constant ($6.67 \times 10^{-11} \, \text{Nm}^2/\text{kg}^2$)
r = distance to the center of the Earth in meters
v_e = escape velocity in units of meters per second.

The equation tells us that the escape velocity depends on the mass of the Earth (M) and on the distance (r) of the object being launched from the center of the Earth. Given that the mass of the Earth is $M = 5.974 \times 10^{24}$ kg, that its radius is $r = 6,378$ km, and that the gravitational constant is $G = 6.67 \times 10^{-11} \, \text{Nm}^2/\text{kg}^2$, the escape velocity of an object on the surface of the Earth is:

$$v_e = \left(\frac{2 \times 6.67 \times 10^{-11} \times 5.974 \times 10^{24}}{6,378 \times 10^3} \right)^{1/2}$$

$$v_e = \left(1.249 \times 10^8 \right)^{1/2} = 1.117 \times 10^4 \, \text{m/s}$$

$$v_e = \frac{1.117 \times 10^4}{10^3} = 11.2 \, \text{km/s}$$

Evidently, the velocity of a rifle bullet (about 1.0 km/s) is much less than the escape velocity, which explains why rifle bullets fired into the air always fall back to the surface of the Earth.

In order to increase the velocity of the projectile, we decide to fire an iron ball out of an old-fashioned cannon as shown in Figure 7.8A. The cannon ball flies a certain distance, but falls back to the surface of the Earth because it does not achieve the necessary escape velocity. If we want the same cannon ball to fly farther, we must increase its velocity by using more black powder. If the force generated by the charge of black powder is large enough, the cannon ball will reach escape velocity and will fly so far that it literally falls off the Earth (i.e., it goes into orbit) as suggested in Figure 7.8B.

Of course, we do not launch satellites into orbit by shooting them out of cannons. Instead, we attach the satellite to a rocket engine which lifts it off the launch pad and carries it to such a height (say 250 km above the surface of the Earth) that its velocity is equal to or exceeds the escape velocity at that distance from the center of the Earth.

We now discover a peculiar property of equation 7.8 because it tells us that the escape velocity decreases the farther the object is located from the center of the Earth. If we increase the value of r in equation 7.8 by 250 km and calculate the escape velocity at that height, we get a value of $v_e = 10.9$ km/s which is less than the escape velocity at the surface of the Earth. These

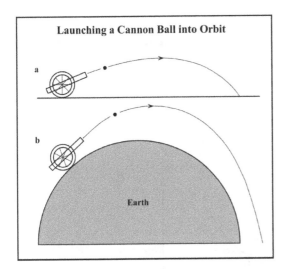

Launching a Cannon Ball into Orbit

a

b

Earth

Figure 7.8. A. Trajectory of an iron sphere fired from a muzzle-loading cannon charged with black powder. B. When the cannon ball reaches escape velocity (11.2 km/s) it falls around the curvature of the Earth and goes into orbit around it

concepts were first mentioned in Section 1.3 and will come up again in later chapters.

When a satellite reaches its destination in space with a velocity that is greater than the velocity required for a circular orbit, it enters an elliptical orbit which is characterized by its *apogee* (point farthest from the planet) and its *perigee* (point closest to the planet). When a satellite in an elliptical orbit is accelerated to a still higher velocity, its orbit may become parabolic in shape, which means that the satellite flies off into interplanetary space and does not return to the Earth. At still higher velocities, the orbit becomes hyperbolic which likewise carries the satellite away from the Earth without return (Hartmann, 2005, Figure 3.4).

7.2.11 Lagrange Points

The gravitational and centrifugal forces acting on a planet in orbit around the Sun (or a satellite orbiting a planet) combine to define five points in space at which another body can maintain its position relative to the other two. The existence of such points was predicted in 1772 by the French mathematician *Joseph Louise Lagrange* (1736–1813) and was confirmed in 1906 by the

discovery of asteroids that occupy two Lagrange points in the orbit of Jupiter. The Lagrange points identified by number in Figure 7.9 revolve around the Sun with the same period as the planet with which they are associated.

Lagrange points 4 and 5 are especially important because objects at these locations tend to remain there indefinitely because the gravitational field at these sites is bowl-shaped. Objects that enter these regions may move around within them like marbles in a bowl, but they cannot leave (Hester et al., 2002). For example, the Lagrange points 4 and 5 associated with Jupiter contain collections of asteroids called the Trojans, which is an allusion to the wooden horse in which soldiers of the Greek armies were hiding during the siege of the city of Troy in the 13th century BC. The Trojan asteroids revolve around the Sun in the same orbit as Jupiter and maintain a constant distance ahead and behind that planet. In addition, Lagrange points 4 and 5 in Figure 7.9 define two equilateral triangles with the Sun and a planet. The other Lagrange points (1,2, and 3) are less effective because solid objects at these locations can drift away or be displaced when another object enters these regions. The reason for this behavior is that the gravitational field at these sites is like an inverted bowl or the summit of a hill. An object can stay there only until it is acted upon by a force that causes it to move away.

Lagrange points are associated with each of the nine planets of the solar system and with the principal satellites that revolve around them. For example, the Lagrange points 4 and 5 of Tethys and Dione (satellites of Saturn) contain several small bodies and the orbit of Mars harbors a Trojan asteroid called 5261 Eureka (Hartmann, 2005, p 56).

The Lagrange points 4 and 5 of the Earth-Moon system are sites where large structures may be constructed in the future with building materials derived from the Moon. The transport of cargo from the Moon to Lagrange points 4 and 5 is facilitated by the fact that these sites maintain a constant position relative to a support base on the lunar surface (O'Neill 1974, 1975, 1977; Johnson and Holbrow, 1977; Heppenheimer, 1977).

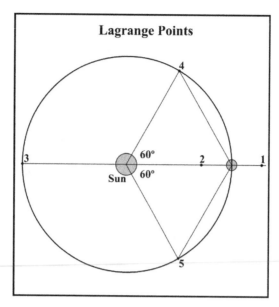

Figure 7.9. The gravitational and centrifugal forces of two bodies in orbit about their center of gravity cause the existence of five Lagrange points identified by number. Objects that enter regions 1,2, and 3 may drift away when acted upon by a force. Objects at sites 4 and 5 behave like marbles in a bowl. They can move within the bowl but they cannot leave it. For example, the Lagrange points 4 and 5 of Jupiter contain collections of solid objects known as the Trojan asteroids which revolve around the Sun in the same orbit as Jupiter and maintain a constant distance in front of and behind Jupiter. Adapted from Hartmann (2005), Freedman and Kaufmann (2002), and Hester et al. (2002)

7.3 Celestial Mechanics of the Earth

The environmental conditions on the surface of the Earth are strongly affected by the inclination of the axis of rotation of the Earth (i.e., its axial obliquity) and much less by the eccentricity of its orbit. These two parameters determine the apparent motions of the Sun as seen from the surface of the Earth and hence the seasonality of the climate in the northern and southern hemispheres.

7.3.1 Rising and Setting of the Sun

The rotation of the Earth causes the Sun to rise in the east and to move in an arc across the southern sky (when viewed from the northern hemisphere) until it sets in the west. The highest position of the Sun in the sky (i.e., noon) depends on the season of the year and on the location of the observer. The time that elapses between successive high-stands of the Sun is defined to be one solar day as discussed in Section 2.3.4 and Science Brief 2.5.4.

The relation between the apparent motions of the Sun and the rotation of the Earth is illustrated in Figure 7.10. Consider that you are standing somewhere on the northern hemisphere of the Earth facing north. Note that east is to the right and west is to the left of you. Now consider that you are standing at point A in Figure 7.10 facing north as before. It is still dark at point A because the first ray of sunlight has not yet reached you. However, as the Earth continues to rotate in a prograde direction, the first ray of sunlight will become visible on the horizon to your right (i.e., in the east). Eventually, your location will be in full daylight at point A' and the Sun will continue to rise in the sky until it reaches its highest position at noon.

Consider next that you are standing at point B in Figure 7.10 facing north. The Sun is still visible in the sky but it is now sinking toward the horizon. As the Earth turns, the Sun sinks below the horizon to your left (i.e., west). By the time the Earth has rotated you to position B', the Sun has set and the night has started.

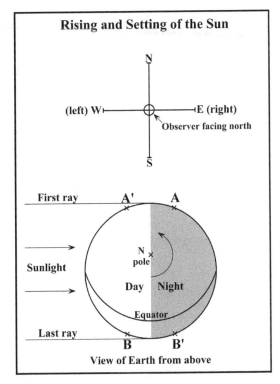

Rising and Setting of the Sun

View of Earth from above

Figure 7.10. The rotation of the Earth between points A and A' explains why the Sun rises in the eastern sky in the morning. Similarly, its rotation from point B and B' causes the Sun to set in the western sky in the evening. The view of the Earth is from above. The North geographic pole is displaced from the center because of the inclination of axis of rotation

7.3.2 Change in the Period of Rotation

The *period* of rotation of the Earth has changed during its history as a result of two opposing processes. It decreased as a result of the growth of the core and increased because of gravitational interactions (tides) with the Moon and neighboring planets. The net result has been that the period of rotation of the Earth has increased from about 15.7 hours at 2.5 billion years ago to 24.0 hours at the present time. In other words, the length of a day has increased because the period of rotation of the Earth increased. The periods of rotation of the other planets and their satellites have also increased for the same reasons. More information on this phenomenon is provided in Science Briefs 7.6.6 and 9.12.2 and in the book by Munk and MacDonald (1960).

7.3.3 Seasons of the Northern Hemisphere

The climate on the surface of the Earth is partly determined by the amount of sunlight it receives, which depends not only on the inclination of its axis of rotation (23.5°) but also on the luminosity of the Sun, the average distance between the Earth and the Sun, the concentration of greenhouse gases in the atmosphere of the Earth (e.g., carbon dioxide, methane, and water vapor), and the fraction of sunlight that is reflected by the surface of the Earth back into interplanetary space called the *albedo* (Graedel and Crutzen, 1993). The ellipticity of the orbit of the Earth and the resulting variation of its distance to the Sun has only a minor effect on the seasonality of the climate because of the low eccentricity of the Earth's orbit (i.e., 0.017).

The key to understanding the seasonality of the climate of the Earth is that the attitude of its axis of rotation in Figure 7.11 remains fixed in space as the Earth revolves around the Sun (except as discussed in Section 7.3.6). The seasons of the northern hemisphere are defined by the summer and winter solstices and by the vernal and autumnal equinoxes, during which the Earth occupies the specific sites in its orbit identified in Figure 7.11. The corresponding seasons of the southern hemisphere are opposite to those of the northern hemisphere:

Northern hemisphere	Southern hemisphere
Summer solstice	Winter solstice
Autumal equinox	Vernal equinox
Winter solstice	Summer solstice
Vernal equinox	Autumnal equinox

7.3.4 The Earth at Summer Solstice

On the day of the summer solstice in the northern hemisphere (June 21) the axis of rotation of the Earth in Figure 7.11 points toward the Sun. Therefore, on this day the northern hemisphere receives a maximum amount of sunlight, whereas the southern hemisphere receives a minimum amount. The vertical cross-section of the Earth in Figure 7.12 demonstrates that during the summer solstice all locations within the arctic circle (including the north pole) have 24 hours of

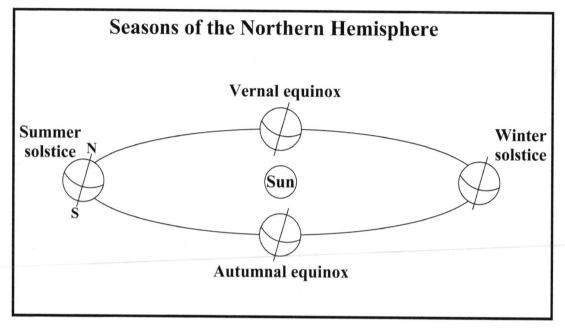

Seasons of the Northern Hemisphere

Vernal equinox

Summer solstice

Winter solstice

Sun

Autumnal equinox

Figure 7.11. The seasons of the northern hemisphere of the Earth are caused by the variation of the amount of sunlight this hemisphere receives because of the 23.5° tilt of the axis of rotation (i.e., the axial obliquity). The difference in the distances to the Sun when the Earth is at perihelion and aphelion is small and does not significantly affect the insolation of the Earth. The amount of sunlight received by the northern hemisphere is a maximum at the summer solstice and a minimum at the winter solstice. During the vernal and autumnal equinoxes both hemispheres get the same amount of sunlight and the length of the day is equal to the length of the night. Adapted from Freedman and Kaufmann (2002)

daylight. At the same time, the area encompassed by the Antarctic circle is in complete darkness for 24 hours on this day. The latitude of the arctic and antarctic circles is at latitude 66.5°N and S (i.e., 90.0–23.5), respectively.

Another significant phenomenon associated with the summer solstice of the northern hemisphere occurs along the *Tropic of Cancer* located at 23.5°N latitude. On this day, the Sun is directly overhead at noon along this circle. In addition, the Tropic of Cancer is the location of the most northerly position on the Earth where the Sun is directly overhead at noon. Similarly, the *Tropic of Capricorn* at 23.5°S latitude is the circle along which the Sun is directly overhead on the day of the summer solstice of the southern hemisphere. As the Earth continues in its orbit past the summer solstice of the northern hemisphere, the location of the circle above which the Sun is directly overhead at noon gradually moves south. On the day of the autumnal equinox (Figure 7.11) the Sun is

directly overhead at noon above the equator of the Earth.

7.3.5 Autumnal and Vernal Equinoxes

On the days of the autumnal and vernal equinox (Figure 7.11) the axis of rotation of the Earth points in the direction in which the Earth is moving in its orbit and therefore is not at all inclined toward the Sun. Therefore, on these two days both hemispheres of the Earth receive the same amount of sunlight, meaning that day and night are equal in length ("equinox" means "equal night"). In addition, on the day of the equinox the Sun is directly overhead along the equator as noted above. As the Earth continues in its orbit toward the winter solstice of the northern hemisphere, the circle along which the Sun is directly overhead at noon moves south and reaches the Tropic of Capricorn at latitude 23.50°S on the day of the winter solstice in the northern hemisphere which coincides with the summer solstice of the southern hemisphere.

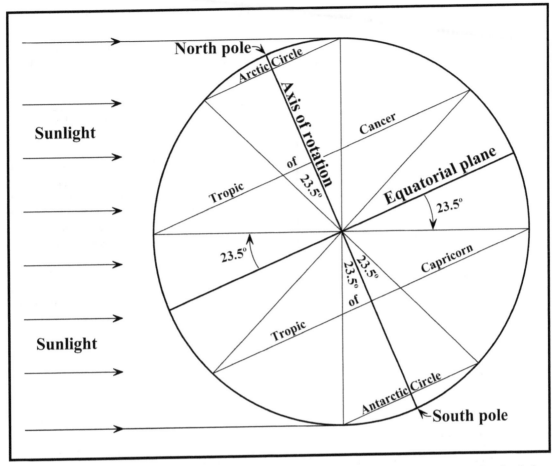

Figure 7.12. The Earth seen in vertical profile to illustrate the sunlight that reaches different parts of the northern hemisphere during the summer solstice. For example, note that all parts of the Earth inside the Arctic Circle have 24 hours of daylight on this day. In addition, the Sun is directly overhead at noon above all points that lie on the great circle of the Tropic of Cancer. The inclination of the axis of rotation of the Earth (i.e., its axial obliquity) determines the positions of the Arctic and Antarctric circles as well as the positions of the Tropic of Cancer and the Tropic of Capricorn

7.3.6 Milankovitch Cycles

The tilt of the axis of the Earth (i.e., its axial obliquity) is not constant over long periods of time but varies from 21.5 to 24.5° with a period of 41,000 years. In addition, the axis of rotation wobbles (precesses) such that it traces out a complete circle in 23,000 years. The effects of the changing inclination and orientation of the axis of rotation of the Earth on the global climate are amplified by the changing elongation (i.e., eccentricity) of its orbit on a timescale of 100,000 years. The result of these variations is that the amount of sunlight received by the northen and southern hemispheres changes systematically on timescales of tens of thousands of years. The Yugoslav astronomer Milutin Milankovitch calculated so-called insolation curves for locations at 45° latitude in the northen hemisphere of the Earth based on the celestial mechanics of its orbit. The results in Figure 7.13 indicate that the insolation (amount of sunlight received) of the northen hemisphere of the Earth has increased and decreased systematically during the past 140,000 years. The work of Milankovitch, which was published in the 1920s and 1930s, was later used by Broecker et al. (1968), Imbrie and Imbrie (1986), and others to

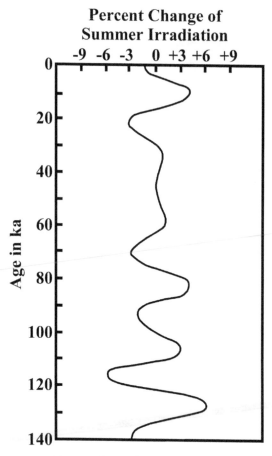

Percent Change of Summer Irradiation

Figure 7.13. Variation of the amount of solar radiation received by the northern hemisphere of the Earth at 45 °N as a result of the celestial mechanics of its orbit. Based on calculations by Milankovitch as adapted by Broecker et al. (1968)

evolved from the calendar used by the Romans who adopted it from the Greeks and from preceding civilizations of the eastern Mediterranean region (i.e., the Egyptians, Babylonians, and Sumerians).

The Roman calendar was originally based on the synodic period of the Moon (29.53059 days) and consisted of 12 months and 355 days. Because the length of the Roman year was inconsistent with the sidereal period of revolution of the Earth (365.256366 days), the Romans had to add ten days to each year in order to keep in step with the seasons, which depend on the sidereal period of the Earth rather than on the synodic period of the Moon. These annual adjustments to the calendar were not always made, which created enough of a problem to prompt the Roman Emperor Julius Caesar in the year 46 BC to invite the Greek astronomer Sosigenes of Alexandria to reform the calendar. Sosigenes started by adding 90 days to the year 46 BC to make up for lost time. In addition, he increased the length of the year to 365.25 days and stipulated that one extra day must be added once every four years (i.e., leap year).

The so-called Julian calendar was in use for almost 1630 years. However, by 1454 AD the calendar had surpassed the movement of the Earth by about 10 days because the Julian year (365.25 days) was longer than the tropical year (365.242199 days) defined as the time between successive vernal equinoxes. After considerable delay, Pope Gregory III proclaimed in 1582 that 10 days would be eliminated from October of that year and that henceforth the length of the year would be 365.2422 days. As a result, the Gregorian calendar differs from the Julian calendar by an amount of time equal to 3.12 days every 400 years.

This discrepancy requires that the calendar must be reduced by three days every 400 years, which is accomplished by specifying that centennial years (i.e, 1900, 1800, etc.) cannot be leap years except when they are divisible by 400. For this reason, the year 2000 was a leap year but 1900 was not. In this way, we have given back the extra three days that had accumulated between 1600 and 2000 AD. A further correction is made by the provision that years divisible by 4000 have only 365 days. As a result, the

explain the cause for the large-scale continental glaciations and the related variations of sealevel during the Pleistocene Epoch that started at 1.81 million years in the past and ended only about 10,000 years ago.

7.4 The Gregorian Calendar

The celestial mechanics of the Earth as manifested by the apparent motions of the Sun and the Moon and by the seasonal appearance of familiar constellations of stars all contribute to the way in which we measure the passage of time (Section 2.3.2). The calendar we use today has

Gregorian calendar only deviates by about one day in 20,000 years (Hester et al., 2002).

The Gregorian calendar was adopted by most Roman Catholic states by 1587 AD, whereas Protestant states delayed their acceptance until the 18th century. For example, England and her colonies did not adopt the "new style" calendar until 1752 and Russia complied only in 1917 after the Bolshevik Revolution.

7.5 Summary

One of the most difficult intellectual problems humans have had to solve is to recognize that we live on a planet that simultaneously revolves around the Sun as it rotates on its axis. These motions regulate our lives by causing the Sun to rise in the morning and to set in the evening. In addition, the inclination of the axis of rotation affects the amount of sunlight the northern and southern hemispheres receive at different times of year, thereby determining the global climate and its seasonal variations.

Celestial mechanics is not an easy subject to master because all parts of the solar system are in continuous motion. Nevertheless, all of us need to understand how we measure time, what causes the Sun to rise and set every day, and why the weather changes in an annual cycle. We also need to grasp how satellites and manned spacecraft are launched from the surface of the Earth into orbits around it.

This knowledge combined with rocket technology and electronic computers has permitted us to land on the Moon, to view the Universe in exquisite detail, and to explore the surfaces of Mars and most of the other planets and their principal satellites. We have also built an orbiting space station where astronauts are learning how to live in space in preparation for the establishment of manned outposts on the Moon and on Mars. In order to achieve such goals, many difficult problems must be investigated and solved. We could not attempt such feats if the pioneers of astronomy had not mastered celestial mechanics.

Therefore, we pay our respects to the pioneers of astronomy who figured out how the solar system works and thereby made possible the exploration of the planets and their satellites. As a result of what we have already achieved, our self-image has changed because, in the words of Archibald McLeish, "We are all of us riders on the Earth together."

7.6 Science Briefs

7.6.1 Newton's Laws of Motion

1. A body at rest remains at rest and a body in motion at a constant speed remains in motion along a straight line unless acted upon by a force.
2. The rate of change of velocity of a body (a) is directly proportional to the force (F) and inversely proportional to the mass (m) of the body (i.e., a = F/m).
3. The actions of two bodies are always equal in magnitude and opposite in direction (Section 1.3).

7.6.2 Equation of an Ellipse

An ellipse is the locus of all points along a closed curve such that the sum of their distances from the two foci remains constant. It is also defined as being the intersection of a right circular cone with a plane that is oriented in such a way that it is *not* parallel to the base or to the axis of a cone. This definition illustrates the relationship between an ellipse and a circle because a circle is the intersection of a right circular cone with a plane oriented *parallel* to the base of the cone.

The equation of an ellipse, oriented such that the semi-major axis is parallel to the x-axis in Cartesian coordinates with the center of the ellipse at (h, k) is:

$$\frac{(x-h)^2}{a^2} + \frac{(y-k)^2}{b^2} = 1 \qquad (7.11)$$

where h = x-coordinate of the center,
k = y-coordinate of the center,
a = semi-major axis,
b = semi-minor axis.
The focal distance (f) of an ellipse is:

$$f = (a^2 - b^2)^{1/2} \qquad (7.12)$$

and the eccentricity (e) is:

$$e = \frac{(a^2 - b^2)^{1/2}}{a} \qquad (7.13)$$

7.6.3 Synodic Periods of Revolution

The synodic period of revolution of a planet (Section 7.2.7) is related to its sidereal period by equations 7.6 and 7.7. If the planet in question is inferior (i.e., has a smaller orbit than the Earth), the rate of revolution of the Earth is $360°/E$, where E is its sidereal period (365.25 days). Similarly, the rate of revolution of the inferior planet is $360°/P$ where P is the sidereal period of revolution of the inferior planet. The angle of rotation of the Earth (A_E) during an interval of time equal to the synodic period (S) of the inferior planet is:

$$A_E = \left(\frac{360°}{E}\right) S$$

The angle of rotation of the inferior planet (A_p) during one synodic period is:

$$A_p = \left(\frac{360°}{P}\right) S$$

Actually, the inferior planet has a higher average orbital velocity than the Earth and therefore has gained a lap on the Earth. Consequently, the total angle of rotation of the inferior planet is equal to the angle of rotation of the Earth plus 360°:

$$\left(\frac{360°}{P}\right) S = \left(\frac{360°}{E}\right) S + 360°$$

Dividing each term by $360°S$ yields:

$$\frac{1}{P} = \frac{1}{E} + \frac{1}{S}$$

which is identical to equation 7.6 in Section 7.2.7.

If the planet in question has a larger orbit than the Earth (i.e., a superior planet), the Earth overtakes the planet. Therefore, the angle of rotation of the Earth during one synodic period of a superior planet is:

$$\left(\frac{360°}{E}\right) S = \left(\frac{360°}{P}\right) S + 360°$$

Dividing each term by $360°S$ yields:

$$\frac{1}{E} = \frac{1}{P} + \frac{1}{S}$$

which is rearranged into the form of equation 7.7:

$$\frac{1}{P} = \frac{1}{E} - \frac{1}{S}$$

7.6.4 Average Orbital Velocity of Mars

Given that the length of the semi-major axis of the orbit of Mars is 1.5236 AU, that the mass of the Sun (M) is 2×10^{30} kg, and that the gravitational constant (G) is 6.67×10^{-11} newton m^2/kg^2, the average orbital velocity of Mars is:

$$v_o = \left(\frac{GM}{r}\right)^{1/2}$$

$$v_o = \left(\frac{6.67 \times 10^{-11} \times 2 \times 10^{30}}{1.5236 \times 149.6 \times 10^6 \times 1000}\right)^{1/2}$$

$$v_o = \left(\frac{13.34 \times 10^{19}}{227.930 \times 10^9}\right)^{1/2} = \left(0.05852 \times 10^{10}\right)^{1/2}$$

$$= 0.2419 \times 10^5 = 24.19 \, \text{km/s}$$

The velocity calculated above is in good agreement with the velocity given in Table 7.1 (24.1 km/s).

7.6.5 Escape Velocity

When a small object on the surface of the Earth (or any other planet or satellite) is acted upon by a force, it is accelerated and thus acquires a certain amount of kinetic energy (KE) depending on its mass (m) and its velocity (v):

$$KE = \frac{1}{2} mv^2$$

As the object rises off the surface of the planet or satellite, it also acquires gravitational potential energy (PE):

$$PE = \frac{GMm}{r}$$

where G = universal gravitational constant,
M = mass of the planet or satellite, and
r = distance between the object and the center the planet or satellite.
In case the kinetic energy of the object is greater than its potential energy, the object will escape from the planet or satellite. The velocity that is required to allow an object to escape from the surface of a planet or satellite is obtained by

setting the kinetic energy of the object equal to its potential energy:

$$KE = PE$$

$$\frac{1}{2}mv^2 = \frac{GMm}{r}$$

Solving for the escape velocity (v_e) yields:

$$v_e = \left(\frac{2GM}{r}\right)^{1/2} \tag{7.14}$$

Since the gravitational constant is expressed in mks units, the mass (M) must be in kilograms, the distance (r) must be in meters, and time is in seconds. Note that the equation for the escape velocity (v_e) is similar in form to the equation for the velocity (v_o) of an object in a stable circular orbit:

$$v_o = \left(\frac{GM}{r}\right)^{1/2}$$

Therefore, we can express the escape velocity by the equation:

$$v_e = 2^{1/2}v_o = 1.41\ v_o$$

In summary, the escape velocity is 1.41 times the velocity of an object in a stable circular orbit (Hartmann, 2005).

7.6.6 Increase of the Period of Rotation of the Earth

The period of rotation of the Earth and hence the length of one day has increased in the course of geologic time because of gravitational interactions of the Earth with the Moon and with neighboring planets in the solar system (Munk and MacDonald, 1960). This tendency was counteracted initially by the growth of the iron core which caused the period of rotation of the Earth to decrease (i.e., the rotation speeded up). However, the period of revolution of the Earth (i.e., the length of one year) was not affected by these processes. Munk and Davies (1964) later combined both effects and demonstrated how the period of rotation of the Earth has increased with time.

For example, the calculations of Munk and Davies (1964) indicate that 2.5 billion years ago one day had only 15.7 hours. Therefore, one year contained a larger number of shorter days compared to the present. The number of days (N) in one year 2.5 billion years ago was:

$$N = \frac{365.25 \times 24}{15.7} = 558.35\ days$$

The geophysical evidence for the increase of the period of rotation was supported by the paleontologist Wells (1963) who reported that some Devonian corals had about 400 daily growth lines per annual layer. This result fits the calculated result of Munk and Davies (1964) and was later corroborated by Clark (1968).

7.7 Problems

1. Calculate the distance from the center of the Sun to the center of the Earth when the Earth is at perihelion (q) and aphelion (Q) of its orbit, given that the eccentricity is 0.017. Express the results in kilometers. (Answer: $q = 147.06 \times 10^6\ km$; $Q = 152.14 \times 10^6\ km$).
2. Calculate the number of jovian days in one jovian year using the data in Table 7.1. (Answer: 1046 jovian days.)
3. Calculate the escape velocity of an object located on the surface of Mercury using data in Appendices I and II and express the results in km/s. (Answer: $v_e = 4.25\ km/s$).
4. Verify by calculation the magnitude of the synodic period of Pluto using data in Table 7.1. (Answer: S = 366.7 d).
5. Use equation 7.3 to calculate the average distance from the surface of the Earth at the equator of a satellite in a "geostationary orbit"(i.e., whose period of revolution is equal to the period of rotation if the Earth). Express the result in kilometers. (Answer: 35,879 km above the surface of the Earth).

7.8 Further Reading

Broecker WS, Thurber DL, Goddard J, Ku TL, Matthews RK, Mesolella KJ (1968) Milankovitch hypothesis supported by precise dating of coral reefs and deep-sea sediment. Science 159:297–300

Clark GR II (1968) Mollusk shell: Daily growth lines. Science 161:800–802

Ferris T (1988) Coming of age in the Milky Way. Doubleday, New York

Freedman RA, Kaufmann III WJ (2002) Universe: The solar system. Freeman, New York

Gingerich O (2004) The book nobody read: Chasing the revolutions of Nicolaus Copernicus. Walker, New York

Graedel TE, Crutzen PJ (1993) Atmospheric change: An Earth-system perspective. Freeman, New York

Hartmann WK (2005) Moons and planets, 5th edn. Brooks/Cole, Belmont, CA

Hester J, Burstein D, Blumenthal G, Greeley R, Smith B, Voss H, Wegner G (2002) 21st century astronomy. Norton, New York

Heppenheimer TA (1977) Colonies in space. Stackpole Books, Harrisburg, PA

Imbrie JD, Imbrie KO (1986) Ice ages: Solving the mystery. Harvard University Press, Cambridge, MA

Johnson RD, Holbrow C (1977) Space settlements – a design study. NASA SP-413, Washington, DC

Munk WH, MacDonald GLF (1960) The rotation of the Earth. Cambridge University Press, Cambridge, UK

Munk WH, Davies D (1964) The relationship between core accretion and the rotation rate of the Earth. In: Craig H, Miller SL, Wasserburg GJ (eds.) Isotopic and Cosmic Chemistry. North-Holland, Amsterdam, pp 342–346

O'Neill GK (1974) The colonization of space. Phys Today Sept.: 32

O'Neill GK (1975) Space colonies and energy supply to the Earth. Science, 10:943

O'Neill GK (1977) The high frontier. Morrow, New York

Wells JW (1963) Coral growth and geochronometry. Nature 197:948–950

Meteorites and Impact Craters

Meteorites are solid objects that fall to the surface of the Earth from the interplanetary space of the solar system. They are named after the county, village, or topographic feature near which they were found or from where their fall was observed. In cases where many specimens are recovered from the same site, all of the specimens are given the same name. The size of meteorite specimens ranges widely from particles smaller than grains of sand to large masses measuring tens of meters in diameter. Even larger meteorites are known to have impacted on the Earth. Most meteorites are composed of silicate and oxide minerals, like those that form igneous rocks of basaltic composition on the Earth, and are therefore classified as "stony" meteorites. In addition, some meteorites consist largely of metallic alloys of iron and nickel, whereas others contain a mixture of silicate minerals in a matrix of metallic iron. The fall of meteorites is accompanied by thunderous explosions of fireballs that streak across the sky trailing dark smoke. Because of their unusual appearance and the terrifying spectacle that accompanies their fall, meteorites have been collected for thousands of years and were considered to be messengers from the gods who resided in the sky. Some meteorite specimens are extremely massive such as the Cape York iron meteorite in Figure 8.1 which weighed about 200 tons before it broke into several fragments as it passed through the atmosphere. Metallic iron from some of these fragments was used by the Inuits to make knife blades and spearpoints. Even a dagger found in King Tutankhamen's burial chamber is suspected of having been forged from meteoritic iron.

The scientific importance of meteorites arises from the fact that they are fragments of larger bodies that formed at the same time as the planets and their satellites very early in the history of the solar system. Accordingly, the study of meteorites has provided a wealth of information about the origin and age of the solar system including the Earth.

When a large meteorite having a diameter of 10 m or more impacts on the Earth, it excavates a crater the diameter of which is roughly ten times larger than the diameter of the impactor. More than 150 such impact craters have been identified on the Earth and tens of thousands occur on the Moon. The impact of large meteorites has serious environmental consequences exemplified in 1908 by the explosion of a meteorite in the atmosphere above the Tunguska River in Siberia.

Before we continue this discourse about meteorites, we need to clarify the terminology that is used to describe them:

Meteors are streaks of light in the sky at night caused by small grains that are heated by friction as they pass through the atmosphere of the Earth.

Meteoroids are solid objects in interplanetary space with diameters ranging from a few centimeters to several meters or even tens of meters. They originated from collisions of asteroids, but the difference between large meteoroids and small asteroids is not defined.

Meteorites are meteoroids that have fallen to the surface of the Earth. In other words, when a meteoroid has passed through the atmosphere and has impacted on the surface of the Earth it is called a meteorite.

The origin of stony meteorites and of the chondrules they contain has been the subject of a growing number of review papers and books including: McCall (1973), Taylor et al. (1983), Grossman (1988), Wasson (1993), Scott et al.

Figure 8.1. Ahnighito (i.e., "tent") is a 34-ton fragment of an iron meteorite (III AB, octahedrite) that fell on the Cape York peninsula of north West Greenland thousands of years before the area was inhabited by Inuits. Ahnighito and two smaller fragments of the so-called Cape York meteorite are on display in the American Museum of Natural History in New York City. Additional specimens can be viewed in the Geological Museum of Copenhagen. The total weight of the Cape York meteorite is approximately 200 tons. The recovery of the fragments of this meteorite was described by Buchwald (1975). The chemical composition and mineralogy of a 20-ton fragment known as Agpalilik in Copenhagen was investigated by Koeberl et al. (1986). Ahnighito and two smaller specimens of the Cape York (i.e., "Woman" and "Dog") were brought to New York City in 1897 by Robert E. Peary, polar explorer and Rear Admiral of the U.S. Navy (Weems, 1967). (Photo: Scot Kaplan)

(1994), Hewins et al. (1996), Hewins (1997), Lipschutz and Schultz (1999), Rubin (2000), Hutchison (2004), Scott and Krot (2004), Davis (2004), Zanda (2004), and Sears (2005).

8.1 Classification and Mineralogy of Meteorites

Meteorites are subdivided on the basis of their composition into stony meteorites, stony – irons, and iron meteorites. Most (but not all) stony meteorites contain small spherical pellets called *chondrules*. Therefore, stony meteorites are further classified into *chondrites* (containing chondrules) and *achondrites* (lacking chondrules). Both types of stony meteorites are subdivided into different classes based primarily on their chemical compositions, mineral compositions, and textures. Many chondrites as well as achondrites are breccias that formed by crushing presumable during collisions in the asteroidal belt. The *stony irons* and *irons* also encompass several different classes defined on the basis of chemical, mineralogical, and textural criteria.

The classification of meteorites in Table 8.1 reveals several interesting facts about them:

1. Ordinary chondrites and achondrites contain minerals that form only at high temperatures and that do not contain hydroxyl ions (OH⁻) or water molecules in their crystal structure.

2. In contrast, carbonaceous chondrites contain minerals that form at comparatively low temperature (e.g., serpentine, sulfates) and they also contain water molecules and solid hydrocarbon compounds.

3. Iron meteorites like Sikhote-Alin in Figure 8.2 contain minerals composed of metallic iron and nickel (e.g., kamacite and taenite).

4. Some stony meteorites are known to have originated from the Moon (Chapter 9) and from Mars (Chapter 12). The martian meteorite Nakhla is pictured in Figure 8.3.

5. A few rare minerals in meteorites do *not* occur naturally on the Earth.

Mason (1962, 1972) listed 80 minerals that occur in stony meteorites, most of which also occur in terrestrial rocks although some are rare (e.g., diamond). Nevertheless, we are reassured that meteorites contain the same chemical elements as terrestrial rocks and that, with a few exceptions, they are also composed of the same minerals as terrestrial rocks.

Table 8.1. Classification of meteorites (Hartmann, 2005; Koeberl, 1998)

Type	Class	Subdivision	Minerals
Stony meteorites			
	Ordinary	E	Enstatite
		H	Bronzite
Chondrites		L	Hypersthene
		LL	Olivine, pyroxene, plagioclase
		Others	
	Carbonaceous	C1 – C2	Serpentine, sulfate, organic compounds
		C3 – C5	Olivine, pyroxene, organic compounds
		Ureilites	Olivine-pigeonite
		Aubrites	Enstatite
Achondrites		Diogenites	Hypersthene
		Howardites	Pyroxene – plag.
		Eucrites	Pigeonite – plag.
		Others	
Stony – irons			
Mesosiderites			Pyroxene, plag., Fe – Ni alloy
Pallasites			Olivine, Fe – Ni alloy
Iron meteorites			
	Octahedrites	IA	Kamacite, taenite
		IIIA	graphite, troilite
Irons		IVA	schreibersite
	Hexahedrites	IIA	daubreelite, phosphate
	Ataxites		
	Others		
	Martian meteorites		
	Lunar meteorites		

Figure 8.2. The Sikhote-Alin iron meteorite fell on February 12 of 1947 in the mountains north of Vladi-vostok, Russia. The original meteoroid weighed between 70 and 100 tons before it exploded about 5.8 km above the surface of the Earth. At that time, the meteorite was no longer moving fast enough to be heated by friction, which explains why some of the recovered fragments have the jagged shapes of shrapnel, whereas others have scalloped surfaces or thumb prints called regmaglypts displayed by the specimen in this image. (Reproduced by permission of GNU Free Documentation License, version 1.2 and retrieved from: "http://en.wikipedia.org/wiki/Image: SikhoteAlinMeteorite.jpg.")

The densities of stony meteorites (except carbonaceous chondrites) are higher than those of olivine ($3.32 \, \text{g/cm}^3$) and pyroxene ($3.3 \, \text{g/cm}^3$) because they contain small grains of metallic iron-nickel alloys. Ordinary chondrites have densities between 3.3 and $3.8 \, \text{g/cm}^3$, whereas the densities of most carbonaceous chondrites range from about 2.2 to $3.5 \, \text{g/cm}^3$ depending on their chemical composition (Mason, 1962), although a few have densities less than $2.2 \, \text{g/cm}^3$ (e.g., the Tagish Lake carbonaceous chondrite). The densities of iron meteorites depend on the concentrations of metallic iron and nickel whose densities in pure form are: iron (Fe) = $7.86 \, \text{g/cm}^3$, nickel (Ni) = $8.90 \, \text{g/cm}^3$. Therefore, the density (d) of the metal phase of an iron meteorite containing 8.0 % Ni and 92 % Fe is:

$$d = 0.08 \times 8.90 + 0.92 \times 7.86 = 7.94 \, \text{g/cm}^3$$

We show in Science Brief 8.11.1 that a spherical specimen of an iron meteorite with a diameter of 30.48 cm has a mass (m) of 117,664 g and therefore is about three times heavier than a granite boulder of the same volume.

8.2 Carbonaceous Chondrites

Nearly 5% of the observed falls of meteorites consist of the so-called carbonaceous chondrites which are characterized by the presence of organic compounds composed of carbon, hydrogen, nitrogen, and sulfur. These compounds formed by chemical reactions in the solar nebula and are not of biogenic origin. They were initially incorporated into the planetesimals and were subsequently transferred into the parent bodies of meteorites and into the planets and their satellites. The organic compounds in carbonaceous chondrites are heat-sensitive and have survived only where the ambient temperature did not exceed 125 °C, as for example in the surface layers of the meteorite parent bodies. In addition, the primordial organic compounds have survived in the ice-rich bodies of the solar system, such as the satellites of the gaseous planets, Kuiper-belt objects, and comets. Organic compounds have also been identified in protoplanetary disks that surround certain young stars and in clouds of gas and dust in interstellar space of the Milky Way galaxy.

The origin and subsequent evolution of chondrules that occur in chondrites was discussed in a book by Hewins et al. (1996) with contributions from nearly 100 meteorit-icists who participated in a conference held in 1994 at the University of New Mexico in Albuquerque. Most of the participants supported the hypothesis that chondrules formed in the solar nebula when small clumps of refractory particles were melted by shock waves and/or by electrical discharge (i.e., lightning). An alternative view is that chondrules are crystallized impact-melt spherules that formed as a result of impact into the regolith that covers the surfaces of asteroids (Sears, 2005).

In accordance with the majority view, chondrules and other kinds of particles in chondrites are older than the meteorites in which they occur and, in some cases, formed inside the ancestral stars prior to or during their terminal explosions (Sections 4.2 and 4.3). These ancient grains include:

Chondrules: Spherical particles, less than 5 mm in diameter, composed primarily of olivine and pyroxene that crystallized from droplets of silicate melt that formed within

Figure 8.3. The stony meteorite Nakhla fell on June 28 of 1911 in Abu Hommos near Alexandria, Egypt. More than 40 specimens of Nakhla weighing about 40 kg were recovered, most of which are still held by the Museum of Cairo. Nakhla is known to be a specimen of basalt that was ejected from the surface of Mars by the impact of a meteorite from the asteroidal belt. The Sm-Nd crystallization age of Nakhla is 1.26 ± 0.07 billion years in contrast to all stony meteorites from the asteroidal belt, which crystallized between 4.50 and 4.60 billion years ago (Nakamura et al., 1982). The stony meteorites, that fell at Shergotty (India), Nakhla (Egypt), and Chassigny (France) are the type examples of the SNC group of meteorites all of which originated from Mars. (Photo No. S98-04014, courtesy of NASA, C. Meyer, 2004: http://www-curator.jsc.nasa.gov.curator/antmet/mmc/mmc.htm)

the solar nebula when small clumps of dust particles were melted by the passage of a shock wave or by lightning (Hewins et al. 1996).

CAIs: Solid particles up to 10 mm in diameter composed of refractory Ca-Al minerals.

Diamonds: Microscopic grains ($\sim 10^{-9}$ m) that formed in stars during supernova explosions and which now occur in interstellar space of the Milky Way galaxy and in carbonaceous chondrites such as Allende which is depicted in Figure 8.4.

Silicon carbide: Microscopic grains (0.5 to 2.6×10^{-6} m) produced by supernovas which ejected them into interstellar space of the Milky Way galaxy. These kinds of grains as well as diamonds have been recovered from the carbonaceous chondrite Murchison (Bernatowicz et al., 2003).

Some of the most widely known carbonaceous chondrites and the dates and locations of their fall are: Allende, 1969, Mexico; Murchison, 1969, Australia; Orgueil, 1864, France; Murray, 1950, Kentucky; Ivuna, 1938, Tanzania; Mighei, 1889, Russia. The fall of Allende was described by Clarke et al. (1971) whose report contains excellent photographs.

An unusually friable carbonaceous chondrite in Figure 8.5 disintegrated over British Columbia on January 18, 2000. The fireball was tracked instrumentally, which established the orbit of this meteorite and indicated that it had a low density, high porosity, and low strength. Fragments of the main mass fell onto the ice of Tagish Lake

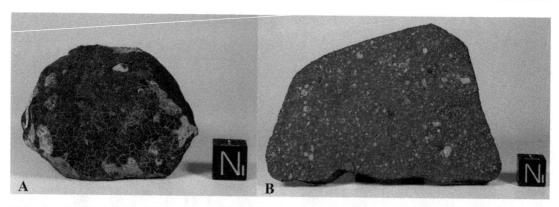

Figure 8.4. A. Allende is a carbonaceous meteorite that fell at 1:05 AM on February 8, 1969, near the village of Pueblito de Allende, Chihuahua, Mexico. The meteorite broke up into a large number of small pieces whose total mass was about two tons. The specimen in part A is covered by a black fusion crust that has flaked off in a few places. The volume of the cube is $1.0 cm^3$. B. The sawed interior surface of Allende reveals the presence of numerous mineral inclusions in a fine-grained, dark grey matrix. The list of mineral that have been identified includes andradite (garnet), anorthite (feldspar), augite (pyroxene), fayalite and forsterite (olivine), melilite, nepheline and sodalite (feldspathoids), perovskite, rhönite, spinel, wollastonite, and other minor constituents. In addition, Allende contains a variety of carbon compounds including fullerene, fullerane, polycyclic aromatic hydrocarboous, benzofluoranthene, and corennulene (Becker and Bunch, 1997). Reproduced by permission of Chinellato Matteo

in British Columbia. Several specimens of the meteorite weighing about 870 g were recovered one week later by Mr. James Brook (a local resident) who wrapped them in plastic bags without touching them and kept them frozen. Additional specimens, found later during spring melt, were saturated with water and had been contaminated. The chemical and mineralogical compositions of the Tagish Lake meteorite were described in a series of reports edited by Brown et al. (2002) including a summary report by Mittlefehldt (2002).

8.3 Tektites

Tektites are small objects composed of black to greenish glass that occur in strewn fields in certain regions of the Earth, including: Australia, Malaysia, Indonesia, Indochina, southern China, Philippines, Africa (Ivory Coast), Czechoslovakia, and North America (Texas and Georgia). The names of tektites are derived from the region in which they were collected (e.g., australites, indochinites, and philippinites). Tektites from Czechoslovakia are known as moldavites (after the Moldau River), those from southern China including the island of Hainan are called Muong Nong tektites, and

the Texas tektites are known as bediasites after the Bedias tribe of native Americans.

The diameters of tektites range from a few millimeters (microtektites) up to 20 cm although most are less than 5 cm across. Their shapes resemble spheres, teardrops, dumbbells, and buttons. The australite tektite in Figure 8.6 is a flanged button. Most tektites appear to have been shaped by aerodynamic processes with concurrent melting and ablation during their passage through the atmosphere. The Muong Nong tektites are layered and weigh up to 12.8 kg. More than 650,000 individual tektite specimens have been collected worldwide weighing between 3 and 4 metric tonnes (Glass, 1982, p. 146–147).

All tektites contain grains of pure silica glass called *lechatelierite* which implies that they reached high temperatures up to 2000 °C. In general, moldavites as well as other tektites have high concentrations of SiO_2 (up to 84.48 %), low concentrations of sodium and potassium ($Na_2O = 0.20\%$, $K_2O = 2.40\%$), and are virtually devoid of water (H_2O) and other volatile compounds (CO_2, CH_4, N_2, etc). In addition, they do not contain crystals of minerals in contrast to volcanic glass (e.g., obsidian) which always contains crystals.

The origin of tektites used to be a controversial subject with several competing hypotheses,

Figure 8.5. The fall of the Tagish Lake carbonaceous chondrite on January 18 of 2000 produced a brilliant fireball and caused thunderous detonations that were observed by many people in northern British Columbia and the Yukon Territory of Canada. Several dozen specimens weighing about 1 kg were collected on January 25 and 26 by Mr. James Brook on the ice of the Taku arm of Tagish Lake located in British Columbia at 59°42′15.7″N and 134°12′4.9″W. About 200 additional specimens weighing between 5 and 10 kg were recovered later between April 20 and May 8 after they had melted into the ice. The strewn field has an area of about 48 square kilometers (3 × 16 km) aligned about S30°E. The Tagish Lake meteorite is a primitive matrix-dominated carbonaceous chondrite containing small chondrules, CAIs, and other grains. The matrix is composed of phyllosilicates, Fe-Ni sulfides, magnetite, abundant Ca-Mg-Fe carbonates, olivine, pyroxene, and about 5.4 % by weight of carbon. Instrumental observations indicate that the meteoroid entered the atmosphere with a velocity of about 16 km/s, that its mass was about 56 tonnes (metric) and that its diameter was 4 meters (Brown et al., 2002). The orbital characteristics of the Tagish Lake meteoroid are: $a = 2.1 \pm 0.2$ AU, $e = 0.57 \pm 0.05$, $q = 0.891 \pm 0.009$ AU, $Q = 3.3 \pm 0.4$ AU, $i = 1.4 \pm 0.9°$, and $p = 3.0 \pm 0.4$ y. The parameters are identified in Chapter 7. (Reproduced by permission of Dr. Peter Brown, Department of Astronomy, University of Western Ontario, London, Ontario)

Figure 8.6. This tektite from Australia is called an australite. It has a diameter of 20 millimeters and weighs 3.5g. Its shape identifies it as a flanged button. Tektites form as a consequence of meteorite impacts into silica-rich soil and rocks on the surface of the Earth. The resulting blobs of molten silicate are thrown high into the atmosphere where they are shaped by aerodynamic forces as they solidify into silica-rich glass. (Courtesy of H. Raab at http://commons.wikimedia.org)

which claimed that they are either obsidian, glassy meteorites, or of lunar origin. Finally, Faul (1966) concluded that tektites formed by melting of terrestrial rocks and soil during impacts of large meteorites upon the Earth. Therefore, each of the major strewn fields is assumed to be genetically associated with a meteorite impact crater. Such a relationship was firmly established between the moldavites and an impact crater at Nördlingen in south-central Germany called Rieskessel (diameter = 24 km) because both formed at the same time about 14.3 million years ago. In addition, the Ivory Coast tektites of Africa originated from the Bosumtwi Crater in Ghana. However, the impact craters that gave rise to the remaining tektite fields have not yet been identified.

According to the current theory, tektites formed by melting of the target rocks during the impacts of large meteorites. The blobs of liquid silicate were propelling into the stratosphere where they solidified before falling back to the surface of the Earth. As a result of their short flight into space, tektites were not irradiated by cosmic rays and therefore do not contain the so-called cosmogenic radionuclides that characterize stony meteorites. Hence, tektites have a terrestrial origin even though they did fall to Earth from the sky (Koeberl, 1986).

8.4 Meteorite Parent Bodies

Most meteorites are fragments of asteroids that occur in large numbers in the space between the orbits of Mars and Jupiter (Chapter 13). Images of individual asteroids recorded by spacecraft indicate that most of the asteroids are themselves fragments of still larger bodies that apparently formed by accretion of planetesimals at the time the solar system formed within the proto-planetary disk. Most of these planetary bodies were subsequently broken up by the gravitational forces of Mars and Jupiter. Therefore, meteoroids in space are fragments of their parent bodies that resembled the Earth and the other terrestrial planets of the solar system. For these reasons, the study of meteorites provides information about the way the planets formed and differentiated into a metallic core, a silicate mantle, and a thin crust (Section 6.1).

Virtually all meteorites contain small amounts of the mineral troilite which is a sulfide of iron with the formula FeS. The shape of the troilite grains indicates that they crystallized from an iron – sulfide liquid that did not mix with the silicate and the iron liquids from which the silicate minerals and the iron – nickel alloys of meteorites crystalized, respectively. The existence of silicate, sulfide, and iron-nickel minerals and their textures indicates that the parent bodies of the meteorites differentiated by the segregation of silicate, sulfide, and iron-nickel liquids which are known to be immiscible in each other (Section 6.2). In addition, this evidence implies that the parent bodies of the meteorites were initially hot enough to melt and then differentiated by the segregation of the immiscible liquids based on their densities and in response to the force of gravity. The segregation of the immiscible liquids took place when large masses of the dense iron liquid sank through the ocean of silicate magma to form the cores of the parent bodies. Some of the sulfide liquid was trapped in the iron liquid and sank with it, whereas the rest formed globules in the silicate liquid. Both eventually crystallized as the parent bodies cooled to form a mantle composed of iron- and magnesium silicate minerals (e.g., pyroxene and olivine), oxide minerals (magnetite, ilmenite, and chromite), and iron sulfide (troilite).

After the parent bodies had solidified, at least some of them were broken up by collisions with other solid bodies that formed within the proto-planetary disk. According to this series of deductions, the different types of meteorites (i.e., stony, iron, and stony iron) originated from different parts of their parent bodies. Chondrites and achondrites represent the material from different levels in the silicate mantle of their parent

bodies, whereas the carbonaceous chondrites formed from the regolith and shallow crust that was subjected to shock waves but maintained a temperature of less than about 125 °C. Some iron meteorites represent the cores of their parent bodies while others may have originated from large masses of metallic iron that were trapped within the mantle of their parent bodies.

8.5 Celestial Mechanics

The origin of meteoroids by collisions of asteroids is supported by the fact that the wave-length spectra of sunlight reflected by asteroids resemble those of different classes of meteorites (e.g., carbonaceous chondrites, ordinary chondrites, iron meteorites, etc.) We therefore have suffi-cient evidence to propose that meteorites originate from the main asteroidal belt between the orbits of Mars and Jupiter and that their orbits were perturbed by collisions among the large fragments of their former parent bodies which presently populate this region of space in the solar system.

The hypothesis that meteoroids are fragments of asteroids is also supported by the observed orbits of several stony meteorites (e.g., Pribram, Farmington, Lost City, and Dhajala). The orbits of these meteorites were determined from photographs of the fireballs that preceded their fall to Earth. The results indicate that the orbits of these meteoroids were highly eccentric and extended from the asteroidal belt at aphelion to the vicinity of the Sun at perihelion. In addition, the orbits of these meteoroids were closely aligned with the plane of the ecliptic as are the orbits of the asteroids. Consequently, these meteoroids crossed the orbits of Mars, Earth, and Venus and could have impacted on any one of these planets or their satellites (e.g., the Moon).

The orbit of Pribram (H chondrite) was one of the first to be established from photographs of its fall in Czechoslovakia during the night of April 7, 1959. Figure 8.7 demonstrates that its orbit was eccentric with an aphelion within the asteroidal belt at about 3.9 AU from the Sun and its perihelion was at a distance of about 0.75 AU outside the orbit of Venus. Therefore, the orbit of Pribram crossed the orbits of Mars and of the Earth-Moon system but not the orbit of Venus. Figure 8.1 illustrates two hypothetical collisions

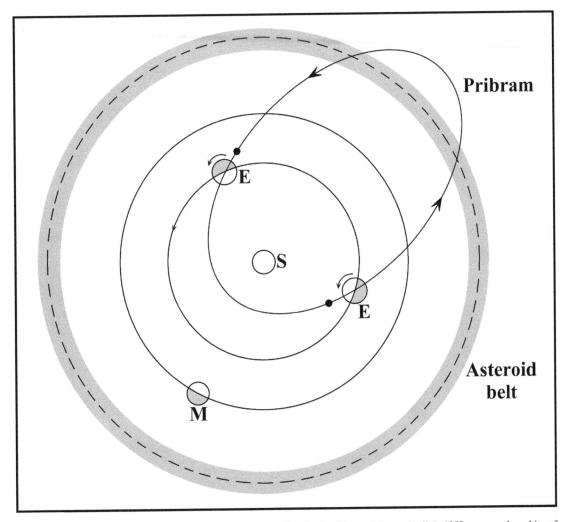

Figure 8.7. The orbit of the Pribram meteorite which fell in Czechoslovakia at night on April 7, 1959, crosses the orbits of Earth and Mars (but not Venus and Mercury) and extends to the outskirts of the asteroidal belt. Therefore, Pribram appears to have originated within the asteroidal belt and was perturbed into an Earth-crossing orbit. During every revolution in its orbit, Pribram had four chances to collide with the Earth and with Mars. The circumstances of its fall indicate that Pribram was approaching the perihelion of its orbit when it caught up to the Earth. (Adapted from Hartmann, 2005, Figure 6.13)

of Pribram with Earth. One of these occurred at night before sunrise as Pribram approached perihelion. The second hypothetical collision occurred in the late afternoon after Pribram had passed the perihelion of its orbit. Since Pribram was reported to have fallen at night, it was on its way to perihelion when it collided with the Earth. The diagram can also be used to illustrate two hypothetical collisions of Pribram with Mars, one at night and the other in the afternoon.

The fall of Pribram makes the point that the time of fall of meteoroids is influenced by celestial mechanics and therefore is not a random event. In fact, Figure 8.8 demonstrates that the frequency of meteorite falls between the years 1790 and 1940 AD reached a peak at 3:00 PM and then declined to a minimum at 3:00 AM of the next day. However, the low frequency of falls at 3 AM is presumably due to the fact that most people are asleep at this time.

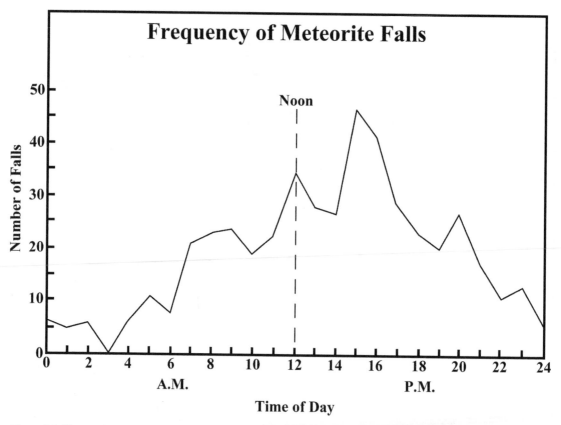

Figure 8.8. The number of observed meteorite falls between 1790 and 1940 AD at different times of day is not constant but rises in the morning to a peak at 3:00 PM and then declines in the late afternoon and evening. The number of recorded meteorite falls between midnight and 4:00 AM may be small because few people are awake at that time. Nevertheless, the large number of falls between noon and midnight suggest that most meteoroids revolve around the Sun in the prograde direction as suggested also by the fall of Pribram in Figure 8.7 (Adapted from Mason, 1962, Figure 8.10)

8.6 Falls and Finds

All meteorites must pass through the atmosphere before they impact on the surface of the Earth. Solid objects that pass through the atmosphere at high velocities (10 to 70 km/s) are heated by friction and are acted upon by mechanical forces that may cause them to break up. Small bodies ranging in diameter from less than 1.0 mm to a few centimeters may be destroyed by the frictional heating and burn up as meteors (shooting stars) in the atmosphere. The same process causes the surface of larger bodies to melt. The molten layer ablates causing the surfaces of meteorites (after they have reached the Earth) to have shallow concave pits (scallops) and a thin fusion crust. The silicate and metallic iron liquids that are stripped from the surface of an incoming meteoroid coalesce into small glass or metallic spherules that form the "smoke" that trails behind the fire ball. The resulting meteorite ablation-spherules have been recovered from glacial ice in Antarctica and Greenland, from deep-sea sediment in the oceans, and from soil on the continents.

When a meteoroid breaks up in the atmosphere, its fall is accompanied by loud thunder and the fragments are scattered in an oval-shaped area called a "strewn field." Consequently, the original meteoroid in space is represented by several smaller meteorite specimens in the strewn field. The fall of such meteorite specimens has only rarely caused significant damage to buildings or injured humans and domestic animals. Freshly fallen meteorite specimens are not hot in most cases because only their outer surface melts while their interiors remain cold. Meteoroids in orbit around the Sun absorb solar radiation depending on the radius and eccentricity of their orbits and on their albedo. (See Glossary). Consequently, the temperature of meteoroids in interplanetary space is well above "absolute zero".

Meteorite specimens that were collected at the time of their fall are especially valuable for research purposes and are designated as "falls", whereas meteorite specimens that were found accidentally long after their fall are called "finds". The abundances of the different types of meteorites whose fall was observed and which were collected a short time later have been recorded in Table 8.2 because this information helps to reconstruct their parent bodies. The results demonstrate that ordinary chondrites are most abundant and constitute 80.6 % of all "falls." Carbonaceous chondrites (4.7%) and achondrite (8.9%) are comparatively rare as are the stony irons (1.2%) and the irons (4.5%).

Meteorites whose fall was not observed and which were later recovered accidentally are rapidly altered by exposure to liquid water and to molecular oxygen (O_2) of the terrestrial atmosphere. As a result, the minerals of meteorites react to form weathering products that are stable in the terrestrial environment (e.g, clay minerals, oxyhydroxides, and carbonates). All meteorite specimens that were found accidentally are suspected of being altered and contaminated by ions and molecules of terrestrial origin and therefore are not suitable for research that depends on the integrity of their chemical and mineralogical compositions.

The abundances of different types of meteorite "finds" are quite different from their abundances among the "falls" in Table 8.2. For example, King (1976, Table I) reported that 54% of the meteorite finds are irons, whereas only 4.5 % of the falls are irons. Conversely, only 40% of the finds are stony meteorites (chondrites and achondrites) compared to 94.2% among the observed falls (Hartmann, 2005, Table 6.2). The reason for the discrepancy is that iron meteorites are found more often because their high density easily distinguishes them from boulders of terrestrial rocks and from stony meteorites which, in

Table 8.2. Abundances of different types of meteorites among "falls" (Hartmann, 2005)

Meteorite classification	Abundance, % by number
Ordinary chondrites	
Enstatite chondrites (E)	1.5
Bronzite chondrites (H)	32.3
Hypersthene chondrites (L)	39.3
Amphoterite chondrites (LL)	7.2
Others	0.3
Sum:	80.6
Carbonaceous chondrites	
C1	0.7
C2	2.0
C3	2.0
Sum:	4.7
Achondrites	
Ureilites	0.4
Aubrites	1.1
Diogenites	1.1
Howardites and eucrites	5.3
Others	1.0
Sum:	8.9
Stony irons	
Mesosiderites	0.9
Pallasites	0.3
Sum:	1.2
Irons	
Octahedrites (IA)	0.8
Octahedrites (IIIA)	1.5
Octahedrites (IVA)	0.4
Hexahedrites (IIA)	0.5
Ataxites, etc.	1.3
Sum:	4.5
Total for all types:	99.9

addition, are more susceptible to weathering than the irons (Science Brief 8.11.1).

Most of the meteorites (about 70 %) that land on the Earth fall into the ocean and are rarely recovered. The remaining 30% are uniformly distributed over the surfaces of the continents without a detectable bias for the polar regions. Nevertheless, more than 10,000 meteorite specimens have been collected on the ice of the East Antarctic ice sheet (Cassidy, 2003). These meteorite specimens originally fell in the accumulation area of the ice sheet, were buried within the ice, and were then transported to the coast by the flow of the ice sheet (Whillans and Cassidy, 1983). In places where the movement of the ice is blocked by the Transantarctic Mountains, the ice sublimates leaving the meteorite specimens stranded on the bare ice. Significant numbers of meteorite specimens have also been collected in the deserts of central Australia, North Africa, and in western South America where meteorites tend to accumulate in the lag gravels of deflation areas. The state of preservation of meteorites collected in Antarctica and in the deserts of the world is generally good and much important research has been done based on these collections (e.g., age determinations).

8.7 Age Determinations

All stony meteorites contain the chemical elements uranium (U) and thorium (Th) whose atoms can decompose spontaneously by emitting subatomic particles from their nuclei. When this happens to a particular atom of U, it is transformed into the atom of a different element. The resulting daughter atom is also unstable and transforms itself into the atom of still another element. This process causes what we call *radioactivity* which manifests itself by the emission of subatomic particles and of penetrating electromagnetic radiation known as gamma rays. The decay of U atoms continues sequentially until at last a stable atom is formed. In the case of the decay of U and Th atoms, the end products are stable atoms of lead (Pb). The result is that the unstable atoms of U and Th are slowly converted to stable atoms of Pb. The unstable atoms of U and Th are the radioactive *parents* and the stable atoms of Pb produced by the decay of U and Th are their radiogenic *daughters* (also called progeny). The rate of decay of radioactive parent atoms to stable daughters is governed by the *halflife* of the parent, which is defined as the time required for one half of the parent atoms to decay. This

simple statement is the basis for measuring the ages not only of meteorites but also of terrestrial rocks (Faure and Mensing, 2005).

For example, consider a one-gram sample of a meteorite that contained 128 radioactive parent atoms of an unspecified element when it formed by crystallization of minerals some time in the past. During a time interval equal to one halflife, half of the initial parent atoms decayed to 64 stable daughter atoms, leaving only 64 parent atoms. During the next halflife more parent atoms decayed to stable daughter atoms leaving only 32 parent atoms at the end of the second halflife. As time continued, the number of parent atoms remaining after three halflives decreased to 16, and then to 8 after four halflives, and so on.

The decay of radioactive parent atoms and the accumulation of stable daughter atoms in the meteorite with time in Figure 8.9A occurs exactly as explained above. As the number of parent atoms decreases, the number of radiogenic daughter atoms increases until in the end all of the parent atoms decay to daughter atoms. Therefore, the number of daughter atoms that have accumulated divided by the number of remaining parent atoms in Figure 8.9B increases from zero (no daughter present) at the beginning to 1.0 after one halflife, 3.0 after two halflives, and so on to 63 after six halflives.

The age of a stony meteorite is determined by measuring the ratio of the number of radiogenic daughter atoms divided by the number of unstable parent atoms remaining (D/P). This measurement is made by skilled analysts working in scrupulously clean chemical laboratories using equipment that can cost \$500,000. Nevertheless, hundreds of meteorites have been dated in this way. For example, if D/P = 15.0, the graph in Figure 8.9B indicates that 4.0 halflives of the parent have elapsed. If the halflife is 1.15×10^9 years, then the age (t) of the meteorite is:

$$t = 4.0 \times 1.15 \times 10^9 = 4.60 \times 10^9 \text{ y}$$

Dates in the past are expressed in terms of mega anna (Ma) or giga anna (Ga) where Ma = 10^6 years ago and Ga = 10^9 years ago, and the word "anna" is the plural form of "annum" which is the Latin word for "year."

Uranium and thorium are not the only elements whose atoms are radioactive. Most of the other

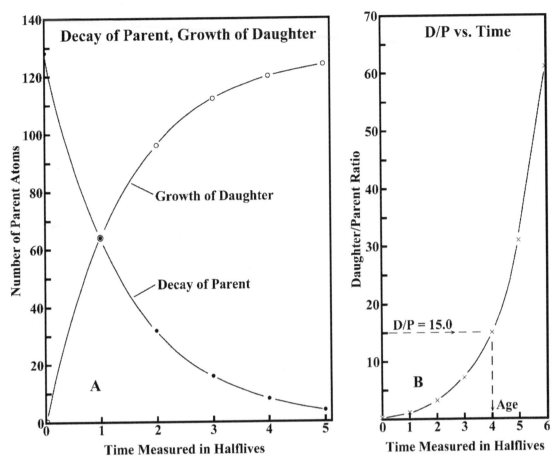

Figure 8.9. A. The radioactive parent atoms decay to stable daughters in accordance with their halflives such that after each halflife only half of the parent atoms remain. Therefore, in this example the number of parent atoms deceases from 128 to 64, 32, 16, 8, 4, 2, etc. At the same time, the number of stable radiogenic daughters increases after each halflife from 0 to 64, 96, 112, 120, 124, 126, etc. For example, after two halflives have elapsed the number of daughters is $64 + 32 = 96$. B. The decay of radioactive parent atoms and the consequent accumulation of stable daughters causes the ratio D/P to increase with time, where D is the number of radiogenic daughters that have accumulated and P is the number of radioactive parents that remain. The time elapsed since crystallization of the minerals of a stony meteorite is determined by measuring the D/P ratio which can be converted into the age of the meteorite by multiplying it by the halflife expressed in terms of billions of years as demonstrated in the text. A more complete explanation of dating is given in Science Brief 8.11.2

elements that have radioactive atoms are listed in Table 8.3 together with their halflives. This table contains nine different parent – daughter pairs each of which has been used to date meteorites. When several of these methods are used to date the same meteorite, the resulting dates are not necessarily identical because each parent – daughter pair responds differently to the events that have affected meteorites during their long histories.

Most stony meteorites that have been dated crystallized within about 100 million years of each other between 4.65 and 4.55 Ga. This result means that the parent bodies of meteorites formed in this interval of time, became hot enough to melt, differentiated by segregation of immiscible liquids, and then cooled and crystallized. According to the theory of the origin of the solar system, the Earth and the other planets of the solar system also formed at this time by

Table 8.3. Elements that have radioactive atoms, their halflives, and the elements of their stable radiogenic daughters. (Faure and Mensing, 2005)

Element of thet radioactive paren	Halflife 10^9 y	Element of the stable daughter
uranium (^{238}U)	4.468	lead (^{206}Pb)
uranium (^{235}U)	0.7038	lead (^{207}Pb)
thorium (^{232}Th)	14.01	lead (^{208}Pb)
potassium (^{40}K)	11.93	argon (^{40}K)
potassium (^{40}K)	1.39	calcium (^{40}Ca)
rubidium (^{87}Rb)	48.8	strontium (^{87}Sr)
samarium (^{147}Sm)	106	neodymium (^{143}Nd)
lutetium (^{176}Lu)	35.7	hafnium (^{176}Hf)
rhenium (^{187}Re)	41.6	osmium (^{187}Os)

the same process as the parent bodies of the meteorites (Sections 5.2, 6.1, and 6.2).

The only exceptions to this generalization are certain meteorites known to have originated from the surfaces of the Moon and Mars. The martian meteorites originally attracted attention because they yielded anomalously low crystallization ages of about 1.3 Ga. Therefore, these meteorites must have originated from a planet on which magmatic activity was still occurring 1.3 billion years ago. Consequently, their home planet must be more massive than the Moon and Mercury both of which have been inactive for more than three billion years. After some hesitation, Mars was identified as the home planet, partly because the abundances of noble gases (neon, argon, krypton, and xenon) in these anomalous meteorites matched the concentrations of these gases in the atmosphere of Mars as reported by the Viking spacecraft that landed on the surface of Mars in 1975.

We return to the lunar and martian meteorites in chapters 9 and 12, respectively. Additional information about the use of radiogenic daughter atoms for dating of meteorites is available in a book by Faure and Mensing (2005). The universal geochronometry equation is derived in Science Brief 8.11.2 and is used to date the achondrite Moama by the slope of the Sm-Nd isochron based on the analytical data of Hamet et al. (1978). The Sm-Nd date of Moama is $4.58 \pm 0.05 \times 10^9$ years.

8.8 Meteorite Impacts

Large meteorites measuring several hundred meters in diameter contain a very large amount of kinetic energy expressed conventionally in terms of the energy released by the explosion of millions of metric tonnes of TNT (dynamite) (Science Brief 8.11.3). As a result, such large meteorites can pass through the atmosphere unscathed and explode on impact with the Earth, thereby forming craters whose diameters are more than ten times larger than the diameters of the impactors. The release of the vast amount of energy and the injection of dust and smoke into the atmosphere both cause environmental damage and sudden climate change on a global scale. Therefore, the impacts of large meteorites in the geologic past have caused global extinction events affecting both animals and plants on the continents and in the oceans (Silver and Schultz, 1982; Melosh, 1989; Koeberl, 1998; Shoemaker and Shoemaker, 1999; Koeberl and MacLeod, 2002; Kenkmann et al. 2005).

8.8.1 Formation of Craters

The physical process by means of which impact craters form has been investigated by the study of:
1. Known impact craters such as Meteor Crater, Arizona (Shoemaker, 1963).
2. Nuclear and chemical explosion craters.
3. Craters formed by hypervelocity projectiles in the laboratory (Gault et al., 1968).

The results of these studies indicate that meteorite impact-craters form in three stages;
1. Compression stage.
2. Excavation stage.
3. Post-formation modification stage (Section 8.8.3).

The compression stage occurs when the kinetic energy of the meteorite is transferred into the target rocks by means of two compressional waves. One of these travels through the meteorite from the front to the back, whereas the second wave moves downward through the target rocks. If the incoming meteorite had a velocity of 10 km/s, the pressure generated in the target rocks by the compressional wave is of the order of several million bars (1 bar = 0.987 atmospheres). These pressures not only crush rocks to a powder but they also cause them to melt and even to vaporize. For example, granite is crushed at about 250,000 bars (250 kilobars) and begins to melt between pressures of 450 and 500 kilobars. When pressures in excess of 600 kilobars are applied to granite, or any other kind of rock, it is vaporized. Therefore, the enormous pressure generated by the impact of a high-velocity meteorite causes the target rocks to form a high-temperature fluid that is expelled at velocities that greatly exceed the velocity of the impactor. The shockwave that passes through the target rocks also causes textural changes in the rocks and crystallographic alteration of minerals. The evidence for this kind of shock metamorphism has been described by French and Short (1968), French (1998), and most recently by Becker (2002). The interested reader may also wish to consult a seminal paper by Dietz (1961) about the history of this branch of planetary science. The compression stage just described lasts only from one thousandth to one tenth of one second depending on the diameter of the impactor between 10 m and 1000 m, respectively.

The *excavation* of the impact crater takes place when a relaxation wave is generated by the decompression of the rocks after the compression wave has passed. At this stage, the fragments of crushed rocks are thrown out of the impact site at comparatively low velocity and form the ejecta blanket that surrounds the crater. The excavation process causes the ejected material to land upside down such that the oldest rocks from the deepest part of the crater fall on top of the youngest rocks derived from shallow depth. Initially, the ejected material forms a cone-shaped curtain which subsequently tears as the cone expands thereby causing the ejected blanket to form rays on the ground. Some of the larger blocks of rocks that may be ejected form small secondary craters in the ejecta blanket. In addition, some of the ejected rock fragments fall back into the crater to form a deposit of fallback breccia. The residual heat within the crater may cause melting of the fallback breccia. When the melt crystallizes, it forms a rock called *impactite* which may be fine-grained and crystalline or glassy depending on the rate of cooling.

The size of the crater that is produced depends on the mass and velocity of the meteorite and

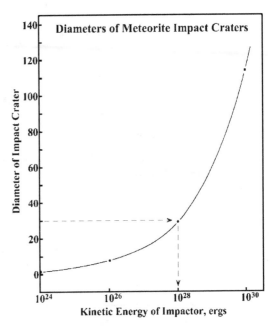

Figure 8.10. The diameters of impact craters increase exponentially with increasing kinetic energy of the impacting meteoroid. The calculations necessary to construct this graph are explained in Science Brief 8.11.4. The results indicate that a crater having a diameter of about 30 km was formed by a meteoroid whose kinetic energy was close to 1×10^{28} ergs

on the mechanical properties of the target rocks. The relation between the diameter of the crater (D) and the kinetic energy (E) of the impactor is expressed graphically in Figure 8.10 based on equation 8.12 stated in Science Brief 8.11.4.

8.8.2 Classification of Craters

The impact craters are classified on the basis of their increasing diameters into several different types:
1. Bowl-shaped craters
2. Central-uplift craters
3. Peak-ring basins
4. Multi-ring basins
The bowl-shaped (or simple) craters on Earth in Figure 8.11A have diameters less than about 5 km (Glass 1982, p. 128). Consequently, they are formed by comparatively small impactors whose kinetic energy is less than 1×10^{24} ergs (based on equation 8.13 in Science Brief 8.11.4).

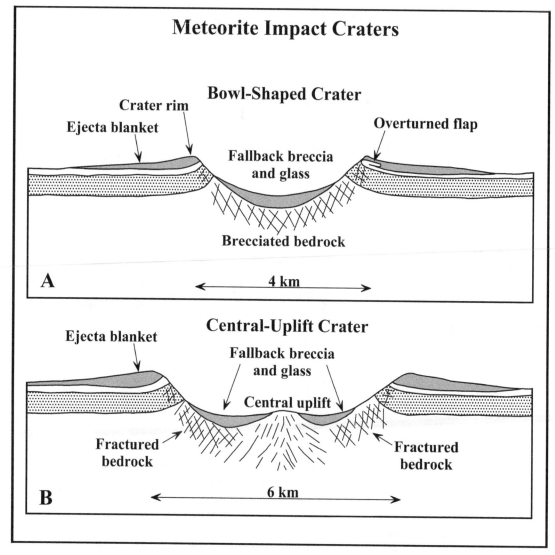

Meteorite Impact Craters

Bowl-Shaped Crater

Figure 8.11. The structure of meteorite impact craters depends on the kinetic energy of the impactor. Bowl-shaped craters (A) form by the release of less than 1×10^{24} ergs, whereas central-uplift craters (B) result from the explosive release of more than 1×10^{25} ergs. Adapted from Glass (1982, p. 128)

Larger craters (e.g., 10 km) have a central uplift that forms by a complex readjustment of stresses suggested in Figure 8.11B. Peak-ring basins and even multi-ring basins probably formed on the Earth during a period of intense bombardment of all terrestrial planets from 4.5 and 3.8 billion year ago. The remnants of such impact basins have been recognized on the Earth, but much of their original structure has been removed by erosion.

8.8.3 Modification of Craters

After impact craters have formed on one of the continents of the Earth, or on another solid planet or satellite elsewhere in the solar system, their shapes are modified by geological processes under the influence of gravity. These processes cause the formation of terraces by sliding of rock masses and the accumulation of talus cones and aprons along the crater walls. In addition, impact craters on the Earth may fill with water which deposits clastic sediment composed of wind-blown dust and locally derived weathering products. If the water eventually evaporates as a result of progressive desertification of the climate, lacustrine evaporite minerals may also be deposited, including calcite ($CaCO_3$), gypsum ($CaSO_4 \cdot 2H_2O$), halite (NaCl), and hematite (Fe_2O_3). In some cases, meteorite impact craters on the near side of the Moon have filled with basalt lava flows until only the raised rims remain visible.

The ultimate fate of impact craters on the continents of the Earth is either to be filled up and buried by sedimentary cover rocks or to be eroded until only the circular outline of the crushed and brecciated rocks that underlie the former craters remain. Consequently, buried or eroded impact craters can only be identified by careful geophysical, mineralogical, and geochemical studies in the field and in the laboratory. The criteria that have been used to detect and to identify impact craters on the Earth include localized geophysical anomalies (e.g., gravity and magnetism), the presence of shatter cones, breccia dikes, breccia bodies containing glass (suevite), and melt rocks (impactites), as well as the occurrence of high-pressure polymorphys of quartz (coesite and stishovite) and of sets of planar partings that identify shocked quartz and feldspar. Remnants of the impactors are found much more rarely (e.g., at Meteor Crater, Arizona), although the dust ejected from meteorite impact craters may have an anomalously high concentration of the element iridium that is much more abundant in meteorites than in terrestrial rocks such as granite.

8.8.4 Frequency of Impacts

Powerful impacts occurred much more frequently during the early history of the Earth when the solar system still contained a large number of solid bodies of various sizes left over from the main phase of planet formation (Section 6.1). For example, Figure 8.12 illustrates the fact that about 4.6 billion years ago the Earth and Moon were hit by meteorites about one million times more often than they are at the present time. The resulting devastation

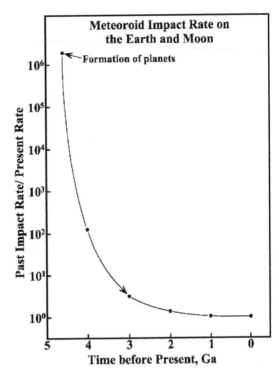

Figure 8.12. Frequency of meteoroid impacts on the Earth and Moon as a function of time in the past. During the time the planets of the solar system formed by impacts of planetesimals (Section 5.2 and 6.1), the impact rate was more than one million times higher than it is at present. By the time the initial period of intense bombardment ended at about 3.8 Ga, the rate had declined to only about 30 times the present rate. Nevertheless, even at the present time the meteorite flux to the Earth is between 10^7 and 10^9 kg/y. (Adapted from Hartmann 2005, Figure 6.6)

of the terrestrial surface may have retarded the development of life on the Earth. Therefore, the time between 4.6 and 3.8 billion years ago is called the Hadean Eon after "hades" the underworld (hell) of Greek mythology. The impact craters, that formed during this time on all planets and their satellites, have been obliterated on the Earth by geological processes, but they have been preserved on the Moon. Nevertheless, more than 150 meteorite impact craters of Precambrian and Phanerozoic age have been identified on the Earth and tens of thousands of craters still exist on the Moon.

Figure 8.12 also demonstrates that the rate of impacts on the Earth and Moon decreased rapidly with time and, at 3.8 Ga, was only about 30 times the present rate. Nevertheless, the Earth and Moon continue to be bombarded by meteorites on a daily basis although most of the particles are small and are therefore classified as *microm-eteorites*. Estimates by Hartmann (2005) based on studies reported in the literature indicate that the Earth receives between 10^7 and 10^9 kg of meteoritic material each year. If the annual amount is taken to be 10^8 kg, the daily meteorite flux (MF) over the whole Earth is:

$$MF = \frac{1 \times 10^8}{10^3 \times 365.256} = 274 \, \text{tonnes/day}$$

The daily deposition of hundreds of tonnes of meteoritic dust over the entire surface of the Earth goes virtually unnoticed primarily because of the small grain size of this material. Larger meteoroids with diameters of 10 m or less (before they enter the atmosphere) impact at hourly intervals, but their fall is observed only when it occurs over densely populated areas and even then is not always reported in the daily news.

Fortunately, meteoroids having diameters of 100 m to 10 km impact on the Earth only very rarely under present circumstances. For example, Figure 8.13 indicates that impacts of meteoroids with diameters of about 100 m occur only about once in 1000 years. However, the impact of such meteoroids produces a crater with a diameter of more than one kilometer. The curve in Figure 8.13 suggests also that still larger meteoroids impact even less frequently and therefore do not pose an imminent threat to the survival of humans on the Earth. For example, meteoroids with diameters of 10 km are expected to strike the Earth only once in several hundred million years.

These predictions apply only to the present time because the large number of impact basins on the Moon having diameters greater than 100 km indicates that impacts of large meteoroids were much more frequent during the early history of the solar system than they are at present.

8.8.5 Environmental Consequences (Tunguska, 1908)

The kinetic energy (KE) of a meteorite approaching the Earth depends on its mass (m)

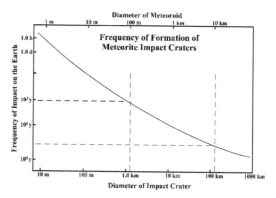

Figure 8.13. The frequency (or probability) of impacts decreases with increasing diameter of meteoroids. For example, meteoroids having a diameter of 100 m impact on the Earth once every 1000 years and make a crater whose diameter is greater than 1.0 km. Large meteoroids (10 km diameter) impact only once in 100 millions (10^8) years, but produce an impact basin with a diameter of more than 100 km. The graph can also be used to estimate the size of the impactor based on the diameter of the crater. For example, a 10 km-crater requires a meteoroid with a diameter of about 1.0 km. Adapted from Hartmann (2005, Figure 6.5, p. 134)

and its velocity (v) in accordance with the formula:

$$KE = \frac{1}{2} m \, v^2$$

where m = mass of the object in grams, v = velocity in centimeters per second, and KE = kinetic energy in ergs. The kinetic energy of a meteoroid having a diameter of 100 m, a density of 3.5 g/cm^3, and a velocity of 15 km/s is equal to 2.06×10^{24} ergs, which is equivalent to the energy released by the explosion of 49 million tonnes of TNT (Science Brief 8.11.3). The amount of energy released by the impact of stony meteoroids with diameters between 10 m and 10 km in Figure 8.14 ranges from less than 10^4 tonnes of TNT (10 m) to more than 10^{12} tonnes of TNT (10 km). The latter exceeds the energy of the most powerful hydrogen-fusion bomb by a factor of about one thousand.

The impact and explosion of large meteoroids in the geologic history of the Earth is recorded by craters some of which have diameters in excess of 100 km (e.g., Vredefort, South Africa, 300 km, 2.02 Ga; Sudbury, Ontario, 250 km, 1.85 Ga; and Chicxulub, Mexico, 180 km, 65.0 Ma). The formation of such craters has been described by Melosh (1989), Koeberl and MacLeod (2002),

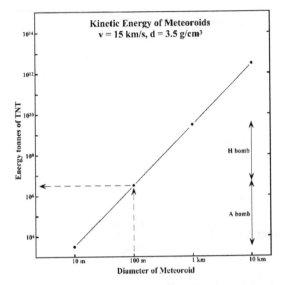

Figure 8.14. Amount of energy released by the impact of meteoroids traveling at 15 km/s. The energy is expressed in terms of the equivalent amount of energy released by the explosion of a certain weight of TNT. Accordingly, the impact of a meteoroid having a diameter of 100 m releases an amount of energy equivalent to about 50 million metric tonnes of TNT or to the explosion of a large nuclear-fission bomb. The instantaneous release of this amount of energy could devastate the surface of a larger continent and exterminate most animals and plants in the affected area. Meteoroids of larger diameter release proportionately more energy and cause damage on a global scale. (Based on 10^3 kg of TNT $= 4.2 \times 10^{16}$ erg)

Grieve and Therriault (2004), and French (2004) among others.

The horrific aftermath of the impact of large meteoroids on the Earth has been extensively studied in order to test the hypothesis that the sudden extinction of the dinosaurs at the end of the Cretaceous Period was caused by the impact of a large meteoroid that excavated the crater at Chicxulub on the northern coast of the Yucatán peninsula of Mexico (Powell, 1998). In addition, the environmental consequences of the impact on the Earth of a large meteoroid resemble the aftermath of the explosion of nuclear weapons in the atmosphere. These matters will be mentioned again in Chapter 13 in connection with the possible impacts of certain asteroids that regularly cross the orbit of the Earth.

A meteoroid with a diameter of 30 to 60 m actually did explode in the atmosphere about

6 km above the Tunguska River, Siberia, in the morning of June 30, 1908 and released energy equivalent to 10 to 15 million tonnes of TNT (Hartmann, 2005). Observers in the area reported seeing a ball of fire crossing the sky followed by very loud thunderous noises. A man who lived 60 km from the explosion site reported that part of the sky was suddenly on fire and that he felt a blast of heat as though his shirt was burning. The heat pulse was followed by a very loud crashing sound and a blast of hot air that threw him off his porch and blew out glass windows in his house.

In the area directly beneath the blast point the trees were stripped of their branches but were left standing. Farther away, about 5 to 15 km from the center, the trees were blown down in a radial pattern pointing away from the blast site as illustrated in Figure 8.15. Reindeer herders camping about 30 km away reported that the forest started to burn and that many reindeer were killed.

This spectacular event in a remote and sparsely populated region of the Earth left no crater and caused little damage to the local communities (Krinov, 1966; Chyba et al., 1993). Impacts of large meteoroids (about 50 ± 20 m in diameter) occur approximately once in 100 years (Figure 8.7). The consequences of such an impact would be catastrophic if it occurred in a more populous region than Siberia.

8.9 Meteor Crater: Memorial to E.M. Shoemaker

A spectacular bowl-shaped crater in Figure 8.16 near the town of Winslow in northern Arizona was originally identified as a volcanic crater by G.K. Gilbert of the U.S. Geological Survey (Gilbert, 1896). However, the mining engineer D.M. Barringer considered it to be a meteorite impact-crater because of the fragments of metallic iron-nickel alloy that had been found in the surrounding area (Barringer, 1905). Consequently, he purchased the crater and attempted to recover the main body of the iron meteorite he expected to find below the floor of the crater. After several years of effort, the project was abandoned without finding the hoped-for bonanza of metal. The crater is presently owned by the Barringer Crater Company and is open to the public. We encourage our readers to

Figure 8.15. Devastation of a spruce forest caused by the explosion on June 30, 1908, of a meteoroid in the atmosphere about 6 km above the ground near the village of Vanavara, Tunguska river of Siberia, Russia. The blastwave knocked down trees in an area of $2150\,km^2$ but did not form a crater (Shoemaker and Shoemaker, 1999). Considering the large amount of energy released by this event, the damage to humans in this sparsely populated area was minor. A fictional but realistic description of such an event in the atmosphere over the city of Vienna by Koeberl (1998, pages 11–16) concluded with the words: "Das gesamte Gebiet ist verbrannt und zerstört." (The entire area was burned and destroyed). (Photo by Kulik, May, 1929, reproduced by courtesy of the Tunguska Page of Bologna University at http://www-th.bo.infn.it/tunguska/)

visit Meteor Crater because it is the best-preserved and most accessible bowl-shaped meteorite impact-crater on the Earth although its name does not conform with the definition of a "meteor".

The failure of D.M. Barringer to find the iron meteorite below the floor of Meteor Crater raised doubts about its origin because geologists had never witnessed the formation of such a crater by the impact of a large chunk of metallic iron that

Figure 8.16. Meteor Crater near Winslow in northern Arizona is 1.1 km wide and 200 m deep. It formed about 50,000 years ago by the explosive impact of an iron meteorite that had a diameter of 30 m, weighed about 100,000 tons, and was traveling at about 20 km/s. The resulting crater was excavated in about 10 seconds. Fragments of the iron meteorite were named Canyon Diablo after a dry streambed (arroyo) in the vicinity of the crater. Meteor Crater is a bowl-shaped crater that was excavated in flat-lying sedimentary rocks of Paleozoic age. The regional pattern of joints of these rocks is reflected by the somewhat rectangular shape of the crater. The rocks that form the ejecta blanket that surrounds the crater are overturned (i.e., the youngest rocks are at the bottom and the oldest rocks are on top). (Courtesy of D. Roddy (U.S. Geological Survey) and Lunar and Planetary Institute in Houston, Texas; http://solarsystem.nasa.gov)

fell out of the sky. Therefore, during the first 50 years of the 20th century most geologists did not accept the idea that meteoroid impacts constitute an important geological process because to do so seemed to violate the Principle of Uniformitarianism (i.e., the present is the key to the past).

A notable exception was Robert Dietz who identified the remnants of numerous meteoroid impact craters he called *astroblemes* on four different continents and referred to them as "geological structures to be recognized as the scars of an agelong and still continuing bombardment of the Earth by rubble from elsewhere in the solar system" (Dietz, 1961). However, the existence of meteoroid impact craters on the Earth was not widely accepted until shatter cones (Dietz, 1947), shocked quartz, and its high-pressure polymorphs coesite and stishovite were demonstrated to be diagnostic

criteria for the impact of large fast-moving meteoroids from interplanetary space. (Chao et al., 1960, 1962).

Meteor Crater has a diameter of 1.2 km and a depth of 180 m. It was formed by the impact of the Canyon Diablo iron meteorite about 40,000 years ago. The crater is located on the southern part of the Colorado Plateau and is underlain by flat-lying sedimentary rocks of Permian to Triassic age. The principal formations are the Coconino Sandstone (Permian), the Kaibab Limestone (Permian), and the Moenkopi Formation (Triassic). The Moenkopi Formation is distinctive because it consists of reddish brown sandstones and siltstones.

When Meteor Crater in Figure 8.16 formed during the time of the Wisconsin glaciation of the Pleistocene Epoch, northern Arizona received considerably more rain than it does at the present time. As a result, a lake formed within the crater causing up to 30 meters of Pleistocene lake beds and debris from the older rock formations to accumulate within it. When the Laurentide ice sheet retreated from North America about 10,000 years ago, the annual amount of precipitation in northern Arizona decreased and the lake in Meteor Crater dried up.

The controversy concerning the origin of Meteor Crater was settled in the late 1950s when Eugene M. Shoemaker mapped it in great detail for his Ph.D. dissertation which he submitted to Princeton University (Shoemaker 1960, 1963; Shoemaker and Kieffer, 1974). His results and their interpretation left no room for doubt that Meteor Crater had in fact been formed by the explosive impact of an iron meteorite whose diameter was about 30 m traveling at 20 km/s. In addition, Shoemaker became convinced that this impact was not an isolated incident but was merely one example of a general geological phenomenon, the evidence for which was most clearly preserved on the Moon. He concluded that the Earth resembled a target in a shooting gallery that had been hit many times before and would be hit again in the future. The projectiles that left their mark on the surface of the Earth and Moon are large meteoroids and asteroids on Earth-crossing orbits.

In the years that followed his investigation of Meteor Crater, Gene Shoemaker became an ardent advocate in favor of the geological exploration of the Moon. He helped to persuade NASA to incorporate the study of the geology of the Moon into the Apollo Program and he trained the astronauts in field geology. Above all, he wanted to be the first geologist to travel to the Moon. Unfortunately, Gene Shoemaker had a medical condition that prevented him from becoming an astronaut. Instead, he became interested in identifying meteoroids and asteroids in Earth-crossing orbits that could someday impact on the Earth. Although he and his wife Carolyn were spectacularly successful in finding near-Earth asteroids, their most famous achievement was the discovery of comet Shoemaker-Levy 9 in 1993. Comets (Chapter 22) are solid objects composed of ice and dust that revolve around the Sun in highly eccentric orbits. The comet Shoemaker-Levy 9 had been broken into a string of 21 fragments in 1992 by the gravi-

tational force of Jupiter. Two years later, in 1994, these fragments sequentially impacted upon Jupiter causing gigantic explosions in full view of astronomers and geologists on Earth. Gene Shoemaker had the good fortune to witness these explosions which confirmed what he and a small group of visionaries had concluded some 30 years before:

1. The Earth has been and continues to be a target of large solid objects (meteoroids, asteroids, and comets) that have been bombarding it since its formation 4.6 billion years ago.
2. The resulting explosions release very large amounts of energy that can cause catastrophic damage to the Earth including the extinction of most forms of life on its surface.
3. Thousands of these projectiles populate the interplanetary space of the solar system and many of them cross the orbit of the Earth.

Gene Shoemaker died in a traffic accident on July 18, 1997, in central Australia where he had been studying a large impact crater. He was 69 years old. His sudden death stunned his many friends and colleagues in the fraternity of planetary scientists. Many people considered what they could do to give meaning to this tragic event. Within a few days, a proposal was developed by Carolyn C. Porco (2000) to place a vial of Gene Shoemaker's ashes on the Lunar Prospector spacecraft which was scheduled to depart for the Moon in the near future. The proposal was approved by NASA and by his wife Carolyn. Accordingly, the Lunar Prospector spacecraft that was launched on January 6, 1998, contained a vial of Gene Shoemaker's ashes and an engraved epitaph. When the spacecraft had completed its mission on July 31, 1999, it was directed to crash into the south-polar region of the Moon. Eugene Shoemaker's wish to go to the Moon was thereby fulfilled.

Gene Shoemaker was widely admired because he was a visionary scientist and a leading explorer of the solar system. His study of Meteor Crater changed forever how we think about the Earth. Science Brief 8.11.5 contains a poem written by Bevan M. French who knew Gene Shoemaker well and who is himself a leader among planetary scientists.

8.10 Summary

Meteorites are an important source of information about the origin and early history of the solar system. They have been used to determine the age of the Sun and the planets of the solar system and have helped us to understand the origin and evolution of the solar nebula. The modern theory of the origin of the solar system and of the internal differentiation of the terrestrial planets is largely based on the study of meteorites.

As a result of the continuing study of meteorites, we now know that they are fragments of parent bodies that formed from planetesimals between the orbits of Mars and Jupiter at about the same time as the other planets of the solar system. Subsequently, the orbits of the parent bodies were perturbed by the gravitational forces of Jupiter and Mars, which caused most of the parent bodies to be broken up by collisions. The resulting fragments are the asteroids that continue to reside in this region of the solar system. When two asteroids collide, the resulting fragments (called meteoroids) may be deflected into eccentric Earth-crossing orbits as they continue to orbit the Sun. Eventually, meteoroids may strike the Earth, the Moon, or one of the other terrestrial planets.

Small grains of meteoritic or cometary dust that hit the Earth glow brightly as meteors because they are heated by friction as they streak through the atmosphere. Larger meteoroids, ranging in diameter up to about 100 m, may break up into a shower of smaller fragments depending on their mechanical properties. Such meteoroid fragments that fall out of the sky are called meteorites. Still larger masses, plough through the atmosphere and explode on impact leaving large impact craters whose radii are more than ten times the diameter of the impactor.

The probability that the Earth will be hit by a meteoroid decreases exponentially with the increasing size of the object because large meteoroids are less numerous in interplanetary space than small ones. Impacts of small objects (less than about one meter) occur daily, whereas impacts of large meteoroids (10 km) occur only once in about 100 million years. However, the impact of a large meteoroid can devastate the surface of the Earth on a global scale by causing short-term changes in the global climate resulting in the extinction of most life forms.

Meteorites have been collected for thousands of years and have been classified on the basis of their mineralogical and chemical composition into stones, irons, and stony irons. They are characterized by their density (e.g., iron meteorites), by the dark brown to black fusion crust, and by their scalloped surface, both of which are the result of heating and ablation during their passage through the atmosphere.

8.11 Science Briefs

8.11.1 The Mass of a Small Iron Meteorite

Calculate the mass of an iron-meteorite specimen having a radius (r) of 15.24 cm (6 inches) and a density (d) equal to 7.94 g/cm^3.

The volume (V) of a sphere of radius (r) is: $V = \frac{4}{3} \pi r^3$ where $\pi = 3.14$ is the circumference of a circle divided by its diameter. Therefore,

$$V = \frac{4 \times 3.14 \times (15.24)^3}{3} = 14,819.1 \text{ cm}^3$$

The mass (m) of the sphere is:

$$m = V \times d = 14,819.1 \times 7.94 = 117,664 \text{ g}$$

$$m = \frac{117,664}{10^3} = 117.664 \text{ kg}$$

Remember that mass is a property of matter which is expressed by its weight when the mass is acted upon by the force of gravity of the Earth or of any other body in the solar system.

8.11.2 Dating Stony Meteorites by the Sm-Nd Method

The Law of Radioactivity states that the rate of decay of a radioactive atom of a chemical element is proportional to the number of atoms remaining. This statement is expressed mathematically by the equation:

$$P = P_0 e^{-\lambda t} \tag{8.1}$$

where P = number of radioactive parent atoms remaining,

P_0 = number of parent atoms present at the time of crystallization of the minerals in the specimen to be dated,

λ = decay constant which is related to the halflife ($T_{1/2}$) by $\lambda = 0.693/T_{1/2}$

e = mathematical constant equal to 2.718,

t = time elapsed since crystallization measured backwards from the present.

The number of radiogenic daughter atoms (D) that have accumulated in a sample of a stony meteorite can be expressed by the equation:

$$D = P_0 - P \qquad (8.2)$$

The number of parent atoms at the time of crystallization (P_0) is obtained by rearranging equation 8.1:

$$P_0 = Pe^{\lambda t} \qquad (8.3)$$

Substituting equation 8.3 into equation 8.2 yields:

$$D = Pe^{\lambda t} - P \qquad (8.4)$$

from which we obtain the basic geochronometry equation (Faure and Mensing, 2005):

$$D = P(e^{\lambda t} - 1) \qquad (8.5)$$

A geochronometer that is especially well suited to dating stony meteorites is based on the decay of an atom of samarium identified as ^{147}Sm. The atom ejects an alpha particle (identical to the nucleus of helium atoms) from its nucleus as it decays to a stable atom of neodymium (^{143}Nd) with a halflife of 106×10^9 years. Therefore, ^{147}Sm is the radioactive parent (P) and ^{143}Nd is the stable radiogenic daughter (D). Substituting these atomic labels into equation 8.5 yields:

$$^{143}Nd = {}^{147}Sm(e^{\lambda t} - 1) \qquad (8.6)$$

For practical reasons we divide both sides of equation 8.6 by the number of stable atoms of Nd identified as ^{144}Nd and thereby obtain the equation:

$$\frac{^{143}Nd}{^{144}Nd} = \frac{^{147}Sm}{^{144}Nd}(e^{\lambda t} - 1) \qquad (8.7)$$

Note that we have not accounted for the presence of atoms of ^{143}Nd that are not produced by decay of ^{147}Sm within the meteorite, but which existed in the solar nebula and were incorporated into the meteorite parent bodies at the time of their formation. Therefore, we now restate equation 8.7 by adding the non-radiogenic ^{143}Nd atoms represented by the ratio $(^{143}Nd/^{144}Nd)_0$:

$$\frac{^{143}Nd}{^{144}Nd} = \left(\frac{^{243}Nd}{^{144}Nd}\right)_0 + \frac{^{147}Sm}{^{144}Nd}(e^{\lambda t} - 1) \qquad (8.8)$$

This equation is used to calculate the age (t) of meteorites from the measured values of the $^{143}Nd/^{144}Nd$ and $^{147}Sm/^{144}Nd$ ratios.

Mineral concentrates prepared from a sample of a stony meteorite all have the same age (t). Therefore, the measured $^{143}Nd/^{144}Nd$ and $^{147}Sm/^{144}Nd$ ratios of minerals separated from the same meteorite define a straight line in coordinates of $^{143}Nd/^{144}Nd$ (y-axis) and $^{147}Sm/^{144}Nd$ (x-axis). The slope of this line is equal to the D/P ratio which is used to calculate the crystallization age of the meteorite.

For example, Hamet et al. (1978) reported measurements of the atomic ratios $^{143}Nd/^{144}Nd$ and $^{147}Sm/^{144}Nd$ for the minerals of the achondrite (eucrite) Moama. The data points in Figure 8.17 define a straight line called an *isochron* because all minerals that define it have the same age t. The slope of the isochron (m) is derivable from the atomic ratios.

$$m = \frac{^{143}Nd/^{144}Nd - \left(^{143}Nd/^{144}Nd\right)_0}{^{147}Sm/^{144}Nd} = \frac{^{143}Nd^*}{^{147}Sm}$$

$$= \frac{D}{P} \qquad (8.9)$$

where $^{143}Nd^*$ is the number of radiogenic ^{143}Nd atoms.

According to equation 8.8, the slope (m) is related to the age (t) of the meteorite by:

$$m = e^{\lambda t} - 1 \qquad (8.10)$$

Solving equation 8.10 for t yields:

$$t = \frac{1}{\lambda}\ln(m + 1) \qquad (8.11)$$

where ln is the logarithm to the base e and λ is the decay constant of ^{147}Sm equal to $6.54 \times 10^{-12}\,y^{-1}$. The slope (m) of the Sm-Nd isochron

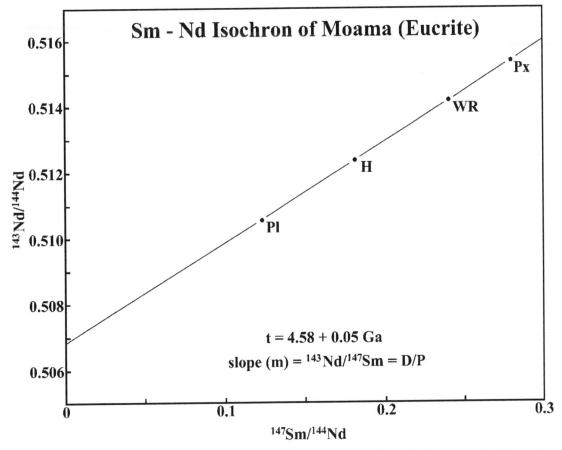

Figure 8.17. Samarium – neodymium isochron of the minerals and of a whole-rock sample of the achondrite Moama. The slope of the straight-line isochron is equal to the D/P ratio which is the number of radiogenic daughters (^{143}Nd) present divided by the number of parent atoms remaining (^{147}Sm) as shown in equation 8.9. The results of this age determination indicate that the minerals of the meteorite Moama crystallized 4.58 ± 0.05 billion years ago. Px = pyroxene, WR = whole rock, H = heavy fraction, density > 3.2 g/cm^3, Pl = plagioclase. Data from Hamet et al. (1978)

of Moama is m = 0.030406. Substituting into equation 8.10 yields:

$$t = \frac{1}{6.54 \times 10^{-12}} \ln(0.030406 + 1)$$

$$t = \frac{0.0299528}{6.54 \times 10^{-12}} = 0.0045799 \times 10^{12}\, y$$

$$t = 4.58 \times 10^9 \text{ years}$$

Almost all of the stony meteorites that have been analyzed for dating have yielded dates between 4.65 and 4.55 Ga thereby defining a time period of about 100 million years when the planets of the solar system formed by impact of planetesimals, differentiated by segregation of immiscible liquids, cooled, and solidified by crystallization of minerals.

8.11.3 Kinetic Energy of a Meteoroid

Assume that a spherical meteoroid with a radius (r) of 50 m, a density (d) of 3.5 g/cm^3, and a velocity (v) of 15 km/s is about to impact on the Earth. The volume (V) of this meteoroid is:

$$V = \frac{4}{3}\pi r^3$$

where the radius $r = 50\,m = 50 \times 100 = 5,000\,cm$

$$V = \frac{4 \times 3.14 \times (5,000)^3}{3} = 5.23 \times 10^{11}\,cm^3$$

The mass (m) of the meteoroid is:

$$m = d \times V$$
$$m = 3.5 \times 5.23 \times 10^{11} = 1.83 \times 10^{12}\,g$$

The kinetic energy (KE) of the meteoroid is:

$$KE = \frac{1}{2}\,m\,v^2$$

where the velocity $v = 15 \times 1000 \times 100 = 1.5 \times 10^6\,cm/s$

$$KE = \frac{1.83 \times 10^{12}\,(1.5 \times 10^6)^2}{2}$$
$$KE = 2.06 \times 10^{24}\,ergs$$

The explosion of one tonne of TNT releases an amount of energy equal to 4.2×10^{16} ergs (Koeberl, 1998, p. 89). Therefore, the kinetic energy of this meteoroid is equivalent to:

$$KE = \frac{2.06 \times 10^{24}}{4.2 \times 10^{16}} = 0.490 \times 10^8$$
$$= 49.0 \times 10^6\,tonnes\ of\ TNT$$

or 49 megatonnes of TNT. Therefore, the impact of this meteoroid has the same explosive energy as a good-sized hydrogen bomb (Figure 8.8). If this meteoroid were to hit a large city like New York, millions of people would die almost instantly. If the meteoroid were to fall into the ocean, the resulting tsunami could flood the adjacent continents, thus killing even more people.

8.11.4 Diameters of Impact Craters

The kinetic energy of a meteoroid that impacts on the Earth determines the diameter of the resulting explosion crater. The relation between the two parameters was expressed by Dence et al. (1977) in the form of an equation:

$$D = 1.96 \times 10^{-5}\,E^{0.294} \tag{8.12}$$

Table 8.4. Diameters of impact craters calculated from equation 8.13 for increasing kinetic energies of the impactor

Diameter of crater, km	Kinetic energy of impactor, ergs
1.95	10^{24}
7.57	10^{26}
2.93	10^{28}
114	10^{30}
444	10^{32}
1700	10^{34}

where D = diameter of the crater in kilometers and
E = energy in joules

Taking logarithms to the base 10 of both sides of the equation yields:

$$\log D = \log(1.96 \times 10^{-5}) + 0.294\log E \tag{8.13}$$

$$\log D = -4.707 + 0.294\log E$$

The unit of energy in the cgs system is the erg which is related to the joule (J) by the relations:

$$1\,J = 10^7\,erg$$
$$10^3\ tonnes\ of\ TNT = 4.2 \times 10^{19}\,erg$$

The diameters of impact craters in Figure 8.10 were plotted from the results of calculations in Table 8.4 derived from equation 8.13.

8.11.5 The Man Passing by on his Way to the Moon

A tribute to Eugene M. Shoemaker
by Bevan M. French

He was born in a Basin
that's now called L.A.
He decided quite early
that he wouldn't stay.
The Moon shone down on him,
there were rocks all around.
And in that combination,
his life's work was found.

He'd lie in his cradle
and smile at his mother,

with a hammer in one hand
and a rock in the other.
And late in the evening,
you might hear him croon,
"I'm just passing by
on my way to the Moon."

He started with field work
like all Survey hands,
but salt and uranium
were not in his plans.
It was Meteor Crater
and all of its kin
that changed our whole view
of the world that we're in.

Then he hooked up with NASA
and worked with Apollo,
'cause where astronauts went,
geologists could follow.
And in conference or meeting,
he'd sing the same tune,
"I'm just passing by
on my way to the Moon."

So all young geologists
who are new on the scene,
if you want to do well,
take your lessons from Gene.
Stay close to your field work,
but leave your mind free,
and don't sit at home when
there's new worlds to see.

For the young are not finished
with the worlds that we know.
They've heard all our stories,
and they are eager to go.
It won't be next August,
or the following June,
but one day they'll pass by
on their way to the Moon.

8.12 Problems

1. Calculate the density (d) of a spherical meteoroid having a diameter of 15 m and a mass of 6×10^9 kg. Express the result in g/cm^3. (Answer: d = 3.39 g/cm^3).

2. Use equation 8.5 to calculate a Rb-Sr date (t) of a lunar meteorite given that its measured daughter/parent (D/P) ratio is 0.04649 and that the decay constant (λ) of ^{87}Rb is 1.42×10^{-11} y^{-1}. (Answer: t = 3.2×10^9 y).

3. The diameter of the meteorite impact crater at Sudbury, Ontario, is approximately 250 km. Calculate the amount of energy (E) released by this impact. (Consult Science Brief 8.11d). Express the energy in terms of tonnes of TNT, where 1 kilotonne of TNT = 4.2×10^{19} erg. (Answer: 3.5×10^{14} tonnes of TNT).

4. Calculate the diameter (D) of the crater formed by the explosive impact of a meteoroid having a mass (m) of 6×10^9 kg and traveling with a velocity (v) of 20 km/s. (Answer: D = 4.05 km).

8.13 Further Reading

Barringer DM (1905) Coon Mountain and its crater. Proceed Acad Nat Sci Philadelphia 57:861–886. Phil Acad Sci, Philadelphia

Becker LA (2002) Repeated blows: Did extraterrestrial collisions capable of causing widespread extinctions pound the Earth? Sci Am Mar.: 77–83

Becker LA Bunch TE (1997) Fullerenes, fulleranes, and polycyclic aromatic hydrocarbons in the Allende meteorite. Meteoritics Planet Sci 32(4):479–487

Bernatowicz TJ, Messenger S, Pravdivtseva O, Swan R, Walker RM (2003) Pristine presolar silicon carbide. Geochim Cosmochim Acta 67(24):4679–4691

Brown PG, ReVelle DO, Tagliaferri E, Hildebrand AR (2002) An entry model for the Tagish Lake fireball using seismic, satellite, and infrasound records. Meteoritics and Planet Sci 37(5):661–675

Buchwald VF (1975) Handbook of iron meteorites. vol. 1–3. University California Press, Berkeley, CA

Cassidy WA (2003) Meteorites, ice, and Antarctica. Cambridge University Press, Cambridge, UK

Chao ECT, Shoemaker EM, Madsen BM (1960) First natural occurrence of coesite. Science, 132:220–222

Chao ECT, Fahey JJ, Littler J, Milton DJ (1962) Stishovite, SiO$_2$, a very high-pressure new mineral from Meteor Crater, Arizona. J Geophys Res 67:419–421

Chyba CF, Thomas PJ, Zahnle KJ (1993) The 1908 Tunguska explosion: Atmospheric disruption of a stony asteroid. Nature, 361: 40–44

Clarke Jr RS, Jarosewich E, Mason B, Nelen J, Gomez M, Hyde JR (1971) The Allende, Mexico, meteorite shower. Smithsonian Contrib Earth Sci 5:1–53

Davis AM, ed. (2004) Meteorites, comets, and planets. Treatise on Geochemistry, Vol. 1, Elsevier, Amsterdam

Dence MR, Grieve RAF, Robertson PB (1977) Terrestrial impact structures: Principal characteristics and energy consideration. In: Roddy D.J., Pepin R.O., Merril R.B. (eds), Impact and Explosion Cratering. Pergamon Press, Oxford, UK, pp 247–275

Dietz RS (1947) Meteorite impact suggested by the orientation of shatter-cones at the Kentland, Indiana, disturbance. Science, 105:42–43

Dietz RS (1961) Astroblemes. Sci Am Aug.: 1–10

Faul H (1966) Tektites are terrestrial. Science 152:1341–1345

Faure G, Mensing TM (2005) Isotopes: Principles and applications. Wiley, Hoboken, NJ

French BM, Short NM (ed) (1968) Shock metamorphism of natural materials. Mono Book Corp., Baltimore, MD

French BM (1998) Traces of catastrophe: A handbook of shock metamorphic effects in terrestrial meteorite impact structures. Contribution 952. Lunar Planet Sci Inst Houston, TX

French BM (1997) The man passing by on his way to the Moon. The Planetary Report 17(6):14–15

French BM (2004) The importance of being cratered: The new role of meteorite impact as a normal geological process. Meteoritics and Planet Sci, 39(2):169–197

Gault DE, Quide WL, Overbeck VR (1968) Impact cratering mechanics and structures. In: French B.M., Short N.M. (eds), Shock Metamorphism of Natural Materials. Mono Book Corp., Baltimore, MD, pp 87–99

Gilbert GK (1896) The origin of hypotheses, illustrated by the discussion of a topographic problem. Science 3:1–13

Glass BP (1982) Introduction to planetary geology. Cambridge University Press, Cambridge, MA

Grieve RAF, Therriault AM (2004) Observations of terrestrial impact structures: Their utility in constraining crater formation. Meteoritics and Planet Sci 39(2): 199–216

Grossman JN (1988) Formation of chondrules. In: J.K. Kerridge, M.S. Matthews (eds), Meteorites and the Early Solar System. University of Arizona Press, Tucson, AZ, pp 680–696

Hamet J, Nakamura N, Unruh DM, Tatsumoto M (1978) Origin and history of the adcumulate eucrite Moama as inferred from REE abundances, Sm-Nd, and U-Pb systematics. Proceedings of the 9th Lunar and Planetary Conference. Geochim Cosmochim Acta Supplement 1115–1136

Hartmann WK (2005) Moons and planets, 5th edn. Brooks/Cole-Thomson, Belmont, CA

Hewins RH, Jones RH, Scott ERD (1996) Chondrules and the protoplanetary disk. Cambridge University Press, Cambridge, UK

Hewins RH (1997) Chondrules. Annual Rev Earth Planet Sci 25:61–83

Hutchison R (2004) Meteorites: A petrologic, chemical, and isotopic synthesis. Cambridge University Press, Cambridge, UK

Kenkmann T, Hörz F, Deutsch A (eds) (2005) Large meteorite impacts III. Geol Soc Amer SPE 384, Boulder, CO

King EA (1976) Space geology: An introduction. Wiley, New York

Koeberl C (1986) Geochemistry of tektites and impact glasses. Ann Rev Earth Planet Sci 14:323–350

Koeberl C, Weinke HH, Kluger F, Kiesl W (1986) Cape York III AB iron meteorite: Trace element distribution in mineral and metallic phases. In: Proceedings of the Tenth Symposium on Antarctic Meteorites, 297–313. Mem Nat Inst Polar Res, Special Issue No. 41

Koeberl C (1998) Impakt; Gefahr aus dem All. VA Bene, Vienna

Koeberl C, MacLeod KG (eds) (2002) Catastrophic events and mass extinctions: Impacts and beyond. SPE 356, Geol Soc Amer, Boulder, CO

Krinov EL (1966) Giant meteorites. Pergamon Press, London

Lipschutz ME, Schultz L (1999) Meteorites. In: P.R. Weissman, L.-A. McFadden, T.V. Johnson (eds), Encyclopedia of the Solar System. Academic Press, San Diego, CA, pp 629–671

Mason B (1962) Meteorites. Wiley, New York

Mason B (1972) The mineralogy of meteorites. Meteoritics 7(3): 309–326

McCall GJH (1973) Meteorites and their orgins. David & Charles, Devon, UK

Melosh HJ (1989) Impact cratering: A geological process. Oxford University Press, New York, 245p

Mittlefehldt DW (2002) Tagish Lake: A meteorite from the far reaches of the asteroid belt. Meteoritics and Planet Sci, 37:703–712

Powell JL (1998) Night comes to the Cretaceous. Harcourt Brace, San Diego, CA

Porco CC (2000) Destination Moon. Astronomy, February, 53–55

Rubin AE (2000) Petrologic, geochemical, and experimental constraints on models of chondrule formation. Earth-Sci Rev, 50:3–27

Scott ERD, Jones RH, Rubin AE (1994) Classification, metamorphic history, and premetamorphic composition of chondrules. Geochim Cosmochim Acta, 58(3): 1203–1209

Scott ERD, Krot AN (2004) Chondrites and their components. In: Davis A.M. (ed.), Meteorites, comets, and planets. Treatise on Geochemistry, vol. 1, Elsevier, Amsterdam, pp 144–200

Sears DWG (2005) The origin of chondrules and chondrites. Cambridge University Press, Cambridge, UK

Shoemaker EM (1960) Penetration mechanics of high velocity meteorites, illustrated by Meteor Crater, Arizona. In: Structure of the Earth's Crust and Deformation of Rocks, 418–434. Internat Geol Congress, XXI Session, Part 18, Copenhagen

Shoemaker EM (1963) Impact mechanics at Meteor Crater, Arizona. In: Middlehurst B.M., Kuiper G.P. (eds), The Moon, Meteorites, and Comets. University of Chicago Press, pp 301–336

Shoemaker EM, Kieffer SW (1974) Guidebook to the geology of Meteor Crater, Arizona. Pub. 17, Center for Meteorite Studies, Arizona State University, Tempe, AZ, 66p

Shoemaker EM, Shoemaker CS (1999) The role of collisions. In: Beatty J.K. Petersen C.C., and Chaikin A., (eds), The New Solar System. Sky Publishing Corp, Cambridge, MA, pp 69–86

Silver LT, Schultz PH (eds) (1982) Geological implications of impacts of large asteroids and comets on the Earth. Geol Soc Amer, SPE 190, Boulder, CO

Taylor GJ, Scott ERD, Keil K (1983) Cosmic setting for chondrule formation. In: E.A. King (eds), Chondrules and their Origins. Lunar Planet Inst, Houston, TX, pp 262–278

Wasson JT (1993) Constraints on chondrule origins. Meteoritics, 28:13–28

Weems JE (1967) Peary: The explorer and the man. St Martin's Press, New York

Whillans IM, Cassidy WA (1983) Catch a falling star: Meteorites and old ice. Science 222:55–27

Zanda B (2004) Chondrules. Earth Planet Sci 224:1–17

The Earth-Moon System

The Moon in Figure 9.1 is the only satellite of the Earth and is usually referred to as "the Moon" even though the ancient Greeks personified the Moon as the goddess Selene and the Romans later knew the Moon as Luna. However, nowadays these names are used only in words like "selenology" and "lunar". Therefore, we will reserve the name "Moon" for the satellite of the Earth and will refer to the satellites of other planets by their proper names (e.g., Io, Europa, Ganymede, and Callisto are the Galilean satellites of the planet Jupiter).

The Moon is a familiar sight in the sky at night. We all know that the face of the Moon has markings that resemble a human face. The markings are approximately circular and a little darker than the surrounding areas. We also know that the Moon "shines" only because it reflects sunlight and that it is a solid body like the Earth. The face of the Moon seems to stay the same, which gives the erroneous impression that the Moon does not rotate on its axis. Actually, the Moon and the Earth carry out a complicated dance of revolutions and rotations that cause the Moon to rise and set, that increase and decrease the area of the Moon's illuminated face, and that cause the rising and falling of the water in the oceans of the Earth. In addition, the periodicity of certain biorythms appears to be synchronized with the period of revolution of the Moon.

Another property of the Moon that is observable from the Earth is that its face is never obscured by lunar clouds, which indicates that the Moon does not have an atmosphere because the Moon lacks water and other volatile compounds such as carbon dioxide and methane. In addition, the mass of the Moon is too small to generate a sufficient gravitational force to retain gases such as helium, nitrogen, oxygen, neon, and argon.

Figures 9.2 and 9.3 record the fact that the Moon is the first body in the solar system to be visited by humans and is scheduled to become the site of the first off-Earth outpost in 2018. The landing of an American spacecraft in the Sea of Tranquillity on July 20, 1969, was a milestone in human history and signaled the start of the era of space exploration (Section 1.1). The names of the astronauts who undertook this daring feat of exploration are: Neil Armstrong, Edwin (Buzz) Aldrin, and Michael Collins. Armstrong and Aldrin descended to the surface of the Moon in the lunar lander (Eagle) while Michael Collins remained in the mothership (Columbia) that returned all three astronauts safely to the Earth, thereby fulfilling the task assigned by President J.F. Kennedy.

The significance of this event was succinctly expressed by Neil Armstrong as he stepped onto the lunar surface:

"That's one small step for a man, one giant leap for mankind."

In the excitement of the moment, Armstrong omitted the letter "a" before the word "man". (Wilhelms, 1993, p. 201). The principle to be observed in the exploration of the Moon and other bodies of the solar system is articulated in a plaque mounted on the descent stage of the lunar lander that still stands at the landing site on the surface of the Moon:

"Here men from the planet Earth first set foot upon the Moon, July 1969 AD.
  We came in peace for all mankind."

Armstrong and Aldrin also planted an American flag on the Moon, but they did not claim possession of the Moon for the United States of America.

Figure 9.1. Full view of the near side of the Moon recorded by the Galileo spacecraft on December 7, 1992. The dark, roughly circular maria are impact basins that later filled with layers of basalt lava. The light-colored highlands are composed of anorthositic gabbro. The maria, are only lightly cratered, whereas the areal density of impact craters on the highlands approaches saturation. This evidence indicates that the highlands are older than the maria. The craters Copernicus and Tycho are surrounded by rays of ejecta that overlie the adjacent mare basins, indicating that they formed after the basins had filled with basalt. The surfaces of the mare basins and the highlands are both covered by a regolith composed of powdered lunar rocks and meteorites including angular boulders ejected from nearby meteorite impact craters (Courtesy of NASA/JPL)

A new spirit of cooperation in the exploration of the solar system was demonstrated on that day.

In subsequent landings, the astronauts of the Apollo program explored five other sites on the surface of the Moon using battery – powered lunar rovers shown in Figure 9.3. The results of these expeditions to the Moon have been described in books and scientific reports starting with a special issue of the journal "Science" (vol. 167, No. 3918, 1970) containing the first results derived from the study of the rock samples collected by Armstrong and Aldrin. Descriptions

Figure 9.2. On July 20, 1969, the spacecraft Eagle carrying Neil Armstrong and Edwin (Buzz) Aldrin landed safely in the southwestern area of Mare Tranquillitatis (Figure 9.1). The astronauts disembarked from their spacecraft and both stepped on the surface of the Moon where they spent a total of 21.36 hours and collected 21.7 kg of rock and soil samples. The Lunar Lander can be seen as a reflection in the visor of Aldrin's space helmet. In the meantime, Michael Collins remained in the spaceship Columbia which continued to orbit the Moon. On that day, the dream of visionaries on the Earth was fulfilled by competent and bold rocket scientists who reached the Moon. The flight of Apollo 11 was followed by five additional trips to the Moon. Only one of the seven attempted flights (i.e., Apollo 13) suffered a malfunction on the way to the Moon; but the astronauts were able to swing around the Moon and returned safely to the Earth. After a long hiatus devoted to the exploration of Venus, Mars, and of the outer reaches of the solar system, a new epoch of lunar exploration will commence in 2018 to prepare the way for a manned flight to Mars. (Courtesy NASA, image # AS 11-40-5903, GRIN Data Base: GPN-2001-000013)

Figure 9.3. Astronaut Charles M. Duke Jr, Lunar Module pilot of the Apollo 16 mission, at station 1 during the first extravehicular activity (EVA) at the Descarte landing site in the lunar highlands at 8.99°S and 15.51°E. Duke is standing at the rim of Plum crater which is 40 m wide and 10 m deep. The Lunar Rover can be seen in the background on the far side of Plum crater. This photograph was taken by astronaut John W. Young. (Courtesy of NASA , image # AS16-114-18423, GRIN Data Base: GPN-2000-001132)

of the Moon landings include books by Collins (1974), Aldrin and McConnell (1991), Wilhelms (1993), Cernan and Davis (1999), Kluger (1999), Harland (2000), Lindsay (2001), Light (2002), Turnill (2003), and Hansen (2005).

9.1 Landforms of the Lunar Surface

The major topographic features of the lunar surface in Figure 9.1 can be seen even with a modest telescope or binoculars and were described hundreds of years ago. Johannes Kepler recognized and named the circular plains and the mountainous highlands. He referred to the plains as the "maria" (plural form of mare) which is the Latin word for "sea" and he named the mountainous highlands "terrae" (plural form of terra, which is the Latin word for land). We now know that the maria are composed of basalt and make up about 30% of the near side of the Moon, whereas the highlands consist of anorthositic gabbro illustrated in Figure 9.4A & B.

9.1.1 Highlands

The mountainous highlands of the Moon are composed of plutonic rocks that range in composition from feldspathic gabbro to light-colored anorthosite both of which consist predominantly of plagioclase feldspar, which causes the lunar highlands to have a higher albedo than the basalts of the lunar maria. Anorthosite also occurs on the Earth in layered gabbro intrusions (e.g., Stillwater complex of Montana) and as large massifs primarily of Precambrian age. For example, Whiteface

Figure 9.4. A. Basalt (sample 15016) collected by the astronauts of Apollo 15 who landed in the Mare Imbrium at the base of the Apennine Mountains (Figure 9.9). The sample is dark grey in color, weighs 923 g, and is highly vesicular. The vesicles are gas bubbles that formed when the basalt lava was extruded at the surface of the Moon. The gas, composed of carbon dioxide and carbon monoxide, escaped from the cooling lava and drifted away into interplanetary space. Basalts are composed primarily of plagioclase and pyroxene with varying amounts of ilmenite ($FeTiO_3$), olivine [$(Fe, Mg)_2SiO_4$], and magnetite (Fe_3O_4). (Courtesy of NASA/Johnson Space Center, photograph S71-45477). B. Anorthosite (sample 60025) collected by the astronauts of Apollo 16 in the Cayley Plains near the crater Descartes of the lunar highlands. The specimen weighs 1.8 kg and is composed primarily of calcium-rich plagioclase. The anorthosites of the lunar highlands formed during the cooling of the magma ocean when plagioclase crystals floated to the surface and formed a thin primordial crust (Section 6.2). The specimen shown here crystallized between 4.44 and 4.51 billion years ago when the Moon was still young. (Courtesy of NASA/ Johnson Space Center, photograph S72-42187)

Mountain in the Adirondacks of upstate New York is composed of light-colored anorthosite.

The dominance of anorthosite in the lunar highlands is significant because, according to the theory of planetary differentiation discussed in Section 6.2, plagioclase-rich rocks formed the original lunar crust as a result of crystallization of a magma ocean like the one we mentioned in connection with the internal differentiation of the Earth. The anorthosite crust of the Earth was destroyed or deeply buried by subsequent geological activity, but it seems to have survived on the Moon where such geological activity did not occur. In fact, the far side of the Moon consists largely of heavily cratered highlands composed of anorthositic gabbro.

9.1.2 Maria

The large circular maria consist of basalt lava that filled giant basins that were carved by the impacts of large meteorites (or asteroids) having diameters that ranged up to hundreds of kilometers. The resulting impact basins subse-quently filled with basalt lava that welled up through deep fractures in the floors of these basins. The maria also contain several kinds of minor topographic features that formed during the initial filling of the basins (e.g., lava drain channels called rilles) and during the subse-quent solidification of the lava lakes (e.g., ridges, domes, cones, pits, and chain craters).

The crater density, defined as the number of impact craters per unit area, of the highlands is much higher than in the basalt plains. This obser-vation indicates that the lunar highlands have been exposed to meteorite impacts for a longer period of time than the basalt plains. Therefore, the topography of the lunar surface indicates the occurrence of three stages in the early history of the Moon:

1. Fractional crystallization of a magma ocean of basaltic composition in accordance with Bowen's reaction series to form an anorthosite crust and a peridotite mantle (Section 6.2).
2. Heavy bombardment of the Moon by large impactors that formed basins some of which have diameters in excess of 1000 km.

3. Eruption of basalt lava through fissures thereby filling the basins and causing some of them to overflow into adjacent impact craters.

For reasons that are still not understood, the surface of the *far side* of the Moon is composed primarily of heavily cratered highland terrain. Lava plains that are so prominent on the *near side* are almost completely absent on the far side except for the Orientale basin, part of which extends to the near side.

9.1.3 Impact Craters

The basalt plains such as the Mare Imbrium in Figure 9.1 as well as the highlands contain innumerable craters first described by Galileo Galilei. The diameters of the craters range from less than one meter to several hundred kilometers. Most of the impact craters are circular in shape, have a raised rim, and a flat floor situated below the elevation of the surrounding terrain. Large craters have a central uplift and the largest craters may contain one or more rings of peaks that rise from their floors (Section 8.8.1,2). Large craters on the Moon are named after famous deceased scientists, scholars, and artists (e.g., Ptolemaeus, Copernicus, Eudoxus, Eratosthenes, Archimedes, etc.). Small craters are given first names (e.g., Shorty and Lara, landing site of Apollo 17, Taurus-Littrow valley).

The rim of the crater Pytheas in Figure 9.5 is elevated because the rocks along the crater rim were tilted so that they dip away from the crater. In addition, craters are surrounded by aprons of rock debris called ejecta blankets. The ejecta blankets on the Moon have rays that can be traced for hundreds of kilometers in some cases (e.g., the crater Tycho in Figure 9.1). Most of the craters are the result of meteorite impacts and only a few are of volcanic origin.

The crater Euler in Figure 9.6 is 27 km wide and has a central uplift. The walls of the crater have been modified by slumping and expose the stratification of the rocks that underlie the surface of the Mare Imbrium. The floor of the crater and the surrounding ejecta are sprinkled with small secondary craters formed by the ejecta of the crater Copernicus.

The craters Archimedes, Aristillus, and Autolycus in the Mare Imbrium are visible in Figure 9.1. Archimedes in Figure 9.7 (80 km wide) was flooded by basalt lava flows which also covered the ejecta deposits along the right side of the crater. The smaller craters Aristillus and Autolycus in the upper right of Figure 9.7 are younger than Archimedes because the lava flow that covers the ejecta of Archimedes did not overflow their ejecta blankets.

9.1.4 Regolith

The ejecta that originated from the explosions that formed the basins on the near side of the Moon originally covered large parts of the lunar surface. This material was subsequently buried under the ejecta blankets associated with smaller impact craters that formed more recently. Consequently, the entire surface of the Moon is now covered by a blanket of regolith that is composed of fine-grained crushed basalt from the maria and of fragments of feldspar-rich gabbro from the highlands. In addition, the lunar regolith contains small beads of glass derived by melting of lunar rocks during meteorite impacts as well as glass of meteoritic origin. The regolith also contains angular boulders of basalt and gabbro that were excavated from the underlying bedrock by the impacts (See Figure 9.4A and 9.4B). The thickness of the regolith layer varies locally from one to several meters. The basalt flows that underlie the maria are exposed in the walls of deep craters (e.g., Figure 9.6) and lava flow-channels such as Hadley Rille at the edge of Mare Imbrium.

The regolith on the lunar surface is sometimes referred to as "lunar soil," even though it does not meet the geological definition of "soil" because it lacks the stratification of terrestrial soil that results from the interaction with rooted plants and from the downward transport of weathering products. The color of the lunar regolith is dark grey to black consistent with its composition.

9.1.5 Water

The absence of an atmosphere mentioned in the introduction also implies that the Moon does not contain water. Petrographic studies of the lunar rocks returned by the astronauts of the Apollo program revealed that crystals containing

Figure 9.5. Oblique view south across the Mare Imbrium toward the crater Copernicus which is clearly visible in Figure 9.1. The Montes Carpatus are located along the south rim of the Mare Imbrium. The bowl-shaped impact crater Pytheas is located near the center of the image. It has a diameter of almost 19 km and a sharply defined raised rim. The ejecta blanket that surrounds Pytheas is rayed and has a hummocky surface. The small craters in this view of the southern area of the Mare Imbrium are aligend along rays of Copernican ejecta that extend north across the plain of Imbrian basalt in Figure 9.1. Both Copernicus and Pytheas formed less than 1.8 billion years ago. (Wilhelms and McCauley, 1971). (Courtesy of NASA, image # AS17-2444-M)

hydroxyl ions (OH⁻) do not occur on the Moon. Therefore, lunar rocks do not contain mica minerals (biotite, phlogopite, and muscovite), amphiboles (hornblende), serpentine, and clay minerals (kaolinite, smectite). Evidently, the basalt magmas that formed the volcanic and plutonic rocks of the lunar crust did not contain water. This insight raises an important question: Why does the Moon lack water when the Earth and the other terrestrial planets apparently did acquire water during their formation? The answer to this question will emerge later in this chapter as part of the theory of the origin of the Moon (Section 9.6).

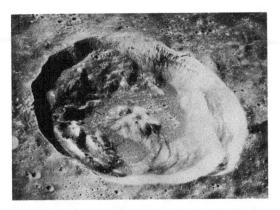

Figure 9.6. Close-up view of the impact crater Euler taken by the astronauts of Apollo 17. This crater is located in the southwestern quadrant of the Mare Imbrium about 400 km northwest of Copernicus. The floor of Euler is lower than the surface of the Mare Imbrium which indicates that the crater has not been appreciably filled with weathering products and ejecta derived from impacts in the neighborhood. Although Euler is a relatively young crater, it is older than Pytheas because Euler was formed **before** the volcanic activity of the mare Imbrium had ended, whereas Pytheas was formed **after** the cessation of volcanic activity. In addition, we can see that Euler formed before Copernicus because Copernican ejecta appear to have cratered Eulerian ejecta. (Courtesy of NASA, image # AS 15-0274, P)

Water in the form of ice may have accumulated in deep craters in the polar regions of the Moon (e.g., the Aitken basin at the South Pole) as a result of impacts of comets that are partly composed of water ice. If confirmed, the availability of water will greatly facilitate the establishment of human outposts on the Moon.

The absence of an atmosphere means that the Moon exists in the vacuum of interplanetary space. As a result, there is no wind and hence there are no dust storms on the surface of the Moon. An entire set of geological processes and their products are eliminated by the combined absence of air and water. For example, it never rains or snows and there are no streams, lakes, and oceans. Consequently, sedimentary rocks that are deposited by wind or water are absent from the Moon such as sandstone, shale, limestone, and evaporites. The only quasi-sedimentary rocks on the Moon are breccias that form when the regolith is compressed by shock waves emanating from meteorite impact sites.

Most important of all is the absence of living organisms of any kind due to the absence of water. Not even bacteria have been found in the lunar regolith because all life forms that we know of require the presence of water and hydrocarbon compounds. Such compounds have not been found because the Moon not only lacks water but is also depleted in hydrogen, carbon, nitrogen, and other volatile elements. Evidently, the chemical composition of the Moon differs from that of the Earth in several important ways even though the Moon is quite close to the Earth (384.4×10^3 km) and presumably formed at the same time and by the same process as the Earth. Here we have another mystery that will be explained later in this chapter.

9.2 Isotopic Dating

The rock samples collected by American astronauts in the course of the Apollo Program between 1969 and 1972 and the soil samples returned by the robotic spacecraft of the USSR (Luna, 1970 to 1976) were dated by means of the radiogenic isotope chronometers described in Section 8.7. The resulting dates based on the Rb-Sr and Sm-Nd methods in Figure 9.8 range from about 4.5 to 3.1 Ga. In addition, the spectrum of isotopic dates of lunar basalts, gabbros, and anorthosites is polymodal which means that the volcanic activity that filled the impact basins with basalt lava was episodic with major events between 4.1 to 3.6 Ga and from 3.5 to 3.1 Ga. The dates also indicate that only a few early-formed rocks that crystallized between 4.6 and 4.2 Ga have survived to the present. The episodic nature of the basin-filling volcanic activity characterizes not only the entire near side of the Moon, but was also recognized by Snyder et al. (1994) in the sequential eruptions of high-titanium basalt in the Sea of Tranquillity at 3.85, 3.71, 3.67, and 3.59 Ga. In all cases, the basalts are younger than the meteorite impacts that formed the basins that later filled with basalt lava.

9.3 Geology of the Near Side

In contrast to the Earth, the rocks on the Moon were not folded and faulted by tectonic forces. The absence of sedimentary rocks deposited by water or wind (Section 9.1.5) and the absence of tectonic activity have prevented the formation of

Figure 9.7. The impact crater Archimedes in the Mare Imbrium (Figure 9.1) has a diameter of 80 km and has been filled by lava flows of the Mare Imbrium. Even the ejecta, visible on the side facing the viewer (south), have been obliterated by lava flows elsewhere around the rim of the crater. For example, east of Archimedes lava flows partially drowned a small crater. These lava flows are only lightly cratered in contrast to the more heavily cratered ejecta blanket south of Archimedes. The craters Aristillus and Autolycus (Figure 9.1) are younger than Archimedes because their ejecta blankets were not covered by Imbrian lava flows. In fact, Aristillus (the northern crater) is younger than Autolycus because the ejecta blanket of Artistillus partly overlaps the ejecta of Autolycus. Wilhelms and McCauley (1971) used these kinds of criteria and the principle of superposition to construct a sequentially correct geologic map of the near side of the Moon. (Courtesy of NASA, image # AS-1541, M)

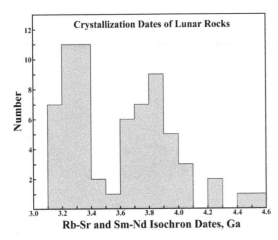

Figure 9.8. The histogram of Rb-Sr and Sm-Nd isochron dates (Science Brief 8.11.3) of lunar rocks is polymodal which suggests that the history of magmatic/volcanic activity on the Moon was episodic. The majority of the dates range from 3.16 to 4.60 Ga. The absence of dates younger than about 3.0 Ga indicates that no magmatic activity has occurred on the Moon during the last three billion years. Data from Turcotte and Kellog (1986)

Table 9.1. Lunar timescale (Adapted from Hartmann, 2005)

Era	Age, Ga	Events
Copernican	< 1.8	Occasional meteoroid impacts.
Eratosthenian	1.8-3.2	End of flooding of mare basins by basalt lava.
Imbrian	3.2-4.0	Mare basins flooded by basalt lava.
Nectarian	4.0-4.6	Intense bombardment, mare basins formed.

complex mountains chains on the Moon. Given these limitations, our present understanding of the geology of the Moon is based on the interpretation of surface topography and of the areal distribution of mappable stratigraphic units contained in the geologic map prepared by Wilhelms and McCauley (1971). The sequence of deposition of these units was determined by means of the principles of *superposition* and of *cross-cutting* relationships. The sequence of meteorite impacts and episodes of volcanic activity was related to a timeframe provided by the isotopic dates of rock samples collected at the landing sites of Apollo astronauts and by the robotic landers of the Luna program of the former USSR (e.g., Luna 16; Mare Fecunditatis; September 1970). The resulting subdivisions of the lunar history in Table 9.1 carry the names of large impact craters whose ejecta are topographic landmarks in the landscape of lunar history and also provide reference points in the timescale of the geologic history of the Moon.

The sequence of meteorite impacts and volcanic activity recorded in the Mare Imbrium in Figures 9.5, 9.6 and 9.7 is representative of the history of the entire near side of the Moon. The pivotal event in the lunar timescale of Table 9.1

is the volcanic activity that resulted in the filling of the Imbrian multi-ring basin which had a diameter of about 1100 km. The mare basalts that filled the Imbrian basin are designated Im (Imbrian mare) in Figure 9.9. Parts of the ejecta deposits outside of the original rim of the basin now form the lunar Carpathian, Apennine, and Caucasus mountains.

In addition, parts of the original ring mountains (shown in black in Figure 9.9) project through the Imbrian basalts (e.g., the Alps in the northern part of Mare Imbrium). While the volcanic activity was still in progress, a meteorite impact formed the crater Archimedes (Figures 9.7 and 9.9) which subsequently filled with Imbrian mare basalt.

Several additional impact basins listed in Table 9.2, that were formed during pre-Imbrian (Nectarian) time, were also filled with basalt lava during the Imbrian Era (3.2 to 4.0 Ga). In other words, the eruption of floods of basalt lava during the Imbrian Era was not restricted to the Imbrian basin but occurred in most, but perhaps not all, of the large impact basins that existed on the near side of the Moon at about 4.0 Ga.

The eruption of mare basalts ended during the Eratosthenian Era (3.2 to 1.8 Ma) named after the impact crater Eratosthenes whose ejecta blanket partly overlaps the Imbrian ejecta that form the lunar Apennine Mountains in Figure 9.9. During this period in the history of the Moon, the frequency of meteorite impacts continued to decline. Nevertheless, the eruption of flood basalt in Mare Imbrium and in other maria listed in Table 9.2 continued. As a result, most of the maria on the near side of the Moon contain basalt flows of Eratosthenian age that overlie the more voluminous Imbrian basalt. The only

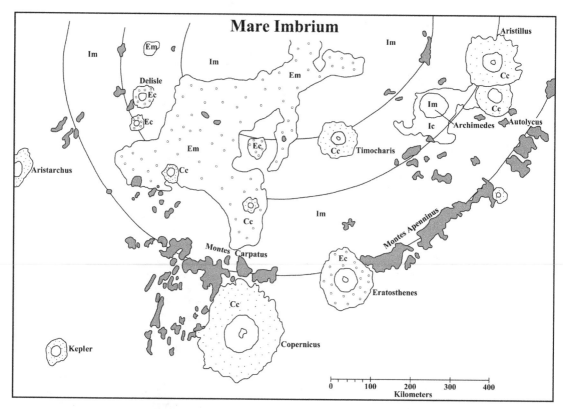

Mare Imbrium

Figure 9.9. Surface deposits of the southern part of the Mare Imbrium. The Imbrian basin formed in pre-Imbrian (Nectarian) time by the impact of a large asteroid. During the subsequent Imbrian Era the resulting multiring basin filled with floods of basalt symbolized by "Im." The ejecta deposits (black) along the original rim of the Imbrian basin now form the Carpathian and Apennine mountain ranges. Meteorites continued to impact on the Moon during the Imbrian Era as indicated by the presence of the crater Archimedes, which filled up with basalt lava from the Imbrian basin. The eruption of flood basalt continued into the Eratosthenian Era that takes its name from the crater Eratosthenes (Ec). Basalt flows of Eratosthenian age (Em) overlie the Imbrian basalts within the Mare Imbrium and in most of the other maria on the near side of the Moon. The most recent era in the history of the Moon is named after the crater Copernicus (Cc). The ejecta blankets of the Copernican craters overlie those the Imbrian and Eratosthenian craters and consist of mixtures of different kinds of rock fragments that originated from both the maria and the highlands. Adapted from the geologic map of the near side of the Moon prepared by Wilhelms and McCauley (1971)

impact basin that is completely filled by Eratosthenian basalt is the Mare Crisium (Figure 9.1).

The most recent history of the Moon is encompassed by the Copernican Era (< 1.8 Ga), named after the crater Copernicus located south of the rim of the Imbrian basin in Figure 9.9. Copernican craters are less numerous and are smaller than the impact basins of pre-Imbrian age. This tendency implies that the frequency of meteorite impacts during the Copernican Era continued to decline and that the impactors were smaller on average than those that carved the mare basins in pre-Imbrian (Nectarian) time.

The Copernican craters Aristillus and Autolycus in Figure 9.9 illustrate the use of the Principle of Superposition in the reconstruction of the history of the Moon. The ejecta of Aristillus overlie those of Autolycus and the ejecta of both overlie those of Archimedes. Consequently, both of these Copernican craters post-date the formation and subsequent filling of Archimedes. The ejecta of these and other Copernican craters overlie those of all previously formed craters. In general, the regolith at the Apollo and Luna landing sites contains rock fragments ejected by several different craters,

Table 9.2. Episodes of basin-filling volcanic activity on the near side of the Moon (Wilhelms and McCauley, 1971)

Mare	Time of Filling	
	Imbrian	Eratosthenian
Imbrium	✓	✓
Oceanus Procel- larum	✓	✓
Tranquillitatis	✓	
Serenitatis	✓	✓
Fecunditatis	✓	
Crisium		✓
Nectaris	✓	
Nubium	✓	✓
Humorum	✓	
Vaporum	✓	
Frigoris	✓	✓

some of which may be located in the highlands. This is the reason why samples of lunar regolith collected at a particular site on the Moon contain rock fragments of several different varieties of mare basalt as well as gabbro and anorthosite clasts from the highlands.

9.4 Lunar Meteorites

The impacts of meteorites on the surface of the Moon dislodged boulders of the bedrock, most of which fell back to the surface of the Moon and formed small secondary craters. Occasionally, boulders of the target rocks were launched into orbit around the Moon or escaped from the Moon and went into orbit around the Sun. Some of these lunar boulders were captured by the Earth or by other terrestrial planets. Such boulders are classified as lunar meteorites after they were captured by the Earth.

The first lunar meteorite pictured in Figure 9.10 was collected on January 18, 1992, by Ian Whillans (glaciologist) and John Schutt (geologist and mountaineer) on the Near Western ice field of the polar plateau adjacent to the Allan Hills, southern Victoria Land, Antarctica. On this day, they collected eleven meteorite specimens including a small breccia fragment about 3 cm in diameter and weighing 31.4 g. The presence of a greenish tan-colored fusion crust on this specimen indicated that its surface had melted

Figure 9.10. The lunar meteorite ALHA 81005 was collected on January 18, 1982, by Ian Whillans and John Schutt on the so-called Middle Western icefield of the East Antarctic icesheet near the Allan Hills of the Transantarctic Mountains. The specimen is about 3 cm wide and has a thin greenish-tan fusion crust. It was considered to be a "strange" meteorite when it was collected. However, the lunar origin of this specimen was immediately recognized by scientists at the Johnson Space Center in Houston, Texas, who had been working with the lunar rock samples collected by the Apollo astronauts. The rock contains fragments of light-colored plagioclase crystals imbedded in a black fine-grained matrix composed of the powdery fraction of the lunar regolith. Therefore, the rock has been classified as an anorthosite breccia from the lunar highlands (Cassidy, 2003). Fragments of this specimen were analyzed in a variety of laboratories around the world and the results were presented at the Fourteenth Lunar and Planetary Science Conference on March 17, 1983 (Keil and Papike, 1983). (Courtesy of NASA/LPI; wikimedia commons)

like those of meteorites that pass through the atmosphere. Therefore, the specimen was very probably a meteorite; but it did not resemble any of the many meteorite specimens John Schutt had previously collected in Antarctica. The lunar origin of this specimen was established only after it had been taken to the Johnson Space Center in Houston, Texas, where it received the designation ALHA 81005 (Cassidy, 2003). This lunar meteorite became the center of attention during the 14th Lunar and Planetary Science Conference on March 17, 1983, in Houston, Texas. The unanimous verdict of the assembled geochemists and meteoriticists was that ALHA 81005 is an anorthositic regolith breccia from the highlands of the Moon and that it was compacted by a shock wave caused by the impact of a meteorite.

Although all geochemical and mineralogical criteria that were considered confirmed the lunar origin of ALHA 81005, the geochemical and mineralogical evidence did not indicate how this specimen had been launched into interplanetary space from the surface of the Moon without being crushed or melted. Some investigators expressed doubt that even a low-angle meteorite impact could cause rock samples to be lifted off the surface of the Moon with enough velocity to escape from its gravitational field. Fortunately, a controversy was averted when H.J. Melosh of the University of Arizona suggested that boulders lying on the surface of the Moon can survive the initial passage of the shock wave and be ejected into space by the subsequent relaxation wave.

The effectiveness of this process is indicated by the fact that lunar meteorites have been found not only in Antarctica, but also in the deserts of Australia, Libya, Oman, and Morocco. The lunar meteorites collected at these sites include anorthositic regolith breccias like ALHA 81005, gabbro and basalt from mare regions, and melt rocks derived from highland anorthosites. In 2004 about 30 specimens of lunar meteorites had been identified, several of which are paired (i.e., they are fragments of the same meteoroid in space). The total weight of lunar meteorites exceeds 5.6 kg and is likely to increase as additional specimens are identified (Eugster, 1989).

Age determinations of the lunar meteorites indicate that they crystallized early in the history of the Moon much like the lunar rocks returned by the Apollo astronauts. After ejection from the Moon by impacts of asteroidal meteorites or comets, individual specimens spent varying intervals of time in space ranging from about ten million to 300 thousand years until they were captured by the Earth. The terrestrial ages of several lunar meteorites collected in Antarctica (i.e., the time elapsed since their arrival on the Earth) range from 170 thousand to 70 thousand years (Eugster, 1989).

Boulders can be ejected by meteorite impacts not only from the surface of the Moon but also from other planets including Mars, the Earth, Venus, and even Mercury. In fact, several stony meteorites collected on the polar plateau of Antarctica and at other sites on the Earth originated on the surface of Mars. Conversely, rocks from Earth could have landed on the Moon, but none were found by the Apollo astronauts. Nevertheless, the transfer of boulders from a terrestrial planet of the solar system to any of its neighbors appears to be feasible and may be an ongoing phenomenon.

9.5 Internal Structure

The composition and structure of the interior of the Moon has been investigated by several indirect methods described by Glass (1982):

1. Changes in the velocities of spacecraft orbiting the Moon.
2. Measurement of the flow of heat from the interior to the surface of the Moon at the landing sites of Apollo 15 and 17.
3. Determination of the strength and orientation of the lunar magnetic field by orbiting spacecraft and by instruments on the surface of the Moon.
4. Recordings of moonquakes by seismometers placed on the surface of the Moon by the Apollo astronauts.

The tracking of spacecraft in lunar orbits indicated that their orbital velocities increased over the maria on the near side and over large craters (greater than about 200 km) that are filled with basalt. The increase in the orbital velocities is attributable to an increase in the force of lunar gravity caused by a local increase of mass. Therefore, the positive gravity anomalies over the maria indicate the presence of mass concentrations referred to as *mascons*. The data also show a decrease of orbital velocities above large unfilled impact basins presumably because of a local deficiency of mass.

The heatflow measurements on the surface of the Moon yielded a value that is only about 20% of the heatflow of the Earth indicating that the interior of the Moon is cooler than the interior of the Earth. This result sets limits on the average concentrations of uranium, thorium, and potassium in the crust of the Moon because the decay of the unstable atoms of these elements is the principal source of heat in the Moon as well as in the Earth (Section 8.7).

The strength of the magnetic field, which extends into the space around the Moon, is only one trillionth (10^{-12}) the strength of the magnetic field of the Earth. However, the rocks of the lunar crust are magnetized, which indicates that the Moon did have a magnetic field at the time the mare basins were flooded by basalt lava between 4.0 and 3.0 billion years ago. The magnetic field of the Moon at that time may have been induced by electrical currents in a liquid core composed of iron and nickel in accordance with the theory of the self-exciting dynamo described in Section 6.4.5. However, several alternative hypotheses have been proposed to explain the magnetization of the rocks in the lunar crust.

The moonquakes recorded by seismometers on the surface of the Moon originate primarily from sources 800 to 1000 km below the lunar surface and less frequently from shallow depths less than 100 km. In addition, the seismometers recorded 1800 impacts of meteorites between 1969 and 1977 when the seismometers were turned off for budgetary reasons. The results indicate that the Moon has far fewer quakes than the Earth, presumably because the Moon is tectonically inactive. The principal cause for the occurrence of deep lunar moonquakes is the stress imposed by tidal forces as the Moon revolves around the Earth.

The seismic data are the best source of information about the structure and composition of the lunar interior. The profile of velocities of compressional (P) waves in Figure 9.11 demonstrates the presence of compositional layers in the lunar *crust* separated by boundary surfaces at depths of about 20 and 60 kilometers. The profile was taken in the eastern part of the Oceanus Procellarum (Ocean of Storms) where the astronauts of Apollo 12 landed on November 14, 1969. The basalt layer is only present in the maria, whereas the anorthositic gabbro forms the crust in the highlands and extends beneath the Oceanus Procellarum. The underlying mantle of the Moon consists of ultramafic rocks (peridotite) composed primarily to the minerals olivine and pyroxene.

The *mantle* of the Moon in Figure 9.12 is subdivided into three layers based on the available seismic data (Glass, 1982, Figure 7.34):

Upper mantle: 60 to 300 km
Middle mantle: 300 to 1000 km
Lower mantle: 1000 to 1475 km
Core (liquid?): 1475 to 1738 km

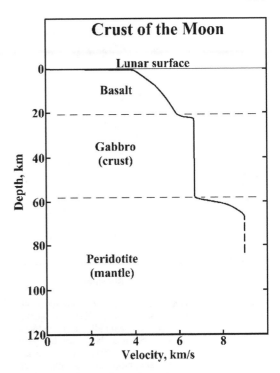

Figure 9.11. Profile of the velocity of seismic P-waves to a depth of about 100 km below the lunar surface in the eastern Oceanus Procellarum (Ocean of Storms). The increases of the velocity at about 20 and 60 km are attributed to transitions from mare basalt to anorthositic gabbro at 20 km and from the gabbro to peridotite (olivine and pyroxene) at 60 km. The basalt layer occurs only in the mare regions and is absent in the highlands. Adapted from Glass (1982, Figure 7.33) and based on data by Toksöz et al. (1973)

The basal part of the *middle mantle* (800 to 1000 km) is the source of most of the deep earthquakes on the Moon. The *lower mantle* attenuates S – waves (shear) suggesting that the rocks in the lower mantle are less rigid because they are partially molten. Consequently, the rigid shell of the lunar crust and mantle extends from the surface to a depth of about 1000 km, thus giving the lunar *lithosphere* (rigid part) a thickness of 1000 km. The underlying lower mantle (1000 to 1475 km) and core (1475 to 1738 km) form the lunar *asthenosphere*, the thickness of which is about 738 km.

According to these interpretations of the available seismic data, the Moon has a much thicker lithosphere than the Earth whose lithosphere is only about 120 km thick. This

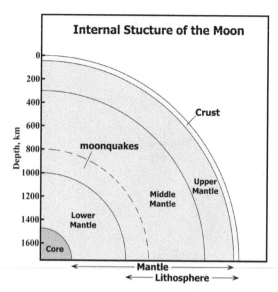

Internal Stucture of the Moon

Figure 9.12. The mantle of the Moon is subdivided into three layers on the basis of seismic data. Deep moonquakes originate from the lower part of the middle mantle between 800 and 1000 km. The core-mantle boundary has been placed here at a depth of about 1470 km. The state and composition of the core are uncertain. The lithosphere (rigid) of the Moon extends from the surface to a depth of about 1000 km. The lower mantle attenuates S-waves presumably because the rocks are partially molten. Accordingly, the asthenophere (not rigid) extends from about 1000 km to the center of the Moon at a depth of 1738 km. Adapted from Figure 7.34 of Glass (1982)

conclusion is consistent with the evidence that the Moon is no longer geologically active because its larger surface-to-volume ratio has allowed it to cool faster than the Earth. However, even though the Moon is tectonically inactive, its interior is still hot. The temperature at a depth of 1000 km is estimated to be 1500 °C, but the thickness of the lunar lithosphere no longer permits magmas to form by decompression melting (Section 6.4.2). The relationship between the cooling rate of a solid body and its surface-to-volume ratio mentioned above means that small bodies cool faster than large bodies because small bodies (e.g., the Moon and Ceres) have larger surface-to-volume ratios than large bodies like Venus and the Earth..

The composition and state of the *core* of the Moon are known only from the seismic signal recorded after the impact of a single meteorite on the far side. The available evidence indicates

that S-waves were not transmitted through the center of the Moon presumably because the seismic waves passed through a molten zone. The hypothetical core implied by the seismic data may have a radius between 170 and 360 km and it may be composed either of iron sulfide or of an alloy of iron and nickel.

9.6 Origin of the Moon

Prior to the landings of the Apollo astronauts in 1969, the experts predicted that the samples of lunar rocks to be collected by the astronauts would reveal the origin of the Moon. However, the information that was subsequently generated by the study of the Moon from orbit and by analysis of lunar rocks did not at first yield a convincing explanation for the origin of the Moon.

The hypotheses that had been proposed prior to the lunar landings included three alternatives:
1. The Moon formed elsewhere in the solar system and was later captured by the Earth.
2. The Moon formed in orbit around the Earth by the accumulation of planetesimals.
3. The Moon formed from material that was torn out of the Earth by one of several alternative processes.

These and other related hypothesis were evaluated by Wood (1986) in relation to certain facts that constrain the origin of the Moon. The facts he considered included:
1. Depletion of the Moon in volatile elements (e.g., potassium, sodium, fluorine, zinc, and 19 others) relative to carbonaceous chondrites.
2. Depletion of the Moon in iron as indicated by its low bulk density ($3.344 \, g/cm^3$) compared to the bulk density of the Earth ($5.520 \, g/cm^3$).
3. The absence of water and other volatile compounds in the igneous rocks of the Moon.

The hypothesis that the Moon formed elsewhere in the solar system and was later captured by the Earth could account for its lower density. In fact, the decrease of the bulk densities of the terrestrial planets with increasing distance from the Sun in Figure 3.6 (Section 3.5) could be used to conclude that the Moon originally formed in the asteroidal belt between the orbits of Mars and Ceres at a distance of 1.8 AU from the Sun. However, in that case we would expect to find hydroxyl ions in some of the minerals that crystallized from lunar magma because water and other volatile compounds become more abundant with increasing distance from the Sun. For the same reason, we would not expect the Moon to be depleted in certain volatile elements. In other words, the absence of water and the depletion of volatile elements of the Moon are both inconsistent with the chemical differentiation of the protoplanetary disk (Section 5.2).

In addition, if the Moon originated from the asteroidal belt or from an even more distant region of the solar system, its velocity relative to the Earth would be too high for capture. Instead, the Moon would pass the Earth and would continue in its eccentric orbit. It would swing around the Sun and then return to the distant region of space from where it originated. If the Moon's orbit prior to capture by the Earth was either parabolic or hyperbolic, the Moon would have made only one pass around the Sun followed by ejection from the solar system (Section 7.2.9).

The hypothesis that the Moon formed in orbit around the Earth by the accretion of planetesimals also seems plausible at first sight. However, it too fails to explain why the Moon lacks water and why its bulk density is much lower than that of the Earth. The special circumstances that are required to reconcile the compositional differences between the Earth and the Moon invalidate this hypothesis.

The third group of hypotheses is based on the idea that the Moon was somehow torn out of the Earth. Wood (1986) described several proposed mechanisms for causing the Earth to fission (e.g., by a rapid rate of rotation). If such fission occurred after the core of the Earth had formed, the deficiency of iron in the Moon might be explained. However, the mechanical problems of fissioning the Earth to form the Earth-Moon system also discredit this hypothesis.

A better case can be made for the hypothesis that the Moon formed from a disk of material that was ejected from the Earth as a result of the impact of a large object having a mass of about 0.1 Earth masses (e.g., approximately the size of Mars). Mathematical modeling has indicated that objects of that magnitude existed within the solar system after the main phase of planetary accretion had ended. Therefore, collisions of such large objects with the Earth were not only possible but probable. The explosion that resulted from such an impact could have ejected sufficient material from the mantle of the Earth to form the Moon.

According to the impact hypothesis, the material ejected from the Earth achieved escape velocity and therefore went into orbit around the Earth. The material initially consisted of vaporized silicates which cooled and condensed rapidly. Elements having low vapor pressures condensed first followed sequentially by other elements and compounds in order of their increasing vapor pressure. The most volatile elements (e.g. potassium, sodium, fluorine, etc.) condensed last and certain fractions of these elements escaped into interplanetary space. Similarly, all of the water and atmospheric gases (e.g., nitrogen, carbon dioxide, etc.) that were initially present also escaped.

This version of the fission hypothesis can explain both the low density of the Moon and the anomalously high density of the Earth (Section 3.5, Figure 3.6A) by the following scenario: The impact occurred after the Earth as well as the impactor had differentiated into an iron-nickel core and a solid silicate mantle composed primarily of pyroxene and olivine or their high-pressure derivatives. The impactor struck the Earth at a low angle of incidence, which expelled a large mass of terrestrial mantle, but left the core of the Earth intact. The impactor itself disintegrated in such a way that its mantle was stripped away allowing its core to be assimilated by the core of the Earth.

Consequently, the Moon formed primarily from mantle rocks derived from the Earth and from the impactor, which explains its low density consistent with the density of peridotite composed of pyroxene and olivine (Table 3.3). On the other hand, the core of the Earth was enlarged by assimilating the iron core of the impactor, which increased the bulk density of the Earth to $5.52 \, g/cm^3$ (Section 3.5). The bulk density of the Earth also increased because the loss of part of the mantle increased its core-to-mantle mass ratio. In this way, the impact hypothesis that was designed to explain the origin of the Moon also explains the anomalously high bulk density of the Earth.

The impactor hypothesis is a compromise that incorporates some aspects of each of the three competing hypotheses because:

1. The impactor did form elsewhere in the solar system.
2. The Moon did form in orbit around the Earth.
3. The Moon did form in large part from material that was ejected by the Earth.

This compromise was reached at a conference in October of 1984 in Kona on the island of Hawaii (i.e., twelve years after the return of the last Apollo mission). The oral presentations were subsequently published in 1986 in the form of a book edited by W.K. Hartmann, R.J. Phillips, and G.J. Taylor. The giant impactor hypothesis continues to be tested and has been modified in minor ways to accommodate geochemical data pertaining to specific elements. None of these modifications have yet invalidated the basic premise of the impactor hypothesis. Consequently, this hypothesis has been accepted in principle by a growing number of planetary scientists and, therefore, it has acquired the stature of a theory.

The current theory concerning the origin of the Moon postulates that it formed in orbit around the Earth by the accretion of solid particles that condensed from the mantle material ejected from the Earth. The impacts of these particles released sufficient heat to cause the newly-formed Moon to melt. As a result, the Moon differentiated by the segregation of immiscible liquids that formed a small core composed of metallic iron and nickel or of iron sulfide surrounded by a solid mantle of silicate minerals (pyroxene and olivine).

9.7 Celestial Mechanics

The physical dimensions of the Moon in Table 9.3 confirm that it is considerably smaller than the Earth with respect to its radius (1738 km), mass (7.349×10^{22} kg), and volume (2.20×10^{19} m^3). In addition, the density of the Moon (3.344 g/cm^3) and its surface gravity (0.17 compared to Earth) are both lower than those of the Earth. The albedo of the Moon (11%) is also less than that of Earth (39%) because the surface of the Moon is covered with regolith composed of fragments of basalt and

Table 9.3. Physical properties of the Moon and the Earth (Hartmann, 2005; Freedman and Kaufman, 2002)

	Moon	Earth
Radius, km	1738	6378
Mass, kg	7.349×10^{22}	5.98×10^{24}
Bulk density, g/cm^3	3.344	5.520
Volume, m^3	2.20×10^{19}	1.086×10^{21}
Visual geometric albedo, %	11	39
Atmosphere	none	nitrogen + oxygen
Surface gravity (Earth = 1.0)	0.17	1.0
Average surface temperature, °C	+130 (day) −180 (night)	+60 to −90 mean: +9
Surface material	regolith composed of fine grained rock fragments and glass spherules.	water, ice, granitic rocks

anorthositic gabbro which do not reflect sunlight as well as the surface of the oceans, ice sheets, and clouds do on the Earth. The low albedo means that, in proportion to its size, the Moon absorbs more sunlight than the Earth. However, the surface temperature of the Moon fluctuates widely (+130 to −180 °C) because the absence of an atmosphere does not allow heat to be transported from the illuminated side of the Moon to its dark side. The surface temperature of the Earth ranges only from +60 to −90 °C with a mean of 15 °C. The high and low temperatures of the surface of the Earth are determined primarily by the seasons and by the latitude of the observer rather than by the day/night cycle as on the Moon.

9.7.1 Spin-Orbit Coupling

Contrary to popular belief, the far side of the Moon is not in perpetual darkness because the

Moon does rotate on its axis. The data in Table 9.4 indicate that the period of rotation of the Moon is equal to its period of revolution in its orbit (i.e., both are 27.32 days). This phenomenon occurs between most planets and their closest satellites as a result of gravitational interactions by means of tides. Figure 9.13 illustrates the well-known fact that the Moon keeps the same face pointed toward the Earth as it revolves around it. The insert in the lower right corner of Figure 9.13 demonstrates that the Moon completes one rotation during each revolution. Therefore, this phenomenon is described as one-to-one spin-orbit coupling. The cause of this phenomenon is explained in more detail in Science Brief 9.12a.

Figure 9.13 also demonstrates which parts of the Moon are illuminated by the Sun as the Moon revolves around the Earth. For example, if the Moon is in position 1 in Figure 9.13, the entire face of the Moon seen from the Earth receives sunlight in the phase called "full Moon". As the Moon continues in its orbit, the area of the illuminated part decreases (wanes) until, at position 5, the face we first saw in position 1 is completely dark. As the Moon continues, the illuminated part of its face increases (waxes). When the Moon returns to point 1 in its orbit 27.32 days later, we again see a full Moon (weather permitting).

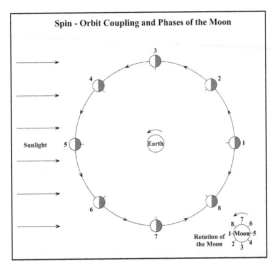

Figure 9.13. As the Moon revolves around the Earth from position 1 to 8 and back to 1, the hatch marks identify the part of the Moon that faces the Earth in position 1. Therefore, the Moon does in fact complete one rotation as it revolves once around the Earth (See inset in lower right). At the same time as the Moon revolves/rotates once, the Earth rotates on its axis 27.32 times, because we measure time by counting the number of rotations (days) of the Earth. The synchronous rotation and revolution of the Moon is described as 1:1 spin-orbit coupling (i.e., one rotation per revolution). The diagram also shows the changing illumination of the Moon by the Sun, which defines the phases of the Moon as seen from the Earth:1. Full Moon, 2. Waning gibbous, 3. Third quarter, 4. Waning crescent, 5. New Moon, 6. Waxing crescent, 7. First quarter, 8. Waxing gibbous

Table 9.4. Orbital parameters of the Moon and the Earth (Hartmann, 2005; Freedman and Kaufman, 2002)

	Moon	Earth
Average distance from the Earth to the Moon, 10^3 km	384.4	–
Average orbital velocity, km/s	1.022	29.79
Eccentricity of the orbit	0.0549	0.017
Sidereal period of revolution, days	27.322	365.256
Synodic period of revolution (new Moon to new Moon) days	29.531	
Period of rotation, days/hours	27.32 d	23.9345 h
Inclination of the orbit to the ecliptic	5.15°	–
Axial obliquity	6.68°	23.45°
Escape velocity km/s	2.38	11.2

9.7.2 Lunar Orbit

The orbit of the Moon is an ellipse in accordance with Kepler's first law. The semimajor axis (a) of its orbit (i.e., the average distance from the center of the Earth to the center of the Moon) in Table 9.4 is 384.4×10^3 km. The eccentricity (e) of the orbit is 0.0549. Since $e = f/a$, where f is the distance between the center of the ellipse and one of the focal points and "a" is the length of the semimajor axis,

$$f = e \times a = 0.0549 \times 384.4 \times 10^3 = 21.10 \times 10^3 \, km$$

Therefore, the distance between the center of the Earth and the center of the Moon in Figure 9.14 increases to $(384.4 + 21.10) \times 10^3 = 405.5 \times 10^3$ km when the Moon is at apogee and decreases to $(384.4 - 21.10) \times 10^3 = 363.3 \times 10^3$ km at perigee. This difference in the distance

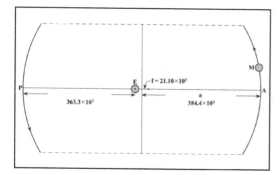

Figure 9.14. Partial view of the orbit of the Moon around the Earth in order to illustrate the average distance (a) between the Earth and the Moon (a = 384.4 × 10³ km) and the distances to the Moon at apogee (A) and perigee (P): A = 405.5 × 10³ km; P = 363.3 × 10³ km. The distance (f) between either one of the foci and the center of the ellipse is f = 21.10 × 10³ km. This distance may appear to be small, but it causes the distance between the Earth and the Moon to vary by 10.9 %

between the Earth and the Moon from apogee to perigee amounts to 10.9 % during each revolution of the Moon.

9.8 Tides

Seafaring peoples along the oceans have known for thousands of years that the water level along the coast rises and falls in daily cycles consisting of two high tides and two low tides. Boat owners know that the safest time to get their sailboats into a harbor is when the tide is rising and the safest time to leave a harbor is after the high tide has peaked and the water is flowing back into the ocean. The height of tides depends not only on astronomical factors, but is also influenced by the shape of the coast and by the local topography of the seafloor. In general, bays and the mouths of estuaries, where harbors are located, have higher tides than places where the coast is straight.

The mathematical analysis of the ocean tides raised by the Moon on the Earth was originally worked out in 1884 by George Darwin (1845–1912), the Plumian Professor of Astronomy and Experimental Philosophy at Cambridge University and second son of Charles Darwin. George Darwin advocated the hypothesis that the Moon formed from matter pulled out of the still-molten Earth by solar tides. Although this hypothesis has been discarded,

Darwin's mathematical analysis of the gravitational interactions between the Earth, the Moon, and the Sun is still the definitive solution of this mathematical problem.

9.8.1 Ocean Tides

In order to illustrate the explanation of tides in the oceans, we imagine that we place three identical billiard balls along a line that extends from the center of the Moon to the center of the Earth and extrapolate that line to the far side of the Earth as shown in Figure 9.15. For the sake of this discussion we say that the gravitational force exerted by the Moon on ball 1, located on the surface of the Earth facing the Moon, is equal to 6 arbitrary units. The force exerted by the Moon on ball 2 at the center of the Earth is only 4 arbitrary units because the distance from ball 2 to the center of the Moon is greater than that of ball 1. Ball 3 on the opposite side of the Earth is acted upon by a still weaker force of only 2 arbitrary units.

The net force (F_1) exerted by the Moon on ball 1 in Figure 9.15 relative to the force it exerts on ball 2 at the center of the Earth is:

$$F_1 = 6 - 4 = +2 \text{ arbitrary units}$$

The plus sign indicates that the force F_1 is directed toward the Moon. Similarly, the net force (F_3) acting on ball 3 relative to the force acting on ball 2 at the center of the Earth is:

$$F_3 = 2 - 4 = -2 \text{ arbitrary units}$$

and the minus sign indicates that the force is directed away from the Moon.

Although this argument is contrived, it provides a plausible explanation for the observed fact that the Moon raises two high tides in the oceans of the Earth: one on the side facing the Moon ($F_1 = +2$ arbitrary units) and one on the opposite side of the Earth ($F_3 = -2$ arbitrary units). Now comes the punch line: The two bulges of water in the ocean remain fixed in their position along the line connecting the center of the Earth to the center of the Moon. Therefore, the daily tidal cycle results from the rotation of the Earth such that a selected spot on the coast in Figure 9.16 experiences

Lunar Tides of the Oceans of the Earth

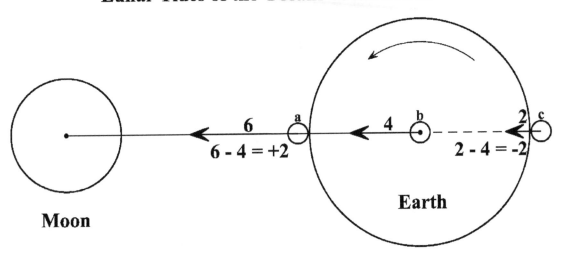

Figure 9.15. View of the Earth and the Moon from above the plane of the ecliptic. The length of the straight arrows suggest the magnitudes of gravitational force (in arbitrary units) exerted by the Moon on three identical billiard balls placed as shown. The diagram provides a plausible explanation for why the net force acting on ball 1 is directed toward the Moon and why the net force acting on ball 3 is directed away from the Moon. As a result, two bulges of water form in the oceans on opposite sides of the Earth. Not drawn to scale

different tidal stages every 6 hours in a 24 hour day. Starting with high tide at $t = 0$ h, low tide at $t = 6$ h, high tide at $t = 12$ h, low tide at $t = 18$ h, and back to high tide at $t = 24$ h. To repeat the punch line: The bulges of water in the oceans remain stationary and the tidal cycle is caused by the rotation of the Earth through these bulges. However, the bulge facing the Moon is actually displaced in the direction of the rotation of the Earth. As a result, gravitational forces have slowed the rate of rotation of the Earth and have caused the Moon to recede from the Earth (Section 7.3.2 and Science Brief 9.12.1).

9.8.2 Body Tides

The tidal forces that affect the water in the oceans also cause bulges to form in the rocks of the crust of the Earth. These bulges maintain their positions on opposite sides of the Earth which means that the rocks of the crust are continually flexed as the Earth rotates. As a result, heat is generated in the rocks of the crust by internal friction. This process is an important source of heat not only for the Earth-Moon system but also for the satellites of other planets

in the solar system (e.g., the Galilean satellites of Jupiter).

The tidal forces the Moon exerts on the Earth act to elongate the body of the Earth along the line that connects the center of the Moon to the center of the Earth. Similarly, the tidal forces exerted by the Earth cause the body of the Moon to deform. The resulting strain on the Moon increases as it approaches the perigee of its orbit and decreases as it recedes toward the apogee because of progressive changes in the distance between the Earth and the Moon. In summary, the body tide caused by the Moon generates heat in the crust of the Earth as it rotates on its axis. The body tide exerted by the Earth on the Moon generates heat because the shape of the Moon is deformed as it approaches the Earth at perigee and relaxes as it recedes to the apogee of its orbit.

Consideration of the ocean and body tides raised by the Moon on the Earth leads to the conclusion stated in Section 9.8.1 that these tides are slowing the rotation of the Earth and are causing the Moon to recede from the Earth (Hartmann, 2005, p. 58). Extrapolating these trends backward in time indicates that the Earth originally rotated faster than it does at present (Section 7.3.2) and that the Moon was originally

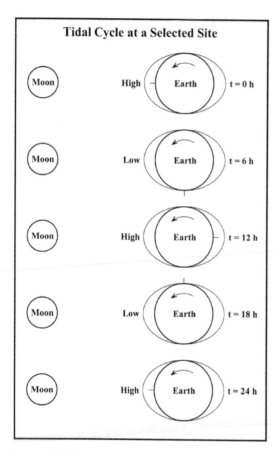

Tidal Cycle at a Selected Site

Figure 9.16. Sequence of high and low tides experienced by an observer at a selected point (hatch mark) along the coast during one 24-hour rotation of the Earth. Not drawn to scale

closer to the Earth than it is at the present time (Eriksson and Simpson, 2000). This is the reason why in Figure 6.2 Chesley Bonestell painted an oversized Moon over the magma ocean of the young Earth. The stress imposed on the crust of the Earth by the body tides may, in some cases, trigger earthquakes and volcanic eruptions. In fact, the intense volcanic activity of Io, the first Galilean satellite of Jupiter, is the result of tidal heating caused by the gravity of Jupiter and the eccentricity of the orbit of Io.

9.9 Research Stations on the Moon

The rapid evolution of our capability to explore the planets and their satellites inevitably raises the questions: Why are we doing it and what benefits do we expect to receive from the efforts we make? Space exploration is an expensive and often dangerous business that should not be undertaken lightly without weighing the risks against the potential benefits (See also Section 7.2.11).

9.9.1 Weighing the Risks and Benefits

The surface of the Moon is not an easy place for humans to live and work. The many hazards that must be controlled include:
1. The absence of an atmosphere which creates the threat of instantaneous and fatal decompression.
2. High intensity of cosmic-ray flux which damages cells.
3. Extreme range of temperatures during the 27.3 lunar day and night which exceeds human tolerance.
4. The low gravity of the Moon can cause bone loss and atrophication of muscles of lunar astronauts.
5. Frequent impacts of small meteorites because of the absence of a protective atmosphere on the Moon.
6. The dust in the lunar soil can damage mechanical and electronic devices.
7. Rescue missions from the Earth will take days to reach the Moon in case of an emergency at a lunar station.

These and other hazards that have not yet been specified must be balanced by the benefits that could be realized by establishing a base on the surface of the Moon. The potential benefits include:
1. Research to prepare for space journeys to distant parts of the solar system:
   a. The development of construction techniques using ceramics manufactured from lunar soil using solar or nuclear energy.
   b. Growth of food crops in lunar soil protected by transparent shelters.
   c. Development of pressurized all-terrain-vehicles for travel and exploration of the lunar surface.
   d. Construction of shelters placed underground for protection against meteorite impacts and cosmic rays.

e. Exploration to detect and recover natural resources on the Moon, including: water, metals, building stone, and fuel (helium-3).

2. The low gravity of the Moon, and hence its lower escape velocity compared to the Earth, requires less energy to launch spacecraft to other planets.

3. Optimal conditions for astronomical research using optical, infrared, and microwave telescopes or receivers placed on the lunar far side for shielding from electronic noise emanating from the Earth and because the absence of an atmosphere improves the clarity of optical images.

4. Repository of important cultural and scientific records to preserve them for posterity.

5. Training of astronauts and citizen volunteers (tourists) whose reports will prepare others to consider moving to space colonies in the future.

6. The most important long-range objective for establishing self-sufficient and viable colonies on the Moon, on Mars, and on any other habitable body in the solar system is to assure the preservation of human life in the solar system in case the Earth becomes uninhabitable as a result of full-scale nuclear war, extreme greenhouse warming, impact of a near-Earth asteroid, or by the uncontrolled spread of an incurable disease.

Therefore, the Moon is not only a challenge to human ingenuity, but it is also our lifeboat that may help humans to survive in case the Earth becomes uninhabitable. For these and other ethical reasons, the Moon should never be used to store nuclear and industrial waste or to launch military missiles at targets on Earth or at spacecraft or satellites of adversaries.

The visionary Russian explorer of space Konstantin Eduardovitch Tsiolkovsky (Section 1.2) saw the future of humans in an even larger perspective. He anticipated that the Earth would become uninhabitable when the Sun begins to expand and eventually becomes a white dwarf star. Therefore, he advocated that humans ultimately must find another habitable planet orbiting a suitable star in our galactic neighborhood. We know that the Sun will, in fact, expand into a red giant in about 5 billion years (Section 23.3.1). We also know that the Sun's surface temperature will start rising long before it enters the red-giant phase of its evolution. The resulting increase in the surface temperature of the Earth may cause the oceans to boil away thus turning the Earth into a hot and dry desert planet like Venus is today.

Tsiolkovsky did not know, as we do, that the death of our Sun is not imminent. Nevertheless, NASA has given high priority to the discovery of potentially habitable planets that may orbit nearby stars and a serious effort is underway to Search for Extra-Terrestrial Intelligence (SETI) because some extrasolar planets in the Milky Way galaxy may already be inhabited by intelligent life forms (Section 24.4 and 24.5). Although the search for extraterrestrial intelligence has been going on for more than 40 years, no life forms of any kind have been discovered anywhere in our solar system and in the Milky Way galaxy. However, we have a saying in science that applies to the search for extraterrestrial intelligence: Absence of evidence is not evidence of absence. Therefore, the search continues.

9.9.2 Utilization of Lunar Resources

The Moon and other solid bodies in the solar system contain certain natural resources that have potential value to colonists or to the crews of spacecraft en route to other destinations (Schrunk et al., 1999). For example, the regolith that covers the surface of the Moon contains atoms of helium-3 that has been embedded in rocks and minerals by the solar wind. These kinds of atoms of helium are rare on the Earth but could be used to fuel commercial fusion reactors, which would reduce our reliance on the combustion of fossil fuel. In addition, fusion reactors could also provide heat and electricity to research stations on the Moon or propel rocket-powered spacecraft that stop to refuel on their way from Earth to Mars and beyond (For more details see Science Briefs 9.12.2 and 3).

The Moon may also contain surface deposits of water ice in deep craters in the polar regions that never receive direct sunlight. Whether such deposits actually exist is not known at the present time (i.e., 2005/2006). Another valuable resource on the surface of the Moon is sunlight because it can be converted directly into electricity and/or heat for industrial processes such as the manufacture of glass by melting of lunar soil and for the extraction of oxygen from silicate or oxide minerals. However, the rotation of the Moon limits the availability of sunlight except for high plateaus and peaks in the polar regions which receive nearly continuous sunlight in spite of the rotation of the Moon.

Therefore, the most desirable sites for a lunar base in Figure 9.17 are in the polar regions of the Moon because of the potential presence of water ice combined with nearly continuous sunlight. In the best-case scenario, a lunar base on a high peak in the Aitken basin at the south pole of the Moon may have access to water needed to sustain human life. In addition, water molecules can be split by electrolysis into oxygen and hydrogen gas, which can be liquefied for use as rocket fuel.

The foregoing examples illustrate the kinds of natural resources that are, or may be, available on the Moon. Guidelines provided by treaties and resolutions of the United Nations encourage international cooperation in the establishment of a lunar base at the most advantageous site and in the recovery and utilization of the natural resources of the Moon. Current plans by NASA call for the establishment of a lunar base in 2018 using a newly designed Crew Exploration Vehicle, a Lunar Surface Access Module, and a Heavy-Lift Vehicle (Reddy, 2006).

9.10 Space Law

The discovery, evaluation, and exploitation of the natural resources of the Moon are the subject of multilateral treaties and resolutions adopted by the United Nations. These international agreements are intended to assure that the people of all countries on Earth share in the benefits to be derived from these resources. The treaties also forbid the storage of nuclear weapons and military exercises on the Moon, and they require that the Moon and all other bodies in the solar system be used only for peaceful purposes such as research and exploration.

9.10.1 United Nations: Laws for States

On December 13 of 1963 the General Assembly of the United Nations adopted a resolution (No. 18) entitled: "Declaration of Legal Principles Governing the Activities

of States in the Exploration and Use of Outer Space." The principles identified in this document were subsequently elaborated in five multilateral treaties:

1. The Treaty on Principles Governing the Activities of States in the Exploration and Use of Outer Space, Including the Moon and Other Celestial Bodies. (Effective date: October 10, 1967). Excerpts of this treaty appear in Science Brief 9.12d.
2. The Agreement on the Rescue of Astronauts and the Return of Objects Launched into Outer Space (Effective date: December 3, 1968)
3. The Convention on International Liability for Damage Caused by Space Objects (Effective date: September 1, 1972).
4. The Convention on Registration of Objects Launched into Outer Space. (Effective date: September 15, 1976).
5. The Agreement Governing the Activities of States on the Moon and Other Celestial Bodies. (Effective date: July 11, 1984).

In addition to resolution 18 mentioned above, the United Nations adopted four resolutions concerning matters arising from the use of space and to encourage international cooperation in space exploration:

The Principles Governing the Use by States of Artificial Earth Satellites for International Direct Television Broadcasting. (Resolution 37/92, adopted December 10,1982)

The Principles Relating to Remote Sensing of the Earth from Outer Space (Resolution 41/65, adopted December 3, 1986).

The Principles Relevant to the Use of Nuclear Power Sources in Outer Space (Resolution 47/68, adopted December 14, 1992).

The Declaration on International Cooperation in the Exploration and Use of Outer Space for the Benefit and in the Interest of All States, Taking into Particular Account the Needs of Developing Countries (Resolution 51/122, adopted December 13, 1996).

These treaties and resolutions of the United Nations are the foundation for future agreements dealing more specifically with human activities on the Moon and on Mars. These agreements are modeled after the Antarctic Treaty that prohibits exploitation of natural resources and does not recognize past claims of ownership advanced

Figure 9.17. A temporary base on the Moon inhabited by two lunar astronauts who are returning to their shelter after an excursion on their lunar rover. The spacecraft which brought them to the Moon and which will return them to the Earth stands ready for take-off in the background. The base shown in this artist's view is located in an impact crater or mare close to the lunar highlands. The regolith at this site consists of ground-up lunar rocks and meteorites containing angular boulders of basalt and anorthositic gabbro that were ejected from nearby meteorite-impact craters. Additional views of a future research station on the Moon appear in an article by Spudis (2003). Courtesy of the Johnson Space Center in Houston (JSC 2004-E-18833, April 2004)

by certain nations based either on active exploration of the continent (e.g., UK, Russia, and Norway) or on territorial proximity, or both (e.g., Australia, New Zealand, Chile, and Argentina). This treaty has been scrupulously observed by all treaty nations in large part because economic considerations have been ruled out. How the issue of economic exploitation of the natural resources of the Moon will be dealt with remains to be seen.

9.10.2 NASA: Code of Conduct for Astronauts

The treaties and resolutions of the United Nations do not provide enough specific guidance for the adjudication of conflicts that may arise when space traffic increases, or when fights occur among the astronauts of a spacecraft on a long mission. In order to regulate the activities of astronauts on the International Space Station (ISS), NASA established a Code of Conduct for the crew in December of 2000 (Tenenbaum, 2002). According to this code, the conduct:

"...shall be such as to maintain a harmonious and cohesive relationship among ISS crewmembers and an appropriate level of mutual confidence and respect through an interactive, participative, and relationship – oriented approach which duly takes into account the international and multicultural nature of the crew and mission."

In addition, the code specifies that crewmembers must not: "act in a manner which results in or creates the appearance of :
1. Giving undue preferential treatment to any person or entity...and/or
2. adversely affecting the confidence of the public in the integrity of, or reflecting unfavorably in a public forum on, any ISS partner, partner state or cooperating agency."

These statements do not specifically prohibit any actions by crewmembers of the ISS except that they are not permitted to speak badly of NASA or of an ISS partner at a public forum. However, the Intergovernmental Agreement concerning the ISS does give jurisdiction for alleged misdeeds to the country of the accused crew member. This provision is similar to the Antarctic Treaty which specifies that violations of its provisions are punished by the country of the alleged perpetrator.

9.11 Summary

The Moon is familiar to all of us and has been closely observed for thousands of years. It is the only body in the solar system on which astronauts have landed. The samples of rocks and soil the astronauts collected have helped to determine the age and geologic history of the Moon. According to the dominant theory, the Moon formed by accretion of solid particles derived from a flat disk (or ring) that originated after the Earth was hit by a large solid body only a few tens of millions of years after the Earth had itself formed by accretion of planetesimals and had differentiated into a core, mantle, and crust.

During the early part of its history, the Moon was bombarded by large bodies that excavated large basins on its near side. These basins were subsequently filled with basalt lava during an episode of volcanic activity that lasted more than 800 million years (4.0 to 3.2 Ga) and affected almost the entire near side of the Moon. Consequently, the present surface of the Moon consists of heavily cratered highlands composed of anorthositic gabbro and sparsely cratered basalt plains (maria). In addition, the entire surface of the Moon is covered by a thin layer of regolith consisting of rock fragments and glass beads derived from meteorite impacts.

Seismic evidence indicates that the lithosphere of the Moon is about 1000 km thick and that the underlying asthenosphere and core are at least partially molten. Even though the interior of the Moon is still hot, the Moon has been geologically and tectonically inactive for about three billion years, presumably because the rigid lithosphere has not allowed magma to form by local decompression of the mantle.

The rate of rotation of the Moon has been slowed by tidal interactions in the Earth-Moon system such that the Moon now has one-to-one spin-orbit coupling, which means that the Moon rotates on its axis only once for each time it orbits the Earth. Gravitational interactions between the Earth and the Moon cause the water in the oceans to form two bulges on opposite sides of the Earth. Consequently, an observer on the coast experiences two high tides and two low tides during a 24-hour day/night rotation of the Earth. Tidal bulges also occur in the rocks of the crust of the Earth, which generate heat by internal friction as the Earth rotates. These gravitational interactions are also slowing the rate of rotation of the Earth and are increasing the radius of the orbit of the Moon.

The exploration of the Moon by astronauts is scheduled to resume in 2018 in preparation for the construction of a research station on the lunar surface. The purpose of such a station will be to prepare for landings of astronauts on Mars. The exploitation of lunar resources in support of a research station on the Moon will require significant advances in technology and will raise legal issues for which there is no precedent.

The natural resources that may be exploited on the Moon include water, helium-3, and sunlight. The existence of water is problematic because the Moon is inherently dry and any ice on its surface would have been deposited by comets. Helium-3, which is implanted by the solar wind in the rock fragments and glass particles of the lunar regolith, could be used in nuclear fusion reactors that could generate large amounts of energy. Sunlight can also be used both as a source of heat and to generate electricity. A few sites in the polar regions receive almost continuous sunshine during the 27.3-day cycle of lunar day and night.

The landing of astronauts on the surface of the Moon during the Apollo Program between 1969 and 1972 was a brilliant achievement of technology, management, and personal courage that initiated the ongoing exploration of the entire solar system. As a result, we recognize now more clearly than ever before that the planet on which we live is the only place where we can live in safety and in comfort.

The poet Archibald McLeish expressed this insight with these eloquent words:

"To see the Earth as it truly is, small and blue and beautiful in that eternal silence where it floats, is to see ourselves as riders on the Earth together, brothers who know now they are truly brothers."

(The Apollo Moon Landings, VHS, Finlay-Holiday, Whitier, CA)

9.12 Science Briefs

9.12.1 Tidal Interactions in the Earth-Moon System

The force of gravity exerted by the Moon on the oceans causes two bulges to form on the side of the Earth facing the Moon and on the opposite side. The resulting elongation of the ocean along the line connecting the center of the Earth to the center of the Moon was discussed in Section 9.8.1 and is illustrated in Figures 9.15 and 9.16.

We now refine the explanation of ocean tides by including the effect of the rotation of the Earth, which causes the bulges in the oceans to be displaced from the line connecting the Earth and the Moon (Hartmann, 2005). This effect in Figure 9.18 causes bulge A on the Earth to pull the Moon forward in its orbit around the Earth. As the Moon speeds up, the radius of its orbit increases. At the same time, the gravitational force exerted by the Moon on bulge A slows the rate of rotation of the Earth. Another way to describe this process is to say that angular momentum is transferred from the Earth to the Moon. As the rate of rotation of the Earth decreases, the radius of the orbit of the Moon increases and angular momentum in the Earth-Moon system is thereby conserved.

The gravitational force exerted by the Earth on the Moon causes the shape of the Moon to be elongated in such a way that the long axis points toward the Earth. The gravitational interaction between the Earth and the tidal bulge on the part of the Moon that faces the Earth has acted as a brake on the rotation of the Moon. There is good reason to assume that the Moon rotated more rapidly after it had formed in orbit around the Earth (Section 9.7.1). The tidal interaction between the Earth and the bulge on the near side of the Moon slowed the rotation of the Moon and forced it into a state of synchronous rotation or one-to-one spin-orbit coupling (Eriksson and Simpson, 2000).

To summarize, tidal interactions in the Earth-Moon system continue to slow the rotation of the Earth and are causing the Moon to recede. Tidal interactions have also reduced the rate of rotation of the Moon and forced it into the present state of one-to-one spin-orbit coupling. The continuing decrease of the rate of rotation of the Earth implies that in the past it rotated more rapidly than it does at the present time. The consequences of a high rotation rate were discussed in Section 7.3.2 and Science Brief 7.6.6.

9.12.2 Energy Released by Hydrogen Fusion

At the very high temperatures and pressures that occur in the cores of stars on the Main Sequence, nuclei of hydrogen atoms (protons) collide with each other with sufficient force to cause them to fuse together to form an atomic version (i.e., an isotope) of hydrogen called deuterium whose nucleus is composed of one proton and one neutron (Section 4.2). The nucleus of deuterium can collide with another proton to form the nucleus of helium-3 which consists of two protons and one neutron. Finally, when two nuclei of helium-3 collide, they form the nucleus of helium-4 (two protons and two neutrons) and release two protons, which return to the hot proton gas in the core of the star. Every time a nucleus of helium-4 is produced by these reactions, a large amount of energy equal to 19.794 million electron volts (MeV) is released.

The energy expressed in units of million electron volts can be converted into ergs by the relation:

$$1\,MeV = 1.60207 \times 10^{-6}\,erg$$

$$19.794\,MeV = 31.711 \times 10^{-6}\,ergs$$

The production of one gram-atomic weight (mole) of helium-4 releases an amount of energy (E) equal to:

$$E = 31.711 \times 10^{-6} \times A\,ergs$$

where $A = 6.022 \times 10^{23}$ atoms/mole (Avogadro's Number).

Therefore, $E = 190.96 \times 10^{17}$ ergs/mole of helium-4.

Given that one mole of helium-4 weighs 4.0026g, the amount of energy released by the production of one gram of helium-4 is:

$$E = \frac{190.96 \times 10^{17}}{4.0026}$$

$$= 47.710 \times 10^{17}\,ergs/gram\ of\ helium\text{-}4.$$

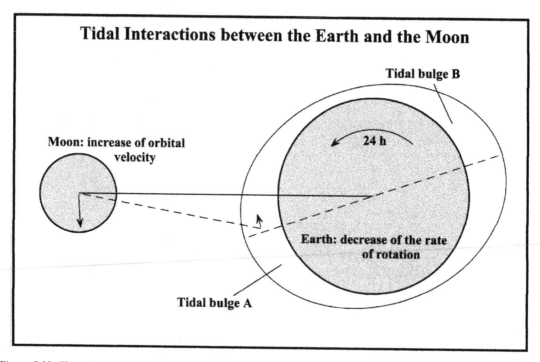

Tidal Interactions between the Earth and the Moon

Figure 9.18. Illustration of the effects of the tidal interactions between tidal bulge A on the Earth and the Moon. The gravitational force exerted by tidal bulge A increase the orbital velocity of the Moon. The gravitational force exerted by tidal bulge B counteracts this effect but does not cancel it because tidal bulge A is closer to the Moon than tidal bulge B. The Moon exerts an opposite gravitational force on tidal bulge A that decreases the rate of rotation of the Earth. The result of these tidal interactions is that the Moon recedes from the Earth and the number of days in one year is decreasing (i.e., the period of rotation of the Earth is increasing) Adapted from Figure 3.11 of Hartmann (2005)

This is a very large amount of energy whose magnitude can be comprehended by converting it to the weight of TNT that releases the same amount of energy as the production of only one gram of helium-4 in the core of a Main-Sequence star. Given that 1 tonne of TNT $= 4.2 \times 10^{16}$ ergs,

$$E = \frac{47.10 \times 10^{17}}{4.2 \times 10^{16}} = 113.59 \text{ tonnes of TNT}$$

The result is that the production of one gram of helium-4 by nuclear fusion reactions in stars releases the same amount of energy as the explosion of about 114 tonnes of TNT. This calculation helps to make the point that nuclear fusion reactors could provide very large amounts of energy.

9.12.3 Fusion Reactors Based on Lunar Helium-3

The solar wind consists of a mixture of the nuclei of hydrogen, deuterium, helium-3, and helium-4. When these particles impact on the surface the Moon, they are trapped in the crystals of the minerals that occur in the lunar soil. This process has been occurring for a very long time (about 4.5 billion years) and substantial quantities of these solar gases have accumulated.

Space engineers have proposed that the helium-3 and other gases can be recovered by heating samples of the lunar soil almost to the melting temperature. Two nuclei of helium-3 recovered from the lunar soil can be fused to produce a nucleus of helium-4 plus two nuclei of hydrogen and a large amount of energy. The desired nuclear fusion reaction can be stated in the form of an equation:

$$_2^3He + _2^3He \rightarrow _2^4He + 2_1^1H + 12.859\,MeV \quad (9.1)$$

where $_2^3He$ = nucleus of helium-3

$_2^4He$ = nucleus of helium-4

$_1^1H$ = nucleus of hydrogen (proton)

The energy originates from the conversion of mass into energy in accordance with Einstein's equation: $E = mc^2$

where E = energy in Joules

m = mass in kilograms

c = speed of light (3×10^8 m/s).

The energy released by this reaction is equivalent to about 74 tonnes of TNT per gram of helium-4 produced.

These kinds of reactions release very large amounts of energy in all Main-Sequence stars like the Sun. Unfortunately, they require an extremely high temperature of about ten million degrees Celsius. Such temperatures can be generated on Earth only for very short periods of time (e.g., in the thermonuclear hydrogen bomb). In spite of continuing efforts, we have not yet succeeded in building a *fusion reactor* that can produce energy by converting hydrogen into helium. The availability of helium-3 on the Moon may offer a shortcut to the production of thermonuclear energy on the Earth as well as on the Moon. Such fusion reactors would have a distinct advantage over U-fission reactors because they produce only stable atoms of hydrogen and helium, that will not contaminate the terrestrial or lunar environment. Nuclear reactors based on *U-fission* produce radioactive waste products that emit energetic radiation for tens of thousands of years.

9.12.4 Treaty on Principles Governing the Activities of States in the Exploration and Use of Outer Space, Including the Moon and Other Celestial Bodies

(Articles 1 to 4 out of a total of 17).

Article 1

The exploration and use of outer space, including the Moon and other celestial bodies, shall be carried out for the benefit and in the interest of all countries, irrespective of their degree of economic or scientific development, and shall be the province of all mankind.

Outer space, including the Moon and other celestial bodies, shall be free for exploration and use by all States without discrimination of any kind, on a basis of equality and in accordance with international law, and there shall be free access to all areas of celestial bodies.

There shall be freedom of scientific investigation in outer space, including the Moon and other celestial bodies, and States shall facilitate and encourage international cooperation in such investigation.

Article 2

Outer space, including the Moon and other celestial bodies, is not subject to national appropriation by claim of sovereignty, by means of use or occupation, or by any other means.

Article 3

States Parties to the Treaty shall carry on activities in the exploration and use of outer space, including the Moon and other celestial bodies, in accordance with international law, including the Charter of the United Nations, in the interest of maintaining international peace and security and promoting international cooperation and understanding.

Article 4

States Parties to the Treaty undertake not to place in orbit around the Earth any objects carrying nuclear weapons or any other kinds of weapons of mass destruction, install such weapons on celestial bodies, or station such weapons in outer space in any other manner.

The Moon and other celestial bodies shall be used by all States Parties to the Treaty exclusively for peaceful purposes. The establishment of military bases, installations and fortifications, the testing of any type of weapons and the conduct of military maneuvers on celestial bodies shall be forbidden. The use of military personnel for scientific research or for any other peaceful purposes shall not be prohibited. The use of any equipment or facility necessary for peaceful exploration of the Moon and other celestial bodies shall also not be prohibited.

9.13 Problems

1. Write an essay to explain why the polar regions of the Moon are the most favorable sites for the establishment of a lunar base.
2. Calculate the escape velocity of the Moon in units of km/s (Appendix 1 and Table 9.3). (Answer: 2.38 km/s).
3. Calculate the distance between the apogee and the perigee of the orbit of the Moon measured along a straight line through the center of the ellipse (Answer: 728.8×10^3 km).
4. Several subsamples of a lunar meteorite collected in Antarctica were analyzed for dating by the Rb-Sr method. The analytical results for two subsamples are:

Subsample	$^{87}Sr/^{86}Sr$	$^{87}Rb/^{86}Sr$
PX1, R	0.703415	0.07779
PL2, R	0.699193	0.001330

The values define two points in the x-y plane where $^{87}Rb/^{86}Sr \equiv x$ and $^{87}Sr/^{86}Sr \equiv y$.

a. Calculate the slope (m) of a straight line connecting the two points. (Answer: m = 0.055218).

b. Calculate a Rb-Sr date from the slope m using equation 8.11 in Science Brief 8.11b:

$$t = \frac{\ln(m+1)}{\lambda}$$

where $\lambda = 1.42 \times 10^{-11} y^{-1}$ (decay constant of ^{87}Rb) and ln is the natural logarithm. (Answer: $t = 3.785 \times 10^9$ y).

9.14 Further Reading

Aldrin B, McConnell M (1991) Men from Earth. Bantam Falcon Books, New York

Cassidy WA (2003) Meteorites, ice, and Antarctica. Cambridge University Press, New York

Cernan E, Davis D (1999) The last man on the Moon. St Martin's Press, New York

Collins M (1974) Carrying the fire: An astronaut's journeys. Ballantine Books, New York

Eriksson KA, Simpson EL (2000) Quantifying the oldest tidal record: The 3.2 Ga Moodies Group, Barberton Greenstone belt, South Afirca. Geol 28:831–834

Eugster O (1989) History of meteorites from the Moon collected in Antarctica. Science 245:1197–1202

Freedman RA, Kaufmann III WJ (2002) Universe: The solar system. Freeman, New York

Glass BP (1982) Introduction to planetary geology. Cambridge University Press, Cambridge, UK

Hansen JR (2005) First man: The life of Neil Armstrong. Simon and Schuster, New York

Harland DM (2000) Exploring the Moon; The Apollo expeditions. Springer-Praxis, New York

Hartmann WK, Phillips RJ, Taylor GJ (eds) (1986) Origin of the Moon. Lunar Planet. Institute, Houston, TX

Hartmann WK (2005) Moons and planets. 5th edn. Brooks/Cole, Belmont CA

Keil K, Papike JJ (1983) Meteorites from the Earth's Moon. Special Session, 14th Lunar and Planetary Science Conference, 17 Mar 1983. Lunar and Planetary Institute Contrib. 501, Houston, TX

Kluger J (1999) Moon hunters: NASA's remarkable expeditions to the ends of the solar system. Simon and Schuster, New York

Light M (2002) Full Moon. Knopf, New York

Lindsay H (2001) Tracking Apollo to the Moon; The story behind the Apollo Program. Springer-Verlag, New York

Reddy F (2006) NASA's next giant leap. Astronomy Feb:62–67

Schrunk D, Sharpe B, Cooper B, Madhu Thangavelu (1999) The Moon: Resources, future development and colonization. Wiley and Praxis Pub., Ltd., Chichester, UK

Snyder GA, Lee D-C, Taylor LA, Halliday AN, Jerde EA (1994) Evolution of the upper mantle of the Earth's Moon: Neodymium and strontium isotopic constraints from high-Ti mare basalt. Geochim Cosmochim Acta 58(21):4795–4808

Spudis PD (2003) Harvest the Moon. Astronomy, June, 42–47

Tenenbaum D (2002) Outer space laws. Astronomy, February, 40–44

Toksöz MN, Dainty AM, Solomon SC, Anderson KR, (1973) Velocity structure and evolution of the Moon. Proceed. Fourth Lunar Sci Conf, Geochim Cosmochim Acta, Supplement 4, vol 3, 2529–2547

Turcotte DL, Kellog LH (1986) Implications of isotopic data for the origin of the Moon. In: Hartmann W.K., Phillips R.J., Taylor G.J. (eds) Origin of the Moon, 311–329. Lunar Planet Sci Inst, Houston, TX, pp 311–329

Turnill R (2003) The Moon landings: An eyewitness account. Cambridge University Press, New York

Wilhelms DE, McCauley JF (1971) Geologic map of the near side of the Moon. Geologic Atlas of the Moon, I-703. U.S. Geological Survey, Washington, DC

Wilhelms DE (1993) To a rocky Moon; A geologist's history of lunar exploration. University Arizona Press, Tucson, AZ

Wood JA (1986) Moon over Mauna Loa: A review of hypotheses of formation of Earth's Moon. In: Hartmann W.K., Phillips R.J., Taylor G.J. (eds) Origin of the Moon. Lunar Planet. Sci. Institute, TX, pp 17–55

Mercury: Too Hot for Comfort

Mercury is the smallest of the terrestrial planets, but it is larger than the Moon. Its surface is covered with impact craters and, like the Moon, it does not have an atmosphere. In spite of these and other similarities, Mercury has many unique properties that distinguish it from the Moon and from all other terrestrial planets.

Although its proximity to the Sun (0.387 AU) makes Mercury difficult to observe from the Earth, it was known to the Sumerians about 5000 years ago. The Greeks called it Apollo when it was visible at sunrise and Hermes when it appeared at sunset. Hermes was the messenger of Zeus (his father) and lived with him on Mount Olympus. The Romans later referred to Hermes as Mercurius because of its rapid motion. We recall that the planet Mercury has the highest average orbital speed (47.9 km/s) of any planet in the solar system (Figure 3.3) in accordance with Kepler's third law (Section 7.1 and Table 7.1).

Very little was known about the surface of Mercury until 1974/75 when the robotic spacecraft Mariner 10 imaged part of its surface during three flybys on March 29 and September 21, 1974, and on March 16, 1975. The resulting mosaic in Figure 10.1 shows that the surface of Mercury is heavily cratered but lacks the large circular mare basins of the Moon. Instead, the surface of Mercury is characterized by sparsely-cratered plains that occur locally. However, only half of the surface of Mercury is visible in the images recorded by Mariner 10.

A new spacecraft called *Messenger* was launched by NASA on May 11, 2004, and is expected to go into orbit around Mercury in 2009. It will continue mapping the surface of Mercury and attempt to determine whether deposits of water ice are present in deep impact craters in the polar regions of Mercury. The European Space Agency (ESA) is also building a spacecraft called Bepi Colombo for launch to Mercury in 2009. This spacecraft was named after the Italian mathematician Giuseppe (Bepi) Colombo who calculated the trajectory that allowed Mariner 10 to reach Mercury in 1974.

Books and articles about Mercury include those by Murray and Burgess (1977), Greeley (1985), Strom (1987, 1999), Vilas et al. (1988, 1999), Harmon (1997), Nelson (1997), and Robinson and Lucey (1997). An atlas of the surface of Mercury was published by Davies et al. (1978).

10.1 Surface Features

The data concerning Mercury sent back by Mariner 10 in 1974 provided information about the abundance of impact craters, the mercurial regolith, the absence of atmospheric gases including water, the extreme temperatures that occur on the surface of this planet, and about the existence of a weak magnetic field.

10.1.1 Impact Craters and Cliffs

The images returned by Mariner 10 indicate that the surface of Mercury is covered with impact craters that resemble those on the surface of the Moon. The craters on Mercury are named after famous composers, painters, poets, etc. For example, Rodin (23°N, 17°W) is a large single-ring impact basin with a diameters of about 240 km. The crater Ibsen (23°S, 37°W; diameter: 160 km) has a smooth uncratered floor suggesting that it was flooded by lava flows. Several other large craters also appear to have been flooded by lava flows (e.g., Homer at 0°,

Figure 10.1. This mosaic of the surface of Mercury was assembled from images recorded in 1974/75 during flybys of Mariner 10. The north pole is at the top and about half of the multi-ring Caloris basin is visible at the left edge of the image. The surface of Mercury is heavily cratered but also includes more sparsely cratered plains formed by local eruptions of basalt lava flows. Strindberg and Van Eyck are old impact craters that were flooded by lava presumably of basaltic composition. Brontë and Degas are comparatively young overlapping craters surrounded by rays of ejecta. The Budh and Tir planitiae are examples of lava plains that are sparsely cratered and are therefore younger than the heavily cratered regions of Mercury located south of the Tir planitia and north of the crater Strindberg. (Courtesy of NASA and JPL/Caltech, Mariner 10, PIA03104: Photomosaic of Mercury)

38 W; Giotto at 13°N, 58°W, Chaikovskij at 8°N, 50°W, and so on). Craters with well preserved central peaks and bowl-shaped craters also occur in abundance on Mercury and are illustrated on the shaded relief maps prepared by the U.S. Geological Survey. These observations indicate that Mercury suffered the same intense bombardment in its early history as the Moon.

The largest impact crater on the surface of Mercury is the Caloris multiring basin parts of which are visible in Figure 10.1. It has a diameter of about 1300 km and the outermost ring forms the Caloris Mountains that are up to 2 km high. The Caloris basin of Mercury resembles the Mare Orientale located partly on the far side of the Moon. The area opposite Caloris basin on the other side of Mercury contains closely spaced hills that may have formed by seismic waves released by the violent impact of the meteoroid that carved out the Caloris basin.

The impact craters on the surface of Mercury are not uniformly distributed because some regions have been resurfaced by the eruption of lava flows. The resulting lava plains are only sparsely cratered, which indicates that the volcanic activity occurred after the period of intense bombardment had ended at about 3.8 Ga. Two of these lava plains (Budh and Tir planitiae) are identified in Figure 10.1. In addition, many of the large early-formed impact craters have smooth and lightly cratered floors because they were partially filled by lava flows. Many such craters are evident in Figure 10.1 including Strindberg and Van Eyck.

The photomosaics of the surface of Mercury also reveal the presence of impact craters that are surrounded by rayed ejecta blankets resembling the crater Tycho on the Moon. This phenomenon is exemplified by the overlapping craters Brontë and Degas in Figure 10.1. Both have central peaks and their floors are lower than the surrounding terrain. Degas is located south of Brontë and overlaps its rim. Both craters are younger than the episode of volcanic activity that caused the formation of the Sobkou planitia visible west of the twin crates in Figure 10.1 (but not labeled).

The most recent episode in the geologic history of Mercury was the formation of scarps called *rupes* that cut across craters as well as the inter-crater lava plains. The Discovery scarp in Figure 10.2 extends for about 350 km and is up to 3 km high. This and other scarps (e.g., the Santa Maria rupes) formed by compression of the crust that may have resulted from the shrinking of the iron core as it cooled and or/by large impacts on the surface of Mercury. Geologically speaking, the scarps or rupes of Mercury are crustal fractures along which there is

vertical displacement. Therefore, the scarps can be classified as a special type of fault that has not been observed on the Moon or on the Earth. The scarps have been preserved on Mercury because of the absence of an atmosphere, which has prevented erosion and redeposition of weathering products. Consequently, the fault scarps of Mercury have remained standing for up to about three billion years.

10.1.2 Regolith

The surface of Mercury is covered by a layer of regolith composed of rock fragments and impact glass of the overlapping ejecta blankets associated with impact craters. The rock fragments are probably composed of basalt and anorthositic gabbro by analogy with the Moon. However, no rock samples from the surface of Mercury have been found on the Earth. In other words, we do not have any mercurial meteorites. The Messenger spacecraft is equipped to measure the chemical composition of the regolith from orbit and the European spacecraft Bepi Colombo will have a lander that may transmit pictures from the surface of Mercury.

10.1.3 Atmosphere

Mercury does not have an atmosphere because the Mariner 10 data indicate that the atmospheric surface pressure is only 2×10^{-12} atmospheres. Therefore, the pressure exerted by the mercurial atmosphere is approximately equal to two trillionths (10^{-12}) of the terrestrial atmosphere. The instruments on Mariner 10 detected the presence of hydrogen and helium in the space around Mercury. These elements originate from the solar wind but cannot accumulate in the space around Mercury because its gravitational force is not strong enough to hold them. Mariner 10 also detected atoms of sodium and potassium that presumably emanate from the regolith because the surface temperature rises to $+350\,°C$ as a result of solar heating.

Mariner 10 did *not* detect the presence of molecular nitrogen, water, and carbon dioxide, all of which may have been present at the time Mercury accreted from planetesimals, but which probably escaped into interplanetary space. The absence of water and carbon dioxide has limited

Figure 10.2. The Discovery scarp (or rupes) is one of several deep fractures in the crust of Mercury with large vertical displacement of up to three kilometers. The trace of this and other scarps on the surface of Mercury is not straight or curved in a consistent direction, but is typically "wavy" or "meandering." The Discovery scarp intersects two impact craters as well as the surrounding lava plain and therefore appears to be the most recent geological feature of this area. Most of the craters in this image have flat and lightly-cratered floors indicating that they were partially filled with lava flows. (Courtesy of NASA and JPL/Caltech, Mariner 10, PIA02446:Discovery Scarp)

the scope of geological activity on the surface of Mercury in the same way as on the Moon. Therefore, sedimentary rocks are not likely to be present on the surface of Mercury, although the igneous rocks of the mercurial crust may contain hydroxyl-bearing minerals such as amphiboles and micas if Mercury acquired water from the impacts of comets and planetesimals at the time of its formation.

10.1.4 Water

Recently-obtained radar images of the north and south poles of Mercury indicate that several craters in these areas contain deposits that strongly reflect radar waves. The bottoms of these craters do not receive sunlight and therefore maintain a uniformly low temperature of about

$-150\,°C$. Therefore, the radar reflections may be caused by water ice that was released by the impacts of comets in the polar regions of Mercury. Another possible explanation is that the radar reflections are caused by deposits of native sulfur which also reflects radar waves. The Messenger spacecraft may solve this mystery after it goes into orbit around Mercury in 2009.

10.1.5 Surface Temperatures

The *average* surface temperatures of Mercury vary between wide limits from $+350\,°C$ on the day side to $-170\,°C$ degrees on the night side. The high day-time temperature arises because Mercury is close to the Sun causing sunlight to be about seven times more intense than on the Earth, because one mercurial day lasts 88 terrestrial days, and because the absence of an atmosphere does not permit heat to be transported from the day side to the night side. Consequently, the daytime temperature along the equator of Mercury rises to +430 °C, which means that the temperature range on Mercury amounts to about 600 degrees. Evidently, the day-time temperatures on the surface of Mercury really are too hot for comfort. In fact, it gets so hot that several industrial metals that occur in solid form on the Earth would melt in the heat of the day on Mercury. These metals and their melting temperatures include: bismuth (271.3 °C), cadmium (320.9 °C), lead (327.502 °C), tin (231.9681 °C), and zinc (419.58 °C). The only metal that is liquid at Earth-surface temperatures is mercury which has a melting temperature of −38.842 °C.

10.2 Internal Structure

The physical dimension of Mercury in Table 10.1 confirm that its radius (2439 km), mass (3.302 × 10^{23}) kg), and volume (6.08 × 10^{19} m^3) are all larger than those of the Moon. In addition, Mercury is a solid body that shines only by reflecting sunlight and it does not have any satellites.

10.2.1 Iron Core

The most noteworthy property of Mercury is its high bulk density of 5.420 g/cm^3, which is much higher than that of the Moon (3.344 g/cm^3) and is exceeded only by the bulk density of the Earth (5.515 g/cm^3). Therefore, the smallest terrestrial planet has a disproportionately large iron core. According to calculations in Science Brief 10.6.1, the iron core of Mercury occupies 42% of its volume whereas the iron core of the Earth makes up only 17% of the terrestrial volume. The radius of the iron core of Mercury can be calculated from the volume of the planet listed in Table 10.1 as demonstrated in Science Brief 10.6.2. The result indicates that the radius of the core of Mercury is 1830 km and that the thickness of the overlying mantle is only 610 km. The cross-section of Mercury in Figure 10.3 illustrates the point that its core is disproportionately large compared to the other terrestrial planets and that its mantle is correspondingly thin.

Table 10.1. Physical properties of Mercury and the Moon (Hartmann, 2005; Freedman and Kaufmann, 2002)

Property	Mercury	Moon
Radius, km	2439	1738
Mass, kg	3.302 × 10^{23}	7.349 × 10^{22}
Bulk density, g/cm^3	5.420	3.344
Volume, km^3	6.08 × 10^{10}	2.20 × 10^{10}
Albedo, %	12	11
Atmosphere	none	none
Surface gravity (Earth = 1.0)	0.38	0.17
Average surface temperature, °C	+350 (day) −170 (night)	+130 (day) −180 (night)
Surface material	igneous rocks of basaltic composition	anorthositic gabbro, basalt, and regolith composition composed of fine grained rock fragments and glass spherules

Consequently, Mercury is more iron-rich than the other terrestrial planets of the solar system.

The large size of the iron core of Mercury can be explained by several alternative hypotheses:

1. The apparent enrichment of Mercury in iron is the result of the chemical differentiation of the protoplanetary disk (Section 5.2).
2. The silicate mantle of Mercury was eroded by the solar wind shortly after the Sun had formed.
3. Soon after Mercury had formed by accretion of planetesimals and had differentiated into an iron core and silicate mantle, it collided with a large body (e.g., a planetesimal). The impact caused part of the silicate mantle of Mercury to be ejected leaving a remnant containing a disproportionate amount of metallic iron. However, the ejected mantle material did not coalesce to form a satellite as in the Earth-Moon system.

Internal Structure of Mercury

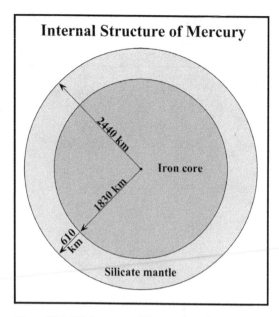

Figure 10.3. The iron core of Mercury occupies about 42% of the volume of this planet (Science Brief, 10.6b). The radius of the core is 1830 km which amounts to 75% of the radius of Mercury. The silicate mantle of Mercury has a thickness of only 610 km. Data from Freedman and Kaufmann (2002, p. 235)

Magnetic Field of Mercury

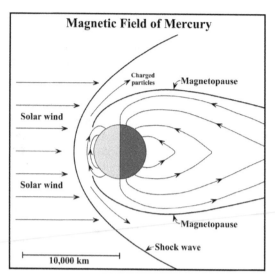

Figure 10.4. The magnetic field of Mercury is deformed by the solar wind causing it to trail in the down-wind direction. The electrically charged particles of the solar wind (primarily protons and electrons) are deflected by the magnetic field and flow around it between the magnetopause and the shock wave, both of which are three-dimensional surfaces in space. Adapted from Figure 10.18 of Freedman and Kaufmann (2002)

10.2.2 Magnetic Field

The Mariner 10 spacecraft detected that Mercury has a magnetic field that is only about 1% of the strength of the magnetic field of the Earth (Section 6.4.5). The magnetic field of Mercury has two poles like the terrestrial magnetic field (Figures 6.5, 6.6 and Science Brief 6.7.4, 5, 6, 7) and the lines of magnetic force in the space around Mercury are deformed by streams of electrically charged particles (primarily protons and electrons) of the solar wind as shown in Figure 10.4.

The existence of even a weak magnetic field in the space surrounding Mercury is surprising because, according to the theory of the self-exciting dynamo, planetary magnetic fields are generated by electrical currents flowing in the liquid iron-nickel cores of rotating planets (Section 6.4.5). This theory does not appear to be applicable to Mercury because its core is thought to be solid and because Mercury rotates very slowly with a period of 58.646 days.

The core of Mercury is assumed to be solid because its surface-to-volume ratio (S/V) in Figure 10.5 and Science Brief 10.6.3 is intermediate between those of the Moon and Mars, both of which do not have magnetic fields presumably because their cores have solidified. In addition, the self-exciting dynamo seems to require stimulation by the rotation of the planet in order to activate the convection currents in the core that cause the induction of the planetary magnetic field. Therefore, Mercury seems to fail both requirements leading to the prediction that it does not have a magnetic field.

The observation that Mercury actually does have a magnetic field, in spite of predictions to the contrary, suggests that some part of the metallic core of Mercury may be liquid because its observed magnetic field cannot be generated by the permanent magnetization of a solid core. The reason is that solid iron that is magnetized at low temperature loses its magnetism when it is heated to 770 °C (i.e., the Curie temperature of iron). Even though the iron core of Mercury is solid, its temperature is higher than 770 °C

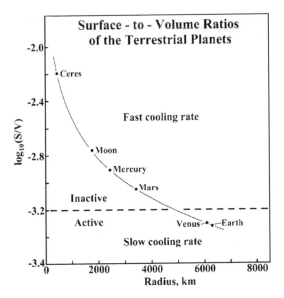

Figure 10.5. The surface-to-volume ratios (S/V) (expressed as logarithms to the base 10) of the asteroid Ceres and the terrestrial planets decrease with increasing values of the planetary radii. These ratios are qualitatively related to the cooling rate because the amount of heat lost by radiation from a planet depends on its surface area per unit volume. Therefore, the comparatively high S/V ratio of Mercury implies a comparatively high cooling rate which has caused the iron core to solidify and has lowered the internal temperature sufficiently to prevent volcanic active at the surface

and therefore the core cannot act as a permanent magnet.

A refinement of the dynamo theory is based on the idea that the convection currents in the liquid part of a planetary core arise because of differences in its temperature. For example, heat is released by the crystallization of liquid iron at the boundary between the solid and liquid parts of the core. The heat increases the temperature of the liquid iron, which expands and thereby reduces its density. Consequently, the iron liquid at the solid-liquid boundary in the core becomes buoyant and rises toward the core-mantle boundary. This mechanism does not depend on the rate of rotation of the planet in question and therefore could explain why Mercury has a weak magnetic field even though it rotates much more slowly than the Earth.

The refinement of the dynamo theory described above, assumes that the iron core of Mercury is partly liquid. Strom (1999) listed

three alternative hypotheses to explain why liquid iron may be present in the core:

1. The iron core of Mercury may be heated by the decay of radioactive atoms of uranium and thorium and their unstable radiogenic daughters (unlikely).
2. The core cooled more slowly than expected because it is unusually well insulated by the mantle of Mercury (questionable).
3. The melting temperature of the iron in the core was lowered by the presence of other elements that melt at lower temperatures than iron (e.g., copper, lead, zinc, sulfur). For example, iron sulfide melts between 800 and 1050 °C compared to 1535 °C for pure iron (unlikely).

Another possible reason why the core of Mercury may be partly molten is that its interior is being heated by tidal friction. The eccentricity of the orbit of Mercury (e = 0.206) around the Sun is much greater than the eccentricities of the other terrestrial planets. Therefore, the shape of Mercury changes as it travels in its orbit from perihelion to aphelion. The constantly changing tidal force exerted by the Sun on Mercury during each revolution and its slow rotation both generate heat by tidal friction. The heat may be concentrated in the region of the core-mantle boundary because of the abrupt change in the physical properties in this transition region. However, none of the alternative hypotheses mentioned above has received sufficient support from direct observations or mathematical models to be accepted as a theory that can explain why Mercury has a magnetic field.

10.2.3 Geological Activity

The surface features of Mercury make clear that this planet is not geologically active. In this regard, Mercury resembles the Moon but differs from the Earth where geological activity is manifested by frequent volcanic eruptions, earthquakes, and the movement of lithospheric plates. The symptoms of geological activity arise partly as a result of melting of solids in the interior of a planet, which depends on several environmental factors including temperature, pressure, and the chemical composition of the solids in the interior of the planet. The absence of an atmosphere on Mercury also excludes chemical weathering, transport of weathering products, and their deposition in layers.

The important point is that the temperature in the interior of a planet is not the only parameter that determines whether melting can take place and hence whether a planet is geologically active. For example, if the pressure in the mantle of the Mercury were to be lowered locally by the impact of a large meteoroid or by some other mechanism, magma could still form as a result of decompression rather than because of an increase in temperature. Nevertheless, the high S/V ratio of Mercury implies that it has cooled more rapidly than the larger terrestrial planets which have lower S/V ratios. Therefore, we return to Figure 10.5 in which the logarithms of the S/V ratios of the terrestrial planets decrease smoothly with increasing numerical values of their radii. The results demonstrate that the S/V ratios of the Moon, Mercury, and Mars, are larger than those of Venus and the Earth. As a result, the Moon, Mercury, and Mars are geologically *inactive*, whereas Venus and the Earth are geologically *active*, although Venus appears to be dormant at the present time. Therefore, the absence of internal geological activity of Mercury is at least partly caused by the loss of internal heat implied by its high S/V ratio, which is similar to the values of this ratio of the Moon and Mars, both of which are also geologically inactive.

10.3 Origin of Mercury

Our understanding of the formation of Mercury is based on its physical and chemical characteristics as interpreted in the context of the theory of the formation of the solar system (Sections 5.1 and 5.2). Accordingly, Mercury formed in the part of the protoplanetary disk that was only 0.387 AU from the center of the Sun. At this close range, all particles that were composed of the ices of water, carbon dioxide, methane, and ammonia sublimated and the resulting gases as well as nitrogen and the noble gases were swept away by the solar wind. Only particles composed of refractory compounds (i.e., not affected by high temperature), such as the oxides of iron, manganese, chromium, titanium, calcium, silicon, and aluminum, survived in close proximity to the Sun. Therefore, we expect that Mercury is enriched in iron and the other elements listed above. In other words, the predicted enrichment of Mercury in these elements may be the result of the *removal* of the volatile elements and their compounds that were converted into gases by the heat that emanated from the young Sun.

The planet Mercury subsequently originated by the accretion of the refractory dust grains and of the planetesimals that formed in the innermost region of the protoplanetary disk. As the mass of the planet continued to grow, occasional ice-bearing planetesimals from the outer regions of the protoplanetary disk could have impacted on Mercury and released water and other volatile compounds. The energy liberated by the impacts of planetesimals of all kinds was sufficient to cause the entire planet to melt. Therefore, when the main accretionary phase ended, Mercury consisted of a magma ocean composed of three kinds of immiscible liquids: molten silicates, metallic iron, and sulfides of iron and other metals. Mercury may also have had an atmosphere at this stage composed of a mixture of gases including water vapor and carbon dioxide.

In the next stage of its evolution, the molten iron and sulfides sank towards the center of Mercury and thereby formed its metallic iron core that may still contain sulfides of iron, zinc, copper, nickel, and cobalt. The growth of the core probably accelerated the rate of rotation of Mercury. In addition, Mercury may have initially had a magnetic field as a result of convection in its molten core.

While the interior of Mercury was differentiating into a core and mantle, gases trapped in its interior escaped into the atmosphere. However, the atmosphere of Mercury was later eroded for three principal reasons:

1. The energy of electrons and protons of the solar wind of the young Sun was sufficient to break up molecules in the atmosphere of Mercury, which allowed the resulting atoms to escape into interplanetary space.
2. The mass of Mercury is too small and its gravitational force is too weak to hold atoms of elements of low atomic number (i.e., the escape velocity of Mercury is only 4.3 km/s).
3. The impact of large planetesimals caused parts of the atmosphere to be ejected.

These three factors combined to allow the early atmosphere of Mercury to disperse and also caused the planet to lose water and other volatile compounds and elements.

After the end of the main accretionary phase, the silicate melt of the magma ocean crystallized in accordance with Bowen's reaction series to form a feldspar-rich crust and a mantle composed of olivine and pyroxene. The iron core cooled more slowly than the magma ocean because it was insulated by the mantle. The early-formed crust continued to be disrupted by frequent impacts of large planetesimals and other kinds of objects from interplanetary space. These impacts formed the craters and impact basins that still characterize the surface of Mercury at the present time.

During and after the period of heavy bombardment, lava of mafic composition flooded pre-existing impact basins and locally covered the surface of Mercury. Although the volcanic rocks on the surface of Mercury have not yet been dated, their ages presumably range from about 4.0 to 3.0 Ga by analogy with the volcanic rocks on the Moon (Section 9.2). When the core of Mercury solidified some time after the period of intense bombardment, the resulting shrinkage of the core compressed the mantle and crust, which caused the formation of scarps by thrust faulting.

In summary, Mercury was initially hot enough to melt, allowing it to differentiate into an iron core and a silicate mantle and primordial crust similar to the Earth. The intense bombardment of Mercury by planetesimals and other objects from the solar system continued after the mantle and crust had solidified as recorded by the abundant impact craters on its surface today. The atmosphere that Mercury may have had at the time of its formation was later lost, which depleted this planet in water, carbon dioxide, methane, and other volatile compounds and elements. Planet-wide volcanic activity occurred toward the end of the period of heavy bombardment and is assumed to have ended at about 3.0 Ga.

These conjectures about the origin of Mercury and its subsequent evolution are extrapolations of the theory of the origin of the solar system. However, they disregard the effects of a possible collision between Mercury and a large planetesimal. Such a collision apparently did occur on the Earth and resulted in the formation of the Moon by accretion of solid particles that formed from matter ejected by the impact (Section 9.6). We mentioned in Section 10.2.1 that a similar collision may have ejected part of the mantle of Mercury except that, in this case, the resulting particles drifted away and did not coalesce to form a satellite of Mercury. Of course, the absence of a satellite may also mean that Mercury did not collide with a large impactor.

10.4 Celestial Mechanics

The planet Mercury orbits the Sun and rotates on its axis in the prograde direction. The average distance of Mercury from the Sun is 0.387 AU and the eccentricity of its orbit is 0.206, which is more than 12 times greater than the eccentricity of the orbit of the Earth (0.017). Therefore, the distance between the center of Mercury and the center of the Sun varies from 0.307 AU at perihelion to 0.466 AU at aphelion or 41% of the average distance. The eccentricity of the orbit of Mercury is exceeded only by the eccentricity of the orbit of Pluto (0.250). In addition, the orbits of both planets are more steeply inclined to the plane of the ecliptic than the orbits of all other planets in the solar system, namely: 7.00° for Mercury and 17.12° for Pluto.

Other noteworthy facts about the celestial mechanics of Mercury listed in Table 10.2 include its high *average* orbital speed of 47.9 km/s compared to 29.79 km/s for the Earth. The orbital speed of Mercury varies widely from 59 km/s at perihelion to 39 km/s at aphelion because of the comparatively large eccentricity of its orbit. Mercury also has an unusually long period of rotation of 58.646 days compared to one day for the Earth. In addition, the axial obliquity of Mercury is only 0.5°, whereas that of the Earth is 23.45°. The significance of the low axial obliquity (inclination of the axis of rotation) of Mercury is that its polar regions receive hardly any sunshine and therefore could contain deposits of ice that have been preserved by the constant low temperatures that prevail in the polar regions (Section 10.1.4). The variation of the orbital speed of Mercury and its slow rate of rotation affect the apparent motions of the Sun across the mercurial sky. Freedman and

Table 10.2. Orbital parameters of Mercury and the Earth (Hartmann, 2005; Freedman and Kaufmann, 2002)

Property	Mercury	Earth
Average distance from the Sun, AU	0.387	1.00
Average orbital speed, km/s	47.9	29.79
Eccentricity of the orbit	0.206	0.017
Sidereal period of revolution, days	87.969	365.256
Synodic period of revolution, days	116	—
Period of rotation, days	58.646	1.0
Inclination of the orbit to the ecliptic	7.00°	—
Axial obliquity	0.5°	23.45°
Escape velocity, km/s	4.3	11.2

Kaufmann (2002, p. 231) reported that the east-to-west motion of the Sun, as seen from the surface of Mercury, actually stops and reverses direction for several Earth days. In addition, when Mercury is at the perihelion of its orbit, the Sun sets in the west and then reappears for several Earth days before setting once again.

10.4.1 Spin-Orbit Coupling

Although Mercury rotates and revolves in the normal prograde direction as noted above, its rotation has been slowed significantly by tidal interactions with the Sun (Section 9.7.1). However, in contrast to the Moon, the period of rotation of Mercury (58.646d) is *not* equal to its period of revolution (87.969d). Instead, the spin/orbit ratio (S/O) is:

$$\frac{S}{O} = \frac{58.646}{87.969} = 0.6666 = \frac{2}{3}$$

It follows that:

$$58.646 \times 3\,d = 87.969 \times 2\,d$$
$$175.938\,d = 175.938\,d$$

In other words, Mercury rotates on its axis three times during two revolutions around the Sun. This unusual relationship, called 3-to-2 spin-orbit coupling, has developed only in the orbit of Mercury and arises because of the gravitational interaction between the tidal bulges that form on Mercury and the significant eccentricity of its orbit (Freedman and Kaufmann, 2002, Figure 10.7).

The 3-to-2 spin-orbit coupling has the practical effect that the Sun does *not* shine constantly on the same side of Mercury. Instead, Figure 10.6 demonstrates that Mercury rotates once on its axis while traversing 240° (or 2/3) of its orbit. The revolution starts when Mercury is at high noon in position 1. When it reaches position 2, it has rotated 90° in the prograde direction while traversing 60° in its orbit around the Sun. Positions 3, 4, and 5 likewise represent 60° increments of the orbit and 90° increments

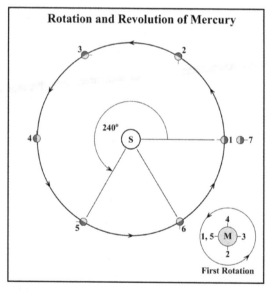

Rotation and Revolution of Mercury

Figure 10.6. Mercury has 3-to-2 spin-orbit coupling which means that it rotates three times during two revolutions. It also means that Mercury rotates once as it revolves through 240° of arc of its orbit. The rotation starts at position 1. At position 2 the planet has traversed 60° of its orbit while rotating 90° in the prograde (counter clockwise) direction. As Mercury continues in its orbit in 60° increments, it rotates through 90° as indicated by the hatch mark. When Mercury reaches position 5, it has completed the first rotation. It continues to rotate as it revolves through position 6 to positions 7. At this point the side of Mercury that was originally in the Sun (point 1) is now in the dark

of the rotation of Mercury. When the planet reaches position 5, it has completed one rotation but has moved only 240° along its orbit. As Mercury completes its first orbit by traveling from position 5 to position 6 and ultimately to position 7, it begins its second rotation. Figure 10.6 demonstrates that the face that was at high noon in position 1 is at midnight when Mercury returns to the same position in its orbit (i.e., position 7).

By tracking the rotation of Mercury around one revolution, we have demonstrated that the time elapsed between high noon (position 1) and midnight (position 7) is equal to 87.969 days (i.e., one period of revolution). When Mercury returns to its starting point after completing one revolution, the side that was dark at the start is now "in the Sun."

The length of time from high noon to midnight on Mercury is virtually identical to the time from sunrise to sunset. Therefore, the part of Mercury that is "in the Sun" is exposed for nearly three months (i.e., 88 days), which causes the temperature of the regolith at the equator to rise to +430°C. The heat acquired by the regolith is radiated into interplanetary space after the day-side of Mercury has rotated to the night-side. As a result, the temperature of the regolith decreases to −170°C during the long night that follows.

10.4.2 The Solar Transit

The inferior planets Mercury and Venus pass regularly between the Earth and the Sun as they approach their respective inferior conjunctions (Section 7.2.6 and Figure 7.5). Their passage in front of the Sun, called a *transit*, is not always visible from the Earth because of the inclination of the orbits of Mercury and Venus relative to the plane of the ecliptic.

Mercury passes in front of the Sun at least three times each year; but transits, when the planet actually moves across the face of the Sun as seen from the Earth, are comparatively rare (i.e., 13 or 14 per century). The most recent transits of Mercury occurred November 13, 1986; November 6, 1993; November 15, 1999; and on May 7, 2003. The next transit will take place on November 8, 2006. A transit of Mercury is only observable during the daytime (i.e., the Sun must be in the sky above the observer) and requires the

use of a telescope. Note that observers must *never* point a telescope directly at the Sun without using an appropriate solar filter. The duration of transits depends on the path the planet takes across the face of the Sun and can last from less than one up to nine hours. For example, the transit of Mercury on November 15, 1999, lasted about 52 minutes (Gordon, 1999; Freedman and Kaufmann, 2002).

10.4.3 Rotation of the Long Axis of the Orbit

Astronomers have known for several centuries that the long axis of the orbit of Mercury rotates in space at the rate of 5600 seconds of arc per century. By the time of the 19th century, calculations had demonstrated that most of this phenomenon is caused by the gravitational effects of the Sun and neighboring planets on the orbit of Mercury. However, a small part of the rotation amounting to only 43 seconds of arc per century could not be attributed to gravitational interactions. Therefore, astronomers proposed that the residual effect is caused by a small undiscovered planet in the space between the Sun and the orbit of Mercury. This hypothetical planet was named Vulcan because its surface temperature was thought to be even higher than that of Mercury. Vulcan was the Roman god of fire, which therefore was an appropriate name for a hypothetical planet that was even closer to the Sun than Mercury (Schultz, 1993; Wagner, 1991).

The problem concerning the rotation of the long axis of the orbit of Mercury was solved in the 20th century after Albert Einstein published his theory of general relativity. One of the predictions of this theory is that the large mass of the Sun actually warps the space around it, which means that the orbit of Mercury around the Sun is not completely describable by Newton's laws of motions. When relativistic effects were included in the calculation, the results agreed exactly with the observed rate of rotation of the long axis of the orbit of Mercury. This result was one of the first tests of Einstein's theory of relativity and it eliminated the need to postulate the existence of planet Vulcan.

10.5 Summary

Mercury is the smallest of the four terrestrial planets, but it is larger and more massive than the Moon. The surface of Mercury is heavily cratered, which implies that it formed at about the same time as the Moon (or perhaps slightly earlier) and was bombarded by large planetesimals and other objects early in the history of the solar system. Following its accretion from planetesimals, Mercury was hot enough to melt and subsequently differentiated by the segregation of iron and sulfide liquids from the silicate melt that formed a magma ocean early in its history. Mercury may have had an atmosphere composed of water vapor, carbon dioxide, and other gases early in its history. This primordial atmosphere was later lost primarily because Mercury does not have enough mass to retain these gases. Consequently, Mercury no longer contains water except perhaps deposits of ice in its polar regions that receive little, if any, sunlight.

One of the most noteworthy characteristics of Mercury is that its iron core is unusually large compared to the volume of the planet (42% of its total volume) and that it has a weak magnetic field. The presence of the magnetic field is unexpected because heat-flow models predict that the core of Mercury is solid and therefore cannot support a magnetic field by means of the planetary dynamo theory.

The orbit of Mercury is more eccentric and its inclination to the plane of the ecliptic is exceeded only by the orbit of the planet Pluto. In addition, Mercury has 3-to-2 spin-orbit coupling which was imposed on it by tidal forces exerted by the Sun. As a result, Mercury rotates very slowly with a period of about 59 days as it revolves in the prograde direction with a period of 88 days. The long exposure to the Sun, caused by the slow rotation and the proximity to the Sun (0.387 AU), allow the "daytime" temperatures of the surface at the equator of Mercury to rise up to 430 °C. The heat received during the mercurial day is radiated into interplanetary space during the long night that follows when the surface temperature drops to −170 °C.

Although Mercury was observed by the Sumerians about 5,000 years ago, very little was known about its surface and interior until the spacecraft Mariner 10 carried out three flybys in 1974/75. A new spacecraft, called Messenger, was launched on May 11, 2004, and is scheduled to go into orbit around Mercury in 2009. It will obtain images of the entire surface of the planet, investigate the possible presence of ice (water and carbon dioxide) in the polar regions, and carry out tests to determine whether part of the core of Mercury is still liquid. However, Messenger is not designed to land on Mercury and will not recover rock samples from its surface.

10.6 Science Briefs

10.6.1 The Core of Mercury

The high bulk density of Mercury is caused by the presence of a large iron core in its interior. The size of this core can be estimated from the densities of metallic iron and of ultramafic rocks that form the mantle of Mercury.

The density (D) of an object is defined in Section 3.5 as the ratio of its mass (M) divided by its volume (V):

$$D = \frac{M}{V} \tag{10.1}$$

Therefore, the mass of the object is:

$$M = D \times V \tag{10.2}$$

The mass of any differentiated planetary body (M_p) is the sum of the masses of its core (M_c) and its mantle (M_m):

$$M = (M_c) + (M_m) \tag{10.3}$$

Expressing the masses by use of equation 10.2 yields:

$$D_p V_p = D_c V_c + D_m V_m \tag{10.4}$$

Dividing by the total volume of the planet (V_p):

$$D_p = D_c \frac{V_c}{V_p} + D_m \frac{V_m}{V_p} \tag{10.5}$$

where (V_c/V_p) and (V_m/V_p) are the volume fractions of the core and mantle, respectively.

The volume fractions must satisfy the relation:

$$\left(\frac{V_c}{V_p}\right) + \left(\frac{V_m}{V_p}\right) = 1.0 \qquad (10.6)$$

Therefore,

$$\left(\frac{V_m}{V_p}\right) = 1 - \left(\frac{V_c}{V_p}\right) \qquad (10.7)$$

Substituting equation 10.7 into equation 10.5 yields:

$$D_p = D_c\left(\frac{V_c}{V_p}\right) + D_m\left(1 - \frac{V_c}{V_p}\right) \qquad (10.8)$$

This equation can be solved for V_c/V_p) which is the volume fraction of the core of the planet:

$$\left(\frac{V_c}{V_p}\right) = \frac{D_p - D_m}{D_c - D_m} \qquad (10.9)$$

In order to calculate the volume fraction of the Mercury, we use its uncompressed density and estimates of the densities of the core and mantle:
$D_p = 5.30\,g/cm^3$ (uncompressed density of Mercury)
$D_m = 3.34\,g/cm^3$ (olivine and pyroxene, Table 3.3)
$D_c = 7.97\,g/cm^3$ (90% iron + 10% nickel as in iron meteorites)
Substituting these values into equation 10.9:

$$\frac{V_c}{V_p} = \frac{5.30 - 3.34}{7.97 - 3.34} = \frac{1.96}{4.63} = 0.423$$

$$\frac{V_m}{V_p} = 1 - 0.428 = 0.577$$

The result of this calculation is that the iron-nickel core occupies about 42% of the volume of Mercury and that the mantle makes up the remaining 58% of the planetary volume.

10.6.2 Radius of the Core of Mercury

The volume (V) of a sphere is (Appendix 1):

$$V = \frac{4}{3}\pi r^3 \qquad (10.10)$$

where r is its radius and $\pi = 3.14$. The volume of the core of Mercury is 42% of the volume of

the planet which is $6.08 \times 10^{10}\,km^3$ (Table 10.1). Therefore, the volume of the core of Mercury (V_c) is:

$$V_c = 6.08 \times 10^{10} \times 0.42 = 2.55 \times 10^{10}\,km^3$$

Substituting this value into equation 10.10 yields:

$$2.55 \times 10^{10} = \frac{4 \times 3.14 \times r_c^3}{3}$$

$$r_c^3 = 0.609 \times 10^{10} = 6.09 \times 10^9\,km^3$$

$$r_c = 1.825 \times 10^3\,km$$

Therefore, the radius of the core of Mercury is: $r_c = 1830\,km$. The thickness d_m of the overlying mantle is: $d_m = 2440 - 1830 = 610\,km$ where the radius of Mercury is 2440 km.

10.6.3 Surface-to-Volume Ratio of Mercury

The surface area (S) of a sphere is (Appendix 1):

$$S = 4\pi r^2$$

where r is the radius of the sphere. Given that the radius of Mercury is 2440 km, the area of its surface is:

$$S = 4 \times 3.14 \times (2440)^2\,km^2$$

$$S = 7.477 \times 10^7\,km^2$$

The volume (V) of Mercury is:

$$V = 6.08 \times 10^{10}\,km^3$$

Therefore, the surface to volume ratio (S/V) of Mercury is:

$$\frac{S}{V} = \frac{7.477 \times 10^7}{6.08 \times 10^{10}} = 1.23 \times 10^{-3}\,km^{-1}$$

The logarithm of the S/V ratio of Mercury in Figure 10.5 is −2.910.

10.7 Problems

1. The map of a small area on Mercury in Figure 10.7 contains numerous impact craters having a range of diameters.

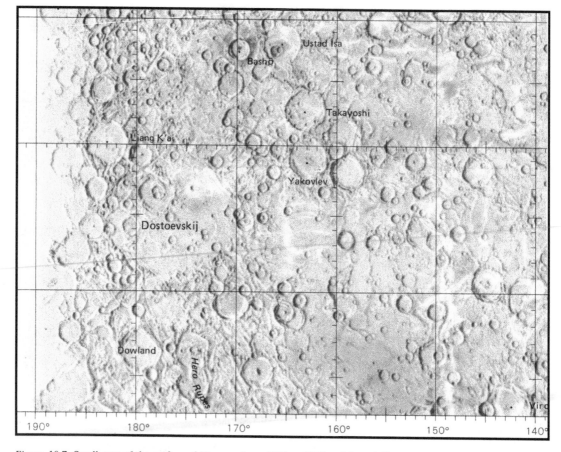

Figure 10.7. Small area of the surface of Mercury from 57°S to 30°S and from 140°W to 190°W. The map is based on images recorded by the spacecraft Mariner 10 and was prepared by the U.S. Geological Survey for NASA (Atlas of Mercury, 1:15 000 000 Topographic Series, H 15M 1RM, 1979, I-1171)

a. Measure the diameter of each crater in this map in millimeters until close to 100 separate craters have been measured.

b. Record these measurements in a table under the headings:

Range	Count	Total	Percentage
0–2			
2–4			
4–6			
6–8			
8–10			
10–12			
12–14			

c. Sum the number of craters in each size range and record these totals in the appropriate column. Add up all of the totals and express the number of craters in each category as a percentage of the total number of craters you measured.

d. Plot the size ranges along the x-axis and the percentage of each category along the y-axis. Fit a smooth curve by eye to the resulting array of points. Make it neat.

e. Interpret the graph you have constructed and explain your conclusions in a short technical report under the subheadings: Objective, Procedure, Presentation of Data, Interpretation, and Conclusions. Compose a cover sheet with the title of your project, your name, and the date. The text must be neatly typed on a high-quality white paper, the histogram must be neatly drafted, and all of the pages must be numbered consecutively and stapled or bound attractively.

10.8 Further Reading

Davies M, Gault D, Dwornik S, Strom R (1978) Atlas of Mercury. NASA SP-423. US Government Printing Office, Washington, DC

Freedman RA, Kaufmann III WJ (2002) The solar system. Freeman, New York

Gordon BB (1999) Adrift on the Sun. Astronomy, Nov, 76–78

Greeley R (1985) Planetary landscapes. Allen and Unwin, London, UK

Harmon JK (1997) Mercury radar studies and lunar comparisons. Adv Space Res 19:1487–1496

Hartman WK (2005) Moons and planets. 5th edn. Brooks/Cole, Belmont, CA

Murray B, Burgess E (1977) Flight to Mercury. Columbia University Press, New York

Murray B, Malin MC, Greeley R (1981) Earthlike planets; Surfaces of Mercury, Venus, Earth, Moon, and Mars. Freeman, San Francisco

Nelson RM (1997) Mercury: The forgotten planet. Sci Am 277(11):56–63

Robinson MS, Lucey PG (1997) Recalibrated Mariner 10 color mosaics: Implications for mercurian volcanism. Science 275:197–200

Schultz L (1993) Planetologie: Eine Einführung. Birkhäuser Verlag, Basel

Strom R (1987) Mercury, The elusive planet. Smithsonian Inst Press, Washington, DC

Strom R (1999) Mercury. In: Weissman P.R., McFadden L.A., Johnson T.V. (eds) Encyclopedia of the Solar System, 123–145. Academic Press, San Diego, CA

Vilas F, Chapman C, Matthews MS (eds) (1988) Mercury. University of Arizona Press, Tucson

Vilas F (1999) Mercury. In: J.K. Beatty, C.C. Petersen, and A. Chaikin, (eds) The new Solar System. 4th edn. Sky Pub. Cambridge, MA, pp 87–96

Wagner JK (1991) Introduction to the solar system. Saunders College Publishing, Philadelphia

Venus; Planetary Evolution Gone Bad

We know Venus as the "star" that shines prominently in the evening and, at other times, in the morning. Of course, Venus is one of the four terrestrial planets of the solar system in Figure 11.1 and is not a star. However, it is the brightest object in the sky at night surpassed only by the Moon. The planet Venus has been associated with the goddess of love since about 3100 BC. It was called Hesper by the Greeks when it appeared in the evening sky and Phosphor (Lucifer) when it was visible in the morning. The Romans named the planet Venus after the goddess of love and beauty. The first telescopic observations of Venus by *Galileo Galilei* around 1610 indicated that Venus has phase changes like the Moon and that its apparent size varies with time. These observations supported the heliocentric model of the solar system and caused Galileo to issue a cautiously worded message: "The mother of Love (i.e., Venus) imitates the form of Cynthia (i.e., the phases the Moon)." However, very little was known about the surface of Venus because it is hidden under a thick and continuous layer of clouds. The presence of an atmosphere was discovered by *Mikhail V. Lomonosov* (1711–1765) who observed a luminous halo around Venus during its solar transit in 1761. The halo results from the scattering of sunlight by the venusian atmosphere when the Sun is directly behind the planet.

Transits of Venus were used in the 18th and 19th century to define the astronomical unit, which established the size of the solar system. A transit of Venus on June 3, 1769, was observed by members of the Royal Society of England who were transported to the island of Tahiti on a sailing ship commanded by James Cook (Section 1.1). Subsequent transits of Venus occurred in 1874 and 1882 and, more recently, in 2004 (Section 11.5.2). The next such event will occur in 2012. In general, the transits of Venus occur in pairs separated by eight years.

The modern era of the exploration of Venus by means of robotic spacecraft started on December 14,1962, when Mariner 2 approached Venus during a flyby and reported that Venus does not have a magnetic field and that its surface temperature is in excess of 430 °C. Present knowledge indicates that the average surface temperature of Venus is 460 °C. Several spacecraft in the Venera series of the USSR subsequently determined that the atmosphere of Venus is largely composed of carbon dioxide. Finally, Venera 7 landed on the surface of Venus on December 15, 1970, and reported that the atmospheric pressure is 90 bars. Two years later on July 22, 1972, Venera 8 carried out a chemical analysis of the rocks on the surface of Venus, and two years after that Mariner 10 recorded closeup images of the clouds of Venus in Figure 11.2 as it flew by on its way to Mercury (Chapter 10). The Venera series continued in 1982 with successful landings of Venera 13 and 14 which sent back the first color pictures of the surface of Venus and obtained improved chemical analyses of samples taken into the spacecraft. The initiatives by the USA included two Pioneer spacecraft that paved the way to the highly successful mission of the Magellan spacecraft, which was launched on May 4, 1989, and subsequently mapped the entire surface of Venus with radar (Young, 1990, Wagner, 1991; Saunders et al., 1992; Ford et al., 1993; Roth and Wall, 1995). The most recent encounter with Venus occurred in 1990 during the flyby of the spacecraft Galileo on its way to the Jupiter system (Johnson et al., 1991) followed

Figure 11.1. The four terrestrial planets: Mercury, Venus, Earth, and Mars reside in the inner region of the solar system. All are solid objects and have high densities because they contain iron cores surrounded by mantles and crusts composed of silicate and oxide minerals. The Earth and Venus are similar in size and chemical composition but their surface environments differ greatly even though both have atmospheres. Mars has only a "thin" atmosphere, whereas Mercury lacks a permanent atmosphere. The Earth has a substantial planetary magnetic field in contrast to Venus and Mars which have none, whereas Mercury has only a weak magnetic field. The distances between the planets are not to scale. (Courtesy of NASA/Lunar and Planetary Institute, Houston, TX)

in 1998 and 1999 by the spacecraft Cassini en route to Saturn. However, we still do not have rock samples from the surface of Venus for isotopic dating and detailed chemical analysis in a terrestrial laboratory.

11.1 Surface Features

The radar images of Magellan revealed that the surface of Venus contains topographic features that are the result of internal geological activity. In this sense, Venus differs from the Moon and Mercury and resembles the Earth. The surface of Venus also contains large impact craters, but they are much less numerous than those of the Moon and Mercury, and they are randomly distributed over the entire surface of the planet. The comparatively low crater density implies that the surface of Venus is younger than the surfaces of the Moon and Mercury, whereas the uniform distribution of the craters tells us that the entire surface of Venus has the same age. In other words, Venus does not have continents composed of ancient rocks similar to the continents of the Earth and the highlands of the Moon. These and other properties of Venus, to

be discussed in this chapter, indicate that the geological evolution of Venus took a different path than the evolution of all other terrestrial planets. Summaries of the available information about Venus have been published by Hunten et al. (1984), Barsukov et al. (1992), Cattermole (1994), Grinspoon (1997), Bougher et al. (1997), Head and Basilevsky (1999), and Saunders (1999).

The principal topographic features of Venus in Figure 11.3 are the highlands (terrae), low plains (planitiae), mountains (montes), and areas of moderate relief (regiones). Most (but not all) of the montes of Venus are volcanoes. The chasmata are deep trenches formed by tectonic processes in the mantle of Venus rather than by erosion. Similarly, coronae are the surface expressions of convection currents in the mantle and appear to be the sources of basalt lava. The International Astronomical Union (IAU) has adopted a system of naming these topographic features on Venus after women, who have been deceased for at least three years and who must be, in some way, notable or worthy of having a planetary feature named after them.

Figure 11.2. The surface of Venus is obscured by a layer of clouds recorded by Mariner 10 during a flyby on February 5, 1974. The clouds are composed of droplets of sulfuric acid (H_2SO_4) and their distribution reflects the circulation of the Venusian atmosphere (Courtesy of NASA Marshall Space Flight Center, MSFC-8915499)

The topography of Venus is dominated by *Ishtar terra* located in the northern hemisphere (60° to 75°N latitude). It is capped by Lakshmi planum and is flanked on the east by the *Maxwell montes*, which consist of tightly folded volcanic rocks. The Lakshmi planum is surrounded by several mountain ranges, including the Fryja montes in the north, the Akna montes in the west,

and the Danu montes in the south. These, in turn, grade into tessera terrains that are characterized by complexly folded volcanic rocks resembling tiles. The major tessera terrains that surround Ishtar terra and the Maxwell montes are the Atropos tessera in the west, the Clotho and Laima tesserae in the south, and the very large Fortuna tessera in the east. The presence of the mountain

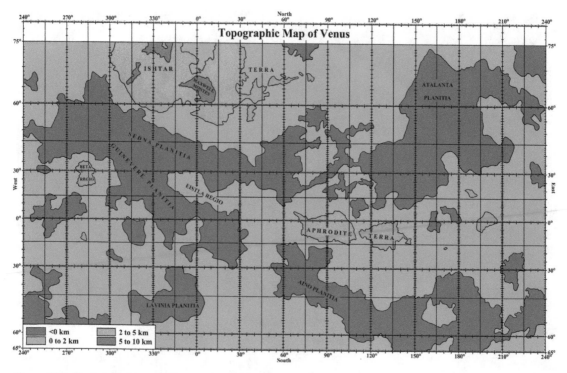

Figure 11.3. Shaded relief map of Venus color coded with respect to elevation of the surface. Adapted from the contour map prepared by the U.S. Geological Survey

ranges and tessera terrains along the margins of Ishtar terra indicates that the crust of Venus has been subjected to large-scale compression.

The northern highlands described above are partly surrounded by low-lying lava plains called planitiae named after mythological heroines: Sedna, Guinevere, Bereghinya, Niobe, and Atalanta. The difference in elevation between the summit of the Maxwell montes and the lowest parts of Sedna and Guinevere planitiae is more than 10,000 meters. If Venus had water on its surface, it would have collected in the planitiae to form a large ocean in the northern hemisphere. However, no evidence has yet been found that the planitiae were ever filled with water.

The equatorial belt of Venus is occupied by another landmass called *Aphrodite terra* that extends in an east-west direction for about 1400 km. Its western end breaks up into Eistla regio which contains several volcanoes including Sif mons and Gula mons. The eastern end of Aphrodite terra is continued by Atla regio, which

contains the volcanoes Maat mons and Ozza mons. The area south of Aphrodite terra is occupied by Guinivere and Aino planitiae.

The southern margin of Aphrodite terra includes several deep trenches (chasmata) some of which extend parallel to Aphrodite terra (i.e., Ovda, Kuanja, Diana, and Dali) in contrast to Artemis chasma which is semicircular. The presence of volcanoes on Aphrodite terra and their location to the north of deep trenches resembles the topography of subduction zones on the Earth (e.g., the Aleutian Islands). However, the resemblance is misleading because there is no compelling evidence that subduction of lithospheric plates has occurred or is occurring on Venus.

The area south of Guinevere and Aino planitiae contains several isolated regiones some of which contain volcanoes (i.e., Alpha, Beta, Themis, and Phoebe). These regiones are comparable to oceanic islands on the Earth. The southern hemisphere also contains another landmass called *Lada terra*. It has a uniformly low elevation compared to Aphrodite and Ishtar terrae and lacks volcanoes or other distinguishing topographic features.

The topographic features of Venus described above clearly demonstrate that Venus is an active planet like the Earth and that its surface is being shaped by tectonic processes occurring in its mantle. The tectonic processes operating on Venus have been discussed by Crumpler et al. (1986), Solomon (1993), Head and Basilevsky (1998), Nimmo and McKenzie (1998), Phillips and Hansen (1998), Leftwich et al. (1999), and Hansen et al. (1999).

11.2 Geological Processes

The images returned by the spacecraft Magellan contain evidence that geological processes have been active on the surface of Venus. Most prominent among these processes are:
1. Volcanic eruptions and intrusions of magma at shallow depth below the surface.
2. Folding of the crust by compression.
3. Rifting of the crust by extension.
4. Uplift of the crust as a result of crustal thickening by compression and thrust faulting.
5. Downwarping of the crust by descending convection currents in the mantle.
6. Chemical and mechanical weathering of surface rocks by interactions with the atmosphere of Venus evident in Figure 11.4.

The Magellan images also demonstrate that water is no longer present on Venus, which means that sedimentary rocks deposited in water are absent. However, eolian (windblown) deposits of volcanic ash and rock particles have been detected in favorable locations. (Kaula, 1990; Stofan, 1993; Price and Suppé, 1994; Basilevsky and Head, 1995.

The absence of water is hardly surprising because it presumably boiled away at the high surface temperature. Moreover, the concentration of water vapor in the atmosphere, expressed in multiples of 10^{-6} g of water per gram of air or parts per million (ppm), is very low and ranges from less than 1.0 ppm at 70 km above the surface to 45 ppm at 40 km above the surface (Hunten, 1999). Another way to show how little water exists in the atmosphere of Venus is to say that, if all the water vapor in the atmosphere were to condense, the resulting layer of liquid water on the surface of Venus would be about one centimeter deep (Wagner, 1991, p. 188). The loss of water from the surface of Venus allowed the carbon dioxide to return into the atmosphere, which has caused its temperature to increase by the greenhouse effect to be discussed in Section 11.3.3.

Another consequence of the absence of water on Venus is that living organisms cannot exist there. Although lifeforms may have developed very early in the history of Venus when water was deposited on its surface by comets and ice-bearing planetesimals, none could have survived after the water evaporated and the atmospheric temperature increased to its present value.

11.2.1 Volcanic Activity

Venus contains a large number of shield volcanoes, some of which appear to have been active in the geologically recent past because the radar images

Figure 11.4. Close-up of the surface of Venus recorded on March 1, 1982, by the Venera 13 lander of the USSR at 7.5 °S, 303 °E located east of Phoebe regio. The lander obtained 14 images during the 2 h and 7 min it survived on the surface of Venus. The image shows flat slabs of volcanic rocks of basaltic composition and weathering products. Part of the spacecraft is visible at the bottom of the image. The object left of center of the image is a lense cap. This image was published by Basilevsky et al. (1985) and is identified as Venera 13 Lander, VG 00261, 262 (Courtesy of NASA)

Figure 11.5. Lava flows extend for several hundred kilometers across a fractured plain of Venus from the crater at the summit of Sapas mons which is located at 8°N and 188°E on the western edge of Atla regio (Figure 11.3). This volcano has a diameter of 400 km and its summit rises to 4,500 m above the mean elevation of Venus. The vertical scale of this image has been exaggerated 10 times. The view is to the southeast toward Maat mons visible on the horizon. The coloration is based on images recorded by Verena 13 and 14 of the USSR. The image (PIA 00107:Venus was produced by the Solar System Visualization project and by the Magellan science team at the Multimission Image Processing Laboratory of JPL/Caltech and was released during a news conference on April 22, 1992.(Courtesy of NASA/JPL/Caltech)

reveal that the lavas that flowed from their summits are bright as a result of surface roughness (e.g., Sapas mons in Figure 11.5 and Maat mons in Figure 11.6). However, none of the volcanoes on Venus erupted during the Magellan mission or during any of the missions that preceded it. (Prinn, 1985; Fegley and Prinn, 1989).

Maat mons at the eastern end of Aphrodite terra is the second highest volcano on Venus with an elevation of about 8 km and a diameter of 395 km at its base. It occupies an area of about 122, 480 km², which is 1.15 times the area of the state of Ohio (106, 760 km²), and the average grade of its slope is 4.0% (i.e., a rise of 4 m

for a horizontal distance of 100 m). The slope of Maat mons is covered by lava flows that reflect radar waves efficiently, suggesting that they are young (i.e., less than 10 million years; Freedman and Kaufmann, 2002, p. 249). The physical dimensions of Maat mons are similar to those of the volcano Mauna Loa on the island of Hawaii (Section 6.4.4). Although the elevation of the summit of Mauna Loa above sealevel is only 4.0 km, its height above the surrounding sea floor is 8.84 km. The slope leading to the summit of Mauna Loa has a grade of 7.3% measured from the coast (Moore and Chadwick, 2000). Therefore, the slope of Maat mons is only

Figure 11.6. Maat mons, in the center of this image, is one of the principal volcanoes on Venus. It is located at 0.9°N and 194.5 E (Figure 11.3). The summit of Maat mons has an elevation of 8,000 m above the mean surface. The vertical scale has been exaggerated 10 times. The lava flows extend for hundreds of kilometers from the summit across the fractured plain. This image is identified as PIA 00254: Venus and was produced by the Solar System Visualization project and by the Magellan science team at the Multimission Image Processing Laboratory of JPL/Caltech and was released during a news conference on April 22, 1992. (Courtesy NASA/JPL/Caltech)

about half as steep as the slope of Mauna Loa. This comparison draws attention to the vertical exaggeration of the images of volcanoes on Venus released by NASA/JPL. For example, the heights of volcanoes are exaggerated by factors of 10 and, in some cases, by 22.5 (Freedman and Kaufmann, 2002, p. 249).

Other volcanic features on the surface of Venus include the *lava domes* in Figure 11.7, also referred to as squeeze-ups or pancake domes. They are about one kilometer high, have nearly flat tops, and are less than about 100 km wide. The lava domes occur in clusters and presumably formed by the extrusion of viscous magma from subsurface magma chambers. The viscosity of lava flows is related to their temperature and to the concentration of silica (SiO_2). For example, rhyolite lava containing about 70% SiO_2 is much more viscous than basalt lava which contains only about 50% SiO_2. The viscous magma that formed the lava domes on Venus differs from the lava that formed the volcanic rocks that are exposed on the low plains (e.g., Sedna and Atalanta planitia), which had low viscosity and appear to be basaltic in composition.

The low viscosity of basalt flows on Venus is also implied by the length of *rilles* or lava channels like those that occur on the Moon

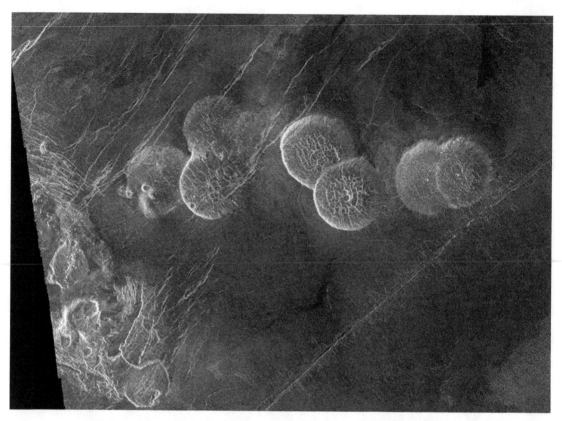

Figure 11.7. Lava domes located along the eastern edge of Alpha regio at 30 °S and 11.8 °E. (North is at the top). The average diameter of the domes in this image is 25 km and their maximum height is 750 m. These domes originated by the extrusion of viscous lava from a subsurface magma reservoir through crustal fractures on a level surface, which allowed the lava to flow radially from the vents. The fractures on the surfaces of the domes formed by extension of the crust that formed by cooling and crystallization as lava continued to be extruded. The lava plain upon which the domes formed contains a set of parallel fractures some of which intersect the domes indicating that the fractures both pre-date and post-date the extrusion of the lava domes. Six of the seven domes in this image are paired which implies that the domes formed sequentially and episodically. The prominent linear feature in the southeastern corner of the image is probably a rift valley. The image is identified as PIA 00215: Venus – Alpha regio. (Courtesy of NASA, JPL/Caltech)

and probably on Mercury. The lava channels on Venus in Figure 11.8 are about 2 km wide and are up to 6800 km long (e.g., Hildr channel, west of the Nokomis montes). These channels were cut into pre-existing volcanic rocks by the flow of large volumes of hot lava that was about as fluid as motor oil. The width of the channels decreases downstream from their source to the end. Some lava channels split into two or more distributaries like stream channels on the Earth. The lava channels occur predominantly on the slopes of Aphrodite terra where they start at coronas and other volcanic landforms. The channels allowed large volumes of lava to spread across the low-lying planitiae on a global scale.

The apparent difference in the viscosity of lava flows suggests the presence of at least two kinds of volcanic rocks on Venus based on their chemical compositions. The chemical analyses of the rocks exposed at the landing sites of the Venera and Vega spacecraft of the USSR can be compared in Table 11.1 to the average chemical composition of basalt in the ocean basins of the Earth. The chemical compositions of rocks composed of silicate minerals (e.g., feldspar, pyroxene, quartz, etc.) are expressed in terms of the oxides of the major elements because oxygen is the most abundant element in such rocks. The differences in the concentrations of some oxides in the venusian rocks (e.g., MgO, CaO, and K_2O) are

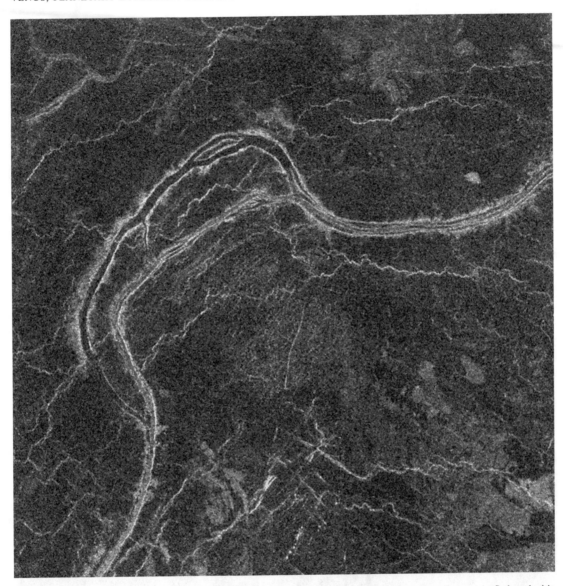

Figure 11.8. This lava channel in Sedna planitia extends from the Fortuna tessera in the north to the eastern Sedna planitia in the south (Figure 11.3). The channel, which is about 2 km wide, branches repeatedly and contains islands indicating that the lava was comparatively fluid. The center of the image is located at 44.96 °N and 19.21 °E and north is at the top. The image is identified as F-MIDR45N019; 1 (framelet 18) and as P-38851. (Courtesy of NASA Goddard Space Flight Center)

caused by variations in the abundances of minerals that contain these elements. In general, the chemical compositions of the rocks at the landing sites on Venus are similar to average terrestrial basalt. The chemical composition of the rocks that form the pancake domes is not yet known.

Volcanic activity is an important mechanism by means of which the terrestrial planets transport heat from the interior to the surface where it can be radiated into interplanetary space. Taylor (1992) cited data from the literature which indicate that on the Earth 18 km^3 of basalt lava are produced annually along midocean ridges and about 2 km^3 at island arcs and intraplate volcanic centers. The production rate of basalt lava on Venus is only about 1 km^3 per year,

Table 11.1. Chemical compositions of basalt on Venus and on the Earth in weight percent (Taylor, 1992)

Component	Venus			Earth, Average oceanic basalt
	1	2	3	
SiO$_2$	46.8	49.8	51.5	51.4
TiO$_2$	1.66	1.28	0.23	1.5
Al$_2$O$_3$	16.4	18.3	18	16.5
FeO	9.7	9.0	8.7	12.24
MgO	11.7	8.3	13	7.56
CaO	7.4	10.5	8.5	9.4
Na$_2$O	(2.1)	(2.5)	–	
K$_2$O	4.2	0.2	0.1	1.0
MnO	–	–	–	0.26

1. Venera 13; 2. Venera 14; 3. Vega 2. O = oxygen, Si = silicon, Ti = titanium, Al = aluminum. Fe = iron, Mg = magnesium, Ca = calcium, Na = sodium, K = potassium, and Mn = manganese.

which means that Venus loses much less heat by volcanic activity than the Earth. Heat loss by conduction is not an efficient process, which therefore raises an *important question*: How does Venus release the heat that continues to be generated in its interior by the decay of the radioactive atoms of uranium, thorium, and potassium?

11.2.2 Tectonic Activity

The landforms on the surface of Venus depicted in Figures 11.9 and 11.10 indicate that the crust has been deformed in various ways by tectonic forces that have caused compression, extension, uplift, and subsidence. The tectonic activity can be explained as a consequence of convection in the form of plumes or diapirs of hot rock that

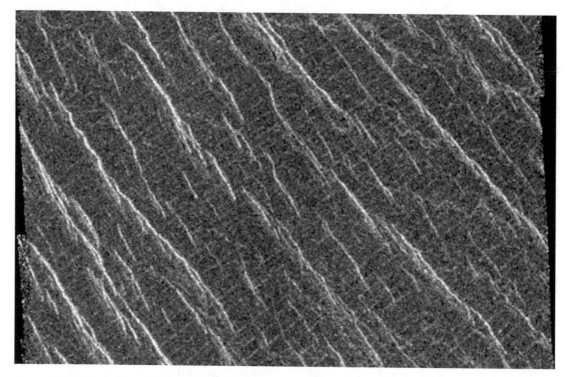

Figure 11.9. Gridded plain at 30°N and 333.3°E between Sedna planitia and Guinevere planitia west of Eistla regio. This lava plain is traversed by two sets of fractures at 90° to each other (Banerdt and Sammis, 1992). The fainter and older fractures (NE/SW) are equally spaced at about 1.0 km. The bright and younger lineations (NW/SE) are more irregular in shape and shorter in length than the preceding fractures. The existence of such fractures on the plains of Venus indicates widespread extensional deformation of the surface. The image is identified as PIA 00085: Venus. (Courtesy of NASA, JPL/Caltech)

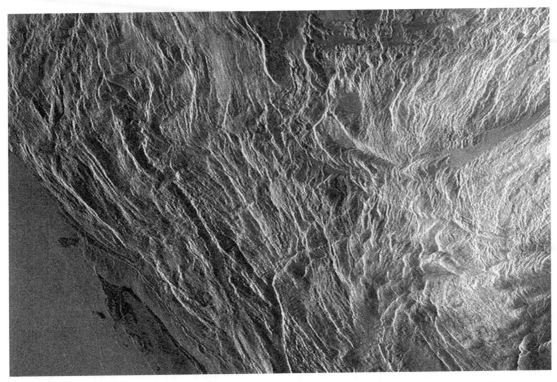

Figure 11.10. The Akna montes of Venus are located west of Lakshmi planum (Figure 11.3). The center of this image is located at 59.89 °N and 291.18 °E with north at the top. These mountains are the result of folding of volcanic rocks caused by compression from the southwest. The image is identified as C1-MIDR60N291; 1, framelet 8. (Courtesy of NASA and Goddard Space Flight Center)

rise in the mantle of Venus until they reach the underside of the lithosphere. The plumes are energized by differences in the density caused by thermal expansion, which generates buoyancy and causes them to rise slowly through the solid mantle. Therefore, the tectonic forces that have deformed the crust of Venus ultimately arise because of temperature differences in the mantle. The heat that is stored in the mantle of Venus originates primarily from the energy released by the decay of the unstable atoms of uranium, thorium, and potassium. Therefore, differences in the temperature of the mantle arise because of the uneven distribution of the heat-producing elements in the mantle and, to a lesser extent, by heat emanating from the iron core, by tidal deformation of the body of Venus, and by other causes.

The existence of plumes in the mantle of Venus is plausible based on their postulated occurrence in the mantle of the Earth (Faure, 2001). The interaction of mantle plumes with the underside of the lithosphere in Figure 11.11 causes uplift, compression and extension of the crust, volcanic activity, as well as localized subsidence where the crust and lithosphere are thinned by extension. The continental crust of the Earth is composed of igneous, metamorphic, and sedimentary rocks whose collective density is about $2.90 \, g/cm^3$ compared to about $3.20 \, g/cm^3$ for the ultramafic rocks of the mantle. Therefore, blocks of continental crust float in the mantle of the Earth and rise above the level of the seafloor by several kilometers depending on the thickness of the crust and on the density contrast between the crust and mantle as shown in Science Brief 6.7.1.

This process is inadequate to explain the elevation of Ishtar terra on Venus because the rocks consist of basalt rather than granite. The difference in densities between the basalt crust and the mantle of Venus is too small to cause

194

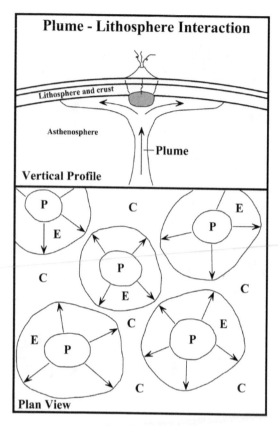

Figure 11.11. Illustration of the *hypothetical* interaction of plumes with the lithospheric shell of Venus (not drawn to scale). In the *vertical profile*, the plume spreads out laterally along the underside of the lithosphere. The resulting extension causes fractures to form which leads to magma formation by decompression melting and volcanic activity on the surface. In the *plan view*, the heads of several plumes cause extension (E) as well as compression (C) where the extensional regimes of individual plumes collide

plates but consists of a single global shell. Plumes that rise through the asthenosphere of Venus until they impact the rigid lithospheric shell spread out laterally and thereby cause extension. If several plumes interact with the lithosphere of Venus in close proximity with each other, the crustal extension caused by each of these plumes results in compression of the crust as suggested in Figure 11.11. In the *compression zones* the crust and lithosphere is thickened by folding and thrust faulting which leads to local uplift. In the *extension zones*, the crust and lithosphere are thinned causing subsidence. The *multi-plume hypothesis* proposed above has not yet been tested to determine whether it can explain some of the topographic features of Venus such as the tessera terrains, chasmata, and coronae. We still have much to learn about the inner workings of Venus.

11.2.3 Impact Craters

The surface of Venus contains about one thousand impact craters most of which have large diameters ranging from several 10s to more than 200 km. Small craters having diameters less than 10 km are rare or absent, presumably because small impactors are destroyed during the passage through the dense atmosphere of Venus. Most of the vinusian impact craters have central uplifts, some have single or even multiple peak rings, and some of the impact craters have a smooth floor indicating that they were partially filled with basalt lava or impact melt (Weitz, 1993).

The large impact craters on Venus are named after women who distinguished themselves in their chosen profession. The largest impact crater is named after the American anthropologist Margaret Meade whose crater has a diameter of 275 km. Other notable craters are named after Halide Adivar, Phillis Wheatley, Liese Meitner, Saskia, Aurelia, and many others. The impact craters in Figure 11.12 and 11.13 are surrounded by ejecta deposits that are bright in the radar images because of the roughness of their surface. The ejecta deposits include impact melts which, in some cases, flowed downslope away from the impact site for several hundred kilometers (Phillips et al., 1991, 1992). The amount of meltrock produced by impacts on Venus is larger

a block of basalt to rise to the observed height of Ishtar terra (i.e., about 3.5 km). Instead, the basaltic crust on Venus may be uplifted locally to form what looks like a continent by plumes rising in the mantle, by crustal thickening caused by folding, and by thrust faulting due to horizontal compression of the crust (Freedman and Kaufmann, 2002).

Another difference between the Earth and Venus is the absence of plate tectonics on Venus. The lithosphere of the Earth consists of about 20 plates that are being moved by convection currents in the asthenosphere (Section 6.4.4). The lithosphere of Venus has not broken up into

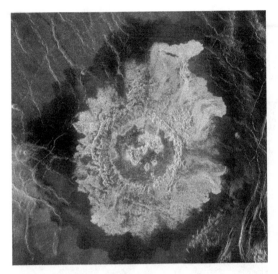

Figure 11.12. The impact crater Yablochkina (48.3°N, 195.3°E) has a diameter of 63 km and contains a partially peak-ring on a flat floor which is about 1.0 km below the surrounding plain of the Atalanta planitia. The ejecta blanket is extensive and includes flows of melt rock. The crater has a dark halo similar to about half the impact craters on Venus (Phillips et al., 1991, 1992). The dark halo was probably formed by the atmospheric shock wave that smoothed the terrane and pulverized rocks during the passage of the meteorite though the atmosphere of Venus (Weitz, 1993). (Project Magellan, courtesy NASA and JPL/Caltech)

than the amounts associated with terrestrial, lunar, and mercurial impact craters, presumably because the heat released by the impacts on Venus caused the temperature of the target rocks to rise from 460°C to higher values than would be the case on the Earth, the Moon, or Mercury where the surface temperatures are much lower.

An unexpected attribute of the impact craters on Venus is their uniform distribution over the entire surface of the planet. The implications of this fact are far-reaching:

1. All parts of the surface of Venus have the same age of about 500 million years.
2. The surface of Venus was rejuvenated by an episode of volcanic activity of global extent that obliterated all previously formed impact craters and other topographic features.
3. Ancient highland regions like those on the Moon and Mercury have not survived on Venus.

4. Although Venus has a young surface, the planet itself has the same age as all other planets in the solar system (i.e., 4.6×10^9 years). Many of the old impact craters on the Earth have also been obliterated by geological processes that have rejuvenated its surface. The difference is that rejuvenation of the surface of the Earth is an ongoing process that affects only small areas at any given time, whereas rejuvenation by resurfacing of Venus seems to occur episodically and on a global scale.

11.3 Atmosphere and Climate

The climate on the surface of Venus was unknown until 1962 when the data recorded by Mariner 2 during a flyby of Venus indicated that its surface was hot and dry. Subsequent landings of some of the Venera spacecraft launched by the Soviet Union confirmed this conclusion and also indicated that the atmospheric pressure on the surface of Venus is 90 times higher than the atmospheric pressure at sealevel on Earth. In other words, the atmospheric pressure on Venus is as high as the water pressure in the oceans at a depth of 1000 meters. The high pressure and high temperature (460°C) of the atmosphere make the climate on the surface of Venus hostile to humans in contrast to the benign conditions on the surface of the Earth. The reasons for the high surface temperature of Venus were briefly discussed in Section 3.4. (Lewis and Prinn, 1984; Lewis, 2004).

11.3.1 Chemical Composition and Structure

The chemical composition of the air in the atmosphere of Venus in Table 11.2 is dominated by carbon dioxide (96.5%) and molecular nitrogen (3.5%). The remaining constituents of the venusian air are sulfur dioxide, carbon monoxide, water vapor, hydrogen sulfide, and the noble gases (helium, neon, argon, and krypton). Molecular oxygen is present only at very low concentrations between 1 and 20 parts per million. The clouds that form a veil over the surface of Venus in Figure 11.2 are composed of tiny (2 μm) droplets of nearly pure sulfuric acid (H_2SO_4).

Figure 11.13. A set of three impact craters centered at 27°S and 339°E in the northwestern portion of Lavinia planitia (Figure 11.3). Howe crater in the foreground has a diameter of 37.3 km, Danilova, located to the left and behind Howe, is 47.6 km wide, and Aglaonice, located to the right and behind Howe is 62.7 km in diameter. The view is toward the northwest. The color in this image was calibrated against the images of Venera 13 and 14. The three craters have central uplifts and extensive ejecta aprons that are bright in this radar image because of the roughness of the surfaces of these features. The image is identified as PIA 00103: Venus and as P39146. It was produced by the Multimission Image Processing Laboratory of JPL/Caltech and was released on May 29, 1991. (Courtesy of NASA and JPL/Caltech)

The temperature of the atmosphere of Venus in Figure 11.14 decreases with increasing altitude above the surface. The resulting temperature profile is used to subdivide the atmosphere into the *troposphere* (0 to 65 km), the *mesophere* (65 to 115 km), and the *ionosphere* (>115 km). Figure 11.14 also indicates that the layer of sulfuric acid clouds occurs at a height of about 50 km where the pressure is about one atmosphere. The droplets of sulfuric acid that form these clouds are produced by chemical reactions based on sulfur dioxide and hydrogen sulfide gas that are emitted during volcanic eruptions on the surface of Venus. The droplets are not lost from the clouds in the form of sulfuric acid rain because they evaporate before they can reach the ground. Nevertheless, the existence of the sulfuric acid clouds is evidence that volcanic activity has occurred in the relatively recent past (i.e., about 30 million years) and may still be occurring at the present time, even though no volcanic activity was detected in the radar images of the surface of Venus by Magellan and preceding spacecraft (e.g., Pioneer Venus and Venera).

Table 11.2. Chemical composition of the atmosphere of Venus at a height of 40 km above its surface (Hunten, 1999) and of the Earth at ground level (Emiliani, 1992)

Compound or element	Venus % or ppm	Earth % or ppm
Carbon dioxide (CO_2)	96.5 %	346 ppm
Nitrogen (N_2)	3.5 %	78.08 %
Oxygen (O_2)	0–20 ppm	20.94 %
Helium (He)	~ 12 ppm	5.2 ppm
Neon (Ne)	7 ppm	18.2 ppm
Argon (Ar)	70 ppm	9340 ppm
Krypton (Kr)	~ 0.2 ppm	1.14 ppm
Carbon monoxide (CO)	45 ppm	0.2 ppm
Water vapor (H_2O)	45 ppm	4% to 40 ppm
Sulfur dioxide (SO_2)	~ 100 ppm	0.004 ppm
Hydrogen sulfide (H_2S)	1 ppm	0.0002 ppm

11.3.2 Circulation

The data returned by the Pioneer Venus multi-probe spacecraft in 1978 revealed that the upper part of the atmosphere of Venus rotates around the planet in only four days and that it does so in the *retrograde* direction. The circulation of the atmosphere around Venus is remarkable because it is much more rapid than the rotation of the planet itself which is also retrograde and has a period of 343.01 days. The corresponding wind speed decreases from about 350 km/h in the upper atmosphere to only about 5 km/h at the surface. The circulation of the atmosphere and the density of the air equalize the surface temperature of Venus from the equatorial region to the north and south poles.

The circulation of the atmosphere of Venus consists of three cells which occur at different elevations above the surface. The *driving cell* in the cloud layer between elevations of 48 and 68 km transports hot air from the equatorial belt toward the poles where it cools, sinks to lower elevations, and then flows back to the equator. The driving cell activates the stratospheric cell above the cloud layer as well as the surface-layer cell below the clouds.

11.3.3 Greenhouse Warming

The high surface temperature of Venus is one of the characteristic properties that distinguish

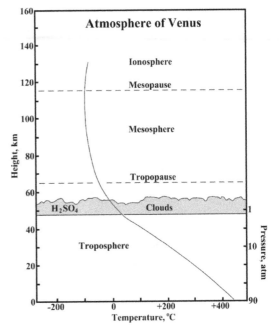

Figure 11.14. Temperature profile of the atmosphere of Venus. The atmosphere is divided into the troposphere (0 to 65 km), the mesophere (65 to 135 km), and the ionosphere (>135 km). The tropopause and mesopause are surfaces that separate the three divisions of the atmosphere. The cloud layer occurs at a height of about 50 km and a pressure of about 1.0 atmospheres. The clouds are composed of droplets of sulfuric acid (H_2SO_4) that has formed from volcanic sulfur-bearing gases. The air of the troposphere consists primarily of carbon dioxide gas (96.5%) with some nitrogen (3.5%). Adapted from Hartmann (2005, Figure 11.4)

this planet from the Earth. Present knowledge indicates that the atmosphere of Venus has heated up because the carbon dioxide that dominates the composition of the atmosphere absorbs infrared radiation emitted by the surface of Venus. The relationship of infrared radiation to the wavelength spectrum of electromagnetic radiation is explained in Science Briefs 2.5.3, 4 and 11.7.1.

About 80% of the sunlight that reaches Venus in Figure 11.15 is reflected back into space by the layer of clouds which cause Venus to have a high albedo 59% compared to only 39% for the Earth. Therefore, the surface of Venus actually receives less sunlight than the surface of the Earth even though Venus is closer to the Sun. The sunlight that does reach the surface of Venus increases the temperature of the exposed rocks, which causes

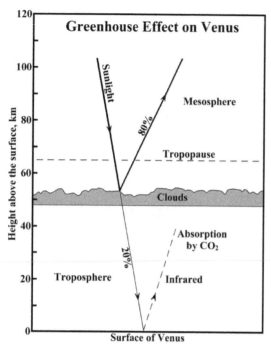

Figure 11.15. Most of the sunlight (80%) that reaches Venus is reflected by the clouds that cover the surface at a height of about 50 km. The sunlight that passes through the clouds (20%) is absorbed by the rocks on the surface of Venus and is re-radiated into the troposphere in the form of infrared radiation described in Science Brief 11.7a. The infrared radiation is absorbed by molecules of carbon dioxide which form 96.5% of the air in the troposphere of Venus. As a result, the temperature of the air has increased to 460 °C compared to only about 45 °C in the absence of the greenhouse effect

through the glass and warms the soil inside. However, the infrared radiation emitted by the soil cannot pass through the glass and is trapped inside the greenhouse. Consequently, the temperature of the air in the greenhouse increases. The magnitude of the greenhouse effect depends on the amount of sunlight the surface receives and on the concentration of the atmospheric gases that absorb infrared radiation. The temperature of the air at groundlevel on Venus (460 °C) is higher than that on the Earth (9 °C) because the concentration of carbon dioxide in the atmosphere of Venus (96.5%) is much higher than in the atmosphere of the Earth (346 ppm). However, greenhouse warming of the terrestrial atmosphere is increased by the presence of water vapor and methane, which together with carbon dioxide raise the concentration of greenhouse gases in the terrestrial atmosphere to about 1.0%. The presence of these gases has caused the average surface temperature of the Earth to increase from −27° to 9 °C (Freedman and Kaufmann, 2002, p. 182), which has made life possible on the Earth, but has given Venus a hellish climate that makes life on its surface impossible.

11.3.4 Evolution of the Surface Environment

The high surface temperature of Venus, the high concentration of carbon dioxide in the atmosphere, and the absence of water are all manifestations of an evolutionary process that has gone bad. The sequence of events has been reconstructed based on the theory of the formation and differentiation of Earth-like planets (Section 6.2 and 6.3) and on the effects of greenhouse warming (Science Brief 11.7.1).

For the sake of this presentation, we assume that Venus originally contained amounts of carbon dioxide and water that were similar to those that existed on the Earth at the time the planets formed by accretion of planetesimals. The heat released by the impacts of planetesimals caused Venus to melt and to differentiate by the segregation of immiscible liquids into an iron core, a silicate mantle, and sulfides of iron and other metals that sank toward the bottom of the mantle before it completely solidified. Any volatile compounds that were present in the planetesimals formed a dense atmosphere

them to emit infrared radiation into the overlying atmosphere. The same phenomenon occurs on the Earth and causes rock surfaces exposed to direct sunlight to be warm to the touch.

Most of the infrared radiation emitted by the surface of Venus is absorbed by the molecules of carbon dioxide in the atmosphere, which allows the energy of this radiation to be converted into heat and thereby increases the temperature of the atmosphere. This phenomenon occurs on all planets and satellites, provided they have atmospheres containing certain compounds that absorb infrared radiation (e.g., carbon dioxide, water vapor, and methane).

The trapping of heat by the atmosphere of a planet resembles the situation that is caused by the glass roof of a greenhouse. Sunlight passes

composed primarily of water vapor, carbon dioxide, and nitrogen, but including also the noble gases and other volatile compounds such as methane and ammonia. At this stage in their evolution, the atmosphere of Venus was similar to the atmosphere of the Earth.

The evolution of the surface environment of the Earth continued as the temperature of the atmosphere decreased sufficiently to cause water vapor to condense. This caused a massive rain-out of the atmosphere to form a global ocean on the surface of the Earth. During this process, most of the carbon dioxide dissolved in the raindrops and was thereby transferred from the atmosphere into the newly-formed ocean. The removal of the two most abundant greenhouse gases from the atmosphere of the Earth greatly reduced the intensity of greenhouse warming. The resulting global climate has remained hospitable to life ever since with perhaps a few short-term interruptions during global ice ages in Late Proterozoic time.

The evolution of the surface environment on Venus proceeded quite differently because the temperature of its early atmosphere did not decrease sufficiently to cause the water vapor to condense. Even in case the water vapor in the early atmosphere of Venus did rain out to form an ocean, the resulting decrease of the global temperature was not sufficient to prevent the water from re-evaporating. The principal reason for the failure of water vapor to condense or for the loss of water from the surface of Venus may be that this planet receives more solar energy because it is closer to the Sun than the Earth. Therefore, the carbon dioxide either remained in the atmosphere of Venus or returned to it and thereby caused the extreme greenhouse warming.

This hypothetical reconstruction of the evolution of the surface environment of Venus does not provide plausible reasons for the absence of water in the present atmosphere. Perhaps, the water molecules in the atmosphere of Venus were broken up by ultraviolet radiation into hydrogen and oxygen both of which escaped into interplanetary space. This explanation for the disappearance of water vapor from the atmosphere of Venus is not altogether satisfying because the dissociation of water molecules takes place primarily in the stratosphere where the concentration of water vapor is low and di-atomic

oxygen molecules are not expected to escape from the atmosphere of Venus. The problems arising from the disappearance of water from the atmosphere of Venus were discussed by Lewis (2004) including the possible fate of oxygen atoms released by the dissociation of molecular water and carbon dioxide. Another hypotheses is implied by Problem 1 in Section 11.8.

11.3.5 Carbon Cycles of Venus and the Earth

We now digress briefly in order to consider whether the extreme greenhouse warming that is occurring on Venus could also occur on the Earth. This question becomes relevant because large amounts of carbon dioxide have been released into the atmosphere of the Earth by the ongoing combustion of fossil fuel (coal, petroleum, and natural gas), by the clearing of forests, and by the decomposition of organic matter in forest soil.

The geochemistry of carbon on the Earth is best considered by use of the carbon cycle, which consists of several large reservoirs that are connected by certain processes that allow carbon compounds to pass among the reservoirs. A simplified representation of the carbon cycle on Earth in Figure 11.16 identifies four reservoirs: the atmosphere, the biosphere, the hydrosphere, and the lithosphere. These reservoirs are connected by arrows that indicate the direction in which carbon compounds can flow from one reservoir to another. Some of the conduits permit two-way traffic while others are unidirectional . Under normal circumstances, the carbon cycle of the Earth maintains a state of dynamic equilibrium such that excess carbon in the atmosphere is transferred into the biosphere or into the hydrosphere and is ultimately sequestered in the lithosphere in the form of limestone (calcium carbonate), or carbon compounds in shale (kerogen), or fossil fuel (coal, petroleum, natural gas).

Consider that a large amount of anthropogenic carbon dioxide has been discharged into the atmosphere of the Earth. Figure 11.16 indicates that the excess carbon dioxide can be removed from the atmosphere by absorption by plants of the biosphere and by dissolution in the water of the hydrosphere (e.g., the oceans). The uptake

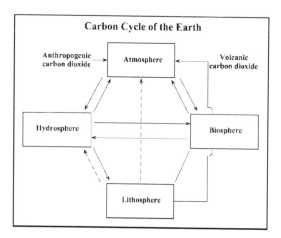

Carbon Cycle of the Earth

Figure 11.16. The carbon cycle of the Earth consists of reservoirs that are connected by means of processes that allow carbon to flow among them. In this diagram only the four major reservoirs are identified in order to illustrate how inputs of carbon dioxide to the atmosphere from anthropologic and volcanic sources are redistributed. Carbon that passes into the biosphere and the hydrosphere can either return to the atmosphere or is deposited in sedimentary rocks and is thereby incorporated into the lithosphere where it is stored for long periods of geologic time. The removal of carbon dioxide from the atmosphere moderates the intensity of greenhouse warming and the subsequent storage of carbon in the sedimentary rocks of the continental crust prevents it from causing the extreme greenhouse warming on the Earth that is occurring on Venus where most of the available carbon dioxide still resides in its atmosphere

of the excess atmospheric carbon dioxide by the biosphere and by the hydrosphere reduces the amount of greenhouse warming that might have been caused by the increase of the concentration of carbon dioxide in the atmosphere.

A large fraction of the carbon dioxide that is absorbed by marine plants is buried in the oceans in the form of complex hydrocarbon molecules (e.g., kerogen) and thereby enters the reservoir of the lithosphere, where it remains until the resulting sedimentary rocks are exposed to chemical weathering on the continents. When plants or animals on the continents die, some of the carbon they contain is recycled into the atmosphere in the form of carbon dioxide and methane depending on whether the organic matter is decomposed by bacteria under aerobic or unaerobic conditions, respectively.

The carbon dioxide that dissolves in the water of the hydrosphere reacts with water molecules to form carbonate ions (HCO_3^- or CO_3^{2-}). Both

ions react with calcium ions in solution to form calcium carbonate (i.e., calcite or aragonite) in the skeletons of corals, clams, snails, etc. When these organisms die, their skeletons and shells accumulate on the bottom of shallow oceans and ultimately form limestone. In this way, the carbon dioxide is removed from the atmosphere and is sequestered in the lithosphere in the form of limestone for tens or even hundreds of millions of years until the limestone dissolves in water on the continents.

The bottom line is that the possible extent of greenhouse warming on the Earth is limited because:

1. Most of the carbon that was present in the early atmosphere of the Earth is now locked up in the sedimentary rocks of the continental crust.

2. A large fraction of the carbon dioxide and methane that have been and are still being released into the atmosphere by anthropological and geological activities has entered the biosphere and the hydrosphere and most of that carbon is being taken out of circulation by entering the lithosphere.

This safety valve that has kept the atmosphere of the Earth cool does not exist on Venus. The failure of water vapor in the early atmosphere of Venus to condense and to form an ocean has allowed the carbon dioxide to remain in the atmosphere. As far as we can tell, the present crust of Venus does not contain sedimentary rocks in which carbon of atmospheric origin is stored. In other words, the carbon cycle of Venus contains only two reservoirs: the atmosphere and the lithosphere. Present knowledge indicates that the lithosphere of Venus is composed of volcanic and plutonic igneous rocks of basaltic composition. The igneous activity that has produced the crust of Venus is a *source* of carbon dioxide rather than a *sink*. Therefore, the intensity of the greenhouse warming of the atmosphere of Venus may have increased during episodes of active volcanic activity. Most of the carbon dioxide that was released by the eruption of volcanoes has remained in the atmosphere and very little has been removed in the form of calcite and/or aragonite during the chemical weathering of basalt exposed at the surface of Venus. Although Venus and the Earth are similar in terms of their masses and bulk chemical compositions, the evolution of their surface environments has made the Earth a haven of life but has turned the surface of Venus into a lifeless inferno.

11.4 Internal Structure

The mass of Venus (4.869×10^{24} kg), radius (6052 km), volume (9.28×10^{11} km^3), and bulk density (5.243 g/cm^3) in Table 11.3 are all less than those of the Earth. Nevertheless, the physical dimensions of Venus and the Earth are similar, which has led to the erroneous conclusion that the two planets also resemble each other in other ways. We have already shown that the surface environment of Venus is completely different from the environmental conditions on the surface of the Earth. We have also concluded that the basaltic crust of Venus is not being subducted into the mantle even though the mantles of Venus and the Earth are both convecting in the form of plumes that rise from depth and cause volcanic activity on the surface.

Table 11.3. Physical properties of Venus and the Earth (Hartmann, 2005; Freedman and Kaufmann, 2002)

Property	Venus	Earth
Radius, km	6,052	6,378
Mass, kg	4.869×10^{24}	5.974×10^{24}
Bulk density g/cm^3	5.243	5.515
Volume; km^3	9.28×10^{11}	10.86×10^{11}
Albedo, %	59	39
Atmosphere	carbon dioxide, nitrogen	nitrogen, oxygen
Surface gravity (Earth = 1.0)	0.91	1.0
Average surface temperature, °C	460	9
Surface material	volcanic rocks, no water	Igneous, sedimentary, and metamorphic rocks. Soil derived from these rocks. Oceans

11.4.1 The Core and Magnetism

The density of Venus in Table 11.3 is 5.243 g/cm^3 which is compelling evidence that Venus contains a large iron core like all of the terrestrial planets of the solar system. In addition, we recall that the uncompressed density of Venus in Table 3.4 is 4.2 g/cm^3. These data can be used to suggest that the core of Venus has a radius of about 3450 km and that its volume is close to 20% of the volume of the planet. The resulting thickness of the mantle of Venus is about 2600 km.

By analogy with the Earth, the core of Venus is at least partly liquid because the surface-to-volume ratios of the two planets in Figure 10.5 are virtually identical, which implies that they have been cooling at about the same rate. If the core of Venus is at least partly liquid, then Venus should have a magnetic field similar in strength to the magnetic field of the Earth as discussed in Section 6.4.5 and in Science Briefs 6.7.3,4,5,6, and 7. However, the Mariner 2 spacecraft determined during a flyby on December 14 of 1962 that Venus does *not* have a planetary magnetic field. The absence of a magnetic field on Venus has been attributed to a variety of causes, such as:

1. The core of Venus is solid (not credible).
2. The rotation of Venus is too slow to activate the convection currents in the liquid part of the core as required by the dynamo theory (possible).
3. Venus lacks a solid inner core like the Earth because the pressure in the core of Venus is not high enough to force liquid iron to solidify (questionable).
4. The magnetic field of Venus has decayed temporarily and may regenerate itself in the future with reversed polarity (questionable).

Venus does rotate even more slowly than Mercury as indicated by their rotation periods: 243.01 days (R) for Venus compared to 58.646 days (P) for Mercury, where R means "retrograde" and P means "prograde". Although Mercury does have a weak magnetic field in spite of its slow rotation, the rotation of Venus may be too slow to activate the dynamo. Therefore, the absence of a magnetic field on Venus has been attributed by many investigators to its slow rate of rotation.

An alternative explanation for the absence of a venusion magnetic field arises from the thermal regime in the mantle of Venus. We recall first of all that planets expel heat from their interiors by means of volcanic activity and that none of the volcanoes on Venus are currently erupting. Consequently, the temperature of the mantle of Venus may be rising. We also recall that heat may be released in the core when liquid iron solidifies along the spherical surface that separates the solid inner core from the liquid outer core. This so-called *latent heat of crystallization* causes the temperature of the liquid iron to increase, which causes it to expand, thereby decreasing its density. If the temperature of the liquid iron at the top of the core along the core-mantle boundary is lower than at the solid core-liquid core boundary, then convection in the liquid part of the core is possible and the dynamo mechanism may generate a planetary magnetic field.

This version of the dynamo theory may not work on Venus because heat from its mantle may have diffused into the liquid upper part of the core. In that case, the temperature gradient in the liquid core may have been eliminated, thereby preventing the liquid part of the core from convecting, which means that a magnetic field cannot be generated. Accordingly, Venus does not have a magnetic field either because it rotates too slowly to stir the liquid part of the core or because thermally driven convection is prevented in the liquid core, or both. In conclusion, we note that the generation and time-dependent variations of planetary magnetic fields are still not well understood. This gap in our understanding will become even more apparent when we try to explain the strong magnetic field of Jupiter and the much weaker magnetic fields of some of its satellites (e.g., Europa).

11.4.2 Episodic Resurfacing

We recall the evidence mentioned in Section 11.2.3 that the impact craters on Venus are uniformly distributed over its entire surface. This means that there are no old and heavily cratered regions on Venus and that, instead, the entire surface of the planet has the same age presumably because it formed at the same time.

In addition, the relatively low crater density indicates that the surface of Venus is only about 500 million years old although the age of Venus is 4.6 billion years as indicated by the crystallization ages of stony meteorites. Therefore, the surface of Venus must have been renewed at about 500 Ma by a process that completely destroyed or at least covered the previous surface.

The apparent domination of basalt and other kinds of volcanic rocks proves that Venus was resurfaced by an outpouring of lava on a global scale. Apparently, the volcanoes on Venus were dormant for hundreds of millions of years thereby allowing the temperature of the mantle to rise until at last a crack formed in the lithosphere. The resulting decompression melting may have been localized at first, but then spread until the entire crust was engulfed by magma on a global scale. The process ended with the formation of a new crust by crystallization of the basalt lava exposed at the surface. During the waning stage of the volcanic upheaval, hundreds of shield volcanoes were active in the low-lying planitiae and in certain volcanic centers such as the regiones of Alpha, Beta, and Phoebe. The resurfacing process ended when these late-blooming shield volcanoes became dormant. According to this hypothesis, the main phase of the volcanic upheaval ended about 500 million years ago and Venus has remained quiescent since then, except for the eruption of minor amounts of lava by a few volcanoes.

This hypothesis does explain the random distribution of impact craters on Venus and the uniformly low age of its surface. However, it raises a question about the thickness of the lithospheric mantle. The volcanic-upheaval hypothesis postulates that Venus has a thick lithosphere (about 300 km) which prevents continuous volcanic activity. On the other hand, the high surface temperature combined with estimates of the temperature gradient in the crust and upper mantle and the evidence of crustal deformation by compression and extension (e.g., tesserae) suggest that Venus has a thin lithosphere. Even the apparent absence of plate tectonics can be attributed to the mechanical weakness of a thin lithosphere that cannot withstand the stresses applied by convection in the asthenospheric part of the mantle.

The lithospheric plates under the ocean basins of the Earth are about 120 km thick and the temperature at that depth is about 1300 °C (Wyllie, 1992). If the temperature at the bottom of the ocean is close to 0 °C, the effective temperature gradient in the depth interval from 0 to 120 km is about 11 °C/km. The same temperature gradient applied to Venus, where the surface temperature is 460 °C, yields a temperature of about 1760 °C at a depth of 120 km. If we assume that the properties of rocks in the mantle of Venus change from lithosphere to asthenophere at a temperature of about 1300 °C, then the lithosphere on Venus is only about 76 km thick. Similar results were obtained by other investigators who used geophysical modeling (e.g., Nimmo and McKenzie, 1998; Hansen et al., 1999; Leftwich et al., 1999). The problem concerning the thickness of the lithosphere of Venus can be resolved by seismic data, which are not yet available because of the difficult environmental conditions on the surface of Venus.

11.5 Celestial Mechanics

The orbit of Venus is very nearly circular with an eccentricity of only 0.0068 and an average distance from the Sun of 0.723 AU. The sidereal period of revolution of Venus in Table 11.4 is 224.70 days and its average orbital speed is 35.0 km/s. By comparison, the orbit of the Earth has a higher eccentricity (0.017), its average distance from the Sun is greater (1.0 AU), and its sidereal period of revolution is longer (365.256 days) giving the Earth a lower average orbital speed of 29.79 km/s. In addition, both Venus and the Earth revolve around the Sun in the prograde direction as do all other planets in the solar system. All of these orbital properties of Venus and the Earth are "normal" and give no hint that Venus rotates "backwards" (i.e., in the retrograde direction).

11.5.1 Retrograde Rotation

The period and direction of rotation of Venus were difficult to determine because the surface of the planet is obscured by clouds that are opaque to visible light. Therefore, the correct period of rotation (243.01 days) and the direction of rotation (retrograde) were eventually determined in the early 1960s by bouncing a beam

Table 11.4. Orbital parameters of Venus and the Earth (Hartmann, 2005; Freedman and Kaufmann, 2002)

Property	Venus	Earth
Average distance from the Sun, AU	0.723	1.00
Average orbital speed, km/s	35.0	29.79
Eccentricity of the orbit	0.0068	0.017
Sidereal period of revolution, days	224.70	365.256
Synodic period of revolution, days	584	—
Period or rotation, days	243.01 R*	1.0
Inclination of the orbit to the ecliptic	3.39°	—
Axial obliquity	177.4°	23.45°
Escape velocity, km/s	10.4	11.2

*R = retrograde

of microwaves off the surface of Venus. The change in the wavelength of the reflected beam was interpreted by means of the Doppler effect to determine the velocity and direction of the rotation (Science Brief 11.7.3).

Another noteworthy property of Venus is that its axis of rotation is inclined relative to the plane of its orbit by an angle of 177.4° (i.e., almost 180°). As a result, the geographic north pole of Venus is pointing "down" and the geographic south pole is pointing "up". In other words, we are saying that Venus has been turned upside down. If that is true, then we can explain why Venus rotates "backwards".

Consider that Venus was originally rightside up and rotated in the prograde direction as seen from a point in space *above* the solar system. Now consider how Venus would look from a point in space *below* the solar system. If Venus appears to rotate in the prograde direction when seen from above, it will appear to rotate in the retrograde direction when seen from below. When Venus is flipped over, the rotation will be in the retrograde direction when viewed from above the solar system.

The problem with this explanation is that the axes of rotation of planets, like those of spinning tops, maintain their orientation in space and cannot be "flipped" easily. In addition, there is no corroborating evidence to support the hypothesis that the axis of Venus was actually flipped. All we can do is to postulate that Venus was impacted by a large planetesimal that caused its axis of rotation to be tipped by nearly 180°. The axes of rotation of Uranus and Pluto also deviate steeply from the vertical direction by 97.86° (Uranus) and 122.52° (Pluto). In fact, Mercury is the only planet with a rotation axis that is inclined by less than 1.0°. Therefore, we could conclude that Mercury is the only planet whose axis of rotation was *not* tipped appreciably by the impact of a large planetesimal.

A more conventional explanation for the retrograde rotation of Venus is that it resulted from gravitational interactions with its neighbors (i.e., the Earth, Mercury, and the Sun). However, these hypothetical interactions have not yet been confirmed by mathematical modeling and therefore remain speculative. Similarly, the inclinations of the axes of rotation of the other planets in the solar system (except Mercury) have also not been explained by gravitational effects. We could therefore regard the inclinations of the axes of rotation of the planets as evidence for the occurrence of celestial accidents or attribute the inclinations to unexplained gravitational effects in the solar system.

11.5.2 Phases, Conjunctions, and Transits

Venus is an *inferior* planet in the astronomical sense of that word, which means that its appearance as seen from the Earth depends on its position relative to the Sun. For example, when Venus is between the Earth and the Sun (i.e., at inferior conjunction), we see only its dark side. Actually, the atmosphere of Venus scatters sunlight from the "sunny" side to the back side. Therefore, the so-called *New phase* of Venus in Figure 11.17 consists of a very faint and diffuse ring of light.

After an inferior conjunction, Venus moves ahead of the Earth because its orbital speed is greater than that of the Earth. Therefore, the illuminated area of Venus as seen from the Earth

increases until the entire face of Venus is illuminated by the Sun in the so-called *Full phase*. As Venus continues in its orbit, it eventually catches up to the Earth again, which explains why the illuminated area of Venus decrease to the *Crescent phase* as Venus approaches the Earth for the next inferior conjunction. The phases of Venus were first observed around 1610 AD by Galileo Galilei who concluded from these observations that Venus orbits the Sun inside the orbit of the Earth, thus confirming the heliocentric model of the solar system postulated by Copernicus in 1545 AD.

The time elapsed between successive inferior conjunctions of Venus is its synodic period of revolution. The numerical value of the synodic period of Venus can be calculated from equation 7.6 stated in Section 7.2.7:

$$\frac{1}{P} = \frac{1}{E} + \frac{1}{S}$$

where P = sidereal period of Venus, E = sidereal period of the Earth, S = synodic period of Venus, and all periods must be stated in the same units of time. By substituting appropriate values from Table 11.4 into the equation we obtain:

$$\frac{1}{224.70.} = \frac{1}{365.256} + \frac{1}{S}$$

from which it follows that S = 583.77 days. Consequently, Venus goes through all of its phases in one synodic period of revolution that lasts 583.77 days or 1.598 Earth years.

When Venus is at *inferior conjunction*, it should pass across the face of the Sun in the phenomenon called a *transit* (Kurtz, 2005). Unfortunately, transits of Venus are rare and none occurred during the entire 20th century because the orbital plane of Venus is inclined by 3.39° relative to the plane of the ecliptic. This arrangement has the effect that Venus moves above or below the face of the Sun as seen from the Earth. However, when a transit of Venus is visible from the Earth, a second transit occurs eight years later. For example, the last transit of Venus observable from the Earth occurred on June 8, 2004, and the next one is predicted to occur eight years later on June 6, 2012. The interval between successive transits is equal to 365.256 × 8 = 2922.048 days. By dividing the

The Phases of Venus

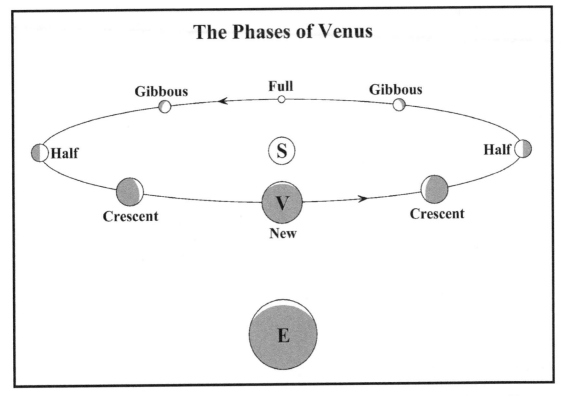

Figure 11.17. The diameter as well as the illumination of Venus (V) depend on its position relative to the Sun (S) as seen from the Earth (E). When Venus is in its Full phase behind the Sun, (i.e., superior conjunction), it is fully illuminated, but its diameter is at a minimum. As Venus moves into its Gibbous phase, more of it is illuminated and its diameter increases because it is approaching the Earth. As Venus continues in its orbit, its diameter continues to increase as it goes through its Half and Crescent phases. When Venus enters into its New phase (inferior conjunction), its diameter reaches a maximum although only a small crescent may be illuminated when viewed from the Earth. Adapted from Figure 4.9 of Freedman and Kaufmann (2002)

number of days between successive transits of Venus by its synodic period of revolution:

$$2922.048/583.77 = 5.00 \text{ synodic periods.}$$

In other words, the second transit occurs after Venus has "lapped" the Earth five times.

11.6 Summary

Venus is far from being the planet of sweetness and light as implied by its name. Instead, Venus disappoints our expectations and confronts us with puzzles that we cannot yet solve. For example, the high albedo that causes Venus to shine brightly in the evening and morning skies is caused by a layer of clouds composed of droplets of highly corrosive sulfuric acid. These clouds are a product of sulfur dioxide gas that was released by volcanic eruptions on the surface of Venus during its long history.

The surface of Venus can only be viewed by means of reflected radar waves because the clouds are opaque to light in the visible part of the electromagnetic spectrum. Nevertheless, we know that the atmosphere of Venus is composed largely of carbon dioxide which absorbs infrared radiation (greenhouse effect) and has caused the temperature at the surface to rise to 460 °C. In addition, the weight of the dense atmosphere has raised the atmospheric pressure at the surface of Venus to 90 times its value on the Earth. The combination of the corrosive clouds, the high surface temperature, and the crushing atmospheric pressure have created an extremely hostile environment on the surface of Venus.

Although Venus probably received about as much water from the planetesimals as the Earth, the water of Venus has vanished without a trace. We can speculate that the water of Venus initially resided in its atmosphere, but its subsequent fate is unknown. The water vapor may have condensed, in which case the surface of Venus was once covered by a deep ocean like the Earth. If so, the water apparently re-evaporated leaving the surface dry, lifeless, and hot. Alternatively, the water vapor in the atmosphere may not have condensed because the high temperature prevented the relative humidity of the atmosphere from reaching 100% (i.e., it never rained on Venus). In either case, the water has disappeared because the present atmosphere contains only a trace of water vapor.

The most likely explanation for the disappearance of water is that the H_2O molecules were dissociated by ultraviolet radiation of the Sun into hydrogen and oxygen gas. The hydrogen gas escaped into interplanetary space because the mass of hydrogen molecules (H_2) is too small to be retained by the gravitational force of Venus. However, we cannot explain what happened to the large amount of oxygen gas (O_2) that was released by the dissociation of water, because oxygen is not expected to escape from the atmosphere of Venus.

But there is more. The surface of Venus contains about 1000 large impact craters whose distribution is entirely random. This means that all parts of the surface of Venus have been exposed to impacts by meteorites for the same amount of time. The exposure age of the surface is about 500 million years. These facts and the resulting conclusions indicate that the entire surface of Venus was resurfaced about 500 million years ago by a global outpouring of basalt lava. In order to explain how such global volcanic eruptions can occur, we note that all of the volcanoes on Venus appear to be dormant at the present time, which implies that the heat generated by the decay of radioactive atoms in the interior of Venus is not being vented into space. Therefore, the temperature in the mantle of Venus increases until magma forms by decompression melting on a global scale. This hypothesis, in turn, requires that Venus has a thick lithosphere (about 300 km) that can

withstand the stresses imposed on it by plumes rising in the asthenosphere. However, there is much evidence that Venus, in fact, has a thin lithosphere of about 70 to 80 km, in direct contradiction to the postulated thick lithosphere.

The physical dimensions (radius, volume, mass, and bulk density) of Venus are similar to those of the Earth, which has led to the expectation that Venus is the twin of the Earth. The composition of the atmosphere and the conditions on the surface of Venus certainly do *not* confirm this relationship. However, the high bulk density does indicate that Venus has an iron core like the Earth, and the core should still be liquid because the two planets have cooled at about the same rate. A planet that contains a liquid iron core should also have a magnetic field like the Earth whose magnetic field is explained by the theory of the self-exciting dynamo. In spite of the similarity of the two planets, Venus does not have a magnetic field. We can resort to special pleading to explain this unexpected situation, but the absence of a magnetic field is a problem we have not yet solved.

One of several alternative explanations that have been proposed to explain the absence of a magnetic field arises from the slow rotation of Venus with a period of 243.01 days compared to 1.0 day for the Earth. In addition, the rotation of Venus is retrograde, which poses yet another problem. Gravitational interactions have been invoked but seem inadequate. A much more drastic solution of this problem is the possibility that the axis of rotation of Venus was flipped, meaning that its south geographic pole is now "up" and its north geographic pole is "down". Something like that has happened to Uranus and Pluto whose axes of rotation are also steeply inclined. However, the axes of rotation of planets and spinning tops are not easily flipped because their orientation in space is stabilized by forces arising from the rotation. Perhaps we should admit that if it happened, it must be possible (i.e., the inverse of Murphy's Law).

11.7 Science Briefs

11.7.1 Greenhouse Warming by Infrared Radiation

The high temperature of the atmosphere of Venus at ground level is caused by the way in which carbon dioxide molecules absorb electromagnetic radiation in the so-called infrared part of the wavelength spectrum (Science Brief 2.5.3,4). The wavelength of electromagnetic radiation in Figure 11.18 ranges widely from less than 10^{-6} nanometers (nm) to more than 10^{12} nm where $1.0\,nm = 10^{-9}\,m$. The wavelength spectrum of electromagnetic radiation is divided into several parts which are identified by the physical phenomena that are caused by specific ranges of the wavelength. For example, Figure 11.18 illustrates the parts of the wavelength spectrum that characterize x-rays, ultraviolet radiation, visible light of different colors from violet to red, infrared radiation, microwaves (radar), and radio waves. The human eye responds only to electromagnetic radiation in a narrow range of wavelengths from about 750 nm (red) to 400 nm (violet). The peak intensity of the radiation emitted by the Sun occurs at about 500 nm, but sunlight actually contains all wavelengths from near zero to 3000 nm and beyond. Electromagnetic radiation whose wavelength is greater than 750 nm is called infrared radiation, whereas radiation whose wavelength is less than 400 nm (but greater than 10 nm) is the ultraviolet part of the spectrum.

Electromagnetic radiation transmits energy whose magnitude increases as the wavelength decreases. In other words, the shorter the wavelength, the higher is the energy of the radiation. Gamma rays that are emitted by the daughter nuclei produced by spontaneous decay of unstable parent nuclei have high energies ranging from 0.1 to 10 million electron volts (MeV). X-rays have longer wavelengths than gamma rays and therefore have lower energies from about 10 kilo electron volts (keV) to 50 keV. Ultraviolet radiation, visible light, infrared radiation, etc. have progressively longer wavelengths and transmit correspondingly smaller amounts of energy.

We are concerned primarily with the infrared part of the wavelength spectrum because carbon dioxide molecules preferentially absorb radiation having wavelengths ranging from greater than 750 nm to less than 10^6 nm. As a result, the energy of infrared radiation is transferred to the carbon dioxide molecules causing the temperature of the carbon dioxide gas to increase. Several other gases (i.e, water vapor and methane) also preferentially absorb infrared

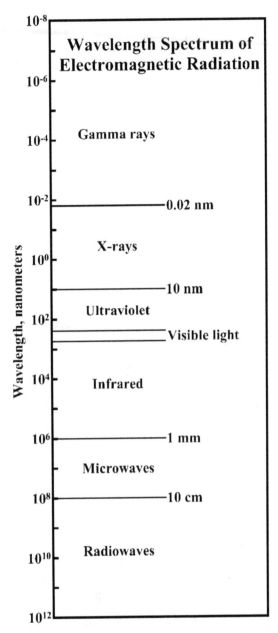

Figure 11.18. Spectrum of the wavelengths of electromagnetic radiation expressed in nanometers (nm), where $1.0\,nm = 10^{-9}\,m$. The human eye can only see radiation whose wavelength is less than 750 and more than 400 nm. Molecules of carbon dioxide, water vapor, and methane selectively absorb infrared radiation and thereby acquire the energy it contains. Consequently, the temperature of air containing these gases increases when it is irradiated with infrared radiation. Adapted from Freedman and Kaufmann (2002, Figure 5.7)

radiation and therefore contribute to the greenhouse warming of planetary atmospheres in which these gases occur.

All forms of electromagnetic radiation are propagated through a vacuum with the speed of light (c):

$$c = 2.99792458 \times 10^8 \, m/s$$

The speed of any wave phenomenon is related to its wavelength (λ) and its period (p) by the equation:

$$c = \frac{\lambda}{p} \qquad (11.1)$$

The reciprocal of the period is called the frequency (f) from which it follows that equation 11.1 can be expressed as:

$$c = \lambda f \qquad (11.2)$$

The energy (E) of electromagnetic radiation of wavelength (λ) and speed (c) is:

$$E = \frac{hc}{\lambda}$$

where h = Planck's constant (6.626176×10^{-34} Js), $1.0 \, J(joules) = 10^7$ ergs and the unit Js is referred to as Joule seconds. (See also Appendix Appendix 1).

11.7.2 Radius of the Core of Venus

According to equation 10.9 in Science Brief 10.6.1:

$$\frac{V_c}{V_p} = \frac{D_p - D_m}{D_c - D_m}$$

where V = volume, D = uncompressed density, and the subscripts mean: c = core, p = planet, and m = mantle. Substituting the uncompressed density of Venus ($D_p = 4.2 \, g.cm^3$, Table 3.4) and the densities of the mantle ($D_m = 3.32 \, g/cm^3$) and core ($D_c = 7.97 \, g/cm^3$) yields:

$$\frac{V_c}{V_p} = \frac{4.2 - 3.32}{7.97 - 3.32} = \frac{0.88}{4.65} = 0.19$$

This result means that the volume of the iron core of Venus is 19 % of the volume of the entire

planet. Consequently, the volume of the mantle of Venus (V_m) is 81% of the total. The data in Table 11.3 indicate that the volume of Venus (V_p) is $9.28 \times 10^{11} \, km^3$. Therefore, the volume of the core of Venus (V_c) is:

$$V_c = 0.19 \times 9.28 \times 10^{11} = 1.76 \times 10^{11} \, km^3$$

and the volume of the mantle of Venus (V_m) is:

$$V_m = 0.81 \times 9.28 \times 10^{11} = 7.51 \times 10^{11} \, km^3$$

We can now calculate the radius (r_c) of the core of Venus using the formula for the volume (V) of a sphere:

$$V = \frac{4}{3} \pi \, r^3$$

$$r_c = \left(\frac{3V_c}{4\pi} \right)^{1/3}$$

$$r_c = \left(\frac{3 \times 1.76 \times 10^{11}}{4 \times 3.14} \right)^{1/3} = (42.03 \times 10^9)^{1/3}$$

$$= 3476 \, km$$

The corresponding thickness of the mantle of Venus is: $6052 - 3476 = 2576 \, km$.

11.7.3 The Doppler Effect

When an object that is receding from an observer with a velocity (v) emits an electromagnetic wave (or a sound wave), the wavelength recorded by the observer is longer than it was when it left its source. In other words, the wave is stretched from its characteristic value (λ) to the observed value (λ') such that λ' is greater than λ. This phenomenon was used by Edwin Hubble to interpret the observed redshift of light emitted by distant galaxies (Sections 4.1). The same principle was used to determine both the rate and the sense of rotation of Venus from the increase and decrease of the wavelength of a beam of microwaves reflected by the surface of the planet (Section 11.5.1). The wavelength of the wave that was reflected by the area on Venus that was receding from the Earth was lengthened (red shift), whereas the wavelength of the wave that was reflected from the surface of Venus that moved toward the Earth was shortened (blue shift).

The equation derived in 1842 by Johann Christian Doppler (1803–1853) of Prague relates the ratio of the wavelength (λ'/λ) to the recessional velocity (v) and to the speed of light (c):

$$\frac{\lambda'}{\lambda} = 1 + \frac{v}{c} \qquad (11.3)$$

For example, if $\lambda'/\lambda = 1.1$, the recessional velocity of the source is $v = 0.1\,c$ or one tenth of the speed of light. Given that $c = 2.997 \times 10^8$ m/s, the recessional velocity in this example is $v = 0.1 \times 2.997 \times 10^8 = 2.997 \times 10^7$ m/s = $29,970$ km/s. (See also Appendix Appendix 1).

11.8 Problems

1. Balance the equation representing the reaction between methane gas (CH_4) and water vapor (H_2O) to form carbon dioxide (CO_2) and molecular hydrogen (H_2)

$$CH_4 + H_2O \rightarrow CO_2 + H_2$$

2. Write a brief essay explaining the potential significance of this reaction for the chemical evolution of the atmosphere of Venus.
3. Explain why Venus rotates in the retrograde direction even though it revolves in the prograde direction.
4. Explain why Venus does not have a magnetic field and consider the consequences for the environmental conditions on the surface of this planet.

11.9 Further Reading

Banerdt WB, Sammis CG (1992) Small-scale fracture of patterns on the volcanic plains of Venus. J Geophys Res 97:16,149–116,166

Barsukov VL, Basilevsky AT, Volkov VP, Zharkov VN (eds) (1992) Venus geology, geochemistry, and geophysics. Research results from the USSR. University Arizona Press, Tucson, AZ

Basilevsky AT, Kuzmin RO, Nikolaeva OV, Pronin AA and six others (1985) The surface of Venus as revealed by the Venera landings: Part II. Geol Soc Amer Bull 96:137–144

Basilevsky AT, Head JW (1995) Regional and global stratigraphy of Venus: A preliminary assessment and implications for the geologic history of Venus. Planet Space Sci, 43:1523–1553

Bougher SW, Hunten DM, Phillips RJ (eds) (1997) Venus II: Geology, geophysics, atmosphere, and solar wind environment. University of Arizona Press, Tucson, AZ

Cattermole P (1994) Venus: The geological story. Johns Hopkins University Press, Baltimore, MD

Crumpler LS, Head JW, Campbell DB (1986) Orogenic belts on Venus. Geol 14:1031–1034

Emiliani C (1992) Planet Earth. Cambridge University Press, Cambridge, UK

Faure G (2001) Origin of igneous rocks: The isotopic evidence. Springer-Verlag, Heidelberg

Fegley B, Prinn Jr RG (1989) Estimation of the rate of volcanism on Venus from reaction rate measurements. Nature, 337:55–57

Ford JP, Plaut JJ, Weitz CM, Farr TG, Senske DA, Stofan ER, Michaels G, Parker TJ (1993) Guide to Magellan image interpretation. Jet Prop Lab, Pub 93-24, Pasadena, CA

Freedman RA, Kaufmann III WJ (2002) The Solar System, 5th edn. Freeman, New York

Grinspoon DH (1997) Venus revealed: A new look below the clouds of our mysterious twin planet. Addison-Wesley, Reading, MA

Head JW, Basilevsky AT (1998) Sequence of tectonic deformation in the history of Venus: Evidence from global stratigraphic relationships. Geol 26(1):35–38

Head JW, Basilevsky AT (1999) Venus: Surface and interior. In: Weissman P.R., McFadden L.-A., Johnson T.V. (eds) Encyclopedia of the Solar System. Academic Press, San Diego, CA, pp 161–190

Hansen VL, Banks BK, Ghent RR (1999) Tessera terrain and crustal plateaus, Venus Geol 27(12): 1071–1074

Hartmann WK (2005) Moons and planets. 5th edn. Brooks/Cole, Belmont, CA

Hunten DM, Colin L, Donahue TM, Moroz VI (eds) (1984) Venus. University Arizona Press, Tucson, AZ

Hunten DM (1999) Venus: Atmosphere. In: Weissman P.R., McFadden L.A., Johnson T.V. (eds) Encyclopedia of the Solar System. Academic Press, San Diego, CA, pp 147–159

Johnson TV, Yeates CM, Young R, Dunne J (1991) The Galileo-Venus encounter. Science 253:1516–1517

Kaula WM (1990) Venus: A contrast in evolution to Earth. Science 247:1191–1196

Kurtz D (eds) (2005) Transits of Venus: New views of the solar system and galaxy. Cambridge University Press, Cambridge, UK

Leftwich TE, von Frese RRB, Kim HR, Noltimier HC, Potts LV, Roman DR, Li Tan (1999) Crustal analysis of Venus from Magellan satellite observations at Atalanta planitia, Beta regio, and Thetis regio. J Geophys Res 104(E4):8441–8462

Lewis JS, Prinn RG (1984) Planets and their atmospheres: Origin and evolution. Academic Press, San Diego, CA

Lewis JS (2004) Physics and chemistry of the solar system. 2nd edn. Elsevier, Amsterdam, The Netherlands

Moore JG, Chadwick Jr WW (2000) Offshore geology of Mauna Loa and adjacent areas, Hawaii. In: Rhodes J.M., Lockwood J.P. (eds) Mauna Loa Revealed. Geophys. Monograph 92, Amer. Geophys. Union, Washington, DC, 21–44

Nimmo F, McKenzie D (1998) Volcanism and tectonics on Venus. Annual Reviews Earth Planet Sci, 26:23–51

Phillips RJ, Arvidson RE, Boyce JM, Campbell DB, Guest JE, Schaber GG, Soderblom LA (1991) Impact craters on Venus: Initial analysis from Magellan. Science 252: 288–297

Phillips RJ, Raubertas RF, Arvidson RE, Sarkar IC, Herrick RR, Izenberg N, Grimm RE (1992) Impact craters and Venus resurfacing history. J Geophys Res 97:15, 923–915, 948

Phillips RJ, Hansen VL (1998) Geological evolution of Venus: Rises, plains, plumes, and plateaus. Science 279:1492–1498

Price M, Suppé J (1994) Mean age of rifting and volcanism on Venus deduced from impact-crater densities. Nature 372:756–759

Prinn RG (1985) The volcanoes and clouds of Venus. Sci Am, 252:46–53

Roth LE, Wall SD (1995) The face of Venus: The Magellan radar-mapping mission. NASA Spec Pub 520

Saunders RS et al., (1992) Magellan mission summary. J Geophys Res 97(E8):13067

Saunders RS (1999) Venus. In: Beatty J.K., Petersen C.C., Chaikin A. (eds) The new Solar System. Sky Pub. Co, Cambridge, MA, pp 97–110

Solomon SC (1993) The geophysics of Venus. Physics Today, 46:49–55

Stofan ER (1993) The new face of Venus. Sky and Telescope 86(8):22–31

Taylor SR (1992) Solar system evolution. A new perspective. Cambridge University Press, Cambridge, UK

Wagner JK (1991) Introduction to the solar system. Saunders College Pub., Philadelphia, PA

Weitz CM (1993) Impact craters. In: Ford J.F. et al. (eds) Guide to Magellan Image Interpretation. JPL Pub. 93–24. NASA and JPL/Caltech, Pasadena, CA, 75–92

Wyllie PJ (1992) Experimental petrology: Earth materials science. In: Brown G.C., Hawkesworth C.J., and Wilson C.R.L (eds) Understanding the Earth; A New Synthesis. Cambridge Univ. Press, Cambridge, UK, 67–87

Young C (eds) (1990) The Magellan Venus explorer's guide. Jet Propulsion Lab (JPL), Publication. Pasadena, CA, pp 90–24

Mars: The Little Planet that Could

The sunlight reflected by Mars has a reddish color, which caused the Romans to name this planet after Mars, the God of War whom they depicted as an invincible warrior in shining armor. The image of the planet Mars in Figure 12.1 reveals a desert world characterized by giant volcanoes and a system of deep canyons. Early telescopic studies of the surface of Mars received a boost in 1877 when its two small satellites Phobos (fear) and Deimos (terror) were discovered by *Asaph Hall* (1829–1907). In addition, the Italian astronomer *Giovanni Schiaparelli* (1835–1910) mapped numerous lineaments on the surface of Mars that had been discovered by *Angelo Sechhi* (1818–1878) who referred to them as "canali". The word was translated into English as "canals" and led to the idea that they were built by the inhabitants of Mars. The American astronomer *Percival Lowell* (1855–1916) actually established an observatory at Flagstaff, Arizona, in order to study these canals of Mars. He concluded (erroneously) that the canals were used to irrigate the crops planted by a dying race of Martians and thereby stirred the imagination of science-fiction writers (e.g., Wells, 1898; Bradbury, 1946; Clarke, 1951).

When the spacecraft Mariner 4 flew by Mars on July 14, 1965, many people on Earth expected that the images it sent back would confirm that intelligent beings had built the irrigation system on the surface of Mars. Instead, the pictures showed that the southern hemisphere of Mars is heavily cratered like the highlands of the Moon, and that the canals mapped by Schiaparelli and Lowell do not exist. Mariner 4 also reported that Mars does not have a magnetic field and therefore lacks radiation belts such as those of the Earth.

Additional images of the surface of Mars were obtained on July 31 and August 5, 1969, by Mariners 6 and 7. A major break-through in the study of Mars occurred on November 13, 1971, when Mariner 9 became the first spacecraft to go into orbit around Mars. The images of the martian surface returned by Mariner 9 revealed the presence of large volcanoes and a system of canyons whose size far exceeds the Grand Canyon of North America. In addition, the images of Mariner 9 included views of long sinuous valleys that were presumably carved by streams at a time in the past when liquid water was present on the surface of Mars.

The unmistakable evidence that liquid water had once flowed on the surface of Mars increased the probability that living organisms had existed on Mars during its early history. Consequently, NASA launched two Viking spacecraft in 1975 that were designed to determine whether microscopic forms of life are still present in the soil of Mars. Viking 1 landed successfully on July 20, 1976, at 23° north latitude in Chryse planitia. A few weeks later, Viking 2 also landed safely on September 3, 1976, at 48° north latitude in Utopia planitia.

The Viking landers remained operational until November of 1982 and returned a great deal of information about Mars. They analyzed the soil and air, searched for life, and recorded the weather conditions at their landing sites. The Viking mission also sent back more than 50,000 images of the surface of Mars and of its two satellites. Although the search for life in the soil of the two landing sites was not successful, this negative result does not exclude the possibility that lifeforms did occur elsewhere in more favorable sites on Mars (e.g., standing bodies of water, hotsprings, polar ice caps, and in the subsurface).

Mars has a thin atmosphere composed primarily of carbon dioxide (CO_2), with minor concentrations of molecular nitrogen (N_2) and water

Figure 12.1. This global view of Mars was recorded in April 1999 by the spacecraft Mars Global Surveyor (MGS) while it was in orbit around the planet. The principal features visible in this image include the north-polar ice cap, the canyons to the Valles Marineris, the three principal volcanoes of the Tharsis plateau, and Olympus mons standing alone to the west of the Tharsis volcanoes. The whispy white clouds reveal that Mars has a thin atmosphere that contains water vapor and carbon dioxide. (PIA 02653, courtesy of NASA/JPL/MSSS)Caltech)

vapor. The atmospheric pressure is less than 10 millibars or about one hundredth of the atmospheric pressure on the surface of the Earth. The air temperature ranges widely from $-33\,°C$ to $-83\,°C$ (night) (Hartmann, 2005, p. 295). The low pressure and temperature of the atmosphere prevent liquid water from occurring on the surface of Mars at the present time. Moreover, the virtual absence of oxygen (O_2) in the atmosphere and

the cold-desert climate impose restrictions on the exploration of the surface of Mars by humans.

The topography of Mars has been shaped both by internal processes as well as by impacts of solid objects. The northern hemisphere includes low-lying plains (planitiae), large shield volcanoes, and tectonic rifts, but contains comparatively few impact craters. The southern hemisphere is heavily cratered and includes several large impact basins and volcanic centers. The difference in the crater density indicates that the surface of the southern hemisphere of Mars is older than the surface of the northern hemisphere.

The technical terms used to describe the topographic features on Mars are similar to the terms used on Venus. For example, a planitia is a low-lying plain, whereas a planum is an elevated plain, mountains are called mons (singular) or montes (plural), valleys are referred to as vallis (singular) or valles (plural). A fossa on Mars is a rift valley (i.e., a graben) and a chasma is a canyon. In addition, the topography of some areas on Mars is described by Latin phrases such as Vastitas Borealis (northern vastness) and Noctis Labyrinthus (labyrinth of the night).

Mars is a much smaller planet than the Earth because its radius is only 3398 km compared to 6378 km for the radius of the Earth, and the mass of Mars (6.42×10^{23} kg) is about one tenth the mass of the Earth (5.978×10^{24} kg). Nevertheless, the former presence of liquid water on the surface of Mars means that it *could* have harbored life in the past and that it *could* be colonized in the future by humans from the Earth. In this sense, Mars is the little planet that "could." The on-going exploration of the surface of Mars has transformed it from a point of light studied by astronomers into a *world* to be explored by geologists. The geology and history of Mars have been described in several excellent books including those by Moore (1977), Greeley (1985), Kieffer et al. (1992), Carr (1996), Sheehan (1996), Sheehan and O'Meara (2001), Morton (2001), Hartmann (2003), Kargel (2004), Hanlon (2004), De Goursac (2004), and Tokano (2005). Each of these books contains references to additional publications about our favorite planet as well as numerous well-chosen images of the landscape.

12.1 Origin and Properties

Mars formed at the same time, from the same chemical elements, and by the same process as all of the planets and their satellites of the solar system. Although the details may vary depending on the distance from the Sun and the amount of mass of a particular body, the theory of the origin of the solar system provides enough guidance to allow us to reconstruct the sequence of events that led to the accumulation and internal differentiation of each of the planets, including Mars.

12.1.1 Origin by Accretion of Planetesimals

Mars accreted from planetesimals within the protoplanetary disk at a distance of 1.52 AU from the Sun. Consequently, it may have contained a greater abundance of volatile elements and compounds than Mercury, Venus, and Earth. Like the other terrestrial planets of the solar system, Mars heated up because of the energy that was released by the impacts of planetesimals and by decay of unstable atoms of uranium, thorium, potassium, and others. As a result, Mars was initially molten and differentiated by the gravitational segregation of immiscible liquids into a core composed primarily of metallic iron, a magma ocean consisting of molten silicates, and a sulfide liquid that may have been trapped within the core of liquid iron and within the magma ocean. The volatile compounds (H_2O, CO_2, CH_4, NH_3, and N_2) were released during the impacts of the planetesimals and formed a dense primordial atmosphere.

As the magma ocean crystallized, an anorthositic crust may have formed by plagioclase crystals that floated in the residual magma. However, this primitive crust has not yet been recognized on Mars, perhaps because it was deeply buried by ejecta during the period of intense bombardment between 4.6 and 3.8 Ga. The present crust of Mars is composed of lava flows interbedded with sedimentary rocks and ejecta deposits derived from large impact basins. The slow cooling of the magma ocean also caused the crystallization of olivine, pyroxene, and other refractory minerals which sank in the residual magma and thus formed the mantle composed of ultramafic rocks (e.g., peridotite). The present mantle of Mars is solid and is divisible into an asthenosphere and a lithosphere on the basis of the temperature-dependent mechanical properties of the ultramafic rocks.

The presence of large shield volcanoes in the northern as well as the southern hemisphere of Mars suggests that its asthenosphere convected by means of plumes that rose from depth until they impinged on the underside of the lithosphere. The basalt magmas that formed by decompression melting of the plume rocks and of the adjacent lithosphere built up shield volcanoes on the surface (e.g., Olympus mons and others). These volcanoes reached great height because the lithosphere of Mars is stationary. In other words, the lithosphere consists of a single plate that encompasses the entire planet, which makes plate tectonics impossible. The absence of plate tectonics also prevented surface rocks from being recycled into the mantle of Mars and thereby prevented the development of igneous rocks of granitic composition. In this respect, Mars resembles Mercury, Venus, and the Moon, which also did not permit rocks of granitic composition to form.

The core of Mars was initially molten and gave rise to a planetary magnetic field that magnetized the oldest crustal rocks of the southern hemisphere. The magnetic field of Mars lasted only a few hundred million years until the core solidified during the Noachian Eon. At the time of the large impact that formed the Hellas basin in the southern hemisphere, the magnetic field of Mars had already ceased to exist because the rocks that underlie that basin were not magnetized as they cooled through their Curie temperature.

12.1.2 Physical and Chemical Properties

The physical dimensions of Mars in Table 12.1 indicate that its radius (3394 km) is only half that of the Earth (6378 km) and that its volume (1.636×10^{11} km^3) is only about 15% of the volume of the Earth. Nevertheless, the surface area of Mars (1.45×10^8 km^2) is similar to the surface of the continents of the Earth (1.43×10^8 km^2). Therefore, to an explorer on the ground, Mars is as large as the Earth. The surface gravity of Mars is 38% of the surface gravity of the Earth because the mass of Mars (6.42×10^{23} kg) is only about 10% of the mass of the Earth (5.974×10^{24} kg). The low gravity on the surface of Mars will someday cause explorers to feel light on their feet. It may also weaken their bones and muscles, which will cause problems when they return to the Earth. The mass and volume of Mars yield its bulk density (D):

$$D = \frac{6.42 \times 10^{23} \times 10^3}{1.636 \times 10^{11} \times 10^{15}} = 3.92 \, \text{g/cm}^3,$$

Table 12.1. Physical properties of Mars and the Earth (Hartmann, 2005; Freedman and Kaufmann, 2002)

Property	Mars	Earth
Radius, km	3394	6378
Mass, kg	6.42×10^{23}	5.974×10^{24}
Bulk density g/cm^3	3.924	5.515
Volume, km^3	1.636×10^{11}	10.86×10^{11}
Albedo, %	15	39
Atmosphere	carbon dioxide, nitrogen	nitrogen, oxygen
Surface gravity (Earth = 1.0)	0.38	1.0
Average surface temperature °C	−53	+9
Maximum	+20	+60
Minimum	−140	−90
Surface material	Igneous and sedimentary rocks and their weathering products	Igneous, sedimentary, and metamorphic rocks, soil, oceans

whereas the uncompressed density of Mars is 3.75 g/cm^3 (Table 3.4 and Figure 3.6A). The albedo of Mars (15%) is less than half that of the Earth (39%) because Mars presently lacks the clouds and oceans that enhance the albedo of the Earth.

The chemical compositions of the mantle + crust and of the core of Mars in Table 12.2 were estimated by Taylor (1992) based on analyses of the martian meteorites (Science Brief 12.13.1). The data show that the mantle and crust are composed primarily of Mg-Fe silicates consistent with the presence of ultramafic rocks in the mantle and of Mg-rich basalts (komatiites?) in the crust. The concentrations of aluminum and calcium are low ($Al_2O_3 = 3.02\%$, $CaO = 2.45\%$), and the concentrations of TiO_2, Na_2O, K_2O, P_2O_5, and Cr_2O_3 add up to only 1.6%.

The chemical composition of the core of Mars in Table 12.2 is dominated by metallic iron (77.8%) which is alloyed with Ni (7.6%), Co (0.36%), and S (14.24 %). Taylor (1992) estimated that the mass of the core amounts to about 21% of the mass of the whole planet (i.e., 1.35×10^{23} kg) and that its radius is close to

Table 12.2. Chemical compositions of Mars and the Earth based on analyses of martian and asteroidal meteorites in percent by weight. Adapted from Table 5.6.2 (Mars) and 5.5.1a (Earth, mantle + crust) of Taylor (1992) and (Earth, core) by Glass (1982)

Elements	Concentrations, %	
	Mars	Earth
Mantle and Crust		
MgO	30.2	35.1
Al_2O_3	3.02	3.64
CaO	2.45	2.89
SiO_2	44.4	49.9
TiO_2	0.14	0.16
FeO	17.9	8.0
Na_2O	0.50	0.34
K_2O	0.038	0.02
P_2O_5	0.16	–
Cr_2O_3	0.76	–
Sum	99.568	100.1
Core		
Fe	77.8	90.6
Ni	7.6	7.9
Co	0.36	0.5
S	14.24	0.7
P		0.2
C		0.4
Sum	100	99.94

50% of the planetary radius (i.e., 1697 km). The high concentration of sulfur implies that the core contains appreciable amounts of sulfides of iron as well as of other metals (Cu, Ni, Mo, Co, and Zn) that form bonds with sulfur.

12.1.3 Evolution of the Atmosphere

The primordial atmosphere of Mars contained most of the water vapor, carbon dioxide, and other gases that were released during the initial accretion of the planet. Additional quantities of volatile compounds and elements were discharged by the frequent volcanic eruptions at the start of the Noachian Eon. The pressure of the primordial atmosphere of Mars was much higher than that of the present atmosphere and may have approached the pressure of the present-day atmosphere of Venus. As the martian atmosphere cooled, the water vapor began to condense to form droplets of liquid water, that accumulated

on the surface of the planet to form a global ocean several kilometers deep. The concurrent removal of carbon dioxide from the atmosphere reduced the greenhouse warming of the atmosphere and caused a decline of the average global temperature of Mars. Nevertheless, during the early Noachian Eon between 4.6 and about 4.0 Ga, the climate on the surface of Mars was similar to the climate that existed on the Earth at that time. There was an abundance of water on the surfaces of both planets and their average global temperatures were well above 0 °C. However, during the period of intense bombardment between 4.6 and 3.8 Ga both planets lost water and part of their atmospheres as a consequence of the explosive impacts of solid objects. The losses suffered by Mars were greater than those of the Earth because Mars has only about one tenth of the mass of the Earth. As a result, the climate of Mars began to deviate from that of the Earth by becoming colder and drier.

12.1.4 Orbit of Mars

Mars revolves around the Sun and rotates on its axis in the prograde direction. In addition the plane of its orbit is inclined by only 1.85° to the plane of the ecliptic. The average distance of the center of Mars from the center of the Sun (i.e., the length of the semi-major axis of its orbit) is 1.524 AU compared to 1.0 AU for the Earth. Therefore, Mars is a superior planet whose sidereal period of revolution is 686.98 days, or 1.880 years by our way of measuring time on the Earth. One day on Mars (called a sol) is 37 minutes and 22 seconds longer than one day on the Earth. Therefore, one sol is equal to 1.0259 Earth days and one year on Mars contains:

$$\frac{686.98}{1.0259} = 669.63 \, sols$$

The sidereal periods of revolution and rotation of Mars in Table 12.3 may someday become the basis for a martian calendar to be used by humans living on Mars.

The slower average orbital speed of Mars causes the Earth to overtake it in its orbit. The time that elapses between successive oppositions or conjunctions of Mars (Sections 7.2.6, 7.2.7) is

Table 12.3. Orbital parameters of Mars and the Earth (Hartmann, 2005; Freedman and Kaufmann, 2002)

Property	Mars	Earth
Average distance from the Sun, AU	1.524	1.00
Average orbital speed, km/s	24.1	29.79
Eccentricity of the orbit	0.093	0.017
Sidereal period of revolution, days	686.98	365.256
Synodic period of revolution, days	780	—
Period of rotation, days, min., s	1.0d, 37 min, 22 s	1.0d
Inclination of the orbit to the ecliptic	1.85°	—
Axial obliquity	25.19°	23.45
Escape velocity, km/s	5.0	11.2

12.2 Surface Features: Northern Hemisphere

The boundary between the northern and southern physiographic regions of Mars does not closely coincide with the planetary equator but ranges up to about 20° north and south of it. The topography of the northern physiographic province is dominated by the Tharsis plateau and by the low plains that merge northward into the Vastitas Borealis which surrounds the large ice cap north of 80°N latitude. The principal planitiae of the northern hemisphere include Amazonis, Acidalia, and Chryse located west of the zero-meridian, whereas Utopia, Elysium, and Isidis occur east of the zero-meridian.

12.2.1 Tharsis Plateau

The Tharsis plateau in Figure 12.2 straddles the equator and extends from about 40°S to about 50°N latitude and from 60°W to 140°W longitude. The main body of this plateau rises to an elevation of 7 km above the zero-elevation contour, but the highest points are the summits of the shield volcanoes Arsia, Pavonis, and Ascraeus which are aligned along a northeast – trending lineament on the top of the plateau. Even more impressive is Olympus mons which stands apart at the western edge of the Tharsis plateau at about 18°N and 134°W and whose summit has an elevation of 21,287 m. The summit elevations of the three volcanoes that rise from the Tharsis plateau are: 17,800 m (Arsia mons), 14,100 m (Pavonis mons), and 18,200 m (Ascraeus mons). The elevation of the summit of Alba patera (40°N, 110°W) at the northern end of the Tharsis plateau is only about 2000 m. Another large volcano called Elysium mons (25°N, 212°W) stands near the edge of Elysium planitia attended by two minor volcanoes named Albor and Hecates. The summit of Elysium mons has an elevation of 14,125 m, which makes it the fourth-highest volcano on Mars.

The surface of the Tharsis plateau in Figure 12.2 contains only a few impact craters indicating that it is younger than the intensely cratered highlands of the southern hemisphere of Mars. In this respect, the surface of the Tharsis plateau resembles the surfaces of the mare regions of the Moon, which suggests that the age

its synodic period whose value is 780 days. The sidereal period of revolution (P) of a superior planet is related to its synodic period (S) and to the sidereal period of the Earth by equation 7.7 that was derived in Science Brief 7.6.2 and was stated in Section 7.2.7:

$$\frac{1}{P} = \frac{1}{E} - \frac{1}{S}$$

Given that E = 365.256 days and that P = 686.98 days, equation 7.7 yields S = 779.93 days or 780 days as stated in Table 12.3.

The axial obliquity of Mars (25.19°) is similar to that of the Earth, (23.45°) which means that the weather on Mars varies seasonally like the weather on the Earth. However, the martian seasons last about 171.7 days each compared to only 91.3 days for the seasons on the Earth. Other factors affecting the seasonality of the weather on Mars are the eccentricity of its orbit (0.093), which is more than five times greater than that of the orbit of the Earth (0.017), and time-dependent changes in the inclination of the axis of rotation of Mars (i.e., its obliquity).

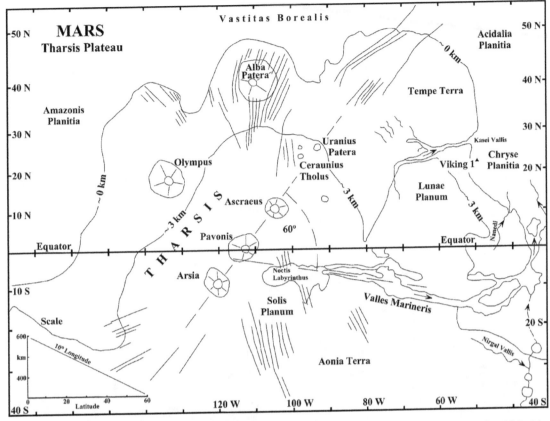

Figure 12.2. Topography of the northern hemisphere of Mars. Adapted from "An Explorer's Guide to Mars" published by the Planetary Society

of the Tharsis plateau is less than about 3.8 Ga. The presence of shield volcanoes implies that the Tharsis plateau is composed of basalt which can form large plateaus because basalt lavas have a low viscosity. Similar basalt plateaus on Earth were described by Faure (2001).

The build-up of the Tharsis plateau is probably attributable to the presence of a large plume at the base of the underlying lithosphere. Such a plume could have formed a dome in the overlying lithosphere and crust as a result of the buoyant force the plume exerted against the underside of the lithosphere and because of the expansion of the rocks in the lithosphere caused by the increase of their temperature. As a result, fractures could have formed that were propagated to the surface to form rift valleys (e.g., fossae), some of which became conduits for large volumes of basalt magma that formed by decompression melting of rocks in the lithospheric

mantle and in the head of the plume. According to this hypothetical scenario, the volcanic rocks that form the surface of the Tharsis plateau were erupted through fissures that were later buried by the resulting layers of flood basalt. Moreover, the three principal shield volcanoes (Arsia, Pavonis, and Ascraeus) could be aligned because the magma that fed them reached the surface by means of the same fracture system. The existence of this fracture system is supported by the positions of several small volcanoes (e.g., Ceraunius tholus and Uranius patera) located along the extension of the chain of the principal volcanoes and by the sets of parallel fossae that occur along strike of the lineament they define.

The Tharsis plateau is located at the center of a radial system of rift valleys in Figure 12.2, some of which extend far into the ancient highlands of the southern hemisphere. Others radiate into the surrounding planitiae of the northern hemisphere and into the north-polar desert (Vastitas Borealis). Although these rift valleys can be traced for hundreds of kilometers in all directions, they are ultimately buried by alluvial and eolian deposits. The largest of these rift valleys that emanate from the Tharsis plateau are the Valles Marineris. The angle between the volcanic lineament in Figure 12.2 and the Valles Marineris is almost exactly 60°, which supports the hypothesis that the two features are conjugate fractures.

12.2.2 Olympus Mons

Olympus mons in Figure 12.3 is the largest volcano in the entire solar system because its summit

has an elevation of 21,287 m above the zero-elevation contour and because it towers 18 km above the surrounding plain. Olympus mons has a diameter of about 600 km and a circumference of 1884 km at its base, which circumscribes an area of 282, 600 km^2 or 2.65 times the area of the state of Ohio. The average gradient of the slope of Olympus mons is about 6%. The physical dimensions of Olympus mons exceed those of Maat mons on Venus as well as those of Mauna Loa on the Earth (Section 11.2.1). The latter rises 8.84 km above the seafloor and the gradient of its slope above sealevel is 7.3%. Olympus mons is so big that it will be difficult for an astronaut standing somewhere in the Tharsis desert

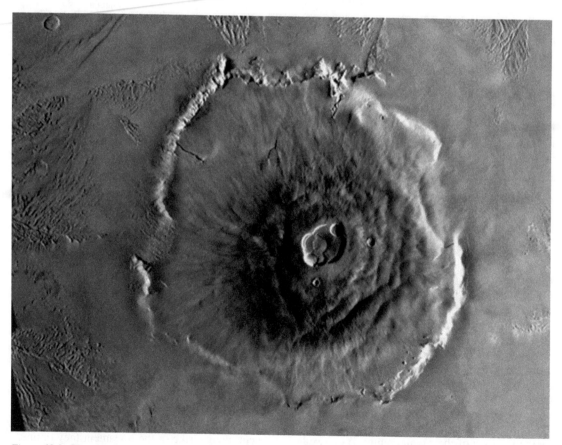

Figure 12.3. Olympus mons is the largest shield volcano in the solar system. Its summit has an elevation of 21,287 m and its base occupies an area that is 2.65 times the area of the state of Ohio. Basalt lava flows were erupted from the large crater at the summit and flowed down the slopes of the mountain. The base of the volcano forms a high cliff, part of which has collapsed. The origin of this cliff is discussed in the text. The lava flows that form the slopes of Olympus mons contain only a few impact craters, which tells us that this volcano may have erupted for the last time less than about ten million years ago. (PIA 02982, Viking Orbiter 1, courtesy of NASA/JPL/Caltech)

to see all of it in one view. However, we have already observed it from space at times when its summit plateau was shrouded by clouds that form as moist air rises until clouds form by condensation of water vapor. On such occasions, Olympus mons resembles Mauna Loa whose summit likewise causes clouds to form on frosty mornings.

The summit plateau of Olympus mons contains a gigantic caldera (70 km wide and 3 km deep) that could accommodate a large city such as New York or Los Angeles. The slopes of the mountain are covered by overlapping lava flows. The flows that are exposed at the surface contain very few if any impact craters and therefore appear to be quite young (i.e., less than 10 million years). As a result, Olympus mons may only be dormant at the present time and could erupt again in the future (Hartmann, 2003, p. 313). Even if Olympus mons has erupted in the past ten million years, the large volume of the volcanic rocks that have been extruded requires that this volcano has been active for hundreds of millions of years or more.

Another noteworthy feature of Olympus mons is the cliff (up to six kilometer high) that surrounds the base of the mountain. This cliff has collapsed in a few places where large rock slides have occurred. Some of the most voluminous lava flows actually reached the top of the cliff and cascaded to the base of the mountain by flowing between their own levees formed of solidified lava. The origin of the cliff at the base of Olympus mons is not known, although various hypotheses have been proposed (e.g., wind erosion, faulting.). The most plausible explanation for the existence of the cliff is that it was either formed by wave erosion (i.e., it is a sea cliff), or that the eruption of lava flows occurred within an ice sheet. These ideas were proposed in 1972 by Henry Faul in an unpublished manuscript mentioned by Kargel (2004, p. 108).

Olympus mons presumably stands above a large mantle plume that caused magma to form by decompression melting in the lower lithosphere. The magma then rose to the surface through fractures in the lithosphere and crust to build up the enormous pile of basalt that forms the mountain. If this is the correct explanation for the origin of Olympus mons, it bears a striking resemblance to the origin of Mauna Loa but with one important difference. The Hawaiian islands are the summits of extinct volcanoes that formed by the eruption of magma above a mantle plume (also called a hotspot) that is presently located beneath Mauna Loa and Mauna Kea on the Big Island. None of the Hawaiian volcanoes have achieved the size of Olympus mons because the movement of the Pacific plate displaced them from the plume and thus deprived them of magma. Therefore, the Hawaiian islands consist of a chain of volcanoes that continue to move with the Pacific plate at a rate of about 11 cm/y (Faure, 2001). In contrast to the Hawaiian islands, Olympus mons achieved its large size because it has remained above the plume that supplies it with magma. In other words, Olympus mons is much larger than Mauna Loa because it has not been displaced from the plume that is the source of the magma it erupts. Or, to put it another way, Mars does not have plate tectonics.

12.2.3 Young Lava Flows

The low abundance of impact craters on some lava flows of Olympus mons and on other volcanoes tells us that these flows "are not very old." Such statements raise two important questions:

1. What are the ages of the youngest lava flows on Mars?
2. Does the presence of "young" lava flows mean that the volcanoes on Mars are not yet extinct and that Mars is still an active planet?

The Amazonis and Elysium planitiae of the northern hemisphere of Mars contain lava flows that appear to be remarkably young. Hartmann (1999) used the images returned by the Mars Global Surveyor to determine that the exposure age of the lava flows of Elyseum mons lies between 600 and 2000 million years. Hartmann (2003) also cited examples of lava flows in Amazonis and Elysium planitiae that may be less than 100 million years old. For example, a lava flow at $9°N, 208°W$ in Elysium planitia is virtually crater-free compared to the heavily cratered surface across which it flowed. The exposure ages of the youngest flows on Olympus mons and on the Amazonis and Elysium planitiae are as low as ten million years. Therefore, dates based on crater counts suggest that magmatic processes have occurred on Mars during the Cenozoic Era. The eruption of young basaltic lava flows on the surface of Mars is confirmed by isotopic age determinations of basaltic shergottites which crystallized between 154 and 474 Ma (Section 12.7 and Science Brief 12.13.1) Apparently, the volcanoes of Mars did erupt more recently than expected and could erupt again, but the probability of witnessing a volcanic eruption on Mars in future centuries is small.

The extrusion of lava flows on the surface of Mars as recently as ten million years ago (if confirmed by isotopic dating of rock samples from Mars) would indicate that this planet is still geologically active. This conclusion is not supported by the high surface-to-volume ratio of Mars in Figure 10.5 which implies that this planet has cooled more rapidly than the Earth and Venus.

12.2.4 Valles Marineris

The Valles Marineris, in Figure 12.2 were named after the spacecraft Mariner 9 and extend east from Noctis Labyrinthus on the eastern slope of the Tharsis plateau for a distance of about 3500 km before turning north for another 1200 km and opening into Chryse planitia. The data of the Mars Global Surveyor Laser Altimeter (MGSLA 1976–2000) indicate that parts of the canyons extend to a depth of 5 km *below* the zero-elevation contour, whereas the adjacent plateau has an elevation of about 4 km *above* the reference contour. Therefore, the lowest part on the martian canyons is about 9 km below the level of the Tharsis plateau adjacent to the canyons. The width of the canyons ranges from about 600 km (Melas, Candor, and Ophir chasmata combined) to about 120 km along most of Coprates chasma.

The length, depth, and width of the Valles Marineris exceed the dimensions of the Grand Canyon of northern Arizona which are: length = 445 km; depth = 1.55 km; width = 0.2 to 29 km. In addition, the Valles Marineris are rift valleys formed as a result of extension of the crust related to the uplift of the Tharsis plateau. The Grand Canyon of Arizona was actually eroded by the Colorado River which transports about 500,000 tons of sediment per day. The rocks exposed in the walls of the Grand Canyon range in age from Archean to Cenozoic and represent one of the best exposures of sedimentary, igneous,

and metamorphic rocks on the Earth. The rocks exposed in the walls of the Valles Marineris consist of layers of sedimentary and volcanic rocks but have not yet been studied in detail.

The Noctis Labyrinthus is located on the Tharsis plateau in Figure 12.4 between the three volcanoes and at the western end of the Valles Marineris. It occupies an area of $350,000 \, km^2$ and consists of a maze of irregular topographic depressions and intersecting canyons resembling collapsed terrain in karst topography on the Earth. This chaotic terrain appears to have been the source of water that flowed through the Valles Marineris and was discharged into the Chryse planitia. The fine grained sediment and boulders transported by these episodic floods were deposited in the planitia that may have contained a body of standing water at that time. Some of the water that flowed through the Valles Marineris could also have been discharged by springs on the valley walls of the canyons where aquifers were exposed.

The nearly vertical walls of the Valles Marineris in Figure 12.5 have collapsed in several places to form huge landslide deposits that have piled up on the floor of the canyons (e.g., Candor and Ganges chasmata). These landslides have caused the canyons to become wider and shallower than they were before the walls began to erode. Some of the fine-grained sediment in the landslide deposits was later redistributed by wind into large dune fields. In many places, the walls of the Valles Marineris are cut by gullies, clearly visible in Figure 12.6, that were eroded by water flowing from the adjacent plateau into the canyons (e.g., Tithonium and Coprates chasmata).

12.2.5 Utopia Planitia

The topographic dichotomy of Mars mentioned in the introduction to this chapter may have been caused by a very large impact that effectively removed the old cratered surface of the northern hemisphere and scattered the ejected rocks all over the southern hemisphere. Such an event could explain why the elevation of the surface of the northern hemisphere of Mars is lower than the elevation of the southern hemisphere. Subsequently, the water on the surface of Mars drained into the depression of the

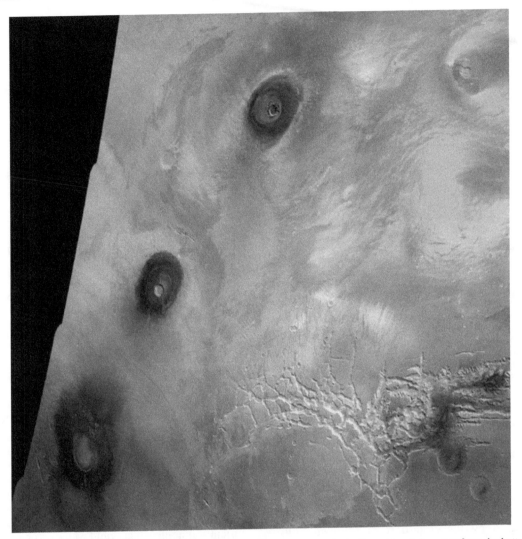

Figure 12.4. The three principal shield volcanoes of the Tharsis plateau are Arsia, Pavonis, and Ascraeus from the bottom to the top of the image. The Noctis Labyrinthus area east of the Tharsis plateau consists of a labyrinth of intersecting box canyons, which were the source of muddy water that flowed into the Valles Marineris depicted in Figure 12.5. The water was ultimately discharged into Chryse planitia identified in Figure 12.2. The canyons of Noctis Labryinthus appear to have formed by caving of the surface when permafrost in the subsurface melted in response to the heat emanating from basalt magma rising from the underlying lithospheric mantle. (Mosaic of images recorded by the orbiter of Viking 1 on February 22 of 1980; MGO 01N104-334S0, courtesy of NASA)

northern hemisphere and deposited sediment that buried impact craters until the climate eventually turned cold and dry and liquid water ceased to occur on the surface of Mars. There is no proof that the northern hemisphere of Mars is, in fact, a giant impact crater, but impacts of this magnitude were certainly possible and probably did occur during the early history of the

solar system (e.g., Section 9.6, the origin of the Moon). The hypothetical impact on the northern hemisphere of Mars would have occurred *before* the formation of the Tharsis plateau and the Valles Marineris.

The principal claim to fame of Utopia planitia is the landing of Viking 2 which touched down on September 3,1976, at 48°N 226°W about

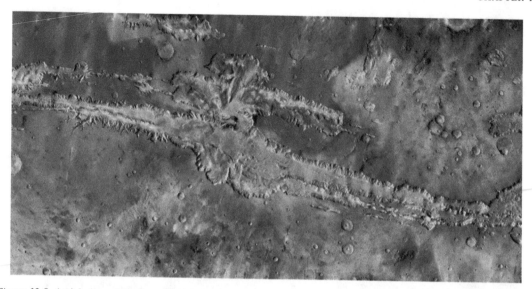

Figure 12.5. Aerial view of the canyons of the Valles Marineris of Mars. The canyons are a system of parallel rift valleys that probably formed as a consequence of the uplift of the crust prior to the eruption of floods of basalt that built up the Tharsis plateau and the large shield volcanoes identified in Figures 12.2 and 12.4. The walls of the rift valleys were later modified by landslides that locally widened the canyons and elevated their floors. In addition, the canyons were modified by mudflows that were released from the subsurface of Noctis Labyrinthus. Additional water drained through gullies from the surface of the Tharsis plateau. The water that flowed through the canyons was discharged into Chryse planitia identified in Figure 12.2. (Mosaic of images (JPG 74MB) recorded by the Thermal Emission Imaging System (THEMIS) of the Mars Odyssey orbiter, courtesy of NASA/JPL/Caltech)

200 km west of a large impact crater called Mie. This site was chosen primarily to assure a safe landing of the spacecraft rather than because it promised scientific rewards not achievable on Chryse planitia where Viking 1 had landed on July 20 of 1976. Therefore, it came as no surprise that the landscape around Viking 2 in Figure 12.7 (like that of Viking 1) is a plain covered by fine-grained reddish-brown "soil" containing a multitude of angular boulders of volcanic rocks some of which appear to be vesicular. Close-up images show that the surface of the fine-grained sediment is marked by wind ripples or "minidunes" indicating that wind has played an important part in shaping the surface of Utopia as well as of Chryse planitia.

12.2.6 Soil of Utopia and Chryse

The reddish brown color of the martian soil is caused by the presence of iron oxides derived by chemical weathering of iron-rich volcanic rocks (e.g., basalt). When basalt weathers on the surface of the Earth, the iron in the silicate minerals (pyroxene and olivine) is first oxidized from the +2 valence state to the +3 state by the action of an oxidizing agent such as molecular oxygen (O_2). The trivalent iron then reacts with water to form ferric hydroxide [$Fe(OH)_3$] which slowly dehydrates to form either goethite (FeOOH, brown) or hematite (Fe_2O_3, brownish red). Therefore, the reddish brown color of the martian soil implies that liquid water was present on the surface of Mars at the time these oxide minerals formed by weathering of iron-bearing volcanic rocks. However, the conversion of divalent iron in the basalt to trivalent iron in the soil of Mars is unclear because the atmosphere of Mars has never contained more than a trace of molecular oxygen.

Iron-oxide minerals that characterize the soils of Mars did not form on Venus because virtually all of the water had disappeared when its surface was rejuvenated by a global volcanic upheaval about 500 million years ago.

The chemical composition of soil in Table 12.4, measured by the Viking landers, indicates that SiO_2 (i.e., silica not quartz) is the most abundant

Figure 12.6. This oblique view of the long axis of the Valles Marineris illustrates how landslides have widened the main branch of the canyon and how the canyon walls have been eroded by gullies cut by water flowing downslope. The landslide deposits that cover the valley floor contain only a few impact craters indicating their comparatively low exposure age. (Mosaic of images recorded by the Thermal Emission Imaging System of the Mars Odyssey spacecraft and assembled in March of 2006. Courtesy of NASA/JPL/Caltech)

constituent of the soil at 43%, followed by Fe_2O_3 (18.0%), Al_2O_3 and SO_3 (7.2% each), MgO (6.0%), CaO (5.8%), and Cl and TiO_2 (0.6% each). In addition, the Vikings detected the presence of carbonate ions $\left(CO_3^{2-}\right)$ and water (H_2O). The analytical equipment of the Viking spacecraft could not determine the concentrations of potassium oxide (K_2O), sodium oxide (Na_2O), phosphorus pentoxide (P_2O_5), manganese oxide (MnO), and chromium oxide (Cr_2O_3). In addition, the Viking landers were not equipped to identify the minerals in the martian soil. However, the results of the chemical analysis in Table 12.4 suggest that the soil contains clay minerals (iron-smectite called nontronite), oxides of iron (goethite and hematite), sulfates of magnesium and calcium ($MgSO_4$ and $CaSO_4$), and chlorides of sodium and potassium (halite and sylvite). The titanium oxide (TiO_2) may reside in grains of the mineral ilmenite ($FeTiO_3$) which is resistant to chemical weathering. The apparent absence of calcium carbonate is puzzling considering that carbon dioxide is the dominant component of martian air.

The presence of evaporite minerals in the martian soil (e.g., $MgSO_4$, $CaSO_4$, NaCl, and KCl) is typical of caliche soil in deserts on the Earth where soilwater is drawn to the surface by capillarity. When the water evaporates, the salts that were dissolved in the water precipitate and form a hard crust at the surface of the soil. This crust prevents water from infiltrating into the soil and interferes with the growth of plants. Consequently, caliche soils on the Earth are not suitable for farming.

12.3 Surface Features: Southern Hemisphere

Compared to the northern hemisphere, the southern hemisphere in Figure 12.8 is mountainous and cratered to such an extent that it resembles the highlands of the Moon. The southern hemisphere of Mars contains the Hellas, Argyre, and Isidis planitiae, all of which are large impact basins. The floors of these basins are

Figure 12.7. The landing site of the Viking 2 robotic spacecraft in the Utopia planitia of Mars is covered by regolith containing angular boulders in a matrix of fine grained reddish-brown sediment. The regolith was formed by chemical weathering of the underlying bedrock and by the deposition of rock debris ejected from meteorite-impact craters following the initial formation of the Utopia basin. In addition, fine-grained sediment and boulders may have been deposited by water during the Noachian Eon more than 3.8 billion years ago. Transport and deposition of fine-grained sediment by wind continues at the present time as indicated by the wind ripples (minidunes). The surface of the regolith recorded on May 18, 1979, is covered by a thin layer of water ice that formed by deposition of ice-covered sediment grains from the atmosphere. (PIA 00571, Viking 2, Utopia planitia, courtesy of NASA/JPL/Caltech)

only sparsely cratered because they are covered by layers of sediment deposited by water and/or by wind. In addition, the southern hemisphere contains several volcanic centers that are much younger than the highlands.

12.3.1 Hellas Impact Basin

The Hellas basin (45°S, 70°E) in Figure 12.8 has a diameter of 1700 km and its lowest point lies about 8.2 km below the surrounding highlands.

The dimensions of this enormous basin indicate that at least 18.6 million cubic kilometers of rocks were removed by the explosive impact of a very large planetesimal. This large volume of rock presumably formed an ejecta blanket that originally surrounded the basin. However, the present rim of the Hellas basin is not elevated and mountainous as expected, but is actually low and flat. Consequently, water was able to flow into the Hellas basin through several winding channels that still extend for hundreds of kilometers in the

Table 12.4. Chemical composition of soil in Chryse and Utopia planitiae of Mars based on analyses by the Viking landers (Carr, 1996; Banin, et al., 1992)

Constituent	Estimated concentration, %
SiO_2	43.0
Al_2O_3	7.2
Fe_2O_3	18.0
MgO	6.0
CaO	5.8
TiO_2	0.6
SO_3	7.2
Cl	0.6
CO_3	<2
H_2O	0 to 1.0
Sum	~91.4

northeastern highlands and continue within the basin. These streams transported sediment that was deposited in layers, some of which were later eroded by flowing water and by wind.

The scarcity of impact craters within Hellas planitia tells us that the sediment was deposited after the period of intense bombardment that occurred between 4.6 and about 3.8 billion years ago. This conjecture is supported by the observation that the basement rocks of the southern highlands of Mars are magnetized, whereas the rocks underlying the Hellas basin itself are not magnetized. Apparently, Mars had a magnetic field when the rocks of the southern highlands crystallized and therefore these rocks were magnetized as they cooled after crystallizing from magma. When the Hellas basin formed, the strength of the magnetic field of Mars had decreased to such an extent that the basement rocks beneath the Hellas impact were not remagnetized as they cooled through their Curie temperature of about 570 °C (Section 10.2.2). Therefore, the Hellas impact occurred after the magnetic field had collapsed, but before the climate cooled and the water on the surface of Mars froze (Hartmann, 2003).

12.3.2 Argyre Impact Basin

The Argyre planitia (50 °S, 45 °W) is located about 2400 km west of the Hellas basin. Its diameter is approximately 750 km and the elevation of its floor is about 4.0 km below the surrounding highlands. The Argyre basin is surrounded by the Nereidum montes in the north and by the Charitum montes in the south, both of which are remnants of the rocks ejected after the impact. The presence of remnants of ejecta can be taken as evidence that the Argyre basin is younger than the Hellas basin where the ejecta close to the basin rim have been eroded.

The southern rim of the Argyre basin was eroded sufficiently to allow a stream to cut a channel across it and to discharge meltwater from the south polar ice cap into the basin (Clifford, 1993). Satellite data also show a stream valley on the northern rim that allowed water to leave the Argyre basin and to flow north into the drainage of the Valles Marineris and thus into Chryse planitia of the northern hemisphere (Clifford and Parker, 2001; Hartmann, 2003; Kargel, 2004).

12.3.3 Isidis Impact Basin

The Isidis planitia (15 °N, 90 °E) in Figure 12.8 is located at the northern edge of the southern highlands which are represented in this area by Terra Tyrrhena whose surface elevation rises up to about 3 km above the zero contour. This basin has a diameter of about 1200 km and a smooth, uncratered floor that is about 3.5 km below the zero-elevation contour. Remnants of the original ejecta deposits form a hummocky surface south and northwest of the basin. The northeast rim of the Isidis basin has been eroded and buried by sediment that covers the surface of the adjacent Elysium planitia of the northern hemisphere. Consequently, the sediment that now covers the floor of the Isidis basin originated in large part from the northern planitiae that may have contained standing water which extended into Isidis. In addition, some sediment or lava flows may have originated from a large shield volcano located adjacent to the western rim of Isidis basin.

12.3.4 Timescale for Mars

The topography reviewed above indicates that the geologic evolution of Mars can be divided

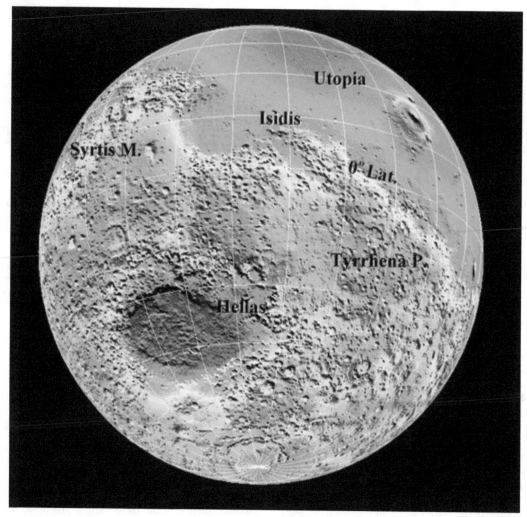

Figure 12.8. The surface of the southern hemisphere of Mars is densely cratered in contrast to the northern hemisphere. The Hellas planitia is the largest impact basin in this hemisphere followed by the Argyre basin (not shown) and the Isidis basin. The large uncratered plane is the Utopia planitia which contains the Elysium volcanic complex. The sparely cratered region adjoining the Isidis basin on the east is the volcano Syrtis Major. (Adapted from Smith et al. 1999. *Science*, 294:1495–1503. Graphics by G.A. Neumann, Massachusetts Institute of Technology and NASA/Goddard Space Flight Center)

into three intervals of time or eons (Tanaka, 1986; Hartmann, 2003):

Noachian Eon (4.6 to about 3.5 Ga).

Frequent impacts cratered the surface and formed several large basins. Volcanic activity occurred on a global scale. Liquid water eroded ejecta deposits and transported sediment into depositional basins located mainly in the northern hemisphere. The atmosphere of Mars was denser

and caused more greenhouse warming than the present atmosphere.

Hesperian Eon (3.5 to about 2.5 Ga)

Volcanic activity became more localized and the volume of lava that was erupted gradually declined. Water was discharged by episodic melting of ice in subsurface deposits as a result of volcanic activity, meteorite impact, and faulting of the crust. The average global temperature declined because of a decease in the amounts of greenhouse gases in the atmosphere. Ice began to form in the polar regions and at high elevations. Large ice sheets may have existed during this time.

Amazonian Eon (2.5 to 0 Ga)

Volcanic activity ceased, except for rare localized eruptions of lava (e.g., Amazonis planitia?). The climate of Mars has remained cold and dry. Liquid water cannot exist on the surface of Mars because of the low atmospheric pressure. However, water ice still occurs in permafrost below the surface and in large deposits in the polar regions, which also contain solid carbon dioxide. The principal geological process that continues on the surface of Mars is the transport and redeposition of sand and silt by dust storms that occasionally engulf the entire planet.

The dates that define the geologic history of Mars are based primarily on the number of craters per unit area and, to a much lesser extent, on the results of isotopic dating of martian meteorites that have been found on the Earth. Hartmann (2003) identified some useful "rules of thumb" that indicate the exposure ages of surfaces on Mars as well as on the Earth and the Moon based on the aerial density of impact craters:

1. Surfaces that are covered by impact craters "shoulder to shoulder" are older than about 3.5 Ga in most cases.
2. Surfaces that contain only a scattering of impact craters have exposure ages that range from about 3.0 to 1.0 Ga.
3. Surfaces on which impact craters are hard to find are generally less than 500 million years old.

These estimates are uncertain by several hundred million years and will be replaced in the future by precise isotopic dates when rock samples, collected on the surface of Mars, become available for analysis in research laboratories on the Earth.

12.4 Volcanoes of the Southern Hemisphere

The exposure of the southern hemisphere to impacts of solid objects from interplanetary space is recorded by the craters that cover its surface in contrast to the northern hemisphere which is dominated by lightly cratered low plains. In addition, the southern hemisphere has also been the site of large-scale volcanic activity whose products have locally buried the ancient basement rocks of Mars.

12.4.1 Syrtis Major Volcano

A large shield volcano called Syrtis Major forms a prominent landmark about 600 km west of the Isidis basin at about 10 °N and 65 °E in Figure 12.8. The summit of this volcano rises to an elevation of 2 km and contains two large calderas: Nili patera (61 km) and Meroe patera (43 km). The slope of Syrtis Major extends east all the way to the edge of Isidis basin, suggesting that lava that was erupted at the summit actually flowed across the western rim into the basin. In addition, the slopes of this volcano contain far fewer impact craters than the highland terrain on which it is located. These observations support the conclusion that Syrtis Major erupted after the Isidis impact had occurred. Although the age of the volcano is unknown, its presence on top of ancient highland terrain confirms that Mars remained geologically active until well past the period of intense bombardment that ended at about 3.8 Ga and suggests that Syrtis Major is of Hesperian age.

12.4.2 Hespera Planum

The southern hemisphere contains additional evidence of relatively recent volcanic activity at Hespera planum (22 °S, 254 °W) in Figure 12.8. This high plain in Terra Tyrrhena consists of lava flows that were extruded by a large volcano called Tyrrhena patera. Close-up images of Tyrrhena patera reveal that the slope of this volcano has been dissected by a radial set of gullies. The origin of these gullies is uncertain because some could be lava channels, whereas others may have been eroded by water that originated as localized rain caused by the discharge of water vapor into the atmosphere during eruptions of the volcano, or by melting of glacial ice on its summit. The lava flows are only lightly cratered like those on Syrtis Major described in the preceding section. In some cases, the lava flows breached the rims of impact craters and partially filled them with lava in a manner that is reminiscent of the Moon.

12.4.3 Hadriaca Patera

The gullies on the slopes of shield volcanoes in the southern highlands of Mars can also be examined on a comparatively small cone-shaped volcano named Hadriaca patera (32°S, 266°W). This mountain is located about halfway between Tyrrhena patera and the northeastern rim of the Hellas impact basin. Close-up images of Hadriaca patera show a set of steep-walled canyons which are connected to the Niger/Dao valley system that extends into the Hellas basin (Section 12.3.1). The formation of the canyons was facilitated because Hadriaca patera is a cinder cone composed of volcanic ash rather than of lava flows (Hartmann, 2003, p. 204). A second valley known as Harmakhis vallis is located about 250 km south of the Niger/Dao vallis and runs parallel to it across the gentle slope of the eastern Hellas basin.

12.5 Stream Valleys

The most plausible explanation for the existence of the valleys mentioned above is that they were carved by running water. In general, the orientation of the valleys in the southern hemisphere indicates that the water flowed north from the southern highlands to the northern lowlands (e.g., Chryse planitia). However, in some cases the water flowed into large impact basins (e.g., Hellas and Argyre planitiae), or flowed downslope from the summits of volcanoes (e.g., Hadriaca and Alba paterae), or flowed into large impact craters from their elevated rims. In many cases, the water originated from collapsed terrain by melting of permafrost or of deposits of subsurface ice. The largest volume of water originated from the collapsed terrain of Noctis Labryrinthus and flowed through the rift valleys of the Valles Marineris and was discharged into the Chryse planitia. Other streams that flowed directly into Chryse include Kasei, Nanedi, and Ares.

12.5.1 Nanedi Vallis

Nanedi vallis in Figure 12.2 is one of several valleys that once carried water from the Valles Marineris and from local chaotic terrains (e.g., Aram chaos) into Chryse planitia. The Nanedi vallis in Figure 12.9 is about 800 km long and

winds its way across a cratered plain although the valley itself received only a few hits. Therefore, Nanedi vallis post-dates the period of intense bombardment and may have contained a river between about 3.5 and 2.5 Ga (Hartmann, 2003, p. 242). Nanedi vallis has only a few short tributaries and was fed by groundwater that undercut the sides of the valley as noted also in the valleys of Nirgal and Vedra in Sections 12.5.2 and 3. In addition, the width of Nanedi valis increases downstream presumably because the stream that

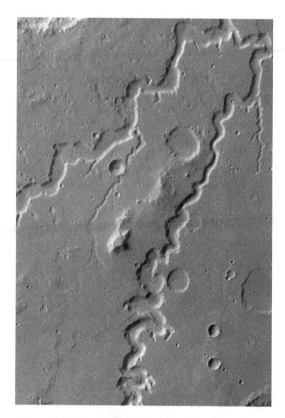

Figure 12.9. The valleys of the Nanedi system are steep-sided and flat-floored, which is typical of former stream channels on Mars. The two principal channels converge prior to entering the Chryse plantia. In addition, the Nanedi valley system contains several tributaries one of which is blocked by an impact crater. Three large impact craters in this view are partially filled with sediment, whereas several smaller craters have preserved their original bowl shape. The area between the principal branches of the Nanedi valley-system contains a flat-topped mesa and low hills both of which may be erosional remnants of the former upland surface. (Mars Express, 25 April 2006, courtesy of ESA/DLR/FU Berlin/G. Neukum)

once flowed in this valley gained water from springs along the valley walls rather than from tributaries.

Nanedi vallis has attracted attention because a close-up image of a segment of this valley in Figure 12.10 indicates that a narrow channel was incised along the valley floor. The presence of this secondary channel indicates that the Nanedi river evolved with time until, near the end of its existence, only a small amount of water remained, which cut the narrow channel along the floor of the much larger valley. This secondary channel is only recognizable where it has not been filled by eolian deposits of sand and silt.

The presence of the narrow secondary channel within the Nanedi vallis indicates that a change in climate occurred, which reduced the amount of water that entered the valley. The morphology of Nanedi vallis may record the transition from warm and humid to cold and dry climatic conditions. We do not yet know when the climate changed and whether the climate changed several times or only once. However, we can guess that

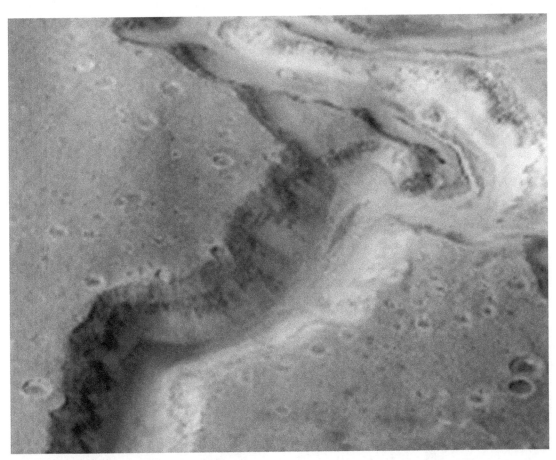

Figure 12.10. This segment of a meander loop of the Nanedi valley in Figure 12.9 reveals that a secondary channel was incised into the bottom of this valley during a late stage in its evolution. The existence of this channel suggests that the amount of water flowing through this valley decreased and became episodic or seasonal. The secondary channel was subsequently buried by eolian deposits except in the area shown in this image. The valley walls are steep and expose flat-lying volcanic rocks overlain by a thick layer of regolith consisting of weathering products and eolian deposits. The Nanedi valley contains only a small number of impact craters compared to the upland surface into which the valley was incised. The valley shown here is about 2.5 km wide except in the upper part of the frame where it widens locally to almost 5 km. The origin of the main terrace in that part of the valley is not yet understood (PIA 01170, Mars Global Surveyor, January 8, 1998, courtesy of NASA/JPL/Caltech/MSSS)

a gradual decrease of the global temperature of Mars caused snow fields to form which could thicken sufficiently to form mountain glaciers and ice caps in the polar regions.

12.5.2 Nirgal Vallis

The Nirgal valley in Figure 12.2 is part of the great waterway that carried water from the former "Lake Argyre" north to Chryse planitia. Nirgal vallis originates in heavily cratered highland terrain and extends east for about 650 km until it joins Uzboi vallis, which was excavated by water that flowed north from the Argyre basin through Hale, Bond, and Holden craters. Nirgal vallis differs from terrestrial stream valleys because it has only a few short tributaries and because the valley walls are unusually steep. These features indicate that the Nirgal river was fed by groundwater that was discharged by springs along the valley floor. Therefore, the valley walls were undercut and steepened by a process called sapping. The Nirgal valley widens downstream indicating that the discharge increased as additional water entered the valley along its length. Recent images such as Figure 12.11 indicate that the floor of Nirgal valley is covered by dunes consisting of windblown sand and silt. The "box canyon" shape of Nirgal vallis and the absence of tributaries is typical of martian stream

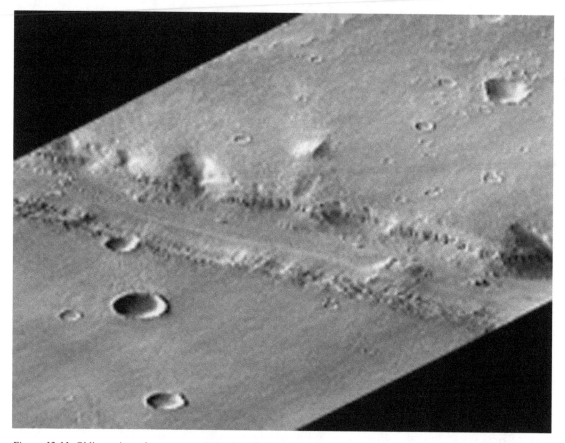

Figure 12.11. Oblique view of a segment of Nirgal vallis in the southern hemisphere of Mars (Figure 12.2). This close-up illustrates the steep walls of this valley and reveals the presence of dunes of wind-blown silt and sand on its floor. The surface of the adjacent terrain contains numerous bowl-shaped impact craters, most of which have been partially filled by silt and sand. However, this segment of Nirgal vallis contains few if any craters because the stream that once occupied this valley obliterated them as soon as they formed. (PIA 00944, Mars Global Surveyor, courtesy of NAS/JPL/Malin Space Science System)

valleys, (e.g, Ares and Nanedi valles), although dendritic stream patterns have also been observed.

12.5.3 Vedra Valles

A dendritic system of streams, collectively known as Vedra valles, occurs at 18°N and 56°W along the eastern slope of Lunae planum. The former streams that flowed in these valleys carried water eastward into the Chryse planita close to the site where the Viking 1 spacecraft landed on July 20, 1976. The valley walls are steep as in Nirgal vallis, but the water eroded numerous channels which converged downstream into two main valleys that carried the water to the western shore of "Lake Chryse." Chryse planitia was chosen as the landing site for Viking 1 not only because it is flat, but also because it is a place that may once have harbored life at a time when it contained water. However, the views transmitted to Earth by Viking 1 revealed that the landing site is a rock-strewn desert whose present soil does not contain organisms of any kind.

12.6 Impact Craters

The surface of Mars bears the scars of the impacts of solid objects of widely varying diameters. The largest impact basins on Mars include the Utopia/Acidalia and Isidis planitiae of the northern hemisphere and the Hellas and Argyre basins of the southern hemisphere, all of which appear to be of Noachian age. Most of the impact craters are smaller than these basins and range in age from Noachian to the most recent past. The oldest craters have been altered by erosion and, in some cases, have been partially filled by basalt flows, by sedimentary rocks deposited in standing bodies of water, and by eolian deposits. The shapes of young craters are well preserved and provide information about the size of the projectile and the composition of the overburden and rocks of the target areas. All of the different shapes of impact craters which we first described in Section 8.8 can be found on Mars (e.g., bowl-shaped and central-uplift craters as well as peak-ring and multi-ring basins).

The unique aspect of many martian impact craters illustrated in Figures 12.12 and 12.13 is that the ejecta blankets that surround them have

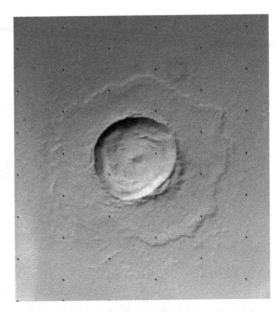

Figure 12.12. Belz impact crater in Chryse planitia is surrounded by a lobate ejecta blanket, which suggests that the ejecta were emplaced as a slurry of rock particles suspended in water. The outer edge of the ejecta lobes is raised forming a "rampart" which is a feature of a military fortification. Such so-called rampart craters formed in cases where subsurface ice was present in the rocks that were impacted by meteorites. Belz crater is bowl-shaped and has a diameter of about 11 km. Its rim is sharply defined and the interior walls of the crater have been modified by slides of rocks and overburden. (Viking 1 Orbiter, image 010A56, courtesy of NASA/Goddard Space Flight Center)

lobate shapes that differ significantly from the rayed ejecta deposits that surround craters on the Moon and on Mercury. In cases where meteorites impacted on surfaces underlain by permafrost, the resulting ejecta consisted of a slurry of mud that splashed down around the rim of the crater. The forward momentum of the ejected slurry caused it to flow radially away from the crater thus giving the deposit a lobate form and pushing up "ramparts" around it.

The presence of *rampart craters* on Mars (Carr, 1996, p. 110) is evidence for the existence of *permafrost* and hence for the presence of water ice on Mars. However, although rampart craters are characteristic of Mars, some craters lack the muddy ejecta blankets because:

1. Permafrost was not present at the impact site or was too far below the surface to be reached by small impact craters.

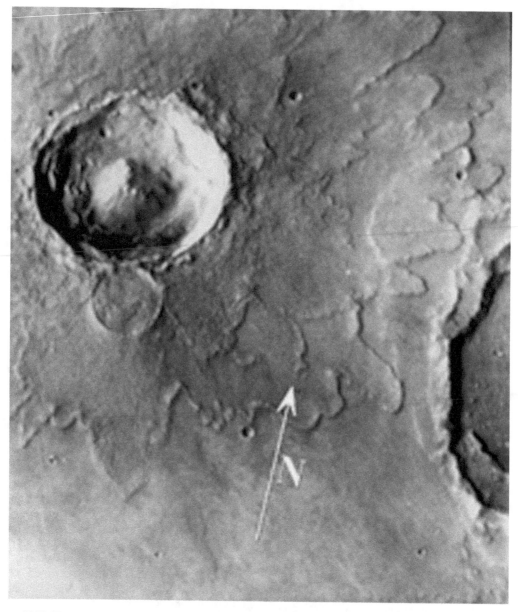

Figure 12.13. Yuty crater is a typical rampart crater on Chryse planitia (22 °N, 34 °W) surrounded by multiple overlapping ejecta lobes. The diameter of this crater is 18 km, it has a central uplift, and the floor of the crater is covered by hummocky landslide deposits. (Viking 1 Orbiter, image 3A07, recorded in 1997, courtesy of NASA)

2. The muddy ejecta were eroded by the action of wind, water, or ice sheets.
3. The ejecta were buried by eolian deposits or lava flows.

The existence of permafrost on Mars is also indicated by *patterned ground* in the form of polygonal fracture patterns in the soil at sites in the planitiae of the northern hemisphere (Carr, 1996, p. 114) and in some impact craters (Hartmann, 2003, p. 141).

The study of the diameters of impact craters on Mars, as well as on the Moon and Mercury, indicates that small craters are more numerous than large ones. We conclude from this fact that small projectiles were more abundant than large ones during the initial period of intense bombardment of planetary surfaces. This size distribution is a natural consequence of the processes that produced the solid objects which collided with the planets and their satellites. For example, the collision of two large objects in the solar system could have produced a large number of small fragments and only a few large ones. When these fragments subsequently impacted on a planet such as Mars or the Moon, the small craters would be more abundant than the large ones.

The craters that formed on a planetary surface during a specified interval of time can be sorted into categories defined by a sequence of numbers obtained by progressively doubling the diameters starting with 1 km (i.e., 1, 2, 4, 8, 16, 32, etc.). Craters with diameters less than 1 km are classified by progressively dividing the diameters by two (i.e., 1, ½, ¼, $^1/_8$, etc.). The number of craters of particular size divided by the area of the surface on which the craters were counted (i.e., number of craters per km^2) and the diameter of the particular size are used to plot a point in Figure 12.14. Additional data points are obtained by counting the craters in each of the categories. The resulting data points define a line called an *isochron* because all of the craters that define the isochron formed during the same interval of time.

The craters of different diameters that define the one-million-year isochron in Figure 12.14 have different surface densities because of the size distribution of the projectiles that formed them. For example, craters having a diameter of 1 km have a higher surface density than craters whose diameter is 16 km because small projectiles were more abundant than large ones. Nevertheless, the surface density of any crater category is consistent with the exposure age of the surface on which the craters were counted.

The crater-count chronometer described above and illustrated in Figure 12.14 can be used to estimate the exposure age of a surface on Mars, or on any other solid body in the solar system. The positions of the resulting data points on a graph containing crater-count isochrons is used to estimate the exposure age of the surface. The validity of such dates depends on several assumptions:

1. All of the craters have been preserved since the formation of the planetary surface being studied.
2. All of the planets were bombarded by objects that had the same size distribution.
3. The frequency of impacts decreased with time in the same way for all planets.
4. The Moon is a reliable indicator of the relationship between the surface density of craters of a certain size and the exposure ages of lunar surfaces determined by isotopic methods.

Planetary surfaces that were exposed during the time of intense bombardment between 4.5 and 3.8 Ga acquired so many craters that all available sites were hit. After a surface has reached a state of saturation in Figure 12.14, the number of craters cannot increase because every new crater destroys one or more previously existing craters. Therefore, planetary surfaces that have achieved saturation have exposure ages between 4.5 and 3.8 Ga. Hartmann (2005, p. 261) showed that such old surfaces exist in the highlands of the Moon and Mercury, on the martian satellite Phobos, as well as on Calisto which is the fourth Galilean satellite of Jupiter. In addition, the surface density of craters in the martian highlands also approaches saturation.

The numbered lines in Figure 12.14 represent different surfaces on Mars.

1. Olympus mons and the Tharsis volcanoes have exposure ages between 10^8 and 10^9 years (i.e., 100 million to one billion years).
2. The highlands in the southern hemisphere of Mars actually approach saturation for crater diameters between 16 and 64 km. Craters having diameters less than 16 km indicate a lower exposure age between 4×10^9 and 1×10^9 years because small craters tend to get buried by eolian deposits.
3. The exposure age of the surface surrounding south pole also ranges from $> 1 \times 10^9$ (large craters) to less than 100 million years (small craters).

The crater counts of the martian highlands (2) and of the south-polar plain (3) both suggest that small craters are buried faster than large ones and that the dates derived from crater counts

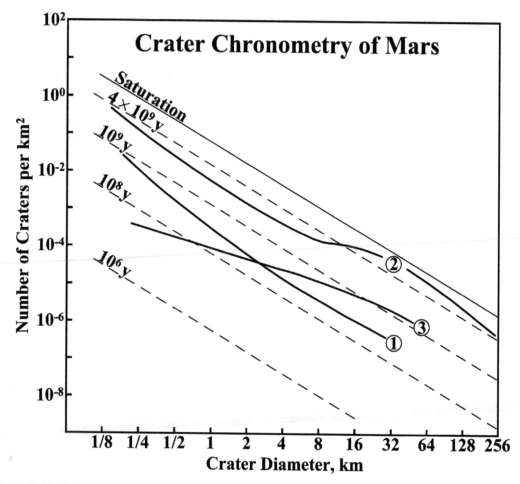

Figure 12.14. The surface densities (number per square kilometer) of craters of specified age decrease with increasing crater diameters from 1/8 to 256 km and beyond. The resulting lines are isochrons because all craters that define them have the same age, such as: 10^6, 10^8, 10^9, and 4×10^9 y. The saturation line is formed by craters from surfaces that were exposed during the period of intense bombardment between 4.6 and 3.8 Ga. The numbered lines are formed by craters from different surfaces on Mars : 1. Olympus mons and central Tharsis volcanics; 2. Martian highlands; 3. South-polar area. Adapted from Figure 9.20f of Hartmann (2005)

are to some extent crater-retention dates rather than estimates of the exposure ages of planetary surfaces.

The dates derived by counting the craters per unit area are currently the best indication we have of the exposure ages of topographic surfaces on Mars. These age estimates will be replaced in the future by precise age determinations of rocks collected by geologists at known locations on the surface of Mars. The new age determinations will be based on measurements of the abundances of nuclear decay products of the radioactive atoms

of certain elements (e.g., potassium, rubidium, samarium, uranium, thorium, and others). The principles of dating rocks by the samarium-neodymium (Sm-Nd) method were explained in Science Brief 8.11.2 and in Section 8.7.

12.7 Martian Meteorites

The dates derived by isotopic methods indicate that most meteorites that have impacted on the surface of the Earth crystallized between 4.50

and 4.60 billion years ago and have thereby established the time of formation of the Sun and of the planets and their satellites of the solar system (Section 8.7). An important exception to this generalization are the lunar meteorites (Section 9.4), which have yielded dates that range from 4.5 to 3.0 Ga (Sections 9.2 and 9.4). Another exception are rock samples that were blasted off the surface of Mars and eventually impacted on the surface of the Earth. Although such martian meteorites are rare, at least 36 specimens have been identified so far (Science Brief 12.13.1).

The martian meteorites were originally classified as achondrites because they are composed of igneous rocks including basalt and several varieties of ultramafic rocks consisting of pyroxene, olivine, and plagioclase feldspar. For example, a shower of about 40 meteorite specimens was observed to fall on June 28, 1911, near Alexandria in Egypt. The total mass of the recovered specimens was 40 kg. This meteorite, known as Nakhla (Figure 8.3), was classified as a diopside-olivine achondrite, but did not attract particular attention for about 63 years until Papanastassiou and Wasserburg (1974) at the California Institute of Technology reported that Nakhla had crystallized only 1.34 ± 0.02 billion years ago. This date was later confirmed by other geochronologists and alerted the scientific community that Nakhla had apparently originated from a planet in the solar system that remained geologically active until about 1.3 billion years ago. The source of Nakhla and other anomalous achondrites was eventually identified as the planet Mars because the composition of noble gases they contain is identical to the composition of the atmosphere of Mars measured by the Viking spacecraft but differs from the composition of the terrestrial atmosphere in Table 12.5.

Another group of martian meteorites yielded even younger dates between 161 ± 11 Ma (Shergotty), 176 ± 4 Ma (Zagami), and 183 ± 12 Ma (ALHA 77005). The Shergotty meteorite contains evidence of shock metamorphism that could have been caused by a meteorite impact on the surface of Mars at 161 Ma. However, the passage of a shock wave emanating from a meteorite impact is not likely to reset the radioactive clock in rocks. Moreover, other

Table 12.5. Concentrations of the noble gases in the atmospheres of Mars and the Earth (Graedel and Crutzen, 1993; Carr, 1996)

Element	Concentration, ppm	
	Mars	Earth
Helium (He)		5
Neon (Ne)	2.5	18
Argon (Ar)	5.3	9300
Krypton (Kr)	0.3	1.1
Xenon (Xe)	0.08	
Ne/Ar	0.47	0.0019
Ne/Kr	8.33	16.3
Ar/Kr	17.6	8454
Ar/Xe	66.2	—
Kr/Xe	3.75	—

evidence indicates that Shergotty, Zagami, ALHA 77005, and other "young" martian rocks were ejected from the surface of Mars only two to three million years ago. Therefore, the dates of about 170 Ma are tentatively accepted as evidence for magmatic activity on Mars at about that time. (See also Science Brief 12.13.1).

The measurement technology that has made these results possible will, in time, be used to study rocks collected from outcrops or recovered by drilling on the surface of Mars. The results will permit a reconstruction of the geologic history of different regions of Mars based on the ages of the heavily cratered highlands of the southern hemisphere, the Tharsis and Elysium volcanic domes, the Hellas and Argyre impact basins, and the numerous volcanic centers of the southern hemisphere.

12.8 Water on Mars

The topographic features described in Sections 12.5 and 12.6 leave no doubt that water in liquid form once existed on the surface of Mars during the Noachian and Early-to-Middle Hesperian Eras (Carr, 1996; Clifford, 1993; Clifford and Parker, 2001). During this time, the atmospheric pressure on Mars was higher than it is today because the atmosphere contained larger amounts of carbon dioxide, nitrogen, and water vapor. Consequently, these gases were more effective in warming the atmosphere and the resulting climate of Mars was warmer and more humid than it is at present.

The effectiveness of the so-called greenhouse gases (e.g., CO_2 and H_2O) in warming the atmosphere of a planet depends not only on the concentration of each gas but also on the total pressure exerted by the atmosphere. For example, the concentrations of carbon dioxide in the atmosphere of Mars (95.32 %) and Venus (96.5 %) in Table 12.6 are nearly identical, but the average global temperatures of Mars is $-53\,°C$, whereas that of Venus is $+460\,°C$. The reason for the discrepancy is that the atmospheric pressure on Mars is less than 10 millibars, whereas the atmospheric pressure on Venus is 90 bars (i.e., 90,000 millibars). Therefore, the pressure due to carbon dioxide at the surface on Mars is only about 9.5 millibars compared to 86,850 millibars at the surface of Venus. In addition, Mars receives less sunlight than Venus even after the effect of the venusian clouds is considered. These factors explain in principle why it is cold on the surface of Mars and hot on the surface of Venus.

The existence of water on Mars still seems to surprise and excite the public perhaps because of a semantic problem. In popular speech the word "water" means the liquid we drink, whereas in science the word signifies the compound whose chemical composition is H_2O regardless of its state of aggregation. Therefore, ever since 1790 AD when the British astronomer William Herschel observed the polar ice caps of Mars through his telescope and considered them to be ice fields, we have known that the compound H_2O in the form of water ice exists there. Later, some observers between 1860 and 1960 questioned whether the ice caps consist of water ice or of frozen carbon dioxide called dry ice. We now know that the ice caps of Mars are composed of permanent deposits of water ice that are covered during the winter by a layer of dry ice which forms by direct condensation of carbon dioxide gas of the atmosphere. The layer of dry ice that forms in the north-polar regions during the martian winter is several meters thick. In the summer, the dry ice in the northern ice cap sublimates into the atmosphere, which causes the barometric pressure of Mars to rise by about 10% as the carbon dioxide migrates to the south pole where it is redeposited during the southern winter. When summer returns to the southern hemisphere, the dry ice sublimates again and migrates back to the north polar ice cap. This seasonal migration of carbon dioxide has occurred annually ever since the winter temperature at the poles decreased sufficiently to cause carbon dioxide gas to condense directly into dry ice.

12.8.1 Hydrologic Cycle

During the period of time when liquid water flowed in streams across the surface of Mars and ponded in large bodies of standing water (e.g., in Chryse, Utopia, Hellas planitiae), some water also infiltrated into the soil and rocks below the surface. Therefore, at this early time in the history of Mars (i.e., Noachian and Early Hesperian) the planet had an active hydrologic cycle. Water evaporated from the surfaces of standing bodies of water into the atmosphere. The circulation of the atmosphere transported the moist airmasses from the equatorial regions to higher latitudes in the northern and southern hemispheres where the air cooled enough to cause the water vapor to condense as rain or snow depending on the temperature. At this time, snowfields may have existed on high plateaus and on the summits of dormant shield volcanoes. Rain and meltwater eroded gullies on slopes and gradually widened them into stream valleys through which surface water flowed into the bodies of standing water.

Table 12.6. Chemical composition of the atmospheres of Mars, Earth, and Venus in percent by volume (Carr, 1996; Owen, 1992). The compositions of the atmospheres of the Earth and Venus are from Table 11.2

Compound or element	Concentration, %		
	Mars	Earth	Venus
Carbon dioxide (CO_2)	95.32	0.0346	96.5
Nitrogen (N_2)	2.70	78.08	3.5
Argon (Ar)	1.60	0.9340	0.0070
Oxygen (O_2)	0.13	20.94	0-0.0020
Carbon monoxide (CO)	0.07	0.00002	0.0045
Water vapor (H_2O)	0.03	variable	0.0045
Methane (CH_4)	~ 0.0000010	0.00017	0

When the average global temperatures began to decline because of decreased greenhouse warming caused by a reduction in the frequency of volcanic eruptions, the water on the surface of Mars remained frozen for increasing periods of time each year depending on the latitude. As a result, the hydrologic cycle ceased to function because the bodies of standing water were frozen, which decreased the rate of evaporation of water. During this period of transition, liquid water existed primarily in subsurface aquifers and was discharged into stream valleys by springs. Eventually, even the groundwater froze causing streams to dry up completely except for intermittent floods caused by the discharge of water from chaotic terrains as a result of melting of frozen groundwater called *permafrost*.

After the hydrologic cycle ceased to function, water ice continued to exist in the polar regions and in permafrost below the former water table. For example, Hartmann (2003, p. 209) cited several lines of evidence that Utopia planitia is underlain by permafrost at a depth of 100 to 150 meters (e.g., patterned ground, rampart impact craters, and the presence of subglacially erupted basalt in Acidalia planitia). The depth to permafrost in the equatorial areas of Mars may be much greater because the former water table was deeper than 150 meters or because of sublimation of permafrost.

The hypothetical scenario outlined above will be altered in the future as more specific data become available. Nevertheless, our present understanding of the former existence of liquid water on Mars justifies the expectation that the water in the pore spaces of rocks in the subsurface is now frozen . We also know that the interior of Mars is still hot because heat is still being produced by the decay of radioactive elements and because not all of the heat that was present initially has escaped into interplanetary space. Therefore, the temperature of the crust of Mars increases with depth below the surface as illustrated in Figure 12.15 and rises to the melting temperature of water at some depth below the surface. Below that depth water may still exist in liquid form underneath the zone of permafrost.

The results of calculations in Science Brief 12.13.2 suggest that the "depth-to-water" may be as much as 5.3 km (i.e., conservative estimate).

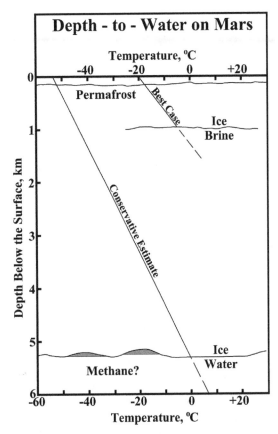

Figure 12.15. The former groundwater on Mars has frozen to form permafrost that starts 100 to 150 meters beneath the surface of the planitiae of the northern hemisphere. As the temperature rises with increasing depth below the surface, a depth is reached below which the groundwater did not freeze. The conservative case (depth-to-water = 5.3 km) assumes a surface temperature of −53 °C, a thermal gradient of 10 °C/km, and a freezing/melting temperature of pure water of 0 °C. The best case (depth to water = 1.0 km) assumes a surface temperature of −20 °C, a thermal gradient of 15 °C/km, and a freezing/melting temperature of salty water of −5 °C. The reported presence of methane in the atmosphere of Mars suggests the possibility that subsurface reservoirs of methane exist beneath the layer of permafrost

Under optimal conditions, the depth to water may only be 1.0 km in places where the average surface temperature is −20 °C (instead of −53 °C), where the thermal gradient of the crust is 15 °C/km (instead of 10 °C/km), and assuming that the water is sufficiently salty to freeze at −5 °C (instead of at 0 °C). Modern equipment available on the Earth is certainly capable of drilling to a depth of five or more kilometers. Whether drilling for water at such depth is feasible on Mars remains to be seen.

In this connection, the presence of methane (CH_4) in the atmosphere of Mars (Table 12.6) becomes significant because this compound is destroyed by energetic ultraviolet radiation of the Sun. Therefore, we may speculate that the crust of Mars may contain, or may have contained, bacteria that manufacture methane which may have accumulated in isolated pockets beneath the layer of permafrost. The existence of potential natural gas deposits that are accessible by drilling would greatly facilitate future human habitation on Mars.

12.8.2 Phase Diagrams of Water and Carbon Dioxide

The construction of the phase diagram for water is explained in Science Brief 12.13.3 and is illustrated in Figure 12.16. The point of intersection of the boiling and freezing curves is called the *triple point* because all three phases of water (ice, liquid, and vapor) can coexist in equilibrium at the pressure and temperature of the triple point (i.e., P = 0.006 atm, T = +0.010 °C). The atmospheric pressure of the Earth (i.e, 1.0 atm) is greater than the pressure of the triple point, which means that liquid water can form on our planet by melting of ice and by condensation of vapor. Liquid water cannot form on the surface of Mars because its atmospheric pressure is less than the pressure of the triple point.

The phases of carbon dioxide are represented by a phase diagram like that for water in Figure 12.16. The principal difference is that the triple point of carbon dioxide occurs at a higher pressure and lower temperature (4.2 atm, −53.2 °C) than the triple point of water. In addition, solid carbon dioxide (dry ice) has a higher density than liquid carbon dioxide, which means that dry ice sinks in liquid carbon dioxide, whereas water ice floats in liquid water. This difference arises because the melting temperature of water ice decreases when it is under pressure because the meltwater occupies a smaller volume than the ice. For example, a person skating on ice at 0 °C is actually skating on a film of water that forms

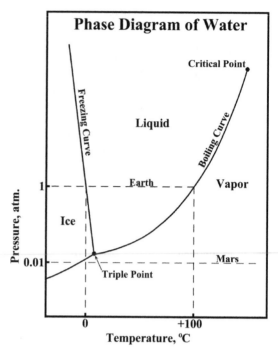

Figure 12.16. Water occurs in three different states called "phases" in physical chemistry: ice, liquid, and vapor. The stability fields of these phases are defined by the boiling and freezing curves of water which intersect in the triple point (T = +0.010 °C, P = 0.006 atm.) where all three phases occur together. The boiling curve ends at the critical point (T = +374 °C, P = 218 atm) where the distinction between liquid and vapor vanishes. On the Earth (P = 1.0 atm), ice melts at 0 °C and water boils at 100 °C because that is how the Celsius scale is defined. The atmospheric pressure on Mars is less than the pressure of the triple point of water. For that reason, liquid water is not stable on the surface of Mars. The diagram was not drawn to scale because it is intended as an illustration. Adapted from Figure 13.1 of Krauskopf (1979)

under the blades of the skates because of the pressure exerted by the weight of the skater. When pressure is exerted on dry ice at its melting temperature it does *not* melt because the liquid carbon dioxide occupies a larger volume than the dry ice from which it formed. In fact, this is the normal behavior of virtually all solids because their melting temperatures increase with increasing pressure (Figure 6.5, Section 6.4.2). The response of volatile compounds such as water, carbon dioxide, methane, ammonia, and nitrogen to changes in pressure and temperature gives rise to the phenomenon of *cryovolcanism* to be presented in Chapter 15.

The phase diagrams of water and carbon dioxide have been combined in Figure 12.17 in order to explain the properties of these compounds on the surfaces of the Earth and Mars. At sealevel on the Earth, indicated by the horizontal dashed line, water and carbon dioxide are both solid at a temperature of less than $-78\,°C$. When the temperature rises, dry ice sublimates to carbon dioxide gas at $T = -78\,°C$, whereas water ice remains solid. When the temperature increases to $0\,°C$, the water ice melts to form liquid water that coexists with carbon dioxide gas. If the temperature rises to $100\,°C$, the liquid water boils and thereby changes to water vapor which mixes with the carbon dioxide gas that formed from dry ice at $-78\,°C$. We note that liquid CO_2 cannot form under natural conditions on the surface of the Earth because the atmospheric pressure is too low. However, liquid carbon dioxide can be made by cooling the gas to about $-40\,°C$ and by compressing it at a pressure of more than 10 bars.

The phase changes of water and carbon dioxide with increasing temperature on the surface of Mars differ from those on the Earth because of the low atmospheric pressure of Mars. First of all, dry ice on Mars sublimates to vapor at about $-127\,°C$ (depending on the pressure), whereas water ice remains stable. When the temperature increases to about $-5\,°C$ (depending on the pressure), the water ice sublimates directly to the vapor phase without melting because liquid water cannot form on the surface of Mars because its atmospheric pressure is below the triple point. Therefore, under present conditions on the surface of Mars, both water vapor and carbon dioxide gas condense directly to their respective solid phases.

12.8.3 Polar Ice Caps

The polar ice caps on Mars, observed by the British astronomer Sir William Herschel around 1790 AD, extend from the poles to latitudes of about $80°$ north and south and consist of water ice interbedded with layers of fine-grained wind-blown sediment. The north-polar ice cap is centered on the north-polar basin whose elevation is close to 5 km *below* the zero-elevation contour based on data from the Mars Orbiter Laser Altimeter (MOLA). The north-polar ice cap in Figure 12.18 is a dome that rises about 3 km above the surrounding plain and has a volume of 1.2 ± 0.2 million cubic kilometers, which amounts to about half the volume of the ice sheet that covers Greenland (Hvidberg, 2005).

The south-polar cap is located on the highly cratered surface of the southern highland that has an elevation of about 1.5 km at this location. The center of the south-polar ice cap is displaced from the south pole by about 200 km and part of the ice sheet covers an old impact basin called Prometheus. The summit of the south-polar ice cap approaches an elevation of 5 km and therefore rises about 3.5 km above the surrounding plain.

The polar ice caps of Mars contain very few craters although the ice sheets may have existed

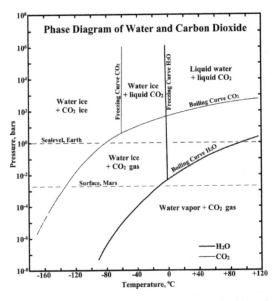

Figure 12.17. The stability fields of solid, liquid, and gas phases of water and carbon dioxide are defined by their boiling-temperature and the freezing-temperature curves. By superimposing the phase diagrams of these two compounds the stability fields can be labeled in terms of the phases of both compounds that can coexist together. The triple point of water ($T = +0.010\,°C$; $P = 0.0060$ atm) permits liquid water to exist on the Earth but not on Mars. The comparatively high pressure of the triple point of CO_2 (4.2 atm, $-53\,°C$) prevents liquid CO_2 from occurring on the Earth and on Mars. Instead, CO_2 ice sublimates to CO_2 gas at $-78\,°C$ on the Earth and at about $-127\,°C$ on Mars. Adapted from Figure 2.3 of Carr (1996)

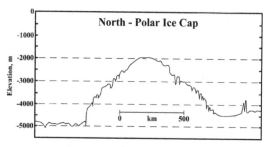

Figure 12.18. Topographic profile of the north-polar ice cap based on a vertical exaggeration of 170. The dome-shaped profile suggests that the ice flows from the center of the ice sheet towards the margins. However, the rate of flow is only about one millimeter per year compared to one to ten meters per year on the Earth. The difference in the flow rates arises because ice on Mars is colder than ice on the Earth, because ice on Mars contains dust layers, and because the force of gravity on the surface of Mars is only 38% of the terrestrial force. The north pole of Mars is located in a basin at an elevation of $-5000\,m$, whereas the south pole has an elevation of $+1000\,m$ above the zero-elevation contour. The difference in the elevation of the poles causes the south pole of Mars to be colder than the north pole. Adapted from Figure 6.1 of Hvidberg (2005)

for billions of years. The scarcity of impact craters is probably caused by the interactions between the ice sheets and the atmosphere and because the craters continue to be obliterated by the flow of the ice. During the winter, each of the polar ice caps is covered by a seasonal layer of carbon dioxide ice because it is cold enough in the polar winters for carbon dioxide gas of the atmosphere to condense directly to dry ice. When the air warms up again in the subsequent spring (e.g., to $-127\,°C$ at the north pole), the dry ice sublimates, whereas the water ice remains behind because the temperature does not rise enough to destabilize it (Figure 12.17). The polar ice caps may still contain layers of dry ice that may have been deposited during colder times in the past when part of the winter layer of dry ice did not sublimate during the summer.

Both ice caps contain spiral valleys that may have formed by selective sublimation of ice and/or by erosion caused by cold and dense air flowing from the poles (i.e., catabatic winds). The valleys curve anticlockwise in the north-polar ice cap and clockwise in the south-polar ice sheet. The steep walls of these polar canyons reveal a complex stratigraphy of layers of ice

and sediment that record changes in the global climate of Mars. Such climate changes may have been caused by variations of the inclination of the axis of rotation (i.e., the axial obliquity) because an increase in the inclination of the rotation axis allows more sunlight to reach the polar regions. The resulting increase in the rate of sublimation of ices of carbon dioxide and water causes the atmospheric pressure to rise and increases greenhouse warming. Therefore, the climate of Mars may have fluctuated between cold and warm period in response to the cyclical variations of its axial obliquity. In this way, the global climate of Mars is controlled by its polar regions.

Ice sheets like those in the polar regions of Mars would flow outward under the influence of gravity if they were on Earth. Terrestrial ice sheets transport sediment which accumulates at their margins in the form of moraines, outwash deposits, eskers, drumlins, and kames. The martian environment differs from conditions on Earth because the temperatures of the polar regions of Mars are lower than the temperatures of North America and northern Europe during the Wisconsin and Würm glaciations, respectively. The low temperature at the martian poles causes the ice to be more rigid and less plastic than it would be on Earth. The resistance to flow of the martian ice sheets is further increased by the presence of dust layers and by the weak gravitational force on the surface of Mars amounting to only 38% of the gravitational force exerted by the Earth.

In spite of these limitations on the plasticity of glacial ice on Mars, the topographic profile of the north-polar ice cap in Figure 12.18 is dome-shaped like the profile of the East Antarctic ice sheet on the Earth. Therefore, the martian ice sheets do appear to flow outward from their centers where the ice is thicker than at the margins. The flow rate of the north-polar ice cap of Mars is about one millimeter per year compared to rates of 1 to 10 meters per year for typical ice caps on the Earth (Hvidberg, 2005). However, the dynamics of martian ice caps may also be affected by differential condensation and sublimation associated with the walls of the spiral canyons that characterize both of the polar ice sheets. Satellite images indicate that the pole-facing walls of the canyons are white, whereas the equator-facing walls are dark. Accordingly, the horizontal and pole-facing surfaces of the ice sheets are interpreted to be areas of accumulation, while the dark equator-facing walls of the canyons are sites of ablation. As a result, the canyons in the polar ice caps of Mars gradually migrate toward the poles (Hvidberg, 2005). The history and present dynamics of the polar ice sheets of Mars may be revealed in the future when glacialogists and glacial geomorphologists are able to study them in the field.

12.8.4 Mountain Glaciers and Continental Ice Sheets

The polar ice caps of Mars may have been larger in the past than they are at present, in which case the area surrounding the present ice sheets may contain geomorphic landforms that resulted from fluvio-glacial processes. However, early reports of the presence of terminal and lateral moraines in the area around the south-polar ice cap later turned out to be erroneous (Kargel, 2004, p. 110). We have already mentioned that the cliff at the base of Olympus mons may record the presence of a continental ice sheet through which the first lavas of that shield volcano were erupted (Faul, 1972, unpublished manuscript). Other indications of the occurrence of ice on Mars include the permafrost and patterned ground of the northern planitiae, as well as the rampart impact craters surrounded by lobes of muddy ejecta. The presence of water ice on Mars at the present time supports the hypothesis that mountain glaciers as well as continental ice sheets occurred on Mars during its long and varied climatic history. The challenge is to recognize the clues provided by satellite images of the martian surface and to interpret them correctly.

The evidence for the former existence of glaciers on Mars has been assembled and interpreted by Kargel (2004). For example, he described sinuous and braided ridges in the southern plains of Argyre planitia north of the Charitum montes (Section 12.3.2). The ridges are 100 to 200 m high, two to three kilometers wide, and occur in a smooth plain over a distance of about 300 km. These ridges may be eskers, which are composed of sediment transported by meltwater streams flowing within an ice sheet (e.g., the Laurentide ice sheet in central North America). In addition, the Charitum montes

contain geomorphic features of alpine valley glaciation including cirques, arêtes, horns, and cols (Kargel, 2004, p. 143). The juxtaposition of eskers on the plain within the Argyre basin and features of alpine glaciation in the adjacent Charitum montes supports the hypothesis that the Argyre basin once contained a large ice sheet that was fed by valley glaciers which originated in the Charitum montes. The case for alpine valley glaciaton of Mars is further strengthened by the existence of U-shaped valleys in the Charitum montes of Mars that resemble the U-shaped valleys of Glacier National Park in Montana.

12.9 Life on Mars

The undeniable presence of liquid water on the surface of Mars during the Noachian and early Hesperian Eons (4.5 to 3.0 Ga) satisfies one of the preconditions for life on a planetary surface (Jakosky and Shock, 1998; Horneck and Baumstark-Khan, 2002). The other conditions include sufficient time for primitive life forms to develop, a source of energy, and the presence of nutrients in the environment such as carbon compounds and various elements required for plant growth (e.g., potassium, nitrogen, phosphorus, iron, etc.). All of these conditions appear to have existed on the surface of the Earth because primitive life forms did arise here in bodies of standing water as early as 3.5 billion years ago and perhaps even earlier. At that time, the surface of Mars was much more similar to the surface of the Earth than it is at the present time. Therefore, we may postulate that life did exist on Mars during its earliest history because we know that life existed on the Earth during that same time.

12.9.1 Origin and Preservation

The development of life in the form of selfreplicating unicellular organisms was initially delayed by the intense bombardment of the surfaces of Earth and Mars by solid objects from interplanetary space. The resulting devastation of the planetary surfaces may have repeatedly interrupted or even terminated the life-forming process (Chyba, 1990). When the period of the most intense bombardment finally ended around 3.8 Ga, organisms appeared on the Earth a few hundred million years later either because they had survived the disruptions, or because life originated spontaneously and evolved rapidly as soon as the environmental conditions permitted it. If it happened on the Earth, it could also have happened on Mars while the environmental conditions were still favorable.

The solid objects (e.g., planetesimals and fragments of disrupted planets) that impacted on the Earth and Mars during the time of the intense bombardment probably contained organic compounds that had formed spontaneously by abiological processes in the solar nebula. When these organic compounds were released into bodies of standing water, they could have inter-acted with each other to form more complex organic compounds that ultimately led to the development of unicellular organism on the Earth. The details of this geochemical process are not yet understood although several hypotheses have been considered (Westall, 2005).

Whether the processes that led to the devel-opment of unicellular organisms on the Earth also produced life on Mars is not certain because:

1. The violent explosions resulting from impacts may have caused large amounts of water on Mars to be ejected into interplanetary space.
2. Bodies of standing water on Mars may have been ice-covered because of the low luminosity of the young Sun prior to 4.0 Ga.
3. The absence of molecular oxygen in the atmosphere of Mars prevented ozone from forming in the stratosphere, which allowed ultraviolet radiation to reach the planetary surface.

These and other potential environmental issues could have prevented the development of lifeforms on Mars even though the same problems did *not* stop the process on the Earth. Success or failure for the development of life on Mars may have depended on timing. The most favorable conditions for lifeforms to develop on Mars occurred as soon as the planet had cooled sufficiently to permit water vapor in its primordial atmosphere to condense to form a global ocean. Calculations by Lazcano and Miller (1994) indicate that the reactions that lead to the beginning of life and to its evolution into cyanobacteria (i.e., blue-green algae capable of photosynthesis) require only about 10 million years.

Therefore, the existence of unicellular organisms on Mars depends on the answers to a series of questions about when they formed and whether they survived the deterioration of the surface environment of Mars:

1. If cyanobacteria *did* form on Mars between 4.5 and 4.0 billion years ago, did they survive the intense bombardment between 4.5 and 3.8 Ga?
2. If cyanobacteria did *not* form during the most favorable time in the history of Mars or if they were destroyed by the explosive impacts, could they have formed later between 4.0 and 3.5 Ga?
3. If cyanobacteria inhabited bodies of standing water on the surface of Mars anytime between 4.5 and 3.5 Ga, how did they adapt to the deteriorating environmental conditions caused by decreasing temperature and by the absence of liquid water?
4. Where are the fossilized remains of martian life *or* colonies of living organisms?

If unicelluar organisms did form on Mars during the most favorable climatic conditions between 4.5 and 4.0 Ga, they presumably inhabited all available bodies of standing water as well as sediment deposited in these bodies of water. In addition, bacteria could have flourished in streams, in groundwater below the surface, and in hotsprings and pools associated with volcanic activity. Other habitats, such as weathering products and snowfields on the surface of the planet, may not have been suitable habitats for bacteria because of exposure to energetic ultra-violet light and the absence of liquid water. On the other hand, lichens and algae are known to occur in porous sedimentary rocks in Antarctica (e.g., endolithic micro-organisms, Friedmann, 1982) and may have found a similar niche on Mars (Friedmann and Koriem, 1989).

The potential habitats of micro-organisms on Mars are also the most likely places where their fossilized remains may be found, such as sedimentary rocks, hotspring deposits, and permafrost (Gilichinsky, 2002). In addition, colonies of living micro-organisms may still exist below the surface of Mars where liquid water is present. Such sites are located at depths of one to five kilometers below the surface where water has remained in the liquid state because of geothermal heat (Section 12.8.1, Figure 12.15).

The recently reported presence of methane gas in the atmosphere of Mars (Table 12.4) is a small clue that reservoirs of bacteriogenic methane may exist below the surface. Although colonies of living bacteria are known to occur in sedimentary rocks several kilometers below the surface of the Earth, the existence of such colonies on Mars after a period of time of at least three billion years seems unlikely.

12.9.2 Search for Life (Vikings)

The principal objective of the two Viking spacecraft that landed in Chryse and Utopia planitiae in 1976 (Figure 12.8) was to search for life (Carr, 1996). Both landers were designed to carry out three experiments to determine whether microscopic organisms inhabit the soil. In addition, each of the Vikings was equipped to detect organic compounds in gases released by soil samples and to identify these gases. This equipment (gas chromatograph/mass spectrometer) could detect such compounds at a concentration of parts per billion. However, the results of the biology experiments were inconclusive and therefore failed to confirm the presence of bacteria or other unicellular organisms in the soil of Mars (Klein, 1978, 1979; Horowitz, 1986).

The absence of living organisms and of organic matter indicates that the present soil on Mars does not support life because (Carr, 1996):

1. The soil consists primarily of wind-blown sediment that is exposed to high doses of energetic ultraviolet radiation which is harmful to life (Jagger, 1985).
2. The soils of Chryse and Utopia planitiae have been dry for two or more billions of years, which means that even dormant bacteria would have died because, in the absence of liquid water, they could not repair the damage caused by the irradiation with ultraviolet radiation.
3. If organic matter once existed in the soil, it would have been converted into carbon dioxide by reacting with oxygen that is bonded to iron and other metals in the soil.

Therefore, the results of the biology experiments carried out by the Viking landers indicate that the soils of Mars are sterile at the present time and that any organic matter that may have existed there has been oxidized to carbon dioxide. However, the results do not rule out the possibility that organic matter has been preserved in hotspring deposits, in porous rocks, in the polar ice caps, and in aquifers where it is less vulnerable to oxidation. The insights gained from the results of the Viking biology experiments have redirected the search for life to sites where organic matter may have been preserved; although the hopes of finding colonies of living organisms at depth below the surface of Mars are fading.

12.9.3 Search for Life (ALH 84001)

The martian meteorites provide another opportunity to search for evidence that life existed on the surface and in the rocks of Mars (Science Brief 12.13a). At least ten of these meteorites have been collected in Antarctica, including ALH 84001, which was found in 1984 on the Far Western Icefield near the Allan Hills of southern Victoria Land (Cassidy, 2003, p. 122). ALH84001 in Figure 12.19 is a brecciated coarse-grained orthopyroxenite and was initially classified as an achondrite (diogenite). Ten years later, the specimen was reclassified as a martian meteorite that had crystallized at $4.5 \pm 0.13\,Ga$ (Mittelfehlt, 1994; Nyquist et al., 1995).

A group of scientists led by Dr. David S. McKay reported in 1996 that fracture surfaces within ALH84001 contain polycyclic aromatic hydrocarbons (PAH) as well as carbonate globules having diameters of about 100 nanometers and tiny crystals of magnetite and iron sulfide, which could have been deposited by anaerobic bacteria. In addition, high-resolution electron-microscope images such as Figure 12.20 revealed the presence of tubular structure that were tentatively identified as fossilized micro-organisms. The origin of the carbonate ovoids is uncertain although bacteria are capable of depositing carbonates having the observed texture. The hydrocarbons that are associated with the carbonate globules have a spectrum of molecular weights between 178 and greater than 450 atomic mass units (1 amu = 1/12 of the mass of carbon -12 atoms) and their chemical formulas range from $C_{14}H_{10}$ (penanthrene) to $C_{22}H_{12}$ (anthanthracene). The authors

Figure 12.19. Sawed face of the martian meteorite ALH 84001 that was collected during the 1984/95 fieldseason on the Far Western Icefield near the Allan Hills of the Transantarctic Mountains in southern Victoria Land. The specimen was found by Roberta Score while she was a member of a group led by William A. Cassidy who initiated the on-going Antarctic Search for Meteorites (ANSMET). The specimen was initially misidentified as a stony meteorite from the asteroidal belt. Technically speaking, it is an achondrite composed of orthopyroxene and was therefore classified as a diogenite (Table 8.1 in Section 8.1). This martian rock formed by the accumulation of pyroxene crystals in a magma chamber 4.5×10^9 years ago. It was ejected from Mars during a meteorite impact 16×10^6 years ago and fell to the surface of the ice in Antarctica 13×10^3 years ago. The specimen emerged on the surface of the ice sheet when the ice in which it was embedded evaporated leaving it stranded on the surface of the East Antarctic ice sheet. (Cassidy, 2003; Whillans and Cassidy, 1983). (Courtesy of NASA, Johnson Space Flight Center)

determined that the PAHs are indigenous to ALH 84001 and concluded:

"Although there are alternative explanations for each of these phenomena taken individually, when they are considered collectively, particularly in view of their spatial association, we conclude that they are evidence for primitive life on early Mars."

McKay et al. (1996)

The announcement of the discovery of possible signs of life on Mars aroused great interest among the people of the world. As a result, the exploration of Mars by NASA was given top priority as was the continuing search for meteorites in Antarctica funded by the National Science Foundation. However, a reassessment of the data reported by McKay et al. (1996) has cast doubt on their conclusion that ALH84001 contains evidence of life on Mars. In the absence of certain proof, the search for life continues by means of remote-controlled rovers on the surface of Mars.

12.9.4 Search for Life (Rovers)

A significant advance in the exploration of Mars occurred on July 4, 1997, when Mars Pathfinder landed safely in Chryse planitia at 19.13 °N and

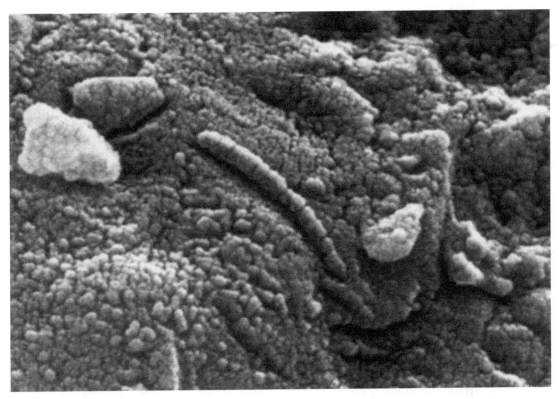

Figure 12.20. High-resolution electron-microscope image of carbonate globules and tube-like structures on the surface of a small fracture in the martian meteorite ALH 84001 (Figure 12.19). The carbonate globules and worm-like structures were tentatively identified as signs of the existence of micro-organisms on Mars (McKay et al., 1996). Although the evidence is not in doubt, its interpretation is not conclusive and does not positively demonstrate that micro-organisms once existed on Mars. (Courtesy of NASA and of the Johnson Space Center)

$33.22\,°W$ and about 1050 km east southeast of the landing site of Viking 1. The Mars Pathfinder was the first spacecraft that landed on inflated airbags and that carried a small rover called Sojourner. After the successful landing, the airbags deflated and the rover rolled cautiously onto the martian surface. For the next 83 sols (martian days) or until September 27 of 1997, Sojourner examined boulders in the vicinity of the landing site and determined their chemical compositions. All of these boulders appear to be dark grey volcanic rocks ranging in composition from basalt to andesite, in contrast to the basaltic composition of the martian meteorites (i.e., the basaltic shergottites, Section 12.7 and Science Brief 12.13.1). The images returned by Pathfinder confirm that Chryse planitia was flooded by muddy water that issued from the Ares and Tiu valleys

(Sections 12.2.4 and 12.5.3). Panoramic views in three dimensions of the landing site of Pathfinder were published in the August-1998-issue of the journal of the National Geographic Society.

The Sojourner traveled about 100 meters in the vicinity of its landing site, carried out 16 chemical analyses of rocks and soil, and sent back more than 17,000 images of the boulder-strewn surface of Chryse planitia. These images confirmed yet again that liquid water had once existed on the surface of Mars; however, the Sojourner did not find fossils or any evidence of life on Mars.

The successful use of inflated airbags to cushion the landing of the Pathfinder spacecraft and the use of a solar-powered robotic rover prepared the way for two larger rovers launched by NASA in 2003. The first of these rovers called *Spirit* landed safely in *Gusev crater* on

January 3, 2004. This rover weighed 174 kg and was equipped with a robotic arm that carried a microscope, an alpha-particle x-ray spectrometer, and a rock-abrasion tool. Gusev crater is 160 km wide and is located at about 15 °S and 175 °E. This site was chosen partly because the crater may have contained a lake in which sedimentary rocks could have been deposited that might contain fossils of organisms that lived in the water of the crater lake. In fact, the floor of Gusev crater turned out to be a plain covered by reddish brown soil containing pebbles and cobbles of volcanic rocks. Several low hills about 100 m high were visible on the horizon about three to four kilometers from the landing site. The landscape projected a sense of tranquillity and solitude seen also in some terrestrial deserts.

While Spirit was parked for the martian winter starting in April of 2006, it was directed to record a detailed panorama of the surrounding area to be known as the McMurdo Panorama. The initial results of this project in Figure 12.21 show the view north toward Husband Hill which Spirit climbed in 2005. The view shows the tracks left by the rover as it traveled to its present location.

Spirit also recorded the image in Figure 12.22 of a scene in the Columbia Hills of Gusev crater. The shiny object in the foreground resembles the iron meteorite the rover Opportunity had previously discovered on January 6 of 2005 on the surface of Meridiani planum (Figure 12.26). The image reproduced in Figure 12.22 contains interesting details that are described in the caption.

An identical rover called *Opportunity* landed on January 25, 2004, in an elongated basin named *Meridiani planum* located at 2 °S, 6 °W. This site was chosen because the soil in this basin contains crystalline hematite (Fe_2O_3) detected from orbit. Fortuitously, the landing site turned out to be a shallow crater about 25 m in diameters which contained outcrops of light-colored thin-bedded sedimentary rocks. The rocks are lacustrine evaporites composed largely of sulfate minerals containing small spherical pellets of specular hematite that have been called "blueberries." The discovery of these sedimentary rocks in Meridiani planum validates the expectation that some of the impact craters on the surface of Mars contained lakes during the Noachian Eon. When water in these lakes evaporated or froze, sulfates and chlorides of the major elements that were originally dissolved in the water precipitated to form thinly bedded evaporite rocks.

The views in Figure 12.23 and 12.24 illustrate the occurrences of thinly bedded sedimentary rocks discovered by the rover Opportunity during its exploration of Meridiani planum. The close-up images in Figure 12.25A and B of the surface of a place called Alamogordo Creek reveal the

Figure 12.21. The so-called McMurdo Panorama was recorded by the rover Spirit during the winter of 2006 in Gusev crater located in the southern highlands of Mars. The surface consists of wind-sculpted fine-grained, reddish-brown sediment with a scattering of angular boulders of black volcanic rocks. Husband Hill is clearly visible on the horizon and Home Plate is located in the right foreground. Spirit actually "climbed" Husband Hill in 2005 and drove across Home Plate on its way to its winter retreat, which it reached in April of 2006. The tracks made by Spirit on this trip can be seen in this image (Courtesy of NASA/JPL/Caltech/Cornell released on June 9, 2006)

Figure 12.22. This image was recorded on April 12, 2006 by Spirit in the Columbia Hills of Gusev crater on Mars. The shiny boulder in the foreground, named Allan Hills, resembles the iron meteorite that was discovered by the rover Opportunity in the Meridiani planum (See Figure 12.26). A second shiny boulder called Zhong Shan is located out of view to the left of this image. The name of the shiny boulder (Allan Hills) is a reference to the site in the Transantarctic Mountains near which many meteorite specimens have been collected on the blue-ice areas of the East Antarctic ice sheet (Cassidy, 2003). Zhong Shan is the name of an Antarctic base established in 1989 by the People's Republic of China. The angular black boulders are vesicular volcanic rocks (e.g., basalt) that were presumably ejected from a nearby impact crater. The fine grained matrix of the regolith at this site is being sculpted and partially eroded by wind. The eolian erosion has uncovered several layers of "soil salts" (e.g., sulfates, carbonates, and chlorides) that have formed by precipitation from aqueous solution. Such salt crusts are a common feature of caliche soils that form under arid and semi-arid conditions in the deserts of the Earth (PIA 08094, courtesy of NASA/JPL/Caltech/Cornell)

presence of hematite pellets in a small area of soil composed of basalt sand adjacent to thinly bedded and light colored evaporite rocks. The pellets may be oöids (diameter < 2 mm) or pisolites (diameter > 2 mm). In terrestrial environments such pellets are concentrically zoned in many cases and may form by the action of micro-organisms (e.g., algae), although

Figure 12.23. The rover Opportunity discovered this outcrop of flat-lying and thin-bedded evaporite rocks on May 12, 2006 during its exploration of Meridiani planum. The crater Victoria is located about 1.1 km from the rover behind the large surface-ripples that confronted the rover at this site. The evaporite rocks were deposited from a saline lake that occupied this site more than three billion years ago (Courtesy of NASA/JPL-Caltech/Cornell)

they may also form by chemical precipitation of minerals around a central nucleus. Terrestrial oöids and pisolites are commonly composed of calcium carbonate (calcite or aragonite). However, sedimentary oolitic iron formations compsoed of hematite are known to occur on the Earth (e.g, Clinton Formation (Silurian), Birmingham, Alabama). (Friedman and Sanders, 1978).

The evaporite rocks in Meridiani planum contain a record of the environmental conditions at the time of deposition. In addition, the rocks may be dated by the decay of longlived radioactive parent atoms and by the accumulation of their stable decay products. In years to come, the evaporite rocks of Mars, like those in Meridiani planum, may become valuable natural resources of iron ore (hematite) and raw materials for a chemical industry (sulfate and chloride minerals) on Mars. Another noteworthy accomplishment of the rover Opportunity recorded in Figure 12.26 was the discovery of a basketball-sized iron meteorite on January 6, 2005. This is the first meteorite from the asteroidal belt to be

Figure 12.24. View of Burns Cliff consisting of flat-lying and thin-bedded sedimentary rocks exposed along the southeast wall of Endurance crater. This crater is located east of Eagle crater where Opportunity first discovered sedimentary rocks including hematite pellets like those in the rocks of Alamogordo Creek in Figure 12.25B. The rocks of Burns Cliff in Endurance crater form a gently curving, continuous surface sloping up and away from the view point. The apparent bulge of this cliff is an artifact of the wide-angle view recorded in this image. (Courtesy of NASA/JPL-Caltech)

Figure 12.26. On January 6, 2005, the rover Opportunity discovered this shiny boulder, which members of the rover science team refer to informally as Heat Shield Rock. This basket-ball sized boulder is composed almost exclusively of metallic iron and nickel and therefore has been identified as an iron meteorite from the asteroidal belt. Two similar shiny boulders named Allan Hills and Zhong Shan were later discovered by the rover Spirit in the Columbia Hills of Gusev crater (Figure 12.22). (PIA 07269, courtesy of NASA/JPL-Caltech/Cornell)

Figure 12.25. A. On May 20, 2006, the rover Opportunity examined a small patch of "soil" called Alamogordo Creek located in Meridiani planum. The sediment in this image has collected in a shallow depression in the rocks shown in Figure 12.23. The sediment contains small hematite pellets and a few small angular rock fragments imbedded in a matrix of silt. B. This microscopic view of the hematite pellets, best described as oöids but colloquially referred to as "blueberries," shows that the pellets are approximately spherical in shape, although a few are irregularly shaped and some are broken. This image does not indicate whether the pellets are zoned concentrically or radially in which case they could be concretions. Nevertheless, the best preliminary interpretation is that the hematite oöids or pisolites in the lacustrine evaporites of Mars actually are concretions. On Earth, similar concretions form as a result of precipitation of minerals by micro-organisms adhering to their surface (Courtesy of NASA/JPL-Caltech/Cornell/US Geological Survey)

discovered on the surface of a planet other than the Earth (Meteoritics and Planetary Science, vol. 40, No. 1, 2005).

Although no signs of life have been found by the rover Opportunity in the sedimentary rocks of Meridiani planum or by Spirit in Gusev crater, the search for life will continue with renewed vigor. The discovery of water-laid sedimentary rocks on Mars is a huge achievement and a turning point in the exploration of this planet that used to be a mere dot of reddish light in the night sky (Squyres, 2005).

12.10 Colonization of Mars

Ever since Giovanni Schiaparelli reported in 1877 that the martian surface is criss-crossed by "canali", people on Earth have speculated that Mars is inhabited by intelligent beings (e.g., Lowell, 1895, 1906, 1908). The modern era of the exploration of Mars, which started on July 14 of 1965 with the flyby of Mariner 4, disposed of these phantasies once and for all and replaced them with an objective assessment of the present state of the surface of Mars. The new view of Mars indicates that its present surface is a cold, dry, and windy place where humans require life-support systems to survive. Nevertheless, preparations are now underway by NASA to land a team of astronauts on the surface of Mars. If the first landing on Mars is successful, it may be followed by the establishment of a permanent research station, which could evolve into a self-sustaining colony of humans on Mars.

The future colonization of Mars is the subject of the science-fiction novels of Robinson (1994, 1994, 1996), who attributed the ultimate success of human colonies to the beneficial consequences of terraforming, by means of which the colonists increased the global atmospheric temperature and pressure of the planet until the ice caps and permafrost melted and liquid water once again existed on the surface of Mars. However, before we can attempt to colonize Mars, we must first develop the technology that will allow us to travel to Mars safely, rapidly, and inexpensively.

12.10.1 Travel to Mars

Current strategies for space travel are based on the use of powerful rockets to lift payloads into orbit around the Earth. Once they have reached orbit, spacecraft use much smaller rockets to escape from the gravitational field of the Earth and then coast toward their destination with minor course corrections as necessary. As a spacecraft approaches its destination, a rocket engine is used to reduce its speed sufficiently to permit it to go into orbit (Section 1.3). In some cases (e.g., Gallileo and Cassini), trajectories were designed to allow the spacecraft to pick up speed by means of gravity assists from other planets along their course. Nevertheless, a one-way trip from Earth to Mars can take six months which means that an expedition to Mars is likely to last about 30 months: Travel time = 12 months; ground time on Mars = 18 months.

The travel time necessary to reach Mars and more distant destinations in the solar system can be reduced by the use of low-energy but continuously acting propulsion systems currently being developed. For example, rockets can generate forward thrust by ejecting charged atoms of a gas through a nozzle at the back (Burnham, 2002). The electric power required to ionize and heat the gas can be provided by solar panels or by nuclear power generators. Such ion rockets powered by nuclear power generators were tested between 1961 and 1973 at Jackass Flat in Nevada. One of the nuclear-powered rocket engines developed 55,000 pounds of thrust for 62 minutes. A spacecraft equipped with an ion drive could reach Mars in much less time (e.g., 40 days) than a spacecraft that merely coasted from Earth orbit. Although nuclear power-generators pose risks especially for space travel by humans, their use is being considered for future robotic missions to the Galilean satellites of Jupiter and for landings on asteroids.

Solar energy cannot generate enough electricity to operate ion drives throughout the solar system because the intensity of sunlight decreases with increasing distance (R) from the Sun by a factor of $1/R^2$. Therefore, the energy needed to operate ion drives outside of the asteroidal belt must be provided either by thermal-electric generators heated by the decay of the radioactive atoms of certain elements such as plutonium or by a conventional nuclear-fission reactor. A review of various propulsion systems with potential applications to spaceflight was published in a set of articles edited by Beardsley (1999).

The most economical form of space transportation is to use a "cycler" which is a spacecraft that orbits the Sun and encounters the Earth and Mars at predictable intervals. Crews and passengers would board the cycler by means of a shuttle craft and would disembark at their destination in the same way. The cycler would not have to be launched from the surface of the Earth for a trip to Mars and it would not even have to slow down as it passes the Earth or Mars. This concept has been advocated by Oberg and Aldrin (2000) and, if implemented, could transport hundreds of people to Mars at regular intervals. This method may be used in the future to transport prospective colonists and business travelers to Mars.

The long duration of the flight to Mars, even with the aid of a technologically advanced propulsion system, poses medical hazards for the crew that may result from the prolonged confinement in a small space, from accidental decompression en route, and from failure to achieve orbit around Mars. The probability that these and other kinds of accidents will occur is small but not zero. In addition, the low gravity within the spacecraft may cause atrophication of muscles and loss of bone mass, which could cause the astronauts to arrive on Mars in a weakened state.

The greatest medical hazard to the astronauts en route to Mars is the exposure to powerful ionizing radiation originating from cosmic rays and solar flares. Parker (2006) estimated that the energy absorbed by an astronaut from cosmic rays in interplanetary space can range from 13 to 25 Rems per year (1 Rem = 100 ergs/g of tissue). At this rate, astronauts traveling in space absorb more radiation in one year than radiation workers do in a lifetime. The radiation dose absorbed in humans living at sealevel on the Earth is only 0.02 to 0.04 Rems per year. The high dose of radiation exposure of astronauts in interplanetary space will damage their DNA. As a result, they may develop cancers of various kinds and other illnesses depending on whether natural repair mechanisms can keep up with the damage. Parker (2006) also described ways of shielding the astronauts from this hazard but concluded that the measures that have been considered add too much mass to the spacecraft and, in some cases, are impractical.

12.10.2 First Steps

The first astronauts to undertake the trip to Mars may either land on the surface of the planet or on one of its two satellites Phobos and Deimos. In either case, they will probably use the landing craft as their habitat and survive on the provisions they brought from Earth. The astronauts will explore the area around the landing site on foot, by means of a rover controlled from inside the habitat, and by means of a pressurized vehicle. The principal objectives of their mission will be to search for evidence of former or present life, to assess the potential for finding and using water and other martian resources, to attempt to grow plants in martian soil, and to return safely to the Earth.

One of the major obstacles of manned missions to Mars is the need to transport the rocket fuel needed for the return trip. This difficulty can be avoided by a strategy advocated by Robert Zubrin (Zubrin, 1996; Zubrin and Wagner, 1996). Zubrin's plan for returning the astronauts from the surface of Mars is to use the carbon dioxide of the martian atmosphere to manufacture methane (CH_4) by reacting it with hydrogen gas shipped to Mars ahead of the arrival of the astronauts. The chemical reaction of this process is well known:

$$CO_2 + 4H_2 \rightarrow CH_4 + 2H_2O \qquad (12.1)$$

The water produced by this reaction is subsequently decomposed into hydrogen and oxygen gas by means of electrolysis:

$$2H_2O \rightarrow 2H_2 + O_2 \qquad (12.2)$$

The oxygen is used to burn the methane in the rocket engine during the return trip. The hydrogen gas is reacted with carbon dioxide of the martian atmosphere to make additional methane. This strategy significantly reduces the cost of the exploration of Mars by human astronauts. After the return vehicle has been fueled on Mars, the astronauts take off from the Earth and land about six months later at a preselected site near the fully fueled Earth-Return Vehicle (ERV).

The main component of the spacecraft that carries the astronauts to Mars is a two-story drum-shaped habitat equipped with a life-support system like that used in the International Space Station and provisions that can sustain a crew of four for three years. If the landing site is in the polar regions (e.g., Vastitas Borealis), the astronauts may be able to melt ice of the north-polar ice cap for their own consumption, to irrigate the martian soil in an experimental greenhouse, or to grow vegetables in their habitat by hydroponics. In addition, the water could provide oxygen for breathing and hydrogen for the production of methane.

After exploring Mars for about 500 days, the astronauts transfer to the ERV and return to Earth for a splashdown in the Pacific or Atlantic Ocean. According to Zubrin (1996), a trip to Mars by four astronauts is achievable with existing technology and would cost about 20 billion dollars. Additional information about the exploration of Mars by astronauts is available in a set of articles published in March of 2000 by the Scientific American.

The techniques and equipment to be used by astronauts on the surface of Mars are already being tested by NASA at the Haughton impact crater on Devon Island in the Nunavut Territory of the Canadian Arctic (Lee, 2002). In addition, members of the Mars Society founded by Robert Zubrin set up a replica of the Mars habitat in the desert near Hanksville, Utah (Schilling, 2003). The participants wore space suits like the astronauts will someday use on Mars, and they maintained pre-arranged schedules on their excursions into the "martian" deserts to collect interesting rocks.

12.10.3 Terraforming

In the long run, the average atmospheric pressure and temperature at the surface of Mars will have to be increased if a self-sustaining human colony is to be established on the Red Planet. The process of improving the climate of a planet in order to make it habitable by humans, called terraforming, will require several centuries. The key to eventual success is to increase the concentration of greenhouse gases in the atmosphere of Mars in order to increase its average global temperature. Several different strategies have been proposed to initiate the process.

1. Release efficient absorbers of infrared radiation such as chloro-fluoro-carbon gases (e.g., freon manufactured on Mars).
2. Increase the rate of sublimation of ices in the polar ice caps by covering them with carbon dust or by focusing sunlight on the ice caps by means of mirrors in stationary orbits above the poles of Mars.
3. Capture several comets and guide them to impact on Mars thereby releasing large quantities of water and other gases.

After some carbon dioxide and water vapor has been released into the atmosphere by these methods, their presence will enhance the greenhouse warming of the atmosphere and thus accelerate the sublimation of ice. As a result, the atmospheric pressure of Mars will rise above the triple point of water and liquid water will form on the surface of Mars.

At this stage in the process, simple plants from the Earth (e.g., algae, lichens, mosses, and liverworts) may grow without the protection of greenhouses. Gradually, the permafrost will melt, water will collect in basins, clouds will form in the martian sky, and the hydrologic cycle will be re-established. Eventually, grasses and shrubs may grow outside and the concentration of oxygen in the atmosphere will begin to increase. As the process of terraforming continues, air-breathing mammals including humans will be able to live on Mars as the planet changes color from red to green and finally to blue (e.g., Robinson, 1993, 1994, 1996).

In 1991 the editors of the popular magazine "Life" published a detailed schedule for terraforming Mars starting in 2015 AD and ending 155 years later in 2170 AD. In the scenario outlined by the editors of Life, large animals were introduced when the air on Mars became breathable, tourism and exports expanded, and Mars became a habitable planet (Darrach et al., 1991). Thousands of colonist moved there to live and, in time, became Martians who adapted to the low gravity and developed their own culture and institutions. Although this may sound like science fiction, we remind our readers that good science fiction should not be taken literally; instead, it should be taken seriously. A case in point is the book by Sagan (1994) "Pale Blue Dot; A Vision of the Human Future in Space."

12.11 Satellites

The two satellites of Mars (Phobos and Deimos) revolve in the prograde direction and have 1:1 spin-orbit coupling. The relevant parameters of their orbits and of their physical properties in Table 12.7 include their sidereal periods of revolution of 0.319 days for Phobos and 1.263 days for Deimos. The sidereal period of revolution of Phobos (7 h and 38 min or 7.65 h) is *shorter* than the period of rotation of Mars (24 h, 37 min, 22 s or 24.622 h. Therefore, Phobos revolves around Mars 3.2 times during each sol. The resulting tidal interaction between Phobos and Mars *increases* the rate of rotation of Mars while causing the orbit of Phobos to decay. As a result, Phobos is expected to crash into Mars in about 50 million years (Freedman and Kaufmann, 2002, p. 278).

The period of revolution of Deimos (30.312 h) is *longer* than the period of rotation of Mars (24.622 h), which means that the tidal interactions of Deimos tend to *slow* the rate of rotation of Mars. As a result, the radius of

Table 12.7. Orbital parameters and physical properties of Phobos and Deimos (Freedman and Kaufmann, 2002; Hartmann, 2005)

Property	Phobos (Fear)	Deimos (Terror)
Average distance from the center of Mars, km	9378	23,460
Sidereal period of revolution, days	0.319 (7.656 h)	1.263 (30.312 h)
Period of rotation, days	0.319	1.263
Eccentricity	0.01	0.00
Diameter, km	$28 \times 23 \times 20$	$16 \times 12 \times 10$
Mass, kg	1.1×10^{16}	1.8×10^{15}
Bulk, density g/cm^3	1.950	1.760

Figure 12.27. Phobos ($27 \times 22 \times 18$ km) is the innermost of the two satellites of Mars, the other being Deimos whose mass is only one sixth that of Phobos (Table 12.7). The surface of Phobos is covered by a layer of regolith and is heavily cratered. The largest impact crater (not shown in this image) was named Stickney by Asaph Hall who discovered both satellites of Mars in 1877. The impact that caused the formation of Stickney appears to have fractured the body of Phobos thereby reducing its bulk density to 1.950 g/cm^3. Traces of these fractures are visible in the form of grooves and aligned series of pits on the surface of Phobos. Both Phobos and Deimos are believed to be former asteroids that were captured by Mars during a close encounter. (PIA 04589, Mars Global Surveyor/Malin Space Science System, courtesy of NASA/JPL-Caltech/MSSS)

the orbit of Deimos is increasing, causing it to recede from Mars. Evidently, the tidal interactions between Deimos and Mars are counteracting those between Phobos and Mars, thereby leaving their ultimate fates in doubt. The tidal interactions between the Earth and the Moon were previously described in Section 9.8.2.

The surfaces of Phobos and Deimos are densely cratered indicating that both are ancient objects whose age is close to 4.6×10^9 years (Hartmann, 2005). Their bulk densities (Phobos: 1.950 g/cm^3; Deimos: 1.760 g/cm^3) are lower than those of Mars (3.934 g/cm^3) and are also lower than those of chondritic meteorites (3.40 to 3.80 g/cm^3), but they are similar to the densities of carbon-rich (Type I) carbonaceous chondrites whose average density is 2.2 g/cm^3 (Glass, 1982). The low densities, dark-colored surfaces, and irregular shapes suggest that the satellites of Mars are captured carbon-rich asteroids. However, their low densities could also indicate that both satellites are either extensively fractured or contain various kinds of ices in their interiors, or both. The existence of fractures in the interior of Phobos is suggested by the presence of parallel grooves on its surface visible in Figure 12.27. In addition, small pits along these grooves may have formed when volatile compounds in its interior vented to the surface (Fanale and Salvail, 1990). Russian investigators associated with the PHOBOS-2 mission actually detected the presence of water molecules at the surface of Phobos. The chemical compositions

and internal structures of the two satellites will undoubtedly be investigate during future missions to Mars.

12.12 Summary

The review of various topographic features in the northern and southern hemisphere of Mars provides the data we need to begin to reconstruct the geologic history of Mars. We can assert that Mars formed by the accretion of planetesimals within the protoplanetary disk at a distance of 1.524 AU from the Sun and that it differentiated by the segregation of immiscible liquids to form a core composed of metallic iron, a mantle composed of silicate minerals (olivine and pyroxene), a crust of plagioclase-rich igneous rocks (i.e., anorthosite), and a dense atmosphere consisting of the oxides of carbon, water vapor, methane, ammonia, nitrogen, and other gases.

As the planet cooled, it continued to be hit by planetesimals at a rapid rate for the first 800 million years after initial accretion. During this time, the Utopia basin of the northern hemisphere as well as the Hellas, Argyre, and Isidis basins were excavated by the explosive impacts of large planetesimals, which disrupted the anorthosite crust and eroded the atmosphere. At the same time, water and carbon dioxide were transferred from the atmosphere to the surface of Mars to form large bodies of standing water in the impact basins of the northern and southern hemispheres.

The impacts of large planetesimals may have triggered the eruption of volcanic rocks that formed a crust composed of basalt flows which continued to be cratered by impacts until about 3.8 Ga. Volcanic activity continued at several centers where asthenospheric plumes interacted with the overlying lithosphere and caused decompression melting of the mantle rocks. The resulting basalt magma was erupted to form basalt plateaus (e.g., the Tharsis plateau), shield volcanoes (e.g., Olympus mons, etc.) and cinder cones (e.g., Hadriaca patera). The surfaces of the plateaus and volcanic mountains contain only a small number of impact craters because they formed after the period of intense bombardment had ended and because some of the flows may have been erupted as recently as ten million years ago. During the main phase of volcanic activity, water flowed in streams that were maintained by springs along the walls of large valleys. At this time, the average global temperature of Mars was sustained by greenhouse warming that resulted from the emission of carbon dioxide by the active volcanoes. As volcanic activity waned, the global surface temperature declined until the hydrologic cycle ceased to function, thereby causing Mars to slip into an ice age. The temperature of the polar regions and on the summits of the highest volcanoes of Mars eventually decreased sufficiently to cause carbon dioxide of the atmosphere to condense to form carbon dioxide ice. The removal of water vapor and carbon dioxide from the atmosphere further decreased the greenhouse effect and thereby caused the climate of Mars to become cold and dry.

Although the present climate on Mars is cold and dry and although the atmosphere is "thin" and lacks oxygen, Mars can be made liveable for humans by terraforming. The transformation of Mars is initiated by increasing its atmospheric pressure, which will stabilize liquid water and will increase its average global temperature by means of the greenhouse effect. Present plans call for a team of astronauts to land on Mars some time after 2020. If terraforming of Mars is initiated during the present century, several hundred years will elapse before Mars becomes habitable for primitive green plants and ultimately for humans. In this way, Mars "could" become habitable for life from the Earth and thereby allow humans to survive in case of a catastrophic impact of an asteroid upon the Earth.

12.13 Science Briefs

12.13.1 Martian Meteorites

The martian meteorites were originally classified as shergottites, nakhlites, and chassignites and were collectively refereed to as the SNC (snik) meteorites. They were considered to be achondrites from the asteroidal belt until several teams of investigators in 1979 and 1980 cautiously hinted that the SNC meteorites may have originated from Mars.

The original shergottite is a stone weighing 5 kg that fell on August 25 of 1865 at the village of Shergotty in India. This stone is composed of fine-grained pyroxene (pigeonite and augite), shock-metamorphosed plagioclase (maskelynite), and accessory Fe-Ti oxides, chromite, as well as sulfates and phosphates. Melt inclusions in the pyroxene contain kaersutite (titaniferous amphibole containing OH^- ions). This stone was classified as a "basaltic shergottite" meteorite because of its unique mineral assemblage and texture. Nearly 100 years later, on October 3 of 1962, another stone weighing 18 kg fell near Zagami in Katsina Province of Nigeria. This stone was also classified as basaltic shergottite because its mineral composition and texture are similar to those of Shergotty. Two additional basaltic shergottites (EETA 79001 and QUE 94201) were subsequently collected on the East Antarctic ice sheet in 1979 and 1994, respectively. Since then, the number of specimens of basaltic shergottites in Table 12.8 has increased to 22 based on finds

Table 12.8. List of the names of shergottites, nakhlites, and chassignites all of which are martian rocks that were ejected from Mars by the impacts of meteorites from the asteroidal belt. Based on Cassidy (2003) and other sources

Name	Mass, g	Comments
Basaltic Shergottites		
Shergotty	5000	age: 185 ± 25 Ma, India
Zagami	18,000	age: 185 ± 25 Ma, Nigeria
QUE 94201	12	age: 327 ± 12 Ma, Antarctica
EETA 79001	7942	age: 185 ± 25 Ma, Antarctica
Dar al Gani 476	2015	age: 474 ± 11, Ma, Sahara desert
Dar al Gani 489	2146	Sahara desert
Dar al Gabu 670	588	Sahara desert
Dar al Gani 876	6.216	Sahara desert
NWA 480	28	Moroccan desert, NW Africa
NWA 856	320	Moroccan desert, NW Africa
NWA 1086	118	Moroccan desert, NW Africa
NWA 1110	576.77	Moroccan desert, NW Africa
Los Angeles 1	452.6	age: 165 ± 11 Ma, Mojave desert
Los Angeles 2	245.4	age: 172 ± 8 Ma, Mojave desert
Sayh al Uhaymir 005*	1344	Oman desert
Sayh al Uhaymir 008*	8579	Oman desert
Sayh al Uhaymi 051*	436	Oman desert
Sayh al Uhaymir 094*	223.3	Oman desert
Sayh al Uhaymir 060	42.28	Oman desert
Sayh al Uhaymir 090	94.84	Oman desert
Dhofar 019	1056	Oman desert
Dhofar 378	15	Oman desert
Lherzolitic Shergottites		
ALHA 77005	482	age: 187 ± 12 Ma, Antarctica 154 ± 6 Ma
Y 793605	18	Yamato Mountains, Antarctica
Lew 88516	13	Lewis Cliff ice tongue, Antarctica
GRV 99027	9.97	Grove Mountain, Antarctica
Nakhlites		
Nakhla	9000	age: 1.3 Ga, Egypt
Lafayette	800	age: 1.3 Ga, Indiana
Governador Valadares	158	age: 1.3 Ga, Brazil
Y000593	13,713	age: 1.53 ± 0.46 Ga, Yamoto Mtns, Antarctica
Y000749	1283	Yamato Mountains, Antarctica
NWA 817	104	Moroccan desert, NW Africa
NW 998		Moroccan desert, NW Africa
MIL 03346	715.2	East Antarctic ice sheet
Chassignites		
Chassigny	4000	age: 1.3 Ga, France
Orthopyroxenite		
ALH 84001	1931	age: 4.5 ± 0.13 Ga, Antarctica

*Fragments of the same meteoroid (Gnos et al. 2002)

in the hot deserts of the world. The sources of water in martian meteorites was discussed by Boctor et al. (2003) and by Karlsson et al. (1992).

In addition, the Antarctic ice sheet has yielded four shergottite specimens that differ from the basaltic shergottites by being composed of coarse grained olivine, diopside (diallage), and orthopyroxene which together define a variety of peridotite called lherzolite. These lherzolite shergottites are plutonic igneous rocks that formed by the accumulation of minerals on the floor of a chamber filled with basalt magma. Therefore, the lherzolitic shergottites are considered to be genetically related to the basaltic shergottites, which originated from lava flows that were erupted at the surface of Mars. Table 12.8 contains 4 specimens of lherzolitic shergottites all of which were collected in the ablation areas of the Earth Antarctic ice sheet.

Nakhla, the original nakhlite, fell on June 28 of 1911 near the village of El Nakhla El Baharia located about 39 km east of Alexandria in Egypt. The meteoroid broke up into a shower of at least 40 stones one of which reportedly killed dog. Nakhla is a clinopyroxenite that probably originated from a shallow intrusion of basalt magma below the surface of Mars. Table 12.8 contains a total of 8 nakhlite specimens. Insofar as age determinations are available, the nakhlites crystallized between 1.3 and 1.5 Ga in contrast to the shergottites which crystallized much more recently between 154 and 474 Ma. The relatively young ages of the shergottites corroborate the evidence derived from crater counts which indicate that some lava flows on Olympus mons and in the northern planitiae of Mars were erupted quite recently (Section 12.2.3 and 12.6).

Chassigny is a dunite (i.e., it is composed of olivine) and, like the nakhlites, it also formed in a shallow intrusion at 1.3 Ga. When ALH 84001 (orthopyroxenite) is included, the total number of meteorite specimens from Mars stands at 36 and their total weight is 81.5 kg.

12.13.2 Depth to Water on Mars

We assume, for the sake of this presentation, that the thermal gradient in the crust of Mars is 10 °C/km. The geothermal gradient in the continental crust of the Earth increases with depth and varies regionally but averages about 15 °C/km.

The assumed value of the thermal gradient (R) of the crust of Mars combined with the average surface temperature (A) in equation 12.3 can be used to estimate the depth below the surface (D) where the temperature (T) rises to 0 °C and where liquid water (in pure form) can exist beneath the layer of permafrost (Figure 12.15).

$$T = A + (R \times D) \tag{12.3}$$

If we want to determine the depth (D) where the temperature (T) is equal to zero, we set $T = 0$ and solve equation 12.3 for D:

$$D = -\frac{A}{R} \tag{12.4}$$

Given that $A = -53\,°C$ and $R = 10\,°C/km$,

$$D = \frac{53}{10} = 5.3\,km$$

The "depth-to-water" (D) will be less than 5.3 km in places where the surface temperature is higher than $-53\,°C$ and/or where the thermal gradient is greater than $10\,°C/km$. In addition, the water in the subsurface of Mars is probably salty because it contains soluble salts produced by chemical weathering of minerals. Salty water freezes at a lower temperature than pure water and therefore can remain liquid at temperatures below 0 °C. We can combine these considerations to obtain the best-case scenario. By setting $T = -5\,°C$, $A = -20\,°C$, and $R = 15\,°C$ we obtain from equation 12.5:

$$D = \frac{T - A}{R} \tag{12.5}$$

$$D = \frac{-5 + 20}{15} = \frac{15}{15} = 1.0\,km$$

Therefore, liquid water may exist about one kilometer below the surface in selected places on Mars where the average surface temperature is $A = -20\,°C$ and where the thermal gradient is $R = 15\,°C/km$, provided also that the water contains enough salt in solution to freeze at $-5\,°C$. Such salty water would have to be purified before it would be suitable for human consumption or for irrigation of crops in greenhouses on Mars.

12.13.3 The Boiling and Freezing Temperatures of Water

Pure water at sealevel on the Earth freezes at 0 °C and boils at 100 °C because that is how the Celsius temperature scale is defined. We also know that the temperature of boiling water remains constant no matter how much heat is applied to the water or how long the water is boiled.

When water is confined in a closed container with an air space above it, water molecules evaporate from the surface until the rate of evaporation becomes equal to the rate of condensation. The pressure exerted by water vapor in equilibrium with liquid water is the *vapor pressure* of water, which increases with rising temperature. Water boils at 100 °C at sealevel because at that temperature the vapor pressure of water is equal to the atmospheric pressure, which is equal to 1.0 atmosphere or 759.94 mm of Hg. At sites where the atmospheric pressure is less than one atmosphere, water boils at less than 100 °C because its vapor pressure becomes equal to the atmospheric pressure at a lower temperature than at sealevel. Since the atmospheric pressure decreases with increasing elevation above sealevel, the boiling temperature of water in Figure 12.28 also decreases with increasing elevation.

On the other hand, when water is placed under pressure, as in a pressure cooker or in the subsurface groundwater reservoir of a geyser, the boiling temperature increases in direct proportion to the applied pressure because the water cannot boil until its vapor pressure is equal to the applied pressure. Therefore, the temperature of water expelled by geysers during an eruption is higher than 100 °C in many cases.

The boiling curve of water in Figure 12.16 specifies the relation between the boiling temperature as a function of the pressure. The boiling curve ends at the critical point located at a temperature of 374 °C and a pressure of 218 atmospheres. At the critical point, the distinction between liquid water and water vapor vanishes.

When pure water at sealevel is cooled to 0 °C, it begins to freeze. The temperature of a mixture of ice and water, remains constant at 0 °C until all of the water has solidified. Similarly, when ice melts to form liquid water, the temperature remains constant until all of the ice has melted.

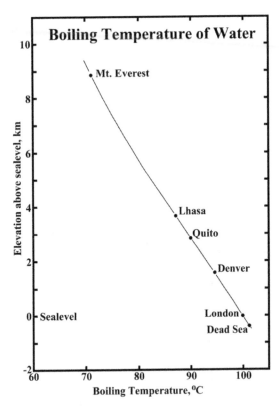

Figure 12.28. The boiling temperature of pure water decreases with increasing elevation from 100 °C at sealevel on the Earth to 71 °C at an elevation of 8.84 km above sealevel on the summit of Mt. Everest. Consequently, potatoes and other food items take longer to cook when they are boiled in water on the summit of Mt. Everest (explorit@ewxplorit.org)

The melting of ice can be accelerated by applying heat, but the temperature of meltwater does not rise above 0 °C until all of the ice has melted. Another surprising property of ice is the fact that it floats in water because the density of ice is lower than the density of water at 0 °C. The low density of ice results from the fact that ice occupies a larger volume than the water from which it formed.

The increase in volume that occurs during freezing of water causes its freezing temperature to decrease when pressure is applied to the water. In other words, the application of pressure prevents water from freezing at 0 °C because the pressure opposes the expansion that accompanies freezing. Therefore, water under pressure must freeze at a lower temperature in order to overcome the effect of the applied pressure. When the freezing temperatures of water are determined at a series of pressures greater than one atmosphere, the resulting data points define a line in coordinates of temperature and pressure. The slope of the resulting freezing-point line is negative in contrast to the boiling-point curve which has a positive slope. When the freezing-point line and the boiling-point curve of water are plotted on the same diagram as in Figure 12.16, the two lines intersect in the triple point whose coordinates are T = +0.010 °C and P = 0.006 atmospheres. At the pressure and temperature of the triple point, the three phases of water: ice, liquid, and vapor occur together in equilibrium.

The boiling-temperature curve and the freezing-temperature line together define the pressures and temperatures that outline the stability fields of the solid, liquid, and vapor phase of water. Therefore, the diagram in Figure 12.16 is known as a *phase diagram*. Similar phase diagrams exist for all volatile compounds including carbon dioxide, methane, and ammonia.

12.14 Problems

1. Calculate the escape speed for Mars in units of km/s (See Appendix I and Table 12.1). Answer: 5.02 km/s
2. Calculate the mass in kilograms of a carbon dioxide molecule whose molecular weight is 44.009 atomic mass units. (Remember that Avogadro's Numbers is 6.022×10^{23} molecules per gram-molecular weight). Answer: 7.308×10^{-26} kg
3. Calculate the average speed of a molecule of carbon dioxide in the atmosphere of Mars assuming an atmospheric temperature of −83 °C. Express the speed in units of km/s. Answer: 0.328 km/s
4. Interpret the results of calculations in problems 2 and 3 above (Consult Section 24 of Appendix 1).
5. Repeat the calculation in problem 3 for a temperature of −33 °C and interpret the result. Answer: 0.368 km/s
6. Plot a graph of the variation of the average speed of CO_2 molecules as a function of temperature between T = 180 K and T = 298 K. Let T be the x-axis and v be the y-axis. Write a descriptive caption for this diagram.

12.15 Further Reading

Banin A, Clark BC, Wänke H (1992) Surface chemistry and mineralogy. In: Kieffer HH. et al. (eds) Mars. Univ. Arizona Press, Tucson, AZ, pp 594–625

Beardsley T (ed.) (1999) The way to go in space. Sci Am February, 80–97

Boctor NZ, Alexander CMO'D, Wang J, Hauri E (2003) The sources of water in martian meteorites: Clues from hydrogen isotopes. Geochim Cosmochim Acta 67(20):3971–3989

Bradbury R (1946) The martian chronicles. Re-issued, September 1979, Bantam, New York

Burnham R (2002) Deep Space 1: The new millenium in spaceflight. The Planetary Report, March/April, 12–17

Carr NH (1996) Water on Mars. Oxford University Press, New York

Cassidy WA (2003) Meteorites, ice, and Antarctica. Cambridge University Press, Cambridge, UK

Chyba CF (1990) Impact delivery and erosion of planetary oceans in the early inner solar system. Nature 343: 129–133

Clarke AC (1951) The sands of Mars. Bantam Books, New York

Clifford SM (1993) A model for the hydrologic and climatic behavior of water on Mars. J Geophys Res, 98: 10,973–11,016

Clifford SM, Parker TJ (2001) The evolution of the martian hydrosphere: Implications for the fate of a primordial ocean and the current state of the northern plains. Icarus 154:40–79

Darrach B, Petranek S, Hollister A (1991) Mars, the planet that once was warm. Life 14(5):24–35

De Goursac O (2004) Visions of Mars. Harry N. Abrams, New York

Fanale F, Salvail J (1990) Evolution of the water regime of Phobos. Icarus 88:383

Faure G (2001) Origin of igneous rocks: The isotopic evidence. Springer-Verlag, Heidelberg

Freedman RA, Kaufmann III WJ (2002) Universe: The solar system. Freeman, New York

Friedmann EI (1982) Endolithic micro-organisms in the Antarctic cold desert. Science 215:1045–1053

Friedmann EI, Koriem AM (1989) Life on Mars: how it disappeared (if it was ever there). Adv Space Res 9(6):167–172

Friedman GM, Sanders JE (1978) Principles of sedimentology. Wiley, NY

Gilichinsky DA (2002) Permafrost model of extraterrestrial habitat. In: Horneck G. and Baumstark-Khan C. (eds) Astrobiology. Springer-Verlag, Berlin, pp 125–142

Glass BP (1982) Introduction to planetary geology. Cambridge University Press, Cambridge, UK

Gnos E, Hofmann B, Franchi IA, Al-Kathiri A, Hauser M, Moser L (2002) Sayh al Uhaymir 094. A new martian meteorite from the Oman desert. Meteoritics Planet Sci 37:835–854

Graedel TE, Crutzen PJ (1993) Atmospheric change: An Earth-system perspective. Freeman, New York

Greeley R (1985) Planetary landscapes. Allen and Unwin, Boston, MA

Hanlon M (2004) The real Mars: Spirit, Opportunity, Mars Express, and the quest to explore the Red Planet. Carroll and Graf, New York

Hartmann WK (1999) Martian cratering VI. Crater count isochrons and evidence for Recent volcanism from Mars Global Surveyor. Meteoritics and Planet Sci 34: 167–177

Hartmann WK (2003) A traveler's guide to Mars. The mysterious landscapes of the Red Planet. Workman Publishing, New York

Hartmann WK (2005) Moons and planets. 5th edn. Brooks/Cole, Belmont, CA

Horneck G, Baumstark-Khan C (eds) (2002) Astrobiology, Springer-Verlag, Berlin

Horowitz NH (1986) To Utopia and back: The search for life in the solar system. W.H. Freeman, New York

Hvidberg CS (2005) Polar caps. In: Tokano T. (ed.) Water on Mars and Life. Springer-Verlag, Heidelberg, pp 129–152

Jagger J (1985) Solar-UV actions on living cells. Praeger, New York

Jakosky BM, Shock EL (1988) The biological potential of Mars, the early Earth, and Europa. J Geophys Res 103(E8):19359–19364

Kargel JS (2004) Mars; A warmer wetter planet. Springer Verlag/Praxis Pub., Heidelberg and Chichester

Karlsson HR, Clayton RN, Gibson Jr EK, Mayeda TK (1992) Water in SNC meteorites: Evidence for a martian hydrosphere. Science 255:1409–1411

Kieffer HH, Jakosky BM, Snyder CW, Matthews MS (eds) (1992) Mars. University of Arizona Press, Tucson, AZ

Krauskopf KB (1979) Introduction to geochemistry, 2nd edn. McGraw-Hill Book Co., New York

Klein HP (1978) The Viking biological experiments on Mars. Icarus, 34:666–674

Klein HP (1979) The Viking mission and the search for life on Mars. Rev Geophys 17:1655–1662

Lazcano A, Miller SL (1994) How long did it take for life to begin and to evolve to cyanobacteria? J Molec Evol 35:546–554

Lee P (2002) From the Earth to Mars: A crater, ice, and life. The Planet. Report, January/February, 12–17

Lowell P (1895) Mars. Houghton Mifflin, Boston MA

Lowell P (1906) Mars and its canals. Macmillan, New York

Lowell P (1908) Mars and the abode of life. Macmillan, New York

McKay DS, Gibson Jr EK, Thomas-Keprta KL, Vali H, Romanek CS, Clemett SJ, Chillier XDF, Maechling CR, Zare RN (1996) Search for past life on Mars: Possible relic biogenic activity in martian meteorite ALH 84001. Science 273:924–930

Mittlefehldt DW (1994) ALH 84001: A cumulate orthopyroxenite member of the martian meteorite clan. Meteoritics 29:2114–2121

Moore P (1977) A guide to Mars. Norton, New York

Morton O (2002) Mapping Mars: Science imagination, and the birth of a world. Picador, New York

Nyquist LE, Bansal BM, Wiesman H, Shih C-Y (1995) "Martians" young and old: Zagami and ALH84001. Proceed Lunar Planet Sci Conf 31: 1065–1066

Oberg J, Aldrin B (2000) A bus between the planets. Sci Am March, 58–60

Owen T (1992) Composition and early history of the atmosphere. In: Kieffer H.H. et al. (eds) Mars. Un. Arizona Press, Tucson, AZ, pp 818–834

Papanastassiou DA, Wasserburg GJ (1974) Evidence for late formation and young metamorphism of the achondrite Nakhla. Geophys Res Letters 1: 23–26

Parker EN (2006) Shielding space travelers. Sci Am 294(3):40–47

Robinson KS (1993) Red Mars. Bantam Books, New York

Robinson KS (1994) Green Mars. Bantam Books, New York

Robinson KS (1996) Blue Mars. Bantam Books, New York

Sagan C (1994) Pale blue dot: A vision of the human future in space. Random House, New York

Schilling G (2003) Mars on Earth. Astronomy, January, 46–51

Sheehan W (1996) The planet Mars: A history of observation and discovery. University of Arizona Press, Tucson, AZ

Sheehan W, O'Meara SJ (2001) Mars: The lure of the Red Planet. Prometheus Books, New York

Smith DE et al. (1999) The global topography of Mars and implcations for surface evolution. Science 284: 1495–1503

Squyres S (2005) Roving Mars. Hyperion, New York

Tanaka KL (1986) The stratigraphy of Mars. Proc. 17th Lunar Planet Sci Conf (Part 1). J Geophys Res 91: 249–252 (supplement)

Taylor SR (1992) Solar system evolution: A new perspective. Cambridge University Press, Cambridge, UK

The Planetary Society (1985) An explorer's guide to Mars. Pasadena, CA

Tokano T (ed.) (2005) Water on Mars and life. Springer-Verlag, Heidelberg

Wagner JK (1991) Introduction to the solar system. Saunders College Pub., Philadelphia, PA

Wells HG (1898) War of the worlds. Reprinted in 1988 by Tor, New York

Westall F (2005) Early life on Earth and analogies to Mars. In: T. Tokano (ed.). Water on Mars and Life. Springer-Verlag, Heidelberg, pp 45–64

Whillans IM, Cassidy WA (1983) Catch a falling star: Meteorites and old ice. Science 222: 55–27

Zubrin R (1996) Mars on a shoestring. Technology Review, November/December, Massachusetts Institute of Technology, Cambridge, MA

Zubrin R, Wagner R (1996) The case for Mars: The plan to settle the Red Planet and why we must. The Free Press, New York

Zubrin R (1999) Entering space: Creating a space-faring civilization. Tarcher-Putnam, New York

Asteroids: Shattered Worlds

The solar system contains hundreds of thousands of irregularly-shaped solid bodies most of which revolve around the Sun in the so-called *Main Asteroid Belt* between the orbits of Mars and Jupiter (Gehrels, 1979; Binzel et al., 1989; Asphaug, 1997; Chapman, 1999; Britt and Lebofsky, 1999). This region of the solar system occupies an area of $1.74 \times 10^{18} \, km^2$ within the plane of the ecliptic. Consequently, the solid objects of the asteroidal belt are spaced far from each other in spite of their large number. As a result, none of the spacecraft that have traveled through this region of the solar system have been damaged or destroyed by collisions with asteroids.

Ceres, the largest asteroid, was discovered in 1801 by the Italian astronomer Giuseppe Piazzi. The length of the semi-major axis of the orbit of *Ceres* was found to be 2.8 AU, exactly as predicted by the Titius-Bode rule (Section 3.2). A second body named *Pallas* was discovered in 1802 at about the same distance from the Sun as Ceres. The existence of two small bodies in close proximity to each other suggested that they were fragments of a larger body that had shattered for some reason. This hypothesis encouraged further searches which led to the discovery of *Juno* in 1804 and *Vesta* in 1807. These bodies were called *asteroids* because they appeared to be star-like points of light. A fifth asteroid, *Astraea*, was discovered in 1830. The number of known asteroids rose to 300 by the year 1890 and has now reached many thousands of specimens (Hartmann, 2005).

13.1 Classification

The continuing search for asteroids during the nineteenth and twentieth centuries yielded a bewildering variety of objects that populate the solar system. For example, asteroids exist not only in the main belt between the orbits of Mars and Jupiter but occur also between the orbits of other planets, at the L4 and L5 Lagrange points of Jupiter (Section 7.2.11), and in orbits around Mars, Jupiter, Saturn, Uranus, and Neptune (e.g., Phobos and Deimos, satellites of Mars). In addition, the spectra of light reflected by the asteroids indicate that most have a range of chemical compositions that match those of stony meteorites (Sections 8.1, 8.2).

13.1.1 Spectral Types

The spectral types of asteroids in Table 13.1 are designated by capital letters and are described in terms of the compositions of meteorites that have similar reflectance spectra. This classification scheme therefore tends to identify the asteroids from which the different kinds of stony and iron meteorites could have originated. The existence of different spectral types also means that the regolith that covers the surfaces of asteroids has a wide range of colors ranging from light tan (Psyche and Dembowska) to reddish brown (Juno, Iris, Hebe, Eunomia, Herculina, and Flora) to varying shades of brownish grey (Pallas) and greyish black (Ceres, Hygiea, Ursula, Davida, Cybele, and many others) (Chapman, 1999).

The abundances of the different spectral types of asteroids in the main belt vary systematically with increasing distance from the Sun between 1.8 and 5.2 AU. For example, the abundance of E-type asteroids in Figure 13.1 deceases from nearly 60% of all known asteroids at 1.8 AU, to only about 2% at 2.15 AU, and finally to 0% at 3.0 AU. The abundance of S-type asteroids increases from about 40% at 1.8 AU to 70% at 2.3 AU and then declines to 20% at 3.0 AU and

Table 13.1. Spectral classification of the asteroids in order of increasing distance from the Sun (Hartmann, 2005)

Spectral type	Albedo %	Meteorite type	Location
E	> 23	Enstatite chondrites?	Inner edge of the main belt
S	7–23	Chondrites?	Inner to central region of main belt
M	7–20	Stony iron or iron	Central part of the main belt
V	38	Basaltic achondrites	Vesta in the main belt
A	~ 25	Pallasite, olivine-rich	Rare in the main belt
C	2–7	Carbonaceous chondrites	Dominant in the outer part of the main belt, > 2.7 AU
P	2–7	Carbonaceous? similar to M type	Outermost part of the main belt
D	2–7	Dark reddish-brown organics, kerogen and carbonaceous material	Outermost part of the main belt and about 2/3 of the Trojan[1] asteroids
Z	4–10?	Organic material, extremely red	Centaurs[2] and Kuiper belt[3]

[1] Trojan asteroids occupy the L4 and L5 Lagrange points of the orbit of Jupiter.
[2] Centaurs have elliptical orbits that extend from about 9 AU at perihelion to about 38 AU at aphelion. (See Table 13.3).
[3] Kuiper belt objects orbit the Sun generally outside of the orbit of Pluto.

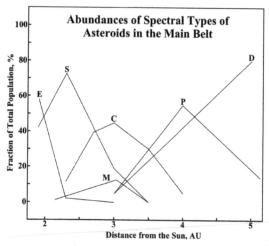

Figure 13.1. The abundances of the asteroids of different spectral types in the main belt between the orbits of Mars and Jupiter vary with distance from the Sun. The chemical and mineralogical compositions of the asteroids are expressed in terms of the meteorites that have similar spectral characteristics: E = enstatite chondrites, S = chondrites, C = carbonaceous chondrites, M = stony iron or iron meteorites, P = similar to M but having low albedo, D = reddish brown, kerogen-rich carbonaceous material. The abundances of B, G, and F spectral types are low and are not shown. Adapted from Chapman (1999)

reaches 0% at 3.25 AU. Asteroids having surface characteristics of carbonaceous chondrites (type C) and asteroids with reddish surfaces (types P and D) are most abundant in the region between 3.0 and 5.0 AU and contain organic material such as kerogen (Hartmann, 2005). The M-type asteroids (iron meteorites) constitute about 10% of all asteroids at a distance of about 3 ± 0.5 AU from the Sun. These asteroids contain not only metallic iron but also nickel, cobalt, gold, and platinum-group metals (Kargel, 1994; Lewis et al., 1993; Lewis, 1996).

The surfaces of the C-bearing asteroids in the outer part of the main belt (spectral types C, P, and D) contain chemically bound water, clay minerals, (montmorillonite), organic polymers,

magnetite, and iron sulfide. The P and D spectal types may contain mixtures of various kinds of ices and organic compounds. The only D-type meteorite known at the present time is the carbonaceous chondrite that fell on January 18 of 2000 on the ice of Tagish Lake in British Columbia, Canada (Hiroi et al. 2001) and Figure 8.5 of Section 8.2. A set of papers describing various aspects of the Tagish Lake meteorite was published in 2002 by Meteoritics and Planetary Science (Vol. 37, number 5) including a paper by Zolensky et al. (2002).

13.1.2 Orbits of the Asteroids

Asteroids are also classified by their location within the solar system and thus by their orbits. This classification in Table 13.2 reveals that asteroids occur not only in the main belt between the orbits of Mars and Jupiter, but that many asteroids also occur elsewhere in the solar system. All asteroids revolve around the Sun in the prograde directions close to the plane of the ecliptic and the eccentricities and inclinations of

Table 13.2. Classification of asteroids by the radii of their orbits around the Sun (Hartmann, 2005, Table 7.2)

Asteroid group	Avg. distance from Sun, AU	Perihelion, AU	Aphelion, AU
Aten	0.83–0.97	0.46–0.79	1.14–1.22
Apollo	1.24–2.39	0.42–0.89	1.66–3.91
Amor	1.46–2.66	1.12–1.48	1.78–4.07
Main belt	2.20–2.45	1.79–2.20	2.57–2.93
Main belt	2.58–3.06	1.78–2.82	2.81–3.60
Main belt	3.13–4.26	2.46–4.12	3.29–4.40
Trojan	5.08–5.28	4.4–5.0	5.2–6.0
Centaur	13.7–24.9	6.6–11.9	19.0–37.7

their orbits are detailed in Appendix 2. The asteroids which are confined to the space within the orbit of Jupiter in Figure 13.2 are classified into groups based on their perihelion distances as well as on their average distances from the Sun and on the relation of their orbits to the orbits of the Earth and Mars.

Aten group: The orbits of Aten asteroids (Appendix 2) lie primarily inside the orbit of the Earth because their perihelion distance (q) is less than 1.0 AU. Members of this group do cross the orbit of the Earth in cases where their aphelion distance (Q) is greater than 1.0 AU. However, we demonstrate in Figure 13.3 that the members of this group do not collide with the Earth in cases where their orbits are steeply inclined relative to the plane of the ecliptic. Collisions between Aten asteroids and the Earth can occur only when an asteroid rises (or descends) through the plane of the ecliptic at a point located on the orbit of the Earth. The diameters of Aten asteroids identified by Hartmann (2005) range from 0.2 to 3.4 km and their reflectance spectra (types C and S) resemble those of stony meteorites. Two members of the Aten group of asteroids and their diameters are: Re-Shalom (3.4 km) and Aten (1.0 km).

Apollo group: The members of the Apollo group (Appendix 2) regularly cross the orbits of the Earth and Mars because their perihelion distances (q) are less than 1.0 AU, whereas their aphelion distances (Q) extend beyond the orbit of Mars (i.e., Q > 1.52 AU). Nevertheless, collisions of Apollo asteroids with the Earth and Mars are prevented in cases where their orbital planes are steeply inclined relative to the plane of the ecliptic (Figure 13.3). A possible

exception is the Apollo asteroid XC (diameter = 3? km) because the inclination of its orbit is only 1° relative to the plane of the ecliptic and its eccentricity is 0.63 (Hartmann, 2005). The diameters of the Apollo asteroids range from 2.0 to 10 km and their reflectance spectra (type S) relate their compositions to stony meteorites. Some of the notable Apollo asteroids and their diameters are: Sisyphus (10 km), Toro (4.8 km), Daedalus (3.4 km), Antonius (3 km), and Geographos (2.0 km).

Amor group: The perihelion distances (q) of the Amor asteroids (Appendix 2) are all greater than 1.0 AU and their aphelion distances (Q) exceed the radius of the orbit of Mars. Therefore, the orbits of the Amor asteroids in Figure 13.2 lie outside of the orbit of the Earth but do intersect the orbit of Mars and, in many cases, extend into the main belt of asteroids. As before, collisions of the Amor asteroids with Mars are prevented by the steep inclination of the planes of their orbit (8° to 36°), but may occur in case an Amor asteroid occupies a point on the orbit of Mars. The diameters of the Amor asteroids listed by Hartmann (2005) range from 6 to 30 km and their compositions resemble those of stony meteorites (spectral types S and C). Some prominent members of the Amor group and their diameters are: Ganymede (30? km), Eros (7 × 19 × 30 km), Beira (11? km), Atami (7? km), and Betula (6 km).

Main Belt group: The main belt of asteroids (Appendix 2) contains by far the largest number of individual objects numbering in the hundreds of thousands. The belt is divided into three regions based on the average radii (a) of the asteroidal orbits: Inner belt: a < 2.5 AU; Middle belt: 2.5 < a < 3.1 AU; Outer belt: 3.1 < a < 4.1 AU.

Inner Main-Belt: The orbits of the asteroids in this group are characterized by q > 1.79 AU and Q < 2.93 AU, which implies that the eccentricities of the orbits of these asteroids (0.09 to 0.23) are lower on average than those of the Aten, Apollo, and Amor asteroids. In addition, the orbits of the Main-Belt asteroids are located closer to the plane of the ecliptic (1° to 15°) than the orbits of the groups named above. The Main-Belt asteroids can impact on one of the terrestrial planets only in case they are ejected from their orbits by collisions with other Main-Belt asteroids or with members of the Apollo and Amor groups whose orbits in

Asteroids of the Inner Solar System

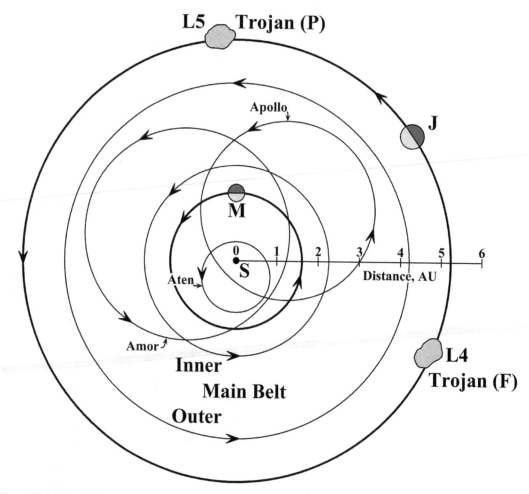

Figure 13.2. The principal types of asteroids in the inner solar system at increasing distance from the Sun are the Aten, Apollo, Amor, Main Belt, and Trojan groups. The orbits of each group are drawn here as circles that encompass the minimum perihelion and the maximum aphelion distances. The orbits of Mars (M) and Jupiter (J) are indicated, but the orbit of the Earth at 1.0 AU from the Sun (S) was omitted. The orbit of the Earth is crossed by members of the Aten and Apollo groups, whereas the orbit of Mars is crossed by members of the Apollo and Amor groups. Note that all asteroids in the inner solar system revolve in the prograde direction, but the inclinations of their orbits relative to the ecliptic range widely from 1° to 68°. In addition, the diameters of the Aten, Apollo, and Amor asteroids are less than 30 km, whereas the diameters of the largest asteroids of the Main Belt and Trojan groups are more than 100 km in many cases. Based on data compiled in Table 7.2 of Hartmann (2005)

Figure 13.2 extend beyond the orbit of Mars. The diameters of the most prominent member of the Inner Main-Belt asteroids range from 115 to 500 km, but many smaller asteroids exist in this group. Most of the largest members of the Inner Main-Belt group have reflectance spectra types C and S, although Vesta has been classified as V (basalt) and Lutetia has been assigned to type M (iron or stony-iron meteorites). The names and diameters of the five largest asteroids in this

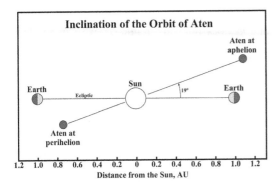

Inclination of the Orbit of Aten

Figure 13.3. The plane of the orbit of the asteroid Aten makes a 19°-angle with the plane of the ecliptic. Therefore, when Aten is at the aphelion of its orbit, it is located in the space above the orbital plane of the Earth and cannot collide with it. This diagram is based on the assumption that the orbit of the Earth (e = 0.017) is virtually circular compared to the orbit of Aten which is distinctly elliptical (e = 0.18). Data from Hartmann (2005)

group are: Vesta (500 km), Iris (204 km), Hebe (192 km), Fortuna (190 km), and Flora (153 km).

Middle Main-Belt: The average distances of the asteroids in the middle of the main belt are greater than 2.5 AU but less than 3.1 AU. The eccentricities of the orbits of the ten largest asteroids in this group range from 0.07 to 0.34 and the inclinations of their orbital planes vary between 3° and 35°. The diameters of the ten largest asteroids in this group range from 214 to 914 km. Most of the largest asteroids in this group have reflectance spectra of types C and S. Only Psyche (264 km) is type M (metallic) among the 10 largest asteroids of this group. The names and diameters of the five largest asteroids (in addition to Psyche) of the middle group of Main-Belt asteroids are: Ceres (914 km), Pallas (522 km), Interamnia (334 km), Eunomia (272 km), and Juno (244 km).

Outer Main-Belt: The orbits of asteroids in the outer part of the main belt (3.1 < a < 4.1 AU), like those of most asteroids in the main belt, have low eccentricities (0.03 to 0.22) and low inclinations relative to the plane of the ecliptic (1° to 26°). The diameters of the 10 largest asteroids in this group range from 202 to 443 km and all but one have C-type reflectance spectra. The one exception is Sylvia (272 km) which has a reddish color like other P-type asteroids. Another asteroid on the outer

fringe of the main belt is Thule (q = 4.12 AU, Q = 4.40 AU), which is also reddish in color and has been assigned to spectral type D. Thule has a diameter of about 60 km and the eccentricity and inclination of its orbit are 0.03 and 2°, respectively. The names and diameters of the five largest regular members (except Sylvia) of the outer group of asteroids in the main belt are: Hygiea (443 km), Davida (336 km), Euphrosyne (248 km), Cybele (246 km), and Camilla (236 km).

In summary, the asteroids that populate the main belt are by far the most numerous, many have diameters greater than 100 km, their orbits have comparatively low eccentricities and low angles of inclination with the plane of the ecliptic, and the abundance of carbonaceous asteroids (spectral types C, P, and D) rises with increasing distance fom the Sun in agreement with the trends in Figure 13.1.

Trojan group: The L4 and L5 Lagrange points, located 60° in front of and behind Jupiter (Section 7.2.11) in its orbit, host swarms of asteroids all of which carry the names of the Greek and Trojan fighters who participated in the siege of the city of Troy (e.g., Agamemnon, Hektor, Achilles, Patroclus, Odysseus, etc.). The name of this group of asteroids is an allusion to the wooden horse that concealed Greek fighters. When the defenders of Troy moved the horse into the city, the Greek fighters emerged from the horse at night and opened the city gate, which led to the downfall of Troy. The configuration of the gravitational field associated with the Lagrange points confines the Trojan asteroids loosely to a comparatively small volume of space partly within and partly outside of the orbit of Jupiter. All of the ten largest Trojan asteroids (Appendix 2) listed by Hartmann (2005) have diameters greater than 100 km, and most have been assigned to spectral types C, D, and P. These types of asteroids contain organic matter which gives them a reddish color. The five largest Trojan asteroids and their diameters are: Hektor (150 × 300 km), Agamemnon (148? km), Patroclus (140 km), Diomedes (130? km), and Aeneas (125 km).

Centaur group: The 11 Centaurs listed in Table 13.3 are large objects having diameters between 40 and 300 km. Their perihelion (6.6 to 11.9 AU) and aphelion (19.0 to 37.7 AU)

Table 13.3. Physical and orbital properties of some of the largest Centaurs that have been discovered. Data from Levison and Weissman (1999) and Trujillo and Brown (2004)

Name	a, AU	e	i, °	Diameter, km
Chiron	13.6	0.38	6.9	300
1997 CU 26	15.7	0.17	23.4	300
Pholus	20.2	0.57	24.7	240
1995 GO	18.0	0.62	17.6	100
1995 DW2	24.9	0.24	4.2	100
Nessus	24.5	0.52	15.7	80
1994 TA	16.9	0.30	5.4	40
2000 EC 98	10.749	0.455	4.3	?
2002 KY 14	12.724	0.137	17	?
2002 PN 34	30.721	0.566	16.7	?
2002 QX 47	18.916	0.049	7.6	?

distances indicate that their orbits are eccentric (0.38 to 0.62), lie outside of the orbit of Jupiter, and extend all the way to the orbit of Pluto which has a perihelion distance of 29.7 AU. The chemical compositions of the Centaurs (spectral types C and D) are similar to those of carbonaceous chondrites but may also contain ices of volatile compounds. In this regard, the Centaurs are transitional between asteroids and the ice worlds of the Edgeworth-Kuiper belt which is a reservoir of comets. For example, Chiron is surrounded by a cloud of gas and dust particles released by its surface as it approaches the perihelion of its orbit. This cloud persists for some time until the dust particles settle back to the surface of Chiron or escape into space as it recedes from the Sun toward the aphelion of its orbit (Levison and Weissman, 1999; Trujillo and Brown, 2004; Hartmann, 2005).

The difficulty of distinguishing between asteroids and comets arises because some ice-rich objects rarely come close enough to the Sun to cause ice to sublimate. In addition, some rocky asteroids may contain ices that would perhaps sublimate and form cometary tails if these asteroids ever approached the Sun.

13.2 Physical Properties and Celestial Mechanics

The compositional zonation of the Main-Belt asteroids in Figure 13.1 supports the conjecture that asteroids in this group formed in the space

where they still reside. However, their distribution does not vary regularly with increasing distance from the Sun because of the existence of the so-called *Kirkwood gaps* illustrated in Figure 13.4. These gaps occur because asteroids that once occupied these positions have been ejected by the effects of resonance between their periods of revolution and that of Jupiter.

For example, the sidereal period of revolution of an asteroid at a distance of 2.5 AU from the Sun is exactly 1/3 of the period of revolution of Jupiter. Therefore, this asteroid completes three revolutions in the same time it takes Jupiter to complete one revolution (i.e., the resonance is 3:1). The repeated gravitational interactions between the asteroids at 2.5 AU and Jupiter increased their orbital velocities and caused them to be ejected from their original orbits (i.e., their orbits became parabolic).

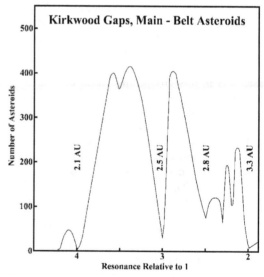

Figure 13.4. The Kirkwood Gaps in the main asteroidal belt were caused by the ejection of asteroids whose orbital periods are fractions of the period of revolution of Jupiter. This condition leads to resonance effects that cause the velocities of asteroids to increase leading to their ejection from their orbits. The resonance scale is expressed as the number of revolutions of an asteroid at a certain distance from the Sun during one revolution of Jupiter. For example, asteroids at a distance of about 2.1 AU make four revolutions around the Sun during one revolution of Jupiter (i.e., the periods of revolution of the asteroid is 1/4 of the period of Jupiter). Adapted from Chapman (1999, Figure 2)

Only seven of the ten Main-Belt asteroids listed Table 13.4 have radii greater than 150 km and only Ceres, the largest asteroid, has a radius that approaches 500 km. A histogram in Figure 13.5 of the radii of the 70 largest Main-Belt asteroids demonstrates that their abundances increase exponentially as their radii decrease. An extrapolation of this relationship indicates that more than 2000 asteroids have radii greater than five kilometers and that as many as 500,000 asteroids have radii greater than 500 meters. Smaller asteroids are even more numerous.

The periods of revolution of the Main-Belt asteroids range from 3.63 to 7.94 y and rise with increasing length of the semi-major axes of their orbits in accordance with Kepler's third law: $a^3 = p^2$, where (a) is the length of the semi-major axis of the orbit in AU and (p) is the sidereal period of revolution in years. For example, the mean distance (a) of Vesta from the Sun is 2.361 AU which yields:

$$p = (2.361)^{3/2} = 3.627 \text{ y}$$

in agreement with the value of 3.63 y listed in Table 13.4.

Most of the large Main-Belt asteroids have short rotation periods of less than 10 hours although Bamberga rotates in 29.43 hours and Nemesis takes 39 hours. Most asteroids have low albedos of less than 10%, which contributes to the difficulty of observing them with telescopes. The intensity of sunlight that is reflected by asteroids as they travel in their orbits fluctuates with time because the albedo of their surface varies as they rotate. The resulting *lightcurves*

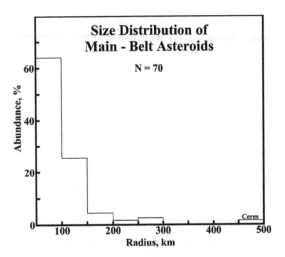

Figure 13.5. The abundance of large asteroids having radii greater than 150 km is much less than the abundance of smaller specimens in this set of 70 of the largest Main-Belt asteroids. The low abundance of large asteroids and the high abundance of small specimens is at least partly attributable to the progressive fragmentation of a small population of parent bodies that may have formed in the space between the orbits of Mars and Jupiter. The trend can be extrapolated to suggest that about 500,000 asteroids have radii greater than about 500 meters. Data from Beatty et al. 1999

are used to determine the periods of rotation and to model the shapes of asteroids.

The low albedo of most stony asteroids is attributable to the presence of organic matter in the regolith that covers their surfaces. The asteroid *Vesta* is an exception because it has a high albedo of 42% causing it to be brighter even than *Ceres*, which is larger than Vesta but has a low albedo of

Table 13.4. Orbital parameters of the ten largest asteroids in the main asteroidal belt (Beatty et al., 1999)

Name	Radius, km	Avg. dist., AU	Orbital period, y	Rotation period, h	Albedo %
Ceres	467	2.768	4.60	9.08	11
Pallas	263	2.774	4.61	7.81	16
Vesta	255	2.361	3.63	5.34	42
Hygiea	204	3.136	5.55	27.6?	7
Davida	163	3.176	5.65	5.13	5
Interamnia	158	3.064	5.36	8.73	7
Europa	151	3.099	5.46	5.63	6
Juno	134	2.669	4.36	7.21	24
Sylvia	130	3.449	6.55	5.18	4
Eunomia	128	2.644	4.30	6.08	21

11%. In addition, Vesta approaches the Earth at certain times due to the eccentricity of its orbit and can be viewed with binoculars at night when it is in opposition. The high albedo of Vesta has been attributed to the presence of a basalt crust. Partly for this reason, Vesta is regarded as a likely source of basaltic achondrite meteorites (Section 8.1) composed of a variety of gabbro known as eucrite (i.e., containing both ortho- and clinopyroxene with Ca-rich plagioclase).

Most asteroids have irregular shapes consistent with their origin by fragmentation of larger parent bodies. Only rocky asteroids with diameters in excess of several hundred kilometers have approximately spherical shapes because their internal pressures are sufficiently large to overcome their mechanical strength. In addition, the elevated temperatures at the centers of large asteroids may have permitted them to deform plastically and to develop spherical shapes (e.g., Ceres, Pallas, Vesta, Hygiea, and Interamnia). However, the high surface-to-volume ratios (Figure 10.3) of even the largest asteroids caused them to cool rapidly which prevented prolonged internal geological activity.

The bulk densities of stony asteroids in the main belt range widely from less than $2.0 \, \mathrm{g/cm^3}$ to greater than $3.3 \, \mathrm{g/cm^3}$. As we have noted before, the bulk density of solid objects in the solar system depends on their chemical compositions and on the extent of internal fragmentation. Asteroids having unusually low bulk densities (i.e., $< 2.6 \, \mathrm{g/cm^3}$) may contain substantial amounts of ices of volatile compounds (i.e., H_2O, CO_2, CH_4, NH_3, etc.) and/or organic matter such as high-molecular-weight hydrocarbons. Alternatively, they may consist of fragments held together loosely by gravity and by a layer of regolith formed by impacts and space weathering. For example, the asteroid *Mathilde* ($66 \times 48 \times 46 \, \mathrm{km}$) in Figure 13.6A has a density of only $1.30 \, \mathrm{g/cm^3}$ because it has been fragmented by multiple impacts during its long history. The largest impact crater on Mathilde is 20 km wide and 10 km deep. The energy released by this impact did not destroy this asteroid because a collection of fragments could absorb the kinetic energy of the impact better than a solid rocky body could have.

The surfaces of all asteroids that have been imaged at close range are covered with impact

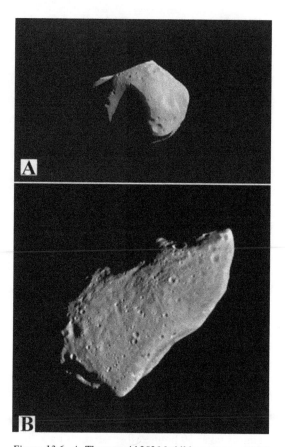

Figure 13.6. A. The asteroid 253 Mathilde was encountered by the NEAR Shoemaker spacecraft on June 27, 1997, en route to a rendezvous with the asteroid Eros. The image, which was recorded at a distance of 2400 km, shows that Mathilde survived several large impacts and many smaller ones. The surface is covered by a layer of regolith which hides the fact that this asteroid consists of a pile of small rock fragments that could absorb the energy released by these impacts without breaking up (Image PIA 02477, courtesy of NASA and Johns Hopkins University/Applied Physics Laboratory) B. The asteroid 951 Gaspra was imaged on October 29, 1991, by the spacecraft Galileo at a distance of 5300 km. Gaspra is an irregularly shaped object whose dimensions are $19 \times 12 \times 11$ km. Its surface is covered by regolith which contains more than 600 craters having diameters between 100 and 500 meters but only a small number of larger craters. The irregular shape of Gaspra identifies it as a fragment of a larger parent body that was broken up by an energetic collision with another asteroid. The faint grooves in the regolith resemble similar grooves on the surface of Phobos and support the hypothesis that the parent body of Gaspra was broken up by a violent collision. The evident scarcity of large impact craters and the comparatively low aerial density of craters suggests that the surface of Gaspra has a low exposure age compared to other main-belt asteroids like Ida. (Image PIA 00118, Galileo orbiter, courtesy of NASA/JPL-Caltech)

craters at varying densities expressed in terms of numbers per square kilometers. The asteroid 951 *Gaspra* in Figure 13.6B, which was imaged by the spacecraft Galileo in 1991, is less densely cratered than the considerably larger asteroid 243 *Ida*, which the spacecraft in encountered two years later in 1993. The surface of Ida in Figure 13.7 is more densely cratered than that of Gaspra suggesting that the surface of Ida is older than the surface of Gaspra. Accordingly, the low exposure age of Gaspra suggests that it is a fragment of a larger body that was broken up relatively recently.

Asteroids may be broken up not only by super-catastrophic collisions with other asteroids but also by tidal effects during close encounters with planets including Jupiter, Mars, Earth, or Venus. For example, the asteroid *Geographos* is 5.1 km long but only 1.8 km wide. The resulting cigar-like shape of this asteroid has been attributed to such a close encounter.

When a stony asteroid is broken up by collisions with another asteroid, the fragments may reassemble to form a pile of rubble like the asteroid Mathilde. In the rare case of a supercatastrophic collision, the fragments scatter to form a collection of small asteroids called a *Hirayama family*, the members of which have similar chemical compositions but travel on separate heliocentric orbits. Several dozen

Figure 13.7. On August 28, 1993, the spacecraft Galileo, on its way to the Jupiter system, encountered the main-belt asteroid 243 Ida at a distance of 10,500 km. The images recorded by the spacecraft on this occasion revealed for the first time that Ida has a satellite that was named Dactyl. Ida is a stony asteroid that is about 58 km long and 22.4 km wide. Its surface is covered by a layer of regolith and is heavily cratered. Dactyl, whose dimensions are 1.2 × 1.4 × 1.6 km, is also heavily cratered including at least one dozen craters that are more than 80 m in diameter. The largest crater on Dactyl has a diameter of about 300 m. At the time of the encounter, Dactyl was about 100 km from the center of Ida. Recent telescopic studies indicate that many asteroids have one satellite and at least one asteroid has two. These observations explain why some impact craters on the Earth and on the Moon occur in pairs. For example, the adjacent Clearwater Lakes in Canada (32 and 22 km wide) are known to be impact craters that formed simultaneously, presumably as a result of the impacts of an asteroid and its satellite. (Project Galileo, courtesy of NASA/JPL-Caltech)

asteroid families have been identified after the Japanese astronomer Kiyotsugu Hirayama first discovered their existence. Examples of Hirayama asteroid families include the Themis, Eos, Koronis, and Flora groups. Quite a few large asteroids have a small satellite that may have been captured or may be a collision product (e.g., Dactyl in Figure 13.7).

13.3 First Landings

The information derived by all available methods indicates that asteroids are ancient objects that contain a record of their evolution during the time of formation of the solar system. In addition, certain kinds of asteroids contain materials that may have value for future space travelers and colonists on Mars and elsewhere in the solar system (Science Brief 13.7.1 and Kargel, 1994; Lewis, 1996). These considerations motivate attempts to study asteroids at close range and/or by landings on their surface. Such missions are not easy to execute because the gravitational field of asteroids is weak and irregular in shape. Only three asteroids have been studied at close range.

13.3.1 Eros

An even more ambitious mission was carried out by the spacecraft named Near Earth Asteroid Rendezvous (NEAR) Shoemaker that was launched on February 17, 1996, and went into orbit around 433 *Eros* in the inner part of the main belt on February 14, 2000. After inserting NEAR into orbit around Eros, the mission controllers reduced the radius of the orbit to 50 km in order to determine its chemical composition. Measurements made from orbit confirmed that Eros (spectral type S) is a solid undifferentiated body whose composition resembles that of stony meteorites. Apparently Eros is *not* a pile of rubble like Mathilde.

Eros in Figure 13.8 is a cigar-shaped object 33 km in length and 8 km in diameter with a bulk density of $2.70 \, g/cm^3$. Its period of rotation is 5.27 h, it is covered by a layer of regolith, and it is heavily cratered. The three largest craters in order of decreasing diameters are: Himeros, Psyche, and Shoemaker. All are several kilometers wide and appear to be the sources of large angular boulders many of which have diameters of more than 30 meters. Smaller craters, having diameters between 75 and 120 m, contain fine-grained sediment that appears to have "ponded" within them. The process that causes fine sediment to move into the craters is not understood because Eros does not have an atmosphere nor is there water that could transport sediment.

On February 12, 2001, NEAR Shoemaker landed gently in the crater Himeros on the surface of Eros. The descent from orbit was recorded by 69 sharply focused images taken by the on-board electronic camera. The gamma-ray spectrometer continued to work for 14 days collecting data at a distance of only 10 centimeters from the surface of Eros. The mission was terminated by NASA on February 28, 2001, by deactivating the spacecraft. It remains parked on the surface of Eros.

13.3.2 Braille

Braille was visited on July 29, 1999, by the spacecraft Deep Space 1 which was propelled by an ion drive and passed this asteroid within about 26 km traveling at a rate of 40,000 km/h. The results of this encounter indicate that Braille is about 2 km long and that the rocks of which it is composed contain pyroxene. The presence of pyroxene-bearing rocks suggests that Braille is a fragment of Vesta or that both asteroids are fragments of the same ancestral body.

Deep Space 1 (DS1) was launched from Cape Canaveral on October 24, 1998, by a conventional chemically-fueled rocket. After it achieved orbit around the Earth, it unfolded its solar panels (11.8 m) and turned on its ion drive. The electricity generated by the solar panels ionized xenon gas and expelled it through a nozzle at more than 27 km/s. This little rocket engine ran for 670 days and increased the speed of DS1 by 4.2 km/s even though its thrust was equivalent to the weight of a sheet of paper on Earth. Although DS1 did encounter the asteroid Braille, problems with the automated navigation system prevented it from obtaining the hoped-for detailed images of this asteroid. Subsequently, DS1 imaged the comet Borelly from a distance of 2,171 km and determined that its potato-shaped nucleus is 8 km long and 4 km wide. In addition, DS1 reported that the surface of the nucleus is jet black and its temperature ranges up to $+72°C$.

Figure 13.8. The asteroid 433 Eros was visited by the spacecraft NEAR Shoemaker (Near-Earth Asteroid Rendezvous) during the year 2000. This view is a mosaic constructed from several images taken at a distance of 200 km. The asteroid is covered by a layer of regolith including large angular boulders having diameter of about 30 m. Eros is the second largest asteroid that passes close to the Earth but is not likely to impact on the Earth. (NEAR Project, courtesy of JHU/APL, NASA/JPL/Caltech)

13.3.3 Itokawa

The asteroid Itokawa in Figure 13.9 passed the Earth in June of 2004 within about 2×10^6 km and was closely observed by ground-based remote-sensing techniques during that flyby. It turned out to be a potato-shaped object only about 550 m long and 310×275 m wide having a chemical composition similar to that of ordinary chondrite meteorites (i.e., it has an S-type optical spectrum). This asteroid, which was named after the Japanese rocket pioneer Hideo Itokawa, is relatively easy to reach from the Earth because its orbit lies close to plane of the ecliptic. On May 9, 2003, the Japan Aerospace Exploration Agency (JAXA) launched a spacecraft named MUSES C

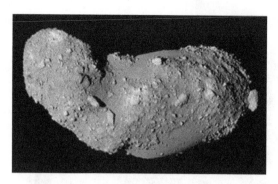

Figure 13.9. Itokawa is a near-Earth asteroid that passed the Earth within about two million kilometers in June of 2004. The images of this asteroid that were recorded by the spacecraft Hayabusa launched by the Japan Aerospace Exploration Agency (JAXA) show that its surface is covered with boulders one of which sticks up prominently in this image. Itkokawa is assumed to be the product of a destructive impact on a larger asteroid because it is a "rubble pile" rather than a solid body . (Website of the Japan Aerospace Exploration Agency)

to intercept Itokawa in order to collect a sample of the regolith from its surface and to return it to Earth for analysis. After launch, the spacecraft was renamed Hayabusa which means "falcon" in Japan.

Hayabusa arrived in the vicinity of Itokawa in September of 2005 after a trip that lasted nearly 2.4 years and included a flyby of the Earth for a gravity assist. At first, the spacecraft mapped the asteroid by remote-sensing techniques and recorded images of its boulder-strewn surface. These images suggest that Itokawa is a rubble pile rather than a solid object and that the escape velocity from its surface is only 0.00014 km/s compared to 11.2 km/s for the Earth. Therefore, a spacecraft that attempts to land on its surface could easily bounce off after impact.

The landing by Hayabusa had to be controlled by the on-board computer because signals from Earth took 16 minutes to reach the spacecraft. The craft practiced the landing several times by approaching the surface of Itokawa to within about 70 m of a dusty plain called the Muses Sea. Finally, on November 19 of 2005 the spacecraft came to rest on the surface of Itokawa for about 30 minutes. Unfortunately, the controllers of the spacecraft were unaware of the fact that

the spacecraft had actually landed and therefore did not order it to collect a sample. However, the spacecraft did release a reflective marker that contained the names of the 880,000 people from 149 countries who had submitted their names to JAXA on the Internet.

During a second attempt to collect a sample on November 25, 2005, the spacecraft again touched the surface of Itokawa; but before the sample could be collected, the spacecraft took off again and subsequently developed problems with the propulsion system that delayed its return to Earth. In spite of the many technical problems that have afflicted the spacecraft Hayabusa during its mission to the asteroid Itokawa, the project demonstrated the feasibility of its ion-propulsion systems which worked continuously for 26,000 hours. In addition, the spacecraft tested an autonomous navigation and guidance system designed by the engineers of JAXA. At a news conference on December 7, 2005, the project manager Junichiro Kawaguchi gave this evaluation of the flight of Hayabusa the falcon (Rayl, 2006): "...we think it is...necessary to take risks and to go forward for space development to progress. If you build a high tower and climb it, you will see a new horizon...."

The ion-drives of Hayabusa are similar to the propulsion system that was used successfully on the American spacecraft DS-1. Although such ion rockets do not generate the thrust necessary to escape from the gravitational field of a planet or a large satellite, they can accelerate spacecraft traveling in free space and thereby shorten the time required to reach distant destinations. In addition, they are well-suited for operations in the low-gravity environment of asteroids. The spacecraft Hayabusa and DS-1 exemplify the design of future spacecraft which will be equipped with computer-controlled navigation and guidance systems that will enable them to function autonomously (i.e., without guidance from the Earth). This innovation in spacecraft design will enhance the chances for success of missions to remote regions of the solar system and will reduce the need for constant ground-based monitoring of missions that may last several years, such as the New Horizons spacecraft that is presently on its way to Pluto and the Edgeworth-Kuiper belt (Rayl, 2006; Yeomans, 2006).

13.4 Near-Earth Objects (NEOs)

Small asteroids in the main belt can escape into the inner region of the solar system as a result of gravitational interactions with Jupiter, by collisions or near misses with other asteroids, and by the *Yarkovsky effect*, which is a weak force derived from the uneven absorption and emission of sunlight by a rotating asteroid (McFadden, 1999). When this force acts over long periods of time (up to one billion years), the orbits of small Main-Belt asteroids (less than 20 km in diameter) can be altered until they are ejected when their periods of revolution come into resonance with the period of Jupiter. As a result, small asteroids from the Main Belt may impact on one of the terrestrial planets including the Moon (Sections 8.8; 8.11.3). The environmental consequences of the explosive impact of a meteoroid or asteroid, measuring from 1 to 10 km in diameter, have repeatedly caused catastrophic climate changes and the sudden extinction of a large number of species on the surface of the Earth (Hallam, 2004).

The occurrence of extinction events in the geologic past (e.g., at the end of the Cretaceous Period 65 million years ago) has motivated efforts to detect Near-Earth Objects (NEOs) whose orbits cross the path of the Earth and which therefore could impact on its surface (Silver and Schultz, 1982; Dressler and Sharpton, 1999; Koeberl and MacLeod, 2002; Kenkmann et al., 2005; Dressler et al., 1994). If a NEO with a diameter of 10 km or more were to impact on the Earth as suggested in Figure 13.10, human civilization would not survive because we are almost as vulnerable now as the dinosaurs were 65 million years ago. However, we do have an advantage over the dinosaurs because we posses the technology to track NEOs in their orbits and to predict whether one of the known objects is on a collision course with the Earth. If we have sufficient warning, the incoming NEO can be deflected and the extinction of the human species can be postponed, even if it is not avoided for all time. In spite of our technological prowess, we remain vulnerable because we cannot track all of the NEOs, and because we are unprepared, at the present time, to deflect the NEOs that we do track.

Wetherill and Shoemaker, writing in 1982, concluded that about 1000 small asteroids of the Aten-Apollo-Amor groups, having diameters greater than one kilometer, regularly cross the orbit of the Earth. They estimated that, on average, three of these bodies impact on the Earth per million years. Impacts of larger bodies, measuring 10 km in diameter, occur once every 40 million years. Even larger bodies up to 20 km in diameter have impacted on the Earth during the last three billion years. The evidence preserved by impact craters indicates that the Earth is a target in a "shooting gallery". It has been hit in the past and will be hit again, but we may be able to deflect the next "bullet" that approaches the Earth.

The International Spaceguard Survey has been discovering about 25 NEOs per month using specially designed telescopes (Gehrels, 1996). As of June 2004, 611 NEOs with diameters greater than one kilometer had been discovered (Durda, 2005). None of these is likely to impact on the Earth in the next 100 years. However, many more NEOs remain to be discovered, especially those having diameter less than one kilometer (Gehrels, 1994; Morrison, 2002; Morrison et al., 2004).

When an approaching NEO has been detected, efforts can be made to deflect it from its collision course with the Earth or to break it into smaller pieces that will cause less damage than the whole object would. (Remember the movies *Armageddon* and *Deep Impact*). This aspect of the study of asteroids has been presented in books by Gehrels (1994), Steel (1995), and Belton et al. (2004). The preparations we can make before the next NEO threatens to impact on the Earth include:

1. Extend the on-going survey of the space in the vicinity of the Earth to include objects having diameters of 200 m or less.
2. Identify large asteroids that could impact in the foreseeable future.
3. Determine the physical properties of all NEOs and their orbits in order to select the most effective countermeasures to prevent their impact upon the Earth.

The possible countermeasures that could be taken to avoid an impact depend on several factors, including: the lead time until impact (the longer the better), the mass of the object and its mechanical strength, its chemical composition

Figure 13.10. During the time of intense bombardment between 4.6 and 3.8 Ga, impacts of large asteroids, having diameters on the order of 500 km, excavated large basins on the planets of the inner solar system. If such a large asteroid impacted on the Earth today, all living organisms would die (i.e, the Earth would be sterilized). Fortunately, projectiles of this magnitude no longer threaten the Earth. However, the impacts of much smaller objects (e.g, diameter = 10 km) can still devastate the surface of the Earth and kill off most of the lifeforms existing at the present time. (Painting by Don Davis, NASA Ames Research Center, no copyright)

and albedo, as well as its orbital parameters such as the velocity relative to the Earth and the inclination of its orbit relative to the plane of the ecliptic.

The kinds of countermeasures that may be brought to bear range from the explosion of nuclear or chemical devices in the immediate vicinity of the incoming NEO to more gentle, but continuously-acting, forces such as those associated with the Yarkovsky effect or gravity (Schweickart et al., 2003). The evidence that a significant number of the known NEOs are rubble piles favors the use of weak forces acting over long periods of time because rubble-pile asteroids can absorb the force released by an explosion without disintegrating. At the present time, none of the available countermeasures could be implemented at short notice because the current "space-faring" nations do not maintain systems for asteroid deflection. The reason is that the expense of building and maintaining such a system is difficult to justify given the low probability that an NEO will impact on the Earth in the next 100 years. Yet, when an impact does occur in a densely populated area of the Earth, millions of people may die, or worse (Morrison et al., 2004).

An additional hazard arises in cases where an incoming asteroid or meteoroid is mistaken for a hostile intercontinental ballistic missile, which may trigger a nuclear response from the target nation. This threat cannot be taken lightly because U.S. satellites have recorded 300 explosions in the atmosphere during a period of eight years caused by meteoroids ranging in diameter from 1 to 10 meters. Explosions of meteoroids 4 m in diameter that release energy equivalent to 5000 tons of TNT occur about once each year. Still larger explosions of meteoroids, measuring 9 m in diameter, release energy equivalent to 50,000 tons of TNT once every 10 years (Brown et al. 2002).

13.5 Torino Scale

The need to quantify the magnitude of the threat posed by impacts of asteroids and comets became apparent in 1992 when the short-period comet Swift-Tuttle passed through the inner solar system in its orbit around the Sun. The information about the orbit of this comet available at that time led to the prediction that Swift-Tuttle may collide with the Earth when it returns in 2126 AD, given that its period of revolution is 134 years (Brandt, 1999). The prospect of this catastrophe has motivated a quantitative assessment of the damage to the surface environment of the Earth and to human civilization (Levy, 2006).

The threat posed by asteroids and comets that approach the Earth increases with the probability that they will actually collide with the Earth and with the amount of kinetic energy that would be released during the impact. These two parameters have been combined in the Torino scale that was designed by R.P. Binzel of M.I.T. in order to assess the magnitude of the hazard posed by asteroids and comets that annually approach the Earth (Binzel, 1999). The scale was adopted in June of 1999 by the participants of an international conference on NEOs that took place in Torino, Italy. It is called the Torino scale (not Turin) in recognition of the spirit of international cooperation in the effort to evaluate the hazard posed by NEOs that was evident at that conference. (<http://en.wikipedia.org/wiki/Torino_Scale>).

The Torino scale in Figure 13.11 assigns numbers between 0 and 10 to potential impactors depending on their kinetic energy (or diameter) and the probability that an impact will occur. In addition, the magnitude of the hazard is color-coded such that white corresponds to zero hazard; green represents hazard-level 1; yellow includes hazard-levels 2, 3, and 4; orange represents hazard-levels 5, 6, and 7 ; and red indicates that a collision is certain and that the resulting damage will be local (8), regional (9), or global (10). In other words, a Torino score of 10 signifies a global catastrophe that is likely to cause the extinction of most living species including humans. More specific predictions of damage and recommendations for actions to be taken by astronomers and by responsible government agencies are included in Science Brief 13.7b. An alternative scale for assessing the hazards posed by impactors called the Palermo Technical Hazard Scale, was published in 2002 by S.R. Chesley of the Jet Propulsion Laboratory in Pasadena, California. This continuous logarithmic scale is intended to be used by astronomers who track NEOs (Levy, 2006).

Meteoroids impact on the Earth virtually on a daily basis, but rarely do any serious damage (Section 8.8). For example, a large meteorite weighing about 70 tonnes fell on February 12 of 1947 in the mountains of Sikhote-Alin along the Tatar Strait and the Sea of Japan. The meteorite broke up in the atmosphere and the fragments excavated several small impact craters without exploding. One of these fragments penetrated the regolith in the target area to a depth of six meters. Approximately 200 tonnes of this meteorite were lost by ablation during the passage through the atmosphere. This fall rates an 8 on the Torino scale because the object did impact (i.e., probability = 1.0) and because the kinetic energy of the meteoroid was less than 100 million tons of TNT. The resulting damage was confined to a small uninhabited area. However, if the Sikhote-Alin meteorite had impacted in a densely populated area, there would have been human casualties.

Another large stony meteorite fell on March 8, 1976, near the city of Kirin in northeast China. The largest fragment weighed 1.7 tonnes and penetrated 6.5 m into clay-rich soil. The total weight of the fragments that were recovered was about 2.7 tonnes. Although tens of thousands of people saw the large fireball (three times the size of the full Moon) and the trailing cloud of smoke and dust and also heard the thunderous echoes of the explosion and fall, no humans

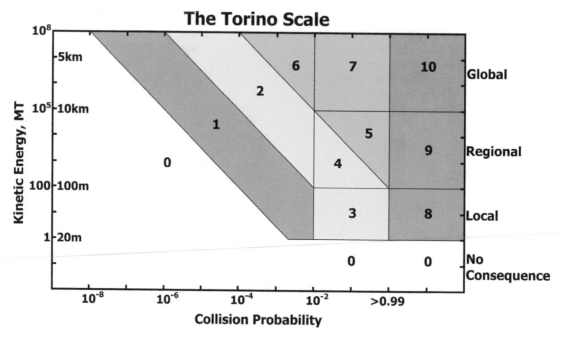

Figure 13.11. The Torino scale invented by Richard Binzel of M.I.T. is used to assess the threat posed by the impacts of near-Earth asteroids and comets. The impact hazard is expressed by a number between 0 and 10 depending on the probability that an impact will occur and by the kinetic energy of the potential impactor. The extent of damage ranges from inconsequential (0) to catastrophic (10) in cases where the probability that an impact will occur is equal to 1.0. The scale is color-coded such that white = no consequences; green = meriting careful monitoring; yellow = meriting concern; orange = threatening events; red = impact is certain. The amount of damage depends on the diameter of the impactor and hence on the area that is affected. For example, the impact of an asteroid or comet having a diameter of 5 km causes catastrophic damage on a global scale. Adapted from Levy (2006). The actions to be taken by astronomers and agencies of the government are stated in Science Brief 13.7b

were hurt and there was no substantial damage (Glass, 1982).

On December 23, 2004, NASA's Office for Near-Earth Objects assigned a rating of 2 on the Torino scale to asteroid 2004 MN4 which has a diameter of 400 m and appeared to be on a collision course with Earth. The threat level was later increased to 4, which implies that an impact would cause regional devastation. However, the subsequent refinement of the orbit of this NEO indicated that it will pass the Earth on April 13 of 2019 without impacting. The official name of this asteroid is 99942 Apophis and its threat-level has been lowered to 1 because its orbit may be altered by gravitational interactions when it returns in 2035/36. Torino scores of 1 have also been given to two other near-Earth asteroids: 1997XR2 and 2004 VD 17. Additional asteroids that may receive a Torino rating of 1 are listed at (< http://neo.jpl.nasa.gov/risk/>).

13.6 Summary

Most asteroids are irregularly shaped solid objects that occur mainly between the orbits of Mars and Jupiter where they move in heliocentric orbits in the prograde direction. Ceres and a few other asteroids having diameters of several hundred kilometers are roughly spherical in shape. The vast majority of Main-Belt asteroids are less than 100 km in diameter and have irregular shapes presumably because most are fragments of small worlds that were shattered by tidal forces of Jupiter and Mars. Groups of small asteroids having diameters less than 30 km occur in the space near the Earth and Mars (i.e., the Aten, Apollo, and Amor asteroids). Another group of large asteroids with diameters greater than 100 km are hiding close to the L4 and L5 Lagrange points of Jupiter. They are known as the Trojan asteroids by analogy with the wooden horse in which Greek fighters were hiding during the siege of Troy. A sixth group of asteroids known as the Centaurs orbit the Sun on highly eccentric orbits that take them to the outskirts of the solar system close to the orbit of Pluto.

The spectra of sunlight reflected by the asteroids and, in some cases, their bulk densities, indicate that the mineral composition of many asteroids resembles the composition of stony meteorites (i.e., achondrites, ordinary chondrites, and carbonaceous chondrites). A few asteroids in the main belt are actually composed of metallic iron (Science Brief 13.7.1) and some asteroids in the outer part of the main belt and among the Centaurs also contain a variety of ices under a crust of refractory dust cemented by organic matter. For this reason, some asteroids (e.g., Chiron) are classifiable as comets when they approach the Sun, but are considered to be asteroids when they are far enough removed from the Sun to avoid the sublimation of ice.

Several asteroids have been imaged at close range during flybys of robotic spacecraft (e.g., Gaspra, Ida, Mathilde, Braille, etc.) and the spacecraft NEAR Shoemaker actually went into orbit around Eros and subsequently landed on its surface. Such close encounters indicate that the surfaces of asteroids are covered by regolith consisting of poorly sorted rock debris formed by impacts of smaller asteroids. The albedo and color of asteroids is determined by the color and composition of the regolith, which has been altered by irradiation with ultraviolet light and nuclear particles of the solar wind and cosmic rays. The impacts of solid objects have left a record of craters on the surfaces of all asteroids from which the exposure ages of fracture surfaces can be determined.

The impacts of small asteroids (1 to 10 km in diameter) have repeatedly caused the sudden extinction of most organisms living on the surface of the Earth in the course of geologic history. The best known example of this phenomenon is the impact of an asteroid on the Yucatán peninsula of Mexico 65 million years ago that caused the extinction of 70% of all species living at that time, including the dinosaurs (i.e., the so-called K-T event). Because of the devastating consequences of asteroid impacts, efforts are now underway to find and document most, if not all, Near-Earth Objects (NEOs). Present knowledge indicates that about 1500 NEOs have diameters of one kilometers or more and that smaller asteroids with diameters between 100 and 1000 m are even more numerous. We now posses the technology to deflect NEOs that are on a collision course with the Earth provided they are detected soon enough (e.g., 100 years before impact). The threat posed by the impact of an asteroid is assessed by the Torino scale of 0 to 10 based on the probability that the impactor will actually collide with the Earth and on its kinetic energy (or diameter). Asteroids may become useful in the future as natural resources for colonists on Mars and as resupply depots for travelers to the outer regions of the solar system.

13.7 Science Briefs

13.7.1 Metallic Asteroid Lutetia

The Main-Belt asteroid Lutetia is a spectral type M that resembles stony-iron and iron meteorites (Table 13.1). This asteroid has a diameter of 115 km (Appendix 2) and its average distance from the Sun is 2.43 AU. If Lutetia is composed primarily of metallic iron similar in composition to iron meteorites, its chemical composition would be: 90.6 % Fe, 7.9 % Ni, 0.5 % Co, 0.7 % S, 0.2 % P, and 0.4 % C. The densities of these elements in grams per cubic centimeter are: Fe = 7.86, Ni = 8.90, Co = 8.90, S = 2.07; P = 1.82, and C = 2.26. The density (d) of a mixture of these elements is:

$$d = 7.86 \times 0.906 + 8.90 \times 0.079 + 8.90 \times 0.005$$
$$+ 2.07 \times 0.007 + 1.82 \times 0.002$$
$$+ 2.26 \times 0.004$$
$$d = 7.90\,\text{g/cm}^3.$$

Assuming that Lutetia is approximately spherical in shape and that its radius (r) is 57.5 km, the volume (V) of Lutetia is:

$$V = \frac{4}{3}\pi r^3 = \frac{4 \times 3.14 \times (57.5)^3}{3}$$
$$= 7.95 \times 10^5 \, km^3 = 7.95 \times 10^{20} \, cm^3$$

The mass (M) of Lutetia is:

$$M = d \times V = 7.90 \times 7.95 \times 10^{20} \, g = 6.28 \times$$
$$10^{18} \, kg = 6.28 \times 10^{15} \text{ metric tonnes}$$

Therefore, the amounts of the elements that make up the asteroid Lutetia are:
Fe $= 6.28 \times 10^{15} \times 0.906 = 5.69 \times 10^{15}$ tonnes
Ni $= 6.28 \times 10^{15} \times 0.079 = 0.496 \times 10^{15}$ tonnes
Co $= 6.28 \times 10^{15} \times 0.005 = 0.0314 \times 10^{15}$ tonnes
S $= 6.28 \times 10^{15} \times 0.007 = 0.0439 \times 10^{15}$ tonnes
C $= 6.28 \times 10^{15} \times 0.004 = 0.0251 \times 10^{15}$ tonnes
$$\text{Sum} = 6.28 \times 10^{15} \text{ tonnes}$$
The monetary value of the asteroid Lutetia depends on the world-market price of the metals it contains. Although the prices of scrap metals on the world market fluctuate, representatives values are:

Iron: $200/tonne (Recycling International, June 2005).
Nickel: $16,000/tonne (Recycling International, June 2005)
Cobalt: $44,092/tonne (US Bur. Mines, 1998)
Based on these prices, the value of the metal contained in the asteroid Lutetia is:
Iron: $5.69 \times 10^{15} \times 200 = \$1.14 \times 10^{18}$
Nickel: $0.496 \times 10^{15} \times 16,000 = \$7.93 \times 10^{18}$
Cobalt: $0.0314 \times 10^{15} \times 44,092 = \$1.38 \times 10^{18}$

yielding a total of about 1×10^{19} or about ten million trillion US dollars.

The estimated value of the metal contained in the asteroid Lutetia depends on the assumptions that:
1. Lutetia is a spherical body with a diameter of 115 km.
2. It is composed only of metal with only minor amounts of stony material.
3. It has the same composition as iron meteorites.
4. The prices of scrap metal stated above are representative in a volatile market.

In reality, none of these assumptions are necessarily correct, the technology for mining any kind of an asteroid does not yet exist, and the space treaties formulated by the United Nations and agreed to by the signatory nations (Section 9.10) do not permit the exploitation of asteroids or any other object in the solar system. In addition, even if recovery of large quantities of iron and other metals takes place over a period of 20 years, the price of the metals being recovered will be depressed thus diminishing their book values. In spite of these and other difficulties, the interest of entrepreneurs has been aroused by the large book value of asteroids like Lutetia.

13.7.2 Actions in Response to Hazard Assessment by the Torino Scale

(Morrison et al., 2004; and < http://neo.jpl. nasa.gov/torino_scale1.html >).

Number	Color	Description
0	White (No hazard)	The likelihood of a collision is zero, or is so low as to be effectively zero. Also applies to small objects such as meteors and bodies that burn up in the atmosphere as well as infrequent meteorite falls that rarely cause damage.
1	Green (Normal)	A routine discovery in which a pass near the Earth is predicted that poses no unusual level of danger. Current calculations show the chance of collision is extremely unlikely with no cause for public attention or concern. New telescopic observations very likely will lead to re-assignment to Level 0.
2	Yellow (Meriting attention by astronomers)	A discovery, which may become routine with expanded searches, of an object making a somewhat close but not highly unusual pass near the Earth. While meriting attention by astronomers, there is no cause for public attention or concern as an actual collision is very unlikely. New telescopic observations very likely will lead to re-assignment to Level 0.

3	Yellow	A close encounter, meriting attention by astronomers. Current calculations give a 1% or greater chance of collision capable of localized destruction .Most likely, new telescopic observations will lead to re-assignment to Level 0. Attention by the public and by public officials is merited if the encounter is less than a decade away.
4	Yellow	A close encounter, meriting attention by astronomers. Current calculations give a 1% or greater chance of collision capable of regional devastation. Most likely, new telescopic observations will lead to re-assignment to Level 0. Attention by the public and by public officials is merited if the encounter is less than a decade away.
5	Orange (Threatening)	A close encounter posing a serious, but still uncertain threat of regional devastation. Critical attention by astronomers is needed to determine conclusively whether or not a collision will occur. If the encounter is less than a decade away, governmental contingency planning may be warranted.
6	Orange	A close encounter by a large object posing a serious but still uncertain threat of a global catastrophe. Critical attention by astronomers is needed to determine conclusively whether or not a collision will occur. If the encounter is less than three decades away, governmental contingency planning may be warranted.
7	Orange	A very close encounter by a large object, which if occurring this century, poses an unprecedented but still uncertain threat of a global catastrophe. For such a threat in this century, international contingency planning is warranted, especially to determine urgently and conclusively whether or not a collision will occur.
8	Red (Certain collision)	A collision is certain, capable to causing localized destruction for an impact over land or possibly a tsunami if close offshore. Such events occur on average between once per 50 years and once per several 1000 years.
9	Red	A collision is certain, capable of causing unprecedented regional devastation for a land impact or the threat of a major tsunami for an ocean impact. Such events occur on average between once per 10,000 years and once per 100,000 years.
10	Red	A collision is certain, capable of causing global climatic catastrophe that may threaten the future civilization as we know it, whether impacting land or ocean. Such events occur on average once per 100,000 years, or less often.

13.8 Problems

1. Estimate the average distance between asteroids that reside between the orbits of Mars and Jupiter. Total area $= 1.74 \times 10^{18}\,\mathrm{km}^2$. Assume that this area is populated by 2×10^6 individual asteroids and that the area surrounding each asteroid is circular in shape. (Answer: r = 526,000 km).

2. The orbit of the Apollo asteroid Geographos has an average radius a = 1.24 AU and an eccentricity e = 0.34. Calculate the distance of this asteroid from the Sun at perihelion (q) and at aphelion (Q). (Answer: q = 0.82 AU; Q = 1.66 AU).

3. Calculate the book value of gold in the iron meteorite Lutetia (Science Brief 13.7.1) based on the following information: Diameter = 115 km, density = $7.90\,\mathrm{g/cm}^3$, concentration of gold $= 0.6\,\mu\mathrm{g/g}$ (ppm), price of gold = \$12,405 per kg (Kargel, 1994). (Answer: \$4.67 \times 10^{16}$ US).

4. Write an essay about the topic: "How should we protect ourselves and our civilization from destruction by the impact of a 10-km asteroid? (Assume that the impactor cannot be deflected).

13.9 Further Reading

Asphaug E (1997) New views of asteroids. Science 278:2070–2071

Beatty JK, Petersen CC, Chaikin A (1999) The new solar system. 4th edn. Sky Publishing, Cambridge, MA

Belton MJS, Morgan TH, Samarasinha N, Yeomans DK (eds) (2004). Mitigation of hazardous comets and asteroids. Cambridge University Press, Cambridge, UK

Binzel RP, Gehrels T, Matthews MS (eds) (1989) Asteroids II. University of Arizona Press, Tucson, AZ

Binzel RP (1999) Assessing the hazard: The development of the Torino scale. Planet Rept Nov/Dec: 6–10

Brandt JC (1999) Comets. In: Beatty, J.K. Petersen, C.C. Chaikin, A. (eds) The New Solar System, 4th edn. Sky Publishing Cambridge, MA, pp 321–336

Britt DT, Lebofsky LA (1999) Asteroids. In: Weissman P.R., McFadden, L.-A. Johnson, T.V. (eds) Encyclopedia of the Solar System. Academic Press, San Diego, CA, pp 585–606

Brown P, Spalding RE, ReVelle DO, Tagliaferri E, Worden SP (2002) The flux of small near-Earth objects colliding with the Earth. Nature 420:294–296

Chapman CR (1999) Asteroids. In: Beatty J.K., Petersen, C.C. Chaikin, A. The New Solar System. Fourth edn. 337–350. Cambridge University Press, Cambridge, UK

Dressler BO, Grieve RAF, Sharpton VL (eds) (1994) Large meteorite impacts and planetary evolution. Geol Soc Amer SPE 293, Boulder, CO

Dressler BO, Sharpton VL (eds) (1999) Large meteorite impacts and planetary evolution II. Geol Soc Amer SPE 339, Boulder, CO

Durda DD (2005) Questions and answers. The Planetary Report, Jan/Feb: 20

Gehrels T (ed) (1979) Asteroids. University of Arizona Press, Tucson, AZ

Gehrels T (ed) (1994) Hazards due to comets and asteroids. University of Arizona Press, Tucson, AZ

Gehrels T (1996) Collisions with comets and asteroids. Sci Am March: 54–59

Glass BP (1982) Introduction to planetary geology. Cambridge University Press, Cambridge, UK

Hallam A (2004) Catastrophes and lesser calamities. Oxford University Press, Oxford, UK

Hartmann WK (2005) Moons and planets, 5th edn. Brooks/Cole, Belmont, CA

Hiroi T, Zolensky M, Pieters C (2001) The Tagish Lake meteorite: A possible sample from a D-type asteroid. Science 293:2234–2236

Kargel JS (1994) Metalliferous asteroids as potential sources of precious metals. J Geophys Res, 99(E10):21, 129–21,144

Kenkmann T, Hörz F, Deutsch A (eds) (2005) Large meteorite impacts III. Geol Soc Amer, SPE 384, Boulder, CO

Koeberl C, MacLeod KG (eds) (2002) Catastrophic events and mass extinctions: Impacts and beyond. Geol Soc Amer SPE 356, Boulder, CO

Levison HF, Weissman PR (1999) The Kuiper belt. In: Weissman P.R., McFadden L.-A., Johnson T.V. (eds) Encyclopedia of the Solar System. Academic Press, San Diego, CA, pp 557–583

Levy DH (2006) Asteroid alerts: A risky business. Sky and Telescope 111(4):90–91

Lewis JS, Matthews MS, Guerrieri ML (eds) (1993) Resources of near-Earth space. University Arizona Press, Tucson, AZ

Lewis JS (1996) Mining the sky, untold riches from the asteroids, comets, and planets. Helix Books/Addison-Wesley, Reading, MA

McFadden L-A (1999) Near-Earth asteroids. In: Weissman P.R., McFadden L.-A., Johnson T.V. (eds) Encyclopedia of the Solar System. Academic Press, San Diego, CA, pp 607–627

Morrison D (2002) Target Earth. Astronomy, February: 47–51

Morrison D, Chapman CR, Steel D, Binzel RP (2004) Impacts and the public: Communicating the nature of the impact hazard. In: Belton M.J.S., Morgan T.H., Samarasinha N.H., Yeomans D.K. (eds) Mitigation of Hazardous Comets and Asteroids. Cambridge University Press, Cambridge, UK

Rayl AJS (2006) Hayabusa: A daring sample-return mission. The Planet Rept 26(1):6–11

Schweickart RL, Lu ET, Hut P, Chapman CR (2003) The asteroid tugboat. Sci Am 289(5):54–61

Silver LT, Schultz PH (eds) (1982) Geological implications of impacts of large asteroids and comets on the Earth. Geol Soc Amer, Special Paper 190, Boulder, CO

Steel D (1995) Rogue asteroids and doomsday comets. Wiley, New York

Trujillo CA, Brown ME (2004) The Caltech wide-area sky survey. In: Davies J.K., and Barrera L.H. (eds) The First Decadal Review of the Edgeworth-Kuiper Belt. Kluwer Academic Publishers, Dordrecht, The Netherlands, pp 99–112

Wetherill GW, Shoemaker EM (1982) Collision of astronomically observable bodies with the Earth. In: Silver L.T. and Schultz, P.H. (eds) Geological Implications of Impacts of Large Asteroids and Comets on the Earth. Geol Soc Amer, Special Paper 190, Boulder, CO, pp 1–3

Yeomans D (2006) Japan visits an asteroid. Astronomy, March: 32–35

Zolensky ME, Nakamura K, Gounelle M, Mikouchi T, Kasama T, Tachikawa O, Tonui E (2002) Mineralogy of Tagish Lake: An ungrouped type 2 carbonaceous chondrite. Meteoritics and Planet Sci 37: 737–761

Jupiter: Heavy-Weight Champion

We now enter the outer regions of the solar system where we encounter Jupiter, the heavy-weight champion among all of the planets. Although the Earth is the most massive and most voluminous of the *terrestrial* planets in the inner solar system, Jupiter in Figure 14.1 is a giant compared to the Earth. For example, its volume is about 1400 times larger than that of the Earth and its mass is about 318 times greater. In addition, Jupiter is 5.2 times farther from the Sun than the Earth and it has an enormously thick atmosphere in which gigantic storms rage with wind speeds of up to 540 km/h. Although Jupiter probably does have a core composed of the same elements that occur in the core and mantle of the Earth, its atmosphere consists largely of hydrogen and helium in about the same proportion as the Sun. The temperature and pressure of the jovian atmosphere both increase with depth below the cloud tops until hydrogen becomes an electrically conducting fluid. However, Jupiter does not have a solid surface at any depth. Convection of the liquid hydrogen in the interior of Jupiter gives rise to a magnetic field, the magnetic moment of which is 20,000 times stronger than that of the magnetic field of the Earth (Van Allen and Bagenal, 1999). The strong gravitational and magnetic fields of Jupiter create a dangerous environment even for robotic spacecraft, which recalls the admonition on the maps used by the early explorers of the Earth about 500 years ago: "There be monsters here."

14.1 Physical and Orbital Properties

The attraction of Jupiter as a target of exploration is based partly on its four satellites that were discovered by Galileo Galilei in 1610 AD following his invention of the telescope. The names of the so-called Galilean satellites in order of increasing distance from Jupiter are: Io, Europa, Ganymede, and Callisto. Many additional satellites were discovered more recently during flybys of the Pioneer and Voyager spacecraft, by the spacecraft Galileo that orbited the planet from 1995 to 2003, and by recent telescopic observations. As a result, the number of satellites of Jupiter has increased from the initial four to a total of 63 in May of 2005. Additional satellites may be discovered in the future. (Gehrels, 1976; Hunt and Moore, 1984; Beebe, 1994; Rogers, 1995; Harland, 2000; and Bagenal et al., 2004).

14.1.1 Physical Properties

The large mass of Jupiter $(1.899 \times 10^{27}\,\text{kg})$ listed in Table 14.1 causes the pressure and temperature of its thick atmosphere to increase with depth from $-108\,°\text{C}$ at the cloud tops to $-13\,°\text{C}$ (260 K) at a depth of 100 km where the pressure is about 4 atmospheres. The upper 60 km of the atmosphere contain three layers of clouds composed of ammonia (NH_3), ammonium hydrosulfide (NH_4HS), and water (H_2O). The temperature and pressure continue to increase with depth all the way to the center of the core where they reach estimated values of 22,000 K and 70 million atmospheres, respectively (Freedman and Kaufmann, 2002). Note that the estimated temperature is far less than the value required by the hydrogen-fusion reaction ($T = 12 \times 10^6\,\text{K}$, Section 4.2). Therefore, Jupiter missed becoming a star by a wide margin although it does emit more radiant energy than it receives from the Sun.

The rapid rate of rotation of Jupiter (9.841 h, Table 14.1) causes its shape to deform

Figure 14.1. Jupiter is the most massive and most voluminous planet of the solar system. Its striped appearance results from the clouds that form in the outer layers of its enormous atmosphere. The dark stripes (or belts) alternate with light-colored stripes (or zones). The winds in the belts and zones blow in opposite directions and cause gigantic cyclonic storms that appear in the image as white ovals. The largest storm is the Great Red Spot (GRS), which occupies a large area and has been observed for at least three centuries. In addition to its large and powerful gravitational field, Jupiter also has the most powerful magnetic field among the planets of the solar system. These two force fields interact with the innermost satellites of Jupiter and with its four large Galilean satellites (Io, Europa, Ganymede, and Callisto presented in Chapter 15). The black spot in this image of Jupiter is the shadow cast by Europa. (PIA 02873, Cassini orbiter, December 7, 2000, courtesy of NASA/JPL/University of Arizona in cooperation with the European and Italian space agencies)

from a sphere to an oblate spheroid whose equatorial diameter (142,984 km) is greater than its polar diameter (133,708 km). The oblateness of Jupiter is expressed as the percent difference between the equatorial and polar diameters:

$$\text{Oblateness} = \frac{(142,984 - 133,708)100}{133,708} = 6.94\%$$

which means that the equatorial diameter is 6.94% longer than the polar diameter. The evident deformation of Jupiter far exceeds the deformation of the Earth which has an oblateness

of only 0.34 %. Jupiter is more strongly deformed than the Earth because it rotates faster and because the gases and liquids in the interior of Jupiter are more easily deformed than the solid crust and mantle of the Earth. This comparison suggests that the oblateness of a rotating body depends partly on the physical properties of the material of which it is composed. This relationship has been used to determine the size of the core of Jupiter whose presence is suggested by the bulk density of the planet (1.326 g/cm^3, Table 14.1). Model calculations referred to by Freedman and Kaufmann (2002, p. 295) suggest

Table 14.1. Physical and orbital properties of Jupiter and the Earth (Freedman and Kaufmann, 2002)

Properties	Jupiter	Earth
Physical Properties		
Diameter, equatorial, km	142,984	12,756
Mass, kg	1.899×10^{27}	5.974×10^{24}
Bulk density, g/cm^3	1.326	5.515
Escape speed, km/s	60.2	11.2
Surface gravity (Earth = 1)	2.36	1.0
Albedo, %	44	39
Avg. temp. at cloudtops, °C	−108	+9
Orbital properties		
Avg. distance from the Sun, AU	5.203	1.000
Perihelion distance, AU	4.950	0.983
Aphelion distance, AU	5.455	1.017
Eccentricity of the orbit	0.048	0.017
Avg. orbital speed, km/s	13.1	29.79
Period of revolution, (sidereal), y	11.86	365.256
Period of rotation, equatorial, h	9.841	23.9345
Axial obliquity	3.12°	23.45°
Inclination of the orbit to the ecliptic	1.30°	0

that the core of Jupiter has a mass of 2.6 % of the mass of the planet:

$$\text{Mass of the core of Jupiter} = 0.026 \times 1.899 \times 10^{27}$$
$$= 4.94 \times 10^{25} \text{kg}$$

This estimate indicates that the core of Jupiter is about eight times more massive than the entire Earth:

$$\frac{\text{Mass of the core of Jupiter}}{\text{Mass of the Earth}} = \frac{4.94 \times 10^{25}}{5.974 \times 10^{24}} = 8.3$$

These calculations support the hypothesis that the massive core of Jupiter initially attracted hydrogen and helium to form a large primordial atmosphere. The composition and mass of the core of Jupiter in Figure 14.2 subsequently

increased because the planet captured asteroids and meteoroids as well as ice planetesimals and comets. The ices of water, ammonia, and methane sank toward the center of Jupiter and now form a layer about 3000 km thick around the rocky core. The ices presumable melted and now exist in liquid form because of the high temperature and in spite of the high pressure in the center of Jupiter. However, the state of aggregation of the rocky core is uncertain because the high pressure may have prevented melting in spite of the high temperature. In fact, the pressure may have compressed the rocky core to a density of about 23 g/cm^3 (Marley, 1999).

The pressure in the interior of Jupiter is so high that hydrogen becomes a conductor of electricity when electrons can move from one atom to its nearest neighbor. As a result, hydrogen gas assumes the properties of a metallic liquid. The transition from non-conducting hydrogen gas to a metallic liquid occurs when the pressure

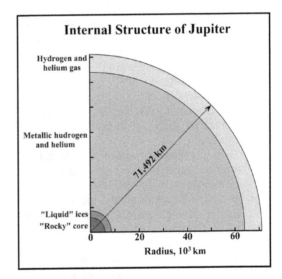

Figure 14.2. The interior of Jupiter contains a core whose mass is about eight times greater than the mass of the Earth. The core is covered by a layer of liquid formed by melting of cometary ices. The bulk of the interior consists of a mixture of hydrogen and helium. The hydrogen has been compressed into a liquid which is metallic in character because it permits electrical currents to flow. The uppermost layer is a mixture of hydrogen and helium gases and contains the colorful clouds we see in images of Jupiter. This layer is the only part of Jupiter that is composed of gases. The bulk of this planet actually consists of liquids. Adapted from Figure 13–16 of Freedman and Kaufmann (2002)

exceeds 1.4×10^6 atmospheres, which occurs at a depth of about 7000 km below the cloud tops of Jupiter in Figure 14.2. The rotation of the planet and the convection of the metallic-hydrogen liquid combine to cause electrons to flow, which induces the strong planetary magnetic field of Jupiter discovered on December 3, 1973, during a flyby by Pioneer 10. The magnetic field of Jupiter was later mapped during flybys of the Voyagers and by the Galileo spacecraft.

Although the magnetosphere of Jupiter is similar in shape to that of the Earth, its dimensions are 1200 times larger and its magnetotail extends at least 650 million kilometers to the orbit of Saturn. If we could see Jupiter's magnetosphere, it would occupy a larger area in the sky at night than the full Moon does. In addition, the pressure of the solar wind on the magnetosphere of Jupiter is only about 4% of the pressure it exerts on the magnetosphere of the Earth (Van Allen and Bagenal, 1999, p. 47). The interactions of Io with the magnetosphere of Jupiter will be presented in Section 15.2.4 of the next chapter.

The electromagnetic activity of Jupiter was originally detected in the 1950s by radioastronomers. The radiation contains three components having different wavelengths:

1. *Decametric* radiation with wavelengths of several meters occurs in sporadic bursts during electrical discharges in Jupiter's ionosphere triggered by interactions with its satellite Io.
2. *Decimetric* radiation, whose wavelength is several decimeters, (i.e., 1 decimeter = 10 centimeters) is emitted continuously by electrons moving through the magnetic field of Jupiter at high speeds approaching the speed of light in a vacuum. The radiation formed by this process, called synchrotron radiation, is an important phenomenon associated with the emission of energy by neutron stars (pulsars), quasars, and entire galaxies.
3. *Thermal* radiation that has a wavelength spectrum consistent with the temperature of Jupiter (i.e. it is blackbody radiation described in Appendix Appendix 1).

In summary, the interior of Jupiter in Figure 14.2 contains a "rocky" core whose mass is equivalent to about eight Earth masses but whose radius is only about 5500 km. The core is surrounded by

a 3000-km layer composed of water, methane, and ammonia that were originally deposited by the impact of ice planetesimals that sank to great depth and may now be liquid. The bulk of the interior of Jupiter below a depth of 7000 km measured downward from the cloud tops is composed of helium and of a metallic-hydrogen liquid. The H_2 – He atmosphere contains several layers of clouds composed of water vapor, ammonium hydosulfide, and ammonia. Therefore, Jupiter is composed primarily of liquids rather than gases and actually is not a "gas planet."

14.1.2 Orbital Properties

The average distance of the center of Jupiter from the center of the Sun in Table 14.1 is 5.203 AU and the eccentricity of its orbit is 0.048 which means that it is about 10 % farther from the Sun at aphelion than it is at perihelion. Jupiter revolves around the Sun and rotates on its axis in the prograde direction. The sidereal period of revolution (11.86 y) and the circumference of the orbit yield an average orbital speed of 13.1 km/s compared to 29.79 km/s for the Earth. The synodic period of revolution can be calculated from equation 7.7:

$$\frac{1}{P} = \frac{1}{E} - \frac{1}{S}$$

where P is the sidereal period of Jupiter, E is the sidereal period of the Earth, and S is the synodic period of Jupiter. Solving equation 7.7 for 1/S and substituting appropriate values for P and E yields S = 398.9 days as demonstrated in Section 7.2.7. The synodic period of Jupiter, illustrated in Figure 14.3, is longer than the sidereal period of the Earth because Jupiter moves a certain distance in its orbit during the time the Earth completes one revolution.

14.2 Atmosphere

Jupiter is the most massive planet in the solar system and therefore has the highest escape velocity (60.2 km/s). This means that Jupiter has retained most of the hydrogen of its primordial atmosphere because even low-mass hydrogen molecules (H_2) rarely achieve the

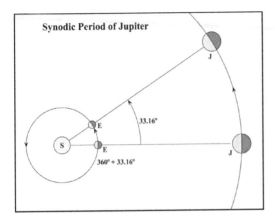

Figure 14.3. The synodic period of Jupiter (J) is the time that elapses between two successive oppositions with the Earth (E) as both revolve around the Sun (S). During one synodic period (398.9 d) the Earth revolves through an angle of 393.16° (360 + 33.16) while Jupiter moves only through an angle of 33.16° in its orbit

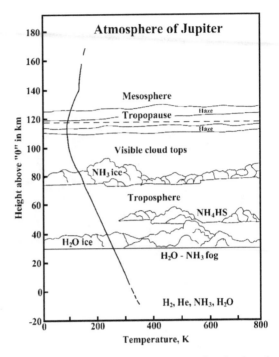

Figure 14.4. Cloud layers that occur at various levels and temperatures in the troposphere of Jupiter where the temperature decreases with increasing height above "zero." The three cloud layers consist of ice crystals of H_2O, NH_4HS, and NH_3 in order of decreasing temperature and increasing altitude. Note that Jupiter does not have a solid surface at the level marked "0" on the diagram. The tropopause is located where the temperature reaches a minimum and begins to increase at higher levels in the mesosphere. Adapted from Hartmann (2005, Figure 11.21)

necessary velocity to escape into interplanetary space (Atreya et al., 1989; West, 1999). The abundances of the chemical components of the atmosphere of Jupiter are expressed in terms of mixing ratios stated in percent by number. Accordingly, the jovian atmosphere consists of 86 % H_2, 13 % He, and 0.07 % CH_4 by number (Hartmann, 2005, Table 11.2). The remainder (0.93 %) includes NH_3, H_2O, H_2S, Ne, Ar, Kr, etc. The cloud layers in the atmosphere of Jupiter in Figure 14.4 consist of water, ammonium hydrosulfide, and ammonia at different levels depending on the temperature. For example, the lowest layer consists of water ice at a temperature of 273 K when water vapor condenses to form ice crystals.

The colorful appearance of Jupiter in Figure 14.1 is caused by the presence of impurities in the cloud layers of the troposphere, whereas the stripes and circular features are evidence of the vigorous circulation of the atmosphere. The stripes, that extend parallel to the equator and encircle the planet, consist of dark *belts* where cool air is sinking and light-colored *zones* where warm air is rising. The wind in adjacent belts and zones is blowing at high speeds in opposite directions as a consequence of the rapid differential rotation of Jupiter. The period of rotation of Jupiter varies with latitude from 9.841 h at the equator to 9.928 h near the

poles. In other words, Jupiter rotates *faster* at the equator than it does at the poles because it is not a solid body like the Earth but consists of liquid hydrogen and other kinds of fluids.

The principal circular feature, the *Great Red Spot* (GRS) in Figure 14.5, is a large storm that was first observed by Robert Hooke in 1664 and has probably existed with only minor changes for more than about 350 years. A second red spot that was discovered in 2006 is smaller than the GRS and appears to have formed by the merger of three small oval storms that were first seen in the 1930s. The resulting new storm was initially white but turned red in February of 2006. The GRS is located in the southern hemisphere of Jupiter and rotates in the counter-clockwise direction with a period of about six days, whereas cyclones (i.e., hurricanes) in the

Figure 14.5. This close-up of the clouds of Jupiter includes the Great Red Spot (GRS) and several smaller storms represented by the white ovals (See also Figure 14.1). The atmosphere of Jupiter is in constant turbulent motion as indicated by the movement of the small storms and by their interaction with the GRS which rotates in the anticlockwise direction. If the GRS is a hurricane (or cyclone), it should rotate in the clockwise direction because it is located in the southern hemisphere of Jupiter. Therefore, the observed anticlockwise rotation means that the GRS is not a hurricane but an anticyclone caused by a high-pressure air mass. The evident dynamical complexity of the atmosphere of Jupiter is primarily attributable to the rotation of the planet, to solar heating, and to the convection of the gases energized by internal heat. However, a complete mathematical description of the circulation of the atmosphere of Jupiter is even more difficult than the mathematical modeling of the atmosphere of the Earth. (Flyby of Voyager 1 on March 5 of 1979, courtesy of NAS/JPL-Caltech and presented by Robert Nemiroff and Jerry Bonnell)

southern hemisphere rotate clockwise. Therefore, the GRS is not a cyclone but an *anticyclone* characterized by the counterclockwise rotation of winds around a dome of high-pressure gas. The area of the GRS has varied over time from a maximum of about $5.6 \times 10^8 \, km^2 (40,000 \times 14,000 \, km)$ to about one third of this area, as for example during the Voyager flybys in 1979. The winds north of the GRS blow from east to west whereas south of the GRS the winds blow from west to east. The wind directions have not changed for several centuries presumably because Jupiter does not have a topographic surface with relief as the Earth does.

The circulation of the atmosphere of Jupiter is energized by direct sunlight, by heat emanating from the interior of the planet, and by its rapid differential rotation. In contrast to the terrestrial planets, Jupiter emits about twice as much energy than it receives from the Sun. The excess energy originated by the compression of gases during the formation of Jupiter 4.6 billion years ago (Freedman and Kaufmann, 2002, p. 280). The increase of the temperature with depth below the cloud tops causes convection cells to form that consist of warm gases rising from depth until they cool and sink, thereby completing the circle. The differential rotation of Jupiter distorts the atmospheric convection cells into narrow zones and belts. The zones are characterized by rising gas that cools and forms light-colored clouds. The dark-colored belts are places where cold gas is sinking and is being heated. The absence of clouds in the dark belts provides a view of the deeper regions of the jovian atmosphere.

14.3 Satellites and Rings

The satellites of Jupiter listed in Table 14.2 are named after mythological characters associated with Zeus. In addition, at least 47 as yet unnamed satellites of Jupiter were discovered by telescopic observations and during flybys of Pioneer (10 and 11), Voyager (1 and 2), and the Galileo spacecraft that orbited Jupiter for eight years from 1995 to 2003. These so-called irregular satellites are located far from Jupiter and their orbits are highly elliptical and steeply inclined to its equatorial plane. In addition, most of the irregular satellites revolve in the retrograde direction, have small diameters ranging down to less than one kilometer, and have irregular shapes. They have low albedos which suggests that some are composed of rocky material while others consist of mixtures of ice and dust similar to comets. Some of these irregular satellites of Jupiter may be fragments of larger objects that were broken up by collisions while others are asteroids captured by Jupiter (Section 13.2).

Table 14.2. The regular satellites of Jupiter (Buratti, 1999; Hartmann, 2005)

Name	Avg. distance from Jupiter, 10^3 km	Period of revolution, days	Diameter, km	Mass, kg	Albedo %
		Inner satellites (Group 1)			
Metis	128	0.295	40	?	low
Adrastea	129	0.298	24×16	?	4
Amalthea	181.5	0.498	270×155	?	5
Thebe	222	0.674	100	?	~ 10
		Galilean satellites (Group 2)			
Io	422	1.769	3630	8.89×10^{22}	63
Europa	671	3.551	3130	4.79×10^{22}	64
Ganymede	1071	7.155	5280	1.48×10^{23}	43
Callisto	1884	16.689	4840	1.0×10^{23}	17
		Outer satellites (Group 3A)			
Leda	11,110	240	~ 16	?	?
Himalia	11,470	250.6	~ 180	?	3
Lysithea	11,710	260	~ 40	?	?
Elara	11,740	260.1	~ 80	?	3
		Outer satellites (Group 3B)			
Ananke	20,700	617R*	~ 30	?	?
Carme	22,350	692 R	~ 44	?	?
Pasiphae	23,300	735 R	~ 35	?	?
Sinope	23,700	758 R	~ 20	?	?

*R = retrograde revolution

14.3.1 Regular Satellites

The regular satellites in Table 14.2 form three groups identified as the "inner", Galilean, and "outer" satellites. Amalthea, the largest of the inner satellites (group 1) of Jupiter in Figure 14.6, was discovered telescopically in 1892. The other satellites in this group (Metis, Adrastea, and Thebe) were discovered in 1979 during the Voyager flybys of Jupiter. All are irregularly shaped rocky objects in prograde orbits between 128×10^3 and 222×10^3 km from the center of Jupiter. The periods of revolution of the inner satellites are less than one day, the orbits are nearly circular in shape, and the orbital planes are closely aligned with the equatorial plane of Jupiter. All of these satellites have low albedos (< 10 %). Amalthea, the largest of the four inner satellites, is 270 km long and 155 km wide and has 1:1 spin-orbit coupling with Jupiter. In spite of problems during the flyby with the onboard computers and with the tape recorder of the spacecraft Galileo, the science team at the Jet Propulsion Laboratory was able to determine the mass of Amalthea and hence its bulk density, which turned out to be slightly less than $1.0\,g/cm^3$ instead of about $3.3\,g/cm^3$ as expected for a rocky object. Apparently, Amalthea is not a large chunk of rock but a pile of rubble covered by a layer of regolith and containing either open spaces or ice in its interior. The low mass of Amalthea also means that its escape velocity is about one centimeter per second. Consequently, grains of dust, pebbles, or astronauts could easily float off its surface. The regolith that covers the surface of Amalthea is reddish in color perhaps because of the presence of sulfur compounds or of organic matter like that of D-type asteroids (Table 13.1). Asteroids of this spectral type occur close to the orbit of Jupiter in the outer regions of the main asteroidal

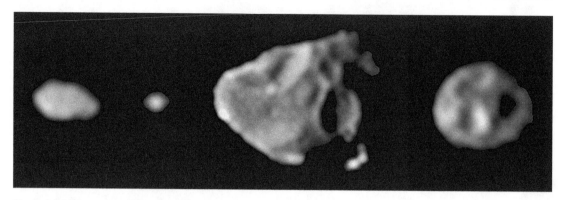

Figure 14.6. The group of inner satellites of Jupiter consists of Metis, Adrastea, Amalthea, and Thebe (Table 14.2). The masses of these satellites are not known but they are comparable to Main-Belt asteroids in terms of their shapes and sizes. Their low albedos (< 10%) imply that the regolith that covers their surface is dark in color because it is composed of silicate and oxide minerals and because it may contain organic matter, sulfur compounds (derived from Io), meteorite debris, and interplanetary dust. Amalthea is the largest member of this group (270 × 155 km). It has a reddish color and an albedo of only 5%. The images of Amalthea recorded by the spacecraft Galileo during flybys in 1996/1997 and by the Voyagers in 1979 indicate that it is heavily cratered. In fact, the body of Amalthea was broken up by these impacts as indicated by its anomalously low bulk density. The sizes of the satellites relative to each other are approximately correct, but the distances between them have been compressed for this family portrait. (PIA 01076, Galileo mission, courtesy of NASA/JPL-Caltech)

belt and at the L4 and L5 Lagrange regions of its orbit (Section 7.2.11).

The Galilean satellites of Jupiter (Io, Europa, Ganymede, and Callisto) were discovered by Galileo Galilei in 1610. They are located between 422×10^3 and 1884×10^3 km from the center of Jupiter and have large diameters (3130 to 5280 km) comparable to the diameter of the Moon of the Earth. Consequently, the Galilean satellites are worlds whose origin, evolution, and present state require consideration (Chapter 15). Although the surface of Io is coated with sulfur, its underlying crust is composed of silicate rocks. In marked contrast to Io, the other satellites in this group (i.e., Europa, Ganymede, and Calisto) have crusts composed of water ice.

The satellites of the group 3A in Table 14.2 (Leda, Himalia, Lysithea, and Elara) were discovered between 1904 (Elara) and 1974 (Leda), which was a remarkable feat in view of their small diameters (16 to 180 km) and their low albedos (< 3%). These satellites may be composed of water ice with a covering of organic matter and grains of refractory particles (Hartmann, 2005). All four members of this group have similar average distances from Jupiter between 11.110×10^6 and 11.740×10^6 km and their periods of revolution range narrowly from 240 to 260 days. In addition, the planes of their orbits are steeply inclined to the equatorial plane of Jupiter (24.8° to 29.0°) and their eccentricities range from 0.107 (Lysithea) to 0.207 (Elara). Nevertheless, all four satellites in this group revolve around Jupiter in the prograde direction.

The satellites in group 3B in Table 14.2 (Ananke, Carme, Pasiphae, and Sinope) were discovered between 1908 and 1951 even though they are about twice as far from Jupiter as the Leda-Elara group and in spite of the fact that they are even smaller than the members of that group. The average distances from Jupiter to satellites in group 3B range from 20.7×10^6 to 23.7×10^6 km, the eccentricities of their orbits are between 0.169 and 0.40, and the planes of their orbits dip from 147° to 163° relative to the equatorial plane of Jupiter. The large angle of inclination of the planes of the orbits of these satellites in Figure 14.7 cause the direction of revolution to change from prograde to retrograde when viewed from a point above the plane of the ecliptic. The periods of revolution of the satellites in this group range from 631 to 758 days (Buratti, 1999). The satellites of the Ananke-Sinope group are thought to be composed of water ice covered by a mixture of organic matter and refractory particles (Hartmann, 2005).

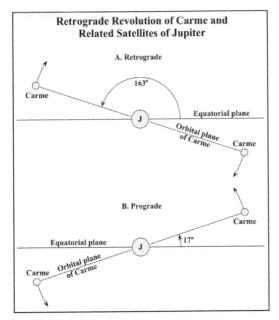

Retrograde Revolution of Carme and Related Satellites of Jupiter

A. Retrograde

163°

Carme

J

Equatorial plane

Orbital plane of Carme

Carme

B. Prograde

Carme

Equatorial plane

J

17°

Carme

Orbital plane of Carme

Figure 14.7. The retrograde revolution of Ananke, Carme, Pasiphae, and Sinope can be explained as a consequence of the large angle of inclination of their orbital planes relative to the equatorial plane of Jupiter. In part A of this diagram, the orbital plane of Carme makes an angle of 163° with the equatorial plane of Jupiter causing the revolution to be retrograde when viewed from above the solar system. In part B, the hypothetical orbital plane of Carme makes an angle of 17° (180°–163°) with the equatorial plane of Jupiter causing the direction of revolution of Carme to be prograde. Data from Buratti (1999)

14.3.2 Rings of Jupiter

All of the large "gas planets" have rings consisting of flat disks composed of solid particles that revolve close to the equatorial planes of their mother planets. In contrast to the giant planets, none of the terrestrial planets have rings at the present time, although they may have had rings shortly after their formation. Although, Galileo Galilei discovered the rings of Saturn in 1610, the modern era of the study of planetary rings started in 1979 after both Voyager 1 and 2 spacecraft passed by Jupiter and Voyager 2 later encountered Saturn in 1981, Uranus in 1986, and Neptune in 1989 (Porco, 1999; Burns, 1999). The observations of Voyager 2 indicated that the ring system of Jupiter consists of three parts:

1. The *Main* ring is about 7000 km wide and less than 30 km thick. It contains the satellites Metis and Adrastea.
2. The *Halo* surrounds the Main ring and attains a thickness of about 20,000 km.
3. The *Gossamer* ring is located outside of the Main ring and the Halo. The satellites Amalthea and Thebe are located within the Gossamer ring and sustain it by shedding small particles released from their surfaces by impact of micrometeoroids (Carroll, 2003).

The particles that form the rings of Jupiter consist of refractory materials and have diameters on a scale of micrometers or less.

In general, the rings of the gas planets are composed of particles whose diameters range from micron-sized dust grains to the dimensions of large houses. Regardless of their diameters, the ring particles are composed of rocky material or ices of volatile compounds. Each particle in a ring revolves around its mother planet in its own orbit independent of neighboring particles. Consequently, the periods of revolution (p) of ring particles increase with increasing distance from the center of the mother planet as required by Newton's version of Kepler's third law. It follows that the periods of revolution of ring particles at certain distances from the center of the planet may resonate with the period of a satellite that orbits the same planet (Section 13.3). For example, the outer limit of the Main ring of Jupiter in Figure 14.8 occurs at a distance of 1.81 radii of Jupiter where the orbital periods of ring particles are in a 3/5 resonance with the period of Amalthea (Science Brief 14.7.1). This means that Amalthea revolves around Jupiter three times in the same time the ring particles orbit five times. In general, the satellites named above and other unseen bodies within the rings of Jupiter are not only the principal sources of particles but also shape the rings and prevent the particles from escaping. Hence the satellites that are located within the system of rings of the gas planets are referred to as *shepherd moons*.

The ring systems are constrained not only by resonance effects with the satellites that orbit a planet but also by the *Roche limit* defined as the distance from the center of a planet within which a liquid satellite that has no tensile strength is disrupted by tidal forces (Science Brief 14.7.2).

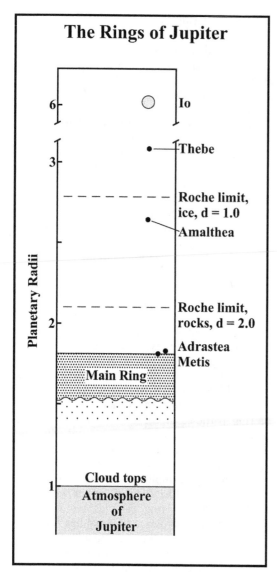

The Rings of Jupiter

Figure 14.8. Relation of the Main ring of Jupiter to the inner satellites: Metis and Adrastea, Amalthera, and Thebe. The outer edge of the Main ring coincides with the orbits of Metis and Adrastea whose periods of revolution resonate with the period of Amalthea (Science Brief 14.7.1). In addition, the Main ring is located inside the Roche limit for objects having a density of $2.0 \, g/cm^3$ in a region where particles cannot accrete to form a solid body because of tidal forces exerted by Jupiter. The satellites of Jupiter are located outside the Roche limit, except Metis and Adrastea, both of which appear to be rocky objects having superior mechanical strength. Adapted from Hartmann (2005, Figure 3.14)

For this reason, rings can exist only *inside* the Roche limit. Particles that orbit a planet *outside* of the Roche limit could, in principle, form a larger solid body in case that body can withstand the stresses caused by planetary tides.

The Roche limit (r_R) is expressed by the equation (Hartmann, 2005, p. 60):

$$r_R = 2.44 \left(\frac{\rho_M}{\rho_S} \right)^{1/3} R \qquad (14.1)$$

where ρ_M is the density of the planet, R is its radius, and ρ_S is the density of the satellite. In case $\rho_M = \rho_S$, the Roche limit is $r_R = 2.44 \, R$.

Note that equation 14.1 implies that the Roche limit for a satellite orbiting a planet such as Jupiter *decreases* with increasing density of the satellite. In other words, satellites composed of rocks can approach more closely than satellites composed of ice before they are broken up by the tidal force of the planet. Therefore, the rings that orbit the large gas planets may have formed when objects composed of rocks or ice came too close to a planet and were broken up by tidal forces. Alternatively, particles that initially formed a disk around a planet at the time of its accretion may have been preserved in the rings by the fact that tidal forces prevented them from coalescing into a satellite.

An additional constraint on ring particles is provided by the *Poynting-Robertson effect* which arises from the way in which centimeter-sized particles in stable orbits absorb and reradiate sunlight (Duncan and Lissauer, 1999). The effect causes particles to lose energy and to spiral toward the mother planet. As ring particles descend into lower orbits, they are further slowed by atmospheric friction and eventually fall into the atmosphere of their mother planet. Consequently, a gap is observed between the top of the atmosphere and the innermost ring of a planet.

14.4 The Spacecraft Galileo

Much of what is known about Jupiter and its satellites was learned during flybys of the robotic spacecraft Pioneers 10 and 11 and by Voyagers 1 and 2. The Voyagers recorded spectacular images of Jupiter and its Galilean satellites, which provided a great deal of information that was succinctly summarized in an issue of Science (volume 204, number 4296, 1979).

The Galileo mission was conceived in the mid-1970s in order to carry out a thorough investigation of Jupiter and its satellites (Johnson, 1995). The spacecraft included a probe that was to be dropped into the atmosphere of Jupiter after Galileo had gone into orbit around the planet. The launch of the spacecraft, originally scheduled for 1982, was repeatedly postponed for various reasons including the loss of the space shuttle Challenger in January of 1986. Galileo was finally launched in October of 1989 by the space shuttle Atlantis. In order to gather momentum for the long flight to the Jupiter system, the spacecraft first flew by Venus and then twice by the Earth.

Two years after launch, the main communications antenna failed to open, leaving only the much smaller back-up antenna in working order. In addition, the tape recorder malfunctioned just before Galileo was to carry out a flyby of Io at a distance of only 900 km. In spite of the delays of the launch and the in-flight malfunctions, the spacecraft did go into orbit around Jupiter in December of 1995 and was expected to explore the planet and its satellites for about two years.

The probe carried by Galileo was released as planned and descended on parachutes into the atmosphere of Jupiter. When it reached a depth of 146 km after 61.4 minutes, communications ended. The probe continued to fall through the atmosphere for another eight hours until its titanium housing melted at about 1000 km below the cloud tops where the pressure was 2000 bars (Johnson, 2000). A summary of the results obtained by the probe was later published in a special section of Science (vol. 272, May 10, 1996, 837–860.)

The Galileo spacecraft continued to orbit Jupiter for a total of eight years until September 21, 2003, when it was directed to plunge into Jupiter's atmosphere. The data it returned have added greatly to our present knowledge about the atmosphere of the planet and its magnetic field. In addition, Galileo sent back close-up images of Jupiter's Galilean satellites (to be described in Chapter 15).

When the Galileo spacecraft finally plunged into the atmosphere of Jupiter it had far exceeded the two-year mission for which it was designed. The information it provided about the volcanic activity of Io and about the ice-covered satellites Europa, Ganymede, and Callisto has strengthened the motivation to explore these bodies in the future by means of a new project called JIMO (Jupiter Icy Moons Orbiter). In addition, the spacecraft Cassini is presently exploring Saturn and its principal satellites.

14.5 Impact of Comet Shoemaker – Levy 9

Impacts on the Earth of asteroids and comets having diameters in excess of one kilometer, though rare, can cause the extinction of most living species including humans. The destructive power of cometary impacts was dramatically demonstrated in July of 1994 when 21 fragments (A to W) of the comet Shoemaker-Levy 9 in Figure 14.9 collided sequentially with Jupiter. The resulting explosions, recorded by the Galileo spacecraft and by the Hubble Space Telescope, confirmed Gene Shoemaker's conclusion that certain craters on the Earth and most of the craters on the Moon are the result of powerful explosions caused by the impacts of asteroids and comets during the long history of the solar system. Gene Shoemaker himself had the unique opportunity to witness such impacts on another planet. This event was widely regarded as a warning that impacts of asteroids and comets on the Earth threaten our very survival (Sections 8.8, 8.9, 9.1, 13.5).

The comet Shoemaker-Levy 9 was discovered by Eugene M. Shoemaker, Carolyn S. Shoemaker, and David H. Levy on May 22, 1993, at the Palomar Observatory in California (Levy et al., 1995). The first image suggested that the newly-discovered comet consisted of a string of 21 separate fragments. Careful observations revealed that the comet was actually in orbit around Jupiter and that it had broken up on July 7, 1992, when it came to within about 20,000 km of the cloud tops of the planet. In addition, Brian G. Marsden, director of the Central Bureau for Astronomical Telegrams at the Harvard-Smithsonian Center for Astrophysics, predicted that the fragments of Shoemaker-Levy 9 would crash into Jupiter in July of 1994.

The first impact on July 16 could not be seen from Earth but was recorded by the Hubble Space Telescope. The impact in Figure 14.10

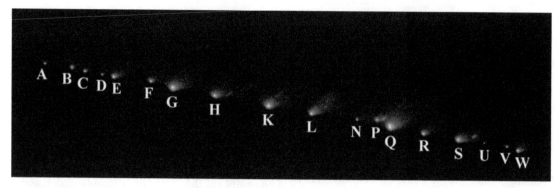

Figure 14.9. The comet Shoemaker-Levy 9 was broken on July 7, 1992, into a string of 21 fragments during a close encounter with Jupiter. These fragments subsequently crashed into Jupiter in July of 1994. The resulting explosions were recorded by the Hubble Space Telescope and by the Galileo spacecraft. This image was recorded on May 17, 1994, by H.A. Weaver and T.E. Smith of the Space telescope Science Institute. (Courtesy of NASA/HST)

released a plume of hot gas that extended about 3000 km above the clouds of Jupiter. When the impact site rotated into view from the Earth, it was clearly visible as a dark area the size of the Earth. The other fragments continued to smash into Jupiter for about one week. Some of these impacts were even more spectacular than the first, depending on the diameter of the object and whether they were solid bodies or consisted of fragments held together weakly by gravity. The scars of the impacts gradually faded in the months that followed but a faint dark band lingered for more than a year.

The story of the crash of comet Shoemaker-Levy 9 into Jupiter was published in a special

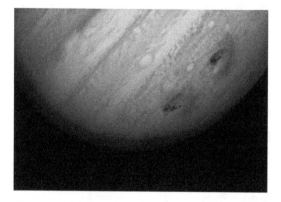

Figure 14.10. Impacts of the fragments of the comet Shoemaker-Levy 9 in the southern hemisphere of Jupiter. This image was recorded by the Hubble Space-Telescope Jupiter-Imaging Team of NASA/JPL and Space Telescope Science Institute. (http://www2.jpl.nasa.gov/s19/image111.html)

issue of Sky and Telescope (vol. 88, N. 4, October 1994), in a section of Science (vol. 267:1277–1323, March 3, 1995), and in a book by Levy (1996).

14.6 Summary

Compared to the Earth, Jupiter is a strange planet. It is 318 times more massive than the Earth even though most of it is composed of hydrogen and helium. It does have a core composed of the same chemical elements that make up the mantle and core of the Earth, although the abundances of the elements in the jovian core may differ from those on the Earth. The mass of the core of Jupiter is about eight times greater than the mass of the Earth, but its density is more than ten times higher than the bulk density of the Earth because of the enormous pressure at the center of the Jupiter.

The core of Jupiter is surrounded by 3000 km of ices of various compounds derived from comets and ice planetesimals that have crashed into Jupiter. The high temperature of the core may have allowed the ice layer to melt in spite of the high pressure which prevails in this region.

The bulk of Jupiter is composed of a mixture of hydrogen and helium that formed an enormous primordial atmosphere held in place by the strong gravitational force exerted by the core. The hydrogen of the primordial atmosphere was compressed to form an electrically conducting fluid. The rapid rotation of Jupiter and thermal convection of the so-called metallic-hydrogen liquid generate electrical currents, which induce the strong planetary magnetic field that was recorded during flybys of the Pioneer and Voyager spacecraft during the 1970s. The magnetic field of Jupiter is deformed by the solar wind, much like that of the Earth, by being compressed on the side facing the Sun and trailing far behind on the dark side. The electromagnetic activity in the interior of Jupiter and its interactions with Io (first of the Galilean satellites) generate "radiowaves" that have been observed by radio-astronomers since the 1950s.

The outermost layer of Jupiter is the atmosphere which is about 7000 km thick and is composed of a mixture of hydrogen and helium gas. The atmosphere also contains layers of clouds near its top composed of ice particles of ammonia (NH_3), ammonium hydrosulfide (NH_4HS), and water (H_2O). When viewed from space, the clouds of Jupiter form dark belts and light-colored zones that extend around the planet parallel to its equator. These linear features reveal aspects of the circulation of the atmosphere which includes thermal convection cells and high winds activated by the rapid rotation of the planet.

The atmosphere of Jupiter contains numerous cyclonic storms, the largest of which is the so-called Great Red Spot (GRS). The GRS is located in the southern hemisphere of Jupiter and has a surface area that could accommodate eight earth-sized planets placed side-by-side. This storm system has apparently persisted for more than 350 years, although its area has varied over time. The circulation of the atmosphere in the GRS is in the counterclockwise direction, which means that it is not a hurricane, because such storms are low pressure airmasses around which the wind blows in the counterclockwise direction when they occur in the northern hemisphere, and clockwise in the southern hemisphere. Therefore, the counterclockwise rotation of the GRS identifies it as an anticyclone consisting of a high-pressure airmass that rotates in the counterclockwise direction because it is located in the southern hemisphere of Jupiter.

The characteristic striped appearance of Jupiter is caused primarily by the rising of gases in the zones and by the sinking of gases in the belts. The light color of the zones is caused by the formation of ice crystals in gases that are cooling as they rise from depth, whereas the dark color of the belts indicates that cold gases are warming as

they sink. When viewed from beneath the clouds of Jupiter, the dark belts are cloudless because condensation of gaseous compounds is prevented by the rising temperature.

The planet Jupiter and its principal satellites and rings were studied by the Galileo spacecraft, which orbited the planet for eight years between 1995 and 2003. The observations returned to Earth by this robotic spacecraft have provided information about the interactions between the magnetic field of Jupiter and the solar wind, about the ring of particles that orbit the planet, and about the sixteen principal satellites including the four large satellites first observed by Galileo Galilei in 1610 AD. The number of satellites of Jupiter has increased to 63 as of May, 2005, and will probably continue to rise.

14.7 Science Briefs

14.7.1 Resonances Among the Inner Satellites

The periods of revolution of the inner satellites of Jupiter are: Metis = 0.295 d, Adrastea = 0.298 d, Amalthea = 0.498 d, and Thebe = 0.674 d. Dividing the periods of Metis and Adrastea by the period of Amalthea:

$$\frac{Metis}{Amalthea} = \frac{0.295}{0.498} = 0.592 \simeq 0.6$$

$$\frac{Adrastea}{Amalthea} = \frac{0.298}{0.498} = 0.598 \simeq 0.6$$

Both ratios are approximately equal to 0.6 or 6/10. Dividing the numerator and denominator by 2 yields 3/5. This result indicates that Metis and Adrastea circle Jupiter five times in the same time Amalthea completes three orbits.

For Metis: $0.295 \times 5 = 1.475$ d
For Adrastea: $0.298 \times 5 = 1.490$ d
For Amalthea: $0.498 \times 3 = 1.494$ d

This result means that every 1.49 days Metis and Adrastea, both of which are smaller than Amalthea, are acted upon by the gravitational force exerted by Amalthea. Ring particles in the immediate vicinity of Metis and Adrastea are similarly affected and may be displaced from their orbits in the same way as are the asteroids that occupy (or once occupied) the Kirkwood gaps (Section 13.3).

14.7.2 Roche Limits

The Roche limit is the smallest distance from the center of a planet or other object at which a second object can be held together by purely gravitational forces (Freedman and Kaufmann, 2002). This concept was formulated in the 19th century by the French mathematician *Edouard Roche* (1820–1883). If a satellite and its planet have similar compositions, the Roche limit (r_R) is about 2.5 times the radius (R) of the planet. The actual value of the Roche limit depends on the size and strength of the smaller body. Strong objects (e.g., Phobos) having diameters less than 60 km may revolve around a planet at a distance less than the Roche limit without breaking up. However, in case the smaller body is on a trajectory which will cause it to impact on a planet that has an atmosphere, aerodynamic stresses may cause the object to disintegrate before reaching the surface (Hartmann, 2005; Melosh, 1981).

The equation that defines the Roche limit is based on the requirement that the gravitational attraction of two particles of equal mass (m) must be equal to the tidal force that acts to separate the particles from each other (Hartmann 2005, p. 60).

$$r_R = 2.5 \left(\frac{\rho_M}{\rho_m} \right)^{1/3} R \qquad (14.2)$$

If the satellite that is orbiting a planet has no mechanical strength (e.g., a rubble pile), the equation takes the form of equation 14.1 in Section 14.3.2:

$$r_R = 2.44 \left(\frac{\rho_M}{\rho_S} \right)^{1/3} R \qquad (14.3)$$

We now use equation 14.2 to calculate the Roche limit of a hypothetical satellite of Jupiter having a density $\rho_S = 1.00 \, \text{g/cm}^3$ (i.e., water ice), assuming that the density of Jupiter (ρ_M) is $1.326 \, \text{g/cm}^3$ and that its radius (R) is 71,492 km. Substituting into equation 14.2:

$$r_R = 2.5 \left(\frac{1.326}{1.000} \right)^{1/3} \times 71,492 = 196.5 \times 10^3 \, \text{km}.$$

This result indicates that a satellite composed of ice would be disrupted by the tides of Jupiter if the radius of its orbit is less than 196.5×10^3 km. The data in Table 14.2 indicate that the Roche limit for a satellites composed of water ice in orbit around Jupiter lies between the orbits of Amalthea (181.5×10^3 km) and Thebe (222×10^3 km). Therefore, Metis and Adrastea whose orbital radii are about 128.5×10^3 km from the center of Jupiter, cannot be composed of ice and therefore probably consist of rocky material.

14.7.3 Masses of Atoms and Molecules

The gram-atomic weight of an element is the atomic weight of the element in grams. According to Avogadro's Law, one gram-atomic weight (or "mole") of an element contains 6.022×10^{23} atoms. The same statements apply to the gram-molecular weight of a compound. Therefore, the mass of one atom (or molecule) of an element (or compound) can be calculated by the following procedure.

The atomic weight of hydrogen is 1.0079 atomic mass units (amu). Therefore, 1 mole of H weighs 1.0079 g, and the mass (m_H) of one H atom is:

$$m_H = \frac{1.0079}{6.022 \times 10^{23}} \text{g} = \frac{1.0079}{6.022 \times 10^{23} \times 10^3} \text{kg}$$

$$m_H = 0.1673 \times 10^{-26} \, \text{kg} = 1.67 \times 10^{-27} \, \text{kg}$$

The mass of one H_2 molecule is

$$m_{H_2} = 2 \times 1.67 \times 10^{-27} = 3.34 \times 10^{-27} \, \text{kg}$$

14.7.4 Average Speed of H_2 Molecules

According to Section 22 of Appendix 1, the average speed (v) of an atom or molecule of mass (m) in a gas at temperature (T) is:

$$v = \left(\frac{3kT}{m} \right)^{1/2}$$

where T is in kelvins, m is in kilograms, k (Boltzmann constant) is 1.38×10^{-23} J/K, and v is expressed in meters per second.

The average speed of one molecule of H_2 at 15 °C (288 K) and mass m = 3.34×10^{-27} kg is:

$$v_{H_2} = \left(\frac{3 \times 1.38 \times 10^{-23} \times 288}{3.34 \times 10^{-27}} \right)^{1/2}$$

$$= \left(356.9 \times 10^4\right)^{1/2}$$

$$v_{H_2} = 18.89 \times 10^2 \, m/s \text{ or } 1.89 \, km/s$$

14.8 Problems

1. Calculate the average speed of H_2 molecules in the atmosphere of Jupiter (Consult Section 22 of Appendix 1), given that the mass of one H_2 molecule is 3.34×10^{-27} kg (Science Brief 14.7.4) and that the temperature of the upper atmosphere of Jupiter is 110 K (Freedman and Kaufmann, 2002, p. 195). (Answer: 1.16 km/s).

2. Calculate the mass of one CO_2 molecules (m_{co_2}) given that the atomic weights of carbon and oxygen are: C = 12.011 amu; O = 15.9994 amu. (Consult Science Brief 14.7.3). (Answer: 7.308×10^{-26} kg).

3. Demonstrate by calculation whether O_2 molecules can escape from the atmosphere of Venus. The mass of O_2 is 5.32×10^{-26} kg, the temperature is 743 K, the escape velocity from the surface of Venus is 10.4 km/s (Section 23, Appendix 1), and the Boltzmann constant is 1.38×10^{-23} J/K. (Consult Science Brief 14.7.4). Answer: v = 0.760 km/s).

4. Calculate the Roche limit for Phobos and explain the significance of this result. (Use equation 14.2 of Science Brief 14.7.2). (Answer: 10,832 km).
   Density of Phobos: 1.90 g/cm³
   Density of Mars: 3.94 g/cm³
   Radius of Mars: 3398 km
   Radius of the orbit of Phobos: 9.38×10^3 km

14.9 Further Reading

Atreya SK, Pollack JB, Matthews MS (1989) Origin and evolution of planetary and satellite atmospheres. University of Arizona Press, Tucson, AZ

Bagenal F, Dowling TE, McKinnon W (eds) (2004) Jupiter: The planet, satellites, and magnetosphere. Cambridge University Press, Cambridge, UK

Beebe R (1994) Jupiter: The giant planet. Smithsonian Inst. Press, Washington, DC

Buratti BJ (1999) Outer planet icy satellites. In: P.R. Weissman, L.-A. McFadden, T.V. Johnson (eds) Encyclopedia of the Solar System. Academic Press. San Diego, CA, pp 435–455

Burns JA (1999) Planetary rings. In: J.K. Beatty, C.C. Petersen, A. Chaikin (eds) 1999. The New Solar System, 4th edn. Cambridge Un. Press, Cambridge, UK, pp 221–240

Carroll M (2003) The long goodbye. Astronomy, October: 37–41

Duncan MJ, Lissauer JJ (1999) Solar system dynamics. In: P.R. Weissman, L.-A. McFadden, T.V. Johnson (eds) Encyclopedia of the Solar System. Academic Press, San Diego, CA, pp 809–824

Freedman RA, Kaufmann III WJ, 2002. Universe: The solar system. Freeman, New York

Gehrels T (ed) (1976) Jupiter. University of Arizona Press, Tucson, AZ

Harland D (2000) Jupiter Odyssey. Springer/Praxis, Chichester, UK

Hartmann WK (2005) Moons and planets, 5th edn. Brooks/Cole, Belmont, CA

Hunt G, Moore P (1984) Jupiter. Rand McNally, New York

Johnson TV (1995) The Galileo mission. Scient Amer December: 44–51

Johnson TV (2000) The Galileo mission to Jupiter and its moons. Sci Am February: 40–49

Levy DH, Shoemaker EM, Shoemaker CS (1995) Comet Shoemaker–Levy 9 meets Jupiter. Scient Amer August: 85–91

Levy DH (1996) Impact Jupiter: The crash of comet Shoemaker-Levy 9. Plenum Press, New York

Marley MS (1999) Interiors of the giant planets. In: P.R. Weissman, L.-A. McFadden, T.V. Johnson (eds.) Encyclopedia of the Solar System. Academic Press, San Diego, CA, pp 339–356

Melosh HJ (1981) Atmospheric breakup of terrestrial impactors. In: P. Schultz, R. Merrill (eds.) Multy-Ring Basins. Pergamon Press, New York

Porco CC (1999) Planetary rings. In: P.R. Weissman, L.-A. McFadden, T.V. Johnson (eds.) Encyclopedia of the Solar System. Academic Press, San Diego, CA, pp 457–475

Rogers JH (1995) The planet Jupiter. Cambridge University Press, New York

Van Allen JA, Bagenal F (1999) Planetary magnetospheres and the interplanetary medium. In: J.K. Beatty, C.C. Petersen, A. Chaikin (eds) The New Solar System, 4th edn. Sky Publishing, Cambridge, MA, pp 39–58

West RA (1999) Atmospheres of the giant planets. In: P.R. Weissman, L.-A. McFadden, T.V. Johnson (eds.) Encyclopedia of the Solar Systems. Academic Press, San Diego, CA, pp 315–338

Galilean Satellites: Jewels of the Solar System

The four Galilean satellites of Jupiter in Figure 15.1 (Io, Europa, Ganymede, and Callisto) were discovered by Galileo Galilei in 1610. He recognized that the four celestial bodies revolve around Jupiter much like the planets of the solar system revolve around the Sun in the heliocentric cosmology of Copernicus. However, for several centuries the Galilean satellites remained mere points of light that accompanied Jupiter in its orbit around the Sun. More recently, optical spectra of sunlight reflected by the Galilean satellites indicated the presence of water ice on their surfaces (Pilcher et al., 1972). The modern era of exploration of the Galilean satellites started in 1979 when Voyagers 1 and 2 recorded spectacular images of all four satellites. Still more information was obtained by the spacecraft Galileo while it orbited Jupiter from December 7, 1995 to September 19, 2003. Summaries of the physical and orbital properties of the Galilean satellites have been published in books by Hunt and Moore (1981), Morrison (1982), Beatty and Chaikin (1990), Beatty et al. (1999), Weissman et al. (1999), McEwen et al. (2000), Harland (2000), Fischer (2001), Hartmann (2005), and Greenberg (2005).

15.1 Physical Properties and Celestial Mechanics

The physical and chemical properties of the Galilean satellites support the hypothesis that they formed in orbit around Jupiter soon after the planet itself had come into existence by accretion of its core and while it was attracting hydrogen, helium, and other volatile compounds that had formed a disk in orbit around it. The subsequent evolution of the Galilean satellites continues to be affected by finely tuned orbital resonances and by the strong body-tides raised by Jupiter.

15.1.1 Physical Properties

The Galilean satellites are spherical objects whose diameters and masses in Table 15.1 exceed those of all other satellites of Jupiter listed in Table 14.2. In addition, their bulk densities in Figure 15.2 decrease with increasing distance from the center of Jupiter starting at $3.530 \, g/cm^3$ for Io to $1.790 \, g/cm^3$ for Callisto compared to $3.34 \, g/cm^3$ for the Moon. The albedos of the Galilean satellites are exceptionally high ranging from 63% for Io to 17% for Callisto compared to only 11% for the Moon. As a result of their comparatively small masses, the escape velocities of the Galilean satellites are between $2.740 \, km/s$ for Ganymede and $2.040 \, km/s$ for Europa, all of which are similar to the escape velocity of the Moon ($2.380 \, km/s$). Although Io has a very tenuous atmosphere of sulfur dioxide (SO_2) released by virtually continuous volcanic eruptions, the surfaces of the other Galilean satellites, like the surface of the Moon, are exposed to the vacuum of interplanetary space.

The high albedo of Io (63%) is caused by the presence of light-colored sulfur and its compounds that cover the underlying basalt lava which continues to be erupted by the numerous active volcanoes that dot its surface. The high albedos of the other Galilean satellites compared to that of the Moon arise because their icy surfaces reflect sunlight much more efficiently than the basalt surface of the Moon. Nevertheless, we note that the albedos of Europa (64%), Ganymede (43%), and Callisto (17%) decrease with increasing distance from Jupiter. The apparent darkening of the ice surfaces of the

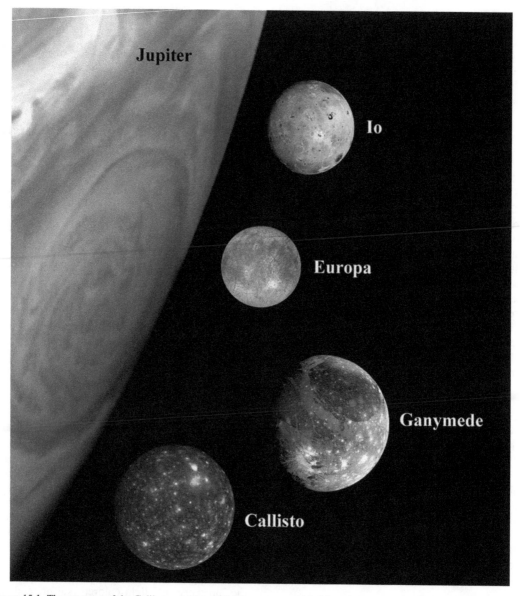

Figure 15.1. The montage of the Galilean satellites of Jupiter was assembled from images recorded by the spacecraft Galileo in 1996 (Io, Europa, and Ganymede) and by the Voyagers in 1979 (Callisto). Each of these satellites is a substantial world characterized by unique physical and chemical properties. Io has the largest number of active volcanoes, Europa has an ice crust underlain by an ocean of salty water, Ganymede is the largest and most massive satellite in the solar system, the interior of Callisto is undifferentiated and its crust is densely cratered. With the exception of Europa, all of the Galilean satellites are larger and more massive than the Moon of the Earth. (Courtesy of NASA)

Galilean satellites also occurs on the surfaces of other ice-covered satellites in the outer part of the solar system. The principal reasons for this phenomenon include:

1. The ice surfaces are covered with regolith resulting from meteorite impacts, by deposition of Interplanetary Dust Particles (IDPs), and by the accumulation of refractory

Table 15.1. Physical and orbital properties of the Galilean satellites and of the Moon (Hartmann, 2005)

Property	Io	Europa	Ganymede	Callisto	Moon
Physical Properties					
Diameter, km	3630	3130	5280	4840	3476
Mass, kg	8.89×10^{22}	4.79×10^{22}	1.48×10^{23}	1.08×10^{23}	7.35×10^{22}
Density, g/cm^3	3.530	3.030	1.930	1.790	3.34
Albedo, %	63	64	43	17	11
Escape velocity, km/s	2.560	2.040	2.740	2.420	2.380
Atmosphere	thin SO_2	none	none	none	none
Orbital Properties					
Semi-major axis, 10^3 km	422	671	1071	1884	384.4
Period of revolution, d	1.769	3.551	7.155	16.689	27.32
Period of rotation, d	1.769	3.551	7.155	16.689	27.32
Orbital eccentricity	0.000	0.000	0.002	0.008	0.055
Inclination of orbit to equat. plane of Jupiter	0.03°	0.46°	0.18°	0.25°	—

dust particles that originally formed in the solar nebula and are being released as the ice sublimates into the vacuum of interplanetary space.

2. The ice contains organic compounds (e.g., CH_4 and other hydrocarbons) that may disintegrate by exposure to ultraviolet radiation and energetic nuclear particles of the solar wind and of cosmic rays leaving a black residue of amorphous carbon and carbonaceous matter of high molecular weight.

Both processes described above can cause the ice surfaces of the Galilean satellites to darken with increasing exposure age as recorded by the number of meteorite-impact craters per unit area. This conjecture is supported by the observation that the crater density of Callisto (albedo = 17%) approaches saturation, whereas the surface of Europa (albedo = 64%) contains only a few craters and the surface of Ganymede (albedo = 43%) includes areas of dark and heavily cratered ice surrounded by light-colored and sparsely cratered terrain. We conclude from these comparisons that the surface of Europa is rejuvenated by an on-going process, which implies that it is still geologically active, whereas Callisto has not been rejuvenating its surface because it is geologically inactive. Once again, Ganymede is the satellite "in the middle" because it is less active than Europa but more active than Callisto. In spite of the differences in the exposure ages of the surfaces, all of the Galilean satellites formed 4.6 billion years ago at the same time as all of the planets of the solar system.

The bulk densities of the Galilean satellites in Figure 15.2 decrease smoothly with increasing

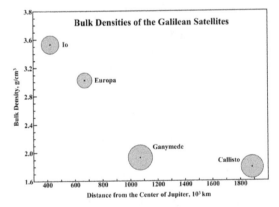

Figure 15.2. The bulk densities of the Galilean satellites decrease with increasing distance from the center of Jupiter. This relationship implies that the amount of water ice relative to rocky material increases similarly, which supports the hypothesis that these satellites formed in orbit around Jupiter from a disk of gas and dust. The diameters of the satellites are approximately to scale

distance from the center of Jupiter. In this regard, the Galilean satellites resemble the terrestrial planets whose bulk densities in Figure 3.6 also decrease with increasing distance from the Sun. The elevated but variable bulk densities imply that the Galilean satellites have different internal structures including iron-cores (Io, Europa, and Ganymede), and mantles composed of silicate minerals (Europa and Ganymede) as shown in Figure 15.3. In spite of these generalizations, the physical properties of each of the Galilean satellites distinguish it from its neighbors. For example, the mantle of Io appears to be molten, the upper mantle of Ganymede consists of a mixture of ice and rocky material, and the interior of Callisto is composed of an ice-rock mixture that has only partially differentiated (i.e, Callisto does not have an iron-rich core). In addition, the outer crust of Io consists of silicate rocks, whereas the crusts of Europa, Ganymede, and Calisto are composed of water ice and the ice crust of Europa is known to float on a global ocean of liquid water. Liquid water may also be present under the ice crusts of Ganymede and Callisto (Johnson, 1999).

The apparent internal homogeneity of Callisto is especially interesting because it indicates that this body has not been heated sufficiently to cause the refractory constituents to sink to the center and for the ice to segregate as it did in the interiors of Ganymede and Europa. Therefore, Callisto may have preserved the initial state of the other Galilean satellites after their formation by accretion of particles composed of ice and refractory materials. In contrast to Callisto, all of the other Galilean satellites evolved by internal differentiation in response to heat provided primarily by on-going tidal heating caused by Jupiter and by the decay of radioactive isotopes of U, Th, and K.

Io is a special case because it no longer contains water or other volatile compounds, except those of sulfur, because of strong heating by tidal friction. The resulting heat-flow to the surface of Io has been estimated at more than 2.5 Watts/m^2, which is about five times greater than the average global heat-flow of the Earth (0.06 Watts/m^2) and even exceeds the heat-flow at the active geothermal area at Wairakei in New Zealand (1.7 Watts/m^2). Evidently, the tidal heating of Io is quite sufficient to account for the

eruptions of basalt lava by the numerous active volcanoes on its surface. For the same reason, the continuing geological activity of Europa and Ganymede is attributable to tidal heating, in the absence of which these bodies would have long ago become inactive like the Moon.

The cores of Io, Europa, and Ganymede are described as being "iron-rich" in composition because the available data do not distinguish between metallic Fe-Ni alloys and sulfides of Fe, Ni, and other chalcophile metals. The composition of the core determines its density and hence its predicted radius. For example, if the core of Io is composed of iron sulfide, its radius is 52% of the planetary radius. If the core consists of metallic iron, its radius is only 36% of the planetary radius (1820 km) (Hartmann, 2005). In addition, questions remain whether the cores are liquid or solid. For example, Johnson (1999) suggested that the iron core of Ganymede may be partly molten. If the cores of Io, Europa, and Ganymede are at least partly liquid, they may permit electrical currents to induce magnetic fields. This matter will be addressed elsewhere in this chapter with respect to each of the Galilean satellites.

15.1.2 Celestial Mechanics

The lengths of the semi-major axes of the orbits of the Galilean satellites in Table 15.1 range from 422×10^3 km (Io) to 1884×10^3 km (Calisto), which means that Callisto is about 4.5 times farther on average from the center of Jupiter than Io. The radii (r) of the orbits of Io and Europa and of Ganymede and Europa are related by a factor of about 1.59 as shown below:

$$\frac{r(\text{Europa})}{r(\text{Io})} = \frac{671 \times 10^3}{422 \times 10^3} = 1.590$$

$$\frac{r(\text{Ganymede})}{r(\text{Europa})} = \frac{1071 \times 10^3}{671 \times 10^3} = 1.596$$

The value of the orbital radius-ratio for Callisto relative to Ganymede rises to 1.759. Nevertheless, these apparent regularities among the orbital radii of the Galilean satellites are the basis for similar regularities of their periods of revolution as required by Kepler's third law.

The data in Table 15.1 indicate that the periods of revolution (p) of the Galilean satellites are

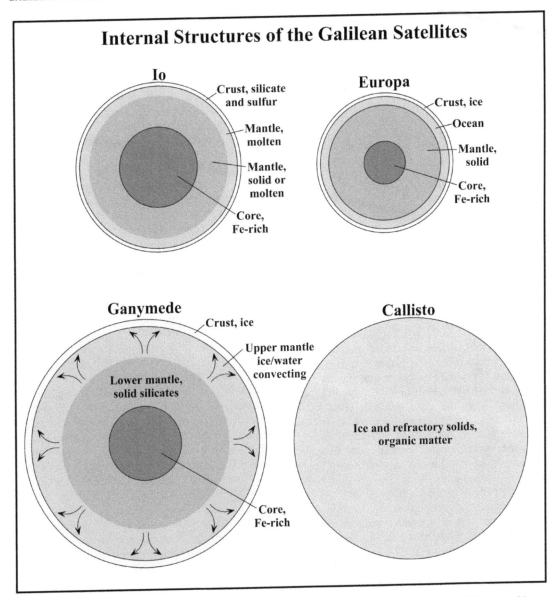

Figure 15.3. Internal structures of the Galilean satellites drawn to scale. Io, Europa, and Ganymede have differentiated into an Fe-rich core, a silicate mantle, and a crust composed of basalt and sulfur (Io) or ice (Europa and Ganymede), whereas Callisto is composed of an undifferentiated mixture of ice and solid particles of refractory compounds and metallic iron. The ice may also contain organic matter similar to that of carbonaceous chondrites. The accumulation of refractory dust particles and of organic matter imbedded in the ice and the deposition of interplanetary dust particles and meteoritic debris cause the surfaces of Europa, Ganymede, and Callisto to darken with increasing exposure age. Adapted from Hartmann (2005, Figure 8.28) with input from Johnson (1999) and Freedman and Kaufmann (2002)

surprisingly short and range from 1.769 days (Io) to 16.689 days (Callisto) compared to 27.32 days for our Moon (Science Brief 15.6.1). The ratios of the periods of neighboring satellites are very nearly whole numbers:

$$\frac{p(\text{Europa})}{p(\text{Io})} = \frac{3.551}{1.769} = 2.007$$

$$\frac{p(\text{Ganymede})}{p(\text{Europa})} = \frac{7.155}{3.441} = 2.014$$

$$\frac{p(\text{Callisto})}{p(\text{Ganymede})} = \frac{16.689}{7.155} = 2.332$$

These results tell us that the period of Europa is twice as long as the period of Io. Therefore, when Io and Europa start their revolutions when they are aligned along a straight line that passes through the center of Jupiter (i.e., when they are in conjunction), the two satellites come back into conjunction after two revolutions for Io and one for Europa:

Io: $1.769 \times 2.007 = 3.550$ days
Europa: $3.551 \times 1.000 = 3.551$ days

In other words, Io and Europa move into conjunction every 3.551 days at very nearly the same location in their orbits and therefore interact gravitationally on the same schedule. This kind of gravitational interaction is referred to as *resonance* and is specified by the fraction 1/2, where the numerator and denominator indicate the number of revolutions completed by two objects in the same length of time. (See also the Kirkwood gaps of asteroids in Section 13.3).

The periods of revolution of Europa and Ganymede are also in resonance because Europa completes 2.014 revolutions in the same time Ganymede completes one. The periodic gravitational interactions of the Galilean satellites at successive conjunctions affect the eccentricities of their orbits in ways that were elucidated by the calculations of the French mathematician *Pierre-Simon Laplace* in the early part of the 19th century. It turns out that the conjunctions of Io and Europa occur when Europa is at the aphelion of its orbit known also as the *apojove*. The conjunctions of Europa with Ganymede occur at the opposite end of Europa's orbit at a point called the *perijove* (i.e., perihelion). Consequently, when Europa and Ganymede are in conjunction on one side of Jupiter, Io is on the other side.

The eccentricities of the orbits of the Galilean satellites that result from the resonance of their periods of revolution cause their distances from the center of Jupiter to vary during each orbit. These variations in turn determine the amount of deformation of the satellites by the tidal forces exerted by Jupiter. The constant changes in the shapes of the Galilean satellites generates heat by internal friction, which explains why Io, Europa, and Ganymede continue to be geologically active. The differential tidal force decreases as the reciprocal of the cube of the distance (Hartmann, 2005, p. 52). Therefore, the amount of tidal heat generated in the Galilean satellites decreases steeply with increasing distance from Jupiter such that Callisto is virtually unaffected by tidal heating and has remained geologically inactive throughout the history of the solar system.

At this point we pause to consider that the eccentricities of the orbits of Io and Europa in Table 15.1 are listed as 0.00 even though we know that the heat produced by tidal friction in these satellites is caused by the eccentricity of their orbits. This discrepancy arises because the orbital eccentricities of the Galilean satellites are the sums of two parts known as the free and the forced components. The *free* component arises from a combination of the distance and velocity of a satellite after it formed in orbit around a planet. In accordance with Kepler's third law, the orbit of a satellite at a specified distance from the center of the planet is circular *only* in case the orbital velocity has exactly the required value. The *forced* component of the eccentricity arises from the gravitational interactions of two satellites whose orbital periods are in resonance. Although the magnitudes of the two components are constant, the direction of the long axis of the orbit corresponding to the forced eccentricity changes with time. Therefore, the free and forced eccentricities are added as though they are vectors. In case the directions are identical, the total eccentricity is the sum of the magnitudes of the free and the forced components. In case the forced component has a different direction than the free component, the total eccentricity is less than the sum of the magnitudes of the two components. The eccentricities of the Galilean satellites in Table 15.1 indicate only the magnitudes of the free component rather than the magnitudes of the actual eccentricities which vary continuously with time because of changes in the direction of the forced component (Greenberg, 2005, Section 4.3).

The data in Table 15.1 also indicate that the periods of rotation of all four Galilean satellites are equal to their respective periods of revolution. In other words, all four Galilean satellites have spin – orbit coupling of 1:1 just like the Moon does as it revolves around the Earth. Consequently the amount of heat generated by tidal friction in a Galilean satellite in Figure 15.4 depends only on the actual eccentricity and on the length of the semi-major axis of its orbit. The rate of rotation does not contribute to the heat production because the satellites do not appear to rotate when viewed from Jupiter.

The periods of revolution of all four Galilean satellites are longer than the period of rotation of Jupiter (9.84 h). Therefore, the gravitational

interactions between Jupiter and the Galilean satellites cause their orbital velocities to increase (Section 9.7.2). As a result, the period of rotation of Jupiter is rising (i.e., its rate of rotation is decreasing) and the lengths of the semi-major axes of the Galilean satellites are increasing (i.e., they are moving farther away from Jupiter). These changes in the relation between Jupiter and its Galilean satellites are based on theoretical considerations and have not actually been observed. In addition, we note that the planes of the orbits of the Galilean satellites are closely aligned with the equatorial plane of Jupiter, whereas the orbits of the two sets of the outer satellites are steeply inclined to the equatorial plane.

15.1.3 Origin: Alternative Hypotheses

The physical and chemical properties of the Galilean satellites and the celestial mechanics of their orbits, to the extent they are known to us, can be used to state alternative hypotheses concerning the origin and subsequent evolution of these satellites of Jupiter. The most plausible hypotheses rely on analogies with the origin of the Sun and the solar system.

The decrease of the densities of the Galilean satellites with increasing radii of their orbits implies that the abundance of water ice increases with distance from Jupiter and that the abundance of silicates of Fe, Ca, and Mg decreases. This kind of differentiation is *inconsistent* with the hypothesis that the Galilean satellites were captured by Jupiter in a sequence of random events. Instead, the systematic differences in the bulk chemical compositions of the Galilean satellites remind us of the differences in the chemical compositions of the terrestrial planets of the solar system which also vary systematically as a function of distance from the Sun (Section 3.5). Therefore, the Galilean satellites may have formed within a disk of gas and dust that was revolving around Jupiter. The dust particles consisted of water ice and refractory materials such as silicates and oxides as well as small grains of metallic iron or iron sulfide. The gas was composed of hydrogen and helium in approximately solar proportions. The Galilean satellites formed only by the accumulation of the solid particles because their gravitational

Tidal Heat Production in a Galilean Satellite

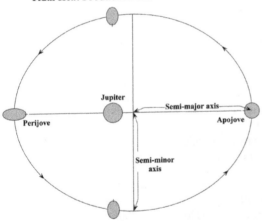

Figure 15.4. Heat generation in a Galilean satellite by tidal friction is caused by the deformation of its spherical shape depending on its distance from Jupiter. The maximum deformation occurs at perijove (perihelion) when the satellite is closest to Jupiter. The least deformation occurs at apojove (aphelion) when it is farthest from Jupiter. The extent of the deformation at perijove is greatest for Io and smallest for Callisto because the size of the tidal bulge decreases by a factory of $1/r^3$ with increasing distance (r) from the center of Jupiter. Consequently, Io is heated sufficiently to be the most volcanically active body in the solar system, whereas Callisto is virtually unaffected by tidal heating and has remained close to its initial state following its accretion. The tick marks on the satellite demonstrate 1:1 spin-orbit coupling

fields were not strong enough to attract and hold atmospheres composed of hydrogen and helium.

The present internal homogeneity of Callisto indicates that the accumulation of dust particles did not release enough energy to cause the satellites to melt. Instead, Io, Europa, and Ganymede were heated by tidal friction depending on their respective average distances from Jupiter. According to this hypothesis, Io lost virtually all of the water ice by melting and direct sublimation as a result of tidal heating, whereas Europa developed an ice crust that covers an ocean of liquid water. Ganymede also has an ice crust that rests on an undifferentiated mixture of ice and rocky material that is convecting in response to temperature-dependent differences in density.

Alternatively, ice crystals that existed in the revolving disk of gas and dust close to Jupiter may have been sublimated by heat radiated from the planet. This heat could have originated by the compression of gases in the atmosphere of Jupiter and from other sources (e.g., decay of radionuclides). In this case, Io may not have contained water or other volatile compounds at the time of its formation, except compounds of sulfur (e.g., FeS). The water vapor released by sublimation of ice particles in the vicinity of Jupiter may have condensed into ice in the colder regions of the disk outside of the orbit of Io and thereby increased the amount of water ice in Europa, Ganymede, and Callisto.

15.2 Io, the Hyperactive Princess

When Voyager 1 approached Io during its flyby in March of 1979, eight plumes were seen on its surface (Smith et al., 1979). When Voyager 2 passed by Io five months later in July 1979, most of these plumes were still active. When the spacecraft Galileo flew by Io again in 1996, about half of the volcanoes that were active in 1979 were still erupting and several others that had been dormant before were then active. The volcanic activity of Io in Figure 15.5 was a surprise because the mass and volume of Io in Table 15.1 are similar to those of our Moon. But instead of being geologically inactive like the Moon, Io has a larger number of active volcanoes than any other planet or satellite in the solar system. The volcanic activity on Io was actually predicted by Peale et al. (1979) in a paper that was published

only one week before the images beamed back to Earth by Voyager 1 confirmed the existence of volcanoes on Io. The authors of this paper based their prediction on the amount of tidal heat generated by the *forced eccentricity* of Io's orbit (0.004) caused by the resonance with Europa. In spite of having noon-time surface temperatures between $-148°$ and $-138°C$ and even though it should have lost most of the original heat it acquired at the time of its formation, Io's upper mantle consists of molten silicate because of the heat produced by tidal friction (Johnson, 1999).

15.2.1 Volcanoes

The images of Io returned by the Voyager spacecraft in 1979 and by the spacecraft Galileo (1995–2003) indicate that its surface contains more than 100 *volcanoes* or "hotspots" where lava flows are being erupted intermittently. A few of the most prominent volcanoes are identified in Table 15.2. The volcanoes are dispersed over the entire surface of Io rather than occurring only in certain kinds of tectonic settings as on the Earth (Faure, 2001). Io also lacks impact craters on its surface because the constant volcanic eruptions are covering them up as soon as they form.

The names of the volcanoes of Io are derived from deities of Volcanoes and Fire in the mythology of different cultures. For example, the large volcano *Pele* was named after the Hawaiian goddess of Fire, *Loki* was a one of the gods in Norse mythology who assisted the more powerful gods Thor and Odin with his clever but devious plans, and *Prometheus* of Greek mythology stole fire from the gods and gave it to humans (Houtzager, 2003).

The colorful surface of Io initially suggested that the volcanoes are erupting liquid sulfur. However, measurements by the spacecraft Galileo indicated that the lava flows are very hot with temperatures that range up to about 1750°C, which means that they are composed of molten silicates rather than of molten sulfur. The high temperature implies that the recently extruded lava flows on Io are composed of Mg-rich basalt similar to komatiite that was erupted during the Early Archean Era (3.8 to 3.5 Ga) on the Earth (Faure, 2001, p. 385). Rocks of this composition were first described in 1969 from the greenstone belts of the Barberton

Figure 15.5. The surface of Io is pockmarked by volcanoes many of which are presently active although some may be dormant. The volcanoes of Io are scattered randomly over its surface, which indicates that the volcanic activity is not a symptom of internal tectonic processes as it is on the Earth. The surface of Io also contains rugged mountains several kilometers in elevation, plateaus composed of layered rocks, and large calderas. However, impact craters are virtually absent because they are buried by lava flows and volcanic ash soon after they form. The surface of Io is colored by deposits of native sulfur that is erupted by the volcanoes together with lava flows of molten silicate at high temperature. Io has 1:1 spin-orbit coupling with Jupiter. The view in this picture is of the side of Io that faces away from Jupiter. (PIA 00583, project Galileo, courtesy of NASA/JPL-Caltech /University of Arizona/LPL)

Mountains in South Africa and are now known to occur in volcano-sedimentary complexes of Archean age on most continents. The high temperature of lava flows erupted on Io also indicates that the interior of Io is hotter than the interior of the Earth at the present time. In fact, recent interpretations illustrated in Figure 15.3 suggest that the crust of Io (100 km thick) is underlain by a magma ocean (800 km deep) in contrast to the mantle of the Earth which is a plastic solid rather

Table 15.2. Major volcanoes and pateras of Io mentioned in the text (Johnson, 1999)

Name of volcano	Location		Height, km	Width, km
	lat.	long.		
Amirani	25°N	116°	95	220
Prometheus	2°S	154°	75	270
Pillan patera[1]	12°S	244°	—	400
Pele	18°S	256°	400	1,200
Loki plume	18°N	303°	200?	400
Loki patera[1]	13°N	309°	—	—
Surt	45°N	338°	300?	1,200

[1] A patera is a volcano with a low profile on Mars and Io

than a liquid. The existence of a molten layer in the upper mantle of Io explains why the volcanoes on its surface are dispersed randomly over its surface instead of occurring above mantle plumes and long subduction zones as on the Earth (Faure, 2001).

The rate of production of lava, volcanic ash, and sulfur compounds on Io is sufficient to cover its entire surface with a layer about one meter thick in 100 years. Therefore, the absence of impact craters does not mean that Io is somehow shielded from solid objects that still populate interplanetary space. Instead, the impact craters that do form on the surface of Io are buried by the material erupted by the active volcanoes and the associated plumes.

Several prominent hotspots on the surface of Io in Figure 15.5 are surrounded by reddish haloes composed of native sulfur in the form of short chains consisting of three and four atoms (i.e., S_3 and S_4) (Spencer et al., 2000). These compounds polymerize to form the stable yellow form of sulfur which consists of eight atoms (S_8). The white material consists of sulfur dioxide snow that forms when the hot gas emitted by plumes crystallizes at the low temperatures of space and falls back to the surface of Io. The lava flows erupted within the calderas of active volcanoes are initially very bright but then turn black as they cool and form a solid crust.

15.2.2 Plumes

The hotspots on Io in Figure 15.6 are associated with plumes that form umbrella-shaped fountains up to 500 km high, aided by the low gravity of Io and by the absence of an atmosphere.

In order to reach such heights, the gas must be emitted at high speeds ranging from 1100 to 3600 km/h which far exceeds the speed of gas emissions by terrestrial volcanoes like Vesuvius, Krakatoa, and Mount St. Helens where gases have reached speeds of only 360 km/h (Freedman and Kaufmann, 2002, p. 307). The first plume was discovered on March 8, 1979, by Linda Morabito in an image recorded by Voyager 1, when the spacecraft approached Io to within 21,000 km of its surface (Matson and Blaney, 1999).

The plumes emit sulfur dioxide gas and varying amounts of particulates composed of solid sulfur dioxide, native sulfur, and silicate minerals derived from the walls of the vent. Accordingly, the deposits that accumulate around the plume vents are mixtures of native sulfur, sulfur dioxide snow, and volcanic ash. Some plumes may be active in the same location for many years while others are short-lived or may change location. In addition plumes may be invisible throughout their existence or become invisible at some time in their lifetime.

The processes that cause the emission of plumes on Io are illustrated in Figure 15.7 (a to d) based on a model developed by Susan Kieffer in 1982 and described by Matson and Blaney (1999). A deposit of sulfur-dioxide snow and/or native sulfur on the surface on the surface of Io in Figure 15.7 a & b is covered by a lava flow. The increase in temperature causes the sulfur dioxide to be vaporized and the resulting gas is discharged through fractures in the lava flow and along its edge as a plume. The plume produced in this way remains active only until the lava flow cools or until the sulfur deposit is exhausted. This mechanism is capable of producing several plumes that may be active at the same time or may form sequentially at different sites in case additional lava flows are extruded. The explosive emission of gas by this mechanism also causes the formation of "rootless" craters which have been observed on Iceland in places where hot lava has covered deposits of water ice (i.e., glaciers and snowfields).

Long-duration plumes are emitted from large subsurface reservoirs depicted in Figures 15.7 c & d. At first, a deposit of sulfur-dioxide snow and native sulfur on the surface of Io is buried deeply by volcanic pyroclastics and

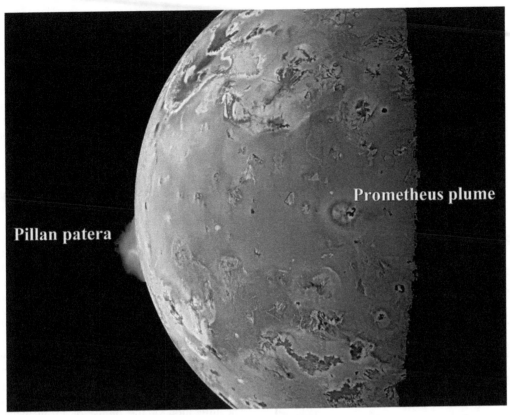

Figure 15.6. The volcanic plume visible along the left edge of Io in this image was erupted by Pillan patera. This plume is 140 km high and consists of sulfur dioxide (SO_2) and fine-grained particulates. A second plume, emitted by the volcano Prometheus, casts a reddish shadow on the surface of Io. The Prometheus plume was first seen in the images recorded by the Voyager spacecraft in 1979 and was still active 18 years later in 1997 when the spacecraft Galileo recorded this image at a distance of more than 600,000 km. (Image PIA 01081, project Galileo, courtesy of NASA/JPL-Caltech/ University of Arizona/LPL)

lava flows. The increase in pressure and temperature converts the solid sulfur-dioxide into a liquid. If the liquid sulfur dioxide is subsequently heated by magma from depth, the resulting gas pressurizes the reservoir which may fracture the roof and allow sulfur-dioxide gas to escape to the surface in the form of a long-lasting and stationary plume. In cases where the material that is emitted consist primarily of sulfur dioxide gas without a significant component of particulates, the resulting plumes are invisible and therefore have been referred to as "stealth" plumes (Matson and Blaney, 1999).

The model developed by Susan Kieffer can explain why some plumes are short-lived whereas others last for years. It can also explain why some plumes appear to migrate and why others are invisible. In addition, the model provides a plausible mechanism for burial of surface deposits of sulfur compounds and the resulting formation of crustal reservoirs of liquid sulfur dioxide and thereby provides a rational basis for the reconstruction of the sulfur-cycle of Io. (Kieffer et al., 2000).

15.2.3 Surface Features

Some of the largest volcanoes on Io (e.g., Prometheus, Loki, Pele, Tvashtar, Pillan, and others) have developed large calderas that contain lakes of cooling lava. In some cases, the lava lakes have formed a solid crust that

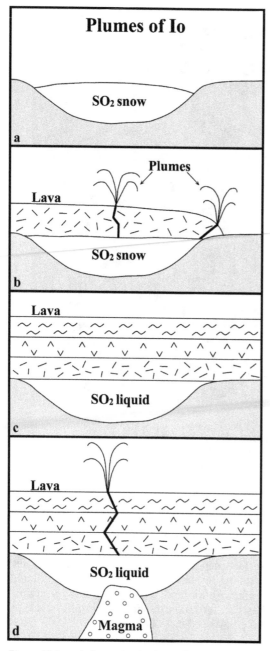

Plumes of Io

a. SO₂ snow

b. Plumes / Lava / SO₂ snow

c. Lava / SO₂ liquid

d. Lava / SO₂ liquid / Magma

insulates the hot lava beneath except along the edges where white-hot lava may be exposed (e.g., Tupan). The lava flows and deposits of sulfur compounds emitted by the volcano Culann patera are displayed in Figure 15.8.

The volcanoes Prometheus, Loki, and Pele are located near the equator of Io and are each associated with prominent plumes. The plume of *Pele* reaches a height of 400 km and has caused the deposition of a red ring of elemental sulfur (S_3 and S_4) with a radius of 700 km from the vent. In addition, the lava lake in the caldera of Pele exposes fresh lava along its shore for a distance of about 10 km. A similar phenomenon has been observed on a much smaller scale in the lava lake of Kilauea on the island of Hawaii.

Loki is the most powerful volcano in the solar system because it releases more heat than all of the volcanoes on Earth combined. The caldera of Loki is filled with lava that may have been erupted shortly before the flyby of Galileo on October 10 of 1999. On this occasion, the spacecraft also observed the plume of *Prometheus* which was first seen in the images returned by Voyager 1 in 1979. To everyone's surprise, the plume (75 km high) was now located about 100 km west of its location in 1979. The migration of the plume of Prometheus is explained by the eruption of a new lava flow onto a deposit of sulfur dioxide snow (Figure 15.7b).

The volcano *Tvashtar*, near the north pole of Io, consists of a series of nested calderas within which fountains of hot lava erupted to heights of 1500 m. Lava fountains on Io can reach such heights because of the low gravity and the low viscosity of the lava (e.g., like olive oil), which allows lava to flow for hundreds of kilometers (Lopes, 2002). For example, the lava emanating from the volcano *Amirani* has flowed up to 250 km. The plume of Tvashtar, like that of Pele, reaches a height of 400 km and is depositing

Figure 15.7. a. A depression in the surface of Io was filled by sulfur-dioxide snow (SO₂) that was erupted by a nearby volcanic plume. b. Part of the solid SO₂ snow is vaporized by the heat emanating from lava that covered the deposit. The resulting SO₂ gas escapes through fracture and/or along the edge of the flow to form one or several short-lived plumes. c. A deposit of solid SO₂ snow was buried by several layers of pyroclatics and lava flows. When their total thickness reached one to two kilometers,the pressure they exerted caused the solid SO₂ to liquefy. d. The resulting subsurface reservoir was intruded by silicate magma from depth and the increase in temperature caused the liquid to boil. The resulting SO₂ gas escapes through fractures in the overlying rocks to form a plume composed of gas and dust. Such plumes may be long-lasting but become invisible (i.e., they are stealth plumes) in cases where the gas that is discharged lacks particulates. Adapted from Figure 12 of Matson and Blaney (1999). Not drawn to scale

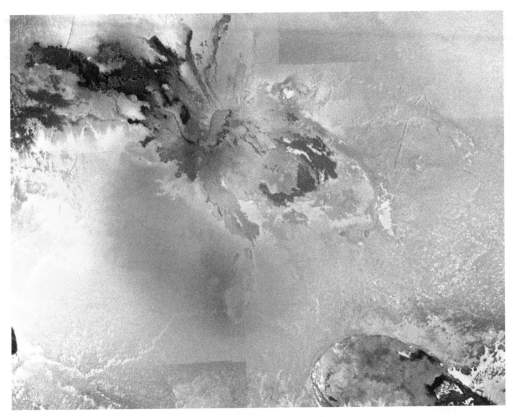

Figure 15.8. Culann patera is one of many active volcanoes on Io. In this image, lava has flowed out of a large caldera in all directions. The diffuse red deposit is native sulfur that was emitted by a plume. This image was recorded in 1999 by the spacecraft Galileo from a distance of 20,000 km. (Image PIA 02535, Galileo project, courtesy of NASA/JPL-Caltech/University of Arizona/LPL)

a reddish ring of sulfur on the area surrounding the vent.

The magnitude and vigor of the Tvashtar plume surprised investigators because all other large plumes occur in the equatorial region of Io. They were even more surprised when the Galileo spacecraft discovered a second large plume about 600 km southwest of Tvashtar during a flyby on August, 6, 2001, when it came within 194 km of the surface of Io. The Tvashtar plume did not even appear in the images taken during this flyby, presumably because it had become a stealth plume, although the Tvashtar volcano was still active. The new plume contained particles composed of sulfur dioxide that rose to a record height of 600 km, thereby exceeding the heights of the plumes of Tvashtar and Pele.

Another unexpected volcanic eruption was observed on February 21, 2002, by the Keck II telescope on the summit of Mauna Kea on the island of Hawaii (Marchis et al., 2002). Without warning, a large hotspot developed in two days near the site of the volcano *Surt* located at about 45°N latitude. The lava erupted at this site had an initial temperature of 1230°C and covered an area 1,900 square kilometers. The heat released by this eruption came close to equaling the amount of heat released by all other volcanoes on Io.

The image of the caldera of Tvashtar in Figure 15.9, recorded by the spacecraft Galileo on February 22, 2000, includes a flat-topped mesa adjacent to the main lava lake. The near-vertical sides of this mesa are scalloped suggesting that a form of mass wasting is

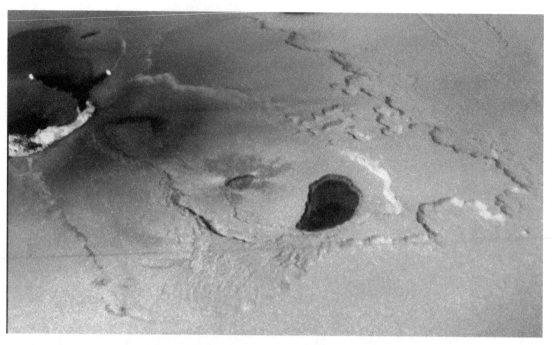

Figure 15.9. Complex of lava lakes in the caldera of the volcano Tvashtar recorded on February 22 of 2000 by the spacecraft Galileo. The caldera, located at 60 °N and 120 °W, was the site of a vigorous eruption in November of 1999 that included lava fountains. The caldera contains several lava lakes which have cooled and appear black in this image. The bright-colored spots are hot lava flows or cracks in the crust of cooling flows. The image indicates that changes are occurring in the topography of this area as a result of the eruption of lava flows and sulfur dioxide. (Image PIA 02550, Galileo project, courtesy of NASA/JPL-Caltech/University of Arizona/HiRise-LPL)

occurring. The erosion of the cliff may be a result of basal sapping caused by a liquid that is discharged by springs at the base of the cliff. The springs may be discharging liquid SO_2 from a subsurface reservoir (Figure 15.7d). The liquid vaporizes instantly in the vacuum at the surface of Io.

The close-up images of the surface of Io also reveal the presence of *mountains* that are not volcanoes but, nevertheless, reach an elevation of 16 km compared to only 8.85 km for Mt. Everest on the Earth. The mountains on Io appear to be the result of uplift of blocks of crust by convection currents in the underlying magma ocean of the upper mantle (McEwen, 1999). The uplifted blocks subside when the pattern of convection in the magma ocean changes and/or because of seismic activity in the crust, which triggers large landslides recorded by ridged debris aprons at the base of the mountains that are visible in images recorded during the Galileo flyby on November 25, 1999.

15.2.4 Interactions with Jupiter

Data collected during flybys of Jupiter by the Pioneer (1973/1974) and Voyager (1979) spacecraft and, most recently, by the Galileo mission (1995 to 2003) indicate that Io interacts significantly with the magnetic field of Jupiter. These interactions cause oscillations in the ambient magnetic field of Jupiter in the vicinity of Io and thereby obscure the evidence concerning the existence of a magnetic field on Io. Therefore, we do not yet know whether Io has a magnetic field (Johnson, 1999; Van Allen and Bagenal, 1999). However, we do know that the orbits of Io and of the other Galilean satellites lie well within the magnetosphere of Jupiter and that Io interacts with Jupiter not only gravitationally but also electromagnetically (Bagenal et al., 2004). For example, the decimetric radiation of Jupiter, mentioned in Section 14.1.1, originates from a toroidal (donut shaped) region that encircles the planet in a plane that dips about 10° to the equatorial plane of Jupiter. The inclination of the torus (donut) matches the tilt of the magnetic moment of Jupiter (9.6°) relative to its axis of rotation. The torus consists of a cloud sodium atoms, positively charged ions of sulfur (S^+) and oxygen (O^+), and electrons (e^-). The sulfur and oxygen ions originate from the decomposition and ionization of sulfur dioxide molecules that are injected into the magnetic field of Jupiter by the plumes of gas and dust discharged by Io. The resulting plasma (ionized gas) is transported all the way around Jupiter by its magnetic field which rotates with the planet at a rate of 9.84 hours per rotation. The electrons are accelerated to high velocities approaching the speed of light (i.e., relativistic velocities) and emit the observed decimetric synchrotron radiation as they move on spiral paths along magnetic field lines in the magnetosphere of Jupiter.

The large number-density of relativistic electrons in the magnetosphere of Jupiter damaged several transistor circuits of the Pioneer 10 and 11 spacecraft as they flew by Jupiter in 1973 and 1974. The dose of radiation they absorbed during their flybys was a thousand times greater than the dose that causes severe radiation sickness or death in humans (Van Allen and Bagenal, 1999, p. 47). The potential for damage of the spacecraft caused by the intense ionizing radiation in the magnetosphere of Jupiter prevented the spacecraft Galileo from attempting close flybys of Io until after the other objectives of the mission had been accomplished.

Measurements by Voyager 1 indicated that the plasma in the torus contains thousands of ions and electrons per cubic centimeter and that the kinetic energies of the ions correspond to a temperature of about 1×10^6 K. In addition, the data reveal that about one tonne of matter per second is being removed from the surface of Io by the magnetic field of Jupiter as it continuously sweeps past the satellite. The bursts of decametric radiation emitted by Jupiter originate from very large electrical currents in the form of current sheets (flux tubes) that connect the plasma of the torus to the ionosphere of Jupiter. The voltage between these terminals is of the order of 400,000 volts and the power generated by the electrical current has been estimated at 2×10^{12} watts.

The energetic plasma derived from Io causes the magnetosphere of Jupiter to expand like a balloon, especially in its equatorial region where the magnetic field of Jupiter is comparatively weak (i.e., 14 times the strength of magnetic field at the equator of the Earth). The magnetosphere is further distorted by the centrifugal force arising from the rapid rotation of Jupiter, which forces the plasma outward into a disk that is aligned with the equatorial plane of the planet. The existence of the so-called *plasma sheet* was directly observed by the Pioneer spacecraft; however, the mechanisms whereby the plasma is heated during its expansion within the magnetosphere of Jupiter is not yet understood but may be related to the rapid rotation of the planet.

The ions and electrons of the plasma that inflate the magnetosphere eventually return to the torus and then cascade into Jupiter's upper atmosphere along lines of magnetic force. The large amount of energy released by this process (10 to 100 trillion watts) energizes the atmosphere in the polar regions of Jupiter and increases its temperature. In addition, this process generates spectacular *auroral displays* of ultraviolet light in the polar regions of Jupiter (Van Allen and Bagenal, 1999, p. 49).

Auroral displays have also been observed in the polar regions of Io when energetic electrons, traveling at high speed along magnetic field lines, collide with molecules in the diffuse atmosphere of Io that is maintained by the eruptions of volcanoes and plumes on its surface. The auroras of Io display a range of colors including bright blue in the equatorial region, red in the polar regions, and green in the night side. The blue glow has been attributed to collisions of energetic electrons with molecules of sulfur dioxide, whereas the red and green auroras are caused by the de-excitation of oxygen and nitrogen atoms, respectively. The auroral displays become weaker when Io passes through the shadow of Jupiter because the resulting decrease of the atmospheric temperature causes the atmospheric gases to condense, thus decreasing the abundance of target atoms and molecules (Geissler et al., 2001).

15.2.5 Summary

Io is the first of the Galilean satellites of Jupiter which dominates it by means of tidal and electromagnetic interactions. Although its orbit appears to be circular, the resonance Io shares with Europa gives rise to a component of forced eccentricity, which enables Jupiter to generate heat within Io by the tidal distortion of its shape. Consequently, Io is the most active body in the solar system with at least 100 volcanoes, several of which are erupting at any given time. The temperature of the lavas erupted by the volcanoes of Io (up to 1750 °C) is higher than the temperature of basalt lava erupted by volcanoes on the Earth. This observation implies that the ionian lavas are similar in composition to Mg-rich komatiites that formed on the Earth during the Early Archean Era by high degrees of partial melting of ultramafic rocks in the mantle of the Earth.

Io formed in orbit around Jupiter by accretion of solid particles in a disk of gas and dust that revolved around the planet at the time of its formation. During this process, ice particles may have evaporated as a result of heat radiated by Jupiter. Alternatively, water that may have been incorporated into Io at the time of its formation later evaporated because of heat generated by tidal friction. In any case, Io has been depleted of water in contrast to the other Galilean satellites which still have thick crusts of ice.

After its formation, Io differentiated into an iron-rich core and a silicate mantle the upper part of which appears to be molten. This reservoir of magma is the source of lava erupted by volcanoes which are randomly distributed over the surface of Io in contrast to the Earth where volcanoes occur in certain well-defined tectonic settings.

The data recorded by the spacecraft Galileo and its predecessors do not indicate unequivocally whether the core of Io is liquid or solid. In any case, the fluctuations of the magnetic field of Jupiter in the vicinity of Io may be caused by electromagnetic phenomena and do not necessarily prove whether or not Io has a magnetic field.

The volcanic activity of Io is accompanied by plumes of sulfur dioxide gas and associated condensation products that rise to great height above the surface. The sulfur dioxide gas condenses in the coldness of space to form snowflakes that are deposited around the vent together with particles of native sulfur and volcanic ash. These deposits account for the coloration of the surface of Io ranging from red to yellow (sulfur) to white (sulfur dioxide snow). Some of the plumes originate when deposits of sulfur dioxide snow are covered by hot lava flows. In cases where deposits of snow are buried deeply (i.e., one to two kilometers), the deposits of solid sulfur dioxide are transformed into subsurface reservoirs of liquid sulfur dioxide. When such a reservoir on Io is heated by the intrusion of hot silicate magma from below, the resulting superheated vapor is expelled in a second kind of plume that resembles a geyser similar to Old Faithful in Yellowstone Park, except that liquid sulfur dioxides takes the place of water.

Io as well as the other Galilean satellites exist within the magnetosphere of Jupiter. Therefore, some of the sulfur dioxide molecules and atoms of other elements are injected into the magnetosphere of Jupiter by the eruption of volcanic plumes on Io. The sulfur dioxide molecules are broken up and the resulting atoms of sulfur and oxygen are converted into positively charged ions by removal of electrons. The resulting energetic plasma is carried away by the magnetic field of Jupiter that continually sweeps past Io as the planet rotates with a period of 9.84 hours. For this reason, Io revolves within a torus (donut) of plasma that surrounds Jupiter. The torus expands to form a plasma sheet close to the equatorial plane of Jupiter. The electromagnetic interactions between Io and Jupiter also cause the emission of the decametric and decimetric radiation and the auroral light displays in the polar regions of Io and Jupiter.

15.3 Europa, the Ice Princess

When the Voyager spacecraft sent back postcards of the Galilean satellites in 1979, Europa appeared to be a cold but serene world whose icy surface in Figure 15.10 gleamed in the reflected light of the Sun. Although first impressions can be misleading, Europa has turned out to be one of the most interesting worlds of the solar system (Greenberg, 2005; Harland, 2000; Kargle et al., 2000; Fischer, 2001; Greeley, 1999).

Europa is the smallest of the Galilean satellites with a radius of 1565 km compared to 1820 km for Io and 2640 km for Ganymede (Hartmann, 2005). Its physical and orbital properties in Table 15.1 indicate that Europa is smaller than

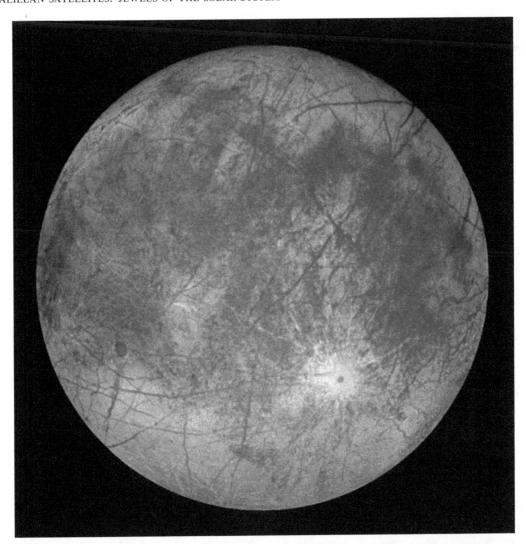

Figure 15.10. Europa is the smallest of the Galilean satellites with a diameter of 3130 km compared to 3476 km for the diameter of the Moon of the Earth. The crust of Europa is composed of water ice and is criss-crossed by innumerable fractures that appear as dark curved lines in this image. The brown discoloration of the surface if caused by deposits of various kinds of salts, meteorite debris, interplanetary dust particles, and organic matter that was embedded in the ice and/or was brought to the surface by water from the ocean beneath the crust. The black spot surrounded by rays of white ejecta is the impact crater Pwyll (diameter = 50 km). Although several other impact craters have been identified, Europa is only sparsely cratered because of the on-going rejuvenation of its surface by the extrusion of water through fractures and melt-throughs. This image shows the trailing hemisphere of Europa which rotates only once during each revolution. (Image PIA 00502, Galileo project, courtesy of NASA/JPL-Caltech. The image was processed by Deutsche Forschungsanstalt für Luft und Raumfahrt, Berlin, Germany)

the Moon and also has a lower mass (4.79×10^{22} kg) and a lower bulk density ($3.030 \, \text{g/cm}^3$). However, the albedo of Europa (64%) is far higher than that of the Moon (11%) which is consistent with the icy surface of Europa. In fact, only a few other ice-covered satellites have higher albedos than Europa (e.g., the satellites of Saturn: Enceladus 95% and Calypso 80%; as well as Triton 75%, satellite of Neptune). The escape velocity of Europa is 2.04 km/s compared to 2.38 km/s for the Moon, 11.2 km/s for the Earth, and 59.5 km/s for Jupiter. The proximity of Europa to Jupiter will become a problem for spacecraft landing on its surface and taking off again during future exploratory missions. Europa does not have an atmosphere and its calculated average surface temperature is −170 °C, whereas the measured temperatures range from −148 °C at daytime to −188 °C at night (Hartmann, 2005, p. 295).

15.3.1 Ice Crust and Ocean

The crust of Europa is composed of pure water ice and contains only a few impact craters because they are covered almost as rapidly as they form. The on-going resurfacing indicates that Europa is still geologically active and that the tidal heat has allowed the lower part of the ice crust of Europa to remain liquid (Pappalardo et al., 1999). This global ocean acts as a reservoir from which liquid water can be extruded to the surface where it freezes instantly to form "lava flows" of water ice. Evidently, on Europa liquid water takes the place of silicate melts in a process called *cryovolcanism* (Greeley, 1999). Several other ice-covered satellites of Jupiter, Saturn, Uranus, and Neptune are also geologically active because of cryovolcanic activity based on melting and freezing or vaporization and condensation of certain volatile compounds that melt or vaporize at very low temperature (e.g., water, methane, nitrogen, etc.).

The topography of the surface of Europa in Figure 15.11 is dominated by sets of intersecting *double ridge systems* that consist of a narrow central valley (crustal fracture) flanked by continuous and low double ridges. The intersections of the double ridges can be used to reconstruct their sequence of formation because the youngest double ridge cuts across all previously formed ridge systems. The fractures are interpreted to form as a consequence of the tidal deformation of the ice crust. The ridges and valleys virtually saturate the surface to form *ridged terrain*.

The ridges and other kinds of topographic features on the surface of Europa are a few hundred meters in elevation. Therefore, the surface of Europa is not as smooth as it appeared in the images returned by the Voyagers.

Instead, the surface is actually quite chaotic and hummocky. There are only a few areas on Europa that are flat and smooth enough to be suitable for the safe landing of a spacecraft. In marked contrast to Io, the surface of Europa does not contain volcanoes and plumes of sulfur dioxide nor are there any fault-block mountains.

In the absence of volcanic eruptions by means of which Io and other geologically active bodies (e.g., the Earth) transport heat from their interiors to the surface, Europa presumably loses heat by convection of water from the bottom of the ocean to the water-ice interface and by conduction through the ice crust. Therefore, the thickness of the crust depends inversely on the amount of heat that needs to be transported to the surface in order to maintain thermal equilibrium. In other words, the more heat needs to be conducted to the surface, the thinner is the crust (Greenberg, 2005). Although the actual thickness of the ice crust of Europa is not yet known, it may vary regionally in case the heat flow from the silicate mantle to the ocean is channeled by hotsprings or by volcanic activity.

The central valleys between the double ridges that criss-cross the surface of Europa contain reddish brown deposits that also cover part of the face of Europa that is never directly exposed to Jupiter (McCord et al., 1998). This material is probably brought to the surface by the ocean water that rises to the surface through fractures in the ice crust. Accordingly, the reddish brown deposits may be composed of salts that form when the ocean water sublimates leaving a residue of *cryogenic evaporites* such as combinations of sulfates, chlorides, and hydroxides of the major metals dissolved in the water (e.g., Na, Mg, K, Ca, Fe, etc.). In addition, the cryogenic evaporites may contain biogenic organic matter provided that organisms actually live in the ocean of Europa. The surface deposits of Europa may also contain accumulations of meteoritic debris and interplanetary dust particles (IDPs).

Although the multitude of intersecting fractures and the associated double ridges have formed in response to the continuous flexing of the ice crust of Europa, the response of the underlying ocean to the fracturing of the crust is still speculative. We assume for the sake of this presentation that the ocean and the overlying ice crust of Europa have interacted with each

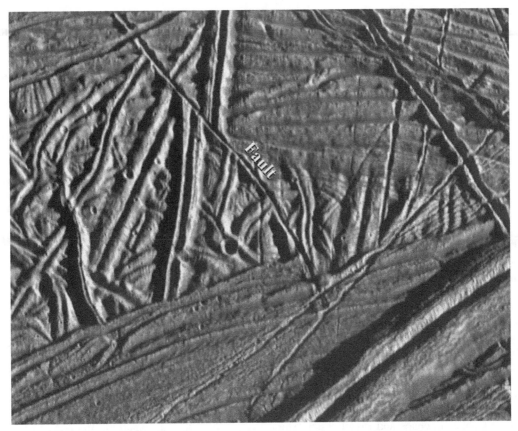

Figure 15.11. The surface of Europa is covered with intersecting sets of double ridges separated by narrow valleys. These ridges form as a result of tidal deformation of the crust of Europa by a process that is not yet clearly understood. The ridged terrain in this image is cut by a right lateral strike-slip fault located at 15 °N and 273 °W. The prominent ridge at the lower right is about 2.5 km wide and is one of the youngest features in this area. (Image PIA 00849, Galileo project, courtesy of NASA/JPL-Caltech/Arizona State University)

other and continue to do so at the present time. If this is true, then the double ridges and certain other surface features visible from space can be attributed to such interactions.

Although several alternative explanations for the formation of the double ridges have been proposed, none is entirely satisfactory. In general, the double ridges have been explained either as pressure ridges, like those that form in the Arctic Ocean of the Earth, or as products of cryovolcanic activity involving the forceful extrusion of geysers of ocean water through tidal cracks followed by condensation of the water vapor to form ice pellets that accumulate on both sides of the tidal fracture.

The model supported by Greenberg (2005) is a variant of the pressure-ridge model in which

water rises to the "float line" of a tidal crack (i.e., about 90% of the thickness of the ice crust). The water boils and freezes in the cold vacuum of space to form a layer of porous and mechanically weak ice that may reach a thickness of about 50 cm in a matter of hours. When the crack begins to close as Europa continues in its orbit, the ice in the crack is crushed. The resulting exposure of water from below increase the thickness of the volume of crushed ice. As the crack closes, some of the crushed ice and interstitial water are squeezed out of the crack and form a single continuous ridge extending along the length of the fracture. As Europa again approaches the perijove of its orbit, the crack reopens, thereby splitting the ice ridge in such a way that deposits of the crushed and frothy ice remain on opposite sides of the crack. This process continues to build up the double-ridges until the fracture eventually becomes inactive as new fractures open elsewhere on Europa or until the crust shifts to change the orientation of the tidal fractures. Greenberg (2005) estimated that this process could form ridges that are about 1 km wide and 100 m high in less than 30,000 years. These hypothetical explanations of the formation of double-ridges need to be tested by direct observation of human observers on the ground or by means of time-lapse photography by robotic explorers.

The intersections of certain prominent fractures in Figure 15.12 provide useful points of reference for other topographic features on the surface of Europa. For example, two east-west trending fractures (*Udaeus and Minos*) intersect in the northern hemisphere of Europa (at 46.5 °N, 216 °W) and form an oblique angle between them. The *Conamara* chaos is located at 12 °N and 273 °W about 1500 km southwest of the Udaeus-Minos cross. About 1000 km south of Conamara is the site of a major impact crater called *Pwyll* (25 °S, 271 °W). A third topographic feature called the *Wedges* region (0° to 20 °S, 180° to 230 °W) is located south of the equator starting about 500 km east of a line drawn between Conamara chaos and Pwyll crater. The Wedges region occupies an area of about 1.16×10^5 km^2 in which the fractures in the crust of Europa are curved and wedge-shaped. Another cross formed by two prominent fractures occurs at 13 °N and 273 °W just north of the Conamara chaos and helps to located that important feature.

15.3.2 Impact Craters

The scarcity of impact craters has led to the conclusion that the surface of Europa is less than 50 million years olds (Zahnle et al., 2003) even though this body formed 4.6 billion years ago during the time of formation of the solar system. Most of the impact craters that did form during the long history of this body have been buried by lava flows composed of ice or have been destroyed by the breakthrough of ocean water to the surface. In addition, craters on the surface of Europa fade by creep and by sublimation of the ice exposed at the surface. The scarcity of impact craters distinguishes Europa from most other

ice-covered satellites of Jupiter, Saturn, Uranus, and Neptune, although Titan (Saturn) and Triton (Neptune) also are only lightly cratered. The craters on Europa are thought to have formed primarily by impacts of comets (Zahnle et al., 2003). However, asteroids, or fragments of asteroids, could also have impacted on Europa (Greenberg, 2005, p. 268).

An important issue in the study of Europa concerns the thickness of the ice that overlies the ocean (Schenk, 2002). Some authors have concluded that the ice is "thick", whereas others maintain that it is "thin". If the ice is thick, the ocean of Europa would be isolated from the surface and topographic features such as the double ridges and chaotic terrain would be the products of the dynamics of the ice crust. In contrast, if the ice is thin, the geomorphic features on the surface of Europa are caused by interactions of the ocean and the ice crust.

The question concerning the thickness of the ice crust of Europa has been investigated by studies of the morphologies of impact craters. Some of the best preserved impact craters on the surface of Europa are identified by name in Table 15.3. Most of these are multi-ring basins (e.g., Tyre, Callanish, Tegid, and Taliesin). The crater Cilix contains a central uplift (or chaotic terrain), whereas *Pwyll* in Figure 15.13 is a recently-formed rayed crater, the ejecta of which were sprayed for more than 1000 km up to and across the Conamara chaos. The multi-ring basin named *Tyre* (33 °N, 147 W) is located on the northern hemisphere of Europa east of the Udaeus-Minos cross and close to the Minos fracture. The diameter of this spectacular feature is about 125 km, which makes Tyre one of the two largest surviving impact craters on Europa. The other one is the multi-ring basin *Callanish* (23 °N, 335 °W) which is located opposite to Tyre on the other side of Europa and the diameter of its outermost ring is similar to that of Tyre. The crater *Amergin* is interesting because its interior appears to contain chaotic terrain similar to a nearby chaos located only about 60 km southeast of the crater. Finally, the crater *Manannán* (3 °N, 240 °W) also contains hummocky terrain that closely resembles an adjacent chaos. In addition to the large impact scars listed in Table 15.3, the surface of Europa contains a large number of small bowl-shaped

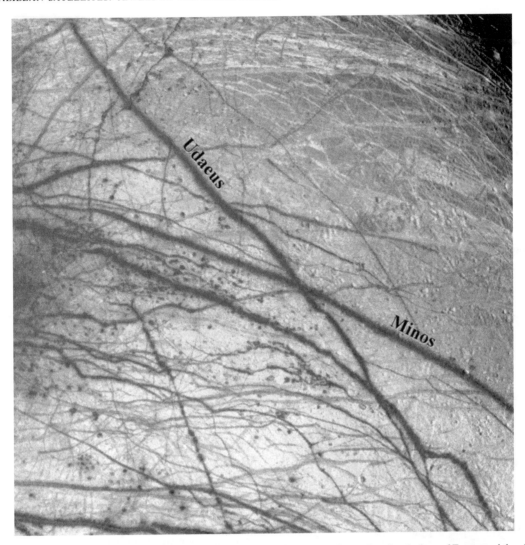

Figure 15.12. Two prominent global fractures (Udaeus and Minos) intersect in the northern hemisphere of Europa and thereby provide a prominent topographic marker. These and the other fractures in this image contain the reddish brown deposit mentioned in the caption to Figure 15.10. The color of the ice ranges from various shades of blue to nearly white based on the size of ice crystals. The region south of the Udaeus-Minos cross is covered with small brown spots or "freckles" (lenticulae). These turn out to be areas of chaotic ice when viewed at higher magnification. The surface of the ice north and south of the Minos fracture is pockmarked with small secondary impact craters that formed when blocks of ice, ejected by the impacts of meteorites and/or comets, fell back to the surface of Europa. (Image PIA 00275, Galileo project, courtesy of NASA/JPL-Caltech/University of Arizona/PIRL)

craters of secondary origin (i.e., they formed by impact of ejecta).

Chaotic terrain is not restricted to the craters Manannán and Amergin but can also be seen inside the ring basins of Tyre and Callanish. The presence of chaos implies that the large impact craters and basins actually penetrated the crust, which allowed ocean water to rise into them and to cause large rafts of crustal ice to be rotated and shifted out of their original positions until the water froze around them to form the rough-textured matrix. Another important property which the impact craters of Europa share with patches of chaotic terrain is that

Table 15.3. Impact craters on Europa (Greenberg, 2005)

Name	Location	Diameter km	Description
Pwyll	25°S, 272°W	24	rayed crater
Callanish	23°S, 335°W	~125	multi-ring
Tegid	1°N, 164°W	~20	multi-ring
Taliesin	?	~20	multi-ring
Tyre	33°N, 147°W	125	multi-ring
Cilix	3°N, 182°W	~50	central uplift or chaotic terrain
Manannán	3°N, 240°W	21	chaotic terrain
Amergin	14°S, 230°W	19	chaotic terrain

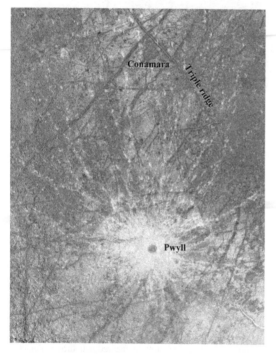

Figure 15.13. The dark spot in this image is the impact crater Pwyll. The rays of white ejecta that surround the crater overlie all older topographic features indicating that the impact that formed Pwyll occurred more recently than the local topography. The rays of ejecta appear to be composed of clean water ice and extend for more than 1000 km from the impact site up to and beyond the Conamara chaos located north of the crater and south of the cross of two prominent tidal fractures. These fractures consist of two dark parallel ridges separated by a light-colored band and are sometimes referred to as "triple ridges." The surface of the ice in this image is discolored by the reddish brown deposits and contains numerous "freckles" caused by the presence of chaotic ice. (Image PIA 01211, Galileo project, courtesy of NASA/JPL-Caltech/PIRL/University of Arizona)

they are both discolored by the presence of the reddish brown surface deposits like those that occur in the central valleys of double ridges. These deposits are attributable to the presence of ocean water at the surface, which implies that the impact craters that contain the reddish brown deposit were flooded by ocean water that entered through deep fractures that penetrated the crust.

For these and other reasons, Greenberg (2005) favored the interpretation that the ice crust of Europa is less than 30 km thick. A lower limit to the thickness of the ice crust is provided by the small bow-shaped craters having diameters less than about 5 km. These craters did not break through the ice and therefore indicate that the thickness of the crust of Europa is more than 5 km. Consequently, the study of impact craters leads to the conclusion that the thickness of the crust of Europa is more than 5 km but less than 30 km. Actually, the thickness of the crust of Europa may vary regionally depending on the amount of heat transported locally by oceanic convection currents to the underside of the crust. Therefore, the variations of the thickness of the crust of Europa may reflect the dynamics of the ocean and the amount of heat escaping locally from the silicate mantle at the bottom of the ocean.

15.3.3 Chaotic Terrains

High resolution images obtained during flybys of the spacecraft Galileo reveal that areas of chaotic terrain are a common geomorphic feature on the surface of Europa. The areas consisting of chaos range in area from about $12, 800 \text{ km}^2 (80 \times 160 \text{ km})$ for *Conamara* (Figure 15.14) to small reddish-brown spots called *lenticulae* (freckles) or *maculae* (Figure 15.12). The designation of small reddish brown spots on the surface of Europa as lenticulae or maculae originated at a time when only low-resolution images were available. High-resolution images of these spots obtained more recently by the spacecraft Galileo reveal that the reddish brown spots consists of chaotic terrain that is not adequately described by names that make sense only at low resolution. All of the examples of chaotic terrain are discolored presumably because they contain the reddish brown deposits that indicate the former presence of ocean water on the surface (Section 15.3.1),

Figure 15.14. The Conamara chaos covers an area of about 35 × 50 km centered on 9 °N and 274 °W (Figure 15.13). It consists of angular blocks of ridged ice that are imbedded in hummocky ice. Several of the blocks (icebergs) have been rotated while others have been displaced from the surrounding crustal ice by widening fractures. Reconstructions of the Conamara chaos demonstrate that much ice evaporated during the formation of this feature. (Image PIA 01403, Galileo project, courtesy of NASA/JPL-Caltech/Arizona State University)

either as a result of melt-through from below or as a result of break-through from above. After a chaotic terrain (e.g, Conemara) or a large impact crater (e.g., Callanish) solidified by freezing of the extruded ocean water, tidal cracks cut through or around the remaining rafts of the original surface because tidal cracks form continuously, even at the present time, whereas melt-throughs and break-throughs occur episodically.

Examples of chaotic terrain typically contain angular blocks of ridged ice that have been displaced and rotated relative to adjacent blocks. These "icebergs" are surrounded by rubbly ice that contains smaller chunks of "brash ice" in a frozen matrix that formed when exposed ocean water boiled and froze in the cold vacuum of space. The Conamara chaos in Figure 15.14 has been partially reconstructed by re-aligning blocks of ice by means of prominent double ridges and by fitting their shapes together like pieces of a jigsaw puzzle. The reconstructions recreate part of the previously existing surface but also

demonstrate that much crustal ice was crushed or vaporized such that large gaps remain (e.g., Figure 15.1b and 16.3 of Greenberg, 2005). Two other noteworthy examples of chaotic terrain are *Thera* (47 °S, 181 °W) and *Thrace* (41° to 48 °S, 170° to 175 °W).

The origin of chaotic terrain has been variously attributed to:

1. Upwelling of ice diapirs with diameters of about 10 km from within a "thick" crust.
2. Extrusion of water-ice slurries through the crust to form dome-shaped lobate deposits on the surface (e.g., the *Mitten*).
3. Local melting of "thin" crust of ice by rising plumes of comparatively warm water of the underlying ocean.

Although each of the three hypotheses has been advocated by certain groups of investigators, the third alternative illustrated in Figure 15.15 requires the smallest number of special assumptions and therefore seems capable of explaining the formation of chaotic topography on Europa in the most plausible way.

The melt-through hypothesis relies on the reasonable assumptions that the ice crust of

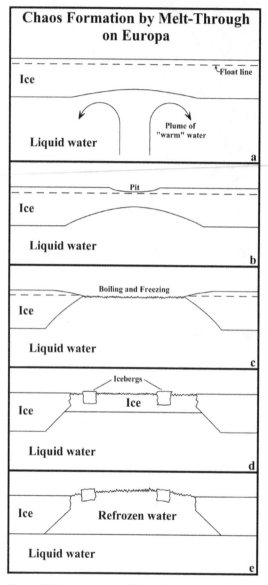

Europa is "thin" (i.e., < 30 km as indicated by crater morphology), that plumes of "warm" water form in the ocean as a result of heat emanating locally from the silicate mantle at the bottom of the ocean (e.g., by hotsprings that discharge ocean water circulating through the upper mantle), and that the ice crust is in buoyant equilibrium with the ocean (i.e., the water in the ocean is not pressurized). The panels (a to e) of Figure 15.15 illustrate the stages of a hypothetical process as result of which chaotic terrain on the surface of Europa can form by melt-through of the ice crust from below . In addition, panels d and e demonstrate that icebergs can break off the edges of the meltwater pool and drift away from their original sites as the water boils and freezes into the rubbly matrix that characterizes chaotic terrain on Europa (e.g., Conamara, the Mitten, Thera, Thrace, and many other examples).

The process of chaos formation by melt-through can also take place in modified form when the ice crust of Europa is fractured locally by the impact of a comet or asteroid. In this case, the resulting impact cavity would fill with ocean water up to the floatline (Figure 15.5c). Rafts of surface ice as well as fragments of crushed ice would drift in the pool until the boiling water freezes around them to form the chaotic surface that is revealed in high-resolution images of impact basins (e.g., Tyre and Callanish). Note that the heat released by an impact on the surface of Europa cannot cause the ice to melt because the ambient pressure is far below the triple point of water (Figure 12.5). Therefore, the ice in the target area is vaporized and the resulting water vapor may either drift away into interplanetary space or recondenses in the form of snowflakes or ice pellets that are then deposited on the surface surrounding the impact site.

The hypothesis that the chaotic terrain is the result of convection of warm ice assumes that the crust of Europa is more than 30 km thick. This

Figure 15.15. a. A plume of "warm" water rises from the bottom of the ocean of Europa and begins to melt the underside of the ice crust above it. The float line is the level to which water rises if the ice crust is in buoyant equilibrium with the underlying ocean. b. Continued melting forms a large cavity on the underside of the ice crust and causes a shallow pit to form on the surface c. The melt cavity expands to the surface and liquid ocean-water boils and freezes to form a thin layer of rubbly matrix ice d. Icebergs break off from the edge of the ice and float in the water as the boiling water freezes around them e. When the plume of warm water subsides, the water that filled the melt cavity in the ice crust refreezes. The surface displays the characteristic properties of chaotic terrain and bulges slightly to accommodate the increase in volume of the refrozen water. Adapted from Figures 16.13 and 16.14 of Greenberg (2005). Not drawn to scale

assumption is not consistent with the morphology of impact craters. The alternative hypothesis that lava composed of water is erupted onto the surface of Europa requires that the ocean is pressurized, perhaps by the tidal distortion of its shape. In that case, the daily opening and closing of tidal fractures should cause cryovolcanic activity in the form of geysers and other kinds of eruptions, but none have been seen in the images of the surface of Europa. Nevertheless, localized melting of the ice crust by warm water can be viewed as a cryovolcanic process in the sense that the warm water that melts the overlying crust is analogous to silicate magma on the Earth intruding the solid crust composed of rocks. When the water (magma) reaches the surface, it cools and freezes (crystallizes) to form solid ice (monomineralic rock). As a result of the exposure of liquid water at the surface, heat from the interior of Europa escapes into space thereby helping to maintain thermal equilibrium.

15.3.4 Life in the Ocean

The view that is emerging from the study of Europa is that it has a thin ice crust ($\sim 10\,km$) which covers a global ocean of liquid water containing dissolved salts. The temperature of the water at the base of the crust depends on the salinity of the water which determines its freezing temperature. For example, seawater on the Earth generally freezes at $-2.2\,°C$. The temperature of the water probably increases with depth, especially close to the bottom of the ocean. The underlying silicate mantle is likely to be a source of heat that locally warms the bottom water, especially where ocean water that has circulated through the upper mantle is discharged by hotsprings. More vigorous volcanic activity, such as eruptions of silicate lava and hot gases, is not indicated by any evidence visible at the surface of Europa. Regardless of whether the crust of Europa is "thick " or "thin", the amount of sunlight that reaches the ocean is very small, except perhaps in the vicinity of melt-throughs or where strike-slip faulting is occurring along tidal fractures (e.g., *Astypalaea* linea, Greenberg 2005, Chapter 12). The ice in these areas is likely to be thin and therefore may allow more sunlight to enter the ocean.

In summary, the ocean of Europa can be described by certain of its properties that can be postulated by reasonable interpretations of existing observations:

1. The ocean consists of liquid water.
2. The water contains salts of common elements derived by chemical weathering of the silicate, oxide, and sulfide minerals in the underlying mantle.
3. The temperature of the water increases with depth and is everywhere above the freezing point of the salty water in the ocean.
4. Heat is entering the ocean through the bottom, perhaps by discharge of hotsprings. Eruption of lava and hot gases is not indicated by presently available evidence.
5. Very little, if any, sunlight enters the ocean except under special short-lived circumstances.
6. Carbon compounds such as carbon monoxide (CO), carbon dioxide (CO_2), methane (CH_4), and complex organic molecules are assumed to be present in the water because of their occurrence in the solar nebula.
7. Minerals whose surfaces contain electrically charged sites exist in the sediment at the bottom of the ocean and in suspension.
8. The water in the ocean is continually stirred by tidal deformation and by thermal convection.

The environmental conditions described above are thought to be favorable for the development of life, although the process by means of which self-replicating cells form is not understood. Nevertheless, these favorable conditions have existed in the ocean of Europa ever since its interior differentiated, and these conditions have remained stable without significant disruption for more than four billion years right up to the present time. Evidently, the environment in the ocean of Europa has been conducive to life for a much longer time than the environment on the surface of Mars (Section 12.8). Even if living organisms did once exist on Mars, there was hardly enough time for evolution to diversify these organisms before increasing aridity and decreasing global temperature caused their extinction or forced them into a long-lasting state of hibernation.

The environmental conditions in the ocean of Europa are at least as favorable to life as the conditions in the primordial ocean of the Earth where life did originate and did evolve in spite of episodes of catastrophic impacts and climate fluctuations (e.g., global glaciations). Therefore, the ocean of Europa may be inhabited by living organisms that have diversified by evolution in order to take advantage of different habitats available to them. Although this prediction is not unreasonable, the existence of life in the ocean of Europa must be confirmed by landing a robotic spacecraft on its surface. Even the reddish brown material that has accumulated on the surface may contain evidence concerning life in the ocean of Europa.

If a future mission to Europa is undertaken, great care will be required to assure that Europa is not contaminated by terrestrial bacteria and viruses that may cause a pandemic among any indigenous organism that may be present in the ocean. Similarly, if any samples from Europa are to be returned to Earth, such materials must be quarantined to prevent alien organisms from infecting humans and/or the rest of the terrestrial biosphere (Sagan and Coleman, 1965; NRC, 2000; Greenberg and Tufts, 2001).

15.3.5 Summary

Europa is the smallest of the Galilean satellites and differs from Io by having an ice crust that covers an ocean of salty water. In addition, Europa has a magnetic field that is induced by electrical currents which arise in the ocean as a result of interactions with the magnetic field of Jupiter. Like Io, Europa is heated by tidal friction caused by the deformation of its shape as it revolves around the Jupiter. The amount of heat released has prevented the water in the ocean from freezing, but is not sufficient to cause volcanic activity.

The ice crust of Europa is virtually covered by intersecting double ridges separated by median valleys that occupy the sites of deep fractures caused by tidal stress. These fractures contain a record of the tidal tectonics that has affected the crust of Europa throughout its history. The double ridges form by freezing of ocean water that rises into the tidal fractures as they open and close at regular intervals.

The surface of Europa is partly covered by reddish brown deposits composed to salts of the major elements that are dissolved in the ocean water. Therefore, the presence of these cryogenic evaporites implies that ocean water does reach the surface of Europa. In addition, the salt deposits may contain organic matter related to the presence of living organisms in the ocean.

Europa has only a small number of impact craters on its surface because they are covered up or destroyed by the continuing geological activity. The largest craters, Tyre and Callanish, are both multi-ring basins about 250 km in diameter. The interiors of most craters with diameters of 20 km or more indicate that ocean water rose into them from below and floated blocks of ice from the rims and floors of the craters thereby creating chaotic terrain. The morphology of the impact craters indicates that the thickness of the crust is about 10 km on average.

Areas of chaotic terrain, ranging in diameter from 160 km (Conamara) to less than 10 km, are widely distributed over the surface of Europa. All of these areas consist of rafts of crustal ice that have been dislodged and rotated by floating in ocean water. In addition, the rubbly ice that surrounds the icebergs is discolored by reddish brown salt deposits, which caused small brown spots on the surface to be referred to as lenticulae (freckles) or maculae (reddish yellow body in the center of the retina) in low-resolution images.

The most plausible explanation for the origin of chaotic terrain is by melt-through of the ice crust by plumes of warm water rising from the floor of the ocean. The end result is that ice-free areas form on the surface of Europa where ocean water is exposed to the cold vacuum of space for short periods of time. The water boils as it freezes to form the matrix of rubbly ice that is characteristic of chaotic terrain.

The environmental conditions in the ocean of Europa permit aquatic life to exist. In addition, these conditions have probably remained stable for about four billion years, which may have allowed these organisms to diversify by evolution to take advantage of all existing habitats. The possibility that the ocean of Europa is inhabited will be investigated in the future by landing one or several robotic spacecraft on its surface. Great care will be required to sterilize the landers and to avoid contaminating the Earth with alien organism that could harm terrestrial plants and animals.

15.4 Ganymede in the Middle

Ganymede, the Galilean satellite, has properties that are intermediate between those of active Europa and those of inactive Callisto. It has the distinction of being the largest of the Galilean satellites in Table 15.1 in terms of volume and mass. In fact, Ganymede is larger than the planets Mercury and Pluto but does not quite measure up to Mars. Among the numerous satellites of Jupiter, Saturn, Uranus, and Neptune only Titan comes close to the volume and mass of Ganymede. . These comparisons drive home the point that Ganymede is a world that is not to be overlooked among the many satellites of the giant planets.

15.4.1 Ice Crust

The images of Ganymede in Figure 15.16 reveals the presence of areas of dark and heavily cratered ice that appear to have been disrupted by bands of light-colored grooved ice. Crater counts confirm that the dark surface is older than four billion years and that the light-colored bands are younger than the dark areas, but nevertheless have exposure ages of the same order of magnitude (Hartmann, 2005, p. 263). The largest area of dark ice, called *Galileo regio*, has a diameter of about 3200 km. Other large areas of dark ice are *Marius* regio, *Perrine* regio, *Barnard* regio, and *Nicholson* regio. Dark ice also occurs in smaller angular areas scattered over the surface of Ganymede. In Figure 15.16 dark ice is cut by narrow bands of the younger light-colored ice that either forced the fragments of old ice to move apart or, more likely, were superimposed on the dark ice. In a global view of Ganymede, the dark areas are the remnants of the original crust that was broken up or rejuvenated, leaving angular remnants of the old ice separated by bands of younger ice.

The diameters of the impact craters that occur both on the dark and the light-colored surfaces of Ganymede range from less than 5 km to more than 100 km. Small craters (5 to 20 km) have central uplifts, raised rims, and recognizable ejecta blankets. Larger craters (20 to 100 km) have a central pit that may be characteristic of craters formed by impacts into the ice crust of Ganymede. Craters having diameters of more than 60 km contain a dome within the central pit. The largest craters have relaxed their original shapes as a result of the creep of the ice. For this reason, they have been called "*palimpsests*" defined as "a reused writing surface on which the original text has been erased" (Hartmann, 2005, p. 302). Large craters having diameters of 100 km relax in about 30 million years, whereas 10-km craters require 30 billion years. The craters that formed by recent impacts on dark as well as light-colored surfaces are surrounded by light-colored and rayed ejecta blankets. This observation is confirmed by the image of Khensu crater in Figure 15.17, which indicates that the dark surfaces on Ganymede have formed on crustal ice that was originally light-colored. The darkening of the ice with increasing exposure age is attributable to the accumulation of sediment such as meteorite debris and interplanetary dust, refractory particles released by sublimation of the crustal ice, and carbon residue derived by decomposition of carbon-bearing compounds by cosmic rays, the solar wind, and ultraviolet light (Section 15.1a).

The large palimpsests on Ganymede are multi-ring basins with diameters of several hundred kilometers (e.g., *Memphis* facula, 350 km). (The word "facula" is also used to describe bright granular structures in the photosphere of the Sun). The large impactors that formed these basins did not break through the ice crust of Ganymede as they did on Europa. Therefore, the ice crust of Ganymede is considerably thicker than the crust of Europa, presumably because the tides of Jupiter generate less heat on Ganymede than they do on Europa. Another indication that Ganymede has a thicker crust than Europa is the absence on Ganymede of chaotic terrain and of the reddish brown salt deposits (Pappalardo, 1999). In addition, Ganymede does not have active or inactive volcanoes, plumes of gas, or geysers of boiling water.

15.4.2 Rejuvenation

The existence of the bright areas implies that parts of the surface of Ganymede were rejuvenated a few hundred million years after its accretion and internal differentiation. The question is: How did it happen? The light-colored bands of ice in Figure 15.18 contain sets of

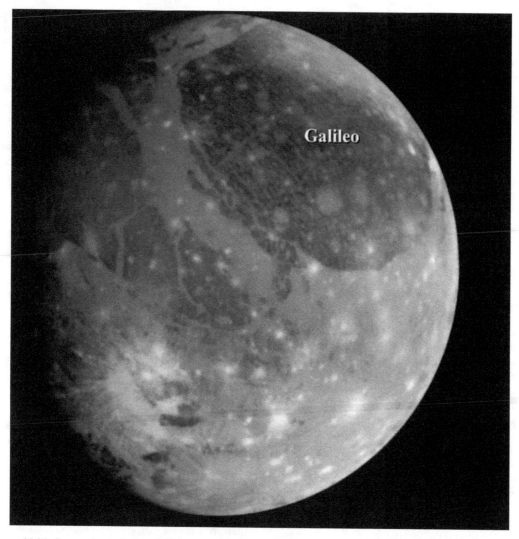

Figure 15.16. Ganymede, the largest Galilean satellite, has a thick crust composed of water ice. The crust contains areas of dark-colored and heavily cratered ice (e.g., Galileo) in contact with areas of light-colored ice that separate what appear to be remnants of the older dark-colored crust. Numerous bright spots in this image are recently-formed impact craters and the associated ejecta. Even though Ganymede is the largest and most massive satellite in the solar system, it does not have an atmosphere and it is no longer geologically active. (Image PIA 00716, Galileo project, courtesy of NASA/JPL-Caltech)

parallel ridges and valleys giving the appearance that the terrain is "grooved". The valleys between the ridges differ from those on Europa because they do not appear to be presently active tidal fractures. Instead, the sets of ridges and valleys on Ganymede may be normal faults (i.e., horsts and grabens) or tilted blocks of the ice crust containing parallel normal faults that are differentially eroded to form the valley-and-ridge topography (Pappalardo, 1999).

Images returned by the Voyagers and by Galileo show that bands of grooved terrain cut across other bands thereby establishing a time series based on the principle of cross-cutting relationships. Some of the bands of grooved terrain of Ganymede are also cut by strike-slip faults. The evident structural complexity of the grooved terrain suggests that it formed during an episode of violent deformation of the ice crust in the early history of Ganymede. The results

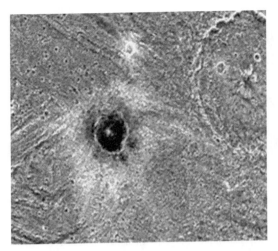

Figure 15.17. The impact crater Khenzu, located at 2°N and 153°W on Ganymede, has a diameter of about 13 km. The surrounding terrain is part of the light-colored Uruk sulcus. The crater has a central uplift projecting upward from a dark floor. Nevertheless, the ejecta blanket consists of white (i.e., clean) ice indicating that the surface of the ice crust of Ganymede has darkened with age. The dark material covering the floor of Khenzu may be a residue of the meteorite that formed the crater. A second crater named El appears in the upper right corner of the image. This crater (diameter = 54 km) is older than Khenzu and has a small pit in the center. Such pits are characteristic of some impact craters on Ganymede. (Image PIA 01090, Galileo project, courtesy of NASA/JPL-Caltech/Brown University)

of the deformation may have been enhanced by the thermal gradient within the thick ice layer. Warm ice at depth could have been soft and plastic, whereas the cold ice at the surface was hart and brittle. Consequently, warm ice from the "lower" crust could have been extruded through the brittle "upper" crust, thereby giving rise to a kind of cryovolcanic activity. However, the available images of Ganymede do not show much evidence of recent cryovolcanic activity (Pappalardo, 1999).

15.4.3 A Conjecture

An alternative explanation for the origin of the grooved terrain on the surface of Ganymede arises from the observation by the Galileo spacecraft that Ganymede has a weak internally-generated magnetic field. According to the theory of the self-exciting dynamo (Section 6.4.5), planetary magnetic fields require the presence

of an electrically conducting fluid which must be stirred by thermal convection and/or by the rotation of the body (e.g., metallic hydrogen in Jupiter). Therefore, the existence of a magnetic field implies either that Ganymede has a molten iron-rich core like the Earth or that a layer of salty water exists beneath its ice crust as in the case of Europa (Figure 15.3). If the magnetic field of Ganymede is actually generated within a layer of salty water, then its crust could have been rejuvenated by the extrusion of ice that formed by freezing of water in tidal fractures as appears to be happening on Europa even at the present time. In order for this process to work on Ganymede, its crust must have been significantly thinner in the past than it is at the present time.

The thickness of the ice crust of Ganymede may have been thinner during its early history (i.e., before four billion years ago) because of the presence of excess heat in its interior arising from compression and the decay of radioactive atoms. The resulting high rate of heat flow caused the crust to be thinner and the layer of water to be thicker than they are at the present time. Consequently, the rejuvenation of the surface of Ganymede occurred during its early history but it gradually ended as the water began to freeze to the underside of the crust causing it to thicken until all geological activity was suppressed. The present thickness of the crust of Ganymede may correspond to the amount of heat that continues to be generated by tidal friction. Accordingly, Ganymede has a thicker ice crust than Europa because the tides of Jupiter are generating less heat in Ganymede than they generate in Europa. The geological histories of the individual Galilean satellites depend primarily on the amounts of tidal heat that are generated within them. The effectiveness of the tides to generate heat in the Galilean satellites decreases with increasing distance from Jupiter. Hence, Ganymede is the satellite in the middle: less active than Europa but more active than Callisto.

15.4.4 Summary

Ganymede is the largest satellite in the solar system exceeding even the planets Mercury and Pluto in volume and mass. Its ice crust contains large remnants of an original crust whose surface

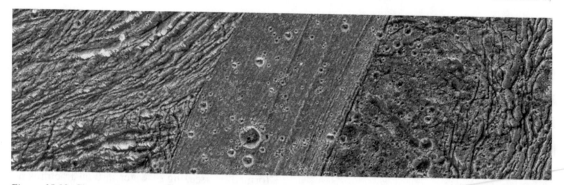

Figure 15.18. Close-up of the surface of Ganymede showing a band of light-colored and grooved ice traversing older and highly corrugated ice. The grooved band is a product of rejuvenation of the surface of Ganymede during an early period in its history when it contained more heat than it does at present. The ejecta deposits of the impact craters in this image have darkened with age. Although the grooved terrain of Ganymede shows some resemblance to the intersecting ridges of Europa, the origin of the topography of Ganymede is not yet understood (Image PIA 02572, Galileo project, courtesy of NASA/JPL-Caltech/DLR (Berlin and Brown University)

has darkened with age, which distinguishes these remnants from younger rejuvenated areas that have a light-colored surface.

Both kinds of surfaces bear the scars of impacts by comets and small asteroids that have left a record of craters ranging in diameter from less than 5 km to more than 100 km. The most recent craters in the dark areas are surrounded by light-colored ejecta blankets indicating that the ice itself is white and clean and that the surface of the dark areas is covered by sediment such as meteorite debris, interplanetary dust, residual sublimation products, and carbon-rich residues that have accumulated for close to 4.6 billion years (Section 15.1). These kinds of deposits are less evident in the light-colored areas because they have been exposed for a shorter interval of time.

The light-colored areas are heavily grooved by sets of parallel ridges separated by valleys. The bands of grooved ice intersect and cross-cut each other indicating that they formed sequentially. The valley-and-ridge topography does not appear to be forming at the present time, presumably because the ice crust of Ganymede has become too thick to be fractured by tides raised by Jupiter. The exposure age of the grooved terrain indicates that it formed a few hundred million years after the accretion and internal differentiation of Ganymede at a time when its ice crust may have been thinner.

The presence of a layer of salty water beneath the crust is suggested by the fact that Ganymede has a weak internally-generated magnetic field. At the time when the grooved terrain rejuvenated parts of the ancient crust, the layer of liquid water may have been thicker and the overlying crust may have been correspondingly thinner than they are at the present time. According to this conjecture, the crust began to thicken when the original ocean started to freeze as Ganymede cooled until its internal heat declined to the level sustained by tidal friction.

15.5 Callisto, Left Out in the Cold

Callisto was a princess who was seduced by Zeus, the chief God of Greek mythology. His wife punished Zeus for his infidelity by transforming Callisto into a bear. Years later, Zeus placed Callisto among the stars as the constellation Ursa Major (Big Dipper, Figure 2.2) and her son Arcas became Ursa Minor (Little Dipper). However, Hera arranged that Ursa Major could never bathe in the water of the Mediterranean Sea (Section 2.1). This tale illustrates the way in which the mythology of the Greeks explained why the constellation Ursa Major is visible in the sky at night from all parts of the Mediterranean region and can therefore be used for navigation by identifying the position of the pole star (Houtzager, 2003).

15.5.1 Physical Properties

The Galilean satellite Callisto in Figure 15.19 does not participate in the orbital resonances and jovian tides that have been so important in the evolution of the other satellites in this group (Section 15.1.2). Callisto is excluded from these phenomena because of the great distance between it and Jupiter (1884×10^3 km). In other words, Callisto was left out in the cold of interplanetary space. Callisto is almost twice as far from Jupiter as Ganymede and almost five times farther than Io (Table 15.1). Its period of revolution (16.689d) is more than

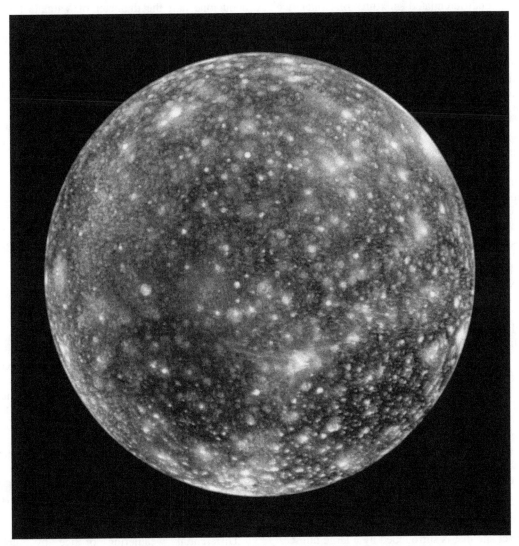

Figure 15.19. Callisto is so far removed from Jupiter that it is not heated by tidal friction and has cooled without interruption since it formed 4.6 billion years ago. As a result, the surface of Callisto is densely covered by impact craters and its interior is undifferentiated (i.e., Callisto lacks a core). As in the case of Ganymede, the surface of Callisto is covered by dark-colored deposits that consist of a mixture of meteorite debris, interplanetary dust particles, and organic matter that has formed by polymerization of methane that was imbedded in the ice. The light-colored ejecta of recent impact craters confirm that the ice beneath the dark surface of Callisto is clean water ice. (Image PIA 03456, Galileo project, courtesy of NASA/JPL/Caltech/DLR, Germany)

two times longer than the period of Ganymede (7.155 d) and almost ten times longer than the period of revolution of Io (1.769d). The diameter of Callisto (4840 km) is smaller than that of Ganymede (5280 km) but larger than those of Io (3630 km) and Europa (3130 km). Callisto has the lowest bulk density $(1.790\,g/cm^3)$ of the Galilean satellites (Figure 15.1) and is composed of an undifferentiated mixture of water ice (density $\sim 1.00\,g/cm^3$) and refractory silicate and oxide particles (density $\sim 3.00\,g/cm^3$). Accordingly, the body of Callisto consists of about 60% ice by volume and 40% refractory particles (Science Brief 15.6.2). Callisto is undifferentiated because the tides of Jupiter, the impacts of planetesimals at the time of accretion, and the decay of radioactive atoms did not release enough heat to cause the ice to separate from the rocky component in its interior.

The surface of Callisto consists of water ice overlain in most places by sediment similar in composition to that which covers the dark areas of Ganymede (Section 15.4.1). However, the low albedo of Callisto (17%) compared to that of Ganymede (43%) indicates that the layer of sediment on the surface of Callisto is probably thicker and more continuous than the sediment layer on the surface of Ganymede. The low albedo of Callisto also indicates that the surface has not been rejuvenated, which means that Callisto has not been active even during its earliest history.

15.5.2 Impact Craters

The absence of cryovolcanism or internal tectonic activity has allowed impact craters to be preserved on the surface of Callisto including large multi-ring basins. The crater density in Figure 15.20 is at saturation, which means that the exposure age of the surface is close to 4.6×10^9 year (Hartmann, 2005). The largest crater on Callisto is a multi-ring basin called *Valhalla* (Science Brief 15.6.3). It consists of a central impact site that is surrounded by a large number of rings spaced about 50 km apart. The rings consist of undulations in the surface formed by energy spreading outward from the impact site with a maximum radius of about 1500 km. The central crater of Valhalla, located at 11°N, 58°W (Beatty and Chaikin, 1990), is a palimpsest

because it was partially erased by plastic flow of the ice which filled the original impact cavity and thereby restored the curvature of Callisto. The center of Valhalla is a light-colored area having a diameter of about 600 km where comparatively clean ice is exposed. A second multi-ring basin in Figure 15.21, located at about 30°N and 140°W, called *Asgard* (Science Brief 15.6.3) has only about one half the diameter of Valhalla but is better preserved.

The color of ejecta blankets that surround recent impacts is white, which confirms that the ice in the outermost crust of Callisto does not contain enough dust particles to darken its color. However, the ice of Callisto, as well as the ice of Europa and Ganymede, may contain carbon-bearing gases including carbon dioxide and methane.

15.5.3 Differential Sublimation

The surface of Callisto contains isolated pinnacles in smooth areas of ice. The resulting knobby landscape in Figure 15.22 appears to be a product of differential sublimation of ice that contains impurities which locally retard the rate of sublimation. The existence of this landscape implies that sublimation of ice is a more effective weathering agent on Callisto than it seems to be on Europa and Ganymede.

When ice is exposed to a vacuum at a temperature close to, but less than, the freezing point of water, the ice sublimates rapidly and leaves behind a residue of the particles and salts it contained. This process, known as freeze drying, is used in industry to remove water, as for example in the manufacture of "instant" coffee. However, the rate of sublimation of ice in a vacuum decreases with decreasing temperature and becomes ineffective at the low temperatures that prevail on the ice-covered satellites of the outer planets. This explains why these ice-covered satellites have survived for billions of years without losing their ice crusts.

Differential weathering of ice by sublimation may be more effective on Callisto than on Europa because the maximum measured noontime temperature of Callisto is 28° warmer than on Europa (i.e., Callisto: −120°C.; Europa: −148°C; Hartmann, 2005). The reason for the comparatively high surface temperature of

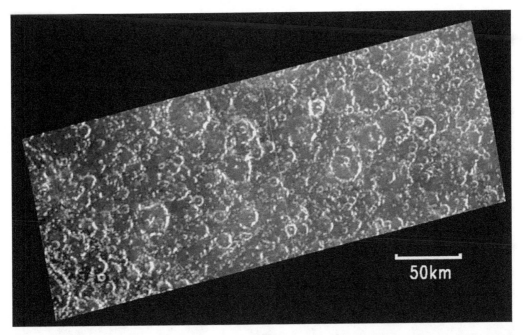

Figure 15.20. The surface of Callisto is completely covered by impact craters packed "shoulder-to shoulder." The area shown here is antipodal to the giant impact crater Valhalla. Nevertheless, the expected hummocky or ridged terrain is *absent*, which implies that the seismic energy from the Valhalla impact was absorbed in the interior of Callisto. This evidence favors the view that Callisto contains liquid water in its interior. A similar conjecture arises from measurements of the magnetic field of Jupiter in the vicinity of Callisto. However, the hypothesis that liquid water is present in the interior of Callisto (and Ganymede) requires confirmation (Image PIA 02593, Galileo project, courtesy of NASA/JPL-Caltech/University of Arizona/HiRise-LPL)

Callisto is that it absorbs more sunlight than Europa (i.e., the albedo of Callisto is lower than that of Europa). The rate at which ice is lost from the surface of Callisto may also be affected locally by the thickness of sediment that covers it. As a result, residual ridges and pinnacles form where the sublimation of ice is retarded. In addition, small impact craters on the surface of Callisto may be erased by sublimation of the ice.

15.5.4 Summary

The gravitational interactions of Callisto with the spacecraft Galileo indicate that the most remote of the Galilean satellites did *not* differentiate internally into a dense iron-rich core, a silicate mantle, and an ice crust. The failure to differentiate tells us that Callisto was not heated sufficiently to allow dense materials (i.e., silicates, oxides, and grains of metallic iron) to sink toward the center thereby concentrating water ice into

an outer shell like the ice crusts of Europa and Ganymede.

One of several reasons why Callisto remained cool is that tidal heating is ineffective because of the great distance between Jupiter and the orbit of Callisto. In addition, Callisto apparently accreted from small particles of ice and dust whose impact did not release enough heat to melt the ice. The other Galilean satellites may have formed similarly without melting and their internal differentiation may have been caused primarily by heat generated by the tides of Jupiter rather than by the impacts of planetesimals or by the decay of radioactive atoms. Nevertheless, after Callisto had accreted without melting, it was hit by comets and asteroids which saturated its surface with craters.

The diameters of craters on the surface of Callisto range from about one to several hundred kilometers. The largest impact crater is the Valhalla multi-ring basin. The rings that surround the impact site are undulations in the surface of

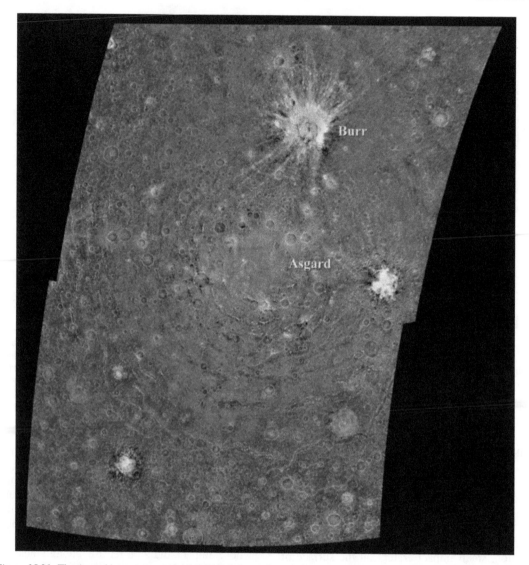

Figure 15.21. The Asgard impact structure of Callisto is located approximately at 30 °N and 142 °W. The structure which has a diameter of about 1700 km consists of a bright central zone surrounded by discontinuous rings. The original impact basin has been filled in by creep of the ice and the rings were formed by the transmission of the impact energy through the ice. The bright-rayed crater Burr north of the center of Asgard was formed during a relatively recent impact and confirms that the ice of Callisto is virtually clean and white. (Image PIA 00517, Galileo project, courtesy NASA/JPL-Caltech/DLR, Germany)

the ice crust that are spaced about 50 km part and have radii up to about 1500 km. A second, but smaller, multi-ring basin is called Asgard.

The surface of Callisto is covered by a layer of sediment composed of debris of various kinds of impactors and of residual dust released by sublimation of ice exposed at the surface. This sediment may be similar in origin to the sediment that covers the areas of old ice on Ganymede but differs from the reddish brown sediment on Europa, which contains salts derived from the ocean beneath the crust.

The landscape of Callisto shows evidence of differential sublimation of ice that results in the development of ridges and pinnacles and causes the shape of local ice plains to be concave up.

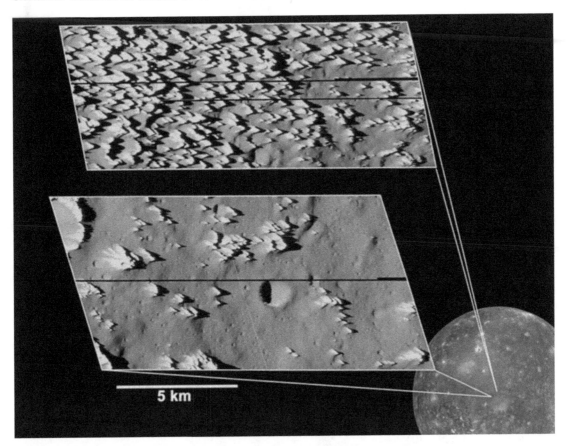

Figure 15.22. An area located south of the Asgard impact basin on Callisto contains clusters of pinnacles that appear to have formed by differential sublimation of ice. The pinnacles are between 80 to 100 m tall and may consist of ejecta from nearby impacts that occurred billions of year ago. The sublimation of ice continues until the layer of accumulated residue becomes too thick for the process to be effective. The pinnacles may eventually diminish in height, as suggested by the lower panel. The smallest object discernable in these images have diameters of about 3 m. These images have the highest resolution ever obtained from orbit of any satellite of Jupiter (Image PIA 03455, Galileo project, courtesy of NASA/JPL-Caltech/Arizona State University)

Sublimation of ice is effective in shaping the surface of Callisto because the surface is not being rejuvenated by cryovolcanic activity and because the surface temperature of Callisto is measurably warmer than the surface temperature of Europa and Ganymede.

15.6 Science Briefs

15.6.1 Period of Revolution of Io

The relation between the period of revolution (p) of Io and the length of the semimajor axis (a) of its orbit is expressed by Newton's version of Kepler's third law:

$$p^2 = \left[\frac{4\pi^2}{G(m_1 + m_2)} \right] a^3 \qquad (15.1)$$

where

$m_1 =$ mass of Jupiter $= 1.90 \times 10^{27}$ kg

$m_2 =$ mass of Io $= 8.89 \times 10^{22}$ kg

$a =$ length of the semi-major axis $= 422 \times 10^6$ m

$p =$ period of revolution of Io in seconds for a circular orbit.

$G = x$ gravitational constant $= 6.67 \times 10^{-11}$ newton m^2/kg^2.

Substituting into equation 15.1:

$$p^2 = \left[\frac{4 \times (3.14)^2}{6.67 \times 10^{-11}(1.90 \times 10^{27} + 8.89 \times 10^{22})} \right]$$
$$\times (422 \times 10^6)^3$$

Note that the mass of $Io(m_2)$ is negligibly small compared to the mass of Jupiter and therefore can be omitted:

$$p^2 = \left[\frac{39.438}{6.67 \times 10^{-11} \times 1.90 \times 10^{27}} \right]$$
$$\times 7.5151 \times 10^{25}$$

$$p = 1.529 \times 10^5\,s = 42.479\,h = 1.769\,d$$

This result agrees exactly with the period of revolution of Io in Table 15.1. Therefore, Io appears to be in a stable orbit because its period of revolution (p), and hence its average orbital velocity (v), have the exact values required by Newton's form of Kepler's third law.

The average orbital velocity of Io is: $v = d/p$, where (d) is the circumference of the orbit: $d = 2a\pi$ and p = period of revolution = 1.529×10^5 s Therefore,

$$v = \frac{2a\pi}{p} = \frac{2 \times 422 \times 10^3 \times 3.14}{1.529 \times 10^5} = 17.33\,km/s$$

15.6.2 Ice-Dust Mixture of Callisto

The bulk density of Callisto (Table 15.1) is $1.790\,g/cm^3$, which means that it cannot be composed entirely of water ice (density $\simeq 1.00\,g/cm^3$) but also contains particles of refractory silicates and oxides (density $\simeq 3.00\,g/cm^3$). These data are used to estimate the volume fractions of ice and refractory particles that make up the body of Callisto.

The equation for this calculation was derived in Science Brief 10.6.1 from the requirement that the total mass of Callisto (M_c) is equal to the mass of ice (M_i) and the mass of the refractory particles (M_r):

The result expressed by equation 10.9:

$$\left(\frac{V_i}{V_c} \right) = \frac{D_c - D_r}{D_i - D_r} \quad (15.2)$$

where

V_i = volume of the ice
V_c = volume of Callisto
D_c = density of Callisto
D_r = density of the silicate and oxide rocks
D_i = density of ice
Substituting values for the densities yields:

$$\frac{V_i}{V_c} = \frac{1.790 - 3.00}{1.00 - 3.00} = \frac{-1.21}{-2.00} = 0.60$$

Conclusion: About 60% of the volume of Callisto consists of water ice. Consequently, the refractory particles make up about 40% of its volume.

15.6.3 Valhalla and Asgard

Valhalla, the largest multi-ring basin on Callisto is named after a hall for slain warriors where, in Norse mythology, they live in comfort under the protection of Odin, one of the principal gods, until Doomsday when they will be called upon to help Odin to fight the Giants.

Valhalla is located in Asgard where the gods of Norse mythology dwell in their own palaces. For example, Valhalla is the home of Odin. Thor lives in Thrudheim, and Balder resides in Breid-ablik.

15.7 Problems

1. Calculate the average distance of Europa from Jupiter using Newton's version of Kepler's third law (See Science Brief 15.6.1 and data in Appendix Appendix 1, Table 15.1, and other sources in this book). (Answer: a = 671 × 10^3 km).

2. Use orbital parameters of Ganymede to calculate the mass of Jupiter. Use data sources available in this book. (Answer: 1.90 × 10^{27} kg).

3. Calculate the escape speed of Ganymede (See Appendix Appendix 1 and Table 15.1). (Answer: 2.73 km/s).

4. Calculate the average speed of N_2 the surface of Ganymede (molecular weight = 28.013 atomic mass units, Avogadro's Number = 6.022×10^{23} molecules/mole, surface temperature T = 107 K). (Answer: $v_m = 0.308$ km/s).

5. Determine whether Ganymede can retain N_2 molecules in the space around it by using the

criterion that N_2 molecules can escape from Ganymede only if their speed multiplied by six is equal to or greater than the escape speed of Ganymede. (Answer: Ganymede can retain N_2 molecules).

6. Suggest one or several plausible explanations why Ganymede does *not* have an atmosphere even though Titan (Chapter 17) does.

15.8 Further Reading

Bagenal F, Dowling T, McKinnon W (2004) Jupiter: The planet, satellites, and magnetosphere. Cambridge University Press, Cambridge, UK

Beatty JK, Chaikin A (eds) (1990) The new solar system. 3rd edn. Sky Publishing Corp, Cambridge, Mass

Beatty JK, Petersen CC, Chaikin A (eds) (1999) The new solar system. 4th edn. Sky Publishing Corp, Cambridge, Mass

Faure G (2001) Origin of igneous rocks. The isotopic evidence. Springer-Verlag, Heidelberg

Fischer D (2001) Mission Jupiter. Copernicus, New York

Freedman RA, Kaufmann III WJ (2002) Universe: The solar system. Freeman, New York

Geissler PE (2003) Volcanic activity on Io during the Galileo era. In: Jeanloz R., Albee A.L., Burke K.C. (eds) Annual Reviews of Earth and Planetary Science, vol. 31

Greeley R (1999) Europa. In: Beatty J.K., Petersen C.C., Chaikin A. (eds) The New Solar System, 4th edn. Sky Publishing, Cambridge, MA, pp 253–262

Greenberg R, Tufts BR (2001) Infecting other worlds. Am Sci 89(4):296–300

Greenberg R (2005) Europa-The ocean moon. Search for an alien biosphere. Springer/Praxis, Chichester, UK

Harland D (2000) Jupiter Odyssey. Springer-Praxis, Chichester, UK

Hartmann WK (2005) Moons and planets, 5th edn. Brooks/Cole, Belmont, CA

Houtzager G (2003) The complete encyclopedia of Greek mythology. Chartwell Books, Edison, NJ

Hunt G, Moore P (1981) Jupiter. Rand McNally, New York

Johnson TV (1999) Io. In: Beatty J.K., Petersen C.C., Chaikin A. (eds) The New Solar System, 4th edn. Sky Publishing Corp., New York, pp 241–252

Kargle JS et al. (2000) Europa's crust and ocean: Origin, composition, and the prospects for life. Icarus, 148:226–265

Kieffer SW, Lopes-Gautier R, McEwen A, Smythe W, Keszthelyi L, Carlson R (2000) Prometheus: Io's wandering plume. Science 288:1204–1208

Lopes R (2002) The rampant volcanoes of Io. Planet Sci Rept Mar/Apr:6–11

Marchis F et al. (2002) High resolution Keck adaptive optics imaging of violent volcanic activity on Io. Icarus, 160:124–131

Matson DL, Blaney DL (1999) Io. In: Weissman P.R., McFadden L.-A., Johnson T.J. (eds) Encyclopedia of the Solar System. Academic Press, San Diego, CA, pp 357–376

McCord TB et al. (1998) Salts on Europa's surface detected by Galileo's near-infrared mapping spectrometer. Science, 280:1242-1245

McEwen AS et al. (2000) Galileo at Io: Results from high-resolution imaging. Science 288:1193–1198

Morrison D (ed) (1982) The satellites of Jupiter. University of Arizona Press, Tucson, AZ

NRC (2000) Preventing the forward contamination of Europa. Nat Acad Sci Press, Washington, DC

Pappalardo RT (1999) Ganymede and Callisto. In: Beatty J.K., Petersen C.C., Chaiken A. (eds), The New Solar System. 4th ed., Sky Publishing, Cambridge, MA, 263–274

Pappalardo RT et al. (1999) Does Europa have a subsurface ocean? J Geophys Res, 104:24,015–24,056

Peale SJ, Cassen P, Reynolds RT (1979) Melting of Io by tidal dissipation. Science 203:892–894

Pilcher CB, Ridgeway ST, McCord TB (1972) Galilean satellites: Identification of water frost. Science 178:1087–1089

Sagan C, Coleman S (1965) Spacecraft sterilization standards and contamination of Mars. J Astronautics and Aeronautics 3(5):22–27

Schenk P (2002) Thickness constraints on the icy shells of the Galilean satellites from a comparison of crater shapes. Nature 417:419–421

Smith BA and the Voyager Imaging Team, (1979) The Jupiter system through the eyes of Voyager 1. Science 204:951–972

Spencer JR, Jessup KL, McGrath MA, Ballester GE, Yelle R (2000) Discovery of gaseous S_2 in Io's Pele plume. Science 288:1208–1210

Van Allen J, Bagenal F (1999) Planetary magnetospheres and the interplanetary medium. In: Beatty J.K., Petersen C.C., Chaikin A. (eds) The New Solar System, 4th edn. Sky Publishing, Cambridge, MA, pp 39–58

Weissman PR, McFadden L-A, Johnson TV (eds) (1999) Encyclopedia of the Solar System. Academic Press, San Diego, CA

Zahnle KL, Schenk P, Levison H, Dones L (2003) Cratering rates in the outer solar system. Icarus, 163: 263–289

Saturn: The Beauty of Rings

Saturn in Figure 16.1 is the second largest planet in the solar system exceeded only by Jupiter in terms of its diameter and mass (Section 14.1). It was discovered in 1610 by Galileo Galilei who noted that the planet appeared to have "handles" on opposite sides. Improvements in the design and construction of telescopes later revealed that the handles of Saturn are actually a system of rings that is much larger than the rings of Jupiter (Section 14.3.2). Saturn has the lowest bulk density $(0.690\,g/cm^3)$ of any planet in the solar system because its rocky core is smaller than that of Jupiter and because the molecular hydrogen and atomic helium of its gas envelope are less strongly compressed. Saturn has 19 regular satellites, five of which are small bodies located within the system of rings: Pan (16 km), Atlas $(25 \times 10\,km)$, Prometheus $(70 \times 40\,km)$, Pandora $(55 \times 35\,km)$, and a small satellite (7 km) discovered by the Cassini spacecraft on May 1, 2005. In addition, Saturn's entourage also includes the noteworthy satellites Mimas (390 km), Enceladus (500 km), Tethys (1048), Dione (1120 km), Rhea (1530 km), Titan (5150 km), Hyperion $(350 \times 200\,km)$, Iapetus (1440 km), and Phoebe (220 km). These satellites are being explored by the Cassini spacecraft that arrived in the saturnian system on July 1, 2004. As of July 2006, Saturn is known to have 31 satellites including 12 irregular ones that have highly eccentric orbits in contrast to the regular satellites whose orbits are nearly circular.

16.1 Physical Properties

The diameter (120,660 km) and mass $(5.69 \times 10^{26}\,kg)$ of Saturn in Table 16.1 are both smaller than those of Jupiter as noted above. Nevertheless, the diameter of Saturn is almost ten times larger than the diameter of the Earth and its mass is nearly one hundred times greater. In addition, Saturn does have a surprisingly low density of $0.690\,g/cm^3$ that deviates from the pattern of decreasing densities in Figure 3.6 of the terrestrial planets including Jupiter. The large mass of Saturn causes it to have a high escape velocity of 35.6 km/s although Saturn, like Jupiter, does not have a solid surface on which a spacecraft could land. The pressure of its atmosphere increases with depth until molecular hydrogen gas converts into a fluid that conducts electricity. Electrical currents that flow within this fluid induce a planetary magnetic field. The dipole moment of the magnetic field of Saturn is 600 times that of the magnetic field of the Earth but is much less than the magnetic moment of Jupiter. (Gehrels and Matthews, 1984; Morrison et al., 1986; Rothery, 1992; McKinnon, 1995).

16.1.1 Atmosphere

The chemical composition of the atmosphere of Saturn in Table 16.2 includes molecular hydrogen (97%), helium (3%), methane (0.2%), and ammonia (0.03 %) where the concentrations are expressed in terms numbers of molecules. The data in Table 16.2 reveal that the atmosphere of Saturn is depleted in helium and enriched in hydrogen compared to the Sun and the other large planets. Molecules of water and hydrogen sulfide have been detected in the atmosphere of Jupiter and the Sun but not in the atmospheres of Saturn, Uranus, and Neptune. The noble gases neon and argon are present in low concentrations but are not listed in Table 16.2 (Ingersoll, 1999).

The clouds in the troposphere of Saturn in Figure 16.2 are stratified like those of Jupiter (Figure 14.4) and consist of condensates of

Figure 16.1. The rings of Saturn are its most remarkable feature. In many other ways, Saturn resembles Jupiter because both planets have very thick atmospheres consisting of hydrogen and helium. The colorful clouds at the top of the atmosphere of Saturn define belts and zones that give this planet a striped appearance. Saturn also has a powerful magnetic field that interacts with the solar wind and gives rise to auroral displays in its polar regions. The rings of Saturn consist of vast number of particles each of which travels in its own orbit in accordance with Kepler's laws. The gaps in the rings, clearly visible in this image, are caused by the selective ejection of particles whose periods of revolution resonate with the periods of revolution of some of the satellites of Saturn. For example, the largest gap, called the Cassini division, is caused by the resonance of ring particles at that distance with the period of revolution of Mimas which is not visible in this image (Hubble Space Telescope, January 24, 2004, courtesy of NASA/ESA/J.Clarke (Boston University) and Z. Levay (STScI)

water, ammonium hydrosulfide, and ammonia in order of decreasing temperature and increasing altitude. The reference level in the atmosphere of Saturn is at a pressure of 100 millibars where the temperature is about $100\,K$ $(-173\,°C)$. The temperature increases exponentially with increasing depth below the reference level and reaches a value of $273\,K$ $(0\,°C)$ at the base of the cloud layer composed of H_2O ice where the atmospheric pressure is greater than 10 bars (Hartmann, 2005, Figure 11.22). The temperature and pressure continue to increase with depth all the way to the center of Saturn where they reach estimated values of $13,000\,K$ and $18\;10^6$ bars, respectively. Saturn, like Jupiter, radiates more

heat into interplanetary space than it receives from the Sun. The total heat flow leaving Saturn is 4.6 watts per square meter (W/m^2) of which $2.6\,W/m^2$ originates from sunlight and $2.0\,W/m^2$ is contributed by the interior (Hubbard, 1999).

The atmosphere of Saturn contains strong winds that blow in opposite direction in light-colored zones and dark belts. The winds in the equatorial zone of Saturn blow at speeds of up to $500\,m/s$ $(1800\,km/h)$ which is considerably faster than the winds of Jupiter. The light coloration of the zones on Saturn and Jupiter is caused by clouds that form as warm gases are rising, whereas in the dark belts cold gases are sinking and clouds are evaporating as the temperature

Table 16.1. Physical and orbital properties of Saturn and Jupiter (Hartmann, 2005)

Property	Saturn	Jupiter
Physical properties		
Diameter, km	120,660	142,800
Mass, kg	5.69×10^{26}	1.90×10^{27}
Bulk density, g/cm^3	0.690	1.314
Albedo,%	46	44
Escape velocity km/s	35.6	59.5
Atmosphere	$H_2 + He$	$H_2 + He$
Magnetic field	yes	yes
Orbital properties		
Semi-major axis, AU	9.539	5.203
Period of revolution, y	29.46	11.86
Period of rotation, d	0.426	0.410
Inclination to the ecliptic	2.49°	1.30°
Eccentricity of the orbit	0.056	0.048
Axial obliquity	26.7°	3.08°

Table 16.2. Chemical composition of the atmospheres of the outer planets in comparison to the Sun (Ingersoll, 1999)

Molecule	Jupiter	Saturn	Uranus	Neptune	Sun
H_2, %	86.4	97	83	79	84
He, %	13.6	3	15	18	16
H_2O, %	(0.1)	—	—	—	0.15
CH_4, %	0.21	0.2	2	3	0.07
NH_3, %	0.07	0.03	—	—	0.02
H_2S, %	0.008	—	—	—	0.003

The concentrations are expressed in terms of percent by number.

rises. Winds blowing in opposite direction in adjacent zones and belts cause eddies to form in the atmosphere of Saturn. However, Saturn does not have a large anticyclonic storm system like the Great Red Spot on Jupiter.

16.1.2 Internal Structure

The rapid rate of rotation of Saturn has caused its shape to deform into an oblate spheroid that is flattened at the poles. As a result, the polar

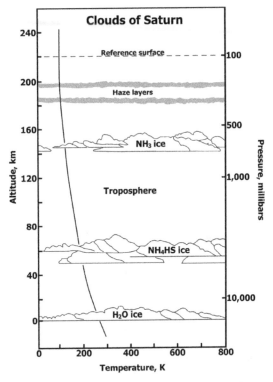

Figure 16.2. The clouds in the atmosphere of Saturn form by condensation of water vapor, ammonium hydrosulfide, and ammonia at decreasing temperature and increasing altitude. The reference surface is at a pressure of 100 millibars. Adapted from Hartmann (2005, Figure 11.22)

radius of Saturn (54,364 km) is shorter than its equatorial radius (60,268 km) by a distance of 5,904 km, which yields an oblateness of $5,904/60,268 = 0.0979$ or 9.8 %. The oblateness of Saturn is greater than that of Jupiter and of all other planets in the solar system.

The internal structure of Saturn in Figure 16.3 has been derived from a combination of the oblateness and the physical chemistry of molecular hydrogen at increasing pressure and temperature (Science Brief 16.7.1). The results indicate that Saturn has a "rocky" core that is surrounded by "ices" followed upward by a layer of electrically-conducting liquid hydrogen commonly referred to as being "metallic". About 26,000 km below the cloud tops of Saturn the liquid hydrogen loses its metallic character and gradually transforms from liquid to a gas with decreasing pressure (Klepeis et al., 1991).

Internal Structure of Saturn

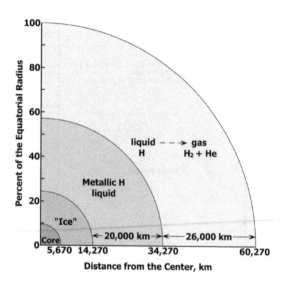

Figure 16.3. The internal structure of Saturn consists of a small core (5670 km) composed of "rocky" material including metallic iron or iron sulfide. The core is overlain by a layer (8600 km) of liquid "ices" composed of water, methane, ammonia, and other compounds released by the impact of comets. The "ice" layer is followed by a region composed of liquid metallic hydrogen (20,000 km) with dissolved atomic helium. The liquid metallic hydrogen is overlain by liquid hydrogen which grades upward into gaseous hydrogen (26,000 km). The depths at which these transitions occur are model dependent (e.g., Klepeis et al., 1991). The stratified layers of clouds in Figure 16.2 occur in the uppermost part of this gas mixture. Adapted from Freedman and Kaufmann (2002, Figure 15.14) and Hartmann (2005, Figure 8.37)

The core of Saturn consists of a mixture of silicon, magnesium, iron, calcium and other elements presumably including oxygen and sulfur. However, these elements are probably not present in the form of minerals because the high pressure and temperature in the core weaken the chemical bonds that form at low pressure and temperature. The density at the center of the core of Saturn is approximately $13\,g/cm^3$ (Marley, 1999). In addition, the material in the core may have mixed with the compounds that form the "ice" in the overlying layer. The core of Saturn, like those of the other giant planets, formed by accretion of refractory dust particles and planetesimals before attracting hydrogen, helium, and other gases.

The so-called ices that overlie the core of Saturn consist primarily of water, methane, and ammonia are actually present in liquid form. Experimental data indicate that these compounds ionize at pressures exceeding about 200 kilobars and that they dissociate at pressures greater than one Megabar (Mbar) (Marley, 1999).

The helium depletion of the outer layer of Saturn (Table 16.2) is attributable to the immiscibility of this element in liquid metallic hydrogen. The phase diagram of hydrogen, presented in Science Brief 16.7.1, contains a triangular field that outlines the P and T conditions for this phenomenon. Therefore, liquid helium may "rain out" in the layer of metallic liquid hydrogen in Figure 16.3. The droplets of helium sink to deeper levels in this layer where they evaporate. This process has caused the upper part of the layer of metallic liquid hydrogen to become depleted in helium, which explains why the concentration of helium in the overlying layer of mixed hydrogen and helium gas is anomalously low (Table 16.2).

The process of helium depletion does not occur in the interior of Jupiter because its P-T profile largely bypasses the field of helium immiscibility (Marley, 1999). Consequently, the atmosphere of Jupiter is only slightly depleted in helium (Table 16.2).

The rapid rate of rotation of Saturn (i.e., 10.23 h/rotation) and thermal convection cause electrical currents to flow in the layer of metallic hydrogen liquid. These electrical currents induce the planetary magnetic field of Saturn, which is weaker and less complex than that of Jupiter because the layer of metallic hydrogen in the interior of Saturn is not as thick as that of Jupiter. The magnetic field of Saturn is inflated by ionized water moleclues that are injected by the geysers of Enceladus. The magnetic field of Saturn rotates at very nearly the same rate as the planet and the orientation of its axis differs by less than one degree from the planetary axis of rotation (Van Allen and Bagenal, 1999).

16.1.3 Celestial Mechanics

The orbital properties of Saturn in Table 16.1 include the length of the semi-major axis of its orbit (9.539AU), its period of prograde revolution (29.46 y), the eccentricity of the orbit

(0.056), and the period of prograde rotation (0.426d). The length of the semi-major axis of Saturn and its period of revolution yield an average orbital speed of 9.64 km/s compared to 13.1 km/s for Jupiter. The inclination of the orbit of Saturn relative to the plane of the ecliptic is 2.49° and the inclination of the axis of rotation of Saturn relative to a line perpendicular to the plane of its orbit is 26.7°, whereas the axis of rotation of Jupiter inclined only 3.08°. The mass of Saturn determines its escape velocity of 35.6 km/s, which indicates indirectly the magnitude of the force of gravity that Saturn exerts on spacecraft that approach it. The bright clouds of the atmosphere of Saturn reflect nearly half of the sunlight it receives as indicted by its albedo of 46%.

16.2 The Rings

Saturn is justly famous for the extensive system of rings in Figure 16.4 that revolve around this planet (Burns, 1999; Porco, 1999; Smith, 1990; Greenberg and Brahic, 1984; and Pollack, 1975). As in the case of the Main ring of Jupiter (Section 14.3.2), the rings of Saturn consist of individual particles that form an extensive flat disk located within the equatorial plane of the planet. Therefore, the rings of Saturn are also inclined by 26.7° relative to the plane of the orbit. As a result, the attitude of the rings relative to an observer on the Earth changes as Saturn moves along its orbit (Freedman and Kaufmann, 2002, Figure 15.3). When the rings dip toward the Earth, the top surface of the rings is visible.

Figure 16.4. The rings of Saturn are a thin disk that is aligned with the equatorial plane of the planet. The image reveals that this disk is composed of a multitude of individual ringlets composed of particles that have closely-spaced orbits. The subdivisions of the disk and its principal gaps are indicated schematically in Figure 16.5 and are discussed in the text. The diameters of the ring particles range from micrometers to boulders that are several meters wide. The particles are either composed of water ice or may consist of refractory materials coated with water ice. (Radio occultation, Cassini project, May 3 of 2005, courtesy of NASA/JPL-Caltech/European and Italian space agencies)

As Saturn continues in its orbit, the view from the Earth is edge-on. Still later, the underside of the rings comes into view.

In accordance with Kepler's third law, the period of revolution of each particle in the rings is positively correlated to its distance from the center of Saturn. In other word, particles close to the planet travel faster than particles at a greater distance from the planet. The particles are about 10 cm in diameter on average but range from 1 to 500 cm. They are either coated with ice or are actually composed of ice as indicated by their high albedo of 80%. Therefore, the rings of Saturn are even brighter than the planet itself, given that its albedo is only 46% (Table 16.1). The particles that form the rings of Saturn may be ancient ice planetesimals that failed to accrete to form a satellite, or they may be fragments of a comet that was captured by Saturn and was broken up because it strayed within the Roche limit (Science Brief 14.7.2). The diameter of the outer edge of the outermost rings in Figure 16.5 is 274, 000 km, whereas the diameter of innermost edge is 149,000 km. Therefore, the rings are about 62,500 km wide.

The rings in Figure 16.5 have been subdivided into three parts labeled A, B, and C from the outside in. Each ring consists of hundreds of narrow bands or *ringlets* of individual particles. The A and B rings are separated from each other by the *Cassini* division. In addition, the A ring contains the *Encke* gap that separates the outer part of the A ring from the inner part (not shown). The space between the inner edge of the C ring and the cloudtops of Saturn contains several very faint ringlets which are collectively referred to by the letter D. Saturn contains three additional rings all of which are located outside of the A ring. The F ring (Figure 16.5) is about 100 km wide and is located about 4000 km from the outer edge of the A ring. At even greater distances, the E and G rings (not shown) are very faint and are located far from the outer edge of the A ring (i.e., G ring = 37,620 km; E ring = 75,240 to 164,490 km). The total mass of the rings is barely sufficient to form a small satellite with a diameter of 100 km. However, such a satellite cannot form because the rings are inside the Roche limit (Section 14.3.2 and Science Brief 14.7.2) (Freedman and Kaufmann, 2002).

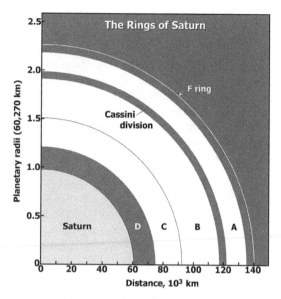

Figure 16.5. The rings of Saturn in Figure 16.4 have been divided into three main parts labeled A, B, and C with two apparent gaps (i.e., the Cassini division and the Encke gap). The Encke gap is 270 km wide and is located within the A ring (not shown). The F ring is only about 100 km wide and is located 4000 km from the outer edge of the A ring. The G and E rings are located far from Saturn outside the F ring. The gap between the inner edge of the C ring and the cloudtops of Saturn is labeled D. Adapted from Figure 15.2 of Freedman and Kaufmann (2002) and drawn to scale

The brightness (i.e., the albedo) of the main rings of Saturn in *reflected light* depends on the size and reflectivity of the particles they contain. The B ring is brighter than the A ring and both are brighter than the C ring, which is actually quite dim and hence difficult to see telescopically. However, when the rings are viewed in light that has passed through the rings (i.e., in *transmitted light*), the A ring is brighter than the B ring, but the Cassini division is even brighter than the A ring. This surprising phenomenon indicates that the Cassini division contains small dust particles that do not reflect much light but which scatter transmitted light. Similarly, light scattered by the F ring indicates that it contains particles whose diameter is approximately equal to one micrometer (μm).

The available evidence suggests that the rings of Saturn formed episodically when small satellites were shattered after being hit by comets or small asteroids. Another possible source of ring particles are comets that were captured by Saturn and were subsequently disrupted by tidal forces inside the Roche limit.

16.3 The Satellites

The regular satellites all have names derived from Greek mythology (Houtzager, 2003). The eighteen regular satellites of Saturn in Table 16.3 form three groups based on the radii of their orbits, the diameters of the satellites , and their masses. All but Phoebe of these satellites revolve around Saturn in the prograde direction and most have spin-orbit coupling of 1:1. Only Hyperion and Phoebe are still tumbling irregularly in their orbit. In addition, the planes of the orbits of all but two (i.e., Iapetus and Phoebe in Group 3) are closely aligned with the equatorial plane of Saturn. The satellites of Group 3 also differ from the others by having more eccentric orbits than the satellites in Group 1

and Group 2. Namely, Hyperion (0.104), Iapetus (0.028), Phoebe (0.163) (Hartmann, 2005).

16.3.1 Ring Satellites

The six satellites of Group 1 are associated with the A and F rings of Saturn. Pan ($\sim 20\,km$) resides in the Encke gap within the A ring, Atlas ($50 \times 20\,km$) and Prometheus ($140 \times 80\,km$) are located in the gap between the A ring and the F ring. Pandora ($110 \times 70\,km$) and the "twin" satellites Epimetheus ($140 \times 100\,km$) and Janus ($220 \times 160\,km$) share an orbit at $151.4 \times 10^3\,km$ outside of the F ring. The newest as yet unnamed member of this group of satellites, discovered by Cassini on May 1, 2005, maintains the narrow *Keeler* gap, which is located between the Encke gap and the outer edge of the A ring.

The satellites of group 1 play important roles in shaping and maintaining the rings of Saturn

Table 16.3. Physical properties and orbital characteristics of the regular satellites of Saturn (Hartmann, 2005)

Name	Semi-major axis, 10^3 km	Period of revolution, d	Diameter, km	Mass, kg	Density, g/cm^3	Albedo, %
Group 1						
Pan	133.6	0.576	16	?	?	?
Atlas	137.6	0.601	50×20	?	?	40
Prometheus	139.4	0.613	140×80	?	?	60
Pandora	141.7	0.629	110×70	?	?	60
Epimetheus	151.4	0.694	140×100	?	?	50
Janus	151.4	0.695	220×160	?	?	50
Group 2						
Mimas	185.54	0.942	390	4.5×10^{19}	1.200	60
Enceladus	238.04	1.370	500	8.4×10^{19}	1.200	95
Tethys	294.67	1.888	1048	7.5×10^{20}	1.210	70
Calypso[1]	294.67 (L$_5$)	1.888	34×22	?	?	80
Telesto[1]	294.67 (L$_4$)	1.888	34×26	?	?	60
Dione	377	2.737	1120	1.05×10^{21}	1.430	50
Helene	377	2.74	36×30	?	?	50
Rhea	527	4.518	1530	2.49×10^{21}	1.330	60
Titan	1222	15.94	5150	2.35×10^{23}	1.880	20
Group 3						
Hyperion	1484	21.28	350×200	?	?	30
Iapetus	3564	79.33	1440	1.88×10^{21}	1.160	5(L) 50 (T)
Phoebe[3]	12,930	550.4	220	?	?	6

[1] Calypso and Telesto are located at Lagrange points in the orbit of Tethys.

[2] The albedos of the leading (L) and trailing (T) hemispheres differ by a factor of 10.

[3] Phoebe is the only satellite of Saturn known *not* to have 1:1 spin-orbit coupling.

(i.e., they are shepherds). For example, the particles of the F ring are being acted upon by Prometheus on the inside and Pandora on the outside. Prometheus travels faster than the particles of the F ring and therefore tends to accelerate them as it passes along the inside of the ring. As a result, Prometheus pushes the ring particles into a slightly higher orbit. Pandora on the outside of the F ring has the opposite effect because it travels more slowly than the ring particles and causes their velocity to decrease as they pass the satellite. As a result, the ring particles drop into a slightly lower orbit. The result is that Prometheus and Pandora hold the particles of the F ring in their orbits.

The satellites of Saturn also affect the rings by resonances with certain satellites. For example, particles that orbit Saturn at the distance of the Cassini division are in resonance with Mimas, the first satellite of group 2 in Table 16.3. The period of revolution (p) of Mimas is 0.942 d or 22.6 h, whereas a hypothetical ring particle in the Cassini division revolves once every 11.3 h. Therefore,

$$\frac{p\ (Mimas)}{p\ (Cassini\ division)} = \frac{22.6}{11.3} = 2.0$$

Accordingly, particles in the Cassini division make two revolutions each time Mimas revolves once, which causes Mimas to exert a gravitational force on the ring particles in the Cassini division once every 22.6 h. The combination of gravitational forces exerted by Saturn and Mimas has caused the ring particles to be ejected from their orbits thereby leaving a gap in the rings of Saturn.

The masses and densities of the ring satellites in Table 16.3 are not yet known. All of them are small compared to the satellites of groups 2 and 3 and have irregular (i.e., nonspherical) shapes which suggests that they may be composed of rocky material. However, their high albedos (40 to 60%) tell us that their surfaces are probably covered by hoar frost or deposits of water ice.

16.3.2 Main Group

Group 2 in Table 16.3 contains nine satellites whose orbits are quasi-circular and have radii that range from 185.54×10^3 km (Mimas) to 1222×10^3 km (Titan) and which place them far outside the rings of Saturn. The periods of revolution of these satellites increase with the radii of their orbits as required by Newton's version of Kepler's third law. Most of the satellites in group 2 have spherical shapes, except Calypso, Telesto, and Helene, which are much smaller than the other satellites in this group and have irregular shapes.

These small satellites are noteworthy because Calypso and Telesto occupy the two Langrangian points L4 and L5 (Section 7.2.11, Figure 7.6) of Tethys. Telesto occupies the L4 position and leads Tethys in its orbit, whereas Calypso resides at L5 and trails behind Tethys. Consequently, both of the Lagrange satellites have the same orbital radius and the same period of revolution as their host. We do not know whether Telesto and Calypso also have spin-orbit coupling because their rotation periods have not yet been measured. In the absence of information about their masses and densities, the irregular shapes and high albedos suggest that Telesto and Calypso are composed of rocky material covered by a layer of water ice.

The status of Helene is somewhat less certain than that of Telesto and Calypso. The orbit of Helene has the same radius as Dione (377×10^3 km) and their periods of revolution appear to be very similar (Helene: 2.74d; Dione: 2.737d). However, the inclinations of the orbits relative to the equatorial plane of Saturn are not identical (Helene: 0.15°; Dione: 0.0°) and the eccentricity of their orbits differ (Helene: 0.005; Dione: 0.002). Hartmann (2005, p. 33) indicated that Helene is located 60° ahead of Dione, which implies that Helen occupies the L4 Lagrange point of Dione.

Another noteworthy feature of the satellites in group 2 is that Mimas (p = 0.942 d) and Tethys (p = 1.888 d) as well as Enceladus (p = 1.370 d) and Dione (p = 2.737 d) are in 2:1 resonance:

$$\frac{p\ (Tethys)}{p\ (Mimas)} = \frac{1.888}{0.942} = 2.00$$

$$\frac{p\ (Dione)}{p\ (Enceladus)} = \frac{2.737}{1.370} = 2.00$$

Titan is by far the largest satellite in group 2 with a diameter of 5150 km and therefore is larger than the planet Mercury. It is also the most massive $(1.35 \times 10^{23}\,\text{kg})$ and has the highest density $(1.880\,\text{g/cm}^3)$ of the saturnian satellites. Titan has an atmosphere primarily composed of molecular nitrogen (N_2) and methane (CH_4), which explains why its surface is obscured by an orange haze. Nevertheless, we have images of the surface of Titan that were recorded by the Huygens probe that was released by spacecraft Cassini during a flyby in December of 2004. More details about Titan are provided in Chapter 17.

16.3.3 Trans-Titanian Satellites

The orbits of the satellites of group 3 in Table 16.3 lie far beyond the orbit of Titan: Hyperion $(1484 \times 10^3\,\text{km})$, Iapetus $(3564 \times 10^3\,\text{km})$, and Phoebe $(12,930 \times 10^3\,\text{km})$. In addition, their periods of revolution range from 21.28 d (Hyperion), to 79.33 d (Iapetus), and finally to 550.4 d (Phoebe). Hyperion is an irregularly shaped object with diameters of $350 \times 200\,\text{km}$ and an albedo of 30%. Iapetus is the third largest of the satellites of Saturn with a diameter of 1440 km compared to 5150 km for the diameter of Titan and 1530 km for Rhea. Iapetus has a bulk density of $1.160\,\text{g/cm}^3$ and the albedo of its leading hemisphere (i.e., the one that faces in the direction of its orbit) is only 5%, whereas the albedo of the trailing hemisphere is 50%. Phoebe is an odd-shaped object composed of rocks and ice located on the very fringe of the saturnian system. Its diameter is only 220 km and its albedo in Table 16.3 is 6%.

16.3.4 Bulk Densities and Diameters

The principal satellites of Saturn include Mimas, Enceladus, Tethys, Dione, Rhea, and Titan. Although Iapetus is the third largest satellite of Saturn, it differs from the others as discussed below. The bulk densities of the principal satellites in Figure 16.6A rise irregularly with increasing distance from the center of the planet and peak at $1.880\,\text{g/cm}^3$ for Titan. The diameters of the same satellites in Figure 16.6B also increase with increasing radii of their orbits. Iapetus deviates from these patterns and is not shown.

The densities of all of the satellites of Saturn in Table 16.3 (insofar as they are known) are

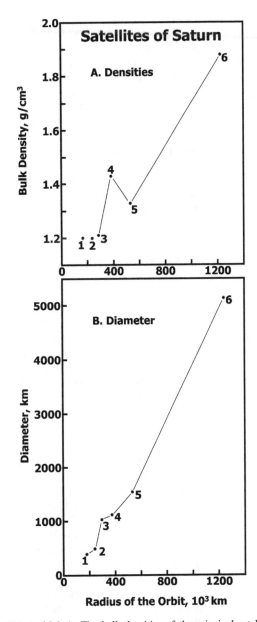

Figure 16.6. A. The bulk densities of the principal satellites of Saturn increase irregularly with increasing radii of their orbits B. The diameters of the principal satellites also increase. The satellites are identified by number: 1 = Mimas; 2 = Enceladus; 3 = Tethys; 4 = Dione; 5 = Rhea; 6 = Titan. Iapetus is not shown because the radius of its orbit is about three times larger than the radius of the orbit of Titan and because the bulk density and diameter of Iapetus deviate from the patterns shown in this diagram. Titan is by far the largest and most massive satellite of Saturn. Data from Table 16.3 and Hartmann, (2005)

greater than $1.0 \, g/cm^3$ (water ice) and less than $3.0 \, g/cm^3$ (Ca-Mg-Fe silicates). Accordingly, we conclude that the satellites for whom densities have been determined consist of a mixture of refractory silicate particles and ice. Although some of these satellites may have differentiated early in the history of the solar system when they were heated by tidal friction, impacts of comets and asteroids, and by the decay of radioactive isotopes of uranium, thorium, and potassium, present knowledge does not indicate the extent of differentiation and whether liquid water still exists in subsurface reservoirs in any of these satellites. (However, see Enceladus, Section 16.4.2).

16.4 Surface Features of the Regular Satellites

All of the regular satellites of Saturn appear to have ice crusts that are variously cratered by impacts of comets and other solid objects (e.g., fragments of asteroids and left-over planetesimals). In addition, the surfaces of several of the principal satellites contain evidence of internal tectonic and cryovolcanic activity.

16.4.1 Mimas

Mimas in Figure 16.7 is a small spherical body (diameter = 390 km) with an ice crust that has been heavily cratered by impacts. The crater density approaches saturation, which indicates that the surface of Mimas has not been rejuvenated since the time of its formation. The largest crater, named Herschel, has a diameter of 130 km, a sharply defined rim, and a central uplift. The hemisphere opposite to the crater Herschel contains parallel grooves that may have originated either as a consequence of the impact that formed this crater, or by tidal forces of Saturn, or by internal tectonic activity (McKinnon, 1999).

The low density of Mimas $(1.200 \, g/cm^3)$ indicates that water ice makes up 90% of its volume. The high albedo(60%) implies that the surface of Mimas is not covered by sediment (e.g., like Ganymede) and that the crust consists of comparatively pure ice. In addition, the low volume fraction of rocky material (10%)

Figure 16.7. Mimas is the smallest of the satellites in Group 2 of Table 16.3. It is composed of a mixture of water ice and refractory particles and its surface is heavily cratered. The largest crater, called Herschel, was formed by an impact that nearly broke up this satellite. Although the orbit of Mimas $(185.54 \times 10^3 \, km)$ is closer to Saturn than the orbit of Io $(422 \times 10^3 \, km)$ is to Jupiter, Mimas receives much less energy by tidal heating than Io (Science Brief 16.7.2). Consequently, Mimas is not geologically active at the present time and has not been affected by cryovolcanic activity since the earliest period of its long history. (Unprocessed image, Cassini project, August 2 of 2005, courtesy of NASA/JPL-Caltech/Space Science Institute)

means that the amounts of radioactive elements (uranium, thorium, and potassium) are also low and are not capable of generating enough heat to cause internal tectonic and cryovolcanic activity. Although Mimas is much closer to Saturn than Io is to Jupiter, the gravitational force exerted by Jupiter on Io is 1270 times greater than the force exerted by Saturn on Mimas (Science Brief 16.7.2). We conclude that Mimas has lost the heat it contained at the time of its formation and that all other potential heat sources are inadequate to sustain any kind of tectonic or cryovolcanic activity.

16.4.2 Enceladus

Enceladus in Figure 16.8 is virtually a twin of Mimas in terms of the radius of its orbit

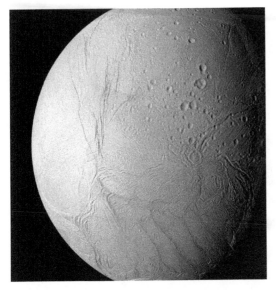

Figure 16.8. Enceladus is about 30% larger and farther from Saturn than Mimas. Considering that Mimas has been geologically inactive since the time of its formation, we might expect that Enceladus should also be internally inactive. Nothing could be farther from the truth! Enceladus has a wrinkled surface that contains an elaborate system of large canyons (the so-called tiger stripes). In addition, large areas of its surface are virtually free of impact craters. These observations leave no room for doubt that Enceladus has been, and continues to be at the present time, a highly active body both in terms of internal tectonic activity and in the form of cryovolcanic activity. (Image PIA 07800, Cassini project, courtesy of NASA/JPL-Caltech/Space Science Institute)

$(238.04 \times 10^3 \, \text{km})$ and its diameter $(500 \, \text{km})$. Although Enceladus $(8.4 \times 10^{19} \, \text{kg})$ is almost twice as massive as Mimas $(4.5 \times 10^{19} \, \text{kg})$, its bulk density $(1.200 \, \text{g/cm}^3)$ is indistinguishable from that of Mimas. The surface of Enceladus has a very high albedo (95%), because the surface has been rejuvenated by the eruption of water from the interior.

Recent images recorded by the spacecraft Cassini and the snapshots of Enceladus returned by the Voyager spacecraft in 1980 and 1981 confirm that some regions are almost free of craters, whereas others approach, but do not reach, the crater density of Mimas. In addition, the surface of Enceladus contains numerous ridges, folds, and fissures that resulted from internal tectonic activity, that appears to have occurred episodically right up to the present.

Therefore, we arrive at the surprising conclusion that Enceladus is still active at the present time (McKinnon, 1999).

The continuing geological activity of Enceladus is attributable to the heat generated by tidal deformation that is caused by the eccentricity of its orbit (0.004). The eccentricities of the orbits of Mimas and Enceladus are both affected by resonance: Mimas with Tethys and Enceladus with Dione (Section 16.3.2). Calculations referred to by Freedman and Kaufmann (2002, p. 336) indicate that enough tidal heat is generated in Enceladus to melt ice. The effect of resonance on the eccentricity of Mimas does not increase the production of tidal heat sufficiently to melt ice, perhaps because Tethys is less massive than Dione (Table 16.3).

Observations made recently by the Cassini spacecraft during flybys on February 17 and March 9, 2005, revealed that Enceladus has a tenuous atmosphere composed of water molecules in spite of the fact that its weak gravity cannot retain them. The water molecules are being ionized by sunlight and are then removed from the atmosphere by the magnetic field of Saturn. The atmosphere of Enceladus is continually replenished with water vapor that is discharged during eruptions of liquid water from a subsurface reservoir by geysers that are located along deep fractures called tiger stripes in the south polar area. The evidence in Figure 16.9 was presented and discussed in a special section of issue 5766 (March 10, 2006) of the journal Science. This issue included contributions from the members of the Cassini science team, including: Kargel (2006), Baker (2006), Porco et al. (2006), Spencer et al. (2006), Hansen et al. (2006), Brown et al. (2006), and others. In addition, images recorded from a distance of only 175 km in mid-July of 2005 contain evidence that the surface of Enceladus is traversed by parallel sets of fissures that intersect (and therefore postdate) both large and small impact craters. The new images also reveal the presence of grooved terrain and of large rift valleys (e.g., Samarkand sulcus) 5 km wide that postdate the grooved terrain similar to the terrain on the surface of Ganymede. The topographic evidence for internal tectonic activity relates Enceladus to Europa and Ganymede rather than to Mimas even though these Galilean satellites

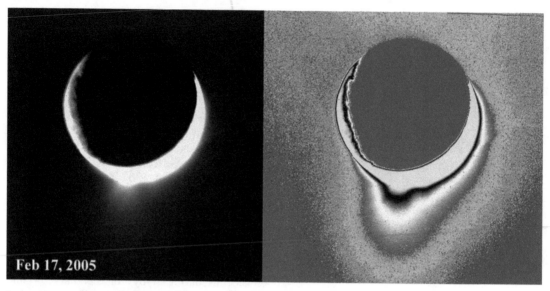

Feb 17, 2005

Figure 16.9. This image of the south pole of Enceladus, recorded on February 17, 2005, reveals that water vapor is being discharged by one or several of the deep fractures (i.e., the tiger stripes) seen in Figure 16.8. Apparently, liquid water at a temperature of 0 °C is being vented into interplanetary space from a pressurized subsurface reservoir on Enceladus. The heat required to allow this to occur appears to be generated by tidal flexing of the body of Enceladus and by the decay of radioactive isotopes of uranium, thorium, and potassium. The existence of cold-water geysers on Enceladus is remarkable because this ice world is about twice as far from the Sun as Io and about ten times as far from the Sun as the Earth. Consequently, the surface temperature of Enceladus is about −190 °C (Hartmann, 2005). (Image PIA 07798, Cassini project, courtesy of NASA/JPL-Caltech/Space Science Institute)

of Jupiter are more massive than Enceladus by factors of 570 (Europa) and 1700 (Ganymede). (See also Kargel and Pozio, 1996).

16.4.3 Tethys and Dione

The second pair of "twins" among the regular satellites of Saturn, Tethys and Dione in Figure 16.10A & B, are about twice as large as Mimas and Enceladus and therefore have higher surface-to-volume ratios than the latter. As a result, Tethys and Dione have cooled more slowly than Mimas and Enceladus (Figure 10.5). Tethys and Dione also have about ten times more mass, higher densities, and lower albedos than the small twin satellites. These properties imply that Tethys and Dione can generate more heat by decay of radionuclides because they contain more of the "rocky" component (i.e., 10% for Tethys and 20% for Dione by volume). The images recorded by the Voyager spacecraft show that some areas of Tethys and Dione are heavily cratered, whereas others are more lightly cratered and contain evidence of internal tectonic activity.

The large impact crater on *Tethys* in Figure 16.10A is a basin called Penelope (i.e., the wife of Odysseus). The floor of this basin has bulged upward to conform to the global curvature of Tethys and its rim has been flattened, leaving only a ghost or palimpsest of the original impact basins. Odysseus, the largest crater on Tethys (not visible in this image) is 40 km wide. The antipode to Odysseus is occupied by a large circular plain that is only lightly cratered and appears to have been resurfaced by the extrusion of watery lava at an early time in the history of the solar system. The craters that now dot this plain were formed by impacts that occurred after resurfacing when the rate of crater formation had declined.

The heavily cratered region of Tethys contains a very large system of rift valleys called Ithaca chasma that extends more than three quarters around the globe and is centered on the Odysseus basin. The Ithaka chasma is up to 100 km wide and several kilometers deep. McKinnon (1999) suggested that this gigantic tectonic feature may have formed as a consequence of the impact that

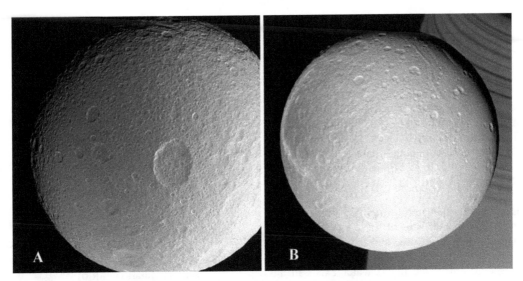

Figure 16.10. Tethys is roughly twice as far from the center of Saturn as Enceladus (Table 16.3) and is about twice as large. The icy surface of Tethys is heavily cratered and includes the Penelope basin visible in this image and a still larger basin called Odysseus (not shown). A plain that is antipodal to Odysseus appears to have been resurfaced by the extrusion of water. Subsequently, the surface of Tethys was extensively cratered which indicates that the cryovolcanic activity, that may have been triggered by the large impacts, occurred during the time of heavy bombardment prior to 3.8 Ga. The surface of Tethys also includes a system of deep canyons visible near the top of this image. (Image PIA 08149, Cassini project, courtesy of NASA/JPL-Caltech/Space Science Institute). B. Dione is similar to Tethys in several respects. Its icy surface is heavily cratered but was rejuvenated in some areas by cryovolcanic activity. The surface of Dione is also cut by a system of canyons that were formed in response to internal tectonics early in its history. However, both Tethys and Dione have been inactive for the major part of their existence. (Cassini project, courtesy of NASA/JPL-Caltech/Space Science Institute)

caused the excavation of the Odysseus basin. Perhaps this impact also triggered the cryovolcanic activity that caused the resurfacing of the large circular area that is antipodal to the Odysseus basin.

The surface of *Dione* in Figure 16.10B also contains a large number of impact craters. In addition, a broad region stretching from pole to pole is only lightly cratered, presumably because it too was resurfaced a long time ago. Dione also contains a system of valleys that extend from the poles to the mid-latitudes. These valley systems may be the result of internal tectonics or they could be "lava drain-channels" like the rilles in the basalt plains (maria) on the Moon (Section 9.1). If the valleys on Dione are flow channels of lava, they were carved by the watery lava that characterizes the icy satellites of Saturn rather than by hot silicate liquids of basaltic composition that were extruded on the Moon and on all of the terrestrial planets.

The surface features of Tethys and Dione described above indicate that both were active early in their histories (i.e., more than about three billion years ago). After the rejuvenation of certain parts of their surfaces, they have continued to cool and have remained inactive to the present time.

16.4.4 Ammonia-Water Lava

We now face up to a complication in the explanation of cryovolcanic activity on Tethys and Dione as well as on Enceladus. Spectroscopic analysis of light reflected by the surfaces of the regular saturnian satellites (except Titan) indicates the presence of pure water ice or hoar frost. Some of these surfaces have been rejuvenated by the extrusion of a fluid from the subsurface. Therefore, the spectroscopic data suggest that the fluid was pure water because, when it froze, it formed pure water ice.

The hypothesis that the surfaces of the icy satellites were rejuvenated by the extrusion of pure liquid water or of a mixture of ice and water (i.e., slush) is flawed for the following reasons:

1. The surface temperatures of the icy satellites of Saturn are less than about $-175\,°C$, whereas pure water ice melts only at a much higher temperature of $0\,°C$ at 1 atm pressure.

2. Tidal friction, which heats Io and Europa, does not generate enough heat to cause pure water ice to melt in Mimas, Tethys, Dione, Rhea, and Iapetus, but it may do so in Enceladus.

3. Even if liquid water formed by melting of ice below the surfaces of these satellites, the water cannot rise to the surface because it is denser than the ice through which it must be extruded, unless the liquid water exists in a pressurized reservoir.

The resurfacing of Tethys and Dione by watery lavas can be attributed to the existence of internal heat that has since escaped into space. Alternatively, the ice in the interiors of these satellites may not be "pure" but contains an ingredient that lowers its melting temperature. The most likely impurity of the ice capable of lowering its melting temperature is ammonia (NH_3) (Lewis, 1971; Croft et al., 1988; Kargel et al., 1991; Kargel, 1992; McKinnon, 1999).

The phase diagram of the ammonia-water system in Figure 16.11 demonstrates that the freezing temperature of mixtures of ammonia and liquid water decreases as the molar concentration of ammonia increases (Lewis, 2004). The freezing temperature reaches its lowest value of $-97\,°C$ when the mixture contains 35.4 mole % of ammonia. The melting temperature of this mixture is almost 100 degrees lower than that of pure water! When ice having this composition melts, it forms ammonia-rich water and pure ice. The ammonia-rich water is less dense than pure ice and is therefore able to rise through the ice to the surface of satellites like Tethys and Dione. In addition, this "watery lava" is viscous and flows like cold honey or basalt lava on the Earth (McKinnon, 1999: Kargel and Pozio, 1996).

One additional question needs to be answered: How did the ammonia get into the ice of satellites like Tethys and Dione? To answer this question we consider the disk-shaped cloud of gas and dust that revolved around Saturn at

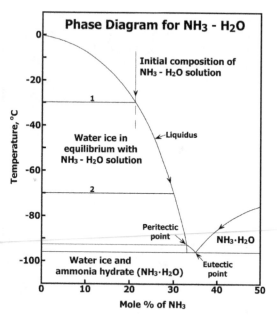

Figure 16.11. When an aqueous solution (vertical line) containing 21.5 mole percent of ammonia (NH_3) in this example is cooled to $-30\,°C$ (tie line 1), crystals of pure water ice begin to form. As the temperature continues to decrease from tie line 1 toward tie line 2, ice crystals continue to form and the concentration of ammonia in the residual water increases along the liquidus curve. When the temperature has decreased to $-70\,°C$ at tie line 2, the remaining water contains 30 mole percent of ammonia. Continued cooling and crystallization of pure water ice raises the ammonia concentration of the liquid to 32.6 mole percent at the peritectic temperature. Additional cooling and crystallization of ice increases the ammonia concentration of the remaining solution to 35.4 mole percent. This solution freezes at the lowest (eutectic) temperature of $-97\,°C$. The temperature cannot decrease to lower values until all of the eutectic liquid has crystalized to form a mixture of ammonia hydrate and water ice. Adapted from Lewis (2004, Figure V.14) and Kargel and Pozio, (1996)

the time of its formation. Most of the regular satellites of Saturn formed within that disk by accretion of solid particles composed primarily of water ice and 10 to 20% of refractory dust particle. The gases in this "protosatellite nebula" included molecules of water, ammonia, methane, and other compounds. Ammonia is much more volatile than water (Figure 12.16) as indicated by the fact that liquid ammonia boils at $-33.35\,°C$ compared to $+100\,°C$ for liquid water when both are at a pressure of one atmosphere. In addition, liquid ammonia freezes at $-77.7\,°C$. When water molecules in the protosatellite nebula condensed to form ice particles at the orbital radius of Tethys and Dione, the ammonia molecules initially remained in the gas phase. As the temperature of the nebula continued to decline to about −123 °C, the ammonia molecules entered into the crystal lattice of water-ice particles to form an ammonia hydrate (McKinnon, 1999). The ammonia hydrate and refractory particles subsequently accumulated into the ice satellites we know as Tethys, Dione, and Rhea. However, the presence of ammonia in the ice of these satellites has not yet been confirmed because the Voyager spacecraft were not equipped to detect this compound. The spacecraft Cassini, that is currently exploring the saturnian system, does carry the necessary equipment and may be able to detect ammonia. However, the ammonia in ice exposed at the surface of Tethys and Dione may be lost by volatilization and/or by molecular decomposition caused by irradiation with ultraviolet light, the solar wind, and cosmic rays.

16.4.5 Rhea and Hyperion

Small bodies cool more rapidly than large ones. Consequently, small bodies should become internally inactive a short time after their formation, whereas large bodies remain active for extended periods of time (relatively speaking). We already know that this rule is violated by Io and Europa both of which continue to be heated by the tides of Jupiter. We also know that Enceladus has remained active to the present time because the resonance with Dione continues to augment the eccentricity of its orbit, which increases the amount of heat generated by tidal friction. In addition, cryovolcanic activity based on melting of water ice (with or without ammonia) does not require nearly as much heat as melting of silicate rocks does. Therefore, the size criterion leads to the expectation that *Rhea*, whose volume is 2.5 times larger than that of Dione and 3.1 times larger than that of Tethys, might have retained enough initial heat to remain active for a longer period of time than Tethys and Dione. It turns out that this expectation is not fulfilled.

The image of Rhea in Figure 16.12 that was recorded by the spacecraft Cassini at a distance of 343,000 km, indicates that its surface is saturated with craters and has not been rejuvenated by cryovolcanic activity. The low bulk density of Rhea (1.330 g/cm^3) tells us that this satellite is largely composed of water ice with about 20% by volume of rocky material that is probably dispersed throughout the ice. The decay of radioactive elements associated with the rocky fraction in the interior of Rhea is its principal source of heat. Tidal heating is not significant because the orbit of Rhea is located 527×10^3 km from the center of Saturn and the orbital eccentricity of Rhea is only 0.001. Moreover, Rhea is not in resonance with any of the neighboring satellites and therefore does not benefit from an increase of its orbital eccentricity (i.e., by a forced component). As a result of these unfavorable circumstances, Rhea became inactive soon after its formation and has remained inactive to the present time.

Nevertheless, the surface of Rhea contains ridges and other topographic features that are attributable to compression of its ice crust. The explanation that has been suggested is that the ice in the interior of Rhea (and of other large cooling satellites) can undergo a phase transformation to form a dense polymorph called ice II. This transformation is accompanied by a decrease in volume of more than 20% and thereby causes the outer layers of ice to be compressed. This process can explain the topographic evidence for compression of the surface ice of Rhea. In addition, the surface of Rhea may not have been rejuvenated because the compression of its ice crust prevented vents from opening through which watery lava could have been erupted (McKinnon, 1999).

Many of the countless craters on the surface of Rhea have diameters of less than 20 km. These craters evidently formed by the impact of large numbers of small bodies that presumably originated from a local (i.e., saturnian) source. A possible explanation of this phenomenon is that *Hyperion* was repeatedly struck by large objects that caused it to have the irregular shape it has today (i.e., $330 \times 260 \times 215$ km). Most of the ejecta from these impacts were captured by Titan, its nearest neighbor and the most massive satellite in the saturnian system. A large number of the remaining projectiles hit Rhea, and some of the debris may have been incorporated into the rings of Saturn. (McKinnon, 1999). Other noteworthy features of Hyperion are its low

Figure 16.12. Rhea is considerably more voluminous than Dione and Tethys and therefore might have cooled more slowly and remained active somewhat longer than the twins in Figure 16.10. This image does not confirm that expectation but indicates instead that Rhea has an old and heavily cratered surface. There is no compelling evidence for cryovolcanic resurfacing of Rhea. Nevertheless, this satellite of Saturn shows topographic signs of compression that has been attributed to the reduction in volume of ice in its interior as a result of a pressure-induced phase transformation from ice I to II. This image contains a relatively fresh impact crater surrounded by rays of light-colored ejecta, which contrast with the darker appearance of this old surface. Evidently, the surface of Rhea, like that of Ganymede and Callisto, has been darkened by the accumulation of organic and particulate matter released by the sublimation of the ice and from interplanetary space (Image PIA 08148, Cassini project, courtesy of NASA/JPL-Caltech/Space Science Institute)

albedo of 30%, compared to 60% for Rhea, and the large eccentricity of its orbit (0.104), which is exceeded only by the eccentricity of Phoebe (0.163). In addition, Hyperion is tumbling irregularly in its orbit and does not have spin-orbit coupling as do most of the saturnian satellites.

The evidence that Hyperion has suffered major damage as a result of impacts by comets and other kinds of solid objects suggests that violent collisions may have been a common occurrence in the saturnian system, especially during the earliest period of its history when the interplanetary space of the solar system contained a large number of left-over planetesimals and fragments of shattered worlds in chaotic orbits around the Sun.

16.4.6 Iapetus and Phoebe

Iapetus has about the same mass and diameter as Rhea but it is located almost seven times farther from Saturn than Rhea at an average distance of 3564×10^3 km. Iapetus was discovered by Jean-Dominique Cassini in 1671, six years after Christiaan Huygens discovered Titan in 1655. The bulk density of Iapetus ($1.160 \, g/cm^3$) is similar to the densities of the other regular satellites of Saturn (except Titan), meaning that it too is composed of a mixture of 90% water ice and 10% refractory dust particles by volume. Therefore, Iapetus should have a large albedo like the other ice satellites of Saturn (e.g., Mimas, Enceladus, Tethys, Dione, and Rhea). Soon after discovering Iapetus, the astronomer Cassini observed that its albedo changed with time and sometimes declined to such an extent that the satellite almost became invisible. The explanation for the apparent disappearance of Iapetus is that the leading hemisphere (i.e., the one that faces in the direction of movement in its orbit) is very low (5%) because it is covered by some kind of dark-colored deposit. The trailing hemisphere (i.e., the backside) of Iapetus is bright and shiny with an albedo of 50% as expected of a satellite composed of ice. The bright hemisphere of Iapetus is as densely cratered as the surface of Rhea and has not been rejuvenated by cryovolcanic activity. The leading hemisphere of Iapetus is so dark that no topographic features are discernible in the Voyager images.

The explanation for the presence of the dark-colored deposit on Iapetus involves *Phoebe*, the outermost of the regular satellites whose orbit has a radius of $12,930 \times 10^3$ km, placing it about 3.6 times farther from Saturn than Iapetus and nearly 25 times farther than Rhea. The images of Phoebe that were recorded by the spacecraft Cassini on its first approach to Saturn on June 11, 2004, show that Phoebe is an ancient world of ice and rocks that has been severely battered by impacts. The plane of Phoebe's orbit is inclined 150° with respect to the equatorial plane of Saturn. Consequently, Phoebe revolves in the retrograde direction when viewed from above the saturnian system, and its orbit is highly eccentric (0.163). In addition, Phoebe has a short rotation period of about 0.4 days (Hartmann, 2005) that does not match its period of revolution. Therefore, Phoebe is the only regular satellite of Saturn that has retrograde revolution, its rotation is not synchronized with its revolution, and the orbit is highly eccentric. These unusual properties of its orbit suggest that Phoebe did not form in the protosatellite disk of gas and dust but was captured by Saturn. This conjecture is supported by the comparatively high bulk density of Phoebe ($1.6 \, g/cm^3$; Talcott, 2004), which resembles the bulk densities of the most massive satellites of Uranus (i.e., Miranda, Ariel, Umbriel, Titania, and Oberon), which have bulk densities ranging from 1.350 to $1.680 \, g/cm^3$.

Most significant for the hypothetical explanation of the dark deposit on the leading hemisphere of Iapetus is the low albedo of Phoebe (6%) caused by the presence of dark carbonaceous material that covers its surface. Impacts on the surface of Phoebe may have dislodged this surface material allowing it to spiral toward Saturn. Subsequently, Iapetus ploughed into this cloud of "Phoebe dust", which was deposited on its leading hemisphere thereby decreasing its albedo. However, optical spectroscopy indicates that the deposit on the leading hemisphere of Iapetus is reddish in color, whereas the surface of Phoebe is black. Perhaps the Phoebe dust was altered in some way to cause its color to change while it was in transit in space or after it was deposited on Iapetus. It is also possible that this hypothesis is incorrect, in which case we need to consider an alternative explanation for the dichotomy of the surface of Iapetus.

Such an alternative is the suggestion that Iapetus formed in a part of the saturnian protosatellite disk that was cold enough to cause methane (CH_4) to condense as methane hydrate ($CH_4 \cdot H_2O$). When the Phoebe dust impacted on the leading hemisphere of Iapetus, the volatile compounds in the ice composed of carbon, hydrogen, nitrogen, oxygen, and sulfur, reacted to form large molecules of organic matter resembling reddish tar similar to the material that coats the surfaces of some asteroids and comets (Section 13.1.1). However, this explanation may also be inadequate and we must await the results of close encounters of the spacecraft Cassini with Iapetus before we can explain how the leading hemisphere of Iapetus came to be covered with a reddish-black deposit (McKinnon, 1999). The exploration of Iapetus by the spacecraft Cassini is scheduled to occur on September 10, 2007, when it will approach the satellite to within 950 km (McEwen, 2004).

16.5 The Cassini-Huygens Mission

A few years after the discovery of Titan by the Dutch astronomer Christiaan Huygens in 1655, the Italian-French astronomer Giovanni Cassini between 1671 and 1684 discovered four of the principal satellites of Saturn: Dione, Iapetus, Rhea, and Tethys. Although telescopic observations of Saturn, its rings, and satellites continued through the centuries, the next major exploration of the saturnian system occurred in 1980 and 1981 when the Voyager spacecraft sent back beautiful pictures that showed these features in unprecedented detail. The interest these images aroused among planetary scientists and in the general public demanded an even more thorough exploration of the new worlds the Voyagers had revealed. Therefore, the managers of NASA and the European Space Agency (ESA) decided to build a large spacecraft for a comprehensive exploration of the planet Saturn, its rings and satellites including Titan, which had attracted attention because of its atmosphere that veils the surface below (Atreya et al., 1989). The mission was to consist of a large spacecraft called Cassini and a smaller probe called Huygens that was to land on the surface of Titan.

The Cassini spacecraft, including the Huygens probe, was launched on October 15, 1997 amid public concern about the 72 pounds of plutonium-238 it carried for use in its three electric generators (Grinspoon, 1999). The nuclear power generators were necessary because solar power cannot produce enough electricity due to the great distance between Saturn and the Sun (i.e., 9.539AU = 1.43 billion kilometers). The plutonium is encapsulated in marshmallow-sized ceramic pellets that are clad in iridium-metal and encased in graphite. Even in case these pellets were accidentally scattered over the surface of the Earth as a result of a launch failure, the plutonium pellets were expected to remain intact. The plutonium in these pellets becomes harmful to humans only in case they are pulverized and the resulting plutonium dust

is inhaled. Plutonium-238 (^{238}Pu) was chosen because its spontaneous decay releases heat which is converted into electric energy by solid-state radioisotope thermoelectric generators (RTGs).

After its successful launch, the spacecraft started a long and winding journey illustrated in Figure 16.13 (McEwen, 2004). First it flew by Venus on April 26, 1998, and again on June 24, 1999, before passing the Earth on August 19,1999. It flew by Jupiter on December 30, 2000, and arrived at Saturn on July 1, 2004. These maneuvers were necessary to gather momentum for the long flight to Saturn because, after its final deep-space maneuver in December of 1998, the spacecraft had to coast the rest of the way.

The objectives of the Cassini-Huygens mission are to study:

1. The magnetic field, the concentration and energy of charged particles, the composition and circulation of the atmosphere, and the structure and composition of the rings of Saturn.

2. To deploy the Huygens probe for a descent and landing on the surface of Titan in order to obtain information about the composition, structure, and dynamics of its atmosphere and to record views of the surface of Titan.

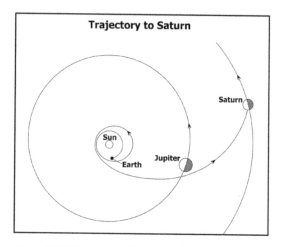

Figure 16.13. The spacecraft Cassini-Huygens was launched from the Earth on October 15,1997 and then circled the Sun twice to gather momentum by two flybys of Venus before being flung on a coasting trajectory past Jupiter and finally to Saturn where it arrived on July 1, 2004. Adapted from McEwen (2004)

3. To obtain images of the surfaces of the regular satellites of Saturn during close flybys and to study their surfaces by remote sensing.

By October of 2005, the Cassini-Huygens mission had achieved several of its major goals. It passed through the jovian system in early January of 2001 at a time when the spacecraft Galileo was still in orbit around Jupiter. On June 11, 2004, Cassini flew by Phoebe and sent back images of its severely cratered surface (Talcott, 2004). Subsequently, the spacecraft fired its main engine for 96 minutes in order to reduce its velocity before going into orbit around Saturn on July 1, 2004. Three days later Cassini-Huygens imaged Iapetus and flew by Titan on October 26 and December 13, 2004. On December 24, 2004, the Huygens probe was released and landed safely on the surface of Titan on January 14, 2005. On February 17 and March 9, 2005, the spacecraft encountered Enceladus and imaged Epimetheus on March 30 at a distance of 74,600 km. After that, Cassini carried out occultation experiments with the rings of Saturn from April to September, 2005, and prepared for close flybys of Dione, Rhea, Hyperion, and Tethys between September and November of 2005.

The mission plan for the time remaining between July 2006 and 2008 calls for mapping and monitoring of the surface of Saturn viewed from distance in the magnetotail. The necessary changes in the orbit of the spacecraft will be accomplished by means of gravity assists during several close flybys of Titan with the opportunity for a slow pass over Iapetus on September 10, 2007. In addition, on December 3, 2007, the spacecraft will view Epimetheus again from a distance of 6,190 km and take a second look at Enceladus on March 12, 2008 at only 995 km above its surface.

16.6 Summary

Our understanding of the inner workings and evolution of Saturn, its rings, and principal satellites is in transition as we await the interpretation of data being recorded by the spacecraft Cassini. Prior to the Cassini mission, most of what we knew about the saturnian system had been deduced from the near-miraculous pictures sent back to Earth by the Voyager spacecraft in 1980 and 1981.

Saturn is the second largest satellite in the solar system in terms of its mass and volume and can be paired with Jupiter. The apparent tendency of planets and satellites to occur in pairs has been noted including Venus and Earth, Jupiter and Saturn, Uranus and Neptune among the planets. Pairs of satellites include Io and Europa, Ganymede and Callisto, Mimas and Enceladus, Tethys and Dione, as well as Rhea and Iapetus. The match is never perfect and some planets and satellites do not have partners (e.g., Mercury and Mars, Titan and Phoebe) which raises doubts about the significance of the pairing of planets and satellites.

Saturn resembles Jupiter not only in volume and mass but also in the composition and structure of its interior, including a central core enclosed in a thick layer of liquid composed of a mixture of volatile compounds (i.e., the "ice layer"), which in turn is overlain by a region consisting of electrically conducting (i.e., metallic) liquid hydrogen. By far the thickest layer (26,000 km) consists of liquid hydrogen that grades upward into a mixture of hydrogen and helium gas. The uppermost part of this layer contains stratified clouds consisting of frozen condensates of water, ammonium hydrosulfide, and ammonia.

The atmosphere of Saturn contains dark belts and light-colored zones where gases sink and rise, respectively. The winds of Saturn blow even harder than those of Jupiter and cause large storm systems to form that do not, however, reach the dimensions or persistence of the Great Red Spot on Jupiter.

The rapid rotation of Saturn and thermal convection cause electrical currents to flow in the layer of the metallic hydrogen liquid. These currents induce the strong planetary magnetic field that interacts with the charged particles of the solar wind in the same way as the magnetic field of Jupiter. However, Saturn does not have a large satellite like Io with which its magnetic field can interact.

Saturn attracts attention among all of the planets of the solar system by having an extensive set of rings composed of small particles of ice that revolve around the planet. The rings include gaps such as the Cassini division and the Encke gap that have formed because of resonance effects with specific satellites. Some of the gaps and the outer edges of the rings are maintained by shepherd moons. The rings are located inside the Roche limit of Saturn and continue to exist because the particles cannot accrete to form large solid bodies.

The regular satellites of Saturn are composed of mixtures of water ice and refractory dust particles. Except for Titan which is discussed in the next chapter, only Enceladus has been recently resurfaced by cryovolcanic activity powered by tidal heat generated in its interior. Tethys and Dione were resurfaced billions of years ago and have been inactive since then. Therefore, the surfaces of Mimas, Tethys, and Dione are heavily cratered as is the surface of Rhea. The leading hemisphere of Iapetus is covered by a dark reddish deposit, whereas the trailing hemisphere exposes an ice surface that has a high albedo. The dark sediment on the leading hemisphere of Iapetus may have originated from Phoebe, which is an ancient heavily cratered body that was probably captured by Saturn and did not form in the protosatellite disk as the other regular satellites did. Data currently being collected by the Cassini spacecraft are expected to shed more light on the past histories and current status of the rings and satellites of Saturn.

16.7 Science Briefs

16.7.1 Phase Diagram of Hydrogen

The phase diagram of hydrogen in Figure 16.14 contains a hypothetical reaction path (dashed line) that illustrates the effects of changes of pressure (P) and temperature (T) on the states of aggregation of hydrogen. When P and T change along the reaction path between points 1 and 2, hydrogen is transformed along the way from a solid at point 1 to a liquid at point 2. Continuing along the reaction path from point 2 to point 3 causes the liquid to change into a gas. The increase in pressure between points 3 and 4 causes hydrogen to become a liquid which becomes electrically conducting at the high pressure of point 4. When this liquid is cooled between points 4 and 5, it solidifies to form solid metallic hydrogen. As the pressure is relaxed between points 5 and 1, the solid hydrogen loses its metallic properties and

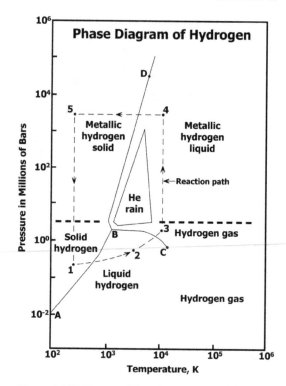

Figure 16.14. Pressure (P) and temperature (T) conditions required to convert hydrogen gas into an electrically conductive fluid in the interiors of Saturn and Jupiter. Adapted from Hubbard (1999, Figure 6)

becomes a non-conducting solid. These transformations take place as the P-T path crosses certain reaction lines: A-B is the melting curve of solid hydrogen; B-C is the boiling curve of liquid hydrogen; C is the critical point at which the distinction between liquid and gaseous hydrogen vanishes; B-D is the freezing curve along which metallic liquid hydrogen freezes to form metallic solid hydrogen. The heavy dashed lines mark the approximate pressure at which hydrogen gas changes into a liquid and where metallic hydrogen solid loses its metallic character.

The triangular area in Figure 16.14 outlines the P-T field where helium is immiscible in liquid metallic hydrogen. The P-T conditions between 14,270 km and 34,270 km measured from the center of Saturn in Figure 16.3 lie within that field. Therefore, droplets of liquid helium may form and subsequently sink to greater depth where they evaporate. This process has caused the upper part of the layer of liquid metallic hydrogen in Saturn to become depleted in helium.

16.7.2 Gravitational Force of Saturn on Mimas and of Jupiter on Io

The magnitude of the gravitational force exerted by Saturn on Mimas can be calculated from Newton's equation:

$$F = \frac{G(M \times m)}{r^2}$$

where M and m are masses in kilograms, r is the distance between the centers of Saturn and Mimas in meters, and $G = 6.67 \times 10^{-11}\,N\,m^2/kg^2$. In the problem at hand: $M = 5.69 \times 10^{26}\,kg$, $m = 4.5 \times 10^{19}\,kg$, and $r = 185.54 \times 10^6\,m$. Therefore:

$$F = \frac{6.67 \times 10^{-11} \times 5.69 \times 10^{26} \times 4.5 \times 10^{19}}{(185.54 \times 10^6)^2}\ \text{newtons}$$

$$F = 0.004962 \times 10^{22} = 4.96 \times 10^{19}\ \text{newtons}$$

The gravitational force exerted by Jupiter on Io is calculated similarly using appropriate values of M, m, and r: $M = 1.90 \times 10^{27}\,kg$, $m = 8.89 \times 10^{22}$, and $r = 422 \times 10^6\,m$:

$$F = \frac{6.67 \times 10^{-11} \times 1.90 \times 10^{27} \times 8.89 \times 10^{22}}{(422 \times 10^6)^2}\ \text{newtons}$$

$$F = 0.0006326 \times 10^{26} = 6.32 \times 10^{22}\ \text{newtons}$$

The comparison indicates that the force exerted by Jupiter on Io is $6.32 \times 10^{22}/4.96 \times 10^{19} = 1.27 \times 10^3$ times stronger than the gravitational force exerted by Saturn on Mimas. The difference in the gravitational forces explains partly why Io is strongly heated by tides, whereas Mimas is not significantly heated.

16.7.3 Phase Diagrams of Ammonia and Water

Ammonia (NH_3) is a volatile compound that forms a gas at the temperature (T) and pressure (P) at the surface of the Earth. The solubility of ammonia in cold water is $0.899\,g/cm^3$, which is equivalent to 52.78 moles per liter of water. The melting temperature of solid ammonia is $-77.7\,°C$ and the boiling temperature of the liquid is $-33.35\,°C$ at a pressure of 1.0 atm (i.e.,

$\log P = 0$). These temperatures have been used in Figure 16.15 to construct a phase diagram for ammonia (heavy lines) based on the assumptions that the melting temperature of solid ammonia and the boiling temperature of the liquid both rise with increasing pressure.

The phase diagram for ammonia has been superimposed on the phase diagram for water (Figure 12.16) by disregarding both the solubility of ammonia in water and the effect which the presence of dissolved ammonia has on the freezing and boiling temperatures of water. Therefore, the combined phase diagrams identify the states of aggregation of ammonia and water that can occur in six fields in the P-T plane. These fields are numbered for convenience and the phases of ammonia and water that are appropriate for each are listed below:

Field Coexisting phases
1. ammonia ice and water ice
2. liquid ammonia and water ice
3. liquid ammonia and liquid water

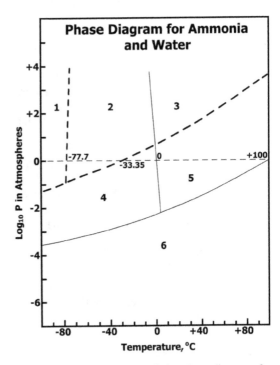

Figure 16.15. Superposition of the phase diagrams for ammonia (NH_3) and water in coordinates of the logarithm of pressure in atmospheres and temperature on the Celsius scale

4. gaseous ammonia and water ice
5. gaseous ammonia and liquid water
6. gaseous ammonia and water vapor

A phase diagram for carbon dioxide and water appears in Figure 12.17. The horizontal dashed line at $\log_{10} P = 0$ represents the pressure on the surface of the Earth. The melting and boiling temperatures of ammonia and water are indicated in the diagram.

16.8 Problems

1. Calculate what fraction expressed in percent of the volume of Rhea is composed of rocky material. Density of Rhea = $1.33 \, \text{g/cm}^3$, density of water ice = $0.919 \, \text{g/cm}^3$, density of rocky material = $3.0 \, \text{g/cm}^3$. (Answer: 19.7 %).

2. Calculate the Roche limit for a comet composed of water ice that is approaching Saturn (density of the comet = $1.10 \, \text{g/cm}^3$, density of Saturn = $0.690 \, \text{g/cm}^3$). Consult Section 14 of Appendix 1 and use the equation for a body of zero strength. (Answer: 125,963 km).

3. Verify that Mimas is outside of the Roche limit for a body consisting of particles that are touching. Density of Mimas = $1.20 \, \text{g/cm}^3$; density of Saturn = $0.690 \, \text{g/cm}^3$. (Answer: 127,425 km).

4. Calculate the average speed of water molecules on the surface of Enceladus and the escape velocity. Interpret the results by concluding whether water molecules can escape from the gravitational field of Enceladus. (Molecular weight of water = 18.0152 amu; mass of Enceladus = 8.4×10^{19} kg; radius of Enceladus = 250 km); T = 100 K.

Escape speed: $\quad v_e = \left(\frac{2GM}{R}\right)^{1/2}$

Average speed of a molecule $v_m = \left(\frac{3kT}{m}\right)^{1/2}$, $k = 1.38 \times 10^{-23}$ J/K, and $G = 6.67 \times 10^{-11}$ Nm^2/kg^2. (Answer: $v_e = 0.2117$ km/s; $v_m = 0.3720$ km/s).

16.9 Further Reading

Atreya SK, Pollack JB, Matthews MS (eds) (1989) Origin and evolution of planetary and satellite atmospheres. University of Arizona Press, Tucson, AZ

Baker J (2006) Tiger, tiger, burning bright. Science 311(5766):1388

Brown RH et al. (2006) Composition and physical properties of Enceladus' surface. Science 311(5766):1425–1428

Burns JA (1999) Planetary rings. In: Beatty J.K., Petersen C.C., Chaikin A. (eds) The New Solar System. Sky Publishing, Cambridge, MA, pp 221–240

Croft SK, Lunine JI, Kargel JS (1988) Equation of state of ammonia-water liquid: Derivation and planetological application. Icarus 73:279–293

Freedman RA, Kaufmann III WJ (2002) Universe: The solar system. Freeman, New York

Gehrels T, Matthews MS (eds) (1984) Saturn. University of Arizona Press, Tucson, AZ

Greenberg R, Brahic A (eds) (1984) Planetary rings. University of Arizona Press, Tucson, AZ

Grinspoon D (1999) Cassini's nuclear risk. Astronomy 29(8):44–47

Hansen CJ et al. (2006) Enceladus' water vapor plume. Science 311(5766):1422–1425

Hartmann WK (2005) Moons and planets, 5th edn. Brooks/Cole, Belmont, CA

Houtzager G (2003) The complete encyclopedia of Greek mythology. Chartwell Books, Edison, NJ

Hubbard WB (1999) Interiors of the giant planets. In: Beatty J.K., Petersen C.C., Chaikin A. (eds) The New Solar System. Sky Publishing., Cambridge, MA, pp 193–200

Ingersoll AP (1999) Atmospheres of the giant planets. In: Beatty J.K., Petersen C.C., and Chaikin A. (eds) The New Solar System. Sky Publishing, Cambridge, MA, pp 201–220

Kargel JS, Croft SK, Lunine JI, Lewis JS (1991) Rheological properties of ammonia-water liquids and crystal-liquid slurries: Planetological applications. Icarus 89:93–112

Kargel JS (1992) Ammonia-water volcanism on icy satellites: Phase relations at 1 atmosphere. Icarus 100:556–574

Kargel JS, Pozio S (1996) The volcanic and tectonic history of Enceladus. Icarus 119:385–404

Kargel JS (2006) Enceladus: Cosmic gymnast, volatile miniworld. Science 311(5766):1389–1391

Klepeis JE et al. (1991) Hydrogen-helium mixtures at megabars pressure: Implications for Jupiter and Saturn. Science 254:986–989

Lewis JS (1971) Satellites of the outer planets. Their chemical and physical nature. Icarus 16:241–252

Lewis JS (2004) Physics and chemistry of the solar system, 2nd edn. Elsevier, Amsterdam, The Netherlands

Marley MS (1999) Interiors of the giant planets. In: Weissman P.R., McFadden L.-A., Johnson T.V. (eds) Encyclopedia of the Solar System. Academic Press, San Diego, CA, pp 339–355

McEwen AS (2004) Journey to Saturn. Astronomy 32(1):34–41

McKinnon WB (1995) Sublime solar system ices. Icarus 375:535–536

McKinnon WB (1999) Mid-size icy satellites. In: Beatty J.K., Petersen C.C., Chaikin A. (eds) The New Solar System, 4th edition. Sky Publishing., Cambridge, MA, pp 297–320

Morrison D et al. (1986) The satellites of Saturn. In: Burns J., Matthews M.S. (eds) Satellites. University of Arizona Press, Tucson, AZ, pp 764–801

Pollack JB (1975) The rings of Saturn. Space Sci Rev 18:3

Porco CC (1999) Planetary rings. In: Weissman PR, McFadden L.-A., Johnson T.V. (eds) Encyclopedia of the Solar System. Academic Press, San Diego, CA, pp 457–475

Porco CC et al. (2006) Cassini observes the active south pole of Enceladus. Science 311(5766):1393–1401

Rothery DA (1992) Satellites of the outer planets: Worlds in their own right. Clarendon Press, NewYork

Smith BA (1999) The Voyager encounters. In: Beatty JK, Petersen C.C., Chaikin A. (eds) The New Solar System, 4th edn. Sky Publishing., Cambridge, MA, pp 107–130

Spencer JR et al. (2006) Cassini encouters Enceladus: Background and the discovery of the south polar hot spot. Science 311(5766):1401–1405

Talcott R (2004) Cassini spies Phoebe. Astronomy 32(9):46–49

Van Allen JA, Bagenal F (1999) Planetary magnetospheres and the interplanetary medium. In: Beatty J.K., Petersen C.C., Chaikin A. (eds) The New Solar System. Sky Publishing, Cambridge, MA, pp 39–58

Titan: An Ancient World in Deep Freeze

Titan, the largest satellite of Saturn, was discovered in 1655 by Christiaan Huygens who initially called it Luna Saturni (the moon of Saturn). Although the orbit of Titan was determined and its brightness and diameter were estimated, not much more was learned about it for nearly300 years. The exploration of Titan received a major boost when Kuiper (1944) reported that this satellites has an atmosphere that contains methane gas (CH_4). This discovery provoked a wide range of speculations about the origin and composition of the atmosphere and about its interaction with the surface of Titan. These conjectures were constrained by measurements of the surface temperature ($-137°$ to $-191°C$) which encompasses the melting point of solid methane ($-182°C$) as well as the boiling point of liquid methane ($-164°C$) at one atmosphere pressure (Hartmann, 2005, p. 295). Therefore, methane can occur on Titan not only in the form of ice or snow but also as a liquid in ponds, lakes, and streams, and as a gas in the atmosphere. In other words, methane behaves on Titan the way water does on the Earth (Owen, 1982, 2005; Hunten et al., 1984; Morrison et al., 1986; Toon, 1988; Lunine, 1994; Coustenis and Lorenz, 1999; Freedman and Kaufmann, 2002; Hartmann, 2005).

The attempts to explore the surface of Titan telescopically have been frustrated by the orange haze of its atmosphere. Even the flyby of the two Voyager spacecraft in 1980/1981 yielded only bland images of an orange-colored sphere in Figure 17.1, but they did confirm the existence of an atmosphere on Titan. The veil was finally lifted by the images recorded by the Cassini/Huygens spacecraft that entered the saturnian system in July of 2004. These images were recorded both optically and by synthetic aperture radar during several flybys of Cassini and during the descent of the Huygens probe in January of 2005 to a safe landing on the surface of Titan.

17.1 Physical and Orbital Properties

Titan is the largest and most massive satellite of Saturn and its diameter and mass are exceeded only by the physical dimensions of Ganymede. Both satellites are larger and more massive than the planet Mercury (Table 10.1). Titan also resembles Callisto in terms of several of its properties including its diameter, mass, and bulk density listed in Table 17.1. The density of Titan ($1.880\,g/cm^3$) is higher than that of the other saturnian satellites presumably because it contains a higher concentration of refractory (i.e., "rocky") material. Assuming that this component has a density of $3.00\,g/cm^3$, its abundance in Titan is about 44%. The estimated abundance of refractory particles in the interior of Titan is greater than that of the other satellites of Saturn but is similar to that of Callisto (Section 15.5.1). Titan also does not have a magnetic field and therefore apparently does not contain an electrically conducting liquid in its interior.

The similarities of the physical properties of Titan and Callisto considered above raise the question whether the gravitational force exerted by Saturn on Titan is similar in magnitude to the gravitational force exerted by Jupiter on Callisto. This question is considered in Science Brief 17.5.1 where we show that the gravitational attraction between Saturn and Titan (3.43×10^{21} newtons) is somewhat lower than the gravitational attraction between Jupiter and Callisto (3.85×10^{21} newtons). We therefore speculate by analogy with Callisto that tidal friction caused by Saturn probably did not generate enough heat

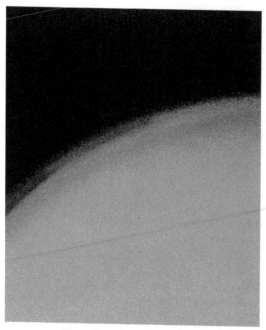

Figure 17.1. The surface of Titan is obscured by the orange-colored photochemical smog in its thick nitrogen-rich atmosphere. The blue haze that envelopes the image of Titan confirms the existence of an atmosphere on this satellite. Titan is the second-largest satellite in the solar system and the only one that has a thick atmosphere. For unknown reasons, Ganymede, which is more massive and more voluminous than Titan, does not have an atmosphere. (Voyager project, courtesy of NASA/JPL-Caltech, http://voyager.jpl.nasa.gov/science/saturn_titan.html)

Table 17.1. Physical and orbital properties of Titan and Callisto (Hartmann, 2005)

Properties	Titan	Callisto
A. Physical Properties		
Diameter, km	5150	4840
Mass, kg	1.35×10^{23}	1.08×10^{23}
Density, g/cm^3	1.880	1.790
Albedo, %	20	17
Escape velocity, km/s	2.64	2.42
Surface comp.	liquid/solid methane	water ice and org. sediment
B. Orbital Properties		
Orbit. radius, 10^3 km	1222	1884
Period of revol., d	15.94	16.689
Period of rotat., d	15.94?	16.689
Eccentricity	0.029	0.008
Obliquity, degrees	∼0?	∼0
Inclinat. of orbit w.r.t. planetary equator, degrees	0.3	0.25

to cause Titan to differentiate internally into a rocky core and a mantle of water ice.

The images of Titan recorded by the spacecraft Cassini, both optically and by synthetic-aperture radar, indicate the presence of a large radar-bright region known as *Xanadu*. Radar images, like the one in Figure 17.2, have shown the presence of impact craters and sinuous channels on the surface of Xanadu. The craters are several tens of kilometers in diameters and appear to have been degraded by erosion. The branching network of channels in Figure 17.2 appears to be part of a drainage system of the area. In addition, other images recorded by the spacecraft Cassini suggest the presence of ice volcanoes on the surface of Titan, as well as of large dune fields.

In conclusion, we note that Titan revolves around Saturn and rotates on its axis in the prograde direction with the same period of 15.94d (i.e., Titan has 1:1 spin-orbit coupling). Its orbit is moderately eccentric (0.029), which means that the distance between Titan and Saturn ranges from 1187×10^3 km at perigee to 1257×10^3 km at apogee, or ±2.9% relative to the average distance of 1222 ± 10^3 km. The albedo of Titan (20%) is low because the haze in its atmosphere absorbs sunlight and because its surface is covered with organic ooze.

17.2 Atmosphere

The data returned by Voyager 1 in 1980 revealed that the atmosphere of Titan is primarily composed of di-atomic nitrogen gas (82 to 99%), whereas the concentration of methane is only 1 to 6%. The total atmospheric pressure on Titan (1.5 bars) is higher than the atmospheric pressure on the Earth.

17.2.1 Chemical Composition

Molecules of nitrogen and methane are broken up in the upper atmosphere of Titan by ultraviolet light and by energetic electrons in the magnetosphere of Saturn. The resulting molecular fragments subsequently recombine to form a wide variety of molecules identified in

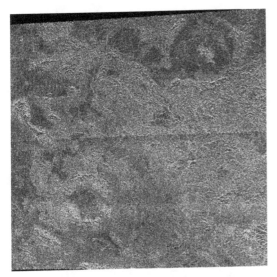

Figure 17.2. The area on the Titan known as Xanadu appears to be an elevated region that contains large impact craters. This image, which was recorded by synthetic aperture radar on April 30, 2006, during a flyby of the spacecraft Cassini, contains two impact craters. The one near the top right of the image has a diameter of 70 km and a central uplift. The smaller crater in the lower left lacks a central uplift but has a dark floor. Both craters appear to have been degraded by erosion. The image also contains several sinuous channels that may be part of the local drainage system (Image PIA 08429, Cassini project (radar mapper), courtesy of NASA/JPL-Caltech)

Table 17.2. Chemical composition of the atmosphere of Titan (Owen, 1999)

Molecular constituents	Chemical formula	Concentration % or ppm
Major constituents		
Nitrogen	N_2	82–99 %
Methane	CH_4	1–6 %
Argon?	Ar	< 1–6%
Minor constituents		
Hydrogen	H_2	2,000 ppm
Hydrocarbon compounds		
Ethane	C_2H_6	20 ppm
Acetylene	C_2H_2	4 ppm
Ethylene	C_2H_4	1 ppm
Propane	C_3H_8	1 ppm
Methylacetylene	C_3H_4	0.03 ppm
Diacetylene	C_4H_2	0.02 ppm
Nitrogen compounds		
Hydrogen cyanide	HCN	1 ppm
Cyanogen	C_2N_2	0.02 ppm
Cyanoacetylene	HC_3N	0.03 ppm
Acetonitrile	CH_3CN	0.003 ppm
Dicyanoacetylene	C_4N_2	condensed
Oxygen compounds		
Carbon monoxide	CO	50 ppm
Carbon dioxide	CO_2	0.01 ppm

Table 17.2. These minor constituents of the atmosphere of Titan include hydrocarbons (e.g., ethane and acetylene), nitrogen compounds (e.g., hydrogen cyanide and cyanogen), and oxygen compounds (e.g., carbon monoxide and carbon dioxide). In addition, the atmosphere of Titan contains hydrogen (H_2) which escapes into interplanetary space and occurs only because it is being produced continuously by the decomposition of methane. Another noteworthy point is that water vapor does not occur in the atmosphere of Titan because water vapor condenses to form ice at the low atmospheric temperature ($-137°$ to $-191°C$). In addition, the atmosphere of Titan does not contain molecular oxygen. Therefore, the hydrocarbon compounds in the atmosphere and on the surface of Titan cannot ignite spontaneously or during collisions with solid objects.

However, if the surface temperature of Titan were similar to that of Mars (e.g., $-35°C$), water ice would sublimate, the resulting water vapor

would be dissociated into oxygen and hydrogen molecules and the oxygen would react with methane to form carbon dioxide:

$$CH_4 + O_2 \rightarrow CO_2 + 2H_2 \qquad (17.1)$$

The hydrogen would escape into interplanetary space and the reaction would continue until all of the methane is consumed. This hypothetical scenario suggests that Titan has preserved a sample of the primordial atmospheres of the terrestrial planets because its low surface temperature has prevented the conversion of the original methane into carbon dioxide.

The orbit of Titan lies within the magnetosphere of Saturn most of the time. Therefore, molecules in the upper atmosphere that are ionized by collisions with energetic ions and electrons are removed by the magnetic field that continuously sweeps past Titan, because the period of rotation of Saturn (10.23 h) is much shorter than the period of revolution of Titan (15.94 d). The resulting loss of methane can deplete the atmosphere of Titan in this constituent within a few million years unless methane is being replenished from other sources (Owen, 1999).

17.2.2 Structure

The temperature of the atmosphere of Titan in Figure 17.3 *decreases* with increasing altitude above the surface and reaches a broad minimum of about $-202\,°C$ at the tropopause located 42 km above the surface. At still greater height, the temperature begins to *increase* with increasing altitude until it reaches a maximum of $-98\,°C$, which is about $80°$ higher than the surface temperature.

The layer of haze on Titan extends to an altitude of about 200 km and is followed upward by a second more tenuous layer at 300 km. Although the number of aerosol particles per

unit volume in the haze is comparatively low, the great thickness of the haze layer increases its opacity and completely obscures the surface for an observer in space (e.g., Figure 17.1). The aerosol particles are composed of the minor atmospheric constituents (except H_2 and CO) that condense at the low temperature of the tropopause (i.e., at 71 K or $-202\,°C$). Consequently, the minor constituents in Table 17.2 condense at the low temperatures that characterize the troposphere of Titan and occur in gaseous form only in the stratosphere. The orange color of the haze indicates that it contains polymers of hydrocarbons (e.g., acetylene) and nitrogen compounds (e.g., hydrogen cyanide) that form by molecular interactions within the haze layer.

The aerosol particles of the atmosphere gradually grow in size until they fall to the surface of Titan where they either re-evaporate, become liquid, or accumulate as solid particles depending on the surface temperature. For example, in accordance with its phase diagram in Figure 17.4, methane forms a liquid on the surface of Titan and therefore can collect in low areas to form ponds and lakes. Methane gas may form by evaporation of the liquid and thereby starts a hydrologic cycle based on methane. Alternatively, ethane (C_2H_6) can also precipitate in the atmosphere and become liquid after it is deposited on the surface of Titan where it can mix with liquid methane. In this case, the ponds and lakes of Titan would be composed of an ethane-methane liquid.

17.3 Huygens Lands on Titan

The Huygens probe, about the size of a Volkswagen beetle, was designed by the European Space Agency (ESA) and was transported to the saturnian system attached to the spacecraft Cassini. It was released on December 24, 2004, and entered the atmosphere of Titan about 21 days later (See also Section 16.5). Its descent through the atmosphere to the surface of Titan was slowed by parachutes and lasted almost 2.5 hours. Finally, in the early morning of Friday, January 14, 2005, Huygens hit the surface of Titan at about 25 km/h. The instruments on board survived the landing and radioed environmental data to the spacecraft Cassini which relayed

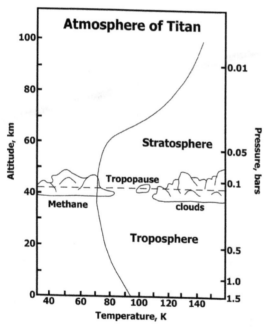

Figure 17.3. The troposphere on Titan extends to an altitude of 42 km above the surface where the temperature reaches a broad minimum of about 71 K ($-203\,°C$). At greater height in the stratosphere, the temperature rises toward a maximum of 175 K ($-98\,°C$). Clouds of condensed methane form a few kilometers below the tropopause where methane gas condenses to form methane ice. The average pressure and temperature at the surface of Titan are 1.5 bars and about 95 K ($-178\,°C$). Adapted from Owen (1999, Figure 5)

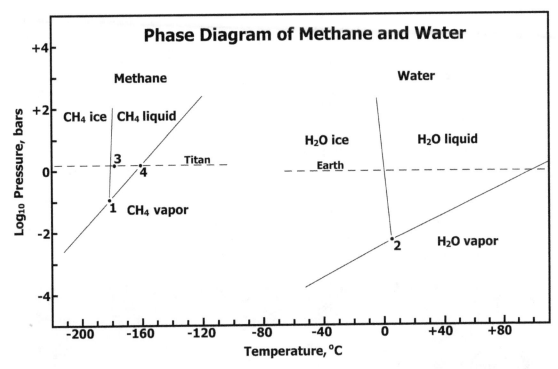

Figure 17.4. The phase diagrams of methane and water in the context of P and T on Titan and the Earth do not overlap. In other words, methane can occur on Titan in the form of ice, liquid, and vapor depending on the temperature, whereas water occurs only in the form of ice. Likewise, water can occur on the Earth in the form of ice, liquid, and vapor, whereas methane occurs only in the form of vapor. The numbered points are: 1. Triple point of methane (−182.3°C, 0.117 bars); 2. Triple point of water (+0.010°C, 0.006 bars); 3. P and T at the landing site of the Huygens probe on the surface of Titan (−179°C, 1.5 bars); 4. Boiling temperature of liquid methane on the surface of Titan (∼ −162°C). Triple-point data from Hartmann (2005)

them to receivers on the Earth. These transmissions lasted for 70 minutes while the spacecraft remained overhead. In addition, the Huygens probe returned 350 images taken during the descent (Smith et al., 1996) and is still parked on the surface of Titan.

The temperature at the landing site was −179°C, which is warm enough for methane to be in the liquid state because it is 3° above its freezing temperature of −182°C (Figure 17.4). However, at −179°C water forms ice that is as hard as rocks are on the Earth. The illumination at the landing site was dim (i.e., about 0.1% of daylight on Earth), partly because of the haze and partly because objects at the distance of Saturn's orbit (9.539 AU) receive much less sunlight than the Earth. The surface on which the probe landed had the consistency of wet clay or sand. The sediment is composed primarily of detrital

particles of water ice and dark hydrocarbon ooze that is washed out of the atmosphere by the rain of liquid methane. The soil at the landing site contained liquid methane that vaporized because of the heat emanating from the spacecraft.

The landscape on the surface of Titan is startlingly familiar. Some of the images recorded during the descent reveal a light-colored elevated surface shrouded locally by patches of white fog and dissected by a dendritic pattern of dark stream valleys that are incised into it. The master streams discharge into dark areas that appear to be large lakes with typical shoreline features. The images do not indicate whether a liquid was actually flowing in the stream valleys and whether the so-called lakes actually contained this liquid at the time the pictures were taken. The stream valleys and the lake basins could have been dry in which case the dark color resulted from the deposits of sediment composed of grains of water ice imbedded in black hydrocarbon ooze.

The image in Figure 17.5 was taken during the descent of the Huygens probe from an altitude of 16.2 km. It shows a branching network of valleys that drain a large light-colored area where bedrock (e.g., water ice) may be exposed. The streams that carved these valleys all appear to flow into a master stream which discharges the liquid into the dark low-lying basin near the top of the image. The environmental conditions on the surface of Titan support the conjecture that the fluid that flows in this and other networks of channels is liquid methane.

The actual landing site of the Huygens probe in Figure 17.6 is a large plain strewn with rounded boulders and cobbles on a substrate of fine grained sand and stilt containing interstitial liquid methane. The round shapes of the boulders indicate that they were abraded during transport in a stream. The unsorted character of the sediment tells us that it was *not* deposited in response to the gradual decrease in the velocity of the liquid in a stream. Instead, the lack of sorting identifies the deposit as an alluvial fan that formed episodically by floods that were discharged by streams flowing into the basins.

The Earth-like appearance of the surface of Titan in Figure 17.7 hides the fact that the ambient temperature is extremely low (i.e., $-179\,°C$) and that the liquid that rains from the sky is not water but a mixture of liquid methane and ethane. Titan is certainly not a world suited

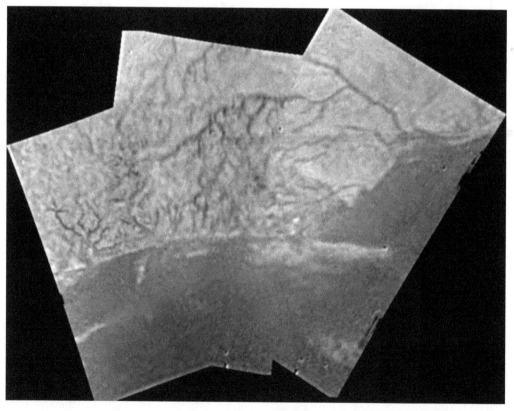

Figure 17.5. This image was recorded by the Huygens probe during its descent to the surface of Titan. It was taken on January 14,2005, at an altitude of 16.2 km. The image shows a branching network of valleys that merge into a master valley, which opens into the dark and low area along the bottom of the picture. The valley floors as well as the floor of the low area are black perhaps because they are coated with black sediment that may consist of grains of solid hydrocarbons or of water-ice particles that are coated with liquid hydrocarbons. (Cassini project, Huygens probe, courtesy of ESA/NASA/JPL-Caltech/University of Arizona)

Figure 17.7. This image from the surface of Titan shows a large lake or basin with bays and islands in a hilly landscape drained by networks of streams. Although this landscape has a familiar appearance, the light-colored rock outcrops consist of water ice, the lake or basin contains a hydrocarbon liquid, the temperature is so low that methane gas condenses to form a liquid which can freeze to from methane ice. Even at noon the surface is only dimly illuminated because the atmosphere contains an orange haze and the sunlight that reaches this distant world is weak. Cassini project, Huygens probe, courtesy of ESA/NASA/JPL-Caltech/University of Arizona). This panorama of the surface of Titan was assembled by Christian Waldvogel. (http://anthony.liekens. net/index.php/Main/Huygens)

Figure 17.6. The landing site of the Huygens probe on Titan is a large plain strewn with boulders and cobbles on a substrate of fine grained sediment. The boulders were probably rounded by abrasion during transport by a fluid in streams. The boulders are composed of water ice which forms the bedrock exposed in the upland surfaces of Titan. The fluid that transported the boulders could be liquid methane because the temperature at the surface of Titan allows methane to exist in the form of ice, as a liquid, and as a gas. In this regard methane on Titan behaves like water does on the Earth. (Cassini project, Huygens probe, courtesy of ESA/NASA/JPL-Caltech/University of Arizona)

for human beings nor is it likely that native organisms of any kind can exist there. We do not know whether cellular metabolism is possible in liquid methane and therefore assume that it is *not* possible. In addition, chemical reaction rates are slowed by the low surface temperature of Titan, which would retard cellular energy production and hence the growth rate of any hypothetical organisms that might exist there.

The atmosphere of Titan has evolved by the rain-out of hydrocarbon compounds that form by progressive polymerization of methane. These compounds appear to have been deposited on the surface of Titan in the form of a dark-colored hydrocarbon ooze. In addition, some methane in the atmosphere is destroyed by ultraviolet light and energetic electrons in the magnetosphere of Saturn. Nevertheless, Titan has preserved a sample of a primitive atmosphere that may resemble the primordial atmospheres of the terrestrial planets including Venus, Earth, and Mars all of which have much higher surface temperatures than Titan. If the Earth once had an atmosphere dominated by nitrogen, methane, and water vapor, the methane may have been converted to carbon dioxide by reacting with molecular oxygen that was released by the photodissociation of water molecules in the stratosphere (Equation 17.1). The global overheating of the Earth that could have occurred was avoided when the carbon dioxide was transferred from the atmosphere into the global ocean that formed when the atmosphere cooled enough to cause water vapor to precipitate as rain (Section 6.3). Venus was less fortunate than the Earth because the carbon dioxide, that may have formed by the oxidation of methane, accumulated in the atmosphere and caused the runaway greenhouse warming that has dominated its climate for billions of years (Section 11.3.3 and 11.3.4).

17.4 Summary

Titan has turned out to be both alien and familiar in appearance. It has an atmosphere of nitrogen and methane at a pressure of about 1.5 bars. However, the temperature of the atmosphere is so low that water vapor and oxygen, derived from the dissociation of water molecules, are absent. The molecules of methane in the atmosphere of Titan have polymerized to form an orange-colored haze that obscures the surface.

The topography of the surface of Titan consists of hilly uplands and local basins that are only sparsely cratered, indicating that Titan has been renewing its surface by cryovolcanic activity and by accumulation of hydrocarbon deposits derived from the atmosphere.

Cryovolcanic activity on Titan may be facilitated by ammonia which can lower the melting temperature of water ice by 100°C. Although an ice volcano or geyser has been seen on Titan, ammonia has not yet been detected in the atmosphere.

Methane gas that condenses in the atmosphere is deposited on the surface of Titan as rain. The downslope flow of this liquid is eroding the water ice exposed at the surface and is transporting the resulting "rock debris" through dendritic stream valleys into basins. Boulders and cobbles of water ice are rounded by abrasion during transport and the fine grained fraction of ice particles constitutes the sediment that covers the water-ice bedrock surface of Titan.

The landing site of the Huygens probe appears to be an alluvial fan composed of unsorted sediment consisting of detrital water-ice particles impregnated with hydrocarbon ooze and liquid methane. However, in spite of the familiar appearance, Titan is not a pleasant place for human beings, primarily because of the low surface temperature and the hydrocarbon ooze that covers its surface. In the absence of evidence of the contrary, it seems unlikely that native organisms exist on Titan. However, the deep-freeze conditions on its surface have preserved an atmosphere whose chemical composition may resemble the composition of the primordial atmospheres of Venus, Earth and Mars.

17.5 Science Briefs

17.5.1 Gravitational Forces on Titan and Callisto

Titan and Callisto have similar masses (i.e., Titan: 1.35×10^{23} kg; Callisto: 1.08×10^{23} kg). However, the mass of Saturn ($5.69 \ 10^{26}$ kg) is less than that of Jupiter (1.90×10^{27} kg) and the radius of the orbit of Titan (1222×10^3 km) is less than the radius of the orbit of Callisto (1884×10^3 km). Therefore, the masses and orbital radii of Saturn and Titan and of Jupiter and Callisto vary in a complementary fashion, which may cause the gravitational forces each planet exerts on its satellite (and vice versa) to be numerically similar.

The force of gravity (F) exerted by a planet of mass (M) on one of its satellites of mass (m) and orbiting at distance (r) is expressed by Newton's equation:

$$F = \frac{GMm}{r^2} \qquad (17.2)$$

where $G = 6.67 \times 10^{-11}$ newton m^2/kg^2 and the distance (r) is expressed in meters.

Therefore, the gravitational force (F_S) exerted by Saturn on Titan is:

$$F_S = \frac{6.67 \times 10^{-11} \times 5.69 \times 10^{26} \times 1.35 \times 10^{23}}{(1.222 \times 10^9)^2}$$

$$= 3.43 \times 10^{21} \, N$$

Similarly, the gravitational force (F_J) exerted by Jupiter on Callisto is:

$$F_J = \frac{6.67 \times 10^{-11} \times 1.90 \times 10^{27} \times 1.08 \times 10^{23}}{(1.884 \times 10^9)^2}$$

$$= 3.85 \times 10^{21} \, N$$

The gravitational force exerted by Jupiter on Callisto is about 12% stronger than the force exerted by Saturn on Titan. In each case, the force exerted by a planet on a satellite in a stable orbit is identical to the force the satellite exerts on the planet.

The results of these calculations support the conjecture developed in the text that both Titan and Callisto have remained undifferentiated because the heat generated by tides is not sufficient to cause melting in the interior.

17.6 Problems

1. Calculate the mass of Saturn based on the orbit of Titan by means of Newton's form of Kepler's third law (Appendix 1, Section 11). Mass of Titan (m) = 1.35×10^{23} kg, average radius of the orbit of Titan (a) = 1222×10^3 km, period of revolution of Titan (p) = 15.94 d, G = 6.67×10^{-11} Nm^2/kg^2. (Answer : 5.688×10^{26} kg).

2. Calculate the average distance of a space-craft in a "stationary" orbit around Titan using Newton's form of Kepler's third law. All data needed for this calculation are provided in Problem 1 above. Express the result in terms of 10^3 km. (Answer: 75.6×10^3 km).

3. Plot two points A and B in Figure 17.4 and determine the states of aggregation (or phases) of water and methane that coexist in each environment.

Point	P (bars)	T(°C)	CH$_4$	H$_2$O
A	3×10^{-2}	15		
B	20	−80		

4. Investigate a possible resonance between Titan and Rhea whose periods of revolution are: Titan = 15.94 d, Rhea = 4.518 d. Express the resonance in terms of whole numbers and explain in words how it works.

5. Investigate the effect of the orbit of Titan on its day/night cycle and on the seasonality of its climate. Present your results in the form of a term paper including neatly drawn diagrams.

17.7 Further Reading

Coustenis A, Lorenz R (1999) Titan. In: Weissman P.R., McFadden L.-A., Johnson T.V. (eds) Encyclopedia of the Solar System. Academic Press, San Diego, CA, pp 405–434

Freedman RA, Kaufmann III WJ (2002) Universe: The solar system. Freeman, New York.

Hartmann WK (2005) Moons and planets, 5th edn. Brooks/Cole, Belmont, CA.

Hunten DM et al. (1984) Titan. In: Gehrels T., Matthews M.S. (eds) Saturn. University of Arizona Press, Tucson, AZ, pp 671–759

Kuiper GP (1944) Titan: A satellite with an atmosphere. Astrophys J 100:378.

Lunine JI (1994) Does Titan have oceans? Am Sci 82:134–143 (Mar/Apr).

Morrison D, Owen T, Soderblom LA (1986) The satellites of Saturn. In: Burns J.A., Matthews M.S. (eds) Satellites. University of Arizona Press, Tucson, AZ, pp 764–801

Owen T (1982) Titan. Sci Am 246(2):98–109.

Owen T (1999) Titan. In: Beatty J.K., Petersen C.C., Chaikin A. (eds) The New Solar System, 4th edn. Sky Publishing., Cambridge, MA, pp 277–284

Owen T (2005) Approaching Xanadu: Cassini-Huygens examines Titan. Planet Report, Jul/Aug:8–13.

Smith PH et al. (1996) Titan's surface revealed by HST imaging. Icarus 119(2):336–349.

Toon OB et al. (1988) Methane rain on Titan. Icarus 75(8):255–284.

Uranus: What Happened Here?

The planet Uranus in Figure 18.1 was observed between 1690 and 1781 by several astronomers all of whom considered it to be a distant star because its apparent motion is much slower than that of Saturn and Jupiter. It was eventually identified as the seventh planet of the solar system by William Herschel who first saw it on March 15, 1781, and subsequently determined that its orbit is approximately circular and that it lies outside of the orbits of Saturn and Jupiter. We now know that the average radius of the orbit of Uranus (19.2 AU) is about twice as large as the radius of the orbit of Saturn (9.5 AU). Therefore, the discovery of Uranus by Herschel in 1781 virtually doubled the diameter of the known solar system. Herschel originally referred to the planet he had discovered as "Georgium Sidus" (i.e., George's star after King George III who was the reigning monarch at the time). The name "Uranus" came into use several decades later.

When viewed through modern telescopes, Uranus is a featureless disk about the size of a golf ball seen from a distance of one kilometer (Freedman and Kaufmann, 2002). However, the observations recorded by Voyager 2 during its flyby in 1986 revealed that the atmosphere of Uranus is in vigorous motion and that Uranus has a magnetic field. In addition, Uranus has a set of narrow rings and 21 satellites including eleven small ones located close to the planet and five regular satellites having diameters between 470 and 1580 km (West, 1999; Bergstralh et al., 1991; Marley, 1999; Ingersoll, 1999; Burns, 1999; Freedman and Kaufmann, 2002; and Hartmann, 2005).

18.1 Physical and Orbital Properties

The most amazing fact about Uranus in Table 18.1 is that its obliquity (i.e., the inclination of its equatorial plane to the plane of its orbit) is 97.86°. In other words, the geographic north pole of Uranus in Figure 18.2 points downward which causes the planet to rotate on its side. Even more amazing is the fact that the satellites of Uranus are closely aligned with the *equatorial plane* rather than with the orbital plane. Therefore, we have to wonder: What happened to Uranus and its satellites?

18.1.1 Physical Properties

The diameter of Uranus (51,118 km) is less than half that of Saturn (120,660 km) but is similar to that of Neptune (49,528 km). We could say that Uranus and Neptune are twins, at least in terms of their size, although they differ in many other ways. The mass of Uranus (8.68×10^{25} kg) is about 85% of that of Neptune (1.02×10^{26} kg); but Uranus is almost 15 times more massive than the Earth (5.98×10^{24} kg). The bulk density (1.290 g/cm^3) of Uranus indicates that it has a rocky core surrounded by layers of liquid and gas composed of hydrogen, helium, and volatile compounds such as methane (CH_4) and its polymerization products (C_4H_2, C_2H_2, C_2H_6, etc.). The albedo of Uranus (56%) is high compared to 44% for Jupiter and 46% for Saturn. The measured "surface" temperature of Uranus at a pressure of 1 bar ($-195\,°C$) is higher than that of Neptune ($-204\,°C$) but lower than the "surface" temperature of Saturn ($-185\,°C$) (Hartmann, 2005).

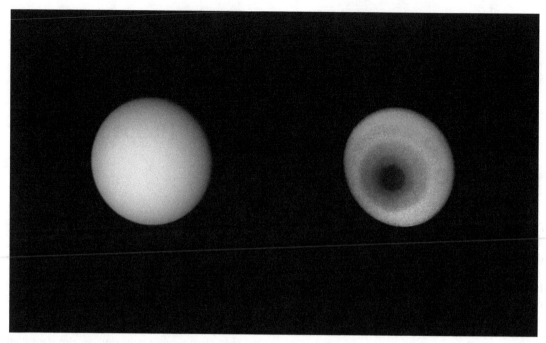

Figure 18.1. The image of Uranus at left displays the planet how it would appear to the human eye. The image at right has been processed to enhance the subtle differences that are caused by convection of the atmosphere of Uranus. Both views are from a point "above" the northern pole of rotation which actually points down because of the large obliquity of the planet. The bluish color of the atmosphere of Uranus is caused by the absorption of red light by methane. The striped appearance of the processed image indicates that the atmosphere of Uranus contains belts and zones where strong winds blow in opposite direction as noted previously in the atmospheres of Jupiter (Figure 16.1) and Saturn (Figure 17.1). The global circulation of the atmosphere of Uranus appears to be dominated by the rotation of the planet in spite of the fact that the polar regions of Saturn receive more sunlight than its equatorial region. (Image PIA 01360, Voyager 2, courtesy of NASA/JPL-Caltech)

The magnetic field of Uranus is generated by induction resulting from electrical currents caused by the flow of plasma in its interior (Van Allen and Bagenal, 1999; Kivelson and Bagenal, 1999). The strength of the magnetic field (i.e., its dipole moment) is 50 times greater than that of the Earth, but about twice the dipole moment of Saturn. In addition, the axis of the magnetic field of Uranus in Figure 18.2 makes an angle of 58.6° with the planetary axis of rotation and is offset from the planetary center. The north magnetic pole of Uranus is located in the northern hemisphere in contrast to the Earth where the south magnetic pole is located in the northern hemisphere (Section 6.4.5). Therefore, the polarity of the magnetic field of Uranus is opposite to that of the Earth. The polarity of the magnetic fields of Jupiter, Saturn, and Neptune are also opposite to that of the Earth (Van Allen and Bagenal, 1999; Kivelson and Bagenal, 1999).

18.1.2 Orbital Properties

Uranus revolves around the Sun in the prograde direction with a period of 84.099 y and the length of its semi-major axis is 19.18 AU. The plane of its orbit is closely aligned with the plane of the ecliptic (0.77°), whereas the angle between the equatorial plane of Uranus and the plane of its own orbit is 97.9°. (i.e., Uranus rotates on its side) and its period of rotation is 18.0 h (0.75d). The severe inclination of the axis of rotation causes the direction of rotation to be retrograde when viewed from a vantage point above the plane of the ecliptic. The shape of the orbit of Uranus is not quite circular with an eccentricity of 0.047 compared to only 0.009 for the orbit of Neptune. The average orbital speed of Uranus is 6.83 km/s whereas that of Neptune is 5.5 km/s.

Table 18.1. Physical and orbital properties of Uranus and Neptune (Hartmann, 2005)

Property	Uranus	Neptune
Physical Properties		
Radius, km	25,559	24,764
Mass, kg	8.68×10^{25}	1.02×10^{26}
Bulk density, g/cm^3	1.290	1.640
Albedo, %	56	51
Temperature, measured at 1 bar, °C	-195	-204
Temperature, mean calculated, °C	-214	-225
Dipole moment of the magnetic field relative to the Earth	50	25
Atmosphere	$H_2 + He$	$H_2 + He$
Orbital Properties		
Semi-major axis, AU	19.18	30.06
Period of revolution, y	84.099	164.86
Period of rotation, d	0.75R	0.80
Average orbital speed, km/s	6.83	5.5
Inclination of the orbit to ecliptic	0.77°	1.77°
Eccentricity of orbit	0.047	0.009
Obliquity	97.9°	29.6°

Dipole moment from Van Allen and Bagenal (1999). The orbital periods and average orbital speeds are from Freedman and Kaufmann (2002).

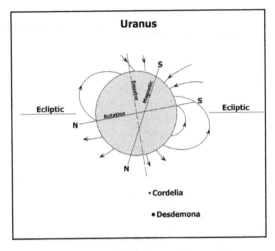

Figure 18.2. The obliquity of Uranus is 97.9°, which causes the northern end of the rotation axis of the planet to point downward and changes the direction of rotation (viewed from above the plane of the ecliptic) from prograde to retrograde. In addition, all of the satellites of Uranus exemplified in this diagram by Cordelia and Desdemona are aligned with the *equatorial plane* of Uranus. The axis of the magnetic field is offset from the center of the planet and makes an angle of 58.6°C with the planetary axis of rotation. The magnetic north pole of Uranus is in the same hemisphere as the geographic north pole. This arrangement is opposite to that of the Earth where the south magnetic pole is located in the northern hemisphere (Section 6.4.5). The magnetic lines of force of Uranus extend from the north magnetic pole to the south magnetic pole as required by the theory of magnetism. Adapted from Van Allen and Bagenal (1999) and Hartmann (2005)

Uranus and Neptune are not only similar in size and composition, but they are also approaching a state of one-to-two resonance of their orbital periods (p):

$$\frac{p(\text{Neptune})}{p(\text{Uranus})} = \frac{164.86}{84.099} = 1.96$$

When these planets are aligned with the Sun (i.e., when they are in conjunction), Uranus races ahead of Neptune and returns to the starting point in 84.099 years. In the same time, Neptune moves in its orbit through an angle (θ) of:

$$\theta_N = \frac{360 \times 84.099}{164.86} = 183.6°,$$

which means that it has traveled a little more than half-way around its orbit. We can also calculate where Uranus will be in its orbit when Neptune has completed its orbit. The angle of revolution (θ) of Uranus 164.86 years after its conjunction with Neptune is:

$$\theta_U = \frac{360 \times 164.86}{84.099} = 705.71°,$$

which means that Uranus will have traversed $705.71 - 360 = 345.71°$ of its second orbit in the time it takes Neptune to complete its first orbit. The synodic period of Neptune (i.e., the time that elapses between successive conjunctions with Uranus) and the gravitational interactions between Uranus and Neptune at conjunction are calculated in Science Briefs 18.8.1 and 18.8.2, respectively.

18.2 Atmosphere

The atmosphere of Uranus is composed of molecular hydrogen (83%), helium (15%), and methane (2%) expressed in percent by number. The noble gases (Ne, Ar, Kr, Xe) and certain volatile compounds O_2, H_2O, NH_3, H_2S, etc.) taken together make up less than 1% of the atmosphere. In general, the chemical composition of the uranian atmosphere resembles that of the Sun translated to low temperatures of about $-200\,°C$. However, the ratio of carbon to hydrogen of the atmosphere of Uranus (and Neptune) is 30 to 40 times higher than that of the Sun. The explanation for the apparent enrichment in carbon is that Uranus and Neptune formed late in the sequence of planets, after the solar wind emitted by the young Sun had selectively blown hydrogen (and helium), but not the heavier atoms of carbon, out of the protoplanetary disk. In contrast to Uranus and Neptune, the C/H ratios of Jupiter and Saturn are only 2.9 and about 3.1 times higher than the solar C/H ratio, respectively. Apparently, these planets formed before large amounts of hydrogen gas were lost from the protoplanetary disk. The delay in the formation of Uranus and Neptune may have been caused by the low density of matter at the outer fringe of the protoplanetary disk, which may have retarded the growth of these planets (Ingersoll, 1999).

The structure of the atmosphere of Uranus is similar to that of Jupiter, Saturn, and Neptune in spite of the inclination of its axis of rotation, which causes the polar regions to receive more sunlight than the equatorial region as the planet passes through the perihelion and aphelion of its orbit. The latitudinal banding, which is prominent in the atmospheres of Jupiter and Saturn, can be made visible on Uranus by computer enhancement of small differences in color. However, small ring-shaped features that appear during computer processing are artifacts caused by the presence of dust particles in the camera (Ingersoll, 1999). The latitudinal bands in the atmosphere of Uranus are maintained by global winds that reach speeds of up to 200 m/s (720 km/h). The direction of the winds enhances and opposes the rotation of the planet depending on the latitude. The images of Uranus recorded by Voyager 2 reveal the existence of plumes of warm air rising and forming white clouds as the air cools. However, large cyclones like those of Jupiter and Saturn have not been observed in the atmosphere of Uranus (Ingersoll, 1999).

The energy that drives the dynamics of the atmosphere of Uranus is provided primarily by sunlight and to a much lesser extent by internal heat. In contrast to Jupiter, Saturn, and even Neptune, Uranus has lost most of the heat that was generated during its formation by the impacts of planetesimals and by the gravitational compression of gases. The temperature of the atmosphere of Uranus in Figure 18.3 increases below the 100 mbar reference surface from $-219\,°C$ to about $-195\,°C$ at 1.0 bar and

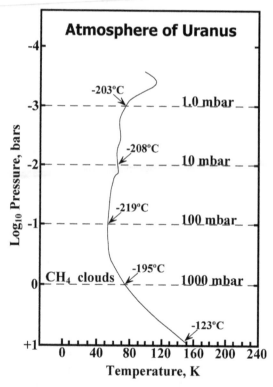

Figure 18.3. Temperature profile of the atmosphere of Uranus above and below the 100 mbar reference surface. The data recorded on January 24, 1986, during the flyby of Voyager 2 indicate that the temperature at the 100 mbar reference surface is 54 K ($-219\,°C$). Clouds composed of frozen methane form at a pressure of 1000 mbar and a temperature of $-195\,°C$. Clouds of ammonia ice have not been observed because they can form only at a temperature $-93\,°C$ which occurs at greater depth where the pressure exceeds 10 bars. Adapted from West (1999) after Gierasch and Conrath (1993)

to about $-123\,°C$ at 10 bars (West, 1999). The atmospheric temperature above the 100 mbar reference surface (i.e., at lower pressure) also increases gradually to $-163\,°C$ at a pressure of 0.4 mbar and then declines at still lower atmospheric pressure (West, 1999).

The atmosphere of Uranus contains clouds of methane that form at a pressure of about 1.0 bar (i.e., below the 100 mbar reference surface) and at a temperature of $-195\,°C$ when CH_4 vapor condenses to CH_4 ice. Clouds of ammonia can only form at greater depth where the pressure exceeds 10 bars and the temperature is about $-93\,°C$. Such clouds have not been observed on Uranus.

18.3 Internal Structure

The rapid rotation of Uranus (Table 18.1) causes its shape to deform into an oblate spheroid such that its equatorial radius (25,559 km) is greater than its polar radius (24,973 km). The resulting oblateness of Saturn:

$$\frac{r_e - r_p}{r_c} = \frac{(25,559 - 24,973)100}{25,559} = 2.29\%$$

is similar to that of Neptune (1.71%), but both are lower than the oblateness of Jupiter (6.49%) and Saturn (9.80%). The lower oblateness of Uranus and Neptune implies that their internal structure differs from that of Jupiter and Saturn (Hubbard, 1999). Another source of information about the internal structure of Uranus is its bulk density of $1.290\,g/cm^3$, which is higher than that of Saturn ($0.690\,g/cm^3$) but similar to the density of Jupiter ($1.314\,g/cm^3$).

These physical parameters together with the comparatively low mass of Uranus (8.68×10^{25} kg) indicate that Uranus contains a "rocky" core at a pressure of 8×10^6 bar and a temperature of 5000 K, both of which are much lower than the pressure (70×10^6 bar) and temperature (22,000 K) of the core of Jupiter (Section 14.1). In addition, the comparatively high density of Uranus indicates that it is enriched in heavy chemical elements compared to the Sun. This conclusion is based on the relation between the radius and the mass of a planet. If Uranus were composed primarily of hydrogen and helium like the Sun, its observed mass would not generate

Table 18.2. Internal structure of Uranus (Hubbard, 1999)

Unit	Depth, km	Thickness, km
Hydrogen and helium gas/liquid	0 – 5110	5110
Liquid ice	5110 – 21,210	16,100
Rocky core	21,210 – 25,559	4,349

enough gravitational force to compress these gases into a body that has the observed radius. Therefore, if Uranus were composed primarily of hydrogen and helium like the Sun, it would have a larger volume than it actually has.

The general conclusion arising from these considerations is that the interior of Uranus consists primarily of "liquid ice" that occurs also in the interior of Jupiter and Saturn where it makes up a much smaller fraction of the whole than it does on Uranus. Accordingly, the interior of Uranus consists of the compositional units listed in Table 18.2 and illustrated in Figure 18.4. The outer layer of Uranus (and Neptune) consists primarily of molecular hydrogen which is gradually liquefied by the increasing temperature and pressure both of which are rising with depth. However, the hydrogen liquid does not convert to the "metallic" state because the pressure and temperature in the outer hydrogen layer do not reach the values required for that transformation (Marley, 1999).

The so-called ice in the interior of Uranus and Neptune is actually a hot and highly compressed aqueous fluid containing methane, ammonia, and other compounds that form by chemical reactions in this environment. This layer may also contain pockets of liquid metallic hydrogen (Hubbard, 1999).

18.4 Rings

Uranus has eleven rings which were discovered by stellar occultation in 1977 and by Voyager 2 in 1986. They are identified in Table 18.3 by letters of the Greek alphabet and by numbers from the inside out: 6, 5, 4, α, β, η, γ, δ, λ, and ε. The Lambda ring (λ), discovered by Voyager 2, is almost invisible because it is very faint. A second ring discovered by Voyager 2, called 1986 U2R, is located 1500 km below ring 6 and extends only

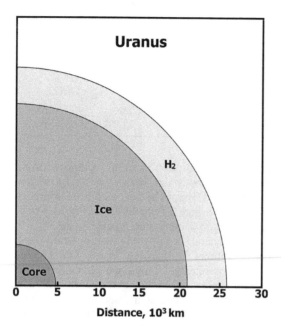

Distance, 10³ km

Figure 18.4. The interior of Saturn consists of a core (radius = 4345 km), composed of elements that form rocks on the terrestrial planets. The core is followed by a layer of "liquid ice" having a thickness of 16,102 km, which is overlain by a mixture of molecular hydrogen and helium (5110 km). The hydrogen changes from gas to liquid with increasing temperature and pressure, but it does not reach the metallic state. The atmosphere in Figure 18.3 is the outermost layer of the H_2 unit of Saturn. Adapted from Hubbard (1999)

Table 18.3. Radii of the orbits of the principal rings of Uranus (Porco, 1999)

Name	Distance from the center of Uranus, km
1986U2R	40, 337
6	41, 837
5	42, 235
4	42, 571
Alpha (α)	44, 718
Beta (β)	45, 661
Eta (η)	47, 176
Gamma (γ)	47, 627
Delta (δ)	48, 300
Lambda (λ)	50, 024
Epsilon (ϵ)	51, 149

to be composed of mixtures of methane and ammonia ice coated with carbon dust and/or organic molecules. In addition, the gaps between the uranian rings contain powdered ice that becomes visible when the gaps are backlighted by the Sun. The thickness of the uranian rings is between 10 and 100 meters, which yields

for about 3000 km. The rings revolve within the equatorial plane of the planet and therefore have the same steep inclination to the orbital plane as the equatorial plane of Saturn. The main Epsilon (ϵ) ring is 60 km wide and is located 51,149 km from the center of Uranus and within the Roche limit for objects having a density of 2.0 g/cm³. It is confined by the satellites Cordelia (inside) and Ophelia (outside) which may also constrain the edges of the Delta (δ), Gamma (γ), and Lambda (λ) rings. The orbits of the uranian rings have eccentricities ranging from 0.001 to 0.01. The highest eccentricity occurs in the orbit of the Epsilon ring.

The rings of Uranus in Figure 18.5 are composed of particles ranging in diameter from 0.1 to 10 m and their albedo is only 1.5% compared to 20 to 80% for the rings of Saturn. The particles that form the uranian rings appear

Figure 18.5. The rings of Uranus form a disk that is closely aligned with the planetary equatorial plane. This image, which was recorded by Voyager 2 from a distance of 236,000 km, reveals the presence of "lanes" of fine-grained dust particles that are not seen from other viewing angles. The short streaks are stars whose images are stretched by the 96-second exposure. (Image PIA 00142, Voyager 2, courtesy of NASA/JPL-Caltech)

an estimated mass between 10^{18} and 10^{19} kg (Hubbard, 1999; Porco, 1999).

The origin of the rings of Uranus has been explained by several alternative hypotheses (Porco, 1999):

1. The ring particles are remnants of the original protoplanetary disk that failed to accrete to form a satellite because of tidal stresses (i.e., they are located within the Roche limit).
2. One or several ice satellites may have been disrupted by tidal forces when the decay of their orbits placed them within the Roche limit.
3. One or several satellites that had strayed past the Roche limit may have been broken up by impacts of comets or asteroids and could not reassemble into a satellite.
4. Ice planetesimals in orbit around the Sun may have been captured by Uranus and broken up as exemplified by the capture of comet Shoemaker-Levy 9 by Jupiter (Section 14.5).

18.5 Satellites

Uranus has 21 satellites, eleven of which are small with diameters of less than 150 km. All but two of the uranian satellites in Table 18.4 were named after characters in Shakespeare's plays. The only exceptions are Belinda and Umbriel which were named after characters in a play of Alexander Pope. The small satellites were discovered in images recorded by Voyager 2 during the flyby in 1986. The so-called classical satellites are large enough to be visible telescopically and were discovered before the Voyager flyby: Miranda (1948), Ariel and Umbriel (1851), Titania and Oberon (1787). The satellites discovered most recently in 1997 were named Caliban, Stephano, Cycorax, Prospero, and Setebos.

18.5.1 Small Satellites

The eleven small satellites (Cordelia to Puck) identified by name in Table 18.4 have circular orbits located close to Uranus. Cordelia and Ophelia shepherd the Epsilon ring, whereas the orbits of all other satellites lie outside of the rings. The semi-major axes of the orbits of the small satellites range from 49.7×10^3 km

(Cordelia) to 86.0×10^3 km (Puck). The periods of revolution of the small satellites are all less than one day, but their periods of rotation are presently unknown (Hartmann, 2005). The masses and bulk densities are also unknown. Their low albedos (< 10) and the low temperature ($-214\,°C$) suggest that the small satellites may be composed of water ice coated with amorphous carbon or organic compounds that formed by decomposition or polymerization of methane.

The periods of revolution (p) of Puck and Ophelia are in resonance:

$$\frac{p(\text{Puck})}{p(\text{Ophelia})} = \frac{0.762}{0.376} = 2.0$$

while Cressida is in resonance with Miranda:

$$\frac{p(\text{Miranda})}{p(\text{Cressida})} = \frac{1.41.4}{0.464} = 3.0$$

These and other resonances may increase the eccentricities of the orbits of the affected satellites and thereby increase the amount of heat generated by tides. However, any tidal heat that may be generated in the small satellites of Uranus is lost by conduction and radiation in accordance with the "law of cooling bodies" (i.e., small bodies cool faster than large bodies). Insofar as we know at this time, none of the small satellites of Uranus (Cordelia to Puck) are geologically active.

18.5.2 Classical Satellites

The diameters of the so-called classical satellites of Uranus (Miranda, Ariel, Umbriel, Titania, and Oberon) in Table 18.4 range from 470 km (Miranda) to 1580 km (Titania). As we have noted before, some of the satellites of Uranus appear to be paired in terms of their sizes: Ariel (1160 km) with Umbriel (1170 km) and Titania (1580 km) with Oberon (1520 km). Only Miranda (470 km) lacks a twin. In addition, Ariel and Umbriel are similar in size to Tethys (1060 km) and Dione (1120 km), whereas Titania and Oberon resemble Rhea (1530 km) and Iapetus (1460 km) while Miranda matches Mimas (392 km) and Enceladus (500 km) among the saturnian satellites. The apparent pairing of

Table 18.4. Physical and orbital properties of the satellites of Uranus (Hartmann, 2005, Freedman and Kaufmann, 2002; Buratti, 1999)

Name and year of discovery	Physical Properties			
	Diameter km	Mass, kg	Density, g/cm³	Albedo, %
Small Satellites				
Cordelia, 1986	~ 30	?	?	<10
Ophelia, 1986	~ 30	?	?	<10
Bianca, 1986	~ 40	?	?	<10
Cressida, 1986	~ 70	?	?	<10
Desdemona, 1986	~ 60	?	?	<10
Juliet, 1986	~ 80	?	?	<10
Portia, 1986	~ 110	?	?	<10
Rosalind, 1986	~ 60	?	?	<10
1986 U1D	40	?	?	<10
Belinda, 1986	~ 70	?	?	<10
Puck, 1986	150	?	?	7
Classical Satellites				
Miranda, 1948	470	6.89×10^{19}	1.350	34
Ariel, 1851	1160	1.26×10^{21}	1.660	40
Umbriel, 1851	1170	1.33×10^{21}	1.510	19
Titania, 1787	1580	3.48×10^{21}	1.680	28
Oberon, 1787	1520	3.03×10^{21}	1.580	24
Exotic Satellites				
Caliban, 1997	60	?	?	?
Stephano	30	?	?	?
Sycorax, 1997	120	?	?	?
Prospero	50	?	?	?
Setebos	40	?	?	?

	Orbital Properties			
	Semi-major axis, 10³ km	Period of revolution, d	Period of rotation, d	Eccentricity
	Small Satellites			
Cordelia	49.7	0.335	?	~ 0
Ophelia	53.8	0.376	?	~ 0.01
Bianca	59.2	0.435	?	~ 0
Cressida	61.8	0.464	?	~ 0
Desdemona	62.7	0.474	?	~ 0
Juliet	64.4	0.493	?	~ 0
Portia	66.1	0.513	?	~ 0
Rosalind	69.9	0.558	?	~ 0
Belinda	75.3	0.624	?	~ 0
Puck	86.0	0.762	?	~ 0
	Classical Satellites			
Miranda	129.8	1.414	1.414	0.00
Ariel	191.2	2.520	2.520	0.00
Umbriel	266.0	4.144	4.144	0.00
Titania	435.8	8.706	8.706	0.00
Oberon	582.6	13.463	13.463	0.00
	Exotic Satellites			
Caliban	7, 168.9	579R	?	0.2
Stephano	7, 942.0	676R	?	?
Cycorax	12, 213.6	1289R	?	0.34
Prospero	16, 114.0	1953R	?	?
Setebos	18, 205.0	2345R	?	?

R signifies retrograde (Freedman and Kaufmann, 2002).

the Galilean satellites of Jupiter (Section 15.1) and of the satellites of Saturn (Section 16.3) and Uranus has given rise to the hypothesis that the paired satellites could have accumulated at the L4 and L5 Lagrange points of their respective planets during the formation of these planets in the protoplanetary disk (Belbruno and Gott, 2005; Howard, 2005).

The bulk densities of the classical satellites of Uranus in Figure 18.6, like those of Saturn (Figure 16.4a), vary irregularly with increasing distance from the center of this planets. The numerical values of the densities of these uranian satellites are greater than $1.0 \, g/cm^3$ and thereby imply that they are composed of mixtures of ice and refractory particles having densities of $3.0 \, g/cm^3$. Miranda, and perhaps the other large satellites of Uranus, may have been heated by tidal friction soon after formation when their orbits were more eccentric than they are at present and before they were forced into 1:1 spin-orbit coupling. Therefore, the large satellites of Uranus may contain rocky cores surrounded by

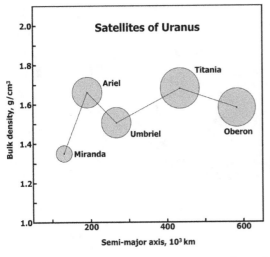

Figure 18.6. The densities of the "classical" satellites of Uranus vary irregularly with increasing distance from the center of the planet. The diameters of the satellites are drawn to scale and illustrate the pairing of Ariel with Umbriel and of Titania with Oberon. The irregular variation of the bulk densities of the uranian satellites differs from the regular decreases of the densities of the Galilean satellites of Jupiter but resembles the relation of density and orbital radius of the saturnian satellites. Data from Hartmann (2005)

thick layers of ice, provided that tidal heating was sufficient to permit internal differentiation to occur.

The ice that forms the mantles of these satellites may be a mixture of water and methane ice. The long-term exposure of methane-bearing ice to energetic electrons of the solar wind causes methane molecules to disintegrate into amorphous carbon and molecular hydrogen gas. The accumulation of residual carbon and hydrocarbon compounds on the surfaces of the large satellites can account for their comparatively low albedos that range from 40% (Ariel) to only 19% (Umbriel).

Miranda in Figure 18.7 is the smallest of the major satellites of Uranus. It contains three large rectangular *coronas* named Arden, Inverness, and Elsinor after places mentioned in Shakespeare's plays. The coronas have straight sides 200 to 300 km in length with rounded corners. The coronas contain grooves and ridges that follow the straight sides and form concentric patterns. The ice in which the coronas are embedded is heavily cratered, whereas the coronas themselves contain only a few craters. Evidently, the coronas have younger exposure ages than the ice in which they are enclosed. All of these features taken together point to the conclusion that Miranda was fragmented by high-energy impacts (perhaps several times) and that the fragments subsequently reassembled to form the satellite we see today (Pappalardo, 2005).

The impact hypothesis is supported by the fact that Uranus attracted solid objects in interplanetary space and accelerated them toward itself. Miranda, being close to Uranus, was therefore hit more frequently by high-energy projectiles than the other major satellites of Uranus. Calculations by Eugene Shoemaker suggest that a small satellite like Miranda, orbiting close to Uranus, should have been smashed to bits several times. The heat released by these impacts could have caused the reformed satellite to differentiate into a sequence of layers which may now be exposed in the form of the coronas that resulted from the last catastrophic impact.

Shoemaker's proposal has been followed up by other planetary scientists who have devised various mechanisms to explain how the internal differentiation of Miranda caused the observed

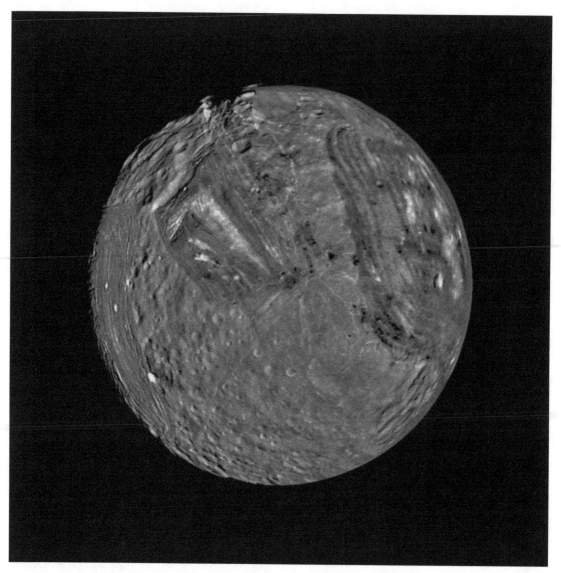

Figure 18.7. View of the south pole of Miranda. This satellite contains three large rectangular coronas, visible in this image. The coronas are embedded in highly cratered ice, whereas the coronas themselves appear to be grooved or layered but have not been intensively cratered. The strange topography of Miranda has been interpreted as evidence that this body was broken up by a powerful impact and that the resulting fragments subsequently re-assembled. Alternative hypotheses concerning the origin and history of Miranda are discussed in the text. (Image PIA 02217, Voyager 2, courtesy of NASA/JPL-Caltech)

layering of the coronas (Pappalardo, 2005). One of the schemes that has been proposed is that the layering resulted from the upward movement of *plumes* of "warm" ice (called risers) from the interior. Some of these plumes may have reached the surface and caused cryovolcanism. The heat required to sustain the differentiation of the interior originated primarily from tidal friction and less by impacts and by residual heat from the time of its formation. At the present time, Miranda's orbit is too circular to cause much tidal heating. In addition, Miranda now has spin-orbit coupling, which also prevents tidal heating. Therefore, the orbit of Miranda is assumed to have been more eccentric, or its rotation may not have been synchronized with its revolution, or both during its early history.

The plume hypothesis does, in fact, explain the origin of the chaotic surface of Miranda well enough to be considered an alternative to the break-up and re-assembly scenario. The two hypotheses are not mutually exclusive because the ancestral Miranda probably was broken up and re-assembled repeatedly during the first few hundred million years after the formation of the solar system. Later, Miranda may have come into resonance with Umbriel which caused Miranda's orbit to become eccentric and resulted in tidal heating of its interior. The area density of impact craters on the surfaces of the coronas suggests that the episode of tidal heating occurred approximately one billion years ago (Pappalardo, 2005).

Accordingly, the coronas are not necessarily fragments of the interior of Miranda that were re-assembled into the present satellite. Instead, the coronas could have formed only about one billion years ago during the internal differentiation of Miranda in response to an episode of tidal heating. The process apparently ended prematurely when the orbit of Miranda became circular again and the tidal heating stopped.

Ariel in Figure 18.8 is the closest neighbor to Miranda and has an ancient heavily cratered surface. Several large craters of recent age expose white ice in contrast to the dark surface of Ariel that is apparently covered by sediment composed of organic matter and amorphous carbon derived from methane embedded in the water ice.

The northern hemisphere of Ariel contains intersecting sets of ridges and grooves as well as several major rift valleys (grabens) with steep sides and flat floors. The presence of these valleys indicates that Ariel was tectonically active in its youth and that lava flows, presumably composed of water-ammonia and water-methane mixtures, were erupted on its surface. The heat for these activities as been attributed to tidal friction at a time when the orbit of Ariel was more eccentric than it is at present. Consequently, Ariel, like Miranda, is assumed to have differentiated into a central rocky core and a thick mantle composed of water ice containing ammonia and methane.

Figure 18.8. Ariel is similar in volume, mass, and bulk density to Umbriel. Ariel has an ancient and heavily cratered surface that has not been resurfaced by cryovolcanic activity. The intersecting flat-floored valleys in the northern hemisphere of Ariel probably formed as a result of tidal heating at an early time in its history before its orbit became circular and before it was forced into 1:1 spin-orbit coupling by the gravity of Uranus. The white ejecta deposits that surround recent impacts indicate that, in spite of its dark surface, the underlying water ice is light-colored and therefore clean. The images from which this mosaic was constructed were recorded by Voyager 2 from a distance of 170,000 km during a flyby of January 24, 1986. (Image PIA 00041, Voyager 2, courtesy of NASA/JPL-Caltech)

Umbriel is one of the darkest satellites in the solar system judging by its low albedo of only 19% compared to 40% for Ariel and 34% for Miranda (Hartmann, 2005). The surface of Umbriel in Figure 18.9 is heavily cratered and has not been resurfaced by cryovolcanic activity. The dark color of Umbriel is caused by sediment composed of organic matter and amorphous carbon that has accumulated on its surface for several billion years. The sediment covers light-colored ice that is exposed in a few places. Although Umbriel may have been tectonically active early in its history, we do not know whether it differentiated into a rocky core and an ice mantle, or whether its interior is composed of a mixture of ice and refractory dust particles similar to the interior of Callisto (Section 15.5).

Titania in Figure 18.10 is the largest of the uranian satellites (diameter = 1580 km). It has a heavily cratered surface that also contains steep-sided rift valleys similar to those on Ariel. Titania is a much brighter satellite than Umbriel as indicated by its albedo of 28%. Apparently, the

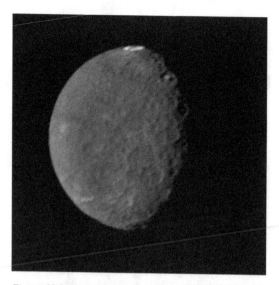

Figure 18.9. The surface of Umbriel is much darker than that of Ariel with which it may be paired. Umbriel, like Ariel, has been tectonically inactive since its early history when its orbit may have been more eccentric than it is at present and before its periods of rotation and revolution were equalized. The crater in the upper right quadrant has a diameter of 110 km and a bright central peak. The diameter of the bright ring at the top of the image is about 140 km, but its origin is presently unknown. This image was recorded during a flyby of Voyager 2 on January 24, 1986, from a distance of 557, 000 km. (Image PIA 00040, Voyager 2, courtesy of NASA/JPL-Caltech)

layer of organic-rich sediment on the surface of Titania is not as thick (or not as dark) as the sediment that covers Umbriel. These observations support the conclusion that Titania was once tectonically active and that parts of its surface were rejuvenated by the eruption of lavas flows composed of water containing ammonia and methane.

Oberon is the "twin" of Titania both in diameter (1520 km) and in the appearance of its surface (Figure 18.10). Therefore, Oberon, like the other regular satellites of Uranus, shows signs of extensional tectonics resulting in the formation of rift valleys and allowing lava flows of water-ammonia mixtures to be erupted on its surface. The geological activity of this satellite, like that of the other large satellites of Uranus, was presumable energized by heat generated by tidal friction during its early history.

The *exotic satellites* of Uranus listed in Table 18.4 were discovered telescopically between 1997 and 1998. They consist of Caliban, Stephano, Sycorax, Prospero, and Setebos. The orbits of these satellites are far removed from the center of Uranus (i.e., 7,168, 900 km for Caliban to 18,250,000 km for Setebos) and their periods of revolution are correspondingly long (i.e., 579 d for Caliban to 2345 d for Setebos). All of these satellites revolve in the retrograde direction and the diameters of the exotic satellites are small and range from approximately 30 km (Stephano) to 120 km (Sycorax). We do not know their masses and therefore cannot calculate their bulk densities. The exotic satellites may be captured ice planetesimals from the outer limits of the solar system. If so, then they are composed of mixtures of ice and dust and have remained unaffected by thermal alteration for the 4.6 billion years that have elapsed since their formation.

18.6 What Happened to Uranus?

The extreme inclination of the axis of rotation of Uranus (i.e., its obliquity of 97.9°) distinguishes it from the other giant planets of the solar system, although Venus and Pluto also have large obliquities (Hartmann, 2005). The explanation of what happened to Uranus starts with the assertion that the protoplanetary disk that revolved around the Sun at the time of its formation was aligned with the plane of the Sun's equator. Therefore, the orbits of the planets were originally aligned with the protoplanetary disk and the rotation axes of the planets that formed in that disk were initially aligned at right angles to the planes of their orbits (i.e., their obliquity was 0°). Rotating bodies in space (like spinning tops or gyros) preserve the attitude of their rotation axes unless they are acted upon by a force that is large enough to change their initial orientation. Therefore, the large obliquity of Uranus implies that the axis of rotation of its core was tipped by a collision with another body while the planet was still forming within the protoplanetary disk. The disk of gas and dust particles that had orbited Uranus prior to the hypothetical impact realigned itself with the equatorial plane of Uranus after its rotation axis had been tipped by the impact. Therefore, the rings and satellites of Uranus system formed by accretion of ice and dust particles within its present steeply dipping

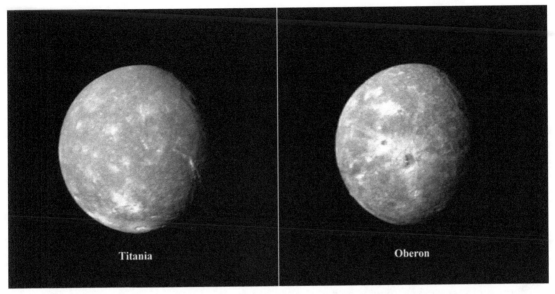

Figure 18.10. Titania (left) and Oberon (right) closely resemble each other both in terms of their physical properties and in the topography and color of their surfaces. Both satellites are heavily cratered and have not rejuvenated their surfaces by cryovolcanic activity. *Titania* shows signs of tectonic activity early in its history when the rift valley along the lower right edge may have formed. In addition, a large impact basin is visible in the upper right quadrant of its image. The surface of *Oberon*, the outermost of the classical satellites of Uranus, is likewise heavily cratered and contains several more recent rayed impact craters. The large crater close to the center of the image has a bright central uplift. A large 11-km high mountain, visible along the lower left edge of Oberon, may be the central peak of another large impact basin that is hidden behind the curvature of this body. The brown discoloration of the surfaces of Titania and Oberon is the result of the darkening of methane-bearing ice that forms the crust of these worlds. Both images were recorded during the flyby of the Voyager 2 on January 24, 1986, when the spacecraft was 500,000 km from Titania and 660,000 km from Oberon. (Images PIA 00036 (Titania) and PIA 00034 (Oberon), Voyager 2, courtesy of NASA/JPL-Caltech)

equatorial plane. Some aspects of this hypothesis resemble the theory concerning the formation of the Moon after an impact of a large object upon the Earth (Section 9.6).

If the core of Saturn was tipped by the impact of a large solid body, the satellites of Uranus may contain elements and compounds derived from the hypothetical impactor and from the rocky core of Uranus. The only evidence in support of this conjecture is the high density of the regular satellites of Uranus (i.e., 1.350 to $1.680\,g/cm^3$) compared to the densities of the satellites of Saturn (i.e., 1.200 to $1.430\,g/cm^3$) except Titan whose density is $1.880\,g/cm^3$. In general, the events that may have tipped the axis of rotation of Uranus and re-oriented its protosatellite disk all reinforce the conclusion that collisions of large bodies occurred frequently during the formation of the solar system. Even at the present time, the state of apparent tranquillity of the solar system can still be interrupted by violent events like

the impacts of the comet Shoemaker-Levy 9 on Jupiter (Section 14.5).

18.7 Summary

Uranus is less massive and has a smaller diameter than Saturn, but resembles Neptune in terms of mass and size. The apparent similarities of Uranus and Neptune, Jupiter and Saturn, as well as Venus and Earth may not be a coincidence and may require a future refinement of the theory of planet formation in the protoplenatry disk. Similar pairings have been noted among the satellites of Jupiter and Saturn as well as among the regular satellites of Uranus (i.e., Ariel with Umbriel and Titania with Oberon).

Uranus revolves around the Sun in the prograde direction but rotates on its axis in the retrograde direction because its obliquity is greater than 90°. The interior of Uranus is differentiated into a rocky core having a radius of about 5000 km, overlain by a mantle of "ice" about 16,000 km thick composed of a hot aqueous fluid containing methane and ammonia in solution. The outermost layer is about 5000 km thick and consists primarily of hydrogen and helium. The hydrogen is compressed into a liquid by pressure which increases with depth beneath the 100 mbar surface. However, the liquid hydrogen does not become electrically conducting. Consequently, the planetary magnetic field of Uranus cannot be attributed to electrical currents in a layer of liquid metallic hydrogen as in the case of Jupiter and Saturn.

The atmosphere of Uranus is composed of molecular hydrogen and helium with a small amount of methane which forms clouds at the 1.0-bar level where the temperature is −195 °C. Clouds composed of frozen water and ammonia can form only at greater depth below the 100 mbar reference surface where the temperature rises to the freezing points of ammonia and water. Therefore, water vapor (and ammonia) do not occur above the 100 mbar surface of the atmosphere of Uranus because they freeze out at lower levels.

The rings of Uranus form a thin disk that lies within the equatorial plane of the planet and consists of eleven discrete bands that are identified by letters of the Greek alphabet and by numbers. The largest and outermost ring is called Epsilon and is confined by two shepherds named Cordelia and Ophelia. The rings of Uranus are much darker than those of Saturn because the ice particles of which they are composed are coated with amorphous carbon and organic compounds.

Uranus has 21 satellites that orbit the planet close to its equatorial plane (i.e., the planes of their orbits have the same steep inclination as the equatorial plane of Uranus). The satellites of Uranus are named after characters in the plays of Shakespeare and Locke rather than after gods and goddesses of Greek mythology. The satellites of Uranus form three groups defined by their diameters and by the radii of their orbits. The eleven satellites that revolve close to Uranus have diameters of less than 150 km and the average radii of their orbits range only from 49.7×10^3 km (Cordelia) to 86.0×10^3 km (Puck). The second group of satellites (Miranda, Ariel, Umbriel, Titania, and Oberon) have large

diameters (470 to 1580 km) and their orbits are much farther removed from Uranus than the orbits of the first group (i.e., 129.8×10^3 km for Miranda to 582.6×10^3 km for Oberon). The surfaces of these satellites are heavily cratered but also contain rift valleys and evidence of cryovolcanic activity that occurred during an early period in their history. The third group of uranian satellites includes five small objects whose orbits are located very far from the center of the planet (i.e., 7.9689×10^6 to 18.2050×10^6 km). These exotic objects may be ice planetesimals that were captured by Uranus and therefore may have preserved their primordial composition and surface characteristics.

18.8 Science Briefs

18.8.1 Successive Conjunctions of Uranus and Neptune

The time that elapses between successive conjunctions of Uranus and Neptune can be calculated from equation 7.7 as stated in Section 7.2.7 and derived in Science Brief 7.6.3:

$$\frac{1}{P} = \frac{1}{E} - \frac{1}{S}$$

where P = sidereal period of revolution of a superior planet (i.e., Neptune)
E = sidereal period of revolution of the Earth (i.e., Uranus)
S = synodic period of Neptune relative to Uranus.

By substituting appropriate values from Table 18.1 we obtain:

$$\frac{1}{164.8} = \frac{1}{84.099} - \frac{1}{S}$$

$$\frac{1}{S} = \frac{1}{84.099} - \frac{1}{164.86} = 0.01189 - 0.006065$$

$$\frac{1}{S} = 0.005825$$

$$S = 171.7\,y$$

This result tells us that Uranus and Neptune are in conjunction at intervals of 171.7 years when the distance between them is approximately equal to $30.06 - 19.18 = 10.88$ AU or 1627.6×10^6 km.

Note that the synodic period is defined as the interval between successive occurrences of the same configuration of a planet. We usually express the synodic period of superior or inferior planets relative to the Earth. However, the relation between sidereal and synodic periods expressed by equation 7.7 is valid for any pair of planets in the solar system.

18.8.2 Gravitational Interactions During Conjunctions

The magnitude of the gravitational force Uranus and Neptune exert on each other is:

$$F = \frac{GM_1 \times M_2}{r^2}$$

By setting $G = 6.67 \times 10^{-11}$ newtons m^2/kg^2 and using the masses of Uranus (8.68×10^{25} kg) and Neptune (1.02×10^{26} kg) after converting the distance (r) from kilometers to meters we obtain:

$$F = \frac{6.67 \times 10^{-11} \times 8.68 \times 10^{25} \times 1.02 \times 10^{26}}{(1627.6 \times 10^9)^2}$$

$$= 0.2229 \times 10^{18} \, N$$

The magnitude of the force exerted by the Sun (1.99×10^{30} kg) on Uranus r = 19.18 AU is:

$$F(Uranus) = \frac{6.67 \times 10^{-11} \times 1.99 \times 10^{30} \times 8.68 \times 10^{25}}{(19.18 \times 149.6 \times 10^9)^2}$$

$$= 1399 \times 10^{18} \, N$$

The magnitude of the force exerted by the Sun on Neptune (r = 30.06 AU) is:

$$F(Neptune) = \frac{6.67 \times 10^{-11} \times 1.99 \times 10^{30} \times 1.02 \times 10^{26}}{(30.06 \times 149.6 \times 10^9)^2}$$

$$= 669.4 \times 10^{18}$$

These results demonstrate that the forces exerted by the Sun on Uranus and Neptune are several thousand times stronger than the gravitational interaction between Uranus and Neptune during conjunctions. Therefore, the periodic gravitational interactions between these planets probably do not significantly affect their orbits.

18.9 Problems

1. Calculate the radius of a hypothetical satellite of Uranus that has a mass of 1.00×10^{19} kg (i.e., equal to the mass of its ring particles) assuming that these particles are composed of water ice having a density of $1.00 \, g/cm^3$. (Answer: r = 134 km)
2. Estimate the radius of the orbit of such a hypothetical satellite of Uranus by calculating the Roche limit for such an object using equation 14.1 of Appendix 1. (Answer: r > 69, 557 km).
3. Calculate the perigee (q) and apogee (Q) distances from the center of Uranus for the "exotic"satellite Caliban (See Table 18.4). (Answer: q = 5, 735 × 10^3 km; Q = 8, 603 × 10^3 km).
4. Investigate whether any two of the known satellites of Uranus in Table 18.4 are "in resonance" with each other.
5. Calculate the Roche limit of the satellite Ophelia assuming that its density is $1.0 \, g/cm^3$ (i.e., water ice) and compare that limit to the radius of its orbit (Use equation 14.1 of Appendix 1). Derive a conclusion from this comparison.

18.10 Further Reading

Belbruno E, Gott JR (2005) Where did the Moon come from? Astron J 129:1724–1745

Bergstralh JT, Miner ED, Matthews MS (eds) (1991) Uranus. University of Arizona Press, Tucson, AZ

Buratti BJ (1999) Outer planet icy satellites. In: Weissman P.R., McFadden L.-A., Johnson T.V. (eds) Encyclopedia of the Solar System, 435–455

Burns JA (1999) Planetary rings. In: Beatty J.K., Petersen C.C., Chaikin A. (eds) The New Solar System, 4th edn. Sky Publishing Cambridge, MA

Freedman RA, Kaufmann III WJ (2002) Universe, The solar system. Freeman, New York

Gierasch PJ, Conrath BJ (1993) Dynamics of the atmosphere of the outer planets: Post-Voyager measurement objectives. J Geophys Res 98:5459–5469

Hartmann WK (2005) Moons and planets, 5th edn. Brooks/Cole, Belmont, CA

Howard E (2005) The effect of Lagrangian L4/L5 on satellite formation. Meteoritics and Planet Sci 40(7):1115

Hubbard WB (1999) Interiors of the giant planets. In: Beatty J.K., Petersen C.C., Chaikin A. (eds) The New Solar System, Fourth edn. Sky Publishing, Cambridge, MA, pp 193–200

Ingersoll AP (1999) Atmospheres of the giant planets. In: Beatty J.K., Petersen C.C., Chaikin A. (eds) The New

Solar System, Fourth edn. Sky Publishing Cambridge, MA, pp 201–220

Kivelson M, Bagenal F (1999) Planetary magnetospheres. In: Weissman P.R., McFadden L.-A., Johnson T.V. (eds) Encyclopedia of the Solar System. Academic Press, San Diego, CA, pp 477–498

Marley MS (1999) Interiors of the giant planets. In: Weissman P.R., McFadden L.A., Johnson T.V. (eds) Encyclopedia of the Solar System. Academic Press, Tucson, AZ, pp 339–355

Pappalardo RT (2005) Miranda: Shattering an image. The Planetary Report, Mar/Apr:14–18

Porco CC (1999) Planetary rings. In: Weissman P.R., McFadden L.-A., Johnson T.V. (eds) Encyclopedia of the Solar System. Academic Press, San Diego, CA, pp 457–475

Van Allen JA, Bagenal F (1999) Planetary magnetospheres and the interplanetary medium. In: Beatty J.K., Petersen C.C., Chaikin A. (eds) The New Solar System, 4th edn. Sky Publishing, Cambridge, MA, pp 39–58

West RA (1999) Atmospheres of the giant planets. In: Weissman P.R., McFadden L.-A., Johnson T.V. (eds) Encyclopedia of the Solar System. Academic Press, San Diego, CA, pp 315–337

Neptune: More Surprises

After Uranus had been discovered in 1781, its position in the sky should have been predictable based on knowledge of its orbit. However, the motion of Uranus did not seem to follow the predictions, which led to the conclusion in the 1830s that the orbit of Uranus is being disturbed by another planet that had not yet been discovered. In the mid 1840s John Adams (1819–1892) of England and Urbain LeVerrier (1811–1877) of France independently determined the position of the undiscovered planet. Adams submitted his result to George Airy (Astronomer Royal of England) who tried but failed to discover the missing planet. Subsequently, LeVerrier approached the German astronomer Johann Galle who found the planet on September 23 of 1846 after searching for about 30 minutes. Urbain LeVerrier and John Adams were both given credit for discovering the new planet which was named Neptune after the Roman God of the Seas because of its greenish color when viewed through a telescope. A search of historical records later revealed that Galileo Galilei saw Neptune during the winter of 1612/1613 and noted its motion but did not pursue the matter. In addition, J.J. Lalande observed Neptune on May 8 of 1795 and noted that it had moved when he looked at it again on May 10. Unfortunately, Lalande concluded that his previous measurement was erroneous (Wagner, 1991). The discovery of Neptune in 1846 raised doubts about the reliability of the Titius-Bode rule (Section 3.2) because the length of the semi-major axis of the orbit of Neptune is 30.06 AU instead of 38.8 AU as predicted. Therefore, the prevailing view among astronomers is that the validity of the Titius-Bode rule is in doubt.

19.1 Surprises Revealed by Voyager 2

As we approach Neptune in Figure 19.1, we enter a region that is virtually on the fringe of the solar system, 30.06 AU from the center of the Sun. In fact, Neptune is 10.88 AU farther from the Sun than Uranus, which itself is 9.64 AU farther than Saturn. Therefore, we expect that Neptune resides in a part of the solar system that is cold and dark because the Sun provides much less energy to Neptune that it does to the Earth. In spite of these dire predictions, the atmosphere of Neptune is more active than the atmospheres of the other giant planets and its satellite Triton continues to rejuvenate its surface with solid nitrogen emitted by geysers that were observed during the flyby of Voyager 2 on August 25, 1989. (Conrath et al., 1989; Lang and Whitney, 1991; Cruikshank, 1995; Beatty et al., 1999; and Ferris, 2002a).

An even bigger surprise is the very existence of a large planet like Neptune so close to the edge of the former protoplanetary disk where the density of matter should have been much lower than it was where Jupiter and Saturn formed. Perhaps the answer to this riddle is that Neptune and Uranus originally formed closer to the Sun and subsequently migrated (or were forced to move) to where their orbits are presently located (Freedman and Kaufmann, 2002, p. 349).

Another surprise to be recorded here is the amazing journey of Voyager 2, which was launched on August 20, 1977, and reached Jupiter two years later on July 9, 1979. Four years after launch, on August 26, 1981, Voyager 2 flew through the saturnian system. Although both planetary systems were later explored in more detail by the Galileo and Cassini missions,

Figure 19.1. Neptune is the last of the great planets of the solar system. Its blue color is caused by the absorption of red ligth by molecules of methane in its atmosphere. The image includes the Great Dark Spot, and wispy clouds of frozen methane, and the Small Dark Spot in the lower right quadrant. The atmosphere contains zones and belts where the wind blows in opposite directions at up to 1260 km/h. This image was recorded from a distance of 7.0 million kilometers as Voyager 2 approached Neptune on August 21, 1989 (Image PIA 01492, Voyager 2, courtesy of NASA/JPL-Caltech)

the Voyager spacecraft recorded images that revealed to us the unsuspected beauty and complexity of these planets and their satellites. Following the encounter with Saturn, Voyager 2 continued to Uranus (January 24, 1986) and finally to Neptune (August 25, 1989). When Voyager 2 arrived in the neptunian system, it had been traveling through interplanetary space for 12 years and therefore had far exceeded its design limits. Nevertheless, Voyager 2 sent back brilliant pictures of Neptune and its principal satellite Triton. As it continued on its journey beyond the orbit of Neptune, Voyager 2 looked back and sent us its last view of Neptune and

Triton receding into darkness. It was a triumph of engineering and human ingenuity.

By the way, both Voyagers are approaching and, in 2012, will pass the outer limit of the solar system where the solar wind collides with the radiation emitted by stars in our galactic neighborhood. After that milestone is passed, Voyager 2 will reach the vicinity of Proxima Centauri in the year 20,319 AD or 18,340 years after launch. Both Voyager spacecraft carry messages consisting of 116 images, greetings in 55 languages, 19 recordings of terrestrial sounds and of 27 musical selections, all of which are inscribed on gold plated copper discs (Ferris, 2002a,b).

The discs also contain a message from President Jimmy Carter including the statement:

"This is a present from a small and distant world, a token of our sounds, our science, our images, our music, our thoughts and our feelings."

If technologically advanced aliens living in the Centauri system decipher these messages, they will be surprised – almost 20,000 years from now. The full text of President Carter's message is presented in Science Brief 19.8.1.

19.2 Physical and Orbital Properties

In many respects, Neptune resembles Uranus (Table 18.1). For example, their diameters, masses, and albedos are similar. However, Neptune differs from Uranus in several important ways. First of all, the obliquity of Neptune is 29.6° compared to 97.9° for Uranus. Therefore, Neptune rotates in the "upright" position unlike Uranus which rotates in the "prone" position. In addition, the bulk density of Neptune $(1.640\,g/cm^3)$ is greater than that of Uranus $(1.290\,g/cm^3)$, presumably because it has a larger rocky core, or because it contains more elements of high atomic number, or both, than Uranus. Neptune also has a larger orbit and therefore has a longer period of revolution (164.8 y) than Uranus (84.01 y). However, the periods of rotation (Neptune = 18.2 h; Uranus = 18.0 h) are both remarkably short. In contrast to Uranus, the rotation of Neptune is prograde because the axis of rotation of Neptune was not tipped as much as that of Uranus. In addition, the

orbit of Neptune has a lower eccentricity (0.009) than the orbit of Uranus (0.047).

The temperature at the one-bar level of Neptune is −225 °C, whereas the temperature at the same pressure level on Uranus is −214 °C consistent with the respective distances from the Sun. However, the measured heat flow of Neptune $(0.28\,J/m^2s)$ is higher than that of Uranus $(< 0.18\,J/m^2s)$. As a result, Neptune radiates 2.7 times more heat than it absorbs from the Sun. In this respect, Neptune exceeds even Saturn and Jupiter which radiate only 2.3 and 2.5 times as much heat as they absorb, respectively. The amount of heat radiated by Uranus is only slightly larger than the heat it receives from the Sun. Therefore, Neptune, Saturn, and Jupiter are still generating heat primarily by the continuing gravitational compression of the gas and liquid in their interiors (Hartmann, 2005, p. 209).

19.3 Atmosphere

In the images returned by Voyager 2 in 1989, Neptune is blue because of the absorption of red light by methane in its atmosphere. The face of Neptune also shows faint latitudinal banding that is more subtle than the banding of Jupiter and Saturn. During the flyby of Voyager 2, the southern hemisphere of Neptune contained a large storm (at 22 °S latitude) called the *Great Dark Spot* which resembled the Great Red Spot of Jupiter both in size and in its anticlockwise rotation. Therefore, both storms are *anticyclones* because low-pressure centers (i.e., cyclones) in planetary atmospheres rotate in the clockwise direction when they occur in the southern hemisphere. The same statement applies to low-pressure centers in the atmosphere of the Earth. The Great Dark Spot on Neptune died out between 1989, when it was observed by Voyager 2, and 1994 when the Hubble space telescope imaged Neptune again. However, a new storm developed in 1995 in the northern hemisphere of Neptune. At the time of the Voyager flyby in 1989 the atmosphere of Neptune also contained a second storm at 55 °S called the *Small Dark Spot* (Atreya et al., 1989; West, 1999; Ferris, 2002a; Freedman and Kaufmann, 2002).

The atmosphere of Neptune consists of molecular hydrogen (79%), helium (18%), and methane (3%) expressed in terms of numbers of molecules. The methane of the atmosphere forms white wispy clouds about 50 km below the 100 mbar surface. The temperature of the atmosphere in Figure 19.2 rises symmetrically both above and below the 100 mbar references surface (0 km) to −138 °C, which it reaches 100 km above and below the reference level (Ingersoll, 1999).

In the polar regions of Neptune the winds reach speeds of up to about 790 km/h and blow in the same direction as the rotation of the planet. Wind speeds in the equatorial zone range up to 1260 km/h but blow in the opposite direction. The winds on Neptune are faster than those of Uranus because the atmosphere of Neptune is energized by the heat that continues to emanate from its interior.

The latitudinal bands in the atmosphere of Neptune consist of *zones* and *belts* that work the same way as they do on the other giant planets. *Zones* are light-colored bands where

warm air is rising until the volatile compounds (e.g., water, ammonia, and methane) condense to form clouds. Only methane clouds can be seen on Neptune (and Saturn) because the other compounds freeze out below the methane clouds. The dark *belts* are bands where cold air is sinking and where condensates that formed at low temperature higher up in the atmosphere are sublimating.

19.4 Internal Structure

The internal structure of Neptune is not completely constrained by the available data, most of which were obtained during the flyby of Voyager 2 (Hubbard 1984). Two of the three possible models of the interior of Neptune described by Marley (1999) include a core that is larger than the core of Uranus. However, in the third model Neptune does not have a core at all. The radius of the core in Figure 19.3 is

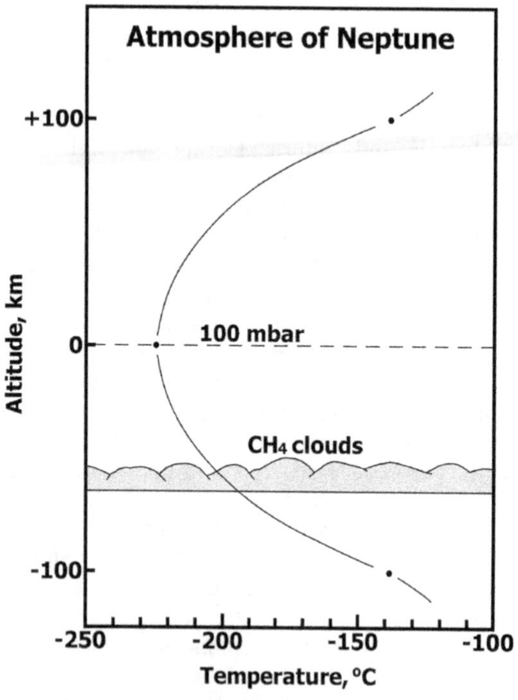

Figure 19.2. The temperature of the atmosphere of Neptune increases both above and below the 100 mbar reference surface where the temperature has a minimum of −225 °C. A layer of methane clouds forms about 50 km below the reference surface. Adapted from Ingersoll (1999)

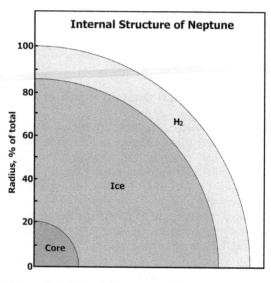

Figure 19.3. Although Neptune resembles Uranus in terms of its volume and mass, Neptune does have a larger core and a thinner atmosphere than its twin. In addition, the bulk density of Neptune (1.640 g/cm³) is greater than that of Uranus (1.290 g/cm³) indicating that Neptune contains larger amounts of elements having high atomic numbers. The "ice" mantle of Neptune consists of a hot aqueous fluid containing ammonia and methane. The molecules of these compounds are dissociated into ions which causes the fluid to be electrically conducting. Thermal convection of the "ice" mantle generates electrical currents that induce the planetary magnetic field. Data from Marley (1999)

about 20% of the planetary radius. The region composed of "ice" extends outward to 85% of the planetary radius where it grades into the atmosphere. According to this interpretation, the core of Neptune occupies a larger fraction of its volume than does the core of Uranus, but the thickness of the Neptunian atmosphere is only 15% of the planetary radius compared to 20% of the radius of Uranus.

The density of the ice mantle of Neptune rises gradually with increasing depth from $1.0\,g/cm^3$ at 85% of the planetary radius to about $4.9\,g/cm^3$ at 20%. At the core-mantle boundary the density increases abruptly to $10.0\,g/cm^3$ and then rises gradually to $12.0\,g/cm^3$ at the center (0%). The variation of the density does not constrain the concentrations of the major constituents of the mantle which include water, ammonia, and methane. If hydrogen is present within the ice layer, it would be in the metallic state. Laboratory experiments have indicated that the molecules in the hot aqueous fluid of the ice mantle of Neptune (and Uranus) are dissociated and that the fluid is therefore electrically conducting. The convection of the fluid causes electrical currents to flow that are capable of inducing the magnetic field of Neptune (Marley, 1999).

The strength (i.e., the magnetic moment) of the resulting magnetic field of Neptune in Figure 19.4 is about 25 times greater than that of the Earth, which makes it only about half as strong as the magnetic field of Uranus. The angle between the axis of rotation and the magnetic axis of Neptune is 47° compared to 59° for Uranus. The north magnetic pole of Neptune is located in the northern hemisphere in contrast to the Earth where the south magnetic pole is located in the northern hemisphere (Section 6.4.3). Therefore, the polarity of the magnetic field of Neptune is opposite to that of the Earth. In addition, the magnetic axis of Neptune is displaced from the geometric center of the planet, but the displacement is less than that of the magnetic axis of Uranus (Kivelson and Bagenal, 1999; Van Allen and Bagenal, 1999).

The chemical composition of the hypothetical core of Neptune is assumed to be dominated by elements of high atomic number which form the silicate and oxide minerals of terrestrial rocks. However, the state of these elements in the core of Neptune (or of any of the giant planets) is

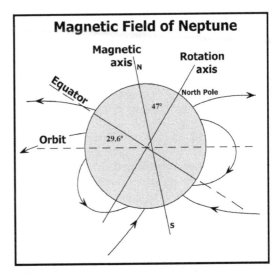

Figure 19.4. The magnetic axis of Neptune is inclined relative to the axis of rotation by 47°. Note that the magnetic axis is displaced from the center of the planet represented by the intersection of the equatorial plane and the rotation axis. The obliquity of Neptune is 29.3° which means that the geographic north pole of Neptune points "up" and that Neptune rotates in the "upright" position and in the prograde direction. The magnetic north pole of Neptune is located in the northern hemisphere. Therefore, the magnetic lines of force extend from the northern hemisphere of Neptune to the southern hemisphere consistent with the theory of magnetism. Adapted from Van Allen and Bagenal (1999, p. 52)

not known. Metallic iron is not expected to be abundant in the cores of Neptune and Uranus because iron-rich planetesimals formed primarily in the inner region of the protoplanetary disk where the terrestrial planets now reside.

The pressure and temperature in the interior of Neptune both rise with increasing depth below 100 mbar reference surface. At the top of the ice mantle (85% of the planetary radius), the pressure is 0.3 million bars (Mbars) and the temperature is 3000 K (2727 °C). Both parameters continue to rise with increasing depth and reach 6 Mbars and 7000 K in the "deep interior" (Marley, 1999). These values are similar to the core pressure of 8 Mbar and a temperature of 5000 K stated by Hubbard (1999).

The rocky cores of Neptune and Uranus formed first by accretion of planetesimals composed of ice and refractory dust particles. The cores of these planets grew more slowly than those of Jupiter and Saturn because of the comparatively low density of matter in the outer regions of the protoplanetary disk. By the time the cores of Neptune and Uranus were sufficiently massive to attract hydrogen and helium, Jupiter and Saturn had already depleted the available supply. Consequently, Uranus and Neptune could not grow as large as Jupiter and Saturn.

An alternative view is that Uranus and Neptune could not have become as large as they are, if they had formed in the outer part of the protoplanetary disk. Therefore, Freedman and Kaufmann (2002) suggested that these two planets formed between 4 and 10 AU from the Sun (i.e., between the present orbits of Jupiter and Saturn) where much more hydrogen and helium was available in the protoplanetary disk. However, before Neptune and Uranus could achieve the large size of Jupiter and Saturn, they were forced out of that region by gravitational interactions with the two large planets. After Neptune and Uranus had been forced to the outer margin of the protoplanetary disk, they stopped growing.

19.5 Rings

Neptune, like all of the giant planets, has a set of rings composed of particles of ice. These particles range in diameter from a few micrometers to about 10 meters. As in the case of Uranus, the ring particles of Neptune are coated with amorphous carbon and organic compounds that have formed by the irradation of methane by electrons that are trapped in the planetary magnetosphere. Therefore, the albedo of the rings of Neptune is about as low as that of bituminous coal. We note in passing that the temperature in the vicinity of Neptune is low enough (i.e., $-225\,°C$ at the 1-bar level) for methane gas to condense to form methane ice.

Because of their low albedo, the rings of Neptune in Figure 19.5 and Table 19.1 were discovered by the way in which they block starlight when Neptune passes in front of a bright star as seen from the Earth. The rings were subsequently observed during the flyby of Voyager 2 on August 25, 1989. As a result of these observations, we know that Neptune has five rings. The outermost ring, called *Adams*, is about 50 km wide and includes three segments where the ring

Figure 19.5. Neptune has five rings the most prominent of which are Adams (outer) and LeVerrier (inner). Three segments of the Adams ring are visibly thicker than the rest of this ring and have been named Fraternité, Egalité, and Liberté. The ring particles originated from the small satellites Naiad, Thalassa, Despina, and Galatea (Table 19.2) whose orbits are located within the Adams ring. The particles are composed of ice (water and methane) and are coated with amorphous carbon and organic compounds. This image was recorded in August of 1989 by Voyager 2 at a distance of 62,400 km from Neptune. (Image PIA 01493 Voyager 2, courtesy of NASA/JPL-Caltech)

particles have clumped together. These thicker segments of the Adams ring have been named Fraternité, Egalité, and Liberté.

The other prominent ring, called *LeVerrier*, is located inside the Adams ring and is also narrow and continuous. The area between Adams and LeVerrier is occupied by the lesser rings *Arago* (narrow) and *Lassell* (broad). The innermost ring (i.e., inside LeVerrier) is called *Galle* and is also faint and broad. All of these rings are revolving around Neptune in the prograde direction close to its equatorial plane. The rings are surrounded by dust that appears to be levitated by the planetary magnetic field because the dust particles are

Table 19.1. The rings of Neptune (Porco, 1999)

Name	Distance from the center of Neptune, km
Adams	62,932
Lassell/Arago	53,200–57,500
LeVerrier	53,200
Galle	41,000–43,000

electrically charged (Porco, 1999). The dust particles may originate by erosion of the small satellites (i.e., Naiad, Thalassa, Despina, and Galatea) that orbit Neptune inside the Adams rings and inside the Roche limit (Burns, 1999).

19.6 Satellites

Neptune has eight-satellites whose physical and orbital properties are listed in Table 19.2. All of the satellites are named after characters in Greek mythology. The small satellites (Naiad to Proteus) were discovered in 1989 during the flyby of Voyager 2. Triton was first observed telescopically in 1847 only one year after Johann Galle had found Neptune, and Nereid was discovered by Gerald P. Kuiper in 1949.

19.6.1 Small Satellites

Four of the six small satellites of Neptune (i.e., Naiad, Thalassa, Despina, and Galatea) orbit within the neptunian rings, whereas the orbits of Larissa and Proteus lie above the outermost ring (Adams). The orbits of the small satellites (Naiad to Proteus) are circular and are closely

aligned with the equatorial plane of Neptune. They revolve in the prograde direction with periods of 0.296 d (Naiad) and 1.121 d (Proteus). The periods (p) of Proteus and Larissa are in resonance:

$$\frac{p(\text{Proteus})}{p(\text{Larissa})} = \frac{1.121}{0.554} = 2.0$$

The periods of rotation of the small satellites are not known but are probably equal to their respective periods of revolution (i.e., they probably have 1:1 spin-orbit coupling). In that case, the circular orbits and synchronous rotation of the small satellites prevent them from being heated appreciably by neptunian tides. Moreover, the small diameters of these satellites (Naiad to Proteus) causes them to cool more rapidly than Triton. Therefore, the small satellites of Neptune are probably inactive at the present time even though they may be largely composed of ices of volatile compounds including water, ammonia, methane, and nitrogen.

19.6.2 Triton

Triton, by far the largest and most massive of the satellites of Neptune in Figure 19.6, has a

Table 19.2. Physical and orbital properties of the satellites of Neptune (Hartmann, 2005)

Name	Physical Properties			
	Radius, km	Mass, kg	Density, g/cm³	Albedo, %
Naiad	~25	?	?	?
Thalassa	~40	?	?	?
Despina	~90	?	?	?
Galatea	~75	?	?	?
Larissa	~95	?	?	?
Proteus	~200	?	?	?
Triton	1350	2.14×10^{22}	2.070	75
Nereid	~170	?	?	14

Name	Orbital Properties			
	Semi-major axis, 10³ km	Revolution period, d	Inclination of orbit, °	Eccentricity
Naiad	48.0	0.296	~0.0	~0
Thalassa	50.0	0.312	~4.5	~0
Despina	52.2	0.333	~0.0	~0
Galatea	62.0	0.429	~0.0	~0
Larissa	73.6	0.554	~0.0	~0
Proteus	117.6	1.121	~0.0	~0
Triton	354.8	5.877	157	0.00
Nereid	5513.4	360.16	29	0.75

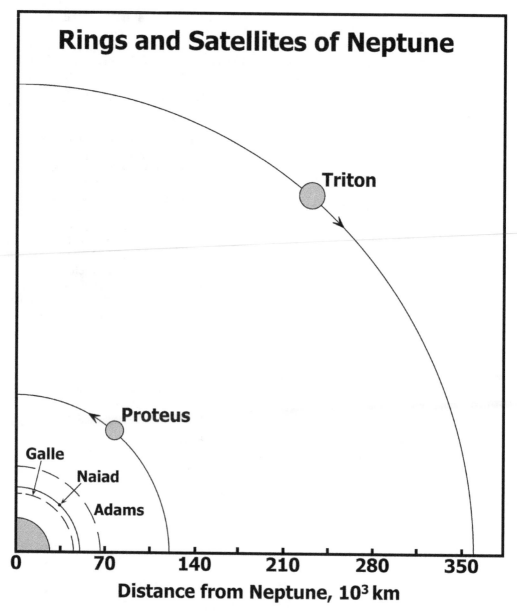

Rings and Satellites of Neptune

Triton

Proteus

Galle

Naiad

Adams

Distance from Neptune, 10³ km

Figure 19.6. The rings of Neptune lie between Adams and Galle indicated by dashed lines drawn to scale. Four of the small satellites, including Naiad, revolve within the system of rings, whereas Larissa (not shown) and Proteus orbit Neptune outside of the system of rings. Triton is much farther from Neptune than Proteus and revolves in the retrograde direction because the inclination of its orbit relative to the equatorial plane of Neptune is 157° (i.e., greater than 90°). Proteus orbits in the prograde direction because the inclination of its orbit relative to the equatorial plane of Neptune is approximately equal to zero. Data from Hartmann (2005) and Porco (1999)

circular orbit with a radius of 354.8×10^3 km which places it far above the Adams ring. It was discovered by William Lassell on October 10 of 1846 only seventeen days after Johann Galle had found Neptune. The newly discovered satellite was named in 1908 after one of the nymphs that attended Neptune in Roman mythology. Triton proved difficult to study by telescope because of its low brightness in spite of its large albedo, which explains why little was known about it until Voyager 2 reached the Neptune system in 1989.

The information recorded by Voyager 2 during its flyby at 39,800 km from Triton indicated that this satellite is larger than Pluto but smaller than the Galilean satellites of Jupiter (Chapter 15) and Titan (Chapter 17) and that it has a large albedo (75%) exceeded only by Enceladus (95%) of the saturnian system (Section 16.4.2). The mass of Triton (2.14×10^{22} kg) and its volume yield a bulk density of 2.070 g/cm^3 which means that it is composed of a mixture of refractory particles and of the ices of certain volatile compounds (Cruikshank et al., 1989).

The images returned by Voyager 2 in Figure 19.7 indicate that Triton has a highly diversified surface that contains only a small number of impact craters. Evidently, Triton has been rejuvenating its surface in spite of its low surface temperatures of $-240\,°C$ when even molecular nitrogen (N_2) is in the solid state given that its melting and boiling temperatures at a pressure of 1.0 atm are $-209.86\,°C$ and $-195.8\,°C$, respectively. All of the other volatile compounds that occur in the satellites of the giant planets (i.e., water, carbon dioxide, ammonia, and methane) are also solid at $-240\,°C$. Therefore, Triton may be composed of a mixture of ices including solid nitrogen. The image of the southern hemisphere of Triton in Figure 19.7 shows the boundary between two different kinds of terrain. The area north of the boundary contains intersecting sets of double ridges that are separated by narrow valleys reminiscent of the ridges on the surface of Europa (Section 15.3.1). The areas between the double ridges contain irregularly shaped hills and plains that resemble the surface markings of a cantaloupe when seen at a distance. This so-called *cantaloupe terrain* is characteristic of Triton and has not been seen on other icy satellites of the giant planets.

The southern physiographic region of Triton contains light-colored deposits that seem to have formed by condensation of volatile compounds (i.e., nitrogen, methane, carbon monoxide, and ammonia). The irregular distribution of these deposits has given the southern terrain a splotchy appearance and has obliterated most impact craters. The virtual absence of impact craters

on the surface of Triton implies that cryovolcanic activity is occurring here even at the low temperature of $-240\,°C$. In fact, several black plumes are visible in the splotchy terrain in Figure 19.7. The plumes rise to a height of 8 km before drifting downwind for more than 100 km in the tenuous atmosphere of Triton. These plumes presumably consist of nitrogen gas that is condensing to form small particles that are ultimately deposited as nitrogen "snow."

The hypothetical phase diagram of nitrogen in Figure 19.8 outlines two alternative pathways for the operation of plumes on Triton. Case 1 involves a reduction of pressure on solid nitrogen in the subsurface of Triton, whereas case 2 considers an increase in temperature at virtually constant temperature. In both cases, nitrogen gas is produced which is expelled through a vent in the crust. The gas expands and cools until it recondenses as solid particles of nitrogen which settle out of the plume to form the observed deposit of nitrogen snow.

The surface of Triton also contains roughly circular plains such as *Ruach* planitia in Figure 19.9 which has a flat floor about 180 km in diameter. The border of Ruach planitia is scalloped and consists of a terraced cliff several hundred meters high. The center of this planitia contains an area of rough terrain consisting of closely-space pits and an impact crater with a central uplift. Cruikshank (1999) concluded that the formation of Ruach and of other planitias on the surface of Triton included flooding by cryovolcanic fluids, freezing, remelting, and collapse. The Voyager images do not indicate whether Ruach planitia is the result of a large impact or whether it is a center of cyovolcanic activity (i.e., a caldera that was subsequently flooded with cryovolcanic fluid).

Mazomba, the largest impact crater on Triton, has a diameter of 27 km and is located south of the equator in the cantaloupe terrain and north of some enigmatic dark areas that are rimmed by light-colored aureoles. The dark patches (*guttae*) and the light-colored *aureoles* constitute topographic features called *maculae*. This term is also used to describe small areas of chaotic terrain on Europa (Section 15.3.3).

The complexity of the topography of Triton implies an unexpected intensity of internal tectonic activity in the relatively recent past.

Figure 19.7. View of the southern hemisphere of Triton that has been superimposed on an image of Neptune in the lower left corner of this picture. The topography of the northern region of Triton resembles the skin of a cantaloupe whereas the southern region looks "splotchy" because it has been shaped by cryovolcanic activity and by sublimation stimulated by wind, even though the temperature is −240 °C. At that temperature all of the volatile compounds that occur naturally in the solar system are in the solid state (e.g., water, carbon dioxide, carbon monoxide, methane, ammonia, and even molecular nitrogen). The cryovolcanic activity on Triton takes the form of plumes of black "smoke" drifting downwind in the tenuous atmosphere of this satellite. The material being vented is assumed to be nitrogen. (Image PIA 00340, Voyager 2, courtesy of NASA/JPL-Caltech)

The heat required to cause this activity could have been generated by tides that were raised when Neptune captured Triton into an elliptical orbit and subsequently forced it into the circular orbit it occupies at the present time. The capture of Triton by Neptune is also indicated by its retrograde revolution and by the steep inclination (i.e.,157°) of its orbit relative to the plane of

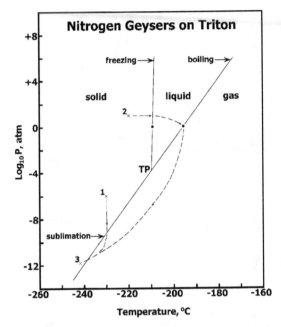

Nitrogen Geysers on Triton

Figure 19.8. This phase diagram is based on the known melting point ($-209.86\,°C$) and boiling point ($-195.8\,°C$) of molecular nitrogen at 1.0 atm pressure (solid dots). The melting-point line is assumed to have a positive slope and the temperature and pressure at the surface of Triton are assumed to be $-240\,°C$ and 10^{-12} atm, respectively. The triple point is identified by the letters TP. Two possible scenarios are outlined by the dashed lines that start from two environments below the surface of Triton identified by crosses labeled 1 and 2, at both of which nitrogen is in the solid state.

1. Solid nitrogen at $-230\,°C$ and 10^{-6} atm is decompressed to $P = 5.0 \times 10^{-10}$ atm. As a result, it converts to the gas phase and is expelled to the surface where it cools to $-240\,°C$ at a pressure of 10^{-12} atm at point 3, which causes it to condense to solid nitrogen that is deposited on the surface of Triton.

2. Solid nitrogen at $-220\,°C$ and 10 atm is warmed to a temperature of $-209\,°C$ which causes it to melt. The temperature of the liquid continues to rise and the pressure drops until the liquid nitrogen reaches the boiling curve at $-195.8\,°C$ at 1.0 atm pressure. The resulting vapor is expelled to the surface where it cools to $-240\,°C$ at 10^{-12} atm and condenses at point 3 on the surface of Triton. Data from the Handbook of Physics and Chemistry, 66th edition

the equator of Neptune. Both of these orbital properties are not expected to occur in satellites that accreted within the disk of gas and dust that surrounded Neptune during its formation. The tidal heat generated during and after the capture of Triton could have caused large-scale melting of the ice in its interior and rejuvenated its surface

either by flooding with cryogenic fluids or by a global meltdown. In either case, the remaining residual heat is still causing the emission of liquid nitrogen by geysers and the extrusion of warm ice or slush on the surface.

Some of the nitrogen gas that is emitted by the geysers and that sublimates from solid nitrogen on the surface of Triton forms a transient atmosphere that exerts a pressure of about 1.6×10^{-5} atm (Freedman and Kaufmann, 2002). This atmosphere sustains the winds that transport the plumes of the geysers downwind and cause the formation of eolian deposits of nitrogen snow and other particles on the surface of Triton.

The retrograde revolution of Triton is causing tidal interactions with Neptune that tend to reduce the orbital speed of Triton. This phenomenon is opposite to the tidal interactions between the Earth and the Moon (Section 9.8.2). Therefore, the orbit of Triton is slowly decaying and when Triton passes the Roche limit in about 100 million years it will be torn apart by tidal forces. The resulting fragments will form a spectacular system of rings around Neptune.

An additional consequence of the capture of Triton by Neptune is the destruction of the native neptunian satellites. The small satellites that presently occupy the space between the clouds of Neptune and the orbit of Triton (i.e., Naiad to Proteus) may be remnants of the original neptunian satellites that were either broken up or ejected by the gravitational forces of Triton and Neptune. The gravitational disturbance of the neptunian satellites caused by the capture of Triton also explains the anomalous orbital properties of Nereid.

19.6.3 Nereid

The semi-major axis of the orbit of Nereid (5513.4×10^3 km) is 15 times larger than that of Triton (354.8×10^3 km) and its orbital eccentricity (0.75) is larger than that of any of the regular satellites in the solar system. In addition, the orbit of Nereid makes a $29°$-angle with the equatorial plane of Neptune. These anomalous orbital properties may be consequences of the postulated capture of Triton by Neptune. However, Nereid revolves in the prograde direction, its radius is about 170 km, and it has a low albedo of 14%, that is similar to the albedo of Proteus (6%) but is much lower than that

Figure 19.9. Oblique view of Ruach planitia which has a diameter of about 180 km and may be either an extinct cryogenic volcano (i.e., a caldera) or the palimpsest of a former impact basin. The image of Neptune appears in the background oriented such that its south pole is located to the left. Note that Ruach planitia has few if any impact craters which suggests that cryogenic resurfacing is still occurring on Triton at the present time. (Image PIA 00344, Voyager 2, courtesy of NAS/JPL-Caltech)

of Triton (75%). The low albedo of Nereid suggests that its surface is covered by dark carbonaceous deposits that characterize many of the icy satellites of Uranus and the small satellites of Neptune.

19.7 Summary

When the Voyager 2 spacecraft arrived at Neptune on August 25, 1989, it recorded images of the planet and of its large satellite Triton that surprised both the experts and the lay people of the Earth. Contrary to expectations and in spite of its remote location, Neptune turned out to be just as active as the other giant planets the Voyagers had visited before. For example, Neptune has a dynamic atmosphere (composed of H_2, He, and CH_4) that exhibits the same kind of latitudinal banding as Jupiter, Saturn, and even Uranus. The energy that drives the circulation of the neptunian atmosphere does not originate from the sunlight that Neptune receives or from solar tides, but is generated by the continuing compression of the gas and fluids that surround the "rocky" core of the planet. The amount of heat emanating from the interior of Neptune is 2.7 times greater than the heat contributed by sunlight. In this regard, Neptune actually surpasses the other giant planets.

The mantle of Neptune consist of a hot fluid composed of water, ammonia, and methane. These compounds are dissociated by the high pressure and temperature causing the fluid to be electrically conducting. Convection activated by the rotation of the planet and by temperature differences cause electrical currents to flow in the interior of Neptune that generate the planetary magnetic field by induction.

Voyager 2 confirmed that Neptune is surrounded by five rings and eight satellites. The rings consist of particles of water ice containing methane, which is decomposed by irradiation with energetic electrons trapped by the magnetic field. As a result, the ring particles are coated with dark-colored organic compounds and amorphous carbon that reduce the albedo of the rings to that of bituminous coal.

Four of the six small satellites (Naiad, Thalassa, Despina, and Galatea) orbit Neptune within the system of rings. The other two (Larissa and Proteus) are located above the outermost ring. Proteus is an irregularly shaped and heavily cratered body of ice with a radius of about 200 km. The other members of this group are smaller but probably resemble Proteus in composition and appearance.

Triton is not only the largest of the neptunian satellites, it is also among the largest satellites in the solar system exceeded only by the Moon, by the Galilean satellites of Jupiter, and by Titan the giant satellite of Saturn. Triton revolves in the *retrograde* direction because the plane of its orbit is inclined by 157° to the equatorial plane of Neptune. Therefore, Triton was probably captured by Neptune into an elliptical orbit that caused extensive tidal heating. The resulting cryovolcanism caused its surface to be rejuvenated and is manifested by the ongoing eruption of geysers, which erupt liquid and gaseous nitrogen that condenses to solid nitrogen on exposure to the low temperature (−225 °C) and pressure of the surface.

Figure 19.10. As Voyager 2 was leaving the solar system on August 28, 1989, it looked back for the last time at Neptune and Triton from a distance of 4.86 million kilometers. The spacecraft is on a parabolic orbit and will not return to the solar system. (Voyager 2, NASA/JPL-Caltech)

The sublimation of nitrogen and other volatile compounds from the surface of Triton forms a diffuse atmosphere. Wind that is generated within this atmosphere deflects the plumes emanating from active geysers and forms eolian deposits of ice particles on its surface.

After Voyager 2 had passed by Neptune in Figure 19.10 it continued to move away from the solar system and, in 2012 AD, it is scheduled to encounter the heliopause where the solar wind collides with the radiation of the interstellar medium. Both Voyager spacecraft carry messages from the people of planet Earth that may be intercepted in about 20,000 years when Voyager 2 will reach the neighborhood of the star Proxima Centauri located about 4.2 lightyears from the Sun. In this way, we have cast a message in a bottle into the cosmic ocean in the hope that it will be read by intelligent beings elsewhere in the Milky Way galaxy.

19.8 Science Briefs

19.8.1 President Carter's Message to Alien Civilizations

"This Voyager spacecraft was constructed by the United States of America. We are a community of 240 million human beings among the more than 4 billion who inhabit the planet Earth. We human beings are still divided into nation states, but these states are rapidly becoming a single global civilization.

We cast this message into the cosmos. It is likely to survive a billion years into the future, when our civilization is profoundly altered and the surface of the Earth may be vastly changed. Of the 200 billion stars in the Milky Way galaxy, some- perhaps many- may have inhabited planets and spacefaring civilizations. If one such civilization intercepts Voyager and can understand these recorded contents, here is our message:

"This is a present from a small distant world, a token of our sounds, our science, our images, our music, our thoughts and our feelings. We are attempting to survive our time so we may live into yours. We hope someday, having solved the problems we face, to join a community of galactic civilizations. This record represents our hope and our determination, and our good will in a vast and awesome universe."

Jimmy Carter
President of the United States of America

The White House
June 16, 1977

The preparation of the recording and the selection of music are described by Ferris (2002b) in an issue of the Planetary Report (vol. 22(5):4–17,2002) devoted to the 25th anniversary of the launch of the Voyager spacecraft.

A plaque depicting a man and a woman, and containing information about the time and place of origin of the spacecraft, was also attached to Pioneers 10 and 11 launched on March 3, 1972 and April 6, 1973, respectively. Both have since left the solar system and continue on their voyage to the stars (Emiliani, 1992; Weissman et al., 1999).

More recently, questions have been raised about the wisdom of advertising our presence on the Earth to alien civilizations because these civilizations may be hostile.

19.9 Problems

1. Calculate the period of revolution (p) of Triton in its orbit around Neptune from Newton's version of Kepler's third law (See Tables 18.1 and 19.2 and refer to Section 11 of Appendix 1 (Answer: p = 5.88 d).
2. Calculate the escape speed from the surface of Triton. (See Section 23 of Appendix 1 and Table 19.2) (Answer: $v_e = 0.459$ km/s).
3. Calculate the average speed of molecules of N_2 in space adjacent to the surface of Triton (T = −240°C, molecular weight of $N_2 = 28.013$ amu). Consult Science

Brief 14.7.3 and use equation 22.3 of Appendix 1. (Answer: 0.171 km/s).
4. Interpret the results of your calculations for problem 2 and 3 above, assuming that significant losses of volatile molecules occur only if the speed of the molecule multiplied by 6 exceeds the escape velocity.
5. Calculate the speed of H_2 molecules at T = 33 K on the surface of Triton. The molecular weight of H_2 is 2.0159 amu. (Answer: v = 0.638 km/s).
6. Describe the variation of molecular speeds as a function of the environmental temperature and the molecular weight.

19.10 Further Reading

Atreya SK, Pollack JB, Matthews MS (eds) (1989) Origin and evolution of planetary and satellite atmospheres. University of Arizona Press, Tucson, AZ

Beatty JK, Petersen CC, Chaikin A (eds) (1999) The new solar system, 4th edn. Sky Publishing Cambridge, MA

Burns JA (1999) Planetary rings. In: J.K. Beatty, C.C. Petersen, A. Chaikin (eds) The New Solar System, 4th edn. Sky Publishing Cambridge, MA, pp 221–240

Conrath B et al. (1989) Infrared observations of the Neptune system. Science 246:1454–1459

Cruikshank DP, Brown R, Giver L, Tokunaga A (1989) Triton: Do we see to the surface? Science 245:283

Cruikshank DP (ed) (1995) Neptune and Triton. University of Arizona Press, Tucson, AZ

Cruikshank DP (1999) Triton, Pluto, and Charon. In: Beatty J.K., Petersen C.C., Chaikin A. (eds) The New Solar System, 4th edn. Sky Publishing Cambridge, MA, pp 285–296

Emiliani C (1992) Planet Earth; cosmology, geology, and the evolution of life and environment. Cambridge University Press, New York, NY

Ferris T (2002a) Seeing in the dark. Simon and Schuster, New York, NY

Ferris T (2002b) Voyager: A message from Earth. The Planet Rept, 22(5):14–17

Freedman RA, Kaufmann III WJ (1999) Universe: The solar system. W.H. Freeman, New York, NY

Hartmann WK (2005) Moons and planets, 5th edn. Brooks/Cole, Belmont, CA

Hubbard WB (1984) Planetary interiors. Van Nostrand Reinhold. New York, NY

Hubbard WB (1999) Interiors of the giant planets. In: Beatty J.K., Petersen C.C., Chaikin A. (eds) The New Solar System, 4th edn. Sky Publishing Cambridge, MA, pp 193–200

Ingersoll AP (1999) Atmospheres of the giant planets. In: Beatty J.K., Petersen C.C., Chaikin A. (eds) The New Solar System, 4th edn. Sky Publishing Cambridge, MA, pp 201–220

Kivelson MG, Bagenal F (1999) Planetary magnetospheres. In: Weissman P.R., McFadden L.-A., Johnson T.V. (eds)

Encyclopedia of the Solar System. Academic Press, San Diego, CA, pp 477–497

Lang KR, Whitney CA (1991) Wanderers in space. Cambridge University Press, Cambridge, UK

Marley MS (1999) Interiors of the giant planets. In: Weissman P.R., McFadden L.-A., Johnson T.V. (eds) Encyclopedia of the Solar Systems. Academic Press, San Diego, CA, pp 339–355

Porco CC (1999) Planetary rings. In: Weissman P.R., McFadden L.-A., Johnson T.V. Encyclopedia of the Solar System. Academic Press, San Diego, CA, pp 457–476

Van Allen JA, Bagenal F (1999) Planetary magnetospheres and the interplanetary medium. In: Beatty J.K., Petersen C.C., Chaikin A. (eds) The New Solar System, 4th edn. Sky Publishing Cambridge, MA, pp 39–58

Wagner JK (1991) Introduction to the solar system. Saunders College Pub., Holt, Rinehart, and Winston, Orlando, FA

Weissman PR, McFadden L-A, Johnson TV (eds) (1999) Encyclopedia of the solar systems. Academic Press, San Diego, CA

West RA (1999) Atmospheres of the giant planets. In: Weissman P.R., McFadden L.A., Johnson T.V. (eds) Encyclopedia of the Solar System. Academic Press, San Diego, CA, pp 315–337

Pluto and Charon: The Odd Couple

Pluto, one of the dwarf planets of the solar system, is a member of a new class of ice worlds that revolve around the Sun in the so-called Edgeworth-Kuiper belt (Section 3.1). Pluto also has much in common with some of the ice satellites of the giant planets (e.g., Callisto, Titan, Oberon, Triton, etc.) and with Chiron the Centaur, which revolves around the Sun between the orbits of Saturn and Uranus (Section 13.1.2). Pluto has a large satellites called Charon, which makes sense because Charon was the ferryman who carried the souls of the dead across the River Styx into the underworld ruled by Pluto. In addition, Pluto has two other satellites named Nix and Hydra that were discovered by the Hubble space telescope on May 15 and 18, 2005. One of these newly-found satellites (S/2005 P1) is located 48,000 km from the center of Pluto and has a diameter of about 56 km. The other satellite (S/2005 P2) orbits Pluto at a distance of 64,000 km and has a diameter of only 48 km (Cowen, 2005).

Pluto and Charon in Figure 20.1 have not yet been visited by a robotic spacecraft and have only been studied remotely by means of the Hubble space telescope and by Earth-based telescopes like those located on the summit of Mauna Kea on the island of Hawaii. This gap in our knowledge of the solar system is scheduled to be closed in 2015 when the New Horizons spacecraft that was launched by NASA in January of 2006 is scheduled to encounter Pluto and Charon. The spacecraft will map the chemical composition of the surfaces of Pluto and Charon, measure their temperatures, study their atmospheres, and search for icy rings and for additional satellites. After flying past Pluto and Charon, the spacecraft will explore the Edgeworth-Kuiper belt, which is expected to

contain many thousands of ice bodies of various sizes (Chapter 21).

Pluto was discovered on February 18 of 1930 by Clyde Tombaugh on photographs taken at the Lowell Observatory in Flagstaff, Arizona. The discovery of Pluto ended a lengthy search by many astronomers who were attempting to explain irregularities in the orbits of Uranus that were not completely accounted for by the gravitational effects of Neptune. Percival Lowell (1855–1916) predicted that the orbital irregularities of Uranus and Neptune were caused by an undiscovered "planet X." However, several searches from 1905 to 1907 and again in 1911 came up empty. The body discovered in 1930 was assumed to be a planet and was named Pluto partly because the first two letters are the initials of Percival Lowell and partly because Pluto, the ruler of the underworld, could make himself invisible as the newly discovered planet had done (Wagner, 1991; Lang and Whitney, 1991).

Charon was discovered in 1978 by James Christy who noticed that Pluto has a bump in photographs taken at the U.S. Naval Observatory and concluded that it was a satellite that orbits Pluto. Ironically, the combined masses of Pluto and Charon are too small to account for the irregularities of the orbits of Uranus and Neptune. Therefore, Pluto is not the "planet X" that Percival Lowell hoped to discover and it was discovered by accident during the unsuccessful search for the hypothetical planet X. In spite of their great distance from the Sun and the limited amount of information that is available, Pluto and Charon are the subjects of several books and many articles, including: Whyte (1980), Tombaugh and Moore (1980), Littmann (1990), Binzel (1990), Stern (1992), Stern and Tholen (1997), Stern and Mitton (1998), Stern and

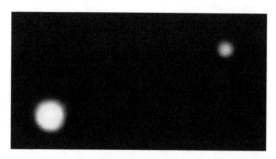

Figure 20.1. Pluto and Charon revolve around their center of gravity at an average distance from the Sun of 39.53 AU. Both bodies are composed of water ice although the surface of Pluto is covered by deposits of frozen methane, carbon monoxide, and nitrogen at a temperature of about −236 °C. Pluto and Charon will be visited in 2015 by the New Horizon spacecraft that was launched in January of 2006. After flying past Pluto and Charon, the spacecraft will continue to investigate the large ice worlds of the Edgeworth-Kuiper belt described in Chapter 21. This image of Pluto and Charon was recorded by NASA's Hubble space telescope on February 21, 1994, when Pluto approached the Earth to within 4.4×10^9 km (29.4 AU). (Courtesy of Dr. R. Albrecht, ESA/ESO Space Telescope European Coordinating Facility; NASA)

Yelle (1999), Cruikshank (1999), Freedman and Kaufmann (2002), and Hartmann (2005).

20.1 Physical Properties

The radius of Charon (595 km) in Table 20.1 is unusually large compared to the radius of Pluto (1150 km). Moreover, the radius of Pluto (1150 km) is smaller than the radii of several satellites of the giant planets (e.g., Triton, 1350 km; Titan, 2575 km; Io, 1820 km; Europa, 1565 km; Ganymede, 2640 km; and Callisto, 2420 km). The small size of Pluto and its anomalous orbital characteristics have caused it to be reclassified as a dwarf planet and it is now regarded as a representative of the ice bodies that orbit the Sun in the Edgeworth-Kuiper belt (Graham, 1999).

When Pluto was discovered in 1930, its mass was estimated to be eleven times larger than the mass of the Earth. However, in subsequent observations the mass of Pluto declined at a rate that would have caused it to go to zero in 1990. That did not actually happen because the mass of Pluto was finally determined to be 1.29×10^{22} kg and that of Charon is 1.77×10^{21} kg. The final irony in

Table 20.1. Physical and orbital properties of Pluto and Charon (Hartmann, 2005)

	Pluto	Charon
Physical Properties		
Radius, km	1150	595
Mass, kg	1.29×10^{22}	1.77×10^{21}
Density, g/cm^3	2.030	∼2.000
Albedo, %	∼60	∼40
Escape velocity, km/s	1.10	∼0.590
Surface	CH$_4$ ice, N$_2$ ice?	H$_2$O ice
Atmosphere	CH$_4$ gas diffuse	no
Orbital Properties		
Semi-major axis	39.53 AU	19.6×10^3 km
Period of revolution	247.7 y	6.39 d
Period of rotation, R	6.39 d	6.39 d
Orbital inclination to ecliptic	17.15°	—
Orbital inclination to planet equator	—	98.8°
Eccentricity of the orbit	0.248	
Obliquity	122.5°	0

R = retrograde

the discovery and subsequent popularity of Pluto occurred in 1993 when E.M. Standish at the Jet Propulsion Laboratory in Pasadena, California, demonstrated that the apparent perturbations of the orbit of Neptune by Pluto are a mathematical artifact. When he used the most reliable estimates of the masses of the giant plants, the irregularities of Neptune's orbit, which Percival Lowell had used to predict the existence of "planet X," vanished.

The bulk density Pluto (2.030 g/cm^3) is similar to that of Charon (2.000 g/cm^3), Triton (2.070 g/cm^3), and Titan (1.880 g/cm^3). All of these bodies are composed of mixtures of ice and refractory particles in different proportions. However, we do not know whether Pluto and Charon are internally differentiated and have rocky cores.

The surface of Pluto consists of water ice overlain by deposits of frozen methane, carbon monoxide, and nitrogen at a temperature of $-236\,°C$. These volatile compounds sublimate from the surface of Pluto and form a thin atmosphere, the density of which varies depending on the position of Pluto in its orbit around the Sun. When Pluto is at the aphelion of its orbit, most of the gases of its atmosphere condense to form deposits of snow on its surface. When Pluto returns to perihelion, its surface temperature rises, which increase the rate of sublimation of the volatile ices and replenishes the atmosphere. After Pluto passed the perihelion of its orbit in 1989, its atmosphere began to diminish, but it will still be detectable when the New Horizons spacecraft encounters the planet in 2015. Images of Pluto taken with the Hubble space telescope show local variations in the albedo of the surface suggesting that Pluto has polar ice caps. However, the resolution of these images is not sufficient to identify impact craters.

The surface of Charon consists primarily of water ice because Charon is not sufficiently massive to retain an atmosphere composed of methane and other volatile compounds. Therefore, these compounds have drifted into space and may have been captured by Pluto, which is separated from Charon by a short distance of only 19,640 km which is 5% of the distance between the Earth and the Moon (Freedman and Kaufmann, 2002).

The similarity of the physical properties and chemical compositions of Pluto and Charon suggests that they are genetically related. For example, Pluto could have collided with a similar ice body, which was partly broken up leaving a large fragment that was captured into orbit by Pluto. Alternatively, Charon may have been captured by Pluto during a close encounter that did not result in a collision. In either case, a collision or close encounter between two similar ice worlds requires that a large number of such bodies originally existed in the outer regions of the solar system between the present orbits of Neptune and Uranus. Therefore, if the space between the orbits of Neptune and Uranus was once populated by 1000 Pluto-like ice bodies, 10 to 50 of these bodies should have been as massive as the Earth because large bodies are not as numerous as small ones. One of

these massive bodies could have collided with Uranus and caused its rotational axis to tip over (Section 18.1.2). The hypothetical ice worlds that may have formed in the vicinity of Uranus and Neptune are no longer there because they were expelled from the solar system by gravitational interactions with the giant planets. Consequently, most of these ice bodies now revolve around the Sun in the space beyond the orbit of Neptune. In fact, Pluto and Charon are the first of these ice bodies to be discovered, which constitute a distinct class of celestial bodies. This line or reasoning leads to the expectation that the outermost region of the solar system contains a large number of ice worlds all of which revolve around the Sun in the so-called Edgeworth-Kuiper belt.

20.2 Orbital Properties

The orbit of Pluto in Table 20.1 has an average radius (semi-major axis) of 39.53 AU, which means that it receives only a small fraction of the sunlight that reaches the Earth. Evidently, Pluto exists in an environment in space that is not only cold but is also quite dark. It revolves around the Sun in the prograde direction with a period of 247.7 years. The plane of the orbit of Pluto is inclined with respect to the plane of the ecliptic at an angle of 17.15°, which is larger than that of any of the planets; although some of the asteroids also have steeply inclined orbits (e.g., Pallas 34.80°, Davida 15.90°, Ceres 10.60°, and Vesta 7.14°). The axial obliquity of Pluto is 122.5°, which means that its axis of rotation has been tipped and that its north pole is pointing "down" like that of Uranus. Consequently, Pluto rotates in the retrograde direction when viewed from a vantage point above the plane of the ecliptic.

The interaction of Pluto with Neptune in Figure 20.2 is affected by:

1. the inclination of the plane of the orbit of Pluto relative to the orbit of Neptune,
2. the higher eccentricity of the orbit of Pluto (0.248) compared to the eccentricity of Neptune's orbit (0.009), and
3. the resonance of their periods (p) of revolution (i.e., p(Pluto)/p(Neptune) = 1.50).

The consequences of these differences are that Pluto moves inside the orbit of Neptune for about 20 years during each revolution. Therefore,

Orbits of Pluto and Neptune

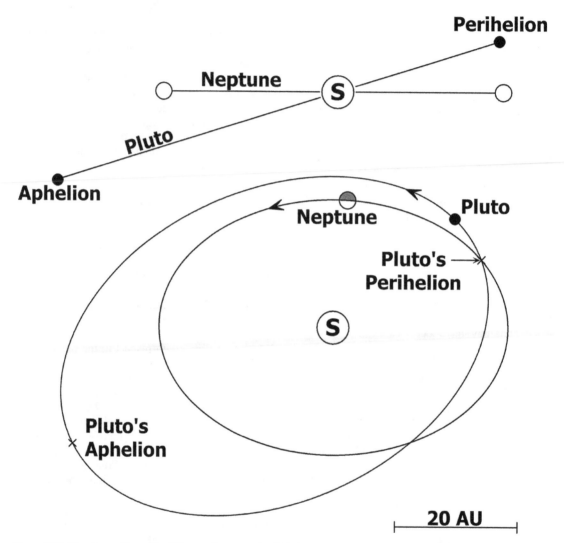

Figure 20.2. The plane of the orbit of Pluto makes an angle of 17.15° with the orbital plane of the Earth. In addition, the orbit of Pluto is more eccentric than the orbit of Neptune. Therefore, Pluto actually moves inside the orbit of Neptune for about 20 years prior to reaching its perihelion and approaches the Sun (S) more closely than Neptune in this part of its orbit. The ratio of the periods of revolution of Pluto to that of Neptune is 1.50 which means that Neptune completes three revolutions in the same time it takes Pluto to complete two revolutions. Adapted from Cruikshank (1999)

during this time Pluto is actually closer to the Sun than Neptune. The last such episode ended in 1989 and Pluto is now on its way toward its aphelion which it will reach in the year 2113 AD. Although Pluto crosses the orbit of Neptune twice during each of its revolutions, the two planets cannot collide because of the difference in the inclinations of their orbital planes illustrated in Figure 20.2.

The interactions of Pluto with Charon are even more interesting because Charon has 1-to-1 spin-orbit coupling at 6.39 days. Therefore, Charon always shows Pluto the same face. The interesting part of this arrangement is that Pluto also rotates with a period of 6.39 days. Therefore, it too shows Charon the same face like two ballroom dancers. Another feature of the Pluto-Charon relationship is that they rotate about a common center of gravity that is located in the space between them because their masses are similar, in contrast to all planet-satellite couples in the solar system. We show in Science Brief 20.4a that the center of gravity of the Pluto-Charon system is located 2369 km from the center of Pluto or 1219 km above its surface along a line that connects the centers of Pluto and Charon.

As a result of these orbital properties Pluto and Charon circle around their common center of gravity with a period of 6.39 days while keeping the same face turned toward each other and while maintaining a separation of 19,636 km. The dance of Pluto and Charon in Figure 20.3 starts at position 1. The two bodies face each other as indicated by hatch marks at a distance of 19,636 km from opposite sides of their center of gravity (G). Pluto moves in the retrograde direction (clockwise) from position 1 to position 2. In the same time, Charon also moves in the retrograde direction from its position 1 to its position 2. The two dancers are still facing each other at a distance of 19,636 km across their common center of gravity. As the two bodies continue to rotate about their center of gravity, they move to positions 3 and 4 before returning to their starting positions. The hatch marks record the fact that both Pluto and Charon have each rotated through 360° in the retrograde direction. Each rotation of Pluto and Charon around their center of gravity lasts 6.39 days. In the meantime, the center of gravity moves a short distance along the orbit of Pluto around the Sun (Science Briefs 20.4.2) Consequently, Pluto orbits the Sun on a spiral path as it continues to rotate with Charon around the common center of gravity.

20.3 Summary

Pluto is one of three presently identified dwarf planets (i.e. September 1, 2006). Like Uranus, Pluto revolves in the prograde direction, but its rotation is retrograde because its rotation axis was tipped (i.e., its axial obliquity is 122.5°). The average distance of Pluto from the Sun (39.53 AU) as well as the eccentricity (0.248) and inclination of its orbit (17°) are all greater than those of any of the regular planets.

Pluto has a large satellite called Charon whose diameter (1190 km) is more than 50% of the diameter of Pluto (2300 km). Pluto and Charon rotate around a common center of gravity in such a way that both always show each other the same face. When Pluto is closest to the Sun, deposits of frozen methane and of other volatile compounds (e.g., N_2 and CO) evaporate and form a diffuse atmosphere. As Pluto approaches the aphelion of its orbit, the surface temperature decreases to −236 °C, which causes these compounds to condense on the surface in the form of snow. Charon does not have an atmosphere because methane and other volatile compounds, which it may have contained, have escaped into space or migrated to Pluto. The bulk densities of both bodies (i.e., about $2.0 \, g/cm^3$) indicate that they are composed of mixtures of water ice and refractory particles.

Pluto appears to have had a complicated history starting with its accretion between the orbits of the giant planets followed by its expulsion to the outer fringe of the solar system. Charon may have formed from fragments that resulted from a collision of Pluto with another ice body. Alternatively, Charon may have been captured by Pluto during a close encounter. The heat generated by these events may have caused differentiation of the interior of Pluto and cryovolcanic activity on its surface . These issues may be resolved when the New Horizons spacecraft reaches the Pluto-Charon system in 2015.

20.4 Science Briefs

20.4.1 Center of Gravity Between Pluto and Charon

The center of gravity (G) of the Pluto-Charon system in Figure 20.4 is located along the straight

Retrograde Rotation of Pluto and Charon

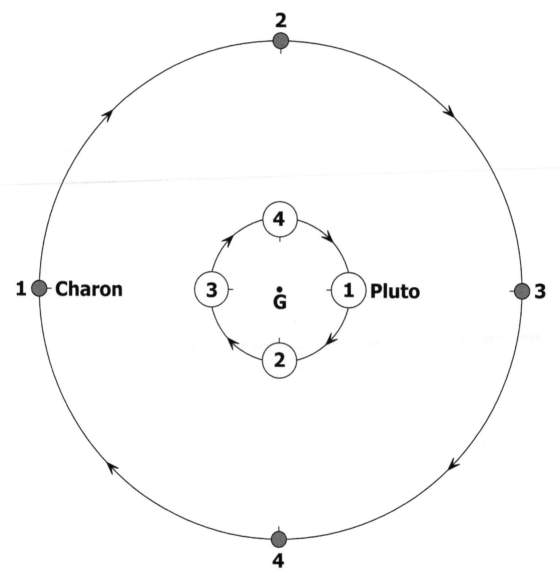

Figure 20.3. Pluto and Charon rotate around their center of gravity (G) in the retrograde direction. The numbers identify the positions of Pluto and Charon at four stages during which each rotates through 360° while continuing to show the same face to their partner. For example, when Pluto is in position 1, Charon is at a distance of 19,636 km on the other side of the center of gravity as explained in the text. The period of rotation of Pluto and Charon is 6.39 days, during which the entire system moves 2.62 million kilometers in the prograde direction along the orbit of Pluto. Data from Hartmann (2005) and Science Brief 20.4b

Center of Gravity of the Pluto-Charon System

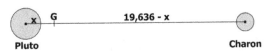

Pluto   Charon

Figure 20.4. The center of gravity (G) of the Pluto-Charon System is located outside of the body of Pluto because of the large mass of Charon amounting to 13.7% of the mass of Pluto. The distance between the two bodies and their radii are drawn to scale

line that connects the center of Pluto to the center of Charon. In addition, we designate the distance between the center of Pluto and G by the letter x. We determine the value of x by taking moments about G where the moment is defined as the product of the mass times the distance.

The masses of Pluto and Charon are:
Pluto $= 1.29 \times 10^{22}$ kg
Charon $= 1.77 \times 10^{21}$ kg
The distances between G and the centers of Pluto and Charon are:
Pluto to G $= (x)$ km
Charon to G $= (19,636 - x)$ km
Therefore,

$$(x) \times 1.29 \times 10^{22} = (19,636 - x) \times 1.77 \times 10^{21}$$

$$12.9\, x = 34,755.72 - 1.77\, x$$

$$x = \frac{34,755.72}{14.67} = 2369 \text{ km}$$

Since the radius of Pluto is 1150 km, the center of gravity (G) of the Pluto-Charon system is $2369 - 1150 = 1219$ km above the surface of Pluto. This results clearly indicates that the center of gravity is located outside of the body of Pluto, in contrast to the Earth-Moon system, the center of which is located within the body of the Earth.

20.4.2 Orbital Speed of Pluto

The orbit of Pluto has a radius (r) of 39.53 AU, which is the length of its semi-major axis. Therefore, the circumference (d) of a circle having that radius is:

$$d = 2r\pi = 2 \times 39.53 \times 149.6 \times 10^6 \times 3.14$$

$$= 3.7137 \times 10^{10} \text{ km}$$

The average orbital speed (v) of Pluto is:

$$v = \frac{d}{t} = \frac{3.7137 \times 10^{10}}{247.7 \times 365.25 \times 24 \times 60 \times 60}$$

$$= 4.75 \text{ km/s}$$

where (t) is the period of revolution.

The orbital distance traveled by Pluto during one rotation, which lasts 6.39d, is:

$$d = v \times t = 4.75 \times 6.39 \times 24 \times 60 \times 60$$

$$= 2.62 \times 10^6 \text{ km}$$

Therefore, the center of gravity of the Pluto-Charon system moves 2.62 million kilometers during each rotation.

20.5 Problems

The problems in this chapter are intended primarily as a review of computations required to solve the problems in preceding chapters. Therefore, the numerical values of the relevant parameters and the answers are not given.

1. Calculate the distance between the center of gravity of the Earth-Moon system and the center of the Earth.
2. Calculate the escape speed from the surface of Charon.
3. Calculate the synodic period of Pluto.
4. Calculate the average speed of molecules of CH_4 (molecular weight $= 1.6043$ amu) on the surface of Charon (T $= 37$ K) and determine whether Charon can retain this compound.

20.6 Further Reading

Binzel RP (1990) Pluto. Sci Am 252(6):50–58
Cowen R (2005) New partners: Hubble finds more moons around Pluto. Science News, 168 (Nov 5): 291
Cruikshank DP (1999) Triton, Pluto, and Charon. In: Beatty J.K., Petersen C.C., Chaikin A. (eds) The New Solar System, 4th edn. Sky Publishing, Cambridge, MA, pp 285–296
Freedman RA, Kaufmann III WJ (2002) Universe: The solar system. Freeman, New York
Graham R (1999) Is Pluto a planet? Astronomy, July: 42–46
Hartmann WK (2005) Moons and planets, 5th edn. Brooks/Cole, Belmont, CA

Lang KR, Whitney CA (1991) Wanderers in space. Cambridge University Press, Cambridge, UK

Littmann M (1990) Planets beyond: Discovering the outer solar system. Wiley, New York

Stern SA (1992) The Pluto-Charon system. Ann Rev Astron Astrophys 30:185–233

Stern SA, Tholen DJ (eds) (1997) Pluto and Charon. University of Arizona Press, Tucson, AZ

Stern SA, Mitton J (1998) Pluto and Charon: Ice dwarfs on the ragged edge of the solar system. Wiley, New York

Stern SA, Yelle RV (1999) Pluto and Charon. In: Weissman P.R., McFadden L.-A., Johnson T.V. (eds) Encyclopedia of the Solar System. Academic Press, San Diego, CA, pp 499–518

Tombaugh CW, Moore P (1980) Out of the darkness: The planet Pluto. Stackpole Books, Harrisburg, PA

Wagner JK (1991) Introduction to the solar system. Saunders College Pub., Philadelphia, PA

Whyte AJ (1978) The planet Pluto. Pergamon Press. Toronto, ON

Ice Worlds at the Outer Limit

The existence of a large number of ice worlds like Pluto and Charon on the outskirts of the solar system was postulated by the Irish astronomer Kenneth Edgeworth and by the American astronomer Gerard Kuiper (Kuiper, 1951). Their prediction was confirmed in 1992 by Luu and Jewitt (1993) who discovered a small ice object (1992 QB1) at a distance of 44 AU from the Sun. They estimated that its diameter is about 200 km and noted its reddish color. In subsequent years, many additional ice objects were discovered that populate a region extending from the orbit of Neptune to beyond 100 AU from the Sun. However, we do not yet have pictures of the surfaces of any ice worlds in the *Edgeworth-Kuiper belt* (Stern, 2000). These objects are called EKOs, EKBOs, KBOs, or TNOs (trans-neptunian objects). They vary in diameter from less than 10 km to more than 1000 km and at least one (Eris) appears to be larger than Pluto. Present estimates of the size distribution suggest that the Edgeworth-Kuiper belt contains about 100,000 objects having diameters of about 100 km and billions of smaller objects with diameters around 10 km. The diameters of the Edgeworth-Kuiper objects (EKOs) have not yet been measured directly because of their small size and great distance from the Earth. Instead, the diameters have been estimated from their observed brightness and their assumed albedo of 4%, which is characteristic of comets. This method overestimates the diameters of those EKOs whose albedos are actually greater than 4 %. In other words, the higher the albedo the smaller is the surface area that is required to account for the observed brightness of an EKO.

The number of EKOs that have been discovered and whose orbits have been determined has risen rapidly since 1992 to almost 1000 and continues to increase. Some of them have nearly circular orbits that are closely aligned with the plane of the ecliptic. These EKOs formed at the outskirts of the protoplanetary disk beyond the present orbit of Neptune. In addition, another group of EKOs has highly inclined and eccentric orbits that form the *Scattered Disk* which extends up to 100 AU from the Sun. The EKOs in the scattered disk probably formed closer to the Sun than their present orbits and were ejected by gravitational interactions with Uranus and Neptune.

The total mass of the EKOs that populate the region between the orbits of Neptune (30 AU) and Pluto (39 AU) has been estimated based on the observed size distribution of EKOs and by assuming that they have a bulk density of $1.0\,g/cm^3$. The results indicate that the total mass of EKOs between 30 and 50 AU is approximately equal to 30 % of the mass of the Earth and about 300 times larger than the combined masses of all asteroids. Clearly, the Edgeworth-Kuiper belt and the scattered disk are substantial parts of the solar system.

The colors of EKOs vary widely from grey to different shades of brown and red depending on the kinds of deposits that cover their surface. All EKOs are composed of mixtures of water ice and refractory particles with admixtures of frozen methane, ammonia, and oxides of carbon. The brownish-red color of some EKOs, like that of the ice satellites of the giant planets, is caused by deposits of hydrocarbon compounds and amorphous carbon that have accumulated on their surfaces. In addition, some of the largest EKOs may have transient atmospheres maintained by sublimation of ices of volatile compounds exposed on their surfaces.

The shapes and sizes of EKOs are being modified by on-going collisions and close encounters. One consequence of this process is that several EKOs are known to have satellites that were either captured or are fragments of bodies that were broken up by collisions. Another consequence of the collisions among the EKOs is that fragments may be ejected from their orbits into highly eccentric orbits that bring them into the inner solar system as short-period comets. A third consequence of the collisions among the EKOs is that the surfaces of the large bodies are altered by the formation of impact craters. In addition, the heat released by impacts, which typically occur at velocities of 0.5 to 1.0 km/s, cause volatile surface ices to vaporize (e.g., methane). Therefore, EKOs that have deposits of methane ice on their surface may be fragments produced during relatively recent collisions, which exposed methane ice that has not yet sublimated.

Many questions about the origin and evolution of some of the largest EKOs will be answered when the New Horizons spacecraft arrives in the Pluto-Charon system in 2015 and continues for a close encounter with at least one EKO (Stern and Spencer, 2004; Spencer et al., 2004). The images to be recorded during such flybys may reveal details that are presently not resolvable even with the Hubble space telescope, such as the presence of impact craters, evidence of cryovolcanism and internal tectonics, as well as the chemical composition of surface deposits and of the underlying ice.

The progress that has already been made in the study of the origin and evolution of the ice worlds that populate the Edgeworth-Kuiper belt was reviewed in March of 2003 at a meeting in Antofagasta, Chile (Davies and Barrera, 2004). This topic was also discussed by Stern and Tholen (1997), Stern and Mitton (1998), Stern (2000), Stern (2002), Levison and Weissman (1999), Fernandez (1999), Weissman (1999), and Brown (2004).

21.1 Principals of the Ice Worlds

A few of the largest ice worlds at the outer limit of the solar system have been described in detail because they reflect enough sunlight to permit spectroscopic analysis (Lemonick, 2005).

Inventories of well-studied EKOs and Centaurs discovered between 1995 and 2003 were published by Trujillo and Brown (2004), whereas Levison and Weissman (1999) listed objects discovered from 1992 to 1997. The diameters of these objects range from several hundred to more than 1000 km, which means that they approach, and may exceed, Pluto and Charon in size. However, the surface characteristics of these EKOs probably differ from those of Pluto and Charon in ways that are yet to be discovered. The largest EKOs in Table 21.1 have been given names drawn from the mytholog of different cultures, but the majority are still only identified by a code consisting of capital letters and numbers including the year of discovery. These designations are awkward and lack human interest. Therefore, the solar-system astronomers who have discovered the EKOs refer to them privately by familiar names. For example, 2003 EL61 has been referred to as "Santa" and its satellite as "Rudolph." These and other informal names (e.g., Buffy) are not likely to be approved by the International Astronomical Union (IAU) because the names of EKOs are supposed to be derived from deities in the creation mythology of human civilizations. The names of all of the principal EKOs listed in Table 21.1 have been approved by the IAU (International Astronomical Union). The only exception was "Xena" (2003 UB 313) which did not have official status in 2005.

21.1.1 Eris (2003 UB 313)

Eris was discovered on October 21, 2003, by M.E. Brown, C.A. Trujillo, and D. Rabinowitz who named it provisionally after the fictional character of the television show: "Xena: Warrior Princess". The diameter of Eris (2384 km) was calculated by means of equation 21.1 (Science Brief 21.5.4) based on its albedo of 86% that was determined from an image of Eris recorded by the Hubble space telescope (Cowen, 2006). The diameter of Eris in Figure 21.1 is about 4% larger than that of Pluto. The surface temperature of Eris is $-243\,°C$, which is cold enough to freeze methane. Therefore, Eris resembles Pluto which also has deposits of frozen methane on its surface when it is close to the aphelion of its orbit. In contrast to several other large EKOs (e.g., Quaoar, Ixion, and Varuna) which are reddish in color, Eris is grey like Pluto.

Table 21.1. Physical and orbital properties of the largest known objects in the Edgeworth-Kuiper belt (EKOs). Data from Hartmann (2005) and Trujillo and Brown (2004) for Quaoar and Ixion

Property	Pluto	Eris[1]	Sedna	Quaoar	Ixion	Varuna[2]
Diameter, km	2300	2397	∼1600	∼1145	∼1020	∼1060
Semi-major axis, AU	39.5	67.7	278	43.249	39.387	43.125
Period of rev., y	247.7	556	4,600	284	247	283.20
Period of rot., d	6.3	44.2°	20?	?	?	0.26432
Eccentricity	0.248	0.44	0.73	0.035	0.243	0.051
Inclination r.t. ecliptic	17°	44.2°	?	8.0°	19.7°	17.2°
Albedo, %	60	86	?	12	9	3.8
Spectral class	?	?	D?	D?	D?	D?
Dist. at perih., AU	29.7	37.8	∼76	41.7	29.8	40.915
Dist. at aph., AU	49.3	97.6	∼480	44.7	49.0	45.335
Discovery date	2/18 /30	9/21/03	3/15/04	2002	5/22/01	11/ 28/00

[1] Previously called Xena

[2] Wikipedia: " http://en.wikipedia.org/wiki/2000_Varuna " updated on December 2, 2005. The diameter and albedo of Eris were reported by Cowen (2006).

In addition, Eris has a satellite discovered on September 10, 2005, which was named Dysnomia after a fictional daughter of Eris. The diameter of Dysnomia is about one tenth that of Eris. Recent observations suggest that many other EKOs also have satellites and that some are members of gravitationally-linked swarms analogous to families of asteroids (Section 13.2).

The existence of satellites in orbit around EKOs is important because the length of the semi-major axis of the orbit and the period of revolution of a satellites can be used to estimate the mass of the primary object (i.e, the EKO they orbit) using Newton's version of Kepler's third law (Appendix 1 and Science Brief 15.6.1).

Eris revolves around the Sun in the prograde direction and presumably rotates in the same sense, although its rotation is not observable telescopically. The orbit of Eris in Figure 21.2 is highly eccentric (e = 0.44, Science Brief 21.5.1) and its inclination is 44°. The orbit extends from 37.8 AU at perihelion to 97.6 AU at aphelion giving it a semi-major axis of 67.7 AU and a period of revolution of 557 y (Reddy, 2005).

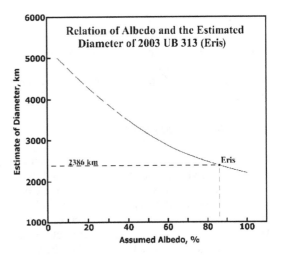

Figure 21.1. The diameter of EKO 2003 UB 313, informally known as Eris, was calculated from its albedo based on equation 21.1 stated in Science Brief 21.5.4. The albedo is known to be 86% which yields a diameter of 2397 km for Eris. Consequently, Eris is larger than Pluto which must now be regarded as the second largest known EKO and also the second largest dwarf planet (Section 3.1). Data from M.E. Brown et al. (2003) and Cowen (2006)

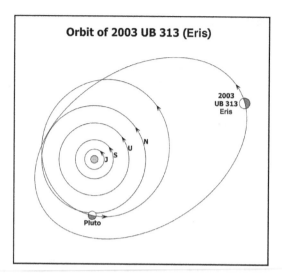

Orbit of 2003 UB 313 (Eris)

Figure 21.2. The orbit of Eris (2003 UB 313) is highly eccentric and extends to 97 AU at aphelion and 38 AU at perihelion. Eris is larger than Pluto and revolves around the Sun in the prograde direction with a period of 556 y. The inclination of its orbit relative to the ecliptic is 44° and its eccentricity is 0.44. The diagram includes the orbits of Pluto, Neptune (N), Uranus (U), Saturn (S), and Jupiter (J). The central body represents the Sun. The orbits of the terrestrial planets are too small to be shown on this scale. Adapted from Brown et al., (2003, internet)

Given that the diameter of Eris appears to be larger than that of Pluto, that its surface is covered with frozen methane like that of Pluto, and that its has a satellite, Eris has been classified as a dwarf planet of the solar system (Section 3.1).

21.1.2 Sedna (2003 VB 12)

The ice world, provisionally known as Sedna, was discovered on March 15, 2004, and was named after a goddess who lives in an ice cave at the bottom of the ocean in the mythology of the Inuits. Sedna has a diameter of about 1600 km or more and its orbit is highly elliptical (i.e., eccentricity ~0.73) presumably because of gravitational interactions with another body in the Edgeworth-Kuiper belt. The high eccentricity causes the orbit of Sedna to be even more elongated than that of Eris such that the aphelion distance of Sedna is about 480 AU compared to 97.6 AU for Eris. Although Eris may be the largest EKO, Sedna has by far the largest orbit and hence the longest period of revolution (~4600 y) of any EKO known in 2005. However, Sedna does not have a satellite and appears to be rotating more slowly than expected (Cowen, 2004). The slow rotation of Sedna, the high orbital eccentricity, and absence of a satellite have been tentatively attributed to a close encounter with a massive object that may have perturbed the orbit and rotation rate of Sedna and may have carried off or destroyed its satellite.

The color of the surface of Sedna is dark brownish-red like that of several other large EKOs (i.e., Quaoar, Ixion, and Varuna) presumably because its surface is covered by deposits of hydrocarbon compounds and amorphous carbon that formed as alteration products of methane. A telescopic search failed to detect the presence of water ice or frozen methane on the surface of Sedna perhaps because the ice is covered by the overburden of organic matter.

21.1.3 Quaoar (2002 LM 60)

The discovery of Quaoar in 2002 by M.E. Brown and C.A. Trujillo aroused a great deal of interest because of its unexpectedly large size. The discoverers named it after a god of the Native American Tongva people who inhabited the Los Angeles Basin of southern California. The diameter of Quaoar is about 1145 km or about one half that of Pluto (2330 km). Its albedo is 12% and it is classified as spectral type D (reddish-brown). The orbit of Quaoar is close to being circular (eccentricity = 0.035), the length of the semi-major axis is 43.249 AU, the period of revolution is 284 y, and the inclination of the plane of the orbit is 8.0° relative to the plane of the ecliptic.

These properties indicate that Quaoar may have formed within the protoplanetary disk beyond the present orbit of Neptune and that it may not have been ejected from the region of the giant planets. If this conjecture is confirmed, Quaoar may be a primitive object that has been preserved since the time of its formation 4.6×10^9 years ago. The New Horizons spacecraft could encounter Quaoar after it has passed the Pluto-Charon system in 2015 because the spacecraft has been designed to remain viable up to a heliocentric distance of 50 AU.

21.1.4 Ixion (2001 KX76)

The EKO identified as 2001KX76 and named Ixion was discovered by Millis et al. (2002) on May 22, 2001. It was named after the king of Thessalon in Greek mythology who killed his father-in-law and thereby disgraced himself in human society. (Houtzager, 2003). The diameter of Ixion (1020 km) was estimated by means of equation 21.1 (Science Brief 21.5.4) based on parameters published by Trujillo and Brown (2004) and using an albedo of 9%. According to this estimate, Ixion is smaller than Xena, Sedna, and Quaoar in Table 21.1 but appears to be similar to Varuna. The eccentricity of the orbit of Ixion (e = 0.243) and the orbital inclination (i = 19.7°) are both similar to those of the orbit of Pluto. In addition, Ixion shares with Pluto the 2:3 resonance with Neptune and therefore has been called a *plutino*. Ixion also resembles Pluto by its period of revolution (247 y) and by its perihelion and aphelion distances from the Sun (29.8 AU and 49.0 AU, respectively). However, the surface of Ixion is brownish red, whereas that of Pluto is grey with only a slight tinge of red.

21.1.5 Varuna (2000 WR106)

Varuna was discovered on November 28 of 2000 by R. McMillan (Spacewatch) and was named after the Hindu god who personifies divine authority. The diameter of Varuna in Table 21.1 is about 1060 km based on its thermal and optical properties. The estimated mass of Varuna of about 5.9×10^{20} kg, combined with its volume, yields a bulk density of about $1 \, g/cm^3$. The semi-major axis of the orbit of Varuna is 43.125 AU and its perihelion and aphelion distances from the Sun are 40.915 and 45.335 AU, respectively. These values yield an orbital eccentricity of 0.051. The period of revolution of Varuna is 283.20 years and its average orbital velocity is 4.53 km/s. Varuna rotates with a period of 3.17 h (0.26432 d). The revolution and rotation of Varuna are both in the prograde direction and its orbital inclination is 17.2°.

The color of the surface of Varuna is reddish brown, presumably because it is covered by a layer of organic matter and amorphous carbon that have accumulated by decomposition and polymerization of frozen methane and other organic compounds that are embedded in the water ice of which this body may be composed. However, spectral analyses of sunlight reflected by the surface have failed to reveal the presence of water ice. Accordingly, Varuna has a low albedo of 3.8% and its surface temperature is −230°C (43 K). The shape of Varuna and the topography and composition of its surface are unknown at the present time but may be revealed in the future because Varuna, like Ixion and Quaoar, is potentially within reach of the Far Horizons spacecraft (Jewitt et al., 2001; Jewitt and Sheppard, 2002).

21.2 Structure of the E-K Belt

The semi-major axes of the orbits of EKOs in Figure 21.3A have a bimodal distribution. The principal modes coincide with specific resonances in relation to Neptune at heliocentric distances of about 39.4 AU (2:3) and at 42.2 AU (3:5) as demonstrated in Science Brief 21.5.2. The EKOs that are represented by the abundance peak between 38 and 40 AU share the same 2:3 resonance as Pluto does with Neptune and have therefore been referred to as the "plutinos." The distribution of the orbits of the EKOs in Figure 21.3A demonstrates that these bodies are *not uniformly distributed* in the space outside of the orbit of Neptune but have been affected by the gravitational forces of their massive neighbor. The "sculpting" of the Edgeworth-Kuiper belt by dynamical interactions with Neptune and by other processes has been discussed by Morbidelli et al. (2004). The subdivisions presented here arise from the observed distributions of the semi-major axes, inclinations, and eccentricities of the orbits of the known EKOs.

The eccentricities of the orbits of EKOs in Figure 21.3B range widely from near zero to 0.848 (e.g., 2001FP185), but the frequency of occurrence decreases with increasing values of the eccentricity. The largest number of EKOs (i.e., 32%) have low eccentricities between 0.0 and 0.05. Consequently, the orbits of many EKOs are nearly circular, but the rest have elliptical orbits of widely varying eccentricity.

The inclinations (i) of the orbits in Figure 21.3C are used to divide the known EKOs into two groups having i < 10° and i > 10°. The percent abundances of the first group

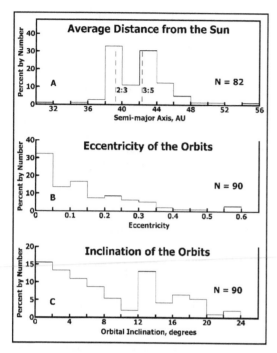

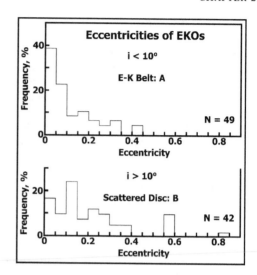

Figure 21.3. A. The average distances of EKOs from the Sun range primarily from about 30 to 50 AU with a gap at 40 to 42 AU. The two principal abundance peaks occur at heliocentric distance of 39.35 AU and 42.24 AU where EKOs are in resonance with Neptune of 3:2 and 5:3, respectively. B. The eccentricities of the orbits of about 62 % of the EKOs are less than 0.15 which means that they are nearly circular. However, the orbits of the remaining 38 % of EKOs are more elliptical with eccentricities that range up to 0.845 in the case of 2003 FX 128. C. The inclination of the orbits of about 54 % of the EKOs range only from 0 to 10° relative to the ecliptic. The orbits of the remaining 46 % of the EKOs range widely from 10 to 35°. Data from Trujillo and Brown (2004) and Levison and Weissman (1999)

Figure 21.4. A. Most of the known EKOs whose orbital inclinations (i) are less than 10° also have low orbital eccentricities (i.e., their orbits are close to being circular in shape). However, some EKOs in this group do have elliptical orbits. B. The eccentricities of EKOs whose orbital inclinations are grater than 10° have a polymodal distribution with peak abundances at e = 0 to 0.05, 0.10 to 0.15, 0.20 to 0.25 and 0.55 to 0.60. Data from Trujillo and Brown (2004) and Levison and Weissman (1999)

(i.e., i < 10°) decrease as their orbital inclinations increase from 0.0° to 10°. The frequency of EKOs in the second group (i.e., i > 10°) varies irregularly with a maximum of 13.3 % at i = 12° to 14° followed by 6.6 % at i = 16° to 17°. The orbits of the EKOs in the first group (i < 10°) lie close to the plane of the ecliptic. Therefore, these EKOs define the *Edgeworth-Kuiper Belt*. The EKOs of the second group (i.e., i > 10°) populate the *Scattered Disk*.

The difference in the orbital characteristics of EKOs in the E-K belt and in the scattered disk are highlighted in Figure 21.4A and B. The orbital eccentricities of about 61 % of the EKOs in the

E-K belt (Figure 21.4A) are less than 0.10 which means that the orbits of these bodies are close to being circular. The eccentricities of the other 39% of the EKOs in this group range from >0.10 to < 0.45, which means that their orbits are strongly elliptical. The spectrum of eccentricities of the EKOs of the second group in Figure 21.4B (i.e., i > 10°) appears to be polymodal with peaks at e = 0.0 to 0.05, 0.10 to 0.15, 0.20 to 0.25, and 0.55 to 0.60. These data indicate that about 26 % of the EKOs in the scattered disk also have circular orbits (e = 0.0 to 0.10), whereas 74 % of the EKOs have elliptical orbits with eccentricities that range up to 0.848 (Trujillo and Brown, 2004).

EKOs that have circular orbits and lie close to the plane of the ecliptic probably formed on the outskirts of the protoplanetary disk by accretion of ice particles (Section 5.2) and their orbits have not been disturbed by gravitational interactions with the giant planets or with massive bodies in the E-K belt. The EKOs whose orbits are inclined more than about 10° relative to the plane of the ecliptic and are elliptical in shape (i.e., e > 0.10) are presumed to have originated from the space between the orbits of Jupiter and Uranus and were subsequently ejected to the region outside of the orbit of Neptune. According to these criteria, most (but not all) of the EKOs that reside in the E-K belt are indigenous to that region, whereas most (but not all) of the EKOs that populate the scattered disk originated from the region now occupied by the giant planets. An additional complication in the dynamical evolution of the E-K belt is that the orbit of Neptune probably increased during the time of formation of the solar system (Gomes, 2004). The outward migration of Neptune caused resonance effects among the KBOs which responded by moving outward as well.

The bulk chemical compositions of EKOs that formed among the giant planets are probably different from those of the indigenous EKOs as a consequence of the differentiation of the protoplanetary disk (Section 5.2). In addition, EKOs that were ejected from the region of the giant planets may have been heated by tidal friction which could have manifested itself by episodic cryovolcanism at the surface and by internal tectonic activity and differentiation. These matters will be addressed by information to be collected when the New Horizons spacecraft reaches the vicinity of Pluto and Charon in 2015.

Although Pluto was originally classified as a planet, it is actually the first EKO to be discovered and is now one of the dwarf planets defined by the IAU in August of 2006. The large angle of inclination ($i = 17°$) and the ellipticity of its orbit ($e = 0.25$) taken together imply that Pluto belongs to the group of EKOs that formed among the giant planets and was subsequently ejected from that region of the solar system. These considerations enhance the importance of determining the bulk chemical composition, internal structure, and evidence for cryovolcanism on the surface of Pluto.

The EKOs that orbit the Sun close to the plane of the ecliptic (i.e., $i < 10°$) can interact with each other by close encounters or collisions that perturb their orbits and may cause them to enter the inner regions of the solar system. As they approach the Sun, the volatile ices on their surface begin to sublimate to form a transient atmosphere and a tail that characterizes comets. Therefore, the E-K belt is the source of short-period comets to be discussed in Chapter 22.

The orbits of EKOs may also be lowered by gravitational interactions causing them to migrate into the region occupied by the giant planets. Ice objects that revolve around the Sun among the orbits of the giant planets are known as the Centaurs (Section 13.1.2, 13.7.1, Table 13.3). Although the Centaurs in Table 13.3 (Section 13.1.2) are classified with the asteroids, they are probably EKOs that were deflected from the outskirts of the solar system. They do not behave like comets (except Chiron) because they stay far enough away from the Sun to avoid being heated. Chiron, which approaches the Sun to within 8.5 AU at perihelion, is an exception because it does develop a tenuous atmosphere during its closest approach to the Sun. The space between the giant planets may contain thousands of Centaurs having diameters of less than 100 km but only a few of the largest have been detected telescopically (e.g., Chiron, diameter = 300 km). In addition, the Caltech wide-area sky survey has recently discovered at least four new Centaurs whose orbits are located between the orbits of the Saturn and Neptune (Trujillo and Brown, 2004).

21.3 The Oort Cloud

In 1950 the Dutch astronomer *Jan Oort* (1900–1992) noticed that the aphelion distances of several long-period comets extend to between 50,000 and 100,000 AU from the Sun (Burnham, 2000). In addition, the orbital inclinations of these comets range widely in contrast to the short-period comets whose orbital inclinations extend only from 0° to about 30°. Oort generalized his observations by postulating that the solar system is surrounded by a very large spherical region of space that contains a large number of ice objects that are gravitationally bound to the Sun even though they are far removed from it.

Oort's hypothetical shell of ice objects was eventually accepted by solar-system astronomers and is now known as the Oort cloud. In reality, the number of objects per unit volume of space in the Oort cloud is so low that this region does not resemble a "cloud" in the conventional meaning of that word. Current estimates indicate that the Oort cloud contains about 6×10^{12} individual objects and that their combined masses are approximately equal to 40 times the mass of the Earth (Burnham, 2000).

The gravitational attraction of the Sun decreases with the reciprocal of the square of the distance and becomes ineffective at a distance of about two light years (18.92×10^{12} km) or 126,470 AU. Therefore, objects on the outer fringe of the Oort cloud (i.e., around 100,000 AU) are only weakly attracted by the Sun and may drift away into interstellar space. As a result, the outer limit of the Oort cloud is not specifically defined. Objects that drift away are replaced by others from the inner regions of the cloud, thereby causing a net loss of matter from the Oort cloud.

As the Sun and the planets move through the Milky Way galaxy, the objects in the Oort cloud may be disturbed by the gravity of passing stars. For example, the red-dwarf star Gliese 710 will pass through the Oort cloud in about 1.4×10^{6} years. Gravitational perturbations can also be caused by clouds of gas and dust in the spiral arms of the Milky Way galaxy. Such gravitational interactions can eject some of the ice objects from the Oort cloud while others are redirected to enter the solar system as a shower of comets (Koeberl, 2004). Some of these comets may impact on one of the planets or collide with one of their satellites, while others fall into the Sun or swing around it at the perihelion of their orbits and then return to the vastness of space that surrounds the solar system.

The fate of ice objects that enter the solar system from the Oort cloud has been investigated by computer simulations, which indicate that most do not survive the trip because they:

are ejected from the solar system by Jupiter

$\sim 63\%$,

break up into small fragments

25%,

lose most of the ice and develop

a dust mantle 7%,

become short-period comets or fall

into the Sun 1%.

Total   96%

Only about 4% of the ice objects of the Oort cloud that enter the solar system survive the trip (Burnham, 2000).

The popular notion that the Oort cloud is disturbed periodically by a dark companion of the Sun or by a massive planet called *Nemesis*

is not supported by any evidence available at this time. Besides, the gravity of the Sun may not be sufficient to hold the hypothetical planet Nemesis in an orbit that has its perihelion within the Oort cloud at a heliocentric distance of 50,000 to 100,000 AU.

The ice bodies of the Oort cloud could not have formed at their present residence because particles move too slowly at great distance from the Sun to allow large objects to form by accumulation of ice and dust particles. In addition, the Oort cloud has the shape of a spherical shell, whereas the protoplanetary disk was flat. Therefore, the ice objects that make up the Oort cloud could not have formed there but may have originated in the space among the giant planets and were later expelled by gravitational interactions with Jupiter and Saturn. In addition, some of these ice objects in the Oort cloud may have originally formed outside of the orbit of Neptune in the E-K belt (i.e., i < 10°) but later re-entered the region of the giant planets and were subsequently ejected by Neptune.

In general, the present locations of KBOs are the result of various gravitational interactions with the giant planets and with each other. Only those KBOs that presently move on circular orbits close to the plane of the ecliptic have avoided celestial accidents that could have heated them sufficiently to cause melting of ice in their interiors which could have led to rejuvenation of their surfaces by cryovolcanic activity.

21.4 Summary

Almost 1000 ice worlds have been discovered in the space outside of the orbit of Neptune and extending primarily from 30 to 50 AU. Some of these objects have highly elliptical orbits whose aphelion distances reach almost to 500 AU. These ice worlds populate the Edgeworth-Kuiper belt and the associated scattered disk. The diameters of the objects that have been discovered range from several hundred to more than one thousand kilometers. The existence of these large bodies implies that billions of smaller ones having diameters less than 100 km also exist in this region.

The distribution of the semi-major axes of the orbits of the known EKOs reveal the effects of orbital resonances with Neptune while the inclinations and eccentricities of their orbits contain a record of violent encounters with the giant planets of the solar system. These clues suggest that most EKOs originally formed by accretion of particles of ice and dust in the space between the orbits of the giant planets. Most of these ice planetesimals collided with the planets that were forming 4.6 billion years ago or were captured by them and thus became their satellites. The survivors were ejected by the slingshot effect and now reside in the scattered disk beyond the orbit of Neptune. The largest EKOs (i.e., Eris, Sedna, Quaoar, Ixion, and Varuna) could not have formed where they are presently located because the amount of matter available in the outer edge of the protoplanetary disk was insufficient and because the ice particles that did exist there moved too slowly to form such large bodies by accretion within the time allowed by the age of the solar system.

A large number of solid bodies that had formed among the giant planets and in the asteroidal belt were ejected by Jupiter in all directions and now form the so-called Oort cloud which extends from the outskirts of the Edgeworth-Kuiper belt to distances of up to about 100,000 AU. Although these objects remain gravitationally bound to the Sun, they move very slowly in their orbits whose eccentricities are virtually equal to 1.0. Most of the bodies in the Oort cloud are probably composed of ice, but some rocky objects (i.e., asteroids) may also be present. Some of these objects may drift away into interstellar space because they are only weakly bound to the Sun. In addition, passing stars may occasionally disturb the gravitational equilibrium of some objects in the Oort cloud and cause them to start on the long journey to the Sun. In the context of the solar system and its dynamics, the Edgeworth-Kuiper belt and the associated scattered disk as well as the Oort cloud are reservoirs of comets that become visible by developing long tails composed of gas and dust as their surfaces are heated by sunlight.

21.5 Science Briefs

21.5.1 Eccentricity of the Orbit of Eris (2003 UB 313)

Perihelion distance $= 37.8$ AU
Aphelion distance $= 97.6$ AU

The length of the semi-major axis (a) of the orbit of Eris is:

$$a = \frac{27.8 + 97.6}{2} = 67.7 \text{ AU}$$

The eccentricity (e) is defined by $e = f/a$ where the focal distance (f) is $f = 67.7 - 37.8 = 29.9$ AU. Therefore, the eccentricity of the orbit of Eris is:

$$e = \frac{29.9}{67.7} = 0.44$$

21.5.2 Semi-major Axis of an EKO in 2:3 Resonance with Neptune (Figure 21.3A)

The period of revolution (p) that yields a 2:3 resonance with Neptune is:

$$\frac{p(\text{Neptune})}{p(\text{EKO})} = \frac{164.8}{p(\text{EKO})} = \frac{2}{3}$$

$$p(\text{EKO}) = \frac{164.8 \times 3}{2} = 247.2 \text{ y}$$

The semi-major axis (a) of the orbit of an EKO that has a period of revolution of 247.2 y is obtained from Kepler's third law:

$$a^3 = p^2$$

Taking logarithms to the base 10:

$$3 \log a = 2 \log p$$

$$\log a = \frac{2 \log 247.2}{3} = 1.5953$$

$$a = 39.4 \text{ AU}$$

21.5.3 Semi-major Axis of an EKO in 3:5 Resonance with Neptune (Figure 21.3A)

$$\frac{p(\text{Neptune})}{p(\text{EKO})} = \frac{164.8}{p(\text{EKO})} = \frac{3}{5}$$

$$p(\text{EKO}) = \frac{164.8 \times 5}{3} = 274.6 \text{ y}$$

$$3 \log a = 2 \log 274.6 = 4.877$$

$$\log a = \frac{4.877}{3} = 1.625$$

$$a = 42.24 \text{ AU}$$

EKOs whose semi-major axes are close to $a = 39.35$ AU and $a = 42.24$ AU occur more frequently in Figure 21.3A than those whose semi-major axes have other values.

21.5.4 Calculation of the Radius of EKOs

The radius (R) of an EKO can be estimated from the equation (Levison and Weissman, 1999):

$$R^2 = \frac{4.53 \times 10^5 \, r^2 \Delta^2}{10^{0.4V} \times p} \tag{21.1}$$

where R = radius in km,

r = heliocentric distance in AU,

Δ = distance from the Earth in AU,

p = geometric albedo expressed as a decimal fraction,

V = visual magnitude of the EKO.

The relevant parameters of 1996 TL66 are (Trujillo and Brown, 2004):

r = 35.03 AU,
Δ = 34.09 AU,
V = 20.76.

Assuming that the albedo (p) = 0.04 and substituting into equation 21.1:

$$R^2 = \frac{4.53 \times 10^5 \times (35.03)^2 (34.09)^2}{10^{0.4 \times 20.76} \times 0.04}$$

$$R^2 = \frac{4.53 \times 10^5 \times 1227.1 \times 1162.1}{2.013 \times 10^8 \times 0.04}$$

$$= \frac{6.4598 \times 10^{11}}{0.08052 \times 10^8}$$

$$R^2 = 80.226 \times 10^3 = 8.0226 \times 10^4$$

$$R = 2.83 \times 10^2 = 283 \, \text{km}, \text{ or } \sim 300 \, \text{km}.$$

21.6 Problems

1. Write a term paper on the theme: "Questions to be answered about the origin and evolution of KBOs by the New Horizons spacecraft."
2. Calculate the radius of "Eris" using equation 21.1, where R = radius in km, r = heliocentric distance (97.6 AU), Δ = distance from the Earth (96.6 AU),

p = albedo expressed as a decimal fraction (0.86), and V = visual magnitude (18.8). (Answer R = 1189 km).
3. Calculate the magnitude of the force of gravity exerted by the Sun on an object in the Oort cloud that has a diameter of 100 km and a density of $1.00 \, \text{g/cm}^3$ at a heliocentric distance of 100,000 AU. Solar mass = 1.99×10^{30} kg. (Answer: F = 3.10×10^4 N).
4. Calculate the period of revolution of an object in the Oort cloud at an average distance of 100,000 AU from the Sun traveling on a circular path. (Answer: p = 31.6×10^6 y).
5. Calculate the average orbital speed of an object on a circular orbit in the Oort cloud at a distance of 100,000 AU (Answer: v = 0.00094 km/s).

21.7 Further Reading

Brown ME (2004) The Kuiper belt. Physics Today April, 49–54

Brown ME, Trujillo CA, Rabinowitz D (2005) The discovery of 2003 UB313, the 10th planet. Internet

Burnham R (2000) Great comets. Cambridge University Press, Cambridge, UK

Cowen R (2006) Tenth planet turns out to be a shiner. Science News 169(16):230

Cowen R (2004) Puzzle on the edge. Science News Apr 24: 262

Davies JK, Barrera LH (eds) 2004. The first decadal review of the Edgeworth-Kuiper belt. Kluwer Academic Publishers, Dordrecht, The Netherlands

Fernandaz JA (1999) Cometary dynamics. In: Weissman P.R., McFadden L.-A., Johnson T.V. (eds) Encyclopedia of the Solar System. Academic Press, San Diego, CA, pp 537–55

Gomes R (2004) The common origin of the high inclination TNOs. In: Davies J.K., Barrera L.H. (eds) The First Decadal Review of the Edgeworth-Kuiper Belt. Kluwer Academic Publishers, Dordrecht, The Netherlands, pp 29–42

Hartmann WK (2005) Moons and planets. 5th edn. Brooks/Cole, Belmont, CA

Houtzager G (2003) The complete encyclopedia of Greek mythology. Chartwell Books, Edison, NJ

Jewitt D, Aussel H, Evans A (2001) The size and albedo of the Kuiper-belt object (20000) Varuna. Nature 411:446–447

Jewitt D, Sheppard S (2002) Physical properties of trans-neptunian object (20000) Varuna. Astronom J 123:2110–2120

Koeberl C (2004) The late heavy bombardment in the inner solar system: Is there any connection to Kuiper belt objects? In: Davies J.K., Barrera L.H. (eds) The First Decadal Review of the Edgeworth-Kuiper Belt.

Kluwer Academic Publishers, Dordrecht, The Netherlands, pp 79–87

Kuiper GP (1951) On the origin of the solar system. In: Hynek J.A. (ed) Astrophysics. McGraw-Hill, New York

Lemonick MD (2005) Meet the new planets. Time, Oct 24: 56–57.

Luu JX, Jewitt DC (1993) Discovery of the candidate Kuiper belt object 1992 QB1. Nature, 362:730–732.

Levison HF, Weissman PR (1999) The Kuiper belt. In: Weisman P.R., McFadden L.-A., Johnson T.V. (eds) Encyclopedia of the Solar System. Academic Press, San Diego, CA, pp 557–583

Millis RL, Buie MW, Wasserman LH, Elliot LH, Kern SD, Wagner RM (2002) The deep ecliptic survey: A search for Kuiper belt objects and centaurs. I. Description of methods and initial results. Astron J 123:2083–2109.

Morbidelli A, Brown ME, Levision HF (2004) The Kuiper belt and its primordial sculpting. In: Davies J.K., Barrera L.H. (eds) The First Decadal Review of the Edgeworth-Kuiper Belt. Kluwer Academic Publishers, Dordrecht, The Netherlands, pp 2–27

Reddy F (2005) The tenth planet. Astronomy Nov: 68–69

Spencer J, Buie M, Young L, Guo Y, Stern SA (2004) Finding KBO flyby targets for New Horizons. In:

Davies J.K., Barrera L.H. (eds) The First Decadal Review of the Edgeworth-Kuiper Belt. Kluwer Academic Publishers Dordrecht. The Netherlands, pp 483–491

Stern SA, Tholen DJ (eds) (1997) Pluto and Charon. University of Arizona Press, Tucson, AZ

Stern SA, Mitton J (1998) Pluto and Charon: Ice dwarfs on the ragged edge of the solar system. Wiley, New York

Stern SA (2000) Into the outer limits. Astronomy Sep: 52–55

Stern SA (2002) Journey to the farthest planet. Sci Am 286:56

Stern SA, Spencer J (2004) New Horizons: The first reconnaissance mission to bodies in the Kuiper belt. In: Davies J.K., Barrera L.H. (eds) The First Decadal Review of the Edgeworth-Kuiper Belt. Kluwer Academic Publishers, Dordrecht, The Netherlands, pp 477–482

Trujillo CA, Brown ME (2004) The Caltech wide-area sky survey. In: Davies J.K., Barrera L.H. (eds) The First Decadal Review of the Edgeworth-Kuiper Belt. Kluwer Academic Publishers, Dordrecht, The Netherlands, pp 99–112

Weissman PR (1999) Cometary reservoirs. In: Beatty J.K., Petersen C.C., Chaikin A. (eds) The New Solar System. Sky Pub., Cambridge, MA, pp 59–68

Comets: Coming Inside from the Cold

Comets have played an important role during and after the accretion of the terrestrial planets because they contributed water, oxides of carbon and sulfur, ammonia, and methane to these bodies (Section 6.1). These compounds were vaporized during impact and accumulated in the primordial atmospheres that enveloped each of the terrestrial planets at the time of their formation. Mercury did not retain its primordial atmosphere, Venus lost the water but retained the carbon dioxide, and the primordial atmospheres of the Earth and Mars were depleted by the effects of explosive impacts and by leakage of molecules into interplanetary space. Comets are, of course, still traversing the solar system and are still impacting on planets (e.g., Shoemaker-Levy 9, Section 14.5), but the frequency of such events has greatly diminished compared to the early history of the solar system.

According to the new classification of celestial objects in the solar system, comets are included among the "small solar-system bodies" along with irregularly shaped asteroids and related meteoroids (Section 3.1).

Comets are composed primarily of water ice and small refractory dust particles and they revolve around the Sun on elliptical orbits. When comets approach the Sun, they develop tails that resemble long hair streaming in the wind, which is why they have been called "hairy stars". Moreover, the word "comet" means "long-haired" in Greek. In contrast to the predictable motions of the Sun, the Moon, and the planets of the solar system, comets appear unexpectedly and without warning, which caused fear among the people of the Middle Ages and gave rise to the popular belief that comets are followed by natural and man-made disasters. (Science Brief 22.9.1).

The recent literature on comets includes books and articles by Hartmann (2005),

Festou et al. (2004), Burnham (2000), Brandt and Chapman (1992, 2004), Brandt (1999), Boice and Huebner (1999), Fernandez (1999), Beatty et al. (1999), Yeomans (1991), Newburn et al. (1991), Whipple (1985), Sagan and Druyan (1985), and Kronk (1984, 2003). The hazards posed by the impact of comets with the Earth are dealt with in books by Gehrels (1994), Spencer and Mitton (1995), Verschuur (1996), Lewis (1998), and Levy (1998), and Koeberl (1998).

22.1 Anatomy of Comets

When a comet approaches the Sun at the perihelion of its orbit, the heat radiated by the Sun causes the ice to sublimate to form a dusty atmosphere called the *coma* that surrounds the body of the comet called the *nucleus*. The nucleus and the coma together form the *head* of the comet. The coma generally begins to form when a comet approaches the Sun to about 3 AU and reaches its largest diameter (10^2 to 10^3 km) at heliocentric distances of 1.5 to 2.0 AU. The coma becomes visible to observers on the Earth primarily because it reflects sunlight. It is surrounded by a large but invisible cloud of hydrogen gas that is released when hydrogen-bearing molecules in the coma are dissociated by ultraviolet sunlight.

The gas and associated dust particles of the coma extend into a long luminous *tail* that is the most characteristic property of comets. The tail of a comet in Figure 22.1 always points away from the Sun such that the tail of a comet that is receding from the Sun actually precedes it in its orbit. In addition, the dust and gas that coexist in the coma, in many cases form separate tails composed of the dust particles and of the gas

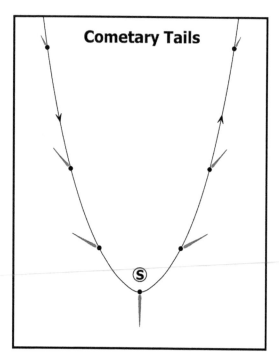

Cometary Tails

Figure 22.1. The tails of comets increase in length as they approach the Sun and always point away from it because they are acted upon by the solar wind and by radiation pressure of sunlight. Note that the tail precedes comets that are moving away from the Sun. The tails of comets split into two components consisting of dust and ionized gas (not shown). Adapted from Figure 17–24 of Freedman and Kaufmann (2002)

molecules which are ionized by the ultraviolet light of the Sun.

The *ion tail* is bluish in color because of the emission of electromagnetic radiation by CO^+, N_2^+, and CO_2^+. It is straight and points away from the Sun because the ions interact with the solar wind which is also composed of charged atoms and nuclear particles. The *dust tail* reflects sunlight and therefore has a yellowish color. It is commonly curved because the radiation pressure exerted by sunlight on the small particles ($\sim 10^{-6}$ m in diameter) decreases their orbital velocities and therefore causes them to lag behind the head of the comet. The curvature of the dust tail arises from the fact that the dust particles are pushed into a "higher" orbit whose semi-major axis is greater than that of the cometary orbit. The masses of the ions in the ion tail are too small to respond gravitationally to the

radiation pressure of sunlight. Therefore, the ion tail is straight, whereas the dust tail is curved.

A major advance in the study of comets occurred in 1705 when the British astronomer *Edmond Halley* (1656–1742) predicted that a comet seen in 1456, 1531, 1607, and 1682 would return in 1758 AD. Although Halley did not live to see it, the comet did return in 1758 as he had predicted. Therefore, the comet was named in his honor. Halley's comet has continued to appear at intervals of about 75 to 76 years. It was last seen in 1986 and will return in 2061 (See also Science Brief 22.9.1).

The orbit of Halley's comet in Figure 22.2 is highly elliptical (e = 0.97) and its aphelion distance of 28.4 AU is close to the orbit of Neptune. In addition, comet Halley revolves in the retrograde direction because the plane of its orbit makes an angle of 162 degrees with the plane of the ecliptic. The large angle of inclination of the plane of the orbit of Halley's comet is not typical for short-period comets, most of which orbit the Sun close to the plane of the ecliptic. However, most short-period comets do have elliptical orbits presumably as a result of gravitational interactions with planets of the solar system (Science Brief 22.9.2). Because of the popularity of Halley's comet, several books have been written about it, including those by Gropman (1985), Metz (1985), Olson (1985), Ottewell and Schaaf (1985), Newburn et al. (1991), and Olson and Pasachoff (1998).

In accordance with the precedent established by the naming of Halley's comet, newly discovered comets continue to be identified by the names of their discoverers. As a result, comet hunting has become a favorite pastime of amateur astronomers who scan the sky with telescopes hoping to find a faint and fuzzy-looking object that moves relative to fixed background stars. About 2000 comets are known at the present time and about 15 to 30 comets are discovered each year. Approximately half of the comets that are discovered annually have been seen before (i.e., they are "recovered"). The remainder are either short-period comets that have previously escaped detection or they are long-period comets that are approaching the Sun either for the first time or after a long absence of 200 years or more (Wagner, 1991).

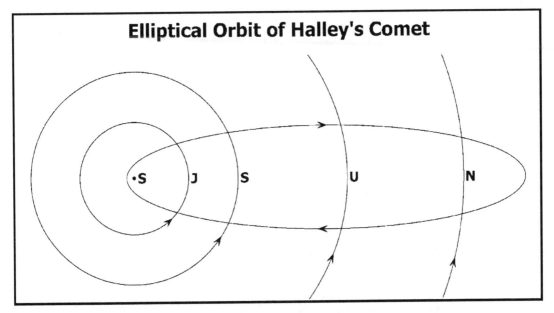

Figure 22.2. The orbit of Halley's comet has an eccentricity of 0.967 and extends from 0.59 AU at perihelion to 28.4 AU at aphelion. The length of its semi-major axis (a) is 17.9 AU. Its period of revolution is 76.1 y and it revolves in the retrograde (clockwise) direction when viewed from above the plane of the ecliptic because the inclination of its orbit is 162.21°. Halley's comet was first sighted by Chinese astronomers in 240 BC and it was probably in the sky at the time of the battle of Hastings in 1066 in southern England. The comet was last seen in 1986 and will return in 2061. Adapted from Figure 26.13 of Wagner (1991)

22.2 Sources of Comets

Comets originally formed as ice planetsimals in the protoplanetary disk between the orbit of Jupiter and the present orbit of Neptune. The ice-planetesimals that survived the period of planet formation 4.6 billion years ago, were later ejected by the giant planets and now reside in the scattered disk of the Edgeworth-Kuiper belt and in the Oort cloud (Sections 21.2 and 21.3). Most of the ice objects that were ejected to the outermost limits of the solar system still remain there.

The orbits of Edgeworth-Kuiper belt objects (EKOs or EKBOs) may be deflected by the gravitational field of Neptune and by low-energy collisions or close encounters with large EKOs. Similarly, ice bodies in the Oort cloud may start to move toward the Sun because of interactions with nearby stars or for other unspecified reasons. As a result, these ice objects leave the cold and dark regions where they may have resided for up to 4.6 billion years and re-enter the solar system where the Sun provides both light

and warmth. As they fall toward the Sun, the ice objects face an uncertain fate that includes:

1. Destruction by impact on one of the nine planets or their satellites, or
2. capture into orbit around one of the giant planets, or
3. deflection from their initial trajectory into an elliptical orbit around the Sun, or
4. return to the Oort cloud in case the orbit of the object is not degraded by interactions with one of the planets, or
5. ejection from the solar system by the sling-shot effect of Jupiter, or
6. annihilation by falling into the Sun.

Ice objects from the E-K belt or from the Oort cloud that impact on a planet or satellite leave a crater and add to the inventory of volatile compounds including water, methane, oxides of carbon, and ammonia on that planet or satellite. In addition, such objects may release a large number of organic compounds that formed previously by abiological processes in the solar nebula. These organic compounds may later facilitate the formation of self-replicating molecules that can evolve into unicellular organism under favorable environmental conditions.

When an EKO or ice object from the Oort cloud is captured into orbit by one of the major planets, it may be torn apart of tidal forces (e.g., comet Shoemaker-Levy 9, Section 14.5), or it may become a satellite of that planet. All of the giant planets have satellites in highly inclined retrograde orbits that may be EKOs or former members of the Oort cloud that were captured. Occasionally, such objects can find refuge in one of the principal Lagrange points (L4 and L5) of a giant planet or of one of its satellites. For example, Calypso (34×22 km) and Telesto (34×26 km) are stowaways in the Lagrange points of Tethys, satellite of Saturn (Section 16.4.3).

In case an EKO or member of the Oort cloud is deflected from its trajectory toward the Sun into an elliptical heliocentric orbit, its ultimate fate depends on its closest approach to the Sun. If the perihelion distance of the new orbit is greater than about 8 AU, the object becomes a *Centaur* that survives as long as it maintains a safe distance from the Sun (Section 13.1.2). However, if the object approaches the Sun at perihelion to within about 6 AU, it becomes a *comet* and loses large amounts of ice by sublimation during each pass around the Sun until it acquires a thick crust of refractory particles cemented by organic matter that covers a small amount of residual ice. Such "dead" comets assume the properties of asteroids (Section 13.2).

Members of the Oort cloud may also escape the gauntlet of gravitational blows exerted by the planets and can return to the cold and dark space of the Oort cloud. Alternatively, any ice object that enters the inner solar system may be ejected by Jupiter without ever getting close to the Sun. The most extreme outcome of a pass through the solar system is to be annihilated by getting too close to the Sun. Such comets are called *sun grazers*. Some comets may actually fall into the Sun.

22.3 Short-Period Comets

A comet in Figure 22.3 that enters the solar system and passes close to and in front of a planet (e.g., Jupiter) may be deflected from its

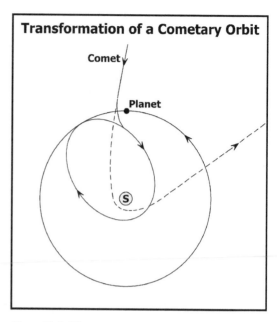

Transformation of a Cometary Orbit

Figure 22.3. When a comet on its way into the inner solar system passes in front of a planet such as Jupiter, the comet may be deflected from a long-period prograde orbit into a short-period retrograde orbit. Note especially that the aphelion of the new cometary orbit is located close to the orbit of the planet that deflected the comet from its path. Adapted from Figure 17.30 of Freedman and Kaufmann (2002)

original path and is forced into a short-period retrograde orbit with an aphelion close to the orbit of the planet that it interacted with. This hypothetical scenario suggests that the aphelion distances of short-period comets may identify the planet that transformed them from long-period to short-period comets. In fact, the aphelion distances of the short-period comets in Figure 22.4 appear to coincide with the orbital radii of the giant planets. According to this criterion, almost 50 % of the short-period comets in Science Brief 22.9.2 have interacted with Jupiter and therefore are members of the Jupiter family of comets. Others appear to have been deflected from their original orbits by Saturn, Uranus, and Neptune.

Short-period comets whose orbits have low inclinations relative to the plane of the ecliptic could have originated in the Edgeworth-Kuiper belt. The data in Science Brief 22.9.2 indicate that nearly 80% of short-period comets have orbital inclinations of less than 20°. The orbital

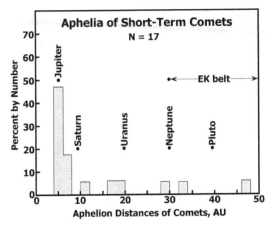

Figure 22.4. The aphelion distances of seventeen short-period comets in Science Brief 22.8c range from 4.2 AU (Encke) to about 47 AU (Swift-Tuttle) and form a multi-modal array. Most of these comets have aphelion distances that scatter above and below the semi-major axis of the orbit of Jupiter. In addition, the aphelion distances of a few comets coincide approximately with the radii of the orbits of Saturn, Uranus, and Neptune. The comet Swift Tuttle (discovered in 1862) actually returns to the Edgeworth-Kuiper belt which may have been its original home. Based on data compiled from Table 7.1 of Hartmann (2005) and from the Appendix of Beatty et al. (1999)

inclinations of the remaining 20% of the short-period comets range from 29.11° (Crommelin) to 162.49° (Tempel-Tuttle). These comets could have originated in the Oort cloud and may have been captured into short-period orbits by gravitational interactions with one of the giant planets as shown in Figure 22.3. The orbital parameters also suggest that Halley's comet, Tempel-Tuttle, Swift-Tuttle, Tuttle, and perhaps even Crommelin originated in the Oort cloud and were each prevented from returning to their source as a result of gravitational interactions with one of the giant planets.

22.4 Long-Period Comets

Certain comets listed in Science Brief 22.9.2 have very long periods of revolution that range from greater than 200 years to infinity. In other words, some long-period comets make only one pass around the Sun and do not return. The eccentricities of the orbits of long-period comets are close to or equal to 1.0 and the inclinations of their orbits range widely from 14°

(Kohoutek) to 160° (Burnham). The aphelion distances of two long-period comets in Science Brief 22.9.2 (Donati and Humason) are of the order of hundreds of astronomical units which places them in the space between the Edgeworth-Kuiper belt and the Oort cloud. The periods of revolution, lengths of semi-major axes, and aphelion distances of most long-period comets are unknown and are described only as being "large."

Several long-period comets were discovered during the second half of the twentieth century including Hyakutake and Hale-Bopp, both of which were seen by many people and were described in detail by Burnham (2000).

22.4.1 Hyakutake

The comet Hyakutake in Figure 22.5 was discovered on January 30, 1996, by the Japanese amateur astronomer Yuji Hyakutake. The period of this comet at the time of its discovery was about 17,000 years, which implies that it has passed through the solar system before and did not come straight from the Oort cloud. The orbit of comet Hyakutake caused it to approach the Earth to within 0.1 AU (about 15×10^6 km) which allowed it to be viewed from the northern hemisphere of the Earth (Fig. 22.5). The comet swung around the Sun at a perihelion distance of 0.231 AU and is now headed back to the aphelion of its orbit that was originally located at about 1300 AU. During the passage through the solar system the orbit of Hyakutake was altered by interactions with the planets it passed along the way. As a result, the distance to aphelion increased to about 3,500 AU and its period of revolution increased to 72,000 years. As the comet approached the Sun, it released large amounts of water vapor even though radar reflections indicated that its nucleus was only about 3 km in diameter. In addition, Hyakutake lost up to 10 tons of dust per second. The water vapor was accompanied by many other compounds including amorphous carbon, carbon monoxide (CO), methane (CH_4), acetylene (C_2H_2), ethane (C_2H_6), and at least ten other compounds of carbon, hydrogen, nitrogen, oxygen, and sulfur. These compounds were discharged by jets that developed on the sunny side of the nucleus which was rotating with a period of 6.23 h. Hyakutake was more active than Halley's comet because it

Figure 22.5. Hyakutake is a long-period comet that was discovered on January 30, 1996, by the Japanese amateur astronomer Yuji Hyakutake. Its period of revolution increased from about 17,000 to 72,000 years as a result of changes in its orbit during its most recent passage through the solar system. Nevertheless, Hyakutake has passed through the solar system before and it will eventually return. As Hyakutake approached the Sun, it developed a large coma and a long tail although the diameter of its nucleus is only about 3 km. In this image, recorded in the spring of 1996 by Rick Scott and Joe Orman from a site in Arizona, the blue ion tail and the yellow dust tail of Hyakutake overlap in this view. (This image was downloaded from "Astronomy Picture of the Day," file: //D:\ Planetary stuff \ Pluto + Comets \ webpages \ APOD July 17, 1998. Reproduced by permission of John Orman and Rick Scott

has not yet developed the thick crust composed of refractory particles and organic matter that covers the surface of Halley's comet.

22.4.2 Hale-Bopp

When Alan Hale and Thomas Bopp independently discovered a small and fuzzy patch of light in the sky during the early hours of July 23, 1995, they suspected that it may be a comet, but they did not anticipate that it would become one of the most active comets on record. It also turned out to have an unusually large elongated nucleus somewhat less than 40 km in length with a rotation period of 11.34 hours. The nucleus emitted bursts of dust and gas that temporarily

brightened the coma. Hale-Bopp in Figure 22.6 was clearly visible in the northern sky during March and April of 1997. During this period, the comet released large amounts of dust at a rate of more than 400 tons per second. The dust particles included silicate minerals such as olivine and pyroxene similar to interplanetary dust particles (IDPs) that enter the atmosphere of the Earth. In spite of the vigorous discharges of gas jets and the release of large quantities of dust, Hale-Bopp probably lost less than 0.1% of its mass during this trip through the solar system (Burnham, 2000, p. 131). Some images recorded by the Hubble space telescope seemed to indicate that several large fragments were ejected from Hale-Bopp during its pass around the Sun. (See Science Brief 22.9.1).

Figure 22.6. Hale-Bopp is a famous long-period comet that was clearly visible from the Earth during March and April of 1997 when this photograph was taken by John Gleason. The tail of the comet has split into the blue ion tail and the yellowish dust tail. The coma is composed of dust and gas that are emitted by the nucleus of the comet. The ion tail contains ionized carbon monoxide which emits blue light when electrons enter vacant molecular orbitals. The coma and the dust tail reflect sunlight but do not themselves generate light. Gravitational interactions with planets of the solar system changed the period of revolution of Hale-Bopp from 4,265 to 2404 years and shortened its aphelion distance from 525 to 358 AU from the Sun. Present knowledge indicates that Hale-Bopp will return to the inner solar system in the years 4401 AD. This image was downloaded from "Astronomy Picture of the Day" at file://D:\ Planetary stuff \ Pluto + Comets \ webpages \ APOD 2005 May 22. Reproduced by permission of John Gleason

The gas emitted by the comet contained a large number of compounds, some of which had not been observed before. These include: sulfur monoxide (SO), sulfur dioxide (SO_2), nitrogen sulfide (NS), formic acid (HCOOH), methyl formate (CH_3COOH), and formamide (NH_2CHO). In addition, many other molecules characteristic of comets were detected in the coma of Hale-Bopp (e.g., H_2O, CO, CO_2, CH_4, and NH_3). The rates of sublimation of these principal components vary with the ambient temperature. For example, carbon monoxide was released as the comets traveled through the asteroidal belt followed by water vapor, which became the dominant component after the comet had approached the Sun to within 3.5 AU.

The analysis of the gases and dust released by comets can identify objects that are indigenous to the solar system (i.e., they are "home-grown") and distinguish them from others that may have escaped from the Oort cloud of another star in our galactic neighborhood. The available evidence suggests that Hale-Bopp is a home-grown body that formed within the protoplanetary disk of the solar system. Comet Hyakutake, on the other hand, may contain matter that did not originate in our solar system (Burnham, 2000, p. 134).

The orbit of Hale-Bopp is highly elongated (e = 0.995) and is oriented nearly at right angles to the ecliptic (i = 89.43°). It passed the Earth on March 22, 1997, at a distance of 1.3 AU and reached perihelion on April 1 when it was 0.914 AU from the Sun. Subsequently, Hale-Bopp descended through the plane of the ecliptic and headed to the aphelion of its orbit which had moved from 525 to 358 AU following a close encounter with Jupiter. In addition, its period of revolution was shortened from 4,265 to 2,404 years. Accordingly, Hale-Bopp was last visible from the Earth in the year 2268 BC (4265-1997), it will reach aphelion in 3199 AD, and will return in 4401 AD (1997 + 2404). We can predict the return of the comet Hale-Bopp but we cannot imagine the environmental conditions on the Earth 2404 years from now.

The ultimate fate of comet Hale-Bopp is difficult to predict because its orbit is *chaotic* (i.e., it is sensitive to perturbations by the planets of the solar system). The most likely outcome is that it will be destroyed by colliding with the Sun, or with a planet, or with one of the many planetary satellites. It may also be ejected to the Oort cloud or even to the interstellar space. The least likely fate of Hale-Bopp is that it will be captured into a stable orbit around one of the giant planets or will enter into a stable orbit that keeps it at a safe distance from the Sun (i.e., it becomes a Centaur). Hale-Bopp may also continue to pass through the inner solar system losing ice each time it approaches the Sun until it develops a thick crust that ultimately prevents the discharge of any residual gases. At this stage in its evolution, Hale-Bopp would resemble an asteroid.

22.5 Meteor Showers and IDPs

During the night of November 12/13 of 1833 the sky over North America was ablaze with meteors also known as "shooting stars" or "falling stars." (See Glossary). The display was so intense that some people became frightened because they thought that the end of the world was at hand. Nevertheless, others remained calm and noted that the meteors appeared to be origi-nating from a single spot within the constellation of Leo. Therefore, these meteors are known as the *Leonids*. Actually, meteors travel on parallel paths and appear to originate from a point, called the *radiant*, only as a result of perspective (Burnham, 2000).

Later work by astronomers demonstrated that meteor showers are a recurring phenomenon that occurs when the Earth travels through streams of small particles whose orbits around the Sun are similar to those of certain comets. In other words, the streams of particles are trails of dust left behind by comets on their way around the Sun. The particles range in size from microscopic grains to sand grains or even small pebbles. When these grains enter the atmosphere of the Earth at speeds of about 35 km/s, they are heated by friction with the air and glow brightly as streaks of light as in Figure 22.7 until they melt and are vaporized. The various meteor showers and the comets that are associated with them are listed in Table 22.1.

The dust shed by comets surrounds the Earth and causes sunlight to diffuse in the form of *zodiacal light* (Grün, 1999). The smallest cometary particles may enter the atmosphere of the Earth without "burning up" and are eventually washed out of the atmosphere by rain or snow. These cometary dust particles constitute a fraction of Interplanetary Dust Particles (IDPs) in Figure 22.8 that have been collected in the stratosphere on air filters by high-flying aircraft (e.g., the U-2) at an altitude of 20 km. The particles are classified as being either "extrater-restrial" or "terrestrial" based on their shape, color, and chemical composition.

Figure 22.7. Fireball recorded in November of 2001 during the Leonid meteor shower. The phenomenon was caused when a small particle, released by the comet Tempel-Tuttle in 1766, entered the atmosphere of the Earth and was heated to incandescence by friction. A second meteor is visible in the background. The stars belong to the constellation Orion. (Reprinted by permission of Kris Asla, Aloha, Oregon, USA)

The *extraterrestrial* particles in Table 22.2 and described by Warren et al. (1997) make up 44% by number of the 458 specimens that were recovered during flights in January/February (L2021) and during June/July (L2036) of 1994. Most of these extraterrestrial particles are less than 40 μm in diameters and consist of irregularly-shaped fluffy aggregates of smaller grains (72.5 %), of equidimensional grains (19.5 %), and of spherical particles (8.0 %). Most of the spherical particles of extraterrestrial origin are composed of iron, nickel, and sulfur and therefore may be meteorite-ablation spherules that formed by frictional heating of stony and iron meteorites passing through the atmosphere of the Earth (Section 8.6). The irregularly shaped

Table 22.1. Meteor showers and their associated comets (Brandt, 1999)

Meteor shower	Date	Rate, per h	comet	Period of revolution, y
Lyrid	April 21	15	Thatcher	410
Eta Aquarid	May 5	35	Halley	75.7
Orionid	October 21	30	Halley	75.7
Perseid	May 12	80	Swift-Tuttle	134
Draconid	October 9	20	Giacobini-Zinner	6.6
Taurid	November 3	10	Encke	3.3
Leonid	November 17	15	Tempel-Tuttle	33.2
Geminid	December 13	90	Phaethon	1.4
Ursid	December 23	10	Tuttle	10

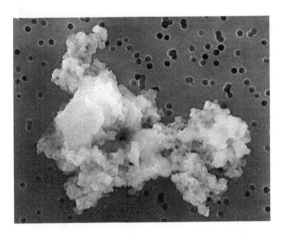

Figure 22.8. Fluffy aggregates of crystalline minerals composed of common lithophile elements are dominant among the Interplanetary Dust Particles (IDPs) and were scattered into space by the dust tails of comets being heated by the Sun. These and other kinds of grains in interplanetary space give rise to meteors when they are vaporized by friction after they enter the atmosphere at high velocities. The minerals of which cometary dust particles are composed form at high temperatures when they occur on the Earth. Crystals of these minerals in IDPs may have formed from vapor either in exploding stars or in interstellar dust clouds that later become "stellar nurseries" such as the solar nebula which later contracted to form the Sun and the planets of the solar system. Courtesy of the Stardust mission, NASA/JPL-Caltech

Table 22.2. Distribution of shapes of 458 particles collected in the stratosphere expressed in percent by number (Warren et al., 1997)

Type	Irregular	Equi-dimensional	Spherical
Extraterrestrial (44 %)	72.5 (145)	19.5 (39)	8.0 (16)
Terrestrial (natural, 25%)	62.6 (72)	29.5 (34)	7.8 (9)
Terrestrial (anthropogenic, 27 %)	69.1 (83)	20.0 (24)	10.8 (13)
Terrestrial (rocket fuel, 4%)	4.3 (1)	0.0	95.6 (22)

The number of specimens in each category is indicated in parentheses. Therefore: "72.5 (145)" means that 72.5 % of the extraterrestrial particles have irregular shapes and the number of such particles is 145.

aggregates in the extraterrestrial component in Figure 22.8 and equidimensional grains are composed primarily of lithophile elements (e.g., silicon, aluminum, magnesium, and calcium) including as yet unspecified concentrations of elements of low atomic number (e.g., carbon). Some, or perhaps most, of the fluffy composite grains may be cometary dust particles.

The *terrestrial* particles among the 458 specimens originated from natural (25%) and anthropogenic (27%) sources including combustion products of solid rocket fuel (4 %). The *natural* terrestrial contaminants consist of mineral grains (e.g., quartz, feldspar, and volcanic glass) that were injected into the stratosphere by volcanic eruptions. This component also includes clay minerals and iron oxide particles derived from soil as well as carbon-rich particles of biogenic origin. The *anthropogenic* particles are characterized by anomalously high concentrations of certain industrial metals (e.g., cadmium, titanium, vanadium, chromium, antimony,

manganese, nickel, copper, and zinc) in addition
to aluminum, iron, and silicon.

The spherical particles (7.8 %) of the natural
terrestrial component recovered by Warren et al.
(1997) appear to be composed of silicate glass
containing silicon, aluminum, magnesium, and
calcium, whereas most of the anthropogenic
spherules appear to consist of metallic copper,
aluminum, cadmium, iron, and antimony. In
some cases, these metals occur with silicon,
sulfur, chlorine, and other elements. Virtually
all of the particles that originated during the
combustion of rocket fuel are spherical in shape
and are composed of aluminum either in metallic
or oxide form with only trace amounts of silicon,
calcium, copper, phosphorus, and chlorine.

22.6 Close Encounters

The gas and dust of the coma obscure the surface
of the nucleus and thereby prevent us from
observing the processes that are occurring there.
We can get a much better view of cometary
surfaces from robotic spacecraft that can fly
through the coma or can intercept comets before
they come close enough to the Sun to develop a
full-blown coma. Missions to comets are difficult
to prepare for and to execute because, in most
cases, comets appear unexpectedly and remain
in the vicinity of the Earth for only a short time.
Halley's comet is exceptional because its orbit
is well known and its appearance is therefore
predictable. Other problems include the steep
orbital inclination of many comets, their high
speeds as they approach perihelion, and their low
gravity which have so far prevented spacecraft
from landing on the surface of a comet.

Several robotic missions have been carried
out which have yielded precise observations
that have greatly advanced our knowledge of
the physical and chemical processes that occur
on surfaces of cometary nuclei and in the
coma. In general, the results of such missions
have confirmed the hypothesis of the American
astronomer Fred Whipple that the nuclei of comets
are analogous to "dirty snowballs" (Rayl, 2004).

The earliest observations of a comet by
a robotic spacecraft occurred in 1970 when
the second Orbiting Astronomical Observatory
(OAO-2) detected a large cloud of hydrogen
(H) and hydroxyl (OH) around the coma of

comet Tago-Sato-Kosaka. This cloud is not
detectable by earth-based telescopes because the
ozone layer of the atmosphere of the Earth
blocks the characteristic ultraviolet radiation
these molecules emit. Similar clouds of
hydrogen were detected in the same year by
the Orbiting Geophysical Observatory (OGO-5)
around comets Bennett and Encke. When the
long-period comet Kohoutek appeared in 1973,
it was observed by the astronauts on Skylab, by
sounding rockets launched from the Earth, by
Mariner 10 on its way to Venus and Mercury
(Chapters 10 and 11), and by OAO-3. The results
of these investigations identified several new
molecules in the coma and clarified the interactions of sunlight and the solar wind with the ion
and dust tails of this comet.

22.6.1 Giacobini-Zinner in 1985

Before Halley's comet returned in 1986, NASA
considered intercepting it and following it around
the Sun. Unfortunately, this plan and several
alternatives had to be abandoned. Instead, NASA
redirected an existing spacecraft to rendezvous
with the comet Giacobini-Zinner in 1985. This
spacecraft, which was renamed International
Comet Explorer (ICE), flew through the tail of
the comet at a speed of 21 km/s, and identified
ionized water and carbon monoxide molecules
in the ion tail. Subsequently, ICE monitored the
solar wind about 32×10^6 km upstream of comet
Halley in order to facilitate the interpretation
of data to be collected by several spacecraft
launched by other countries that were converging
on Halley. The mission of ICE has not yet
ended because the spacecraft may be recovered
by NASA in 2014 from the vicinity of the Moon
in order to collect the cometary particles that
may be embedded in its solar panels and metal
shrouds. When its journey finally ends and the
dust study has been concluded, the ICE spacecraft will be donated to the Smithsonian Institution in Washington, D.C..

22.6.2 Halley in 1986

The observations recorded by the five spacecraft
that encountered comet Halley in 1986 made a
huge contribution to the study of comets. Two
of the spacecraft were launched by the USSR

(i.e., Vega 1 and Vega 2), two were sent up from Japan (i.e., Suisei and Sakigake), and one was dispatched by the European Space Agency (i.e., Giotto). The two Vega spacecraft imaged the nucleus of the comet at distances of 8,900 km (Vega 1) and 8000 km (Vega 2) and reported that its surface is covered by a layer of carbonaceous dust which is heated by sunlight to room temperature or higher (i.e., > 25 °C).

The spacecraft Giotto was named after the Italian painter *Giotto di Bondone* (1266–1337) who may have witnessed the reappearance of Halley's comet in 1301 and therefore painted the star of Bethlehem as a comet (Burnham, 2000, p. 147). On March 13, 1986, the spacecraft Giotto passed in front of the nucleus of comet Halley at a distance of only 569 km and at a speed of 68 km/s (i.e., 244,800 km/h). It determined that the nucleus is "potato-shaped" rather than spherical and that its dimensions are 16 × 8.4 × 8.2 km. The sunny side of the nucleus contained seven active jets that spewed gas and dust particles into the coma. A small particle (~ 1 mm in diameter) did impact on the spacecraft and caused it to spin out of control for a few moments. In addition, Giotto identified a large number of compounds some of which formed by on-going chemical reactions in the coma. Six years after its encounter with Halley, Giotto flew by the comet Grigg-Skjellerup in July of 1992.

22.6.3 Borelly in 1999

The mission of spacecraft Deep Space 1 (DS-1), which was launched in October of 1998, was to test the reliability of its ion-propulsion system (Section 13.3.2) and other innovative technologies. When its primary mission ended in September of 1999, NASA redirected the spacecraft to rendezvous with the comet Borelly, which was named after the Italian physiologist, physicist, and astronomer *Giovanni Alfonso Borelli* (1608 to 1679). During the encounter on September 22, 2001, the comet in Figure 22.9 passed the spacecraft at a distance of about 2200 km while the comet was traveling at a speed of 59,330 km/h. The images recorded during the flyby indicate that the potato-shaped nucleus of Borelly is about 8 km long and 3 km wide. The orbital inclination of the comet is 30° and its period of revolution is 6.86 years

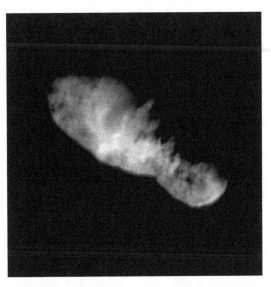

Figure 22.9. The nucleus of the comet Borelly was observed at close range by the spacecraft Deep Space 1 during a flyby on September 22, 2001. The nucleus of this comet is about 8 km long and is covered by a black crust composed of solid particles cemented by organic compounds. Several jets appear to be active in the midsection of the nucleus where white ice is exposed. Evidently, the gas and dust of the coma are ejected from the interior of the nucleus through vents in its black crust. (Image PIA 03500, Deep Space 1, courtesy of NASA/JPL-Caltech)

making it a short-period comet that presumably migrated into the inner solar system from the Edgeworth-Kuiper belt.

The surface of Borelly is covered by a black crust (average albedo = 3%) that completely covers the ice and dust remaining in its interior. Several jets of gas and dust, emanating from deep holes in the crust, were active during the flyby which occurred shortly after the comet had passed the perihelion of its orbit. The force exerted by such jets can change the orbits of comets and can affect the shape of the coma. The surface of Borelly contains rolling plains in the vicinity of active jets and rugged terrain elsewhere including deep fractures, ridges, mesas, and exceptionally dark areas. However, impact craters are not abundant indicating that Borelly has a geologically young surface. After this successful encounter with the comet, the extended mission of DS-1 was terminated and the spacecraft was turned off on December 18, 2001 (Talcott, 2002).

22.6.4 Wild 2 in 2004

More recently, there have been other missions to comets including Stardust launched in February of 1999 and Deep Impact in January of 2004. The Stardust spacecraft obtained close-up images of comet Wild 2 in Figure 22.10 during a flyby on January 6 of 2004 at a distance of only 236 km. The period of revolution of this comet is 6.2 years and its perihelion distance from the Sun is slightly greater than 1.5 AU. The nucleus has a diameter of 5 km and is approximately spherical in shape, presumably because Wild 2 was deflected into its present orbit by Jupiter in 1974 and therefore it had made only five passes around the Sun when it was studied by the Stardust spacecraft in 2004. The encounter revealed that the surface of the nucleus of Wild 2 contains impact craters, spires, and mesas, as well as jets of gas and dust that turn on and off as the nucleus rotates and different parts of the

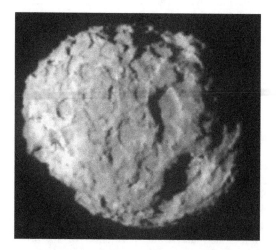

Figure 22.10. The Stardust spacecraft achieved a close encounter with the nucleus of comet Wild 2 on January 6 of 2004 at a distance of only 236 km. During this encounter, the spacecraft collected close to on million solid particles in a panel of aerogel, including very small crystals of olivine and other high-temperature minerals. The surface of Wild 2 contains several large circular depressions, which may be impact craters or may have formed by the collapse of the roofs of subsurface cavities from which water vapor and dust had vented to the surface. Comet Wild 2 was deflected into its present orbit in 1974 and therefore had completed only about five orbits around the Sun prior to the encounter with Stardust. (Image PIA 05579, Stardust, courtesy of NASA/JPL-Caltech)

surface are exposed to the Sun. The presence of these topographic features indicates that the nucleus has been hit by small objects and that its crust has collapsed locally where cavities have formed by the sublimation of the ice and by the discharge of the resulting mixture of gases through jets. The Stardust spacecraft also detected abundant organic matter and collected dust particles as well as molecules of the gas that were successfully returned to Earth on January 15 of 2006 (Moomaw, 2004). The first reports indicate that the spacecraft collected more than one million microscopic dust particles during flybys of asteroid Annefrank in November 2002 and of comet Wild 2 in January of 2004.

22.6.5 Tempel 1 in 2005

The return of the Tempel 1 comet in 2005 provided the opportunity for the most recent encounter. This comet was originally discovered in 1867 by Ernst W.L. Tempel but was lost in 1881 because the orbit was altered during a close encounter with Jupiter. Fortunately, comet Tempel 1 was rediscovered in 1972 by B.G. Marsden. Its current orbit extends from 4.7 AU at aphelion to 1.5 AU at perihelion (e = 0.52) and its period of revolution is 5.5 years.

The Deep Impact Spacecraft was launched on January 12 of 2005 in order to intercept Tempel 1 in Figure 22.11A as it passed through the perihelion of its orbit. During the encounter, the spacecraft released a projectile made of copper (372 kg), which slammed into the "fist-shaped" cometary nucleus whose dimensions are about 5 × 11 km. Actually, the impactor separated from its mother ship about 24 hours before the comet arrived and moved to a predetermined point on the comet's orbital path. When the comet arrived on schedule traveling at a speed of about 37,000 km/h, it collided with the projectile that had been waiting for it.

The projectile carried a camera that recorded a series of images of the surface of the approaching cometary nucleus, whereas the impact and its aftermath were recorded by the spacecraft from a distance of about 800 km and by the Hubble space telescope. The images of the cometary nucleus that were recorded by the camera of the projectile a few seconds before the impact reveal a variety of topographic features including

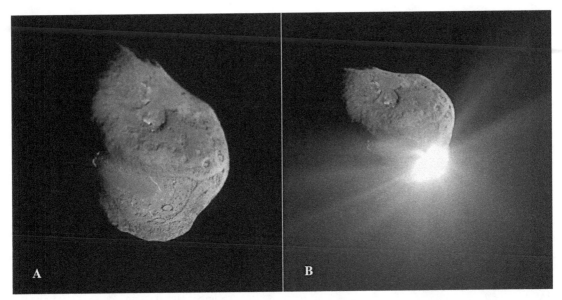

Figure 22.11. A. The nucleus of comet Tempel 1 is a "fist-shaped" object about 11 km in diameter. The surface is covered by a black crust and contains numerous pits, basins, and plains that are the scars of its long life in space. (Image PIA 02142, Deep Impact, courtesy of NASA/JPL-Caltech/UMD). B. The impact of the copper projectile caused an explosion that resulted in the ejection of a cone-shaped plume of water vapor and dust particles derived from a depth of up to about 30 meters. The response to the impact indicates that the nucleus of comet Tempel 1 is not just a "dirty iceberg" but actually seems to consist of a fluffy aggregate of small ice crystals that accumulated gently and were not compressed or melted. (Image PIA 02137, Deep Impact, courtesy of NASA/JPL-Caltech/UMD)

impact craters, smooth plains, a line of cliffs about 20 m high, and evidence of layering on a scale of 20 to 30 m. The topography of the surface of Tempel 1 implies that the comet has been hit many times during its long history in space and that parts of the surface have collapsed into cavities that formed by the sublimation of the underlying ice. The resulting vapor of water and other volatile compounds may have been redeposited on the surface after being ejected through vents to form the observed plains on the surface of the comet (i.e., cometary cryovolcanism).

The collision of the projectile with the cometary nucleus caused an explosion that ejected a cone-shaped plume of debris in Figure 22.11B traveling at about 180,000 km/h and released an amount of energy equivalent to 4.8 tons of TNT. The debris cloud was composed of hot gas and droplets of molten silicate at a temperature of about 3500 °C both of which originated from about 20 to 30 meters below the surface of the comet. The glow of the debris cloud remained visible for more than one hour and the dust cloud did not settle for about two days.

The data recorded during this experiment suggest that comet Tempel 1 is composed of a fluffy mixture of dust particles and ice crystals and that its interior is crudely stratified. The unexpected abundance of dust particles having grain diameters like talcum powder (i.e., < 100 micrometers) gave rise to the suggestion that this comet resembles an icy dirt ball rather than a dirty snowball. Although the impact crater is not visible in the images recorded by the Deep Impact spacecraft, it is probably about 100 m wide and 30 m deep. The depth of the crater implies that the material ejected from the bottom of the crater had not been heated by repeated exposures to sunlight and therefore is representative of the material from which this and other comets formed in the protoplanetary disk 4.6 billion years ago (Sections 5.2 and 6.1). The bulk density of the nucleus, estimated from its mass and volume, is about 0.6 g/cm^3.

Evidently, the material of which the nucleus is composed accumulated slowly and gently during millions of years and was never subjected to pressures and temperatures that are required to stabilize liquid water (Section 12.8.2). If the pore spaces between ice crystals and dust particles in the cometary nucleus had ever been filled with water, which subsequently solidified to ice, the bulk density of the nucleus would have increased to values approaching or perhaps even exceeding $1.0\,g/cm^3$ and the impact crater would be less than 30 m deep.

The Deep Impact spacecraft will return to Earth in December 2007 and may then be redirected to rendezvous with another short-period comet known as Boethin. Although the spacecraft does not carry a second impactor, its cameras and spectrometers may reveal information obtainable only by close encounters during flybys (Talcott, 2005; Cowen, 2005; Lakdawalla, 2005; McFadden, 2004; Weissman, 2006).

22.6.6 Churyumov-Gerasimenko in 2014

The low bulk density of comet Tempel 1 (i.e., $0.6 \pm 0.3\,g/cm^3$) is remarkable, but it does not yet indicate unequivocally whether this comet is composed of fluffy ice crystals or whether it is a rubble pile consisting of chunks of ice with open spaces between them. The internal structure and composition of the periodic comet Churyumov-Gerasimenko will be determined definitively by the Rosetta spacecraft that was launched on March 2, 2004, by the European Space Agency (ESA). The spacecraft will first encounter the comet at a distance of 4 AU from the Sun (i.e., in the main asteroid belt) and will fly with it to its perihelion at 1.29 AU. In November of 2014 Rosetta will release a small lander called *Philae* which will take 10 scientific instruments to the surface of the comet. If this maneuver is successful, the ground-penetrating radar of the lander will determine whether the interior of the comet is a fluffy aggregate or a rubble pile.

The results of this mission will suggest how comets originally formed in the protoplanetary disk and to what extent they were altered by gravitational compression and by collisions with other bodies composed of ice or refractory materials. The results will also indicate how best to deflect a comet from its orbit in case one is on a collision course with the Earth.

22.7 Chemical Composition

The chemical composition of cometary nuclei is an important clue concerning their origin by accretion of particles of ice and dust in the solar nebula and in the protoplanetary disk. Much of the information that is available to us was obtained during the close encounters with comet Halley in 1986 and during other flyby missions described in Section 22.6. The chemical composition of the ice of cometary nuclei may vary in systematic ways arising from the chemical differentiation of the protoplanetary disk (Sections 5.2 and 6.1) and depending on the place of origin of different comets. However, such differences may also result from the fact that information about the chemical composition of comets is derived from the gases that are emitted by jets whose composition does not necessarily represent the entire nucleus. The data in Table 22.3 indicate that water molecules make up 85% by number of the gas in the coma. Several other molecules occur in measurable concentrations including compounds of carbon (CO, CO_2, H_2CO, CH_3OH) and diatomic nitrogen (N_2). In addition, many other neutral molecules and positively charged ions have been identified in the mixture of gases that surround cometary nuclei. An even greater variety of molecules and ions of hydrogen, carbon, nitrogen, oxygen, fluorine, magnesium, aluminum, silicon, phosphorus, sulfur, sodium, and potassium occur in interstellar space (Boice and Huebner, 1999).

The elemental concentrations of cometary ice and of dust particles in Table 22.4 indicate that hydrogen (59.29 %) and oxygen (33.47 %)

Table 22.3. Chemical compositions of ice in cometary nuclei (Boice and Huebner, 1999)

Molecules	Concentration, % by number
H_2O	85
CO	4
CO_2	3
H_2CO	2
CH_3OH	2
N_2	1
Others	3
Total	100

Others: H_2S, HCN, NH_3, CH_4, CS_2, C_2H_2, H_2CO_2, C_2H_4, etc.

Table 22.4. Elemental concentrations of cometary ice and dust (Boice and Huebner, 1999)

Element	Ice	Dust
	Percent by number	
Hydrogen	59.29	48.10
Carbon	5.70	19.34
Nitrogen	1.54	1.00
Oxygen	33.47	21.14
Sodium	—	0.24
Magnesium	—	2.38
Aluminum	—	0.16
Silicon	—	4.39
Sulfur	—	1.71
Calcium	—	0.15
Chromium	—	0.02
Iron	—	1.24
Nickel	—	0.10
Totals	100	99.97

are by far the most abundant elements in the mixture of gases followed by carbon (5.70 %) and nitrogen (1.54 %). The hydrogen is bonded to oxygen in molecules of water and to a lesser extent to carbon and nitrogen, whereas oxygen is not only bonded to hydrogen but also to carbon as indicated in Table 22.3. The dust particles in comets contain significant concentrations of elements whose atomic numbers range from 11 (sodium) to 28 (nickel). These elements form a variety of refractory compounds such as oxides, carbonates, sulfides, and silicates that originated by condensation from the gas phase in the solar nebula and in the protoplanetary disk. Even though the particles may have the crystal structure of minerals (e.g., olivine, feldspar, and quartz), they did not crystallize from silicate melts but formed by condensation of vapor at low temperature. The dust particles in Table 22.4 also contain hydrogen (48.10 %), oxygen (21.14 %), and carbon (19.34 %) presumably because they are coated with water ice and carbon compounds of various kinds.

The organic compounds that coat the dust grains and are contained in the ice of cometary nuclei are dispersed into the interplanetary space of the inner solar system every time a comet passes close to the Sun. These compounds are deposited on the Earth as part of the "cosmic dust" that is continuously settling out of the atmosphere (Section 22.5). In addition, the organic compounds can survive the collision of a comet with the Earth, especially in cases where the impact occurs in the oceans. The organic molecules that were deposited on the Earth, after it had cooled sufficiently to permit liquid water to accumulate on its surface, may have evolved in various ways to form self-replicating molecules and hence unicellular organism known to have existed more than 3.5 billion years ago.

22.8 Summary

Comets are irregularly-shaped objects that orbit the Sun and are composed of ice and refractory particles. In most cases, their long axes are less than 100 km and some have diameters of only a few kilometers. Comets are classified depending on whether their periods of revolution (p) are greater or less than 200 years. The orbits of most short-periods comets (p < 200 y) lie close to the plane of the ecliptic and are contained within the orbit of Jupiter. However, Halley's comet (p = 76 y) has an eccentric and steeply dipping orbit (162.22°), which causes it to revolve around the Sun in the retrograde direction with an aphelion distance of 28.4 AU (i.e., close to the orbit of Neptune). Long-period comets (p > 200 y) have highly elliptical orbits (e ~ 1.0) and their aphelion distances are far from the Sun in the Edgeworth-Kuiper belt and in the Oort cloud. Recent examples of long-period comets are Hyakutake (1996) and Hale-Bopp (1997).

When a comet approaches the Sun, ice begins to sublimate from its sunny side and the resulting gas escapes from the nucleus into the coma, which is a transient atmosphere maintained by jets of gas and dust. As a result of this process, short-period comets acquire a crust composed of the refractory particles cemented by organic matter present in the ice. The coma reflects sunlight, which causes comets to shine brightly as they pass the Sun even though the dark crust of the nucleus has a low albedo (i.e., ~ 4%). Some comets get so close to the Sun at perihelion that they seem to graze it (i.e., Sun grazers). These comets may be broken up by the gravitational force of the Sun and may even fall into the Sun. Short-period comets that survive their periodic close encounters with the Sun eventually build up a thick black crust that prevents further sublimation of ice and turns them into inert objects that resemble asteroids (e.g., the Aten group, Section 13.1b).

Impacts of comets on the growing planets and their satellites originally deposited water and other volatile compounds on their surfaces and thereby contributed to the formation of their atmospheres. In addition, comets released organic molecules that may have enabled self-replicating molecules to form in bodies of standing water such as the primordial ocean of the Earth (Chapter 23).

The gases and particles of the coma extend into the spectacular tails that characterize comets as they pass the Sun. Cometary tails typically split into the ion tail and dust tail. The ion tail is straight and points away from the Sun because the ions interact with the solar wind. The dust tail is curved because the particles are retarded by the pressure exerted by sunlight and therefore shift into a higher orbit where they travel at a lower velocity than the nucleus of the comet. Both tails extend away from the Sun, which means that the tails precede comets that are moving away from the Sun. The tails gradually shrink as the distance to the Sun increases and the tails as well as the comae eventually dissipate when the comets move out of range. When comets are in the distant part of their orbits, they are only chunks of ice (i.e., their nuclei) and are not visible from the Earth.

The dusty trails that mark the orbits of short-period comets are a permanent fixture of the interplanetary space occupied by the Earth. When the Earth travels through these cometary dust trails, the particles enter the atmosphere and are vaporized by friction. The resulting streaks of light in the night sky form meteor showers that appear to originate from a small region in the celestial sphere called the radian. The meteor showers caused by the dust trails left by different comets are identified by the constellation within which the radian of the meteor shower is located (e.g, the radian of the Leonid shower on November 17 is located in the constellation Leo and the dust trail is that of the comet Tempel-Tuttle). Cometary dust particles are one of several components of interplanetary dust particles (IDPs) that rain down upon the Earth

and have been collected by high-flying aircraft (e.g., the U-2).

22.9 Science Briefs

22.9.1 Comets as Messengers of Doom

Until quite recently, comets were not a welcome sight in the sky because their appearance was unpredictable and not understood (Burnham, 2000). The first recorded observation of a comet by Chinese naturalists in the eleventh century BC was regarded with apprehension. Other human societies have reacted similarly. For example, comets were believed to forecast impending wars or battles because their curved tails resemble sabers or flames. When a comet appeared in the sky over Mexico, the Aztecs feared that their gods had turned against them. The subsequent invasion by the Spanish conquistadors confirmed their fear.

Perhaps best known to us is the presence of a comet at the battle of Hastings on October 14 of 1066 in southern England between the armies of Duke William of Normandy and of Harold II, the king of the Anglo-Saxons. Duke William was confident that his cause was just and considered the presence of the comet depicted in the Bayeux Tapestry to be a good omen, whereas King Harold was worried about it because the had perjured himself.

In the previous year, King Edward of England had sent Harold to Normandy to confirm that William was to inherit the throne of Anglo-Saxons after Edward had died. Before Harold returned to England, he swore an oath of allegiance to William and received arms and horses in return. However, when King Edward died on January 5 of 1066, Harold arranged to be crowned King of the Anglo-Saxons, thus violating his oath to support Duke William of Normandy.

In any case, the Normans defeated the Saxons at the Battle of Hastings and Harold himself was killed. This was a pivotal event in the history of the English-speaking peoples because for almost 500 years England was ruled by French kings who left a strong imprint in the English language and culture (Lawson, 2002; Burnham, 2000; Churchill, 1961).

Comets were eventually demystified by Edmond Halley who predicted in 1705 that a comet that had been seen in 1456, 1531, 1607, and 1682 would return in 1758. When the comet appeared on schedule, it received much popular attention. We know now that Halley's comet has been returning regularly for 2250 years since 240 BC when it was first observed by Chinese astronomers. We also know that it did appear in the sky in 1066 during the Battle of Hastings. The next appearance of Halley's comet at perihelion will be on July 28, 2061 (Lang and Whitney, 1991).

The return of comet Halley in 1910 was widely anticipated and coincided with the death of King Edward VII of England (Burnham, 2000). News about the comet appeared in the newspapers and aroused a great deal of public attention. As a result, sales of telescopes and binoculars increased dramatically and astronomical observatories opened their doors to let people see the comet. However, the mood of the people changed when reports circulated that the tail of comet Halley contains cyanide gas in the form of cyanogen (C_2N_2) which is a deadly poison. The astronomers of that time reassured the public that the concentration of cyanogen in the tail of Halley's comet is exceedingly low and poses no threat whatever, even though the Earth would pass through the comet's tail. Nevertheless, newspaper reporters drew attention to the large amount of cyanogen that existed in Halley's tail in spite of its low concentration. In addition, the well known science writer Camille Flammarion was quoted as saying that the cyanogen could infiltrate the atmosphere of the Earth and, in that cause, could destroy all life forms.

The contradictory statements by the experts confused the public and caused mass hysteria. During the night of May 18/19 of 1910, when the Earth passed through the tail of comet Halley, some people took precautions by sealing the chimneys, windows, and doors of their houses. Others confessed to crimes they had committed because they did not expect to survive the night, and a few panic-stricken people actually committed suicide. Enterprising merchants sold comet pills and oxygen bottles, church services were held for overflow crowds, and people in the countryside took to their storm shelters.

A strangely frivolous mood caused thousands of people to gather in restaurants, coffee houses, parks, and on the rooftops of apartment buildings to await their doom in the company of fellow humans.

Eventually, the night ended when the Sun rose as it always does in the morning,. The people who had expected the worst were glad that nothing bad had happened and a little embarrassed at having lost their heads. They went home relieved and had their comet breakfast before resuming their every-day lives (Burnham, 2000). Surely, nobody would ever panic again because of the apparition of a comet. However, there is more.

On November 14 of 1996 an amateur astronomer in Houston recorded an image of the comet Hale-Bopp that revealed the presence of a fuzzy-looking object close to the comet (Burnham, 2000). This observer considered that the object was following the comet and announced his conclusion during an appearance on a national radio talk-show. People who were predisposed to believe in paranormal phenomena quickly concluded that the object following the comet was a large spaceship. However, when the image was published, professional astronomers identified the object as an ordinary star that had no connection to Hale-Bopp. Unfortunately, this explanation was rejected by some people who considered it to be a cover-up by the government. This view was shared by 39 members of a religious group called Heaven's Gate who met in a mansion in San Diego during the month of March of 1997 and committed suicide. They left behind video taped messages in which they explained that the spaceship following Hale-Bopp was taking them to a higher level of existence. The members of the Heaven's Gate society considered the comet to be a signal from heaven whose arrival, in their opinion, was fraught with theological significance.

22.9.2 Orbital Parameters of Comets

Some authors cite only the orbital period, and the perihelion distance of comets. This information can be used to calculate the length of the semi-major axis (a) and the aphelion distance (Q) using Kepler's third law and the geometry of ellipses. For example, the comet Swift-Tuttle has a period or revolution (p) of 135 years and a perihelion distance (q) of a 0.968 AU (Brandt, 1999). According to Kepler's third law: $p^2 = a^3$, where p is expressed in years and a in astronomical units.

$$\log a = \frac{2 \log p}{3} = \frac{2 \log 135}{3} = 1.4202$$

$$a = 26.3 \text{ AU}$$

The aphelion distance $(Q = 2a - q)$ of Swift-Tuttle is:

$$Q = 2 \times 26.3 - 0.968 = 51.7 \text{ AU}$$

We conclude that the comet Swift-Tuttle originates from the Edgeworth-Kuiper belt.

22.9.3 Listing of the Orbital Properties of Comets

22.10 Problems

1. Calculate the area of the Edgeworth-Kuiper belt located between the orbit of Neptune (30 UA) and 50 AU. Express the result as a fraction of the total area of the solar system extending to 50 AU. (Answer: $A_{30\text{-}50}/A_{50} = 0.64$ or 64 %).
2. The orbital parameters of the short-period comet Crommelin (Science Brief 22.9.3) are $p = 27.89\,y$ and $q = 0.743$ AU. Use these data to calculate the values of a, Q, and e. (Answer: $a = 9.196$ AU; $Q = 17.64$ AU; $e = 0.919$).
3. The orbital parameters of the long-period comet Hale-Bopp *after* its last pass through the solar system include $Q = 358$ AU and $q = 0.914$ AU. Calculate the corresponding values of p and e of its new orbit. (Answer: $p = 2404\,y$; $e = 0.995$).

Name	a, AU	e	i, (°)	p, years	q, AU	Q, AU	Year of discovery	Last perihelion	Ref.
				Short Period					
Encke	2.21	0.847	11.8	3.30	0.340	4.2	1786	2003	2,3
Biela	3.52	0.756	12.55	6.62	0.861	6.18	1772	lost	2
Schwassmann–Wachmann 1	6.1	0.046	9.38	14.9	5.729	6.7	1908	2004	1,2
S-W2	3.4	0.39	4	6.4	2.1	4.8	—	2006	1
S-W3	3.06	0.694	11.4	5.36	0.937	5.183	—	2006	3
Grigg-Skjellerup	2.96	0.664	21.1	5.09	0.989	6.0	—	2002	1,3
Brorsen-Metcalf	16.86	0.972	19.28	69.5	0.479	33.25	1847	1989	2
d'Arrest	3.49	0.614	19.5	6.51	1.346	5.5	1851	2001	1,2,3
Crommelin	9.20	0.919	29.0	27.89	0.743	17.47	1818	1984	2,3
Swift-Tuttle	~24	0.963	113.47	~135	0.968	~47	1862	1992	1,2
Tempel-Tuttle	10.33	0.906	162.5	33.92	0.982	19.72	1866	1998	1,2,3
Tempel 1	3.11	0.517	10.5	5.5	1.500	4.7	—	2005	1,3
Tuttle	5.69	0.819	54.94	13.6	1.034	10.34	1858	1994	1,2
Arend-Rigaux	3.5	0.60	18	6.8	1.4	4.9	—	—	1
Wild 2	3.44	0.540	3.24	6.39	1.583	5.30	1978	1997	1,2,3
Giacobini-Zinner	3.52	0.706	31.86	6.52	0.996	6.1	1900	1998	2,3
Halley	17.94	0.967	162.22	76.1	0.587	28.4	239 BC	1986	2,3
Wirtanen	3.12	0.652	11.7	5.46	1.063	5.13	1954	1997	1,2,3
Borelly	3.61	0.624	30.3	6.86	1.358	5.862	—	2001	3
Honda-Mrkos-Pajdusakova	3.02	0.825	4.3	5.29	0.528	5.512	—	2005	3

	a	e	i	p	q	Q			
Churyumov-Gerasimenko	3.51	0.632	7.1	6.57	1.292	5.728	—	2002	3
West-Kohoutek-Ikemura	3.45	0.540	30.5	6.46	1.596	5.304	—	2000	3
Wilson-Harrington	2.64	0.623	2.8	4.29	1.000	4.28	—	2005	3
Macholz I	3.016	0.959	59.96	5.24	0.125	5.907	1986	1996	1,2
Long Period									
Donati	~157	0.996	117	~2000	0.578	~313	1858	—	2
Humason	~204	0.990	153	~2900	2.13	~400	1962	—	1
Morehouse	Large	1.00	140	Large	0.95	Large	1908	—	1
Burnham	Large	1.00	160	Large	0.95	Large	1960	—	1
Kohoutek*	Large	1.00	14	Large	0.14	Large	1973	—	1
Arend-Roland	—	1.00	119.95	—	0.316	—	1956	—	2
Bennett	—	0.996	90.04	—	0.538	—	1970	—	2
Hale-Bopp*	263	0.995	89.43	4265	0.914	525	1995	1997	2,3
Hyakutake*	~660	0.999	124.95	~17,000	0.231	~1,300	1996	1996	2,3
Ikeya-Seki	—	1.000	141.86	—	0.008	—	1965	—	2
IRAS-Araki-Alcock	—	0.990	73.25	—	0.991	—	1983	—	2

* Before encounters with Jupiter
a = semi-major axis of the orbit
e = eccentricity of the orbit
i = inclination of the orbit relative to the plane of the ecliptic
p = period of revolution
q = distance from the Sun at perihelion
Q = distance from the Sun at aphelion

References: 1 = Hartmann (2005); 2 = Beatty et al. (1999); 3 = Williams (2005).

4. Explain why Kepler's third law applies to the orbits of comets but not to the orbits of satellites.
5. Does Newton's version of Kepler's third law apply to comets? Make a calculation for the comet Encke and find out.

22.11 Further Reading

Beatty JK, Petersen CC, Chaikin A (eds) (1999) The new solar system, 4th edn. Sky Publishing, Cambridge, MA

Boice DC, Huebner W (1999) Physics and chemistry of comets. In: Weissman P.R., McFadden L.-A., Johnsen T.V. (eds) Encyclopedia of the Solar System. Academic Press, San Diego, CA, pp 519–536

Brandt JC (1999) Comets. In, Beatty J.K., Petersen C.C., Chaikin A. (eds) The New Solar System, 4th edn. Sky Publishing, Cambridge, MA, pp 321–336

Brandt JC, Chapman RD (1992) Rendezvous in space. Freeman, New York

Brandt JC, Chapman RD (2004) Introduction to comets, 2nd edn. Cambridge University Press, New York

Burnham R (2000) Great comets. Cambridge University Press, Cambridge, UK

Churchill WS (1956) The birth of Britain. A history of the English-speaking peoples. vol 1. Dodd, Mead, and Co., New York

Cowen R (2005) Deep impact: A spacecraft breaks open a comet's secret. Science News Sep 10: 168–180

Fernandez JA (1999) Cometary dynamics. In: Weissman P.R., McFadden L.-A., Johnsen T.V. (eds) Encyclopedia of the Solar System. Academic Press, San Diego, CA, pp 537–556

Festou MC, Keller HU, Weaver HA (eds) (2004) Comets II. University of Arizona Press, Tucson, AZ

Freedman RA, Kaufmann III WJ (2002) Universe: The solar system. Freeman, New York

Gehrels T (ed) (1994) Hazards due to comets and asteroids. University of Arizona Press, Tucson, AZ

Gropman D (1985) Comet fever. Fireside Books, New York

Grün E (1999) Interplanetary dust and the zodiacal cloud. In: Weissman P.R., McFadden L.-A., Johnson T.V. (eds) Encyclopedia of the Solar System. Academic Press, San Diego, CA, pp 673–696

Hartmann WK (2005) Moons and planets, 5th edn. Brooks/Cole, Belmont, CA

Koeberl C (1998) Impakt, Gefahr aus dem All: Das Ende unserer Zivilisation. Va Bene, Vienna, Austria

Kronk GW (1984) Comets: A descriptive catalog. Enslow Publishers, Hillside, NJ

Kronk GW (2003) Cometography: A catalog of comets. vol 2 (1800–1899). Cambridge University Press, Cambridge, UK

Lang KR, Whitney CA (1991) Wanderers in space. Cambridge University Press, Cambridge, UK

Lakdawalla E (2005) A smashing success. Planet Rept Jul/Aug: 21

Lawson MK (2002) The battle of Hastings 1066. Tempus Publishers Ltd., Brimscombe Port Stroud, Gloucestershire, UK

Levy DH (1998) Comets: Creators and destroyers. Touchstone/Simon and Schuster, New York

Lewis JS (1998) Rain of iron and ice. Addision-Wesley, Reading, MA

McFadden L (2004) Deep impact: Our first look inside a comet. Planet Rept Nov/Dec: 12–17

Metz J (1985) Halley's comet, 1910: Fire in the sky. Singing Bone Press, New York

Moomaw B (2004) Stardust collects bits of comet Wild 2. Astronomy, April: 24

Newburn RL, Neugebauer MM, Rahe J (eds.) (1991) Comets in the post-Halley era. Kluwer, Dordrecht, The Netherlands

Olson RJM (1985) Fire and ice. Walker and Co., New York

Olson RJM, Pasachoff JM (1998) Fire in the sky. Cambridge University Press, Cambridge, UK

Ottewell G, Schaaf F (1985) Mankind's comet. Astronomical Workshop, Greenville, SC

Rayl AJS (2004) In memoriam-Fred Whipple. Planet Rept Nov/Dec: 15

Sagan C, Druyan A (1985) Comet. Random House, New York

Spencer JR, Mitton J (eds) (1995) The great comet crash. Cambridge University Press, Cambridge, UK

Talcott R (2002) Comet Borelly's dark nature. Astronomy April: 42–45

Talcott R (2005) Blasting the past. Astronomy October: 72–75

Verschuur G (1996) Impact! Oxford University Press, New York

Wagner JK (1991) Introduction to the solar system. Saunders College Publishers, Philadelphia, PA

Warren JL, Zolensky ME, Thomas K, Dodson AL, Watts LA, Wentworth S (1997) Cosmic dust catalog. vol 15, National Aeronautics and Space Administration, Houston, TX, 477 p

Whipple FL (1974) The nature of comets. Sci Am, 230: 48–57

Weissman PR (2006) A comet tale. Sky and Telescope. 111(2):36–41

Williams DR (2005) Comet fact sheet. NASA Goddard Space Flight Center, Mail Code 690.1, Greenbelt, MD 20771 (< dave.williams@gsfs.nasa.gov >)

Yeomans DK (1991) Comets. Wiley, New York

Earth: The Cradle of Humans

Now that we have explored the planets and satellites of the solar system, we know that the Earth is unique because:

1. a large portion of its surface is covered by liquid water,
2. it has an atmosphere that contains oxygen gas, and
3. it is inhabited by living organisms that together form the biosphere.

It is possible that primitive organisms once existed on Mars after the planet had cooled sufficiently to allow water vapor to condense on its surface and after the frequency of impacts had diminished enough to allow life forms to survive. It is also possible that the ocean of Europa harbors organisms that may have evolved into a variety of forms in the course of time. Perhaps life exists even on Enceladus or Titan, although the low temperature of their surfaces makes that quite improbable. The existence of life elsewhere in the solar system is possible but not probable given the harsh environmental conditions that exist on all other planets and satellites, except on the Earth. Possibilities and probabilities of finding life on other planets and their satellites motivate our exploration of the solar system. However, in spite of our best efforts to find living organisms in the solar system, the Earth is still the only planet that is known to be inhabited (Chyba and McDonald, 1995; Soffen, 1999).

23.1 Origin of Life

Life on the Earth is based on the chemical properties of carbon, which can bond with itself and with hydrogen, nitrogen, oxygen, sulfur, and phosphorus. Therefore, the presence of these and a few other elements (e.g., iron, cobalt, nickel, copper, zinc, etc.) is one of the prerequisite conditions that make life possible. In addition, liquid water must be present because chemical reactions are facilitated in an aqueous medium, although complex organic molecules did form in the solar nebula which consisted of a very dilute gas. The organic molecules in the solar nebula were initially sorbed to the surfaces of microscopic ice particles that subsequently accreted to form larger bodies having diameters that ranged from less than 10 m to more than 1000 km (i.e., the planetesimals). These bodies could form only in the cold outer regions of the protoplanetary disk at a distance of more than 5 AU from the Sun. Nevertheless, a large number of these bodies entered the inner part of the solar system and impacted on the terrestrial planets as they were forming. As a result, all of the terrestrial planets initially acquired water and other volatile compounds as well as large amounts of the organic matter that was embedded in the ice planetesimals (Kasting and Ackerman, 1986; Hunten, 1993).

Exactly how life in the form of self-replicating molecules arose in the primordial ocean of the Earth is still unknown. None of the several hypotheses that have been proposed has gained the stature of a scientific theory and no life forms of any kind have been synthesized by chemical reactions under controlled conditions. The failure of these hypotheses and laboratory experiments implies that the synthesis of life by abiotic chemical reactions is a highly improbable process that can succeed only in the course of millions or even tens of millions of years. The subsequent evolution of sentient beings from primitive unicellular organisms is even more improbable and requires billions of years without guaranty that the process of evolution will be successful. These considerations provide perspective on the uniqueness of the existence of human beings on the Earth and raise doubts that all earthlike planets in the Milky Way galaxy are necessarily inhabited by sentient beings (Chang et al., 1983; Davis and McKay, 1996; McKay and Davis, 1999; Ward and Brownlee, 2000).

23.1.1 Panspermia

Micro-organisms could have formed elsewhere in the solar system (e.g., Mars) and were then transported to the Earth inside meteorites that fell into the oceans of the Earth (Warmflash and Weiss, 2005; Mileikowsky et al., 2000). Mars is a possible source of micro-organisms because it had standing bodies of water on its surface during its earliest history, because it contains the essential elements, and because more than 30 rocks that originated from Mars have been found on the Earth (Sections 12.7, 12.9). However, panspermia does not explain the origin of micro-organisms or of complex organic molecules on other planets or their satellites. In addition, it is not yet certain that micro-organisms could survive the trip and the impact on the Earth.

Panspermia becomes more probable when it is expanded to include the transport of organic molecules from Mars, or from some other body in the solar system, to the Earth. For example, carbonaceous chondrites and comets contain abiogenic organic matter that is released into the terrestrial environment when these objects impact upon the Earth. In this view of panspermia, life arose on the Earth by as yet unspecified reactions of extraterrestrial organic molecules.

The validity of panspermia hinges on the feasibility of dislodging rock samples from the surface of a superior planet (e.g., Mars) and transporting them to the Earth. The presence on the Earth of rocks that originated from the Moon (Section 9.4) and from Mars (Section 12.7) demonstrates not only that such interplanetary transfers are possible but that large amounts of material have successfully made this trip in the course of geologic time. Transport in the opposite direction (e.g., from Earth to Mars or to the Moon) is much less likely because it is opposed by the gravity of the Sun.

The transfer of organic molecules or micro-organisms from Mars to the Earth begins when a large meteorite or comet impacts on Mars and ejects rocks and dust from the surface into interplanetary space. A small fraction of this material has the necessary trajectory to reach the Earth in less than one year, which minimizes the radiation dose to which microbes and abiogenic organic matter are exposed. Hypothetical martian micro-organisms similar to the terrestrial species *Deinococcus radiodurans* can survive a short trip from Mars to Earth if they are protected from ionizing radiation inside fist-sized rock fragments whose interiors remain cool during the passage through the atmosphere of the Earth (Section 8.6). Warmflash and Weiss (2005) estimated that about 1 ton of martian rock fragments is deposited on the Earth every year. Most of this material has been in transit for several million years depending on the size of the fragments. In general, large pieces take longer than small pieces (pebbles and dust). As a result, on average ten fist-sized samples of martian rocks arrive on the Earth within three years after a large impact on Mars.

Rocks from the surface of Mars that spent only about one year in transit could transfer micro-organisms from the surface of Mars to the Earth, provided the organisms were not injured by exposure to ionizing radiation in space and by the heat during passage through the terrestrial atmosphere. However, no martian micro-organisms have been discovered in any of the martian rocks that have been studied and no evidence has been found that micro-organisms once existed on the surface of Mars (e.g., ALH 84001, Section 12.9.3). Therefore, panspermia remains an interesting hypothesis that has not yet been supported by direct observation.

23.1.2 Miller-Urey Experiment

An alternative to panspermia is the hypothesis proposed by the Russian geochemist Alexandr I. Oparin that the first living organisms developed spontaneously from simple organic compounds in the primordial ocean of the Earth (Oparin, 1938). He and other geochemists (e.g., the British geneticist John B.S. Haldane) considered that the organic compounds were synthesized in the primordial atmosphere of the Earth composed of methane, ammonia, water vapor, and hydrogen but lacking oxygen (Haldane, 1954).

The hypothesis of A.I. Oparin was tested by S.L. Miller and H.C. Urey in the 1950s by means of a famous experiment in which they generated organic compounds by electrical discharges in a mixture of the gases listed above (Miller and Urey, 1959; Miller 1953, 1955, 1957). The compounds that were produced in this experiment included amino acids and sugars that could serve as building blocks for the assembly of ribonucleic acid (RNA), which is a polymer that carries genetic information and can act as an enzyme (Ferris, 2005). When the Miller-Urey experiment was later repeated with a different mixture of atmospheric gases consisting of carbon dioxide, nitrogen, and water vapor, a similar mixture of organic compounds was produced. These results demonstrate that organic compounds could have formed by lightning in the primordial atmosphere of the Earth. In other words, the origin of life on the Earth does not necessarily depend on the importation of organic compounds from extraterrestrial sources.

23.1.3 Mineral Catalysts

The concept that the primordial ocean was a "soup" of organic compounds is misleading. The concentration of organic molecules in the ocean was quite low because of the large volume of water and the slow rate of production of such molecules in the atmosphere (Hazen, 2005). Therefore, the simple organic compounds that were dissolved in the water had to be concentrated in some way so that they could combine to form more complex molecules. Lasaga et al. (1971) proposed that organic compounds were concentrated in the form of an oil slick on the surface of the primordial ocean. Another suggestion, advanced by Lahav et al. (1978), concentrated organic matter in evaporating tidal pools. Organic molecules are also concentrated by sorption to the crystalline surfaces of minerals. For example, Sowerby et al. (1996, 1998) demonstrated that molecules of adenine ($C_5H_5N_5$) and other biologically interesting compounds are sorbed to cleavage planes of graphite (C) and molybdenite (MoS_2) and that the sorbed molecules organize themselves into two-dimensional structures. Hazen (2005) considered that the formation of such structures by the assembly of smaller organic molecules on the surfaces of minerals may be an "early step in the emergence of life."

Mineral surfaces also sorb organic molecules selectively. For example, calcite ($CaCO_3$) and quartz (SiO_2) sorb specific amino acids and distinguish between left-handed and right-handed versions of the same molecule (Churchill et al., 2004). This ability of mineral surfaces may explain why living organisms on the Earth use only about 20 of the more than 70 different amino acids and almost exclusively prefer the left-handed molecular forms (Hazen et al., 2001).

An important group of minerals that can form in reducing environments (i.e., lacking oxygen) are the sulfides of transition metals (iron, cobalt, nickel, copper, zinc, etc.) that form in the ocean where superheated aqueous solutions are discharged by hotsprings along spreading ridges (Cody, 2005). In this environment iron monosulfide (FeS) reacts with hydrogen sulfide (H_2S) to form the mineral pyrite (FeS_2) and molecular hydrogen (H_2) (Wächtershäuser, 1988, 1990):

$$FeS + H_2S \rightarrow FeS_2 + H_2 \qquad (23.1)$$

The energy released by this reaction provides the driving force for a reaction that transforms carbon dioxide (CO_2) into formic acid (HCOOH):

$$CO_2(aq) + FeS + H_2S \rightarrow HCOOH + FeS_2 + H_2O \qquad (23.2)$$

In this way, the precipitation of pyrite catalyzes the formation of an organic molecule (formic acid). Wächtershäuser (1992) postulated that organic films on pyrite crystals developed the ability to synthesize "food" from molecules in the environment by means of these kinds reactions. When the organic membranes subsequently detached from the pyrite surface and formed cells, they continued to convert carbon dioxide and iron monosulfide into formic acid by means of the metabolic process they had "learned" on the pyrite surface.

These conjectures are supported by the fact that micro-organisms alive today utilize enzymes that contain atoms of transition metals and sulfur. For example, *nitrogenase* which helps to reduce N_2 to NH_3 contains molybdenum, iron, and sulfur; *hydrogenase* which is used to remove electrons from molecular hydrogen to form the ion H^+ contains iron, nickel, and sulfur; and *aldehyde oxidoreductase* enzymes use tungsten and molybdenum to transform organic aldehydes into acids. The existence of such organo-metallic compounds in living organisms is consistent with the hypothesis that life arose in close association with sulfide minerals (Smith, 2005).

23.1.4 Cradle of Life

It is tempting to imagine that the process which allowed self-replicating molecules to form took place on the surface of the Earth in shallow lagoons filled with warm water containing dissolved organic compounds. Actually, the surface of the Earth during the Hadean Eon (4.6 to 3.8 billion years ago) was disrupted by frequent impacts of asteroids and comets as well as by continuous eruptions of lava through fissures and volcanoes (Maher and Stevenson, 1988). In addition, the surface of the Earth was exposed to high-energy ultraviolet (UV) radiation, which is harmful to organisms because it breaks the chemical bonds between the atoms of large molecules of organic compounds. Ultraviolet radiation emitted by the Sun could reach the surface of the Earth during the Hadean Eon because the atmosphere of the Earth did not contain sufficient oxygen to sustain an ozone layer like the one that presently shields the surface of the Earth from UV radiation.

The cradle of life had to be a place that was undisturbed for long periods of geologic time (i.e., millions to tens of millions of years), that was protected from UV radiation, and that contained sufficient heat to facilitate the synthesis of complex molecules from simple organic compounds. A further condition is that the environment had to contain transition metals and sulfur which are required in the synthesis of enzymes. Contrary to our intuitive expectation, the cradle of life did not contain molecular oxygen (O_2) because organic compounds are inherently unstable in the presence of oxygen and the cradle of life was not exposed to direct sunlight because of the problem with UV radiation.

These requirements are met by hotsprings on the bottom of the ocean adjacent to submarine volcanoes and fissures where superheated aqueous fluids are discharged. These fluids form precipitates of sulfides of iron and other transition metals as they cool and mix with the ambient seawater. The precipitates consist of small black particles that discolor the water and make it look like black smoke. For this reason, the hotsprings associated with volcanic activity on the bottom of the ocean are referred to as *black smokers*. These sites are inhabited by bacteria that thrive in this seemingly hostile environment. The bacteria in turn support communities of mollusks and fish which feed on the bacteria. The unexpected existence of isolated oases of life at the bottom of the present oceans of the Earth supports the view that these places could be the cradle of life (Freedman and Kaufmann, 2002: Ballard, 1995; Baross and Hoffman, 1985; Baross and Deming, 1993; Karl, 1995; Huber et al., 1990; Miller and Bada, 1988; Ohmoto and Felder, 1987).

This line of reasoning raises the possibility that life-forming geochemical processes are still occurring on the Earth in the vicinity of black smokers. It also increases the probability that the global ocean of Europa contains organisms that originated near hotsprings which discharge brines that were heated by contact with the rocks that underlie the ocean (Sections 15.3.1 to 15.3.4). If the ocean of Europa does contain communities of living organisms, the hypothesis concerning the origin of life in the primordial ocean of the Earth would be supported and mankind would take another giant leap forward (Greenberg, 2005).

23.2 Importance of the Biosphere

The fossils preserved in sedimentary rocks of Early Archean age indicate unequivocally that micro-organisms existed in the oceans of the Earth about 3.5 billion years ago and perhaps as early as 3.8 billion years ago (Section 27.2a to 27.2b, Faure and Mensing, 2005). These simple life-forms evolved slowly during the passage of geologic time until eventually all available habitats were occupied and living organisms became important participants in the geological processes that take place on the surface of the Earth. The evolutionary history of the biosphere and its effect on the environment on the surface of the Earth are the subjects of a growing body of the scientific literature including books by Gould (1994), Lovelock (1979), Schopf (1983, 1992, 1994), Holland (1984), Berner and Berner (1987), Emiliani (1992), Holland and Petersen (1995), Hunten (1993), Schidlowski (1993), Chang (1994), and others.

The importance of the biosphere to the geological processes on the surface of the Earth was recognized a long time ago by the visionary Russian geochemist Vladimir Ivanovich Vernadsky (1863–1945) who, among many other activities, taught mineralogy and crystallography at the University of Moscow (Ivanov, 1997). Vernadsky extended the limits of the biosphere to include the lower atmosphere to an altitude of 30 km and the upper part of the crust to a depth of two to three kilometers. He considered that the biosphere is an active part of the Earth and that it has existed since the earliest period in the history of the Earth. The papers that contained Vernadsky's doctrine of the biosphere were published in the 1920s, but his ideas were not widely accepted by contemporary European and North American scientists because they were written in several different languages and because they deviated from conventional views prevalent at the time. In later life, Vernadsky came to believe that humans will ultimately control the biosphere and thereby create a utopian society in which malnutrition, hunger, misery and disease will be eliminated (Ivanov, 1997, p. 18).

Vernadsky's doctrine of the biosphere was popularized in the 1960s by John E. Lovelock who developed the concept that the biosphere acts to maintain environmental conditions on the surface of the Earth that are required for its own survival. For example, the well-being of the biosphere, called Gaia by Lovelock, is presently endangered by the anthropogenic contamination of the atmosphere, oceans, and surface of the Earth. We can only hope that the biosphere will respond effectively and thus prevent environmental changes that could endanger the survival of humans and other life-forms on the Earth.

23.2.1 Atmosphere

The primordial atmosphere contained only trace amounts of molecular oxygen that formed by the dissociation of water molecules by UV radiation (Holland, 1984; Kasting and Ackerman, 1986; Hunten, 1993). Consequently, the micro-organisms that formed in the dark places of the Earth were *anaerobic* for whom molecular oxygen was poisonous. In the course of time, certain kinds of unicellular organisms began to manufacture a magnesium-bearing metallo-organic compound called *chlorophyll* that enabled these organisms to synthesize carbohydrates by combining carbon dioxide and water. This process is energized by sunlight and produces molecular oxygen as a byproduct:

$$6CO_2 + 6H_2O \overset{\text{sunlight}}{\rightarrow} \underset{\text{glucose}}{C_6H_{12}O_6} + O_2 \qquad (23.3)$$

The ability to synthesize carbohydrate molecules by means of photosynthesis enabled micro-organisms to live at the surface of the ocean and to tolerate oxygen, which they released into the atmosphere. Consequently, the concentration of oxygen in the atmosphere of the Earth began to increase during the Archean Eon. The change was slow at first and more than one billion years elapsed before the mass of photosynthetic plants in the oceans had increased sufficiently to cause the concentration of oxygen in the atmosphere to increase to about 20 %. The change in the composition of the atmosphere of the Earth occurred during the Early Proterozoic Era between 2.5 and 2.0 billion years ago. The presence of oxygen in the atmosphere revolutionized the biosphere of the Earth and allowed it ultimately to spread from the oceans to the surface of the continents and to diversify into plants (oxygen producers) and animals (oxygen consumers).

23.2.2 Sedimentary Rocks

All sedimentary rocks contain organic matter and some are formed from the shells and skeletons of organisms that inhabit the oceans. The organic matter of *shales* occurs in the form of complex organic molecules called *kerogen* and particles of amorphous carbon, both of which are alteration products of the tissue of plants that live in the marine or nonmarine basins where fine-grained sediment accumulates in layers. *Limestones*, in most cases, are composed of the exoskeletons of corals and crinoids as well as of the shells of mollusks which form calcareous ooze that recrystallizes after it has been buried by overlying layers of sediment.

In many cases, sedimentary rocks (e.g., shale and limestone) contain fossils of the plants and animals that lived in the oceans or in lakes at the time and place where the sedimentary rocks were deposited. The study of such fossils in layered sequences of sedimentary rocks has provided direct evidence for the evolution of life-forms on the Earth and has been used to define the geological timescale (Gould, 1994).

No morphologically preserved fossils have yet been recognized in the martian rocks that have been collected on the Earth (Section 12.9.3) or in the close-up views of rock outcrops on the surface of Mars. The only hypothetical exception are the small hematite pellets (i.e., the so-called blueberries), which may have formed by the action of bacteria in the shallow lakes that long ago filled the craters in Meridiani planum of Mars (Section 12.9.4) (Davis and McKay, 1996; Goldsmith, 1997; McKay, 1997).

23.2.3 Fossil Fuels

Some of the sedimentary rocks that have formed on the Earth are so rich in organic matter (i.e., kerogen) that they will burn when lit with a match. These kinds of organic-rich rocks include oil shale, tar sand, and different varieties of coal (i.e., brown coal, bituminous coal, and anthracite). In addition, the crust of the Earth contains reservoirs of liquid hydrocarbons and hydrocarbon gas (i.e., petroleum and natural gas). These materials are collectively referred to as fossil fuels because they formed by the alteration of organic matter derived from plants on land (coal) and from organisms in the oceans (petroleum and natural gas). The availability of deposits of fossil fuels made possible the industrial revolution during the 19th century and the concurrent development of technologies which have increased the standard of living of the human population of the Earth.

23.2.4 Humans

Perhaps as much as six million years ago, primitive anthropoids emerged within the biosphere. They developed the ability to walk upright and to grasp objects with their hands which had opposable thumbs. The early humans lived in small groups and sustained themselves by hunting small animals and by gathering the roots and seeds of plants. About 1.8 million years ago a more advanced species called *Homo erectus* migrated from East Africa into Europe, Asia, and Indonesia. Toward the end of the Pleistocene Epoch humans of the species *Homo neanderthalensis* survived the repeated advances of continental ice sheets into northern Europe by living in caves and by hunting mammoth and other large animals. The Neanderthal people were eventually absorbed or displaced by *Homo sapiens* who, about 10,000 years ago, began to grow grain and to domesticate animals. By doing so, humans became the masters of the biosphere who favored plants and animals that were useful to them and tried to eradicate those that were not useful.

Farming improved nutrition and increased the life expectancy of humans. Farming also led to ownership of land which caused disputes about borders and led to warfare. Starting about 10,000 years ago, the population of humans began to grow and the rate of growth increased with time, especially after the Industrial Revolution in the 19th century AD (Musser, 2005; Cohen, 1995, 2005; Livi-Bacci, 2001). The human population in Figure 23.1 increased exponentially from about 1.0 billion at 1800 AD, to 1.5 billion at 1900 AD, and to 6.0 billion at 2000 AD. Holland and Petersen (1995) predicted that the population will grow to 40 billion at 2100 AD, whereas Musser (2005) placed the upper limit at 14 billion. Although we are relative late-comers on the Earth, we have achieved dominance over the biosphere and have become significant agents of change on the surface of the Earth.

The expansion of the human population and the increase of the standard of living has been accompanied by an increase in the rate of consumption of non-renewable and finite natural resources including fossil fuels and ore deposits of metals. The combustion of fossil fuel has increased the concentration of carbon dioxide and decreased the concentration of oxygen in the atmosphere. The contamination of the atmosphere by the release of carbon dioxide, methane, sulfur dioxide, and chlorofluorocarbon gases is causing global warming and

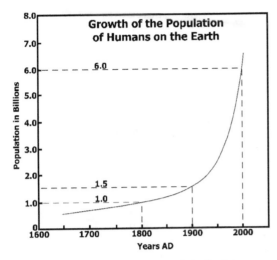

Figure 23.1. The human population of the Earth began to increase slowly after our species started to cultivate crops and to domesticate animals about 10,000 years ago. Even as recently as 1800 AD the population amounted to only about one billion. Two hundred years later, in the year 2000 AD, the population reached six billion and is projected to reach 40 billion in 2100 AD. Most of the increase will occur in developing countries, whereas the population of the industrialized nations is expected to stabilize at one billion by the middle of the 21th century. Adapted from Figure 5.7 of Holland and Petersen (1995)

climate change (Kasting and Ackerman, 1986). The presence of chlorine in the stratosphere is destroying the ozone layer which has been shielding the surface of the Earth from harmful UV radiation. Global warming is increasing the rate of evaporation of water from the surface of the ocean and may increase the amount of snow that forms in the polar regions. Therefore, global warming may lead to global glaciation. Alternatively, global warming of the atmosphere may cause mountain glaciers and the ice sheets of Antarctica and Greenland to melt leading to widespread coastal flooding because of the increase in sealevel. The most disturbing aspect of this conandrum is that we are unable to predict the consequences of the global experiment we are performing on the Earth.

The mining and smelting of the ore deposits of industrial metals requires large amounts of energy and contaminates the atmosphere by the release of gases (e.g., carbon dioxide and sulfur dioxide) and dust. In addition, chemical weathering of ore and gangue minerals contami-nates streams, water reservoirs, and groundwater. Mining operations also disrupt the surface of the Earth and make it unattractive or even unsuitable for human habitation. Attempts to supplement energy production by means of uranium-fueled nuclear-fission reactors and by building dams across stream valleys for the production of electric power also have undesirable side effects. Mining and processing uranium ore is hazardous and the spent uranium fuel remains dangerously radioactive for up to 100,000 years. The dams across major rivers interrupt the transport of sediment from the continents into the oceans, interfere with the migration of fish, and store potentially hazardous sediment.

The amount of arable land is declining as a result of increasing urban sprawl and may decline further in case sealevel begins to rise appreciably as a consequence of global warming. Additional land area will be lost because of the need to dispose of increasing amounts of municipal and industrial waste. These and other consequences of human activities diminish the capacity of the Earth to sustain the growing human population at the level of the present standard of living. If the population of the Earth does rise to 40 billion by the end of the present century, living conditions will become difficult. Linear extrapolations of present trends inevitably lead to predictions of doom.

Fortunately, there are signs that the worst-case scenarios can be avoided. The population of the industrial nations are expected to stabilize at about one billion by the year 2050. The impli-cation of this information is that the population growth of developing nations will also slow as the living standard of the people rises. Although the areas devoted to forest and pastures have been declining, the area of cropland has been increasing from 1030 million hectares in 1850 to 3002 million hectares in 1980 (1 hectare = $10,000 \, m^2$). In addition, the yields of grain crops have been rising; slowly at first, but more rapidly after 1950. For example, in the United Kingdom wheat yields rose from 4 kg/ha/yr prior to 1950 to 78 kg/ha/y after that year. In the same time period, wheat yields in the USA rose from 3 kg/ha/y (prior to 1950) to 50 kg/ha/y (post 1950). In general, the number of nutritional calories per capita per year has been rising slowly throughout the world (Repetto, 1987; Holland and Petersen, 1995).

Global warming and the resulting increase in sealevel can be slowed by reducing emissions of carbon dioxide and other greenhouse gases world-wide. Organic waste can be fermented to produce methane fuel. Industrial waste (metals, plastics, glass, and paper) can be sorted and recycled. These and other strategies can be employed to sustain a stable population of humans during the millennia of the future. The implementation of these strategies must be adopted world-wide, which requires that people must recognize the need to do so and therefore be willing to make the necessary changes in their lifestyle. One of the lessons we have learned from the exploration of the solar system is that the Earth is the only place in the solar system where humans can live without the aid of life-support. Our future depends on the effort we make to maintain a healthy environment on our home planet (Renner et al., 2005). If we fail, the human population is likely to decline catastrophically and without warning.

23.3 Tales of Doom

Konstantin Tsiolkovsky (Section 1.2) believed that humans will have to leave the Earth before the Sun runs out of fuel and can no longer provide heat and light to the Earth (Section 4.2). When that happens, the Earth will become uninhabitable and the human race will become extinct unless humans emigrate to a planet orbiting another star in the Milky Way galaxy. These convictions may have motivated Tsiolkovsky to design rocket-propelled spaceships that can transport humans from the Earth to a new home among the stars. He also thought about the practical problems of living in space for long periods of time and developed plans for doing so. Tsiolkovsky became convinced that space travel beyond the solar system is feasible and would become necessary to assure the survival of the human race. Therefore, in Tsiokovsky's view, space travel is our destiny because we have an obligation to preserve our species. He expressed his urgent plea to our descendants by the famous statement:

"Earth is the cradle of humanity, but we cannot live in the cradle forever." (Hartman et al., 1984)

23.3.1 Evolution of the Sun

The survival of the biosphere of the Earth is closely related to the future of the Sun as predicted by the theory of stellar evolution (Section 4.2 and 5.3.5). According to this theory, the Sun will begin to contract because of the increasing density of its core caused by the conversion of hydrogen to helium. As a result, the core temperature will rise, which will increase the amount of heat generated by hydrogen fusion and therefore will lead to an increase in the solar luminosity. In this way, the Sun will raise the surface temperature of the Earth and will eventually cause the oceans to evaporate. The resulting greenhouse effect will cause the Earth to become hot and dry like Venus is today (Section 11.3.3). These changes will take place gradually during many hundreds of millions of years. As climatic conditions on the Earth become increasingly difficult, the biosphere will adapt at first until it is overwhelmed and life on the Earth is eventually extinguished.

The Sun will ultimately evolve into a red giant when its diameter will expand beyond the orbits of Mercury and Venus and will come close to the present orbit of the Earth. This transformation will occur in about four to five *billion* years and may cause the crust of the Earth to melt. The expansion of the Sun into a red giant will also increase the surface temperatures of the superior planets and their satellites (i.e., Mars, Jupiter, Saturn, etc.). Exactly how this will change them is uncertain and is still in the domain of science fiction.

The perspective we gain from a consideration of the evolution of the Sun is that it will remain *stable* for at least another billion years and that any changes in the solar luminosity are so far in the future as to be beyond human comprehension. Although Tsiolkovsky was correct in principle, we have much less to fear from the eventual death of the Sun than we do from other celestial and anthropogenic threats to our existence on the Earth.

23.3.2 Impacts

The Earth is continually being bombarded by objects in interplanetary space whose heliocentric orbits cross the orbit of the Earth (e.g., meteoroids, comets, and asteroids). Most of the impactors have diameters less than 10 m and cause only local damage on the surface of the Earth. Impacts of larger objects having diameters between 10 and 1000 m are less common but can destroy a city and can cause millions of human casualties. These kinds of impacts are regrettable, but they do not endanger the survival of the human race or of our civilization. We are concerned here with the impacts of asteroids and comets having diameters greater than 10 km that cause the extinction of large fractions of the biosphere and therefore do endanger the survival of the human population of the Earth (Kring, 2000).

The biosphere of the Earth has sustained many such extinction events in the past and has actually been reinvigorated by the opportunities they provided for the evolution of new species (Gould, 1994). The best-documented extinction event occurred 65 million years ago when a small asteroid 10 km in diameter impacted on the Yucatán peninsula of Mexico and formed the Chicxulub crater that is 180 km wide. About 70% of all species of plants and animals became extinct in the aftermath of this event, including the dinosaurs which had dominated the biosphere during the Mesozoic Era for up to about 185 million years (Silver and Schultz, 1982; Emiliani et al., 1981; Dressler et al., 1994; Powell, 1998; Koeberl and MacLeod, 2002).

The Chicxulub impact caused a global fire storm and injected large amounts of dust, soot, and smoke into the atmosphere, which caused short-term darkness and a decrease of the global temperature. In addition, the nitrogen of the atmosphere at the impact site reacted with oxygen to form large quantities of oxides of nitrogen which subsequently reacted with water vapor to form nitric acid:

$$N_2 + \frac{5}{2}O_2 \rightarrow N_2O_5 (\text{nitrogen pentoxide})$$
$$(23.4)$$

$$N_2O_5 + H_2O \rightarrow 2HNO_3 (\text{nitric acid}) \quad (23.5)$$

The target rocks on the Yucatán peninsula became another source of acid because they contain evaporite rocks including the minerals anhydrite ($CaSO_4$) and gypsum ($CaSO_4 \cdot 2H_2O$). These minerals formed sulfur dioxide gas which subsequently reacted with oxygen and water to form molecules of sulfuric acid:

$$SO_2 + \frac{1}{2}O_2 + H_2O \rightarrow H_2SO_4 (\text{sulfuric acid})$$
$$(23.6)$$

The deposition of acid rain all over the Earth and the global climatic cooling was followed by global warming caused by the release of water vapor and carbon dioxide into the atmosphere. The profile of temperature fluctuations in Figure 23.2 shows a strong heat pulse that lasted for several days after the impact, followed by global cooling for several decades, and later by global greenhouse warming that abated slowly in the course of the following century. Plants on land were killed by the initial blast wave, by the fires that broke out all over the Earth, by the month-long interruption of photosynthesis, and by the acidification of the environment. Animals that could not retreat into subsurface shelters were also killed by the sudden devastation of the surface of the Earth, regardless of whether they were herbivores or carnivores. Even plants and animal in the ocean were affected by the acidification of the surface water and because photosynthesis by planktonic organisms was prevented for several months after the impact. All of the large reptiles perished within a matter of days or less, but small reptiles survived, presumably because they found shelter below the surface of the Earth and their populations eventually rebounded. However, no new species of dinosaurs evolved from these survivors. Apparently, extinction is forever.

If another asteroid with a diameter of 10 km or larger were to impact on the Earth, most humans would be killed and our technological civilization would end abruptly. However, the frequency of such events is less than one in 100×10^6 years (Hartmann, 2005, Figure 6.5). In other words, global extinction events are very rare and the next such event is not likely to occur in the foreseeable future. Therefore, the possibility that humans will become extinct as a result of the impact of an asteroid does not require us to abandon the Earth to seek a new life on another

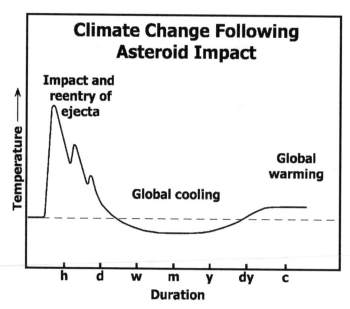

Figure 23.2. The aftermath of the impact of an asteroid or comet with a diameter of about 10 km (e.g., Chicxulub) is depicted here by the effect on the surface temperature of the whole Earth. After the initial heat pulse that lasts from hours (h) to days (d), the temperature declines due to the presence of dust, soot, and smoke in the atmosphere that linger for up to one decade (dy). After the dust has settled and the smoke has cleared, the water vapor and carbon dioxide injected into the atmosphere by the impact cause global warming that continues for centuries (c) until the water vapor condenses and carbon dioxide is absorbed by the biosphere and by the ocean. Adapted from Figure 3 of Kring (2000)

planet. It does, however, demand our continued vigilance to detect incoming asteroids decades before they actually reach the Earth so that they can be deflected from their collision course. We do have the knowledge and the means to prevent large-scale impacts of asteroids on the Earth and therefore have an obligation to use both in order to safeguard the future of our species.

23.3.3 Suicide

In the course of the 20th century we have developed weapons based on the energy that is released by the uncontrolled fission of the nuclei of uranium and plutonium atoms and by the fusion of hydrogen atoms. These weapons are so powerful that they can be used to destroy all human civilizations and most plants and animals on the surface of the Earth. Although initially only the USA and the USSR possessed these weapons, the technology has spread to other countries. The nuclear weapons can be delivered to selected targets by intercontinental ballistic

missiles that are propelled by the same kinds of rockets that have enabled us to land astronauts on the Moon and to explore the solar system. In other words, the rapid growth of technology is both a triumph of human ingenuity and a potential cause for our self-destruction.

The damage that nuclear weapons can inflict arises from the explosive expansion of the high-temperature fireball and from the dispersal of the radioactive atoms of many chemical elements that are produced in the process. Therefore, survivors of the initial blast are threatened by the harmful effects of the nuclear radiation emitted by the radioactive atoms that are produced.

The threat to the survival of the human race posed by nuclear weapons is widely recognized and efforts are being made to reduce the stock-piles of such weapons and to prevent their proliferation. In addition, the long-range goal of the "space-faring" nations is to establish human colonies in satellites orbiting the Earth as well as on the surface of Mars. If such colonies can become self-sufficient, the human race may survive even if the Earth is devastated by global nuclear war. When viewed from this perspective, colonization of Mars becomes an urgent survival strategy especially if the climate of Mars can be altered by terraforming (Section 12.10.3).

The most prudent course of action is to avoid nuclear war and to minimize the contamination of the surface of the Earth. The future well-being of humans on the Earth also requires a source of abundant energy that does not release waste products into the atmosphere and therefore avoids the undesirable effects of global warming. We have become the masters of our own fate. If we act to prevent our own extinction caused by the impact of an asteroid or by global nuclear war, we may reap the benefits of exploring the solar system and of establishing colonies elsewhere in the solar system.

23.4 Summary

The Earth is the only planet in the solar system that is inhabited by life-forms which together form the biosphere. Life originated after the Earth had cooled enough to allow liquid water to collect on its surface. The organic molecules that served as the building blocks in the assembly of self-replicating molecules may have originated from extraterrestrial sources (panspermia) or they formed in the primordial atmosphere of the Earth by lightning (Miller-Urey experiment), or both. The assembly of self-replicating molecules may have occurred on the surfaces of minerals being deposited by hotsprings associated with volcanic activity at the bottom of the primordial ocean.

The first micro-organisms in the ocean eventually developed the ability to manufacture chlorophyll which enabled them to produce carbohydrates and molecular oxygen from carbon dioxide and water. Therefore, the first global impact of the biosphere on the Earth was to increase the concentration of molecular oxygen in the atmosphere from near zero during the Early Archean Era to about 20% in Early Proterozoic time between 2.5 to 2.0 billion years ago.

The second effect of life on the Earth was to sequester most of the carbon dioxide in sedimentary rocks and in fossil fuel deposited during the Paleozoic and Mesozoic Eras. The storage of carbon in these reservoirs ensures that run-away greenhouse warming cannot occur on the Earth.

The appearance within the biosphere of Homo sapiens during the most recent past is a triumph of evolution by natural selection and adaptation. However, modern humans are also causing greenhouse warming of the Earth by combusting fossil fuel which has begun to increase the concentration of carbon dioxide in the atmosphere. In addition, humans are causing significant environmental degradation of the surface of the Earth and have acquired the ability of destroy the biosphere by means of nuclear weapons. However, humans also have the ability to identify asteroids that cross the orbit of the Earth and have the means to deflect them from their collision course.

The future of the human race depends on the choices we make. We can avoid extinction by disavowing the use of nuclear weapons and by remaining vigilant against the impacts of asteroids. We can also improve the chances of surviving a man-made or natural disaster by setting up and supporting human colonies in large orbiting satellites and on the surface of Mars.

23.5 Further Reading

Ballard RD (1995) Explorations: My quest for adventure and discovery under the sea. Hyperion, New York

Baross JA, Hoffman SE (1985) Submarine hydrothermal vents and associated gradient environments as sites for the origin and evolution of life. Origins of Life 15: 327–345

Baross JA, Deming JW (1993) Deep-sea smokers: Windows to a subsurface biosphere? Geochim Cosmochim Acta 57:3219–3230

Berner EK, Berner RA (1987) The global water cycle: Geochemistry and environment. Prentice Hall, Englewood Cliffs, NJ

Chang S, DesMarais D, Mack R, Miller SL, Strathearn GE (1983) Prebiotic organic syntheses and the origin of life. In: Schopf J.W. (ed) Earth's Earliest Biosphere; Its Origin and Evolution. Princeton University Press, Princeton, NJ, pp 53–92

Chang S (1994) The planetary setting of prebiotic evolution. In: Bengston S. (eds) Early Life on Earth. Nobel symposium No. 84. Columbia University Press, New York, pp 10–23

Churchill H, Teng H, Hazen RM (2004) Correlation of pH-dependent surface interaction forces to amino acid adsorption: Implications for the origin of life. Am Mineral 89:1048–1055

Chyba CF, McDonald GD (1995) The origin of life in the solar system: Current issues. Ann Rev Earth Planet Sci. 23:215–249

Cody GD (2005) Geochemical connections to primitive metabolism. Elements 1:139–143

Cohen JE (1995) How many people can the Earth support? Norton, New York

Cohen JE (2005) Human population grows up. Sci Am 293(3):48–55

Davis WL, McKay CP (1996) Origins of life: A comparison of theories and application to Mars. Orig. Life Evol Biosph 25:61–73

Dressler BO, Grieve RAF, Sharpton VL (eds.) (1994) Large meteorite impacts and planetary evolution. Geol Soc Amer, Special Paper 293. Boulder, CO

Emiliani C, Kraus EB, Shoemaker EM (1981) Sudden death at the end of the Mesozoic. Earth Planet Sci Letters 55:317–334

Emiliani C (1992) Planet Earth; Cosmology, geology and the evolution of life and environment. Cambridge University Press, Cambridge, UK

Faure G, Mensing TM (2005) Isotopes: Principles and applications. Wiley, Hoboken, NJ

Ferris JP (2005) Mineral catalysis and prebiotic synthesis: Montmorillonite-cataliyzed formation of RNA. Elements 1:145–149

Freedman RA, Kaufmann III WJ (2002) Universe: The solar system. Freeman, New York

Goldsmith D (1997) The hunt for life on Mars. Penguin, Baltimore, MD

Gould SJ (1994) The evolution of life on Earth. Sci Am 271:85–91

Greenberg R (2005) Europa-The ocean moon. Springer/Praxis, Chichester, UK

Haldane JBS (1954) The origins of life. New Biology 16:12–27

Hartmann WK, Miller R, Lee P (1984) Out of the cradle: Exploring the frontiers beyond Earth. Workman Publications, New York

Hartmann WK (2005) Moons and planets, 5th edn. Brooks/Cole, Belmont, CA

Hazen RM, Filley T, Goodfriend GA (2001) Selective adsorption of L- and D-amino acids on calcite: Implications for biochemical homochirality. Proc Nat Acad Sci USA 98:5487–5490

Hazen RM (2005) Genesis: Rocks, minerals, and the geochemical origin of life. Elements, 1:135–137

Holland HD (1984) The chemical evolution of the atmosphere and the oceans. Princeton University Press, Princeton, NJ

Holland HD, Petersen U (1995). Living dangerously. Princeton University Press, Princeton, NJ

Huber R, Stoffers P, Hohenhaus S, Rachel R, Burggraf S, Jannasch HW, Stetter KO (1990) Hyperthermophilic archaeabacteria within the crater and open-sea plume of erupting MacDonald seamount. Nature 345:179–182

Hunten DM (1993) Atmospheric evolution of the terrestrial planets. Science 259:915–920

Ivanov SS (1997) (ed.) Vernadsky V.I: Scientific thought as a planetary phenomenon. Ecological Vernadsky V.I. Foundation, Moscow, Russia

Karl DM (ed.) (1995) The microbiology of deep-sea hydrothermal vent habitats. CRC Press, Boca Raton, FL

Kasting JF, Ackerman TP (1986) Climatic consequences of very high carbon dioxide levels in the Earth's early atmosphere. Science 234:1383–1385

Koeberl C, MacLeod KG (eds.) (2002) Catastrophic events and mass extinctions: Impacts and beyond. Geol Soc Am SPE 356. Boulder, CO

Kring DA (2000) Impact events and their effect on the origin, evolution, and distribution of life. GSA Today 10(8):1–6

Lahav N, White D, Chang S (1978) Peptide formation in the prebiotic era: Thermal condensation of glycine in fluctuating clay environments. Science 201:67–69

Lasaga AC, Holland HD, Dwyer MJ (1971) Primordial oil slick. Science 174:53–55

Livi-Bacci M (2001) A concise history of world population: An introduction to population processes. Blackwell, Oxford, UK

Lovelock JE (1979) Gaia; A new look at life on Earth. Oxford University Press, Oxford, UK

Maher KA, Stevenson DJ (1988) Impact frustration of the origin of life. Nature 331:612–614

McKay CP (1997) The search for life on Mars. Orig Life Evol Biosph 27:263–289

McKay CP, Davis WL (1999) Planets and the origin of life. In: Weissman P.R., McFadden L.-A., Johnson T.V. (eds) Encyclopedia of the Solar System. Academic Press, San Diego, CA, pp 899–922

Mileikowsky C, Cucinotta FA, Wilson JW, Gladman B, Horneck G, Lindegren L, Melosh HJ, Rickman H, Valtonen M, Zheng JQ (2000) Risks threatening viable transfer of microbes between bodies in our solar system. Planet Space Sci 48(11):1107–1115

Miller SL (1953) Production of amino acids under possible primitive Earth conditions. Science 17:528–529

Miller SL (1955) Production of some organic compounds under possible primitive-Earth conditions. J Amer Chem Soc 77:2351

Miller SL (1957) The mechanism of synthesis of amino acids by electrical discharges. Biochim Biophys Acta 23:480

Miller SL, Urey HC (1959) Organic compounds syntheses on the primitive Earth. Science 130:245–251

Miller SL, Bada JL (1988) Submarine hotsprings and the origin of life. Nature 334:609–611

Musser G (2005) The climax of humanity. Sci Am 293(3):44–47

Ohmoto H, Felder RP (1987) Bacterial activity in the warmer, sulphate-bearing Archean oceans. Nature 328:224–246

Oparin AI (1938) The origin of life on Earth. Macmillan, New York

Powell JL (1998) Night comes to the Cretaceous. Harcourt Brace, San Diego, CA

Renner M, French H, Assadourin E (eds.) (2005) State of the world: A Worldwatch Institute report on progress toward a sustainable society. W.W. Norton, New York

Repetto R (1987) Population, resources, environment: An uncertain future. Population Bulletin 42(2):1–94

Schidlowski M (1993) The initiation of biological processes on Earth: Summary of empirical evidence. In: Engel

M.H., Macko S.A. (eds) Org Geochem. Plenum Press, New York, pp 639–655

Schopf JW (ed.) (1983) Earth's earliest biosphere: Its origin and evolution. Princeton University Press, Princeton, NJ

Schopf JW (1992) Major events in the history of life. Jones and Bartlett, Boston, MA

Schopf JW (1994) The oldest known records of life: Early Archean stromatolites, microfossils, and organic matter. In: Bengston S. (eds.) Early Life on Earth. Nobel Symposium No. 84, Columbia University Press, New York, pp 193–206

Silver LT, Schultz PH (eds.) (1982) Geological implications of impacts of large asteroids and comets on the Earth. Geol Soc Amer Special Paper 190, Boulder, CO

Smith JV (2005) Geochemical influences on life's origins and evolution. Elements 1:151–156

Soffen GA (1999) Life in the solar system. In: Beatty J.K., Petersen C.C., Chaikin A.(eds.) The New Solar System, 4th edn. Sky Pub., Cambridge, MA, pp 365–376

Sowerby SJ, Heckl WM, Petersen GB (1996) Chiral symmetry breaking during self-assembly of monolayers from achiral purine molecules. J Mol Evol 43: 419–424

Sowerby SJ, Edelwirth M, Heckl WM (1998) Self-assembly at the prebiotic solid -liquid interface: Structure of self-assembled monolayers of adenine and guanine bases formed on inorganic surfaces. J Phys Chem B102: 5914–5922

Wächtershäuser G (1988) Before enzymes and templates: Theory of surface metabolism. Microbiol Reviews 52:452–484

Wächtershäuser G (1990) Evolution of the first metabolic cycles. Proc Nat Acad Sci USA 87:200–204

Wächtershäuser G (1992) Groundworks for an evolutionary biochemistry: The iron-sulfur world. Progress in Biophys and Molecular Biol 58:85–201

Ward PD, Brownlee D (2000) Rare Earth. Copernicus Books, New York

Warmflash D, Weiss B (2005) Did life come from another world? Sci Am Nov: 64–71

Brown-Dwarf Stars and Extrasolar Planets

The presence of sentient life on the Earth inevitably raises the question whether planets like the Earth orbit other stars in the Milk Way galaxy and whether some of these planets are inhabited by intelligent lifeforms that have achieved a level of technological expertise similar to ours (Taylor, 1998). These questions have been explored by science-fiction writers who have generally assumed that Earth-like planets do exist in our galaxy and that they are inhabited by carbon-based flora and fauna including intelligent lifeforms. Scientists have remained skeptical because for a long time we did not know whether planets of any kind are associated with nearby stars in our galactic neighborhood. We now know that, in fact, planets do orbit other stars in the galaxy and we have reason to predict that some of these planets are similar to the terrestrial planets of our solar system. However, we still have not actually identified any Earth-like planets among the close-to-160 stellar companions that have been discovered by astronomers. The probability that alien life exists on the extrasolar planets that have been discovered is *small* because most of them are massive gas giants like Jupiter. In addition, the orbits of many of these Jupiter-like planets are even closer to their stars than Mercury is to our Sun.

The search for extrasolar planets and the probability that they are inhabited are the subjects of many books and articles in popular science journals: Ward and Brownlee (2000), Ward (2005), Villard and Cook (2005), Butler (1999), Black (1999), Sagan and Drake (1975), (Sagan 1993, 1994), McKay (1996), McKay and Davis (1999), Soffen (1999), Lemonick (1998), Clark (1998), Zuckerman and Hart (1995), Goldsmith and Owen (1992), Drake and Sobel (1992), and Shklovskii and Sagan (1966).

24.1 Methods of Discovery

The theory of star formation leads to the conclusion that young stars in the process of formation are surrounded by disks composed of gas and dust. The theory also predicts that the circum-stellar disks may spontaneously develop gravitational centers that can grow by sweeping up solid particles until their masses increase sufficiently to attract atmospheres of hydrogen and helium. Therefore, this theory indicates that stars may have systems of planets similar to the planets that revolve around the Sun. Unfortunately, the existence of planets orbiting nearby stars could not be confirmed by telescopic observations until quite recently because planets "shine" only by reflecting the light of their central star. In addition, the light reflected by planets is overpowered by the light emitted by the star they orbit and the separation angle between the star and its planets is not resolvable in many cases by conventional telescopes. These difficulties were overcome in the 1990s by two complementary techniques based on the celestial mechanics of star-planet systems.

24.1.1 Doppler Spectroscopy

When a massive planet revolves around a star having a mass comparable to that of the Sun, the star appears to "wobble" because the two bodies are actually revolving around their common center of gravity which is displaced from the geometric center of the star. As a result, the position of the star relative to the Earth is not fixed but moves back and forth.

A similar phenomenon occurs in the gravitational interactions between Pluto and Charon (Section 20.2). The motions of a star that has an unseen companion can be detected and quantified by Doppler spectroscopy (Science Brief 11.7.3) based on the measurable displacement of absorption lines in the wavelength spectrum of the light emitted by the star. When the star moves *away* from the Earth, the wavelengths of the absorption lines increase thereby shifting them toward the red color in the spectrum. Conversely, when the star moves *toward* the Earth, the wavelengths of the absorption lines are shortened and the spectrum is shifted toward the blue color. The periodic shifts of the absorption lines are used to calculate the velocity of the star as it moves toward and away from the Earth. Therefore, this technique is also called the *radial velocity* method (Marcy and Butler, 1998).

The observed oscillations of the star reveal the gravitational effect of an unseen companion, while the frequency of the oscillations indicates the period of revolution of the companion, and the amplitude of the wobble is used to estimate the mass of the companion. The trace of the oscillations recorded during a period of time measured in years also reveals the shape of the orbit of the companion. If the trace is sinusoidal, the orbit of the companion is circular. If the trace is skewed, the orbit of the companion is elliptical. In that case, the semi-major axis of the elliptical orbit can be calculated from the observed period of revolution using Kepler's third law.

24.1.2 Occultation

In cases where the plane of the orbit of a stellar companion is aligned with the line of sight from the Earth to the star, the companion passes in front of the star and thereby reduces the amount of light that reaches the Earth. Therefore, stars whose luminosity varies periodically by a small amount are assumed to have a companion. The magnitude of the reduction is used to estimate the diameter of the companion and the period of the light curve yields the period of revolution of the companion from which the semi-major axis of its orbit can be calculated as before.

24.2 Brown-Dwarf Stars

The masses of the stellar companions that have been detected by Doppler spectroscopy and by occultation range widely from less than the mass of Jupiter (M_j) to more than 75 (M_j). Consequently, many of these objects are classified as *brown-dwarf stars* although the light they emit is closer to being red than brown (Black, 1999). The companion of the star Gliese 299, located only 19 light years from the Earth, is known to be a brown dwarf that revolves around the star at a distance of 32 AU. The mass of this brown dwarf, which was discovered telescopically, lies between 30 to 55 (M_j), but its radius is similar to that of Jupiter and its surface temperature is about 1000 K (Freedman and Kaufmann, 2002, Figure 7.22).

Brown dwarfs are composed of hydrogen and helium and their masses range from 12 to 75 (M_j). Stars in this mass range can generate energy only by fusion of deuterium nuclei ($^2_1$H) because their core temperatures are not high enough to sustain fusion of hydrogen nuclei ($^1_1$H). Therefore, brown dwarfs have low luminosities and are difficult to detect by means of optical telescopes, which explains why very few brown-dwarf stars were known prior to the mid-1990s when the Doppler-spectroscopy method began to be used (Henry, 1996; Hartmann, 2005). More recently, brown dwarfs have been discovered by the Hubble space telescope and by infrared-detectors because brown dwarfs radiate primarily in the infrared part of the spectrum. The estimated masses of the brown dwarfs in Figure 24.1 have a unimodal Gaussian distribution centered on 35 to 40 (M_j).

In accordance with the theory of stellar evolution (Section 4.2), brown dwarfs form as companions to larger stars in the course of the contraction of nebulae in interstellar space. Both the principal star and the brown-dwarf companion may acquire disks of dust and gas that revolve around them. These protoplanetary disks are the sites where planets can form by accretion of solid particles. Therefore, brown dwarfs may be associated with planets just like red dwarfs and solar-mass stars. However, planets that orbit brown-dwarf stars receive less radiant energy than the planets of our solar system because brown dwarfs typically have much lower luminosities than the Sun. These conjectures, if

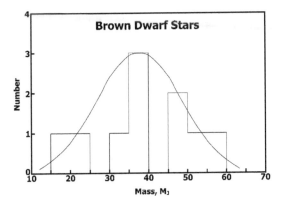

Figure 24.1. The masses of the brown-dwarf stars in Table 24.1 appear to have a normal or Gaussian distribution ranging from greater than 12 to less than 80 M$_j$. Brown dwarfs having masses between 35 and 40 M$_j$ have the highest frequency although the number of individuals is probably too small to be definitive. Based on data of Butler (1999) and Black (1999)

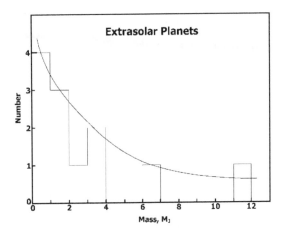

Figure 24.2. The distribution of masses of extrasolar planets appears to be log-normal which implies that many additional planets having masses less than about 0.5 M$_j$ may exist but have not yet been discovered. The mass distribution of the extrasolar planets appears to differ from the range of masses of brown-dwarf stars although their masses may overlap at about 12 M$_j$. Data from Butler (1999) and Black (1999)

Table 24.1. Properties of brown dwarfs that orbit stars in our galactic neighborhood (Butler, 1999; Black, 1999)

Name	Mass, M$_j$	Period, d	a AU	e
HD 110833	> 17	270	0.817	0.69
BD −04 : 782	> 21	240.92	0.757	0.28
HD 112758	> 35	103.22	0.430	0.16
HD 98230	> 35	3.98	0.049	0.00
HD 18445	> 39	554.67	1.32	0.54
HD 29587	> 40	1471.70	2.53	0.37
HD 140913	> 46	147.94	0.547	0.61
BD +26 : 730	> 50	1.79	0.0288	0.02
HD 89707	> 54	298.25	0.873	0.95
HD 217580	> 60	454.66	1.15	0.52

true, suggest that brown dwarfs bridge the gap between stars and planets (Kulkarni, 1997).

The orbits of the brown dwarfs in Table 24.1 are highly elliptical in most cases with orbital eccentricities (e) that range from e = 0.00 to e = 0.95. However, the semi-major axes (a) of the orbits of the brown dwarfs are surprisingly short (a = 0.0288 AU to a = 2.53 AU), which means that they closely approach the stars they orbit and presumably are heated by them both by direct irradiation and by tidal friction. For example, the semi-major axis (a) and eccentricity (e) of the orbit of brown dwarf HD 110833 in Table 24.1 are: a = 0.817 AU and e = 0.69

(Black, 1999). The perihelion distance (q) is:

$$q = a - f = a(1 - e) \qquad (24.1)$$

where the focal length f = e × a. The aphelion distance (Q) is:

$$Q = 2a - q \qquad (24.2)$$

Accordingly, the perihelion distance of the orbit of HD 110833 is q = 0.253 AU and its aphelion distance is Q = 1.38 AU. These results indicate that this brown dwarf, having at least seventeen times more mass than Jupiter, has a perihelion distance that is less than the orbital radius of Mercury. As a result, the parent star exerts a strong tide that deforms this brown dwarf as it moves in its orbit at an average speed of 32.8 km/s (118,440 km/h). The stars may also strip hydrogen from a brown dwarf by blasting it with protons and electrons as it passes through perihelion. The violent gravitational interactions between the central star and its brown-dwarf companion may cause the companion to be ejected into interstellar space where it may drift aimlessly forever or until it is captured by another star.

The apparent ejection of a brown dwarf by its parent stars may be exemplified by a mysterious object identified as TMR-1C that was imaged on August, 4, 1997, by the Hubble space telescope. This object is located 450 light years from the Earth in the constellation Taurus and was about 1400 AU from its central star. If the ejection of stellar companions and associated planets is a common event, a large number of such bodies may exist in the interstellar space of the Milky Way galaxy.

Recent surveys of stellar populations by means of the Hubble space telescope have revealed that brown dwarfs occur in large numbers among clusters of young stars (e.g., IC 348 in Perseus and in the Trapezium cluster of the Orion nebula). Most of the brown dwarfs in IC 348 are free-floating and are not orbiting around larger stars. Young star clusters are a good place to search for brown dwarfs because the luminosity of brown dwarfs decreases with age as they cool after the deuterium in their cores has been consumed leaving only the residual heat of their initial compression. In general, brown dwarf stars are nearly as common as ordinary stars and many are less than one billion years old (Basri, 2000). Older brown dwarfs exist but are more difficult to find because their luminosities have decreased as a result of cooling.

Current research concerning brown-dwarf stars seeks to determine their numbers, masses, and distribution in the Milky Way galaxy. In addition, questions yet to be answered concern their origin in stellar "nurseries" and their subsequent internal evolution as they exhaust their deuterium fuel and cool by radiating energy into space. Brown dwarfs that form in orbit around a solar-mass star may derive energy by tidal interactions and therefore may cool more slowly than floaters. In a few cases, binary systems of free-floating brown dwarfs that orbit each other have been discovered (Mohanty and Jayawardhana, 2006).

24.3 Extrasolar Planets

The extrasolar planets in Figure 24.2 are more massive than the Earth and their orbits are more elliptical in most cases than the orbits of the planets in the solar system. The apparent absence of low-mass planets in orbit around other stars does not necessarily indicate that Earth-like planets do not exist. The most likely reason why

small planets have not yet been found is that the methods of detecting extrasolar planets work best for massive planets and brown-dwarf stars.

Another interesting property of the extrasolar planets is that their orbits take them very close to their respective parent stars. For example, five planets in Table 24.2 have nearly circular orbits whose radii (a) are all less than 0.25 AU. The stars that have such planets are : 51 Pegasi (a = 0.051 AU), Tau Boötis (a = 0.045 AU), 55 Cnc (a = 0.117 AU), Rho Cancri (a = 0.11 AU), Upselon Andromedae (a = 0.054 AU), and Rho Coronae Borealis (a = 0.23). The masses of these planets range from 0.45 to 3.7 M_j making them more massive than the Earth by factors of 145 to almost 1200. These massive planets revolve barely above the surfaces of their respective parent stars which have probably stripped them of their atmospheres and heated their rocky surfaces close to the melting point of silicate minerals. These so-called *hot Jupiters* are not likely to be inhabited by alien life forms.

A second set of four extrasolar planets in Table 24.2 has elliptical orbits with perihelion distances (q) that also come close to the surfaces of their receptive planet stars. This group includes planets that revolve around 16 Cygnis B (q = 0.56 AU), 70 Virginis (q = 0.28 AU), HD 114762 (q = 0.24 AU), and Gliese 876 (q = 0.13 AU). These planets are heated not only by irradiation during their perihelion passage but also by tidal deformation caused by the gravity of the star and the ellipticity of the planetary orbits. The surfaces of these planets may not be quite as hot as those of the hot Jupiters, partly because they may have retained residual atmospheres.

Both groups of extrasolar planets are likely to have 1:1 spin-orbit coupling which means that the side facing their central star is irradiated continuously and is very hot, while the opposite side is never irradiated and therefore is cold and perpetually dark. The difference in surface temperature is especially severe for planets that lack an atmosphere that could redistribute the heat globally. However, neither the hot Jupiters nor the planets that have highly elliptical orbits are likely to be habitable.

The bodies that orbit the stars 14 Herculis and 47 Ursae Majoris in Table 24.2 represent a third class of planets whose surface environments may be less hostile to life than those of the two groups

Table 24.2. Properties of extrasolar planets that orbit stars in our galactic neighborhood (Butler, 1999; Black, 1999)

| Name | Star | | Extrasolar Planets | | | |
	Distance, ly	Mass, (solar)	Mass, M_j	Period, d	a, AU	e
51 Pegasi	50	1.0	> 0.45	4.23	0.051	0.01
Tau Boötis	49	1.25	> 3.7	3.31	0.045	0.006
55 Cnc	—	—	> 0.84	14.65	0.117	0.05
14 Herculis	59	0.80	> 3.3	1,619	2.5	0.35
47 Ursa Major.	46	1.1	> 2.4	1,098	2.1	0.10
16 Cygni B	72	1.0	> 1.7	802	1.7	0.67
70 Virginis	59	0.95	> 6.8	116.6	0.47	0.40
HD 114762	90	1.15	>11.6	83.9	0.36	0.34
Gliese 876	15	0.32	> 1.89	61.1	0.30	0.37
Rho Cor. Bor.	57	1.0	> 1.1	39.6	0.23	0.04
Rho Cancri	44	0.85	> 0.93	14.64	0.11	0.03
Upsilon Andro.	57	1.25	> 0.65	4.61	0.054	0.10

described above. The planet that orbits Herculis has a mass of 3.3 M_j, the eccentricity (e) of its orbit is 0.35, and its semi-major axis (a) is 2.5 AU. The perihelion distance of this planet is $q = a(1-e) = 1.6$ AU and the aphelion distance is $Q = 2a - q = 3.4$ AU. Therefore, this planet approaches its star to within 1.6 AU (similar to the orbital radius of Mars) and recedes from it as far as 3.4 AU (less than the orbital radius of Jupiter). Since the mass of 14 Herculis is only 0.80 times the mass of the Sun, its satellite at perihelion receives less radiant energy than Mars receives from the Sun. The large mass of the planet orbiting Herculis (3.3 Mj) indicates that it has an even thicker atmosphere than Jupiter and that the atmosphere is composed primarily of hydrogen and helium. The apparent similarity of this planet to Jupiter implies that it is not likely to harbor alien life. However, the large radius of its orbit means that its rotation has not been synchronized with its revolution and that this planet is therefore heated uniformly by stellar radiation. In addition, the large average radius of its orbit (a = 2.5 AU) leaves room for several small and Earth-like planets to orbit close to Herculis. If such Earth-like planets exist, they would be protected by the exoplanet from annihilation by impacts of large solid objects.

The exoplanet in orbit around 47 Ursa Major also has a large mass (2.4 Mj). The semi-major axis of its orbit is 2.1 AU, and the orbital eccentricity is 0.10. The mass of Ursa Majoris is 1.10 solar masses giving it a higher luminosity than

the Sun. However, the planet in orbit around this star is similar to the one described above and therefore may shelter one or more small and Earth-like planets within the large radius of its orbit (a = 2.1 AU). However, these conjectures have not been supported by any direct observational evidence because Earth-like planets that may orbit nearby stars cannot be detected at the present time because of their small mass.

The existence of hot Jupiters challenges the theory of the origin of our solar system because planets as massive as Jupiter cannot form so close to the central star. A possible explanation for this apparent deviation from the theory is that the hot Jupiters originally formed at greater distances from their stars (e.g., more than 5 AU). The subsequent inward migration of planets may have been caused by the flow of gas and dust in the protoplanetary disk toward the growing star and by resonance of their orbital periods with spiral waves in the rotating disk (Butler, 1999, p. 383). Even if such inward migration is an inherent property of planets that form in the protoplanetary disk, two questions remain:

1. Why did the hot Jupiters stop their inward migration before falling into their parent star?
2. Why did the giant gas planets of *our* solar system remain close to the region in the proto-planetary disk where they formed?

All of the extrasolar planets in Table 24.2 are associated with stars in our galactic neighborhood (i.e., 15 to 90 ly from the Sun) and the masses of these stars are similar to the mass of the Sun (i.e., 0.32 to 1.25 solar masses). Consequently, these stars may be surrounded by sets of planets resembling the solar system. In most cases, only the largest planets of such hypothetical "planetary systems" have been detected although in a few cases two or even three planets have been identified. For example, the pulsar $1275 + 12$ in the constellation Virgo appears to have three planets whose masses relative to the Earth (M_E), orbital radii (a), and periods of revolution (p) are (Butler, 1999):

    Planet 1: > 0.015 M_E, a $= 0.19$ AU, p $= 23$ d
    Planet 2: > 3.4 M_E, a $= 0.36$ AU, p $= 67$ d
    Planet 3: > 28 M_E, a $= 0.47$ AU, p $= 95$ d

Of course, pulsars (i.e, rapidly rotating neutron stars) are not suitable hosts for habitable planets; however, apparently even neutron stars, which are remnants of supernovae, can have planets.

24.4 Voyage to the Stars

The exploration of the solar system is preparing us to undertake voyages to nearby star systems. The Voyager and Pioneer spacecraft are our first probes of this vast and hostile region of space. Although the physical and medical sciences as well as technology have taken giant leaps forward during the twentieth century, several more such leaps will be required before we can hope to meet the challenges we will face when we try to voyage to the stars.

Science-fiction writers have painted a rosy picture of spacecraft from Earth visiting planets where intelligent aliens have evolved their own cultures based on science and technology. Such daydreams help us to imagine the range of possibilities that may confront us in interstellar space. However, authors of science fiction gloss over the fact that we still lack the necessary propulsion and life-support systems for voyages by human crews lasting several centuries. The awesome difficulties of such voyages require exceptional motivation arising from dire necessity. One of the ironies of seeking salvation among the stars is that *when conditions on the Earth have deteriorated sufficiently to force us to leave, we will no longer possess the resources to attempt voyages to the stars.*

A much more prudent course of action would be to preserve a livable environment on the Earth and to proceed with terraforming Mars. Our Sun will not act up anytime in the foreseeable future and we can avoid catastrophic impacts because we have the ability to deflect asteroids and comets from a collision course with the Earth. The greatest danger we face is that we can destroy our planet in a frenzy of nuclear explosions. If that happens, the life of the human race on the Earth will be brilliant but short and our window of opportunity to voyage to the stars will close forever.

24.4.1 Nearby Stars

The galactic neighborhood of the Sun is populated by 34 stars which exist in a spherical region of space extending out to 13.0 light years (ly). Twelve of these stars are listed in Table 24.3 together with their spectral types and luminosities. Three of the twelve nearest stars are binaries designated by the letters A and B: Alpha Centauri A and B, L 726-8 A and B, and Sirius A and B. The number of stars in the vicinity of the Sun in Figure 24.3 increases with increasing distance extending to 13.05 ly. Proxima Centauri, the two

Table 24.3. The nearest stars located in a spherical region of space extending about 10 ly from the Sun (Freedman and Kaufmann, 2002, Appendix 4)

Name	Distance, ly	Spectral, type	Luminosity (Sun = 1)
Proxima Centauri	4.22	M	8.2×10^{-4}
Alpha[1] Centauri A	4.40	G	1.77
Alpha Centauri B	4.40	K	0.55
Barnard's star	5.94	M	3.6×10^{-3}
Wolf 359	7.80	M	3.5×10^{-4}
Lalande 21185	8.32	M	0.023
L 726-8 A	8.56	M	9.4×10^{-4}
L 726-8 B	8.56	M	5.6×10^{-4}
Sirius A	8.61	A	26.1
Sirius B	8.61	white dwarf	2.4×10^{-3}
Ross 154	9.71	M	4.1×10^{-3}
Ross 248	10.32	M	1.5×10^{-3}

[1] Stars labeled A and B are members of a binary star system.

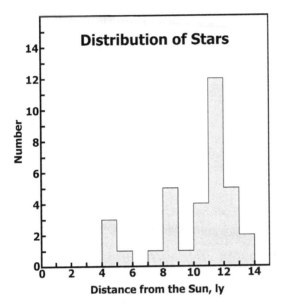

Distribution of Stars

Number

Distance from the Sun, ly

Figure 24.3. The Sun is surrounded by 34 stars that occupy the space around it within about 13 light years. The distribution of the distances to these stars is polymodal with peaks between 4 to 5, 8 to 9, and 11 to 12 lightyears. The closest group of stars consists of Proxima Centauri (4.22 ly), Alpha Centauri A and B (4.40 ly), and Barnard's star (5.94 ly). Proxima Centuri is an M-type and metal-rich star that emits reddish light and has a surface temperature of about 3500 K. The Sun is a G-type star whose surface temperature is about 5500 K. Although Proxima Centauri is the closest star, it is not within range of presently available propulsion systems. Data from Freedman and Kaufmann (2002) and Protheroe et al. (1989)

Alpha Centauries, and Barnard's star form a small group that is located less than 6 ly from the Sun followed by Wolf 359 at 7.80 ly. The four closests stars are separated from the more distant group by a gap of almost two lightyears. All of the stars in Figure 24.3 are moving either toward the Sun or away from it at different velocities. For example, the four closest stars mentioned above are moving toward the Sun at speeds ranging from 21 km/s (Alpha Centauri B) to 111 km/s (Barnard's star). Three of the closest stars are red dwarfs (Proxima Centauri, Barnard's star, and Wolf 359) that are known to flare up occasionally. The stellar flares can scorch the surfaces of nearby planets and therefore constitute a threat to life inhabiting such planets. For example, the flare emitted by Barnard's star on July 17, 1998, had a temperature of 8000 K compared to only 3100 K for the surface of this star. The flare-up

of Barnard's star was unexpected because it is about twice as old as the Sun (i.e., 11 to 12 billion years) and its period of rotation (130d) is much longer than that of young red-dwarf stars (e.g., Proxima Centauri and Wolf 359) (Croswell, 2006).

Proxima Centauri and Alpha Centauri A and B are of special interest to us, not only because of their proximity to the Sun, but also because they may have formed within the same solar nebula as the Sun and therefore may have similar ages and similar chemical compositions. Three of the stars in the closest group are classified as M-type stars (Proxima Centauri, Barnard's star, and Wolf 359), Alpha Centauri (A) is a type G star like the Sun, and Alpha Centauri B is a type K star. The approximate surface temperatures of the star in these spectral types are: $G = 5,500$ K, $K = 4,500$ K, and $M = 3000$ K (Seeds, 1997). The low surface temperatures of the M-type stars in Table 24.3 cause them to have much lower luminosities than the Sun by factors of 10^{-3} and 10^{-4}. The low luminosities of these and other red dwarf stars require that the radii of the orbits of habitable planets must be less than 1.0 AU in order to prevent water on their surfaces from freezing. The luminosities of the Alpha Centauri stars A and B are similar to that of the Sun (e.g., 0.55 and 1.77 times solar). In principle, the space surrounding each of these stars contains a zone in which a terrestrial planet could be habitable (Doyle, 1996; Kasting, 1996; Ward and Brownlee, 2000). However, no evidence exists at this time that any of the closest stars actually have planetary systems.

24.4.2 Habitable Planets

None of the extrasolar planets that have been discovered to date are likely to be inhabited by alien lifeforms. What we are looking for, but have not yet found, are planets that reside in the habitable zones of their central stars and therefore meet certain other criteria required for the initiation, evolution, and preservation of life. The properties of habitable planets have been identified and discussed by Ward and Brownlee (2000) in order to determine whether such planets are common or rare in the Milky Way and other galaxies. Some of the prerequisites for the existence of life on the surfaces of extrasolar planets are listed below.

1. The distance of the planet from the star must yield a surface temperature that allows water to exist in the liquid state but must be large enough to avoid spin-orbit coupling.
2. The mass of the star must be low enough to allow for a long residence on the Main Sequence (i.e., about 4 to 5 billion years).
3. The orbit of the planet must be stable which requires the absence of external gravitational interactions with nearby giant planets.
4. The mass of the planet must be sufficient to retain an atmosphere and an ocean of liquid water.
5. The planet must also have sufficient internal heat (both residual and radiogenic) to permit volcanic activity and plate tectonics as well as a liquid core which can generate a magnetic field that shields the planet from high-energy nuclear particles.
6. The planetary system should contain giant planets at a safe distance which can deflect incoming comets and asteroids and thus prevent impacts that could sterilize the planet of interest.
7. The obliquity of the axis of rotation of the planet should allow moderate seasonal variations of the climate.
8. The planet must have a sufficient inventory of carbon to sustain an effective biosphere and an effective carbon cycle that can prevent runaway greenhouse warming.
9. The evolution of the biosphere must include photosynthesis resulting in the release of oxygen into the atmosphere.
10. The central star must have formed from a nebula that contained heavy elements and it must be located in the habitable zone of its galaxy (i.e., outside of the galactic center and inside of its outer region).

These and additional requirements identified by Ward and Brownlee (2000) probably exclude most of the planets that may have formed around other stars in the Milky Way galaxy. Therefore, these authors concluded that habitable planets like the Earth are rare.

24.4.3 Voyage to Proxima Centauri

Regardless of whether Proxima Centauri has a habitable planet, it is likely to be the target of the first attempt to reach the vicinity of a neighboring star because it is closer to the Sun than any other star in the galaxy. The distance (d) to Proxima Centauri is 4.22 ly which is equivalent to:

$$d = 4.22 \times 9.46 \times 10^{12} = 39.9 \times 10^{12} \, \text{km}$$
$$= 266,000 \, \text{AU}$$

where $1 \, \text{ly} = 9.46 \times 10^{12} \, \text{km}$ and $1 \, \text{AU} = 149.6 \times 10^6 \, \text{km}$. Two hundred sixty six thousand astronomical units is a long way. If the speed of a spacecraft is 100 km/s or $3.15 \times 10^9 \, \text{km/y}$, a one-way trip to Proxima Centauri would take $39.9 \times 10^{12} / 3.15 \times 10^9 = 12,670$ years. That is way too long. If we want to reach Proxima Centauri in a reasonable amount of time, the speed of the spacecraft has to be a lot faster. However, even at a speed of 1000 km/s, a one-way trip to Proxima Centauri would take 1267 years. Even at the unheard of speed of 1000 km/s the length of time required to reach the nearest star is prohibitive. We cannot get there with presently available technology.

The problem gets even worse if we consider traveling at speeds that are comparable to the speed of light ($c = 2.9979 \times 10^5 \, \text{km/s}$). According to Einstein's special theory of relativity, the mass of a particle traveling at a certain velocity (v) increases to infinity as v approaches c. The relevant equation is:

$$m_v = \frac{m_o}{\left(1 - \left(\frac{v}{c}\right)^2\right)^{1/2}} \qquad (24.3)$$

where m_v = mass of a particle at velocity v
       m_o = mass of the same particle at rest
       c = speed of light.

The graph in Figure 24.4 indicates that the ratio m_v/m_o rises imperceptibly as the v/c ratio increases from $v/c = 0$ to $v/c = 0.2$. At higher values of the v/c ratio the mass ratio rises steeply and goes to infinity as $v/c \rightarrow 1.0$.

Another way to interpret these results is to recognize that $v/c = 0.2$ means that the velocity of the spacecraft is:

$$v = 0.2 \times 2.9979 \times 10^5 = 59,958 \, \text{km/s}$$

If the spacecraft travels at this speed, it can reach Proxima Centauri in:

$$t = \frac{39.9 \times 10^{12}}{59,958 \times 60 \times 60 \times 24 \times 365.25} = 21.0 \, \text{y}$$

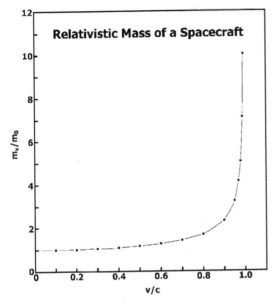

Figure 24.4. Einstein's theory of special relativity leads to the conclusions that the mass (m_v) of an object traveling at velocity (v) increases as its velocity approaches the speed of light (c). Therefore, the mass ratio (m_v/m_0) rises exponentially as the velocity ratio (v/c) increases and approaches $v/c = 1.0$, where m_0 is the mass of the body at rest. The practical consequences of this relationship is that the velocity of a body must remain less than the speed of light. Einstein's equation also limits the velocity of a spacecraft such that a one-way trip to Proxima Centauri will always take a lot longer than 4.22 years

At that speed, the mass ratio increases to 1.020, which means that the mass of the spacecraft has increased by 2% relative to its rest mass. Consequently, more fuel will be required to complete the trip. If the speed of the spacecraft is doubled such that $v/c = 0.4$, it will be traveling at a speed of 119,916 km/s and it will require only 10.5 y to reach Proxima Centauri. However, the mass of the spacecraft will be 9.1 % greater than the rest mass. Therefore, even more energy will be required.

The shape of the curve in Figure 24.4 indicates that the relativistic mass of the spacecraft increases the faster it travels. A high speed reduces the travel time but requires increasing amounts of fuel to provide the energy to push the increasing mass. Here we encounter the second irony of interstellar travel. The stars are only reachable by traveling at very high speeds, *but the faster we travel the more fuel is required.*

The laws of nature make it impossible for us to reach Proxima Centauri in only 4.22 years.

If flybys of robotic spacecraft in the distant future were to find a habitable planet in orbit around Proxima Centauri that met all of the requirements for carbon-based life including liquid water and an atmosphere that contained close to 20 % molecular oxygen, the third irony of interstellar travel would become apparent: *Habitable planets are already inhabited.* The lifeforms on such a planet would not necessarily welcome an invasion by humans, and may resist by force of arms or by infectious diseases caused by micro-organisms and viruses (Wells, 1898). Therefore, traveling among the stars and inhabiting other planets in our galaxy is not our destiny. We have to continue living on the Earth and elsewhere in our own solar system.

24.5 SETI and the Drake Equation

All efforts to find life on the planets and satellites of our solar system have failed although Mars and Europa still need to be re-examined to ascertain whether organisms once existed on Mars and may still be present on Europa (Klein, 1979; Greenberg, 2005). Life did arise on the Earth more than four billion years ago and has evolved without interruption to the present time. The various sensors on the robotic spacecraft we use to search for signs of life elsewhere in the solar system have, in fact, detected the presence of life on the Earth (Sagan, 1993, 1994).

When the spacecraft Galileo flew by the Earth on its way to the Jupiter system, its sensors detected the presence of:

1. molecular oxygen in the atmosphere,
2. chlorophyll on the continents of the Earth,
3. trace amounts of methane gas, and
4. pulsed, amplitude-modulated, narrow-band radio waves emanating from sources on the surface of the Earth.

These results are reassuring because they imply that our instruments are capable of detecting life elsewhere in the solar system. The radio waves detected by the Galileo spacecraft reveal the presence of our technological civilization on the Earth. However, if an alien spacecraft had scanned the Earth 200 years ago, it would not have detected such signals because radio and television came into wide use only about 100 years ago.

24.5.1 SETI

Pulsed electromagnetic radiation should also be generated by alien civilizations on planets orbiting other stars in the Milky Way galaxy. These waves are propagated at the speed of light and may reveal the presence of technologically advanced civilizations elsewhere in our galaxy. The first systematic search for extraterrestrial intelligence (*SETI*) was initiated on October 12, 1992, by NASA. Although the Congress of the USA cancelled the project after one year, privately funded searches by the SETI Institute in Mountain View, California, have continued. Since we cannot yet detect the existence of lifeforms by sending spacecraft to the planets of other stars, project SETI searches for signs of *intelligent* life on planets that are out of range for our spacecraft (Tarter and Chyba, 1999; Sagan, 1994).

To date, increasingly sophisticated searches of large parts of the sky have failed to detect signals that could have originated from an alien civilization. Although the absence of a positive result does not mean that such civilizations do not exist, it does reinforce the view that only a small number of the planets in the Milky Way galaxy are habitable and that only a few of these planets are actually inhabited by techno-logical civilizations. However, before we give up on SETI we need to remember that we have only been listening for a short time and that the transmissions of some civilizations may not have reached us yet. The alternative scenarios indicated in Figure 24.5 include:

1. An alien civilization on a planet in the Milky Way galaxy is emitting a beam of radio waves that reaches the Earth and is detected by us. This is the ideal case.
2. We are detecting a pulsed radio signal emitted by an alien civilization that has already become extinct.
3. An alien civilization emitted a signal that has not yet reached us but the civilization became extinct a short time after it emitted the signal.
4. An alien civilization emitted a signal that has not yet reached the Earth but the emissions are continuing.

5. A potential source is not emitting radio waves because the lifeforms have not yet become technically able to do so. Alternatively, the alien civilization may have emitted signals while we were not listening and the transmis-sions stopped arriving on the Earth before our receivers were turned on.

Among the five alternatives considered here only one results from the presence of a planet that is inhabited by an alien civilization that we could communicate with because that civilization still exists (1). The other hypothetical civilizations have already died out (2) and (3), or their signal has not yet reached us (4). Planet (5) is either too young to be inhabited by a technological civilization or it is so old that the civilization became extinct a long time ago and its signal is no longer detectable on the Earth.

24.5.2 The Drake Equation

In the absence of definitive information about the existence or absence of intelligent life on other planets in the Milky Way galaxy, we turn to an equation proposed in the 1950s by the astronomer Frank Drake. This equation is an attempt to estimate the number of intelligent civilizations (N) in the Milky Way galaxy with which we could communicate (Ward and Brownlee, 2000). The original version of Drake's equation is:

$$N = N^* \times fs \times fp \times ne \times fi \times fc \times fl \qquad (24.4)$$

where N^* = number of stars in the Milky Way galaxy,

fs = fraction of Sun-like stars,
fp = fraction of the Sun-like stars that have planets,
ne = number of planets in the habitable zone of Sun-like stars,
fi = fraction of the habitable planets where life actually exists,
fc = fraction of the habitable planets that are inhabited by intelligent beings,
fl = fraction of the lifetime of a planet during which a communicative civilization exists.

The total number of stars in the Milky Way galaxy has been estimated to be about 300×10^9 based on observational evidence. Most of these stars are not "Sun-like" because they are either more massive

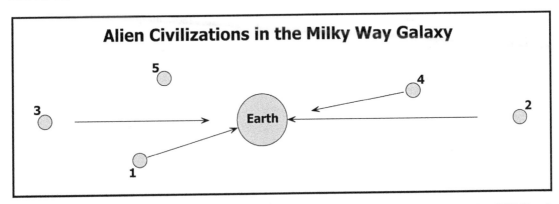

Figure 24.5. An alien civilization on a distant planet in the Milky Way galaxy can only be detected by project SETI (Search for Extraterrestrial Intelligence) when the electromagnetic radiation the aliens radiate into space reaches the Earth. Five possible cases are considered in this diagram: 1. The signal is emitted and reaches the Earth; 2. The signal is still reaching the Earth but the alien civilization that sent it has ceased to transmit it; 3. A signal was sent but has not yet reached the Earth but the transmission has already ended; 4. A young civilization has started to emit a signal that has not yet reached us; 5. No signal is emanating from a habitable planet either because the sentient beings that live there have not yet developed the necessary technology, or because the alien race became extinct long ago after broadcasting briefly at a time we were not yet listening

or less massive than the Sun. Stars having masses greater than the Sun do not remain on the Main Sequence long enough to allow intelligent beings to evolve. Stars whose masses are less than that of the Sun have low luminosities. Therefore, planets that revolve around small stars (i.e., red dwarfs) at a distance of 1 AU have low global surface temperatures that cause water to freeze. Planets whose orbits lie closer than 1 AU from a small star receive more "stellar" energy, but they run the risk of being locked into one-to-one spin-orbit coupling. When that happens, the side of the planet that faces the red-dwarf star is continuously irradiated and becomes hot while the far side is in permanent darkness and becomes very cold. Planets may also occur in binary star systems and therefore need to be included in the census of Earth-like planets.

Sun-like stars must also contain elements that are heavier than helium. These elements are considered to be "metals" by astronomers and the concentration of these elements is called the "metallicity" of the star. The Sun has a high metallicity of 2 or 3%, which permitted the planets of the solar system to form. In order to be Sun-like a star must also be located in the life zone of the galaxy at a safe distance from the violent upheavals of the galactic nucleus and from the exposed outer fringe of the galaxy. Evidently, the properties of Sun-like stars are narrowly constrained which means that such stars

constitute only a small fraction of the multitude of stars in the Milky Way galaxy. For that reason we tentatively set $fs = 10^{-8}$ and thereby limit the number of Sun-like stars in the Milky Way galaxy to about 3000. We also stipulate that all of the Sun-like stars are associated with planetary systems (i.e., $fp = 1$). Furthermore, we assume that the orbit of at least one of the planets that orbit each Sun-like star is located within the life zone of each Sun-like star (i.e., $ne = 1$). Accordingly, we postulate that the Milky Way galaxy contains approximately 3000 planets that revolve within the habitable zone of the Sun-like star they orbit.

Planets in the habitable zone of their star must have masses similar to the mass of the Earth. Planets as massive as Jupiter have thick atmospheres composed primarily of hydrogen and helium and have high surface pressures. Planets that are significantly less massive than the Earth do not retain atmospheres and therefore cannot stabilize liquid water on their surfaces. In addition, planets less massive than the Earth may not have magnetic fields and therefore are not shielded from cosmic rays. These constraints limit the number of planets that can sustain life even though their orbits are located in the life zone of their star. For example, the Earth is the only planet in our solar system that is sufficiently massive to have retained an atmosphere and to have a core that is capable of generating a magnetic field. Therefore, in our solar system fi = 1/9 = 0.1 (including Pluto). This criterion reduces the number of habitable planets in the Milky Way galaxy to 300.

What fraction of these habitable planets is actually inhabited by intelligent beings who have developed a technological civilization? The answer to this question depends partly on the ages of the planets and partly on the favorable outcomes of random events in the histories of these planets. In our solar system, it took more than 3.5 billion years for unicellular organisms to evolve into a technological civilization. Similar amounts of time may be required on other habitable planets that orbit Sun-like stars. However, even if sufficient time was available, the process may have been terminated prematurely by catastrophic impacts of large asteroids that killed all organisms and may have destroyed the planet itself. Nevertheless, we tentatively conclude that the destruction of inhabited planets is a rare event and therefore set fc = 0.9. By doing so, we say that 10% of the habitable planets in the Milky Way galaxy fail to be inhabited by technological civilizations because they are destroyed by celestial accidents.

The only civilizations with whom we can communicate must exist on planets whose age is similar to that of the Earth. Planets that revolve around stars that formed more recently than the Sun may be inhabited, but the inhabitants have not yet achieved the technology required to communicate with us at the present time. Similarly, civilizations on planets that revolve around stars that are significantly older than our Sun may have ceased to exist for a variety of reasons and therefore are no longer broadcasting signals at the present time.

These considerations lead to the final parameter (fl) of the Drake equation that expresses the fraction of the lifetime of a planet during which a communicative civilization exists. Although our civilization has been "communicative' only for about 100 years, we assume that such civilizations on other planets may last for as long as the dinosaurs dominated the Earth from the Early Jurassic to the Late Cretaceous Epochs or about 140×10^6 yeas. Since the age of the Earth is 4.6×10^9 y, the parameter fl = $140 \times 10^6/4.6 \times 10^9 = 3 \times 10^{-2}$.

Therefore, the number of technologically advanced civilizations in the Milky Way galaxy that may be able to communicate by broadcasting signals that our SETI project could intercept is:

$$N = 300 \times 10^9 \times 10^{-8} \times 1 \times 1 \times 0.1 \times 0.9$$
$$\times 3 \times 10^{-2} = 8$$

This result probably underestimates the number of planets that are inhabited by technologically advanced civilizations depending primarily on the number of Sun-like stars and Earth-like planets that exist in the Milky Way galaxy and on how long such civilizations survive on average. Therefore, we conclude that planets that harbor civilizations of intelligent beings are rare, but we are probably not alone in the Milky Way galaxy.

24.6 Summary

Close examination of stars in the Milky Way galaxy primarily by Doppler spectroscopy has revealed that at least 150 of them have unseen companions. Many of these companions are actually brown-dwarf stars with masses that range from 12 to 75 times the mass of Jupiter. Brown dwarfs are composed of hydrogen and helium and generate energy in their cores only by fusion of deuterium (heavy isotope of hydrogen) to helium. Consequently, brown-dwarf stars have low luminosities which make them difficult to see with telescopes. The orbits of many brown dwarfs are elliptical and their perihelion distances are very close to their central stars. The resulting tidal interactions may cause brown dwarfs to be ejected from their orbits and to become "floaters" in interstellar space. Surveys of star-forming nebulae suggest that brown-dwarf stars are about as numerous as red dwarfs and other kinds of stars. Some brown-dwarf stars may have companions of their own including planets.

Some stars in the Milky Way galaxy have unseen companions whose masses are less than 12 Jupiters. These bodies are classified as extrasolar planets although even the smallest of these "exoplanets" are much more massive than any of the terrestrial planets of the solar system. Some of the exoplanets have nearly-circular orbits located at distances of less than 0.25 AU from their respective central stars. These planets are therefore known as hot Jupiters. Another group of exoplanets has eccentric orbits that have short perihelion distances. The hot Jupiters and the "eccentric" planets are not habitable because of their high surface temperatures. The orbits of a third group of exoplanets have average radii (semi-major axes) that are large enough (2 to 3 AU) to avoid extreme heating. The large masses of these kinds of planets (i.e., more than twice the mass of Jupiter) presumable cause them to have thick atmospheres composed of hydrogen and helium, thus making them uninhabitable as well. However, these kinds of planets may shield small Earth-like planets that orbit close to the star from being hit by solid projectiles whose impacts could interfere with the evolution of organisms living on their surfaces.

The nearest star, Proxima Centauri, is located at a distance of 4.22 ly from the Sun. Proxima Centauri is a red dwarf star whose luminosity is about 82,000 times weaker than the luminosity of the Sun. Even if this star has an Earth-like planet, that planet has to be closer to Proxima Centauri than the Earth is to the Sun in order to permit water to remain liquid on its surface. The propulsion systems available at the present time do not yet permit even a robotic spacecraft to reach Proxima Centauri in less than many centuries of travel time.

Earth-like planets that revolve around Sun-like stars in the Milky Way galaxy may be inhabited by intelligent beings who are generating electromagnetic signals that are radiated into interstellar space. However, the on-going search for extraterrestrial intelligence (SETI) has not yielded any credible evidence that such civilizations exist in our galaxy. An evaluation of the factors that make up the Drake equation suggests that Earth-like planets in the Milky Way galaxy are rare but permits the optimistic assessment that we are probably not alone in our galaxy.

24.7 Further Reading

Basri G (2000) The discovery of brown dwarfs. Sci Am 288(4):76–83

Black DC (1999) Extra-solar planets: Searching for other planetary systems. In: Weissman P.R., McFadden L.-A., Johnson T.V. (eds) Encyclopedia of the Solar System. Academic Press, San Diego, CA, pp 941–955

Butler RP (1999) Other planetary systems. In: Beatty J.K., Petersen C.C., Chaikin A. (eds) The New Solar System, 4th edn. Sky Publishing, Cambridge, MA, pp 377–386

Clark S (1998) Extrasolar planets: The search for new worlds. Wiley-Praxis, Chichester, UK

Crosswell K (2006) A flare for Barnard's star. Astronomy 34(3):24

Doyle L (ed) (1996) Circumstellar habitable zones. Travis House, Menlo Park, CA

Drake F, Sobel D (1992) Is anyone out there? The scientific search for extraterrestrial intelligence. Delacorte Press

Freedman RA, Kaufmann III WJ (2002) Universe: The Solar System. Freeman, New York

Goldsmith D, Owen T (1992) The search for life in the Universe. University Science Books, New York

Greenberg R (2005) Europa-The ocean moon: Search for an alien biosphere. Springer, Chichester, UK

Hartmann WR (2005) Moons and planets, 5th edn. Brooks/Cole, Belmont, CA, pp 91–95

Henry R (1996) Brown dwarfs revealed at last! Sky and Telescope, 91:24–28

Kasting J (1996) Habitable zones around main-sequence stars: An update. In: Doyle L. (ed) Circumstellar Habitable Zones. Travis House, Menlo Park, CA, pp 17–28

Klein HP (1979) The Viking mission and the search for life on Mars. Rev Geophys Space Phys 17:1655–1662

Kulkarni SR (1997) Brown dwarfs: A possible missing link between stars and planets. Science 276 (May 30):1350–1354

Lemonick MD (1998) Other worlds: The search for life in the Universe. Simon and Schuster, New York

Marcy G, Butler RP (1998) Detection of extrasolar planets. Ann Rev Astron Astrophys 36:57–97

McKay C (1996) Time for intelligence on other planets. In: Doyle L. (ed) Circumstellar Habitable Zones. Travis House, Menlo Park, CA, pp 405–409

McKay CP, Davis WL (1999) Planets and the origin of life. In: Weissman P.R., McFadden L.-A., Johnson T.V. (eds) Encyclopedia of the Solar System. Academic Press, San Diego, CA, pp 899–922

Mohanty S, Jayawardhana R (2006) The mystery of brown-dwarf origins. Sci Am 294(1):38–45.

Protheroe WM, Capriotti ER, Newsom GH (1989) Exploring the Universe, 4th edn. Merrill Publishing Co, Columbus, OH

Sagan C, Drake F (1975) The search for extraterrestrial intelligence. Sci Am 232:80–89

Sagan C (1993) A search for life on Earth from the Galileo spacecraft. Nature 365:715–721

Sagan C (1994) The search for extraterrestrial life. Sci Am October: 93–99

Seeds MA (1997) Foundations of astronomy, 4th edn. Wadsworth Publishers, Belmont, CA

Shklowskii D, Sagan C (1966) Intelligent life in the Universe. Holden-Day, New York

Soffen GA (1999) Life in the solar system. In: Beatty J. K., Petersen C.C., Chaikin A. (eds) The New Solar System, 4th edn. Sky Publishing Corp., Cambridge, MA, pp 365–376

Tarter JC, Chyba CF (1999) Is there life elsewhere in the Universe? Sci Am Dec:118–123

Taylor SR (1998) Destiny or chance: Our solar system and its place in the cosmos. Cambridge University Press, New York

Villard R, Cook LR (2005) Infinite worlds: An illustrated voyage to planets beyond our Sun. University of California Press, Los Angeles, CA

Ward PD, Brownlee D (2000) Rare Earth; Why complex life is uncommon in the Universe. Copernicus Books, New York

Ward P (2005) Life as we do not know it: The NASA search for and synthesis of alien life. Viking Adult, New York

Wells HG (1898) The war of the worlds. Reissued in 1988 by Tom Doherty Associates, New York

Zuckerman B, Hart MH (eds) (1995) Extraterrestrials – where are they? Cambridge University Press, Cambridge, UK

Appendix 1

Mathematical Equations Used in Astronomy

(Freedman and Kaufmann, 2002; Hartmann, 2005)

A1.1 Astronomical Constants

Astronomical unit	AU	$= 149.60 \times 10^6 \, \text{km}$
		$= 1.4960 \times 10^{11} \, \text{m}$
Light year	ly	$= 9.4605 \times 10^{12} \, \text{km}$
		$= 9.4605 \times 10^{15} \, \text{m}$
		$= 63{,}240 \, \text{AU}$
Year	y	$= 365.2564 \, \text{d}$
		$= 3.156 \times 10^7 \, \text{s}$
Solar mass	$M\odot$	$= 1989 \times 10^{30} \, \text{kg}$
Solar radius	$R\odot$	$= 6.9599 \times 10^8 \, \text{m}$
		$= 6.9599 \times 10^5 \, \text{km}$
Solar luminosity	$L\odot$	$= 3.90 \times 10^{26} \, \text{W}$
Speed of light	c	$= 2.9979 \times 10^8 \, \text{m/s}$
		$= 2.9979 \times 10^{10} \, \text{cm/s}$
		$= 2.9979 \times 10^5 \, \text{km/s}$

A1.2 Physical Constants

Gravitational constant	G	$= 6.6726 \times 10^{-11} \, \text{Nm}^2/\text{kg}^2$
Planck's constant	h	$= 6.6261 \times 10^{-34} \, \text{Js}$
		$= 4.1357 \times 10^{-15} \, \text{eVs}$
Boltzmann constant	k	$= 1.3807 \times 10^{-23} \, \text{J/K}$
		$= 8.6174 \times 10^{-5} \, \text{eV/K}$
Stefan-Boltzmann constant	σ	$= 5.6705 \times 10^{-8} \, \text{Wm}^{-2}\text{k}^{-4}$
Mass of an electron	m_e	$= 9.1094 \times 10^{-31} \, \text{kg}$
Mass of a proton	m_p	$= 1.6726 \times 10^{-27} \, \text{kg}$
Mass of a neutron	m_n	$= 1.6749 \times 10^{-27} \, \text{kg}$
Mass of a hydrogen atom	m_H	$= 1.6735 \times 10^{-27} \, \text{kg}$
Rydberg constant	R	$= 1.0968 \times 10^7 \, \text{m}^{-1}$
Electron volt	$1 \, \text{eV}$	$= 1.6022 \times 10^{-19} \, \text{J}$

A1.3 Geometrical Relations

Area of a rectangle	$A = a \times b$
Volume of rectangular solid	$V = a \times b \times c$
Hypotenuse of a right triangle	$c = (a^2 + b^2)^{1/2}$
Circumference of a circle	$C = 2r\pi$
Area of a circle	$A = r^2\pi$
Surface of a sphere	$A = 4\pi r^2$
Volume of a sphere	$V = \frac{4}{3}\pi r^3$

$$\pi = \frac{C}{2r} = 3.1415926536,$$

$$r = \text{radius of a circle}$$

A1.4 Units of Distance, Time, and Mass

In Science the unit of length or distance is the meter (m), the unit of time is the second (s), and the unit of mass is the kilogram (kg).

A1.4.1 Distance (m)

$$10^{-2}\,m = 1 \text{ centimeter (cm)}$$
$$10^{-3}\,m = 1 \text{ millimeter (mm)}$$
$$10^{-6}\,m = 1 \text{ micrometer } (\mu m)$$
$$10^{-9}\,m = \text{nanometer (nm)}$$
$$10^{3}\,m = 1 \text{ kilometer (km)}$$
$$149.6 \times 10^{6}\,km = 1 \text{ astronomical unit (AU)}$$
$$9.46 \times 10^{12}\,km = 1 \text{ lightyear (ly)} = 63,240 \text{ (AU)}$$

A1.4.2 Time (s)

$$60\,s = 1 \text{ minute (min)}$$
$$3600\,s = 1 \text{ hour (h)} = 60 \text{ min}$$
$$86,400\,s = 1 \text{ day (d)} = 24\,h$$
$$3.156 \times 10^{7}\,s = 1 \text{ year(y)} = 365.256\,d$$

A1.4.3 Mass (kg)

$$1\,kg = 10^{3} \text{ gram(g)}$$
$$1000\,kg = 1 \text{ metric tonne (t)}$$
$$1.99 \times 10^{30}\,kg = \text{solar mass } (M_{\odot})$$

A1.5 Speed, Velocity, and Acceleration

Speed is a measure of how fast a body is moving expressed in terms of the ratio:

$$\text{speed} = \frac{\text{distance}}{\text{time}}$$

$$10^{3}\,m/s = 1\,km/s$$
$$1.609 \times 10^{3}\,m/s = 1\,mi/s$$

Velocity differs from speed because it includes the direction in which a body is moving. For example, a car moving north at 100 km/h and a car moving south at 100 km/h have the same speed but different velocities.

Acceleration is the rate at which the velocity of a body in motion is changing either because of a change in its speed or because of a change in its direction. Acceleration occurs when the speed of a body in motion increases or decreases, or when the direction of motion is changing. When an apple falls out of a tree, its speed increase from zero to 9.8 m/s during the first second, to 19.6 m/s during the next second, and to 29.4 m/s during the third second. In other words, the speed of the apple in free fall on the Earth increases by 9.8 m/s for each second that elapses. Therefore, the acceleration of the apple is 9.8 m/s/s or 9.8 m/s².

A planet on a circular orbit around the Sun has a constant speed but it is continually accelerated because of the change in direction.

A1.6 Newton's Laws of Motion

1. A body at rest remains at rest, a body in motion continues to move in a straight line at a constant speed unless it is acted upon by a net outside force.
2. The acceleration of an object is proportional to the net outside force that is acting on it.

If a net force (F) acts on body of mass of mass (m), the body will experience an acceleration (a) such that:

$$F = ma$$

F = net outside force acting on the body (newtons),
m = mass of the object (kg),
a = acceleration(m/s²).

3. Whenever one body exerts a force on a second body, the second body exerts an equal and opposite force on the first body.z

A1.7 Mass and Weight

The *mass* of a body depends on the amount of material it contains and is expressed in grams or kilograms. The mass of a body does not depend on its location in the Universe.

The *weight* of a body is the magnitude of the gravitational force that acts on it. Since weight is a force, it is expressed in newtons or pounds where:

$$1 \text{ newton} = 0.225 \text{ pounds}$$

A1.8 Sidereal and Synodic Periods

The *sidereal* period of a planet is the time required for the planet to complete one orbit. The *synodic* period of a planet is the time between two successive identical configurations.

A1.8a Inferior Planet

$$\frac{1}{P} = \frac{1}{E} + \frac{1}{S} \tag{A1.1}$$

P = sidereal period of an inferior planet (i.e., the radius of its orbit is smaller than that of the Earth),
E = sidereal period of the Earth (1 year),
S = synodic period of an inferior planet.

A1.8b Superior Planet

$$\frac{1}{P} = \frac{1}{E} - \frac{1}{S} \qquad (A1.2)$$

P = sidereal period of a superior planet
E = sidereal period of the Earth (1 year)
S = synodic period of a superior planet
Note that all periods are expressed in multiples of sidereal Earth years.

A1.9 Kepler's Third Law

$$p^2 = a^3$$

p = period expressed in sidereal Earth years,
a = semi-major axis of the orbit of a planet in the solar system expressed in astronomical units (AU)
This form of Kepler's third law applies only to objects that orbit the Sun. It does not apply to satellites that orbit planets in the solar system or planets that orbit a star other than the Sun.

A1.10 Newton's Law of Universal Gravitation

$$F = G\left(\frac{m_1 m_2}{r^2}\right)$$

F = gravitational force between two objects in newtons,
m_1 = the mass of the first object in kilograms,
m_2 = the mass of the second object in kilograms,
r = distance between the centers of the objects in meters,
G = universal constant of gravitation = 6.67×10^{-11} newton m^2/kg^2.

A1.11 Newton's Form of Kepler's Third Law

$$p^2 = \frac{4\pi^2}{G(m_1 + m_2)} a^3$$

p = sidereal period in seconds,
a = semi-major axis of the orbit in meters,
m_1 = mass of the first object in kilograms,
m_2 = mass of the second object in kilograms,
G = universal constant of gravitation (See above).

A1.12 Tidal Force Exerted by the Earth on the Moon

$$F_{tidal} = \frac{2GM\,md}{r^3}$$

F_{tidal} = tidal force in newtons (N)
G = $6.67 \times 10^{-11}\,Nm^2/kg^2$
M = mass of the Earth = $5.974 \times 10^{24}\,kg$
m = small mass of 1 kg
d = diameter of the Moon in meters
= $3.476 \times 10^6\,m$
r = distance between the center of the Earth and the center of the Moon in meters
= $3.844 \times 10^8\,m$

$$F_{tidal} = \frac{2(6.67 \times 10^{-11})(5.974 \times 10^{24})(1)(3.476 \times 10^6)}{(3.844 \times 10^8)^3}$$
$$= 4.88 \times 10^{-5}\,N\,(\text{Freedman and Kaufmann, 2002, p.215}).$$

A1.13 Tidal Bulge

The mass of the tidal bulge (m_b) is:

$$m_b = \frac{A}{r^3}$$

where A depends on the mechanical properties of the body that is being deformed.

By substituting $m_b = A/r^3$ into the formula for the tidal force, we obtain the net tidal force that acts on the Moon:

$$F_{tidal-net} = \frac{2GMd}{r^3} \times \frac{A}{r^3} = \frac{2GMdA}{r^6}$$

The ratio of the net tidal forces acting on the Moon at perigee and apogee is:

$$\frac{F_{per}}{F_{apo}} = \frac{2GMAd\,(r_{apo})^6}{(r_{peri})^6\,2GMAd} = \left(\frac{r_{apo}}{r_{peri}}\right)^6$$

$$= \left(\frac{405,500}{363,300}\right)^6 = 1.93\,(\text{Freedman and}$$

Kaufmann, 2002, p. 216).

A1.14 Roche Limits

Two touching particles:

$$r_R = 2.5 \left(\frac{\rho_M}{\rho_m} \right)^{1/3} R \qquad (A1.3)$$

Liquid (or zero-strength) spherical body:

$$r_R = 2.44 \left(\frac{\rho_M}{\rho_m} \right)^{1/3} R \qquad (A1.4)$$

r_R = Roche limit
ρ_m = density of a body of mass M
ρ_m = density of body of mass m
R = radius of the body of mass M (km)
(Hartmann, 2005, p.60)

A1.15 Frequency and Wavelength of Electromagnetic Radiation

$$\nu = \frac{c}{\lambda} \qquad (A1.5)$$

ν = frequency of an electromagnetic wave in hertz (Hz)
c = speed of light in meters per second
λ = wavelength of the wave in meters
The frequency of a wave of any kind is the number of wave crests that pass a given point in one second.

$$\nu = \frac{1}{p} \qquad (A1.6)$$

where p = the time between successive wave crests that pass a given point. The speed of light, or the speed of any wave, is equal to the wavelength divided by the period.

$$c = \frac{\lambda}{p} \qquad (A1.7)$$

A1.16 Temperature Scales

$$C = K - 273.15 \qquad (A1.8)$$

$$K = C + 273.15 \qquad (A1.9)$$

$$C = \frac{(F-32)\,5}{9} \qquad (A1.10)$$

$$F = \frac{9C}{5} + 32 \qquad (A1.11)$$

$$F = \frac{(K-273.15)\,9}{5} + 32 \qquad (A1.12)$$

$$K = \frac{(F-32)\,5}{9} + 273.15 \qquad (A1.13)$$

C = temperature on the Celsius scale,
K = temperature on the Kelvin scale,
f = temperature on the Fahrenheit scale.

A1.17 Wien's Law for Blackbody Radiation

The relation between the temperature of a radiating body and the wavelength of the most intense radiation it emits is:

$$\lambda_{max} = \frac{0.0029}{T}$$

λ_{max} = wavelength of the most intense radiation in meters,
T = temperature of the object in kelvins.

A1.18 Stefan-Boltzmann Law for Blackbody Radiation

$$F = \sigma T^4$$

F = flux of energy in units of $J/m^2/s$ at the surface of the radiating body,
$\sigma = 5.67 \times 10^{-8}$ $W/m^2/T^4$ (or W m^{-2} T^{-4}),
T = temperature of the object in kelvins.

A1.19 Surface Temperature of the Sun

The wavelength of the most intense radiation emitted by the Sun is 500 nm (Freedman and Kaufmann, 2002, p. 103). Therefore, Wien's Law yields a temperature of:

$$T = \frac{0.0029}{\lambda_{max}} = \frac{0.0029}{500 \times 10^{-9}} = 5800\,K$$

The luminosity (L) of the Sun is $L = 3.90 \times 10^{26}$ W. Given that 1 Watt $= 1\,\text{J/s}$, the Sun radiates 3.90×10^{26} J every second. The energy flux (F) is obtained by dividing the total luminosity by surface area (A) of the Sun where $A = 4\pi R^2$ and R is the radius of the Sun (Freedman and Kaufmann, 2002, p. 103):

$$F = \frac{L}{4\pi R^2} = \frac{3.90 \times 10^{26}}{4 \times 3.14(6.96 \times 10^8)^2}$$

$$= 6.41 \times 10^7\,\text{W/m}^2$$

We can also calculate the surface temperature of the Sun from the Stefan-Boltzmann Law:

$$T^4 = \frac{F}{\sigma} = \frac{6.41 \times 10^7}{5.67 \times 10^{-8}} = 1.13 \times 10^{15}$$

$$\log T = \frac{\log 1.13 \times 10^{15}}{4} = \frac{15.0530}{4} = 3.7632$$

$$T = 5797\,\text{K}$$

which rounds to 5800 K.

A1.20 Energy of Photons

$$E = \frac{hc}{\lambda} \qquad (A1.14)$$

E = energy in electron volt (eV) where 1 eV
  $= 1.60210^{-19}$ J,
h = Planck's constant $= 6.625 \times 10^{-34}$ Js,
c = speed of light,
λ = wavelength of light.
One photon of red light having a wavelength $\lambda = 633$ nm carries an energy of:

$$E = \frac{6.625 \times 10^{-34} \times 3.00 \times 10^8}{633 \times 10^{-9}} = 3.14 \times 10^{-19}\,\text{J}$$

The energy of a photon can also be expressed in terms of the frequency (ν) where

$$\nu = \frac{c}{\lambda} \qquad (A1.15)$$

$$E = h\nu \qquad (A1.16)$$

A1.21 Doppler Effect

$$\frac{\Delta\lambda}{\lambda_0} = \frac{\nu}{c}$$

$\Delta\lambda$ = measured change in wavelength in nanometers,
λ = wavelength in nanometers when the source is not moving,
ν = velocity of the source measured along the line of sight (i.e., the radial velocity) in km/s,
c = speed of light $(3.0 \times 10^5\,\text{km/s})$.
If $\Delta\lambda$ is negative, the source is approaching the Earth (i.e., blue shift). If $\Delta\lambda$ is positive, the source is receding from the Earth (i.e., red shift).

A1.22 Kinetic Energy and Speed of Atoms and Molecules in a Gas

The kinetic energy of an object is:

$$KE = \frac{1}{2} m \nu^2 \qquad (A1.17)$$

m = mass of the object in kilograms,
ν = speed in meters per second,
KE = kinetic energy in Joules (J).
The *average* kinetic energy of an atom or molecule in a gas is:

$$KE = \frac{3}{2} kT \qquad (A1.18)$$

k = Boltzmann constant $= 1.38 \times 10^{-23}$ J/K,
T = temperature in kelvins,
KE = average kinetic energy of a single atom or molecule in Joules.
The average speed of an atom or molecule in a gas depends on its mass.

$$\frac{1}{2} m\nu^2 = \frac{3}{2} kT$$

$$\nu = \left(\frac{3kT}{m}\right)^{1/2} \qquad (A1.19)$$

ν = average speed of an atom or molecule in a gas in m/s, k $= 1.38 \times 10^{-23}$ J/K,

T $=$ temperature of the gas in kelvins,

m $=$ mass of the atom or molecule in kilograms.

A1.23 Escape Speed from Planetary Surfaces

The escape speed is the minimum speed an object at the surface of a planet must have in order to permanently leave the planet:

$$\nu_e = \left(\frac{2GM}{r}\right)^{1/2}$$

The escape speed of planets in km/s is:

Mercury	4.3
Venus	10.4
Earth	11.2
Moon	2.4
Mars	5.0
Jupiter	59.5
Saturn	35.5
Uranus	21.5
Neptune	23.4
Pluto	1.3

A1.24 Retention of Molecular Oxygen in the Atmosphere of the Earth

The mass of a molecule of oxygen (O_2) is 5.32×10^{-26} kg. Therefore the average speed of oxygen molecules at $20\,°C$ (293 K) is (Freedman and Kaufmann, 2002, p. 160):

$$\nu = \left(\frac{3\left(1.38 \times 10^{-23}\right)(293)}{5.32 \times 10^{-26}}\right)^{1/2} = 478\,m/s$$

$$= 0.478\,km/s$$

Since the escape speed from the Earth is 11.2 km/s, oxygen cannot escape from the atmosphere of the Earth. In general, atoms and molecules cannot escape in cases where the escape speed from the surface of the planet is more than six times higher than the average speed of the atom or molecule under consideration.

A1.25 References

Freedman RA, Kaufmann III WJ (2002) Universe: The solar system. Freeman, New York

Hartman WK (2005) Moons and planets, 5th edn. Brooks/Cole, Belmont, CA

Appendix 2

Summaries of Physical and Orbital Parameters

A2.1 The Planets and Satellites of the Solar System

Planet	Satellite	m kg	d km	a AU/10^3 km	p	e	i (°)
Mercury		3.30 (23)	4878	0.387 AU	87.97 d	0.206	7.00
Venus		4.87 (24)	12,104	0.723 AU	224.70 d	0.007	3.39
Earth		5.98 (24)	12,756	1.000 AU	365.256 d	0.017	0.00
	Moon	7.35 (22)	3,476	384.4	27.32 d	0.055	18 to 29
Mars		6.42(23)	6,787	1.524 AU	686.98 d	0.093	1.85
	Phobos	9.6 (15)	27 × 19	9.38	0.319 d	0.018	1.0
	Deimos	2.0 (15)	15 × 11	23.50	1.262 d	0.002	2.8
Jupiter		1.90 (27)	142,800	5.203 AU	11.86 y	0.048	1.30
	Metis	?	40	128	0.295 d	0.0	0.0
	Adrastea	?	24 × 16	129	0.298 d	0.0	0.0
	Amalthea	?	270 × 155	181.5	0.498 d	0.003	0.45
	Thebe	?	100	222	0.674 d	0.01	0.8?
	Io	8.89 (22)	3630	422	1.769 d	0.000	0.03
	Europa	4.79 (22)	3130	671	3.551 d	0.000	0.46
	Ganymede	1.48 (23)	5280	1071	7.155 d	0.002	0.18
	Callisto	1.08 (23)	4840	1884	16.689 d	0.008	0.25
	Leda	?	∼ 16	11,110	240 d	0.146	26.7
	Himalaya	?	∼ 180	11,470	250.6 d	0.158	27.6
	Lysithea	?	∼ 40	11,710	260 d	0.12	29.0
	Elara	?	∼ 80	11,740	260.1 d	0.207	24.8
	Ananke	?	∼ 30	20,700	617 R	0.169	147
	Carme	?	∼ 44	22,350	692 R	0.207	163
	Pasiphae	?	∼ 35	23,300	735 R	0.40	147
	Sinope	?	∼ 20	23,700	758 R	0.275	156
Saturn		5.69 (26)	120,660	9.539 AU	29.46 y	0.056	2.49
	Pan	?	∼ 20	133.6	0.576 d	0	0
	Atlas	?	50 × 20	137.6	0.601 d	0.002	0.3

(*Continued*)

(*Continued*)

Planet	Satellite	m kg	d km	a AU/10³ km	p	e	i (°)
	Prometheus	?	140 × 80	139.4	0.613 d	0.003	0.0
	Pandora	?	110 × 70	141.7	0.629 d	0.004	0.05
	Epimetheus	?	140 × 100	151.4	0.694 d	0.009	0.34
	Janus	?	220 × 160	151.4	0.695 d	0.007	0.14
	Mimas	4.5 (19)	390	185.54	0.942 d	0.020	1.5
	Enceladus	8.4 (19)	500	238.04	1.370 d	0.004	0.0
	Tethys	7.5 (20)	1048	294.67	1.888 d	0.000	1.1
	Calypso	?	34 × 22	294.67 (T)	1.888 d	0.0	~ 1?
	Telesto	?	34 × 22	294.67 (L)	1.888 d	0.0	~ 1?
	Dione	1.05 (21)	1120	377	2.737 d	0.002	0.0
	Helene	?	36 × 30	377	2.74 d	0.005	0.15
	Rhea	2.49 (21)	1530	527	4.518 d	0.001	0.4
	Titan	1.35 (23)	5150	1222	15.94 d	0.029	0.3
	Hyperion	?	350 × 200	1484	21.28 d	0.104	~ 0.5
	Iapetus	?	1440	3564	79.33 d	0.028	14.72
	Phoebe	?	220	12,930	550.4 dR	0.163	150
Uranus		8.68 (25)	51,118	19.18 AU	84.01 y	0.047	0.77
	Cordelia	?	~ 30	49.7	0.335 d	~ 0	~ 0.14
	Ophelia	?	~ 30	53.8	0.376 d	~ 0.01	~ 0.09
	Bianca	?	~ 40	59.2	0.435 d	~ 0	~ 0.16
	Cressida	?	~ 70	61.8	0.464 d	~ 0	~ 0.04
	Desdemona	?	~ 60	62.7	0.474 d	~ 0	~ 0.16
	Jnliet	?	~ 80	64.4	0.493 d	~ 0	~ 0.06
	Portia	?	~ 110	66.1	0.513 d	~ 0	~ 0.09
	Rosalind	?	~ 60	69.9	0.558 d	~ 0	~ 0.28
	Belinda	?	~ 70	75.3	0.624 d	~ 0	~ 0.03
	Puck	?	150	86.0	0.762 d	~ 0	~ 0.31
	Miranda	6.89 (19)	470	129.8	1.414 d	0.00	3.40
	Ariel	1.26 (21)	1160	191.2	2.520 d	0.00	0.00
	Umbriel	1.33 (21)	1170	266.0	4.144 d	0.00	0.00
	Titania	3.48 (21)	1580	435.8	8.706 d	0.00	0.00
	Oberon	3.03 (21)	1520	582.6	13,463 d	0.00	0.00
Neptune		1.02 (26)	49,528	30.06 AU	164.8 y	0.009	1.77
	Naiad	?	~ 50	48.0	0.296 d	~ 0	~ 0.0
	Thalassa	?	~ 80	50.0	0.312 d	~ 0	~ 4.5
	Despina	?	~ 180	52.2	0.333 d	~ 0	~ 0.0
	Galatea	?	~ 150	62.0	0.429 d	~ 0	~ 0.0
	Larissa	?	~ 190	73.6	0.554 d	~ 0	~ 0.0
	Peoteus	?	~ 400	117.6	1.121 d	~ 0	~ 0.0
	Triton	2.14 (22)	2700	354.8	5.877 dR	0.00	157
	Nereid	?	~ 340	5513.4	360.16 d	0.75	29
Pluto		1.29 (22)	2300	39.53 AU	247.7 y	0.248	17.15
	Charon	1.77 (21)	1190	19.6	6.39 d	98.8	

m = mass in kg, d = diameter in km, a = semi-major axis in AU (planets) or in 10^3 km (satellites), p = period of revolution in years (planets) or days (satellites), R = retrograde, e = eccentricity, i = orbital inclination relative to the ecliptic plane for planets and relative to the equatorial plane of planets for their satellites. Adapted from Hartmann (2005).

A2.2 Orbital Properties Asteroids

Name	a AU	q AU	Q AU	e	i	Diameter, km
Aten (Earth crossers)						
Re-Shalom	0.83	0.47	1.20	0.44	16°	3.4
Aten	0.97	0.79	1.14	0.18	19°	1.0
Hathor	0.84	0.46	1.22	0.45	6°	0.2
Apollo (Earth and Mars crossers)						
Sisyphus	1.89	0.87	2.92	0.53	41°	10
Toro	1.37	0.77	1.96	0.44	9°	4.8
MA	1.78	0.42	3.13	0.76	38°	6?
NA	2.39	0.88	3.91	0.63	68°	6
Antonius	2.26	0.89	3.63	0.61	18°	3
XC	2.25	0.83	3.67	0.63	1°	3?
Daedalus	1.46	0.56	2.36	0.62	22°	3.4
Geographos	1.24	0.83	1.66	0.34	13°	2.0
Amor (Mars crossers)						
Ganymede	2.66	1.22	4.10	0.54	26°	30?
Eros	1.46	1.13	1.78	0.22	11°	$7 \times 19 \times 30$
Beira	2.73	1.39	4.07	0.49	27°	11?
UB	2.12	1.36	2.89	0.36	36°	8?
AD	2.37	1.48	3.26	0.38	20°	8?
Atami	1.95	1.45	2.44	0.26	13°	7?
RH	2.38	1.48	3.28	0.38	21°	7?
Betulia	2.20	1.12	3.27	0.49	52°	6
Ivar	1.86	1.12	2.60	0.40	8°	6?
Main belt (a < 2.5 AU)						
Vesta	2.36	2.15	2.57	0.09	7°	500
Iris	2.39	1.84	2.93	0.23	6°	204
Hebe	2.43	1.94	2.92	0.20	15°	192
Fortuna	2.44	2.05	2.83	0.16	2°	190
Flora	2.20	1.85	2.55	0.16	6°	153
Parthenope	2.45	2.20	2.70	0.10	5°	150
Melpomene	2.30	1.79	2.81	0.22	10°	150
Massalia	2.41	2.07	2.75	0.14	1°	131
Victoria	2.34	1.83	2.85	0.22	8°	126
Lutetia	2.43	2.04	2.82	0.16	3°	115
Main belt (2.5 < a < 3.1 AU)						
Ceres	2.77	2.55	2.99	0.08	11°	914
Pallas	2.77	2.13	3.41	0.23	35°	522
Interamnia	3.06	2.57	3.55	0.16	17°	334
Eunomia	2.64	2.14	3.14	0.19	12°	272
Psyche	2.92	2.51	3.32	0.14	3°	264
Juno	2.67	1.98	3.60	0.26	13°	244
Bamberga	2.69	1.78	3.60	0.34	11°	242
Patientia	3.06	2.82	3.30	0.07	15°	230
Ejeria	2.58	2.35	2.81	0.09	16°	214
Eugenia	2.72	2.50	2.94	0.08	7°	214

(Continued)

(*Continued*)

Name	a AU	q AU	Q AU	e	i	Diameter, km
Main belt (3.1 < a < 4.1 AU)						
Hygiea	3.15	2.84	3.46	0.10	4°	443
Davida	3.18	2.65	3.73	0.17	16°	336
Sylvia	3.48	3.13	3.83	0.10	11°	272
Euphrosyne	3.15	2.46	3.84	0.22	26°	248?
Cybele	3.43	3.02	3.84	0.12	4°	246
Camilla	3.49	3.25	3.73	0.07	10°	236
Themis	3.14	2.76	3.52	0.12	1°	228
Aurora	3.15	2.84	3.47	0.09	8°	212
Alauda	3.19	3.09	3.29	0.03	21°	202
Loreley	3.13	2.88	3.38	0.08	11°	202
Main belt (4.1 < a < 5.1)						
Thule	4.26	4.12	4.40	0.03	2°	60?
Trojans (L4 and L5 Lagrange points) (P = preceding Jupiter, F = following Jupiter)						
Hektor (P)	5.15	5.0	5.2	0.02	18°	150 × 300
Agamemnon(P)	5.19	4.8	5.5	0.07	22°	148?
Patroclus (F)	5.21	4.5	5.9	0.14	22°	140
Diomedes (P)	5.08	4.8	5.3	0.05	21°	130?
Aeneas (F)	5.17	4.4	5.9	0.10	17°	125
Achilles (P)	5.17	4.4	6.0	0.15	10°	118?
Odysseus (P)	5.21	4.7	5.7	0.09	3°	118?
Nestor (P)	5.26	4.7	5.8	0.11	4°	102?
Troilus (F)	5.17	4.7	5.6	0.09	34°	98?
Antilochus (P)	5.28	5.0	5.6	0.05	28°	98?

a = average distance from the Sun; q = distance from the Sun at perihelion; Q = distance from the Sun at aphelion, e = eccentricity; i = inclination of the orbit relative to the ecliptic. Adapted from Hartmann (2005).

A2.3 Orbital Properties of Selected Comets

Name	a AU	q AU	Q AU	e	i (°)	p y
A2.3.1. Short Period						
3D Biela	3.52	0.861	6.18	0.756	12.55	6.62
23P Brorsen-Metcalf	16.9	0.479	33.3	0.972	19.28	69.5
27 P Crommelin	9.11	0.745	17.5	0.918	29.11	27.5
6 P d'Arrest	3.4	1.347	5.5	0.614	19.52	6.52
2 P Encke	2.2	0.333	4.1	0.849	11.91	3.29
21 P Giacobini-Zinner	3.5	1.034	6.0	0.706	31.86	6.61
1 P Halley	17.9	0.596	35.2	0.967	162.22	75.8 (R)
96 P Machholz	3.02	0.125	5.91	0.959	59.96	5.24
29 P Schwassmann-W.1	6.1	5.729	6.5	0.046	9.38	14.9
109 P Swift-Tuttle	∼ 24	0.968	∼ 47	0.963	113.47	135 (R)
55 P Tempel-Tuttle	10.3	0.977	19.6	0.906	162.49	33.3 (R)
8 P Tuttle	5.69	1.034	10.3	0.819	54.94	13.6
81 P Wild 2	3.44	1.583	5.30	0.540	3.24	6.39

46 P Wirtanen	3.09	1.063	5.11	0.657	11.72	5.45

A2.3.2. Long Period

					Discovery date
R1 Arend-Roland		0.316	1.000	119.95	1956
Y1 Bennett		0.538	0.996	90.04	1970
L1 Donati	∼ 157	0.578	0.996	116.96	1858
O1 Hale-Bopp		0.914	0.995	89.43	1995
B2 Hyakutake		0.231	0.999	124.95	1996
S1 Ikeya-Seki		0.008	1.000	141.86	1965
H1 IRAS-Araki-A.		0.991	0.990	73.25	1983

a = average distance from the Sun; q = distance from the Sun at perihelion; Q = distance fro the Sun at aphelion; e = eccentricity; i = inclination of the orbit relative to the ecliptic; p = period of revolution; R = retrograde. Adapted from Beatty et al. (1999) and Hartmann (2005).

A2.4 References

Beatty JK, Petersen CC, Chaikin A (eds) (1999) The new solar system, 4th edn. Sky Publishing Corp. Cambridge, MA

Hartmann WK (2005) Moons and planets, 5th edn. Brooks/Cole, Belmont, CA

Glossary

Geographic places and the names of satellites and meteorites are referenced in the Index.

A

Absolute zero A temperature of $-273.15\,°C$, or $-459.67\,°F$, or $0\,K$. It is the lowest temperature when all molecular motion stops.

Absorption lines Dark lines superimposed on a continuous wavelength spectrum of star light caused by the absorption of certain wavelengths of light by atoms in the photospheres of stars.

Acceleration The rate at which the velocity of an object changes as a result of changes in its speed or direction, or both.

Accretion The gradual accumulation of small particles under the influence of gravity to form a large body such as a planet.

Accretion disk A disk of material in orbit around a star in which particles are accreting to form planets.

Achondrite A stony meteorite that lacks chondrules (in most cases). (See also chondrules).

Aerosol Microscopic droplets of liquid or solid particles dispersed in a gas or mixture of gases like air.

Albedo The fraction of sunlight that is reflected by planets, satellites, asteroids, and comets. It is expressed either as a decimal fraction or as percent.

Algae Primitive aquatic single-celled or colonial micro-organisms capable of photosynthesis ranging in age from $3.5 \times 10^9\,y$ to the present time.

Alluvial fan Fan-shaped body of sediment deposited at the base of a steep slope by a flowing liquid such as water or methane depending on the temperature.

Alpine glacier A glacier whose flow is restricted by the walls of a valley in the mountains. Valleys eroded by an alpine glacier have a characteristic U-shaped cross-section.

Amazonian Era The latest era of time on Mars that lasted from 1.8×10^9 years ago to the present time.

Amino acids A group of organic compounds containing both amine (NH_2) and carboxyl (COOH) components. They are the building blocks of proteins and are therefore essential to life processes.

Ammonia A chemical compound (NH_3) that occurs as a gas on the surface of the Earth but may form a solid at the low temperatures in the outer regions of the solar system.

Ammonium hydrosulfide (NH_4HS) a compound that occurs in the atmospheres of the giant planets of the solar system.

Amor asteroids Asteroids whose orbits intersect the orbit of Mars and whose perihelion distances lie between 1.017 and 1.3 AU.

Amplitude The height of a wave. The intensity of light increases as the amplitude of light waves increase.

Angstrom $1\text{Å} = 10^{-10}\,\text{m}$

Angular momentum A physical property of a rotation body. It is analogous to the linear momentum of a body moving in a straight line. (See also momentum).

Annular eclipse An eclipse of the Sun in which the Moon is too distant (when viewed from the Earth) to cover the entire disk of the Sun allowing a ring of sunlight to be seen around the Moon at mid-eclipse.

Anorthosite Plutonic igneous rock composed primarily of Ca-rich plagioclase feldspar. It occurs both on the Earth and in the highlands of the Moon.

Antarctic circle The latitude line at 66.5° south (i.e., 23.5° north of the South Pole of the Earth).

Aphelion The point on the orbit of a planet that is located at the greatest distance from the center of the Sun.

Apogee The point on the orbit of the Moon (or of any other satellite) that is farthest from the center of the Earth (or from another planet).

Apollo American space program of the 1960s and 1970s that achieved the objective, proclaimed by President John F. Kennedy on May 25, 1961, of landing an astronaut on the Moon and bringing him back alive.

Apollo asteroids A group of asteroids whose orbits intersect the orbit of the Earth and whose semi-major axes are greater than 1 AU.

Arctic circle The latitude line at 66.5 23.5° north (i.e., south of the North Pole of the Earth).

Asteroid One of tens of thousands small rocky bodies that orbit the Sun.

Asteroidal belt The main asteroidal belt located between the orbits of Mars and Jupiter.

Asthenosphere The lower part of the rocky mantle of a planet that deforms plastically at the elevated temperature that prevails in that region.

Astronomical unit The average distance between the center of the Earth and the center of the Sun ($1\,\text{AU} = 149.6 \times 10^6\,\text{km}$).

Astrobiology The study of the life in Universe.

Astronomy The science that studies the Universe and the stars it contains.

Aten asteroids Asteroids whose orbits intersect the orbit of the Earth and whose average distance from the center of the Sun is less than 1.0 AU.

Atmosphere A unit of pressure equal to the atmospheric pressure at sealevel along the equator of the Earth. **Also** A layer of gas that surrounds a solid body orbiting a star or planet.

Atom The basic unit of matter composed of a central positively-charged nucleus surrounded by a cloud of negatively-charged electrons to form an electrically neutral atom.

Atomic number The number of protons in the nucleus of an atom of a particular chemical element. It is equal to the number of extra-nuclear electrons of a neutral atom.

Aurora (borealis or australis) A glow in the upper atmosphere of a planet caused by the emission of radiant energy by excited atoms of nitrogen and oxygen.

Autumnal equinox The day in a year on Earth when the Sun is exactly above the equator at noon such that day and night are of equal length. The autumnal equinox occurs on September 22 in the northern hemisphere.

B

Bar A unit of pressure equal to 1×10^6 dynes per square centimeter.

Barycenter The common center of gravity of two objects that orbit around their barycenter.

Basalt Volcanic igneous rock composed of pyroxene, plagioclase feldspar, olivine, and magnetite with varying amounts of volcanic glass. It is fine grained, black, and may be vesicular.

Big Bang The explosion of all space that started the expansion of the Universe about 14 billion years ago.

Billion 1,000,000,000 or 10^9.

Binary star system Two stars that orbit a center of gravity located between them.

Biosphere The total of all living organisms in the atmosphere, surface, soil, and upper crust of the Earth.

Black hole Objects at the centers of galaxies that exert extremely strong gravitational forces such that even photons of electromagnetic radiation cannot escape from their surfaces.

Blue shift A decrease in the wavelength of light waves that are emitted by a source that is approaching the Earth.

Bode's rule A sequence of numbers that matches the average radii of the orbits of the planets of the solar system including the asteroid Ceres but excluding Neptune and Pluto. Also called the Titius-Bode rule.

Bolide An exploding fireball that can occur when a meteoroid passes through the atmosphere of the Earth prior to its impact.

Breccia A rock that is composed of angular fragments of rocks or minerals cemented by a fine-grained matrix or mineral cement.

Brown dwarf A starlike object composed of hydrogen and helium that is not sufficiently massive to sustain fusion of hydrogen in its core. Instead, such objects can fuse only deuterium.

Bulk density The mass of a body in space divided by its volume (kg/m^3 or g/cm^3).

C

CAI Calcium-aluminum inclusion in carbonaceous chondrite meteorites.

Caldera A large depression in the summit of an extinct or dormant volcano that forms when part of the volcano collapses into an empty magma chamber.

Canals Linear markings on Mars first seen in telescopic images during the 19th century. They are now known to be optical illusions.

Capture theory Explanation of the origin of the Moon by capture from orbit by Earth. This theory has been discredited in favor of the impact theory.

Carbonaceous chondrites Stony meteorites containing chondrules and abiogenic organic matter of large molecular weight that have not been altered by heating since their formation 4.6×10^9 years ago.

Carbon dioxide An oxide of carbon (CO_2) that occurs as a gas in the atmospheres of Venus, Earth, and Mars but that can condense at low temperature to form "dry ice."

Cassini division The most conspicuous gap in the system of rings of Saturn.

Celestial sphere An imaginary sphere of very large radius centered on the observer and encompassing to sky overhead.

Celsius scale A temperature scale defined by the freezing temperature of water ($0\,°C$) and by the boiling temperature of water ($100\,°C$) at 1 atm pressure. Absolute zero is at $-273.15\,°C$.

Cenozoic Era Geologic time from the end of the Cretaceous Period at 65 million years ago to the present time.

Center of mass A point along the line between the centers of two stars, or a star and a planet, or a planet and its satellite around which both objects revolve. (See also barycenter).

Centigrade scale See Celsius scale.

Chemical composition A statement of the concentrations of all chemical elements that exist within a sample of matter.

Chemical differentiation A process that results in the enrichment of some chemical elements and in the depletion of others in different parts of a body such as a star, a planet, or a satellite.

Chondrules Spherical particles composed of silicate and sulfide minerals that crystallized from small blobs of glass in some kinds of stony meteorites.

Chromosphere A layer in the atmosphere of the Sun between the photosphere and the corona.

CNO cycle A series of nuclear reactions in which ^{12}C acts as a catalyst to form the nucleus of a helium atom from four protons.

Coma The diffuse gaseous envelope that surrounds the nucleus of a comet and constitutes part of its head.

Comet A body composed of ice and dust that revolves around the Sun and develops a coma and a long tail as it approaches the Sun.

Condensation A process that causes a compound in the gas state to form a solid at a certain temperature and pressure.

Conduction A process by means of which heat flows through matter by the passage of energy between atoms and molecules. Also refers to the passage of an electrical current through a conductor.

Cone section A section through a cone-shaped solid that yields different kinds of curves ranging from circles to ellipses, parabolas, and hyperbolas.

Conjunction An alignment of a planet with the Sun and the Earth. (See inferior conjunction and superior conjunction).

Conservation of angular momentum A law of physics stating that in an isolated system the amount of angular momentum (a measure of the amount of rotation) remains constant.

Constellation A set of stars in a specified region of the sky.

Convection A process whereby energy is transported by moving currents of fluid or gas in response to a difference in the temperature of different parts of a body of solids, liquids, or gases.

Core The central region of a star or planet that is distinguished by differences in its chemical composition, temperature, and pressure from the shells of matter surrounding it.

Corona The outer atmosphere of the Sun having a low density but a high temperature.

Cosmic rays Nuclear particles and atomic nuclei that enter the atmosphere of the Earth at very high energies. They may originate from solar flares, supernova explosions of distant stars, and other sources.

Cosmology Study of the properties, origin, and evolution of the Universe.

Crater A circular excavation on the surface of a planet or satellite formed either by the explosive impact of a solid body or by the eruption of lava from a volcano.

Critical point The temperature and pressure at which the vapor and liquid phase of a compound have the same density which causes the liquid to change to vapor without boiling.

Crust The surface layer of a terrestrial planet or satellite that differs in its chemical composition from the underlying mantle.

Crystal A solid object formed by the orderly arrangements of charged atoms (ions).

D

Dark matter An invisible kind of matter in the Universe.

Decametric radiation Radio waves having a wavelength of about 10 m emitted by Jupiter.

Decimetric radiation Radio waves having a wavelength of about 0.1 m emitted by Jupiter.

Deuterium An isotope of hydrogen containing one proton and one neutron in its nucleus.

Differential gravitational force A force caused by the unequal attraction of gravity on different parts of an object.

Differential rotation Rotation of a non-solid body in which different latitude regions of the body rotate at different rates.

Differentation The separation of distinct layers in the interior of a star, planet, or satellite based on differences in density.

Diffraction The dispersion of light into a spectrum of wavelengths.

Dike A tabular body of igneous rocks that cuts across the structure of a pre-existing rocks.

Dinosaur Large reptiles that lived on the Earth during the Mesozoic Era.

Dipole field A magnetic field emanating from a planet that has two magnetic poles of opposite polarity.

Dirty snowball model Description of the composition of comets.

DNA Deoxyribonucleic acid. Organic compound that stores the genetic information of all organisms on the Earth.

Doppler effect Change in the wavelength of electromagnetic radiation and sound waves caused by the motion of their sources away from or toward the observer. (See also blue shift and red shift).

Dormant volcano A volcano that is inactive but has the potential of erupting again in the future.

Drake equation Attempt to estimate the number of technological civilizations in the Milky Way galaxy with whom we may be able communicate.

Dry ice Carbon dioxide ice.

Dune A deposit of wind-blown sand having a range of characteristic shapes.

Dust devil Small whirlwind that can inject dust into the air in the deserts of the Earth and Mars.

Dust tail The part of the tail of a comet that is composed of dust particles that are blown outward by the pressure of sunlight.

Dynamo theory Explains the origin of the magnetic field of the Earth as a result of induction by electrical currents in the liquid part of the core.

Dyne The force necessary to give an acceleration of $1\,cm/s^2$ to 1 gram of mass.

E

Earthshine The faint illumination of the crescent Moon caused by sunlight reflected by the Earth.

Eccentricity The ratio of the focal distance (f) divided by the semi-major axis (a) of the orbit of a stellar companion, a planet, or satellite: $e = f/a$.

Eclipse The occultation of the Sun by the Moon as seen from the Earth.

Ecliptic plane Plane defined by the orbit of the Earth.

Edgeworth-Kuiper belt A region that extends from the orbit of Pluto to a distance of about 500 AU from the Sun where thousands of ice bodies of various diameters orbit the Sun.

Ejecta Fractured and pulverized rocks ejected from meteorite impact craters.

Electromagnetic radiation Waves of electromagnetic energy having a wide range of wavelengths from gamma rays to radiowaves including visible light (400 to 750 nm).

Electron volt (eV) The energy that is produced when an electron is accelerated by a voltage difference of 1 volt.

Electrons Particles or waves carrying a unit electrical charge that exist in the space surrounding the nucleus of an atom.

Electrostatic forces Opposite electrical charges attract each other while charges of equal polarity repel each other.

Element A form of matter composed of atoms all of which contain the same number of protons in their nuclei.

Ellipse (1) A closed curve enclosing two foci such that the total distance from one focus to any point on the curve and back to the other focus is constant regardless of the position of the point of the ellipse.

Ellipse (2) A section of a cone at an angle to its base in the form of a closed geometrical figure having an eccentricity between 0 and 1.

Elongation The angle subtended by a planet in its orbit, the center of the Sun, and the center of the Earth.

Emission spectrum A dispersion of light into separate wavelengths in the form of bright lines produced by the electromagnetic radiation emitted by atoms of hot gases.

Encke gap A narrow gap in the A ring of Saturn.

Energy The ability to do work.

Energy flux The rate of energy flow expressed in units of Joules per square meter per second $(J\,m^{-2}\,s^{-1})$.

Enstatite chondrite Stony meteorite containing chondrules and composed primarily of the magnesium pyroxene enstatite ($MgSiO_3$).

Enzymes Special proteins that control various processes in an organism.

Equator The line of latitude located exactly midway between the north and south geographic poles of a spherical body in space.

Equilibrium A state of balance between two opposing forces (e.g., gravitational and centrifugal forces) or processes (e.g., evaporation and condensation).

Eratosthenian Era Time in the history of the Moon between about 3.2 and 1.0 billion years ago.

Erg Energy expended when a force of 1 dyne acts over a distance of 1 cm ($1\,J = 10^7$ ergs).

Erosion Removal of weathering products by flowing water, moving glaciers, wind, and by mass wasting occurring on the surfaces of terrestrial planets.

Escape velocity The velocity required for an object of any mass to escape completely from the surface of a planet or satellite.

Eucrite A basaltic achondrite meteorite that may have originated from the asteroid Vesta.

Evaporation Change in state from liquid to vapor of a volatile compound.

Evolution A theory that proposes that the form and function of adult living organisms change gradually over many generations.

Exosphere The outermost region of the thermosphere in the atmosphere of the Earth.

Extinct volcano A volcano that can no longer erupt in contrast to "dormant" volcanoes.

Extrasolar planet (exoplanet) A planet orbiting a star other than the Sun.

Extremophiles Micro-organisms that live in extreme environments such as sea ice, black smokers, hotsprings, brine lakes, etc.

F

F ring Narrow ring of Saturn located outside of the A ring.

Facula Small, bright areas above the photosphere of the Sun.

Fahrenheit Temperature scale in which water freezes at 32 °F and boils 212 °F and where absolute zero occurs at −459 °F.

Fall A meteorite that was observed to fall and was recovered shortly thereafter.

Far side of the Moon The side of the Moon that always faces away from the Earth.

Fault A fracture in the rocks of the crust of terrestrial planets and satellites along which the two sides have been displaced relative to each other by vertical, horizontal, and oblique motions.

Feldspar A common aluminosilicate mineral (e.g., $KAlSi_3O_8$).

Find A meteorite that was found but was not seen to fall.

Fission The splitting of the nucleus of an atom (e.g., uranium-235) into two fragments of unequal mass, accompanied by the release of neutrons and energy.

Flare A violent eruption on the surface of the Sun or star.

Focus One of two points inside an ellipse that satisfy the condition that the distance from one focus to any point on the ellipse and from that point to the other focus is constant.

Force A push or pull that acts on an object.

Fossil A remnant of ancient life or a trace of its activity that has been preserved in sedimentary rocks deposited throughout geologic time.

Frequency The number of wave crests that pass a fixed point per second.

Fusion crust Thin layer of glass that coats the surface of stony meteorites caused by frictional heating during their passage through the atmosphere of the Earth.

Full moon A phase of the Moon when its entire hemisphere that faces the Earth is illuminated by the Sun.

Fusion reaction A nuclear reaction in the cores of stars on the main sequence that releases the energy stars radiate into space. (See also nuclear fusion and CNO cycle).

G

Galaxy An aggregate of hundreds of billions of stars that interact gravitationally and form a coherent body in intergalactic space (e.g., the Milky Way galaxy).

Galilean satellites The four large satellites of Jupiter that were discovered by Galileo Galilei in 1610: Io, Europa, Ganymede, and Callisto.

Gamma rays Extremely short wavelength and highly energetic electromagnetic radiation emitted by atomic nuclei that are in an excited state.

Gas tail The part of the tail of a comet that consists of gas and is blows outward by the solar wind.

Gauss (G) A unit used to measure the strength of a magnetic field.

Geocentric concept The theory that the Earth is located at the center of the Universe and that the Sun and all of the stars revolve around it. This theory was discredited by Nicolaus Copernicus in 1543 AD.

Geology The science of the Earth including rocks of which it is composed, the topography of its surface, the structure and composition of its interior, its origin, and its history.

Geosynchronous orbit The orbit of an object whose period of revolution is equal to the period of rotation of the Earth. Such an object remains stationary above a point on the Earth.

Giant planet Term used with reference to Jupiter, Saturn, Uranus, and Neptune all of which have thick atmospheres composed of hydrogen and helium.

Gibbous A phase of the Moon between the quarter and full phase.

Glacier A body of ice that flows under the influence of gravity.

Global warming The increase of the average global temperature of the atmosphere of the Earth caused by the greenhouse effect and arising from the discharge of gases that absorb infrared radiation.

Gram The basic unit of mass in the cgs system of units.

Gravitational contraction The force of gravity of a cloud of gas and dust or of a star until the gravitation force is balanced by their internal pressure.

Gravity A force, first described by Sir Isaac Newton, that is exerted by all bodies that have mass on other massive bodies in their vicinity.

Great dark spot A large storm in the atmosphere of Neptune.

Great red spot (GRS) A large storm in the atmosphere of Jupiter that has been observed for several centuries.

Greatest elongation The point at which an inferior planet subtends the maximum angle with the center of the Sun and the Earth and therefore is easiest to observe from the Earth.

Greenhouse effect The absorption of infrared radiation by certain gases (e.g., CO_2, CH_4, and H_2O) in the atmosphere of the Earth or other celestial body (e.g., Venus) which causes the temperature of the atmosphere to rise.

Greenhouse gas A compound that absorbs infrared radiation and therefore contributes to global warming by means of the greenhouse effect.

Gregorian calendar The modern calendar initiated by Pope Gregory XIII in 1582 AD.

Groundwater Water that fills pore spaces and fractures in bedrock and overburden.

H

Habitable zone The volume of space around a star where a planet would have a surface temperature that would allow water to exist in the liquid state.

Halflife The time required for one half of an initial number of radioactive atoms to decay.

Halley's comet The comet whose return in 1758 AD was predicted by Sir Edmond Halley.

Halo The large envelope of diffuse hydrogen gas that surrounds the head of a comet.

Harmonic law Kepler's third law which states that the square of the period of revolution is proportional to the cube of the semi-major axis of the orbit.

Head The nucleus and coma of a comet.

Heat Thermal energy of a body consisting of the motion or vibration of atoms and molecules.

Heat flow The movement of heat caused by the existence of a temperature gradient. For example, heat flows from the hot interior of a planet or satellite to its cold surface.

Heavily cratered terrain The surface topography that occurs in some parts of Mercury, the Moon, Callisto, and other celestial objects.

Heliocentric concept The theory that the planets of the solar system (including the Earth) revolve around the Sun. This theory was first proposed by Greek astronomers and was reinstated by Nicolaus Copernicus.

Helium Chemical element (He) composed of atoms that contain two protons in their nuclei. Helium is a noble gas and is the second-most abundant element in the Sun and in stars of the Universe.

Heliocentric Centered on the Sun.

Heliopause The outer boundary of the heliosphere where the pressure of the solar wind equals that of the interstellar medium.

Heliosphere The spherical region of space that is permeated by the gases and by the magnetic field of the Sun.

Hertz Unit of frequency of waves. 1 Htz = 1 wave crest/s.

Hertzsprung-Russell diagram A plot of absolute magnitudes (or luminosities) of stars versus their spectral types (or surface temperatures).

Hesperian Era Time period in the history of Mars from 3.5 to 1.8 billion years ago.

Hirayama family A group of asteroids that appear to be fragments of a larger body that was broken up by collision with another asteroid.

Hubble's constant The proportionality constant (H) relating the recessional velocity (v) of a galaxy to its distance (d) from the Earth ($v = H \times d$).

Hubble's law States that the farther a galaxy is from the Earth the faster it is moving away from it.

Hydrocarbons Organic compounds composed of hydrogen and carbon atoms such as asphalt, petroleum or natural gas. Solid hydrocarbon compounds are generally brown or black in color.

Hydrogen Chemical element (H) composed of atoms that contain only one proton in their nuclei. Hydrogen is by far the most abundant element in the Sun, the solar system, and the stars in the Universe.

Hydroxyl The chemical radical OH^-.

Hyperbola A cone section having an eccentricity greater than one. The orbital velocity of an object on a hyperbolic orbit is greater than the escape velocity which means that the object is not gravitationally bound.

Hypothesis A proposed solution to a scientific problem that must be verified by experiment, observation, or mathematical modeling.

I

Ice Solid phase of volatile compounds such as water, carbon dioxide, methane, and ammonia.

Ice age Period of time in the past when large parts of a planet or satellite were covered by ice.

Ice Planet Planet that has a thick crust and mantle of ice surrounding, in many cases, a core composed of silicates, oxides, or even metallic iron.

Igneous rock Originated by crystallization of a silicate melt (magma or lava) either within the crust or on the surface of a planet or satellite.

Imbrian Era An interval of time in the history of the Moon from 3.85 to 3.2 billion years ago.

Impact crater Circular cavity in the surface of a planet, satellite, asteroid, or comet that was excavated by the explosive impact of a meteoroid, asteroid, or comet.

Impact melt rock A rock that formed by the cooling and crystallization of minerals from a melt produced by the impact of a meteoroid, asteroid, or comet.

Inclination The angle between the plane of the orbit of an object revolving around the Sun and the plane of the ecliptic (i.e., the plane of the orbit of the Earth).

Inertia The property of a body to remain at rest (if it is at rest) or to remain in motion (if it is in motion).

Inferior conjunction The alignment of an inferior planet (i.e., Venus or Mercury) between the Earth and the Sun.

Inferior planet A planet whose orbit lies between the orbit of the Earth and the Sun.

Infrared light Electromagnetic radiation having a wavelength that is slightly longer than the wavelength of red light.

Inner core The solid central portion of the core of the Earth composed of an alloy of iron, nickel, and other elements.

Intelligence The characteristic ability of certain living organisms to be aware of their own existence and the capacity for abstract thought.

Interplanetary dust particles (IDPs) Microscopic dust particles including fluffy aggregates of mineral matter as well as micrometeorites that are deposited on the surface of the Earth from interplanetary space.

Interplanetary medium The gas and particles that pervade the space between the planets of a solar system.

Ion An atom or molecule that has an electrical charge (positive or negative) caused by the loss or gain of electrons.

Ionic bond A chemical bond that arises from the electrostatic attraction of opposite electrical charges of ions in a crystal.

Ionosphere A region in the stratosphere of the Earth in which ionized atoms and molecules are present.

Iridium A platinum-group metal that is more abundant in meteorites than it is in the rocks of the crust of the Earth. The atomic number of iridium is 77.

Iron meteorite A meteorite composed of metallic iron and nickel.

Isostacy The theory that the crust of the Earth is in buoyant equilibrium which is maintained by vertical motions in the crust and mantle due to density differences between them.

Isotopes Atoms of an element whose nuclei contain different numbers of neutrons but the same number of protons.

Isotopic dating (also mistakenly called "radiometric" dating) A method of measuring the time elapsed since crystallization of minerals that contain naturally occurring long-lived radionuclides.

Isotropic Material whose physical and chemical properties are the same in all directions.

J

Joule A unit of energy in the mks system. The Joule is the work done when a force of one newton acts over a distance of 1 meter. 1 Joule $= 10^7$ ergs $= 1$ Watt second.

Julian calendar A calendar initiated by Julius Caesar in 45 BC with the aid of the astronomer Sosigines.

Jupiter-family comet A comet with a period of revolution of less than 20 years.

K

Kamacite A mineral composed of metallic iron alloyed with nickel that occurs in iron meteorites.

Karst topography A landscape that develops from the weathering of limestone under humid climatic conditions. Karst topography is characterized by sinkholes and caverns.

Kelvin (K) A unit of temperature on the Kelvin scale equivalent to one degree on the Celsius scale.

Kelvin scale A temperature scale that starts with absolute zero (0 K), the freezing temperatures of water at 273.15 K, and the boiling temperature of water at 373.15 K at a pressure of 1 atm.

Kepler's laws Mathematical laws that govern the motions of planets and their satellites everywhere in the Universe.

Keplerian motion Orbital motion in accord with Kepler's laws of planetary motion.

Kerogen Organic matter of high molecular weight that occurs in terrestrial sedimentary rocks (e.g., shale) and carbonaceous chondrites.

Kinetic energy The energy a body in motion possesses. $KE = \frac{1}{2} m\, v^2$ where m = mass and v = velocity.

Kirkwood gaps Ring-shaped areas (centered on the Sun) in the asteroidal belt containing few asteroids because the asteroids that occurred in these gaps were ejected by resonance effects with Jupiter.

Komatiite An ultramafic lava with $SiO_2 < 53\%$, $Na_2O + K_2O < 1\%$, $MgO > 18\%$, and $TiO_2 < 1\%$. The name is derived from the Komati River, Barberton Mountain Land, Transvaal, South Africa.

KREEP A product of fractional crystallization of basalt magma on the Moon that is enriched in potassium (K), rare earth elements (REE), and phosphorus (P).

Kuiper belt See Edgeworth-Kuiper belt.

L

Lagrangian point Positions with respect to an orbiting object where another object can get a free ride (e.g., two Lagrangian points exist 60° preceding and 60° behind the primary orbiting object).

Lagrangian satellite A satellite that occupies a stable Lagrangian point in the orbit of a planet or satellite (e.g., Calypso and Telesto are Lagrangian satellites of Tethys one of the large satellites of Saturn).

Lapse rate The rate of change of temperature as a function of elevation in the atmosphere of a planet or satellite.

Latitude Lines that run parallel north and south of the equator and are used to specify locations on the surface of a spherical body. (See also longitude and meridian).

Leading hemisphere The hemisphere of a 1:1 spin-orbit coupled satellite that is facing in the direction of orbital motion.

Leap year Every four years one day is added to the year (i.e., 366 days per year) to correct for the fact that the year has 365.25 days.

Least-energy orbit A spacecraft traveling from one planet to another planet is placed in an elliptical orbit in order to minimize the energy required to reach its destination.

Libration The variation of the Moon's orientation as seen from the Earth. As a result of libration and other effects we can actually see 59% of the surface of the Moon in spite of its 1 to 1 spin-orbit coupling.

Life The system of complex biochemical reactions that occur in any living organism.

Life zone A region around a star within which a planet can have surface temperatures that permit liquid water to exist.

Light curve A plot of the brightness of a celestial object versus time.

Lightyear The distance that light travels in one year which is about 9.46×10^{12} km.

Lithosphere Upper mantle and crust of a terrestrial planet that are cool enough to respond to stress by brittle failure. The lithosphere of the Earth is about 100 km thick.

Longitude Lines that run from the north geographic pole to the south geographic pole. The longitude line that passes through Greenwich, England, is the prime meridian.

Long-period comet A comet whose period of revolution around the Sun is more than 200 years.

Low-Earth orbit An orbit that lies about 160 km above the surface of the Earth.

Luminosity The rate at which a star emits electromagnetic radiation.

Lunar eclipse A passage of the Moon through the shadow of the Earth.

Lunar orbiter Five American spacecraft that orbited the Moon during the 1960s in order to photograph its surface.

M

Mafic rocks Igneous or metamorphic rocks that contain silicate minerals of iron and magnesium (i.e., olivine, pyroxene, amphiboles).

Magellanic clouds The two irregular galaxies that are satellites of the Milky Way galaxy.

Magnetic field The invisible pattern of forces generated by a magnet.

Magnetic storm A strong, temporary disturbance in the magnetic field of the Earth caused by the ejection of ionized particles by the Sun in the form of a solar flare. (See also flare).

Magnetopause The outer boundary of the magnetosphere of the Earth or planet that has a magnetic field.

Magnetosheath The transition region between the magnetopause and the solar wind.

Magnitude A numerical scale that expresses the brightness of a star, planet, or other celestial object.

Main-belt asteroids Asteroids that revolve around the Sun between the orbits of Mars and Jupiter.

Main sequence An array of data points in coordinates of the luminosity and surface temperature of the stars in a galaxy. (See Hertzsprung-Russell diagram).

Major axis The longest line that can be drawn across an ellipse that intersects both focal points.

Mantle The interior part of a planet that lies between the top of its core and the base of its crust.

Mare Large impact basins on the Moon that have been filled with basalt lava flows.

Mariner American spacecraft: 2, 5, and 10 to Venus, 4, 6, 7, and 9 to Mars, and 10 to Mercury.

Mascon Acronym for "mass concentration" implied by positive gravity anomalies that are associated with the mare basins on the Moon.

Mass A property of all matter. Also used to refer to the amount of matter contained within an object.

Maunder minimum The period of time between 1645 and 1715 when very few sunspots appeared and when the normal sunspot cycle did not occur.

Melting point The temperature a which a substances changes from a solid to a liquid depending on the pressure.

Mesosphere Part of the atmosphere of the Earth that is located between the stratosphere and the thermosphere.

Mesozoic Era Geologic time between 250 and 65 million years ago. Includes the Triassic, Jurassic, and Cretaceous Periods.

Metabolism The reactions that occur in living organisms that consume materials taken into the body, emit wastes, and produce energy for growth and reproduction.

Metallic hydrogen A form of hydrogen at high pressure and temperature that conducts electricity like a metal.

Metallicity The concentration in stars of elements having atomic numbers greater than 2.

Metamorphic rocks Igneous and sedimentary rocks that have been altered by high pressure and temperature acting over long periods of time.

Meteor A streak of light in the sky caused by a small meteoroid being heated by friction as it passes through the atmosphere of a planet or satellite.

Meteor shower An unusually large number of meteors that occur when the Earth moves through a cloud of particles in the orbital track of a comet.

Meteorite A fragment of a meteoroid (or a rock from the surface of the Moon and Mars) that has survived the passage through the atmosphere and has landed on the surface of the Earth.

Meteoroid A solid object in orbit around the Sun that may be a fragment of an asteroid (or originated from the surface of the Moon and Mars) and could eventually impact on the Earth.

Meter Basic unit of length in the mks system of units. ($1\,m = 100\,cm$).

Methane A hydrocarbon compound (CH_4) that forms a gas at the temperature and pressure of the terrestrial atmosphere, but occurs in liquid and solid form on the ice-covered satellites of Saturn Uranus, Neptune, and Pluto (e.g., Titan).

Micrometeorite Small fragment of a meteorite generally less than 1 mm in diameter.

Microwaves Part of the spectrum of electromagnetic radiation between infrared radiation and radio waves.

Midoceans ridge Continuous chain of submarine volcanoes along a rift in the floor of the ocean where the volcanic activity results in seafloor spreading (e.g., Mid-Atlantic ridge).

Milankovitch theory The theory that changes in the celestial mechanics of the Earth cause variations in the amount of solar energy received by the northern (and southern) hemisphere of the Earth.

Milky Way galaxy The galaxy in which the Sun and the system of planets including the Earth are located.

Miller-Urey experiment Synthesis of complex organic molecules by electrical discharges in a mixture of atmosphere gases.

Million 1,000,000 or 10^6.

Mineral A naturally occurring, crystalline solid form of a chemical compound.

Minor axis The shortest line that can be drawn across an ellipse through the center between the focal points and at right angles to the long axis.

Minute Both an angular measurement equal to 1/60 of one degree and an interval of time equal to 60 seconds.

Mixing ratio The ratio of the mass of a specified gas to the mass of all the gases in a unit volume of the atmosphere.

Mohorovicic discontinuity The surface that separates the bottom of the continental and oceanic crust from the underlying rocks of the upper mantle of the Earth.

Molecular hydrogen Diatomic hydrogen (H_2). (See also metallic hydrogen and hydrogen).

Molecule Covalent bonding between two or more atoms to form a compound. (e.g., H_2O, N_2, CO_2, NH_3, H_2).

Moment of inertia A measure of the inertia of a rotating body. It is controlled by the distribution of mass within the body. (See also inertia).

Moment-of-inertia-coefficient A number that describes the distribution of mass within a rotating body.

Momentum The tendency of a moving object to keep moving. Expressed mathematically as the product of mass and velocity.

Multi-ring basin A large impact basin that contains several concentric rings produced by the energy released during the impact of a small asteroid or comet.

Multiple star system A set of three or more stars that are gravitationally bound to each other. (See also binary star system).

Mutation A random error in the replication of DNA during the reproductive process of an organism.

N

Nadir The point on the celestial sphere lying directly beneath the observer or located 180° from the zenith. (See also celestial sphere and zenith).

Nanometer 10^{-9} m or one billionth of a meter (nm).

Natural selection The process by means of which the best traits of an organism are passed on to the descendents thereby allowing the organisms that are best adapted to their environment to survive.

Neap tide A tide in the oceans of the Earth that occurs when the Moon is close to the first quarter or the third quarter phase (i.e., when the Sun-Earth-Moon form a 90° angle causing a low tidal effect).

Near-Earth object (NEO) An asteroid whose orbit lies wholly or partly within the orbit of Mars.

Nebula A large cloud of gas and dust in interstellar space.

Nectarian Era An interval of time in the history of the Moon between 3.90 and 3.85 billion years ago.

Neutrinos Subatomic and electrically neutral particles released during nuclear fusion reactions in stars. Neutrinos interact sparingly with matter and are known to have mass when they are at rest.

Neutron Electrically neutral nuclear particle. Extra-nuclear neutrons are unstable and decay to form a proton and an electron.

Neutron star A small but very dense star composed on neutrons that are bound by gravity.

Newton (N) A basic unit of weight in the mks system of units.

Noachian Era An interval of time in the history of Mars that lasted from 4.6 to 3.5 billion years ago.

Noble gas An element whose atoms do not form bonds with the atoms of other elements (i.e., helium, neon, argon, krypton, xenon, and radon).

Node A point at which the orbit of a planet in the solar system intersects the plane of the ecliptic.

North celestial pole A point on the celestial sphere that is located directly above the geographic north pole of the Earth. (See also Polaris).

Northern lights Also known as aurora borealis (See aurora).

Nova A star close to the end of its evolution that suddenly increases in brightness by factors of 10 to 10^4 and then fades to its original luminosity.

Nuclear fission A reaction that causes the nucleus of an atom (e.g, uranium -235) to split into two fragments accompanied by the release of neutrons and energy. (See also fission).

Nuclear fusion A nuclear reaction that causes the nuclei of two atoms to merge into one nucleus thereby releasing a large amount of energy (e.g., fusion of four hydrogen nuclei to form one nucleus of helium).

Nucleosynthesis A complex of nuclear reactions in stars by means of which atomic nuclei are produced of all of the chemical elements.

Nucleus 1. The central particle of an atom where most of its mass is located; 2. The body of a comet that is composed of ice and dust; 3. The central body of a cell that contains the genetic material.

Nutation Periodic variation in the precession of the Earth caused by irregularities in the orbit of the Moon. (See also precession).

O

Oblateness The extent of polar flattening of a planet or satellite. It is expressed by the difference in the equatorial and polar radii divided by the equatorial radius (i.e, oblateness $= r_e - r_p)/r_e$).

Oblate spheroid A sphere that is flattened at the poles such that is polar diameter is shorter than its equatorial diameter (e.g., the Earth and Mars are oblate spheroids but Mercury and Venus are not because of their slow rate to rotation).

Obliquity The angle between the axis of rotation of a celestial body and the direction perpendicular to the plane of the orbit. Alternatively it is definable as the angle between the equatorial plane and the plane of the orbit of a celestial body.

Occultation An event that occurs when one celestial body obscures the view of another body that has a smaller apparent diameter (e.g., the Moon blocking the view of a planet or star).

Occam's razor A statement recommending that if two or more competing hypotheses attempt to explain a phenomenon, the one that makes the fewest assumptions should be preferred.

Oceanic rift A fracture in the oceanic crust and lithosphere causing decompression melting, magma formation, and extrusion of basalt lava along the resulting mid-ocean ridge.

Olivine A mineral composed of magnesium-iron silicate ($(Mg, Fe)_2 SiO_4$).

Oort cloud A large spherical region of space up to 100,000 AU from the Sun that surrounds the solar system within which a large number of ice bodies exist that turn into comets when they approach the Sun.

Opacity The property of certain materials to block light or other electromagnetic radiation.

Opposition The alignment of a superior planet with the Earth and the Sun such that the Earth lies between the superior planet and the Sun.

Optical depth A measure of the radiation absorbed as it passes through a medium: If the opacity is zero, the medium is transparent, if it is less than 1, it is optically thin, and if the opacity is greater than 1, the medium is optically thick.

Orbit The path in space taken by a celestial object moving under the gravitational influence of another object or set of objects.

Orbital resonance A relationship between the periods of revolution of two celestial bodies that revolve around the same primary body such that the ratio of the periods of revolution is a simple numerical fraction.

Orbital velocity The velocity of a body in orbit around another body.

Ordinary chondrite A group of stony meteorites that contain chondrules and are most abundant among falls. (See also chondrule, chondrite, and fall).

Organic molecule A molecule composed of atoms of carbon and hydrogen bonded by covalent bonds. Formed by living organisms as well as by abiogenic reactions in the solar nebula.

Organism any member of the biosphere that is alive. (See also biosphere and life).

Outer core The outer shell of the core of the Earth that is composed of a liquid iron-nickel alloy (See also core and dynamo theory).

Outflow channel Channels on the surface of Mars through which large volumes of water and sediment were discharged into the northern lowlands. (See also Mars).

Outgassing The release of gases from the hot interior of a planet.

Ozone hole The stratosphere over Antarctica and the Arctic region of the Earth where the concentration of ozone during the early spring is low because of its destruction by chlorine atoms of chlorofluorocarbon (CFC) gas.

Ozone layer A region in the stratosphere of the Earth that contains ozone (O_3) which absorbs harmful ultraviolet radiation emitted by the Sun.

P

P wave A seismic wave that is transmitted by solid as well as liquid materials.

Paleomagnetism The study of the past magnetic field of the Earth based on remanent magnetism that is best preserved in fine-grained volcanic rocks that cooled quickly on the surface of the Earth.

Palimpsest An impact crater that lacks topographic relief because of isostatic rebound (e.g. impact craters on the Galilean satellites Europa, Ganymede, and Callisto).

Pangaea The supercontinent that existed in late Paleozoic time (300 to 200 million years ago) and contained all of the continents of the present Earth. Pangaea split into Laurasia and Gondwana at about 200 million years ago and Gondwana later broke apart to form South America, Africa, India, Antarctica, and Australia.

Panspermia The hypothesis that life on the Earth originated from dormant spores and hardy bacteria that were deposited on the Earth by meteorites and interplanetary dust particles (IDPs). (See also IDPs).

Parabola A cone section having an eccentricity e = 1.0. The velocities of objects on parabolic orbits exceed the escape velocity of the object at the focal point and therefore are not gravitationally bound (i.e., they do not come back).

Parallax The shift in the apparent position of an object caused by differences in the location of the observer.

Parent atom The radioactive atom that decays to form a radiogenic daughter which is an atom of a different chemical element than its parent.

Parent bodies The objects that were broken up to form most of the present asteroids and meteoroids which are fragments of their respective parent bodies.

Parsec A unit of distance equal to 206,265 AU. It is the distance at which an object would have a parallax of 1 second of arc. (See also parallax).

Patera A shield volcano of low relief that occurs on Mars.

Penumbra 1. The lighter region of a sunspot; 2. A region of a two-part shadow in which the light is not completely blocked.

Perigee The point on the orbit of the Moon (or any other satellite) that is closest to the Earth (or to the planet being orbited by a satellite). (See also apogee).

Perihelion The point in the orbit of a planet, asteroid or comet at its closest approach to the Sun.

Period The time required by a celestial body to complete one rotation on its axis or one revolution around the primary object (i.e., the Sun or a planet).

Periodic table A listing of the chemical elements in order of their increasing atomic numbers which control their chemical and physical properties.

Permafrost Regolith that contains water that is permanently frozen. Also called "ground ice".

Photodissociation The break-up of molecules in the atmosphere by irradiation with sunlight.

Photoelectric effect The ejection of electrons from a surface being irradiated by photons. (See also photon).

Photon A virtual particle of light equivalent to electromagnetic waves.

Photosphere The region of the Sun that emits the light we see on the Earth.

Photosynthesis The process whereby green plants produce molecules of glucose by combining six molecules of carbon dioxide and six molecules of water aided by chlorophyl and powered by sunlight.

Pioneer American program of planetary exploration. Pioneer 10 explored Jupiter, Pioneer 11 went to Jupiter and Saturn, and Pioneer Venus orbited Venus and released a probe into its atmosphere.

Plagioclase A common mineral in mafic igneous rocks (e.g., basalt) composed of a solid solution of albite ($NaAlSi_3O_8$) and anorthite ($CaAl_2Si_2O_8$).

Planck's constant (h) The proportionality constant in the relation between the energy (E) and the frequency (ν) of a photon: $E = h\nu$ where $h = 6.626176 \times 10^{-34}$ Js.

Plane of the ecliptic The plane of the orbit of the Earth around the Sun.

Planet One of the nine large bodies that orbit the Sun.

Planetary science The science of the solar system including its origin, celestial mechanics, geology, geophysics, aeronomy, and biology.

Planetesimal Solid objects of wide-ranging size that formed within the protoplanetary disc and which ultimately accreted to form the planets, satellites, asteroids, comets, and Kuiper-belt objects.

Planetoid Term used to describe asteroids (i.e., they are "planet-like").

Plasma Hot gas in which all atoms have electrical charges (i.e., they are ionized).

Plasma tail The ion tail of comets that consists of ionized atoms and molecules pushed away from the Sun by the solar wind. (See also comets and solar wind).

Plate tectonics A theory concerning the movement and interaction of lithospheric plates of the Earth that are driven by convective motions in the underlying asthenosphere.

Polar ice-caps The ice sheets that exist in the polar regions of Mars (as well as of the Earth). The ice caps of Mars are composed of water ice and carbon-dioxide ice interbedded with layers of windblown dust.

Polaris The star that is currently located directly above the geographic North Pole on a line that extends the axis of rotation of the Earth to the celestial sphere. (See also north celestial pole).

Polarization of light An alignment of electromagnetic radiation such that all vibrations are parallel in direction.

Polymer A chain of molecules linked together by covalent bonds.

Positron A positively-charged electron that is ejected from the nucleus of certain radioactive atoms during their decay.

Potential energy The energy that a body possesses because of its location.

Poynting-Robertson effect The drag on a small orbiting particle due to the asymmetrical re-radiation of absorbed sunlight or starlight.

Precambrian Eon The interval of geologic time that lasted about 3.2 billion years from the start of the Early Archean Era at 3.8 billion years ago to the start of the Cambrian Period at 545 million years ago.

Precession The slow "wobble" of the orientation in space of the axis of rotation of the Earth and of other rotating bodies. It is caused by tidal interaction with other celestial bodies.

Precession of the equinoxes The slow westward motion of the equinoxes along the ecliptic due to precession of the axis of rotation of the Earth.

Precipitation The formation of solid particles or liquid droplets from aqueous solutions and atmospheric gases.

Pre-Nectarian Eon The interval of time from the formation of the Moon to 3.85 billion years ago. It coincides with the Hadean Eon in the geologic history of the Earth.

Pressure A measure of the force exerted per unit area.

Primates The group of mammals that includes humans.

Primordial An adjective referring to conditions at the time of formation of the solar system 4.6 billion years ago.

Principle of uniformity of process Also called Uniformitarianism, is the guiding principle of geology stating that geological processes act slowly over long periods of time and that they have not changed in kind or energy (i.e., The present is the key to the past).

Prograde Rotation and revolution in the anticlockwise direction when viewed from above the plane of the orbit of a celestial body. (See also retrograde).

Prolate spheriod A sphere that has been stretched along its polar axis causing its polar diameter to be great than its equatorial diameter. (See also oblate spheroid).

Proton A nuclear particle that has a mass of about one atomic mass units and an electrical charge of +1 electronic units.

Protoplanet A large solid object that has formed by accretion of planetesimals that may continue to grow to become one of the planets that revolve around a star like the Sun.

Protostar A youthful star that has begun to emit electromagnetic radiation as a result of internal heat generated by compression but before the onset of hydrogen fusion in the core.

Pulsar A pulsating source of radiowaves in a galaxy that is thought to be a rapidly spinning neutron star. The periods of most pulsars range from 0.001 to 5 seconds.

Pyroclastic material Composed of hot (or even molten) fragments of lava and fractured rocks emitted by volcanoes that erupt lavas of andesitic to rhyolitic composition.

Pyroxene A common silicate mineral in mafic rocks (e.g., basalt) containing Fe, Mg, Ca and minor amounts of Al and Na.

Q

QSO Abbreviation of quasi-stellar object.

Quantum mechanics The branch of physics dealing with the structure and behavior of atoms and their constituents as well as with the interaction of atoms and their constituents with light.

Quasar A starlike object with a very large red shift. (See also red shift and Doppler effect).

R

Radar Acronym derived from "radio detecting and ranging". It is an electronic system used to detect and locate objects at distances and under conditions of obscuration.

Radial velocity The speed at which an object is either moving away or toward an observer.

Radian The point in the sky from which meteors appear to radiate outward during a meteor shower. (See also meteor shower).

Radiation A method of heat transport based on energy being carried by photons. (See also photon).

Radiation darkening The darkening of methane ice by impacts of electrons. (See also solar wind).

Radiation pressure Pressure exerted on a surface by photons during an irradiation.

Radiative diffusion The random migration of photons from the center of a star toward the surface.

Radio telescope A telescope designed to detect radio waves.

Radio waves Electromagnetic radiation of long wavelength.

Radio window The range of wavelengths of radio waves that can pass through the atmosphere of the Earth.

Radioactivity A process of nuclear decay that is accompanied by the emission of particles and electromagnetic radiation from the nucleus. (See also gamma rays and halflife).

Radiogenic An adjective referring to the products of radioactive decay (e.g., radiogenic daughter atoms and radiogenic heat).

Radionuclide An atomic species that is radioactive (i.e., its nucleus is unstable and decays spontaneously).

Ranger American space program of the 1960s that photographed the surface of the Moon at close range prior to impacting.

Rays Streaks of ejecta radiating outward from recently formed impact craters (e.g., on Mercury and the Moon). (See also impact crater and ejecta).

Recovery The first observation of a short-period comet as it approaches the Sun each time it returns to the inner solar system.

Red dwarf A cool low-mass star residing at the lower end of the main sequence on the Hertzsprung-Russell diagram.

Red giant Very large and luminous star in a late stage of stellar evolution. (See also main sequence).

Red shift The increase in the wavelength of light emitted by a source that is moving away from the Earth. (See also Doppler effect).

Refractory element or compound A chemical element or compound that has a high melting and boiling temperature and a low vapor pressure depending on the temperature.

Regmaglypts Thumbprint-like depressions in the surface of a meteorite caused by ablation during passage through the atmosphere.

Regolith Unconsolidated rock debris and weathering products that cover the underlying bedrock. "Regolith" is not synonymous with "soil" and should not be referred to as "dirt."

Remanent magnetism Magnetism that is embedded in igneous and clastic sedimentary rocks at the time of their formation.

Remote sensing Observing a celestial object from a distance by recording the electromagnetic radiation it reflects and/or emits at different wavelengths.

Respiration A biological process that produces energy by consuming oxygen and releasing carbon dioxide.

Retrograde Revolution and rotation of a celestial object in the clockwise direction when viewed from a point above the plane of the ecliptic. (See also prograde and plane of the ecliptic).

Rhyolite Volcanic igneous rock, commonly porphyritic and exhibiting flow banding, containing crystals of quartz and alkali feldspar in a glassy to cryptocrystalline matrix.

Rift valley A tectonic valley on the surface of a terrestrial planet or satellite that forms by extension of the crust which causes fracturing and the formation of normal faults (e.g., East African rift valleys).

Rille Long, narrow, and sinuous valleys on the Moon believed to be lava-drain channels.

Ring particles Particles that orbit one of the giant planets of the solar system composed of rocky material or ice and coated by abiogenic organic matter in some cases.

Ring system Disks composed of solid particles that revolve in the equatorial planes around each the giant planets of the solar system.

Roche limit The closest distance to which an orbiting body can approach another body without being broken up by tidal forces.

Rocket A propulsion system that is based on Newton's third law of motion (to every action there is an equal and opposite reaction).

Rotation The spinning of a celestial body about an axis.

Runaway greenhouse effect A case of global warming by the greenhouse effect when the temperature rises to high values (e.g., Venus).

Runaway icehouse effect A case of global climatic cooling in which the oceans freeze and the entire Earth is covered by ice (e.g., Snowball Earth in the Late Proterozoic Era).

S

S wave A seismic shear wave produced during earthquakes that is absorbed by liquid media.

Satellite A celestial body that orbits another celestial body of equal or greater mass (e.g., the Moon is a satellite of the Earth).

Saturation A state that occurs when the concentration of an element or compound in an aqueous solution or in a mixture of gases reaches a maximum at a specified temperature. A further increase of the concentration causes precipitation. (See also precipitation).

Science The explanation of natural phenomena based on observations, the rational interpretation of data, and testing of conclusions by experiment, additional observation, or mathematical modeling.

Scientific method A procedure that governs scientific research. It consists of reproducible measurements and observations, the formulation of multiple hypotheses, and testing of these hypotheses in an attempt to prove them wrong. A hypothesis that withstands the tests becomes a theory that its deemed to be the best available explanation of the phenomenon under study.

Scientific notation Use of powers of ten to express both large and small numbers (e.g., $1000 = 10^3$ and $0.001 = 10^{-3}$).

Seafloor spreading The movement of plates away from spreading ridges in the oceans where oceanic crust is generated by submarine volcanic activity.

Seasons Periodic changes in climate that result from the annual changes in the amount of sunlight received at a selected place on the Earth. The seasonality of the weather results from the inclination of the axis of rotation of the Earth.

Second The basic unit in which time is measured. It is also the angle that is 1/60th of 1 minute of arc.

Secondary crater Small impact crater formed by the impact of a mass of rock ejected from a large meteorite impact crater.

Sediment Particles of rocks and minerals or their weathering products that are being moved (or were moved) by gravity, flowing water, or wind to a place of deposition (e.g., the ocean basins).

Sedimentary rocks Layers of sediment that have been cemented or compressed and chemical precipitates that form in the oceans or in lakes by biological activity or by evaporation.

Seismic tomography A technique used to generate three-dimensional images of the interior of the Earth from seismic waves recorded by seismographs.

Seismic wave Compressional waves and shear waves that are generated by sudden motions of rocks below the surface that cause the phenomenon of earthquakes.

Semi-major axis Half of the long axis of an ellipse. It is also the average distance of points on an ellipse to one of the two focal points.

SETI Search for extratrestrial intelligence.

Shatter cone Cone-shaped fracture pattern that develops in rocks by the passage of the shockwave emanating from the site of a meteorite impact.

Shepherd satellite A satellite whose gravity restricts the motions of particles in a planetary ring thereby preventing the dispersion of the particles. (See also ring).

Shield volcano Large volcanic mountain with gently sloping sides and a broad summit plateau that forms by the extrusion of basalt lavas that have a low viscosity.

Shock wave A fast-moving and highly energetic wave emanating from the explosive impact of a meteorite and spreading in a hemispherical pattern through the target rocks.

Short-period comet Comets that have periods of revolution of less than 200 years and that therefore return to the inner solar system at regular intervals of less than 200 years.

SI units The International System of Units (Système international des unités) including the meter (m), kilogram (kg), and second (s).

Sidereal day The time it takes the Earth to complete one rotation relative to fixed background stars.

Sidereal month The period of revolution of the Moon with respect to fixed background stars.

Sidereal period of revolution The period of revolution of the Earth relative to fixed background stars. The sidereal period of revolution is one year by definition.

Siderophile elements Chemical elements that associate with metallic iron.

Silicate A salt of silicic acid (H_4SiO_4). Naturally occurring silicates are minerals that form by crystallization from cooling silicate melts or by metamorphic recrystallization of mixtures of oxides and silicates.

Sill A tabular body of igneous rocks that formed by the intrusion of magma along bedding planes of sedimentary rocks. In most cases, sills are flat-lying and conformable but they may also be transgressive.

Sinkhole A depression in the surface of the Earth caused by the collapse of a limestone cavern in the subsurface. (See also karst topography).

Slingshot effect The use of the gravitational field of a celestial body to increase the velocity of a spacecraft during a flyby of that body.

SNC meteorites (Shergotty, Nakhla, Chassigny) all of which originated from Mars. The term also applies to other meteorites from Mars whose chemical and mineral compositions resemble those of shergottites and nakhlites.

Snowball Earth A somewhat controversial theory based on evidence that the entire Earth was twice covered by ice during the Late Proterozoic Era between 800 and 650 million years ago, each glaciation lasting less than one million years.

Solar constant The average amount of solar energy received by the Earth above its atmosphere and expressed in watts per meter squared per second.

Solar corona Hot, faintly glowing gases seen around the Sun during a total eclipse. It is the uppermost region of the atmosphere of the Sun.

Solar day The length of time required for the Earth (or another body) to complete a rotation measured with respect to the Sun. It is the length of time between successive appearances of the Sun at the highest point in the sky. (See also sidereal day).

Solar eclipse A transit of the Moon between the Earth and the Sun causing sunlight to be blocked temporarily.

Solar mass A unit of mass where the mass of the Sun is equal to 1.

Solar system The Sun and all of the bodies that revolve around it including the Oort cloud of ice objects.

Solar wind The continuous stream of gas emitted by the Sun. The principal component of the solar wind are protons and electrons derived from hydrogen atoms.

Solid-fuel rocket A rocket engine that uses a mixture of solid fuel and oxidizer.

Solstice The date when the Earth passes through the perihelion or aphelion of its orbit.

Sounding rocket A rocket that is launched vertically into the stratosphere and that returns its payload to the Earth (i.e.,it does not go into orbit).

South celestial pole The point on the celestial sphere that is located directly above the geographic South Pole of the Earth.

Specific gravity Is the ratio of the density of a solid or liquid substance divided by the density of a standard substance such as pure water at $4\,^\circ$C which has a density of $1.000\,g/cm^3$. Therefore, the specific gravity of a solid or liquid substance is numerically identical to its density, but specific gravity is a dimensionless number because it is the ratio of densities. The specific gravity of gases is the ratio of the density of a gas at a specified temperature and pressure divided by the density of dry air at $0\,^\circ$C and 1 atm pressure (STP) which is $1.29\,g/L$. For example, the density of carbon dioxide at STP is $1.976\,g/L$, but its specific gravity is $1.976/1.29 = 1.53$ (no units). (See also density).

Spectral classification of stars A classification based on decreasing surface temperatures of stars indicated by the letters: O, B, A, F, G, K, M, L.

Spectroscopy The analysis of light emitted by a body. The light is dispersed into its wavelength spectrum which is used to determine the chemical composition of the body.

Speed Rate of motion of a body regardless of the direction. (See also velocity).

Speed of light The speed with which light travels in a vacuum. According to Einstein's theory of special relativity, the speed of light cannot be exceeded ($c = 2.99 \times 10^8\,m/s$).

Spherule Spherical particles of silicate glass, or metallic iron, or a mixture of iron oxide and glass that form from molten material that is stripped from the surfaces of meteorites passing through the atmosphere.

Spin-orbit coupling The condition that the periods of revolution and rotation of an orbiting body are either equal to each other or related by simple ratios such as 1:2, 1:3, or 2:3, etc..

Spiral galaxy A galaxy, such as the Milky Way, in which most of the stars are distributed in the arms of a pinwheel that forms a flat disc in space.

Spring tide Ocean tides of high amplitude that occur at new-Moon and full-Moon phases (See also tides).

S-type asteroid A common type of asteroid in the inner part of the main belt of asteroids. The optical spectrum of these asteroids resemble those of stony meteorites.

Star Large, hot, spherical mass of hydrogen and helium gas that radiates energy in the form of electromagnetic waves. The energy of stars is generated primarily by a nuclear fusion reaction in their cores and by compression caused by gravity.

Stefan-Boltzmann law A mathematical relationship between the temperature of a blackbody and the rate at which it radiates energy. (See Appendix 1).

Stishovite A dense polymorph of the mineral quartz (SiO_2) produced by the shock wave generated by the impact of a large meteorite (e.g., Meteor Crater, Winslow, Arizona).

Stony meteorite A meteorite composed primarily of silicate and oxide minerals as well as small grains of metallic iron. (See also chondrites).

Stony-iron meteorite A meteorite composed of both metallic iron-nickel alloy and silicate minerals (e.g., pallasites).

Stratosphere The second of four layers of the atmosphere in which the temperature increases with height above the surface of the Earth and which contains the ozone layer. (See also ozone).

Strewn field The oval-shaped area defined by fragments of a meteorite that exploded in the atmosphere.

Subduction The return of oceanic crust and underlying lithosphere into the mantle along deep-sea trenches as a result of convective motions in the asthenosphere. (See also lithosphere and asthenosphere).

Sulfuric acid A strong inorganic acid whose chemical formula is H_2SO_4.

Sun-grazing comets Long-period comets that come very close to the Sun at perihelion and may, in some cases, fall into the Sun.

Sunspot Dark spots in the photosphere of the Sun that are about 1500 K cooler than the surroundings. Sun spots are caused by changes in the Sun's magnetic field.

Sunspot cycle An 11-year cycle during which both the number and the location of sunspots vary.

Super-rotating atmosphere A condition where the atmosphere rotates faster than the body of the planet beneath it.

Superior conjunction The alignment of a planet on the side of the Sun that is opposite to the Earth.

Superior planet a planet whose orbit has a larger average radius than the orbit of the Earth (e.g., Mars, Jupiter, etc.).

Supernova A massive star that explodes at the end of its evolution and ejects the chemical elements it synthesized during its life cycle into interstellar space. (See also nucleosythesis).

Surface wave A type of seismic wave that travels only over the surface of the Earth.

Surveyor An American space program during the 1960s that landed spacecraft on the surface of the Moon.

Synchronous rotation The rotation of a celestial body with a period equal to its period of revolution. (See also spin-orbit-coupling).

Synchrotron radiation A type of nonthermal radiation emitted by charged particles moving through a magnetic field.

Synodic period The time required between successive oppositions or conjunctions of a planet (See also sidereal period).

T

T Tauri star A stage in the early evolution of stars during which a strong solar wind blows away remnants of the solar nebula.

Tektites Small glassy objects formed from molten rocks ejected from meteorite impact craters into the stratosphere where they solidify before falling back to the Earth.

Terminator A line on the surface of a celestial body orbiting the Sun along which sunrise or sunset is occurring.

Terrestrial planets Celestial bodies in orbit around a star that are comparatively small, are composed of solid rocks, may have an atmosphere and satellites, but lack rings.

Theory The best available explanation of a natural phenomenon that has been confirmed by testing and is therefore accepted by scientists.

Thermosphere The outermost of the four layers of the atmosphere of the Earth.

Tholus A small, steep-sided volcanic mountain on Mars.

Tholeiite A silica-oversaturated basalt containing orthopyroxene, calcic plagioclase, and olivine.

Tidal forces The differential forces exerted by two celestial bodies on each other due to gravity depending on the distance between them.

Tides in the ocean The daily rise and fall of the surface of the ocean caused by the tidal forces exerted by the Moon and the Sun.

Time zones Regions of the Earth where, by agreement, all clocks have the same time.

Titius-Bode rule See Bode rule.

Torino scale A numerical scale from zero to ten that expresses the magnitude of the risk of the impact of an asteroid or comet.

Torus A three-dimensional geometric body resembling a doughnut. The sodium atoms removed from Io by the magnetic field of Jupiter form a torus around it.

Trailing hemisphere The hemisphere of a satellite with 1:1 spin-orbit coupling that faces away from the direction in which the satellite moves in its orbit.

Transfer orbit An orbit that carries a spacecraft from one planet in the solar system to another.

Transit The passage of an inferior planet across the face of the Sun as seen from the Earth.

Trans-Neptunian objects (TNO) Celestial objects that orbit the Sun beyond the orbit of Neptune (e.g., Pluto and the objects of the Edgeworth-Kuiper belt).

Trillion $1,000,000,000,000$ or 10^{12}.

Triple-alpha process The process whereby atoms of carbon-12 are formed by nuclear reactions that result from the merger of three helium nuclei (alpha particles) to form one nucleus of carbon-12.

Trojan asteroids Asteroids located at the two Lagrange points (4 and 5) on the orbit of Jupiter. (See also Lagrange points).

Tropic of Cancer A circle at 23.5° latitude north of the equator of the Earth.

Tropic of Capricorn A circle at 23.5° latitude south of the equator of the Earth.

Tropical year The time between the beginnings of a terrestrial season in successive years.

Tropopause The boundary between the troposphere and the overlying stratosphere of the Earth.

Troposphere The lowermost layer of the atmosphere of the Earth.

Tuff A rock composed of pyroclastic materials that have been welded together or cemented by minerals deposited by aqueous solutions.

Tunguska event Violent explosion in the atmosphere over the Tunguska river in Siberia in 1908 caused by the entry of a comet or meteoroid into the atmosphere.

U

Ultraviolet light Electromagnetic radiation that has shorter wavelengths and is more energetic than visible light.

Umbra Latin word for shadow 1. Dark central portion of sunspots; 2. Darkest part of a shadow cast by a celestial body by blocking sunlight.

Uncompressed density The bulk density of a celestial body that has been adjusted by decompressing its interior.

Universe All of the matter and energy and the space they occupy.

Utopia planitia A low plain in the northern hemisphere of Mars.

V

Van Allen radiation belts Two belts of subatomic particles that are trapped by the magnetic field of the Earth and that are located above its equator.

Velocity A measure of the speed of an object and the direction in which it is moving.

Ventifact A pebble or cobble that has been faceted by wind-blown sand and silt.

Vernal equinox March 21 when the Sun at noon is directly above the equator of the Earth. (See also autumnal equinox).

Viking An American program that successfully landed two spacecraft in 1976 on Chryse and Utopia planitia of Mars.

Volatiles Compounds that have low melting and boiling temperatures and high vapor pressures depending on the temperature.

Voyagers 1 and 2 Two spacecraft that visited the giant planets of the solar system (Jupiter and Saturn by Voyager 1; Jupiter, Saturn, Uranus, and Neptune by Voyager 2).

W

Waning Lunar phases when the illuminated portion of the Moon is decreasing.

Wavelength The distance between successive crests (or troughs) of a wave.

Waxing Lunar phases when the illuminated portion of the lunar surface is increasing.

Weathering Decomposition of rocks and minerals by chemical reactions at Earth-surface temperatures.

Weight The force exerted by gravity on a body that has mass.

White dwarf A small, dense star at the end of its evolution in which little or no energy is being generated and which has contracted until no further contraction is possible.

Widmanstätten texture A pattern of interlocking crystals of minerals composed of metallic iron and nickel (kamacite and taenite). Made visible by polishing and etching a cut surface of an iron meteorite.

Wien's law The relationship between the temperature of a light-emitting body at the wave-length at which it emits the maximum amount of light.

Watt A unit of power equal to on joule per second.

X

X-rays Electromagnetic radiation that is intermediate between ultraviolet light and gamma rays in terms of its wavelength and energy.

Y

Yardangs Long, irregular, sharp-crested ridges between round-bottom troughs carved on a plateau or unsheltered plain by wind erosion in desert regions.

Z

Zenith The point on the celestial sphere directly above the observer. (See also nadir and celestial sphere).

Zodiac The 12 constellations that become visible in the night sky in the course of one revolution of the Earth around the Sun.

Zodiacal light Faint light visible near the horizon after sunset and before sunrise caused by sunlight scattered by cometary dust in the plane of the ecliptic.

Zonal winds The pattern of alternating eastward and westward winds in the atmospheres of Jupiter and Saturn.

Zone A light-colored band in the atmosphere of Jupiter.

Sources

Beatty JK, Petersen CC, Chaikin A (eds.) (1999) The new solar system. Sky Publishers, Cambridge, MA
Freedman RA, Kaufmann III WJ (2002) Universe: The solar system. Freeman, New York
Hartmann WK (2005) Moons and planets, 5th edn. Brooks/Cole, Belmont, CA
Jackson J (ed.) (1997) Glossary of geology, 4th edn., Amer Geol Inst, Alexandria, VA
Protheroe WM, Capriotti ER, Newson GH (1989) Exploring the Universe, 4th edn. Merrill Pub. Co., Columbus, Oh
Seeds MA (1997) Foundations of astronomy, 4th edn. Wadsworth, Belmont, CA
Wagner JK (1991) Introduction to the solar system. Saunders, Philadelphia, PA
Weast RC, Astle MJ, Beyer WH (1986) CRC handbook of chemistry and physics, 66 edn. CRC Press, Boca Raton, FL

Author Index

Subject Index